T0213956

Lecture Notes in Computer Science   10573

Commenced Publication in 1973
Founding and Former Series Editors:
Gerhard Goos, Juris Hartmanis, and Jan van Leeuwen

More information about this series at http://www.springer.com/series/7408

Hervé Panetto · Christophe Debruyne
Walid Gaaloul · Mike Papazoglou
Adrian Paschke · Claudio Agostino Ardagna
Robert Meersman (Eds.)

On the Move to Meaningful Internet Systems

OTM 2017 Conferences

Confederated International Conferences:
CoopIS, C&TC, and ODBASE 2017
Rhodes, Greece, October 23–27, 2017
Proceedings, Part I

 Springer

Editors

Hervé Panetto
University of Lorraine
Nancy
France

Christophe Debruyne
Odisee University College
Brussels
Belgium

Walid Gaaloul
Télécom SudParis
Évry
France

Mike Papazoglou
Tilburg University
Tilburg
The Netherlands

Adrian Paschke
Freie Universität Berlin and Fraunhofer
  FOKUS
Berlin
Germany

Claudio Agostino Ardagna
Università degli Studi di Milano
Crema
Italy

Robert Meersman
TU Graz
Graz
Austria

ISSN 0302-9743   ISSN 1611-3349 (electronic)
Lecture Notes in Computer Science
ISBN 978-3-319-69461-0   ISBN 978-3-319-69462-7   (eBook)
https://doi.org/10.1007/978-3-319-69462-7

Library of Congress Control Number: 2017956721

LNCS Sublibrary: SL2 – Programming and Software Engineering

Printed on acid-free paper

This Springer imprint is published by Springer Nature
The registered company is Springer International Publishing AG
The registered company address is: Gewerbestrasse 11, 6330 Cham, Switzerland

Hervé Panetto · Christophe Debruyne
Walid Gaaloul · Mike Papazoglou
Adrian Paschke · Claudio Agostino Ardagna
Robert Meersman (Eds.)

On the Move to Meaningful Internet Systems

OTM 2017 Conferences

Confederated International Conferences:
CoopIS, C&TC, and ODBASE 2017
Rhodes, Greece, October 23–27, 2017
Proceedings, Part I

 Springer

Editors
Hervé Panetto
University of Lorraine
Nancy
France

Christophe Debruyne
Odisee University College
Brussels
Belgium

Walid Gaaloul
Télécom SudParis
Évry
France

Mike Papazoglou
Tilburg University
Tilburg
The Netherlands

Adrian Paschke
Freie Universität Berlin and Fraunhofer
  FOKUS
Berlin
Germany

Claudio Agostino Ardagna
Università degli Studi di Milano
Crema
Italy

Robert Meersman
TU Graz
Graz
Austria

ISSN 0302-9743   ISSN 1611-3349 (electronic)
Lecture Notes in Computer Science
ISBN 978-3-319-69461-0   ISBN 978-3-319-69462-7   (eBook)
https://doi.org/10.1007/978-3-319-69462-7

Library of Congress Control Number: 2017956721

LNCS Sublibrary: SL2 – Programming and Software Engineering

Printed on acid-free paper

This Springer imprint is published by Springer Nature
The registered company is Springer International Publishing AG
The registered company address is: Gewerbestrasse 11, 6330 Cham, Switzerland

General Co-chairs and Editors' Message for OnTheMove 2017

The OnTheMove 2017 event held October 23–27 in Rhodes, Greece, further consolidated the importance of the series of annual conferences that was started in 2002 in Irvine, California. It then moved to Catania, Sicily in 2003, to Cyprus in 2004 and 2005, Montpellier in 2006, Vilamoura in 2007 and 2009, in 2008 to Monterrey, Mexico, to Heraklion, Crete in 2010 and 2011, Rome 2012, Graz in 2013, Amantea, Italy in 2014 and lastly in Rhodes in 2015 and 2016 as well.

This prime event continues to attract a diverse and relevant selection of today's research worldwide on the scientific concepts underlying new computing paradigms, which of necessity must be distributed, heterogeneous and supporting an environment of resources that are autonomous yet must meaningfully cooperate. Indeed, as such large, complex and networked intelligent information systems become the focus and norm for computing, there continues to be an acute and even increasing need to address the respective software, system, and enterprise issues and discuss them face to face in an integrated forum that covers methodological, semantic, theoretical, and application issues as well. As we all realize, e-mail, the Internet, and even video conferences are not by themselves optimal or even sufficient for effective and efficient scientific exchange.

The OnTheMove (OTM) International Federated Conference series has been created precisely to cover the scientific exchange needs of the communities that work in the broad yet closely connected fundamental technological spectrum of Web-based distributed computing. The OTM program every year covers data and Web semantics, distributed objects, Web services, databases, information systems, enterprise workflow and collaboration, ubiquity, interoperability, mobility, and grid and high-performance computing.

OnTheMove is proud to give meaning to the "federated" aspect in its full title: It aspires to be a primary scientific meeting place where all aspects of research and development of Internet- and intranet-based systems in organizations and for e-business are discussed in a scientifically motivated way, in a forum of interconnected workshops and conferences. This year's 15th edition of the OTM Federated Conferences event therefore once more provided an opportunity for researchers and practitioners to understand, discuss, and publish these developments within the broader context of distributed, ubiquitous computing. To further promote synergy and coherence, the main conferences of OTM 2017 were conceived against a background of their three interlocking global themes:

- Trusted Cloud Computing Infrastructures Emphasizing Security and Privacy
- Technology and Methodology for Data and Knowledge Resources on the (Semantic) Web

– Deployment of Collaborative and Social Computing for and in an Enterprise Context

Originally the federative structure of OTM was formed by the co-location of three related, complementary, and successful main conference series: DOA (Distributed Objects and Applications, held since 1999), covering the relevant infrastructure-enabling technologies, ODBASE (Ontologies, DataBases and Applications of SEmantics, since 2002) covering Web semantics, XML databases and ontologies, and of course CoopIS (Cooperative Information Systems, held since 1993) which studies the application of these technologies in an enterprise context through, e.g., workflow systems and knowledge management. In the 2011 edition security aspects issues, originally started as topics of the IS workshop in OTM 2006, became the focus of DOA as secure virtual infrastructures, further broadened to cover aspects of trust and privacy in so-called Cloud-based systems. As this latter aspect came to dominate agendas in this and overlapping research communities, we decided in 2014 to rename the event as the Cloud and Trusted Computing (C&TC) conference, and originally launched in a workshop format.

These three main conferences specifically seek high-quality, contributions of a more mature nature and encourage researchers to treat their respective topics within a framework that simultaneously incorporates (a) theory, (b) conceptual design and development, (c) methodology and pragmatics, and (d) application in particular case studies and industrial solutions.

As in previous years we again solicited and selected additional quality workshop proposals to complement the more mature and "archival" nature of the main conferences. Our workshops are intended to serve as "incubators" for emergent research results in selected areas related, or becoming related, to the general domain of Web-based distributed computing. This year this difficult and time-consuming job of selecting and coordinating the workshops was brought to a successful end by Ioana Ciuciu, and we were very glad to see that our earlier successful workshops (EI2N, META4eS, FBM) re-appeared in 2017, in some cases in alliance with other older or newly emerging workshops. The Fact Based Modeling (FBM) workshop in 2015 succeeded and expanded the scope of the successful earlier ORM workshop. The Industry Case Studies Program, started in 2011 under the leadership of Hervé Panetto and OMG's Richard Mark Soley, further gained momentum and visibility in its 7th edition this year.

The OTM registration format ("one workshop resp. conference buys all workshops resp. conferences") actively intends to promote synergy between related areas in the field of distributed computing and to stimulate workshop audiences to productively mingle with each other and, optionally, with those of the main conferences. In particular EI2N continues to so create and exploit a visible cross-pollination with CoopIS.

We were very happy to see that in 2017 the number of quality submissions for the OnTheMove Academy (OTMA) noticeably increased. OTMA implements our unique, actively coached and therefore very time- and effort-intensive formula to bring PhD students together, and aims to carry our "vision for the future" in research in the areas covered by OTM. Its 2017 edition was organized and managed by a dedicated team of collaborators and faculty, Peter Spyns, Maria-Esther Vidal, inspired as always by OTMA Dean, Erich Neuhold.

In the OTM Academy, PhD research proposals are submitted by students for peer review; selected submissions and their approaches are to be presented by the students in front of a wider audience at the conference, and are independently and extensively analyzed and discussed in front of this audience by a panel of senior professors. One may readily appreciate the time, effort, and funds invested in this by OnTheMove and especially by the OTMA Faculty.

As the three main conferences and the associated workshops all share the distributed aspects of modern computing systems, they experience the application pull created by the Internet and by the so-called Semantic Web, in particular developments of big data, increased importance of security issues, and the globalization of mobile-based technologies. For ODBASE 2017, the focus somewhat shifted from knowledge bases and methods required for enabling the use of formal semantics in Web-based databases and information systems to applications, especially those within IT-driven communities. For CoopIS 2017, the focus as before was on the interaction of such technologies and methods with business process issues, such as occur in networked organizations and enterprises. These subject areas overlap in a scientifically natural and fascinating fashion and many submissions in fact also covered and exploited the mutual impact among them. For our event C&TC 2017, the primary emphasis was again squarely put on the virtual and security aspects of Web-based computing in the broadest sense. As with the earlier OnTheMove editions, the organizers wanted to stimulate this cross-pollination by a program of engaging keynote speakers from academia and industry and shared by all OTM component events. We are quite proud to list for this year:

- Stephen Mellor, Industrial Internet Consortium, Needham, USA
- Markus Lanthaler, Google, Switzerland

The general downturn in submissions observed in recent years for almost all conferences in computer science and IT has also affected OnTheMove, but this year the harvest again stabilized at a total of 180 submissions for the three main conferences and 40 submissions in total for the workshops. Not only may we indeed again claim success in attracting a representative volume of scientific papers, many from the USA and Asia, but these numbers of course allow the respective Program Committees to again compose a high-quality cross-section of current research in the areas covered by OTM. Acceptance rates vary but the aim was to stay consistently at about one accepted full paper for three submitted, yet as always these rates are subject to professional peer assessment of proper scientific quality.

As usual we separated the proceedings into two volumes with their own titles, one for the main conferences and one for the workshops and posters. But in a different approach to previous years, we decided the latter should appear after the event and thus allow workshop authors to improve their peer-reviewed papers based on the critiques by the Program Committees and on the live interaction at OTM. The resulting additional complexity and effort of editing the proceedings was professionally shouldered by our leading editor, Christophe Debruyne, with the general chairs for the conference volume, and with Ioana Ciuciu and Hervé Panetto for the workshop volume. We are again most grateful to the Springer LNCS team in Heidelberg for their professional support, suggestions, and meticulous collaboration in producing the files and indexes ready for downloading on the USB sticks. It is a pleasure to work with staff that so deeply understands the scientific context at large and the specific logistics of conference proceedings publication.

The reviewing process by the respective OTM Program Committees was performed to professional quality standards: Each paper review in the main conferences was assigned to at least three referees, with arbitrated e-mail discussions in the case of strongly diverging evaluations. It may be worth emphasizing once more that it is an explicit OnTheMove policy that all conference Program Committees and chairs make their selections in a completely sovereign manner, autonomous and independent from any OTM organizational considerations. As in recent years, proceedings in paper form are now only available to be ordered separately.

The general chairs are once more especially grateful to the many people directly or indirectly involved in the set-up of these federated conferences. Not everyone realizes the large number of qualified persons that need to be involved, and the huge amount of work, commitment, and financial risk in the uncertain economic and funding climate of 2017 that is entailed by the organization of an event like OTM. Apart from the persons in their roles mentioned earlier, we therefore wish to thank in particular explicitly our main conference Program Committee chairs:

- CoopIS 2017: Mike Papazoglou, Walid Gaaloul, and Liang Zhang
- ODBASE 2017: Declan O'Sullivan, Joseph Davis, and Satya Sahoo
- C&TC 2017: Adrian Paschke, Hans Weigand, and Nick Bassiliades

And similarly we thank the Program Committee (Co-)chairs of the 2017 ICSP, OTMA and Workshops (in their order of appearance on the website): Peter Spyns, Maria-Esther Vidal, Mario Lezoche, Wided Guédria, Qing Li, Georg Weichhart, Peter Bollen, Hans Mulder, Maurice Nijssen, Anna Fensel, and Ioana Ciuciu. Together with their many Program Committee members, they performed a superb and professional job in managing the difficult yet existential process of peer review and selection of the best papers from the harvest of submissions. We all also owe a significant debt of gratitude to our supremely competent and experienced conference secretariat and technical admin staff in Guadalajara and Dublin, respectively, Daniel Meersman and Christophe Debruyne.

The general conference and workshop co-chairs also thankfully acknowledge the academic freedom, logistic support, and facilities they enjoy from their respective institutions — Technical University of Graz, Austria; Université de Lorraine, Nancy, France; Latrobe University, Melbourne, Australia; and Babes-Bolyai University, Cluj, Romania — without which such a project quite simply would not be feasible. Reader, we do hope that the results of this federated scientific enterprise contribute to your research and your place in the scientific network... and we hope to welcome you at next year's event!

September 2017

Robert Meersman
Hervé Panetto
Christophe Debruyne

Organization

OTM (On The Move) is a federated event involving a series of major international conferences and workshops. These proceedings contain the papers presented at the OTM 2017 Federated conferences, consisting of CoopIS 2017 (Cooperative Information Systems), C&TC 2017 (Cloud and Trusted Computing), and ODBASE 2017 (Ontologies, Databases, and Applications of Semantics).

Executive Committee

General Co-chairs

Robert Meersman	TU Graz, Austria
Tharam Dillon	La Trobe University, Melbourne, Australia
Hervé Panetto	University of Lorraine, France
Ernesto Damiani	Politecnico di Milano, Italy

OnTheMove Academy Dean

Erich Neuhold	University of Vienna, Austria

Industry Case Studies Program Chair

Hervé Panetto	University of Lorraine, France

CoopIS 2017 PC Co-chairs

Mike Papazoglou	European Research Institute in Service Science, Tilburg University, The Netherlands
Walid Gaaloul	TELECOM SudParis, France
Liang Zhang	Fudan University, China

ODBASE 2017 PC Co-chairs

Adrian Paschke	Freie Universität Berlin and Fraunhofer FOKUS, Germany
Nick Bassiliades	Aristotle University of Thessaloniki, Greece
Hans Weigand	Tilburg School of Economics and Management, The Netherlands

C&TC 2017 PC Co-chairs

Claudio Ardagna	Università degli Studi di Milano, Italy
Adrian Belmonte	European Union Agency for Network and Information Security (ENISA), Greece

Konstantinos Markantonakis	Royal Holloway, University of London, UK

Local Organization Chair

Stefanos Gritzalis	University of the Aegean, Greece

Publication Chair

Christophe Debruyne	Odisee University College, Belgium

Logistics Team

Daniel Meersman

CoopIS 2017 Program Committee

Aditya Ghose
Akhil Kumar
Alex Norta
Alfredo Cuzzocrea
Aly Megahed
Amal Elgammal
Amel Bouzeghoub
Amel Mammar
Andreas Andreou
Andreas Oberweis
Andreas Opdahl
Antonio Ruiz Cortés
Arturo Molina
Athman Bouguettaya
Barbara Pernici
Barbara Weber
Beatrice Finance
Bruno Defude
Carlo Combi
Cesare Pautasso
Chengzheng Sun
Chihab Hanachi
Chirine Ghedira
Christian Huemer
Claude Godart
Claudia Diamantini
Daniel Florian
Daniela Grigori
David Carlos Romero Díaz

Djamal Benslimane
Djamel Belaid
Elisabettta di Nitto
Epaminondas Kapetanios
Ernesto Exposito
Eva Kühn
Faiez Gargouri
Farouk Toumani
Francois Charoy
Frank Leymann
Frank-Walter Jäkel
Georg Weichhart
George Samaras
Gerald Oster
Giancarlo Guizzardi
Guido Wirtz
Heiko Ludwig
Heinrich Mayr
Hongji Yang
Imen Grida Ben Yahia
Ivona Brandic
Jan Mendling
Jian Yang
Jiang Cao
Jianwen Su
John Miller
Joonsoo Bae
Jörg Niemöller
Jose Luis Garrido

José Palazzo Moreira de Oliveira
Joyce El Haddad
Juan Manuel Murillo Rodríguez
Juan Manuel Vara Mesa
Julius Köpke
Kais Klai
Karim Baina
Khalid Belhajjame
Khalil Drira
Kostas Magoutis
Lakshmish Ramaswamy
Layth Sliman
Leandro Krug Wives
Liang Zhang
Lijie Wen
Lin Liu
Lucinéia Heloisa Thom
Mahmoud Barhamgi
Manfred Jeusfeld
Manfred Reichert
Marcelo Fantinato
Marco Aiello
Maristella Matera
Marouane Kessentini
Martin Gaedke
Martine Collard
Massimo Mecella
Matthias Klusch
Maurizio Lenzerini
Mehdi Ahmed-Nacer
Michael Mrissa
Michael Rosemann
Michele Missikoff
Mike Papazoglou
Mohamed Graiet
Mohamed Jmaiel
Mohamed Mohamed
Mohamed Sellami

Mohammed Ouzzif
Mohand-Said Hacid
Mourad Kmimech
Narjes Bellamine-Ben Saoud
Nizar Messai
Nour Assy
Oktay Turetken
Olivier Perrin
Oscar Pastor
Pablo Villarreal
Paolo Giorgini
Peter Forbrig
Philippe Merle
Richard Chbeir
Rik Eshuis
Salima Benbernou
Sami Bhiri
Sami Yangui
Samir Tata
Sanjay K. Madria
Selmin Nurcan
Shazia Sadiq
Sherif Sakr
Slim Kallel
Sonia Bergamaschi
Sotiris Koussouris
Stefan Jablonski
Tiziana Catarci
Vassilios Andrikopoulos
Wil M.P. van der Aalst
Walid Gaaloul
Willem-Jan van den Heuvel
Yehia Taher
Youcef Baghdadi
Zakaria Maamar
Zhangbing Zhou
Zohra Bellahsene

ODBASE 2017 Program Committee

Adrian Paschke
Alessandra Mileo
Alexander Artikis
Anastasios Gounaris

Anna Fensel
Annika Hinze
Asuncion Gomez Perez
Athanasios Tsadiras

Bernd Neumayr
Charalampos Bratsas
Christian Kop
Christophe Debruyne
Costin Badica
Danh Le Phuoc
Dietrich Rebholz
Dimitris Plexousakis
Dumitru Roman
Efstratios Kontopoulos
Fotios Kokkoras
Georg Rehm
George Vouros
Georgios Meditskos
Gines Moreno
Giorgos Giannopoulos
Giorgos Stamou
Giorgos Stoilos
Gokhan Coskun
Grigoris Antoniou
Grzegorz J. Nalepa
Hans Weigand
Harald Sack

Harry Halpin
Heiko Paulheim
Ioannis Katakis
Irlán Grangel-González
Kalliopi Kravari
Kia Teymourian
Manolis Koubarakis
Marcin Wylot
Markus Luczak-Roesch
Naouel Karam
Nick Bassiliades
Olga Streibel
Oscar Corcho
Ralph Schäfermeier
Rolf Fricke
Ruben Verborgh
Soren Auer
Sotiris Batsakis
Stefania Costantini
Vadim Ermolayev
Vassilios Peristeras
Witold Abramowicz

C&TC 2017 Program Committee

Marco Anisetti
Claudio A. Ardagna
Rasool Asal
Ioannis Askoxylakis
Adrian Belmonte
Michele Bezzi
David Chadwick
Mauro Conti
Ernesto Damiani
Francesco Di Cerbo
Scharam Dustdar
Nabil El Ioini
Stefanos Gritzalis
Marit Hansen
Sotiris Ioannidis
Martin Jaatun

Meiko Jensen
Gwanggil Jeon
George Karabatis
Antonio Mana
Konstantinos Markantonakis
Raja Naeem Akram
Eugenia Nikolouzou
Claus Pahl
Konstantinos Rantos
Damien Sauveron
Stefan Schulte
Julian Schutte
Daniele Sgandurra
Miguel Vargas Martin
Luca Viganò
Christos Xenakis

OnTheMove 2017 Keynotes

Pragmatic Semantics at Web Scale

Markus Lanthaler

Google, Switzerland

Short Bio

Dr. Markus Lanthaler is a software engineer and tech lead at Google where he currently works on YouTube. He received his Ph.D. in Computer Science from the Graz University of Technology in 2014 for his research on Web APIs and Linked Data. Dr. Lanthaler is one of the core designers of JSON-LD and the inventor of Hydra. He has published several scientific articles, is a frequent speaker at conferences, and chairs the Hydra W3C Community Group.

Talk

Despite huge investments, the traditional Semantic Web stack failed to gain widespread adoption and deliver on its promises. The proposed solutions focused almost exclusively on theoretical purity at the expense of their usability. Both academia and industry ignored for a long time the fact that the Web is more a social creation than a technical one. After a long period of disillusionment, we see a renewed interest in the problems the Semantic Web set out to solve and first practical approaches delivering promising results. More than 30% of all websites contain structured information now. Initiatives such as Schema.org allow, e.g., search engines to extract and understand such data, integrate it, and create knowledge graphs to improve their services.

This talk analyzes the problems that hindered the adoption of the Semantic Web, present new, promising technologies and shows how they might be used to build the foundation of the longstanding vision of a Semantic Web of Services.

Evolution of the Industrial Internet of Things: Preparing for Change

Stephen Mellor

Industrial Internet Consortium, Needham, MA 02492, USA

Short Bio

Stephen Mellor is the Chief Technical Officer for the Industrial Internet Consortium, where he directs the standards requirements and technology & security priorities for the Industrial Internet. In that role, he coordinates the activities of the several engineering, architecture, security and testbed working groups and teams. He also co-chairs both the Definitions, Taxonomy and Reference Architecture workgroup and the Use Cases workgroup for the NIST CPS PWG (National Institute for Standards and Technology Cyberphysical System Public Working Group).

He is a well-known technology consultant on methods for the construction of real-time and embedded systems, a signatory to the Agile Manifesto, and adjunct professor at the Australian National University in Canberra, ACT, Australia. Stephen is the author of Structured Development for Real-Time Systems, Object Lifecycles, Executable UML, and MDA Distilled.

Until recently, he was Chief Scientist of the Embedded Software Division at Mentor Graphics, and founder and past president of Project Technology, Inc., before its acquisition. He participated in multiple UML/modeling-related activities at the Object Management Group (OMG), and was a member of the OMG Architecture Board, which is the final technical gateway for all OMG standards. Stephen was the Chairman of the Advisory Board to IEEE Software for ten years and a two-time Guest Editor of the magazine, most recently for an issue on Model-Driven Development.

Talk

The fundamental technological trends presently are more connectivity and more capability to analyze large quantities of data cheaply. But no one knows where those technological trends will take us, so we need to prepare for change.

Prediction is difficult, especially about the future, as several people are reputed to have said. But this keynote will peer ahead into several areas that we can see need attention, such as:

– Security for everything
– Innovation and funding
– Learning, deployment and competitiveness

We need strategies to prepare for evolution in these areas, and we also need to understand longer term trends. Already we see improvements in operational efficiency, and changes in the economy from pay-per-asset to pay-per-use. More changes are likely, towards pay-per-outcome and direct consumer access to "pull" products autonomously.

These changes will fundamentally change the economy and drive technological innovation. The industrial internet is only at the beginning of perhaps forty more years of change.

Contents – Part I

Contents – Part II

International Conference on Cooperative Information Systems (CoopIS) 2017

CoopIS 2017 PC Co-chairs' Message

Walid Gaaloul, Mike Papazoglou, and Liang Zhang

The venue of CoopIS 2017 (the 25th CoopIS) is Rhodes, Greece. In this edition, CoopIS celebrates a quarter of century of scientific excellence, academic and industrial success. The longevity of CoopIS demonstrates its undeniable success in the scientific and academic community. CoopIS represents one of one of the most qualified and well-established international conferences positioned in a scientific area focusing on the engineering of Cooperative Information Systems (CIS).

CIS paradigm integrates the research results from many related computing areas, such as: distributed systems, coordination technologies, process management, knowledge management, collective decision making, and systems integration technologies. In recent years, several innovative technologies have emerged: Cloud Computing, Service Oriented Computing, Internet of Things, Linked Open Data, mash-ups, Semantic Systems, Collective Awareness Platforms, Processes as a Service, etc. New technologies have fuelled the need for new forms of large-scale social computing, an even tighter integration of data and knowledge with large-scale collaboration platforms, by crowd-sourcing and community-centric cooperation.

Thanks to the many high-quality submissions we were able to put a strong program to celebrate this 25th anniversary. A total of 119 papers were submitted, of which 36 were accepted as full papers (30% acceptance), and 11 as short papers. The authors of the accepted papers are from numerous different countries around the world, thus keeping with the tradition of the international nature of CoopIS. Each paper received at least three, in a number of cases even five, independent peer reviews.

We thank everyone who contributed to the success of CoopIS 2017. We would especially like to thank: (1) the 121 program committee members who have reviewed submissions in a timely manner, providing valuable and constructive feedback (and their expertise) to the authors and assured the quality of the selected papers; (2) the authors who contributed their papers on their research to COOPIS 2017; (3) the Publicity Chair: Zhangbing Zhou; (4) the Business Program Chair: Hervé Panetto, and (7) the local organizing committee.

Characterizing Regulatory Documents and Guidelines Based on Text Mining

Karolin Winter[1][✉], Stefanie Rinderle-Ma[1], Wilfried Grossmann[1],
Ingo Feinerer[2], and Zhendong Ma[3]

[1] Faculty of Computer Science, University of Vienna, Vienna, Austria
{karolin.winter,stefanie.rinderle-ma,wilfried.grossmann}@univie.ac.at
[2] University of Applied Sciences Wiener Neustadt, Wiener Neustadt, Austria
ingo.feinerer@fhwn.ac.at
[3] Center for Digital Safety and Security,
Austrian Institute of Technology, Vienna, Austria
zhendong.ma@ait.ac.at

Abstract. Implementing rules, constraints, and requirements contained in regulatory documents such as standards or guidelines constitutes a mandatory task for organizations and institutions across several domains. Due to the amount of domain-specific information and actions encoded in these documents, organizations often need to establish cooperations between several departments and consulting experts to guide managers and employees in eliciting compliance requirements. Providing computer-based guidance and support for this often costly and tedious compliance task is the aim of this paper. The presented methodology utilizes well-known text mining techniques and clustering algorithms to classify (families) of documents according to topics and to derive significant sentences which support users in understanding and implementing compliance-related documents. Applying the approach to collections of documents from the security and the medical domain demonstrates that text mining is a promising domain-independent mean to provide support to the understanding, extraction, and analysis of regulatory documents.

Keywords: Compliance · Regulatory documents · Requirements extraction · Text mining

1 Introduction

Eliciting and implementing requirements from textual sources, i.e., regulations such as Basel III [13], represents a major challenge for todays businesses and requires a significant effort in terms of time and cost (cf., e.g., [18]). For example, for a company the average cost (respectively duration) to implement the ISO 27001 standard is estimated between $6500 and $26000 (respectively between 6 and 12 months) [16]. It is often a manual and tedious process to interpret, adapt, and implement the clauses from standards and guidelines into appropriate technologies, processes, and actions, supported by consultants. Moreover, the

© Springer International Publishing AG 2017
H. Panetto et al. (Eds.): OTM 2017 Conferences, Part I, LNCS 10573, pp. 3–20, 2017.
https://doi.org/10.1007/978-3-319-69462-7_1

cooperation between consultants and several departments is required in order to correctly interpret and adapt these regulations. Most employees and managers are experts in their field but dealing with these cumbersome documents is a challenge. The sheer breadth, intentional neutrality, and lack of actionable details are the main obstacles to understand and apply the information in a meaningful way. On the other hand, since many of these documents went through a very rigorous drafting and voting process, the use of terminologies and words is carefully thought aiming for maximal precision. This raises an interesting question i.e., *what if computers can be used to assist to understand, interpret, and implement regulatory documents and guidelines and to extract the salient features?* A comprehensive approach towards this question has not been provided yet [21].

In order to access this problem the paper aims at answering the following research questions:

RQ1 *How can standard text mining tools help to understand the topics and content of regulatory documents and guidelines?*
RQ2 *How to improve the results produced by these methods?*
RQ3 *Is it possible to extract sentences which are relevant for implementing such documents?*

For this purpose, a methodology is proposed that constitutes a first step towards covering RQ1, RQ2, and RQ3. One idea of this methodology is the fragmentation of (larger) documents into subdocuments in order to exploit their logical structure and to improve the results of text mining techniques. Another idea is the further analysis of the documents using clustering to group document fragments by topics. The proposed methodology is evaluated based on three case studies whereupon two of them are described in the paper and the third one is added as supplementary material[1]. The first case study features a selection of ISO 27000 standard documents. Contrary, for the second case study only one medical document was chosen to demonstrate the usage of the document fragmentation contained in the methodology. Therefore, it will be possible to outline the feasibility and applicability of the approach to a variety of domains and documents. Privacy documents are the subject of the third case study.

The paper is structured as follows. First, in Sect. 2, the methodology of the approach is presented. Afterwards, the document collections used for the case studies are described in Sects. 3.1 and 3.2. The evaluation of these case studies is issued in Sect. 4 while in Sect. 5 use cases and limitations are discussed. Related work is presented in Sect. 6 and the paper terminates with a conclusion and outline of future work in Sect. 7.

2   Methodology

The methodology presented in this section is depicted in Fig. 1 and was implemented in R^2. The starting point is the collection of selected documents which

---

[1] http://www.wst.univie.ac.at/projects/sprint/index.php?t=tm.
[2] https://www.R-project.org/.

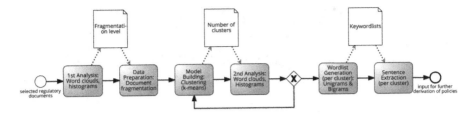

Fig. 1. Characterization of regulatory documents – methodology

is loaded into one corpus. Standard corpus transformations as implemented in the *tm-package*[3] are applied on this basic corpus, for example converting all characters to lower case, removing the numbering, punctuation, stop words as well as customized (stop) words depending on the content and structure of the document (e.g., copyright labels).

To acquire an overview of the topics covered in this set of documents, frequent terms are resolved and represented by word clouds and histograms, applying on the one hand *weightTf* (a weighting scheme using term frequencies) and on the other hand *weightTfIdf* (term frequency — inverse document frequency) [23] (**1st Analysis**). Using *weightTfIdf* is especially suited for document sets of different length as frequencies are normalized [17]. In our context *weightTfIdf* is appropriate due to the wide length spread of the documents within the first case study (cf. Sect. 3.1; ↦ RQ1). Nevertheless, these methods will in general not lead to an in-depth understanding of each document or parts of documents thus further analysis is necessary.

In order to achieve improved analytical results, the idea is to divide the documents into logical units or fragments and to perform analysis based on clustered fragments (**Data Preparation**; ↦ RQ2). It is well-known that text mining techniques deliver better results when applied to a large number of short documents (with respect to the number of words), e.g., tweets, than on few but large documents [11] with a dense information value. Figure 2 provides a categorization of the selected document collections for the case studies (cf. Sect. 3) compared to tweets with respect to two parameters, i.e., number of words per document and number of documents (per collection). The intention is to design the fragmentation in such a way that the resulting partial documents imitate the characteristics of tweets, i.e., few words, but many single documents in order to thin out the information density of the original document. So the goal of the data preparation step is to fragment the documents in such a way that the outcome, i.e., all partial documents considered together is located in the upper left (grey) corner of the figure. Consequently, the fragmentation should be fine-granule enough, but without leading to overly short fragments.

Additionally, it seems to be meaningful to operate on "connected" fragments that reflect the structure of the document. Normally this is suggested by the structuring of a text into sections and subsections. It depends on the documents

---

[3] https://CRAN.R-project.org/package=tm.

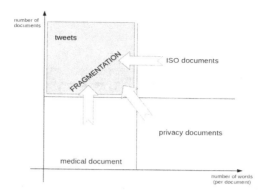

Fig. 2. Categorization of selected (collections of) documents based on their number and size, i.e., number of words per document

which fragmentation level should be chosen. For implementation purposes it can be helpful if the documents have a clear structure and a table of contents, like the ISO documents (cf. R-Script in Sect. 4.1). In this case it is possible to extract (sub-) sections automatically based on the table of contents and suitable regular expressions.

After the fragmentation an optional further step is to merge all partial documents into one corpus on which the above mentioned transformations are applied once more. Now, frequent terms can be computed again in order to check if the splitting has led to a better and more detailed understanding of the documents' topics. Since as a next step clustering methods are used for resolving fragments treating similar topics (**Model building**), this could be helpful when a decision on the number of clusters has to be made.

Why is clustering of the fragments in this case feasible? Families of documents containing guidelines often have content overlaps so it is possible that fragments of different documents may treat the same topic and should therefore be considered together. The opposite is also imaginable and consequently fragments dealing with totally different issues should be separated. Additionally, not much informaion about the documents is available and so clustering is the right choice for this setting [6]. There are multiple clustering algorithms available. The most popular is *k-means*. Choosing the appropriate number of clusters k is not a trivial task since the results of k-means strongly depend on the initial selection of seeds [6]. In the evaluation, the number of clusters k is determined by using the elbow method (cf. [24]) whereupon for each k the average variability (within sum of squares) is computed ten times and the arithmetic mean over these ten runs is used in the elbow plot. In combination with the information from the previous analysis steps a reliable decision on the number of clusters is thus possible.

The evolved clusters define the new corpora, for example, if the number of clusters is 10, also 10 corpora are set up. In the **2nd analysis** word clouds, histograms or dendrograms (not displayed in the paper) can be computed per cluster. These last two steps (Model building & 2nd Analysis) can of course be iterated per cluster if the number of document fragments is very large.

At the end, the topic(s) of document fragments in one cluster can be derived based on which a characterization of the fragmented documents becomes possible. As a final step (**Wordlist generation, Sentence extraction**) wordlists are built (per cluster) and used to extract significant sentences from the fragmented documents in one cluster (cf. Algorithm 1; ↦ RQ3). The sentence extraction enables the outline of potential requirements and implementation guidelines.

Data: clustered fragmented documents
Result: significant sentences per cluster
for *each cluster* **do**

> build a corpus consisting of all documents in the cluster;
> determine (unique) uni- and bigram keywordlists based on frequent terms;
> optional: let user edit these lists;
> perform POS tagging per sentence for documents in the corpus;
> extract sentences containing at least one word (unigram or bigram) present in the keywordlists

end

Algorithm 1. Determine significant sentences for each cluster

---

Algorithm 1 computes per cluster unigram and bigram wordlists. These consist of frequent terms using both *weightTf* and *weightTfIdf*. A user could refine these lists or add terms that might be of importance. Then sentences are tagged using the function *Maxent_Sent_Token_Annotator()* contained in the *R*-package *OpenNLP*[4]. If a sentence contains a word present in the wordlists it is saved for output and the user can view all extracted sentences per cluster in a .txt file.

Overall, the presented methodology combines existing text mining and analytical techniques with a novel document fragmentation approach. The following case studies will illustrate the applicability of the methodology to documents from different domains and show that it faciliates a sound understanding of the fragmented documents.

3   Description of Documents

3.1   Description of ISO/IEC Documents

For the first case study, 13 documents from the ISO 27000 security standard family, i.e., a document composition treating a similar topic (in this case IT security), were selected. This selection consists of ISO 27000_2014, ISO 27001 – ISO 27005, ISO 27010, ISO 27011, ISO 27013, and ISO 27032 – ISO 27035. Document ISO 27000_2014 is an overview document that guides the reader through the more specific topics of the following documents, e.g., guidelines for cybersecurity in

---

[4] https://CRAN.R-project.org/package=openNLP.

ISO 27032. It also contains a glossary with important general terms. Moreover, every document also encloses a collection of terms specifically important for this document. The documents contain between 42 and 136 pages.

A qualitative assessment of the documents was conducted based on an expert interview. The results are summarized in the following:

1. ISO 27001: Overview and vocabulary: $\frac{1}{3}$ is management-related, $\frac{2}{3}$ technical; document has a special role, i.e., it provides a general overview for the management; document contains description of overall process; the components are defined in the other documents
2. ISO 27002: Code of practice; describes actors and roles as well as different aspects of organization (cf. roles)/management (cf. information processing)
3. ISO 27003: Guidance of implementation
4. ISO 27004: Information security management system (ISMS) measurement
5. ISO 27005: Risk assessment, breakdown of risk assessment
6. ISO 27010: Information exchange inter-organizational, inter-sector/inter-organizational communication
7. ISO 27011: Instantiation for telecommunication
8. ISO 27013: Integrated implementation (document per se not so relevant)
9. ISO 27032: Cybersecurity
10. ISO 27033: Network Security
11. ISO 27034: Application Security
12. ISO 27035: Information Security, Incident Management

3.2 Description of Medical Document

The document "Diagnosis and treatment of melanoma: European consensus-based interdisciplinary guideline" [8] consists of 14 pages and contains instructions on cutaneous melanoma diagnosis and treatment. Within a previous case study (cf. [3]) process models and constraints were manually resolved from the document. Therefore it will be possible to compare this manual outcome with the results of the application of the presented methodology. Selecting this document for a second case study is reasonable since it illustrates the variety of domains and document collections the methodology can be applied to (cf. Fig. 2).

4 Evaluation

The methodology outlined in Sect. 2 is applied within three case studies. First, to security documents from the ISO 27000 family (cf. Sect. 3.1) and secondly on a medical document (cf. Sect. 3.2). The third case study as well as all results and figures can be downloaded (see Footnote 1).

4.1   Case Study 1: ISO Documents

As described in Sect. 3.1, 13 documents were chosen for the first case study. All documents are merged into one corpus *securityAll* and the previously described corpus transformations are applied. Following the methodology, depicted in Fig. 1 in a **1st analysis** the most frequent terms are visualized using a word cloud with a maximum of 100 words and a histogram (taking the distribution over the documents into account), always applying *weightTf* as well as *weight-TfIdf*. The results are displayed in Fig. 3a and b.

(a) Word cloud for *securityAll, weightTf* (b) Word cloud for *securityAll, weightTfIdf*

Fig. 3. Word clouds for *securityAll*

What can be recognized but is not surprising is that the terms `information` and `security` are frequent. Also the terms `management`, `iso/iec`, `isms`, `organization`, `risk` as well as `measurement`, `cyberspace`, `cybersecurity` and `telecommunications` occur quite often. Figure 4 shows the distribution of frequent terms (computed for *weightTf*) in each of the selected documents (one color per document). What can be observed is that `telecommunications` only shows up in ISO 27011 whereas `information` and `security` are present in each document. In Fig. 5 *weightTfIdf* was used and the frequent terms are even more correlated to a specific document.

After this first analysis step, all documents are (semi-) automatically split into sections. For this **data preparation** step a R-Script was implemented which (per document)

1. extracts the table of contents via regular expressions; each section headline corresponds to one entry in a vector
2. searches the main part of the document for the section headlines
3. extracts the part in between the section headlines (inclusive headlines)
4. saves the fragments to separate txt files.

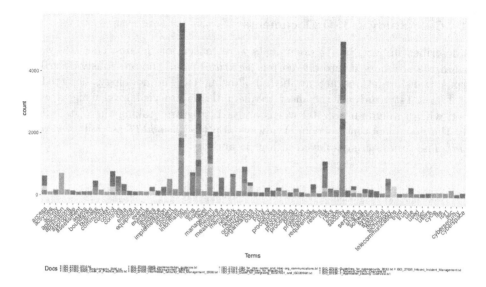

Fig. 4. Histogram for *securityAll, weightTf*

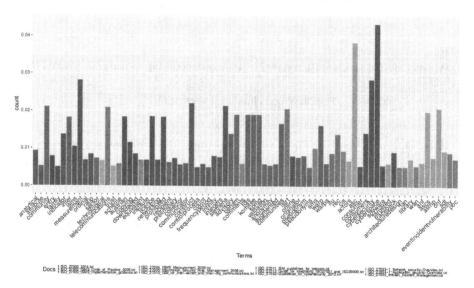

Fig. 5. Histogram for *securityAll, weightTfIdf*

This script is tailored to the structure of the security documents, but its main idea (to perform fragmentation according to the table of contents) can be adopted to fit other document structures. Section headlines are saved twice since they are considered to contain important terms. The fragmentation level was set at section level for all documents in this case study, so in the end 202 fragments are obtained. For the subsequent steps, the introduction, as well as the scope and terms and definitions sections of each document were not included since these fragments would increase the frequency of already frequent terms and would therefore cause noise which should be avoided. Thus the second corpus *securitySections* only contains 129 files.

After preprocessing the second corpus, word clouds and histograms are computed in order to see if the results have improved compared to the first analysis. Only when using *weightTfIdf* a change was noticed. In this case the terms ISMS, inter-organizational, inter-sector, risk and community are more recurrent than for corpus *securityAll*.

For **model building**, the number of clusters for k-means is determined. Therefore an elbow plot is used. For the document-term matrix *weightTfIdf* is applied and the distance measure is the cosine distance. The initial centers are selected randomly and as mentioned in Sect. 2 the average variability (within sum of squares) is computed ten times for each k but the plot (cf. Fig. 6) includes only the arithmetic mean over these ten runs.

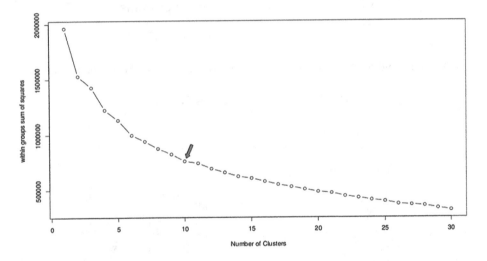

Fig. 6. Elbow plot for *securitySections*

According to Fig. 6, 10 clusters are reasonable, so k-means is performed for $k = 10$. The resulting clusters contain between 7 and 25, in average 12.9 fragments. In a **2nd analysis** for each of these clusters word clouds and histograms are computed. Additionally, the **wordlist determination** is performed in order to **derive significant sentences**.

For showing the feasibility of the methodology the results for one randomly picked cluster are given in the following.

Example Cluster (Security) ($\mathcal{ECS}$): All 9 fragments of this cluster are put into a corpus *corpusECS* and after preprocessing *corpusECS* it can be recognized that the following terms stand out

- unigram word clouds (cf. Fig. 7a and b): `measurement, information, security, isms, management, results, criteria, data, organization, effectiveness, program, base, improvements, assurance, integration, verification, risk`
- bigram word clouds (cf. Fig. 8a and b): `security measurement, measurement results, derived measure, base measure, ensure measurement, data analysis` and `data collection`.

(a) Word cloud for $\mathcal{ECS}$, weightTf

(b) Word cloud for $\mathcal{ECS}$, weightTfIdf

Fig. 7. Unigram word clouds for $\mathcal{ECS}$

(a) Word cloud for $\mathcal{ECS}$, weightTf

(b) Word cloud for $\mathcal{ECS}$, weightTfIdf

Fig. 8. Bigram word clouds for $\mathcal{ECS}$

Based on these results it can be concluded that the documents contained in this cluster treat the measurement and evaluation of ISMS, as well as topics on data collection, storage and handling of ISMS in general. Responsibility assignments are also covered in the fragments. This is checked by inspecting the fragments which are 5. `information security measurement overview`, 6. `management responsibilities`, 7. `measures and measurement development`, 8. `measurement operation`, 9. `data analysis and measurement results reporting`, 10. `information security measurement programme evaluation and improvement`, `annex a`, `annex b` all contained in ISO 27004 and `annex e` from ISO 27003. The observed terms and deduced topics fit the description of ISO 27004 and 27003 in Sect. 3.1.

Now the overall topic of the fragments is known and two specific wordlists (one for unigrams and another for bigrams) can be acquired. This is done by resolving frequent unigrams respectively bigrams with function *freqTerms* implemented in the tm-package. Additionally, a domain expert could extend and refine these wordlists which can now be applied in order to figure out relevant phrases and sentences.

The wordlists for the selected cluster are

- Unigrams: `base, construct, control, criteria, data, decision, indicator, information, isms, management, measure, measurement, method, number, organization, reporting, reserved, results, review, rights, security, specification`.
- Bigrams: `base measure, decision criteria, derived measure, frequency data, information security, management review, measure specification, measurement construct, measurement method, measurement results, object measurement, rights reserved, security measurement, siemens ag, third party`.

Three examples of the derived sentences determined by Algorithm 1 are

- "An organization should develop and implement measurement constructs in order to obtain repeatable, objective and useful results of measurement based on the information security measurement model."
- "The information security measurement programme and the developed measurement construct should ensure that an organization effectively achieves objective and repeatable measurement and provides measurement results for relevant stakeholders to identify needs for improving the implemented isms, including its scope, policies, objectives, controls, processes and procedures."
- "All relevant measures applied to an implemented isms, controls or groups of controls should be implemented based on the selected information needs".

Implementation instructions are clearly contained in these sentences. So it is possible to figure out requirements (semi-) automatically by combining standard text mining tools with fragmentation and clustering of documents.

The ISO documents have significant section titles and so the advantage of the approach concentrates more on the automatic grouping of fragments, the (semi-) automatic deduction of wordlists and the extraction of relevant sentences than giving a first insight of the content.

4.2   Case Study 2: Medical Document

As described in Sect. 3.2, the medical document differs from the ISO document collection. There is only one short document (14 pages) available compared to the ISO documents ($\geq$42 pages). Moreover, it has a different structure, e.g., no table of contents is included. So, fragmentation based on the table of contents is not possible here. Instead, one has to figure out appropriate regular expressions for being able to perform an automatic fragmentation.

After importing the document into the corpus *corpusMedical*, transformations like converting all words to lower case and removing customized words are applied. Performing the **1st analysis** step of the methodology generated the word clouds depicted in Fig. 9. Applying *weightTfIdf* is not reasonable here, since *corpusMedical* contains only one document. Examples of frequent unigrams and bigrams are `melanoma`, `patients`, `metastases`, `treatment`, `cancer`,`therapy`, `et al.`, `lymph node`, `malignant melanoma`, `cutaneous melanoma`.

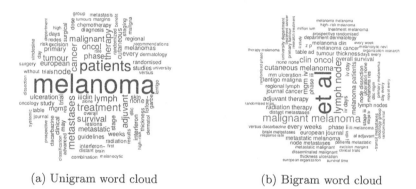

(a) Unigram word cloud   (b) Bigram word cloud

Fig. 9. Word clouds for the medical document

Having a closer look at the remaining terms gives a good impression of the content of the document. One could verify this by e.g., reading the abstract.

For the **data preparation** the document is fragmented. As mentioned before, the document does not contain a table of contents, so the fragmentation is issued via suitable regular expressions. Splitting the document into subsections results in 35 partial documents while fragmentation on section level produces just 7 partial documents. For this analysis the subsection level is chosen in order to produce a fine-granule fragmentation.

As in the first case study an elbow plot for determining the number of clusters is used (cf. Fig. 10; **model building**). Based on this plot $k = 6$ is chosen for k−means clustering and the clusters contain between 3 and 9 fragments. Like before, the results of one randomly picked cluster are given below.

Example Cluster (Medical) ($\mathcal{ECM}$): Following the methodology outlined in Sect. 2 a **2nd analysis** is issued on the five documents contained in the

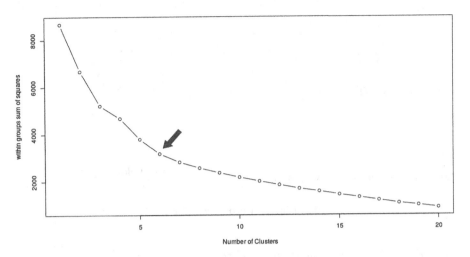

Fig. 10. Elbow plot for the medical document

cluster. For this purpose the documents are merged into a corpus *corpusECM*. Subsequently, corpus transformations are again applied on this corpus and word clouds are computed (cf. Figs. 11 and 12). Here, *weightTfIdf* can be used again since the corpus contains more than one document.

In contrast to the primary word clouds a more detailed description of the documents in $\mathcal{ECM}$ seems to be possible. Here, the frequent terms are not only **metastases** and **therapy** but also different types of metastases like **skin metastases, distant metastases, bone metastases, brain metastases**. The documents in the cluster also seem to contain therapy and treatment suggestions (cf. **patients, survival, radiation, surgical, rate, indications, pain, stability, treatment choice, effectively palliated**).

(a) Unigram word cloud, weightTf   (b) Unigram word cloud, weightTfIdf

Fig. 11. Unigram word clouds for $\mathcal{ECM}$

(a) Bigram word cloud, weightTf

(b) Bigram word cloud, weightTfIdf

Fig. 12. Bigram word clouds for $\mathcal{ECM}$

Keyword lists are generated and used to *extract sentences* resulting in

- "For multiple lesions on a limb, isolated limb perfusion with melphalan ± tumour necrosis factor (TNF) has palliative value."
- "In stage iii patients with satellite/in-transit metastases the procedure can be curative, as indicated by the reported 5- and 10-year survival rates of 40% and 30%, respectively."
- "Even though excision is the treatment of choice for lentigo maligna, radiation therapy may achieve adequate tumour in-transit metastases, which are too extensive for a surgical approach, may be effectively controlled by radiation therapy alone."

Without an expert interview a qualitative assessment of the derived terms and sentences is not possible right away, but reading through the partial documents in the cluster (`3.9 skin metastases`, `3.10 distant metastases`, `4.1 primary melanoma`, `4.4 bone metastases`, `4.5 brain metastases`)   can provide an evaluation for the feasibility of the clustering, i.e., if the topic of the documents in the cluster is derived correctly. This is the case for $\mathcal{ECM}$.

Nevertheless, in order to be able to compare the significance of the derived sentences per cluster, we searched for the cluster with the partial document containing a manually derived subprocess (cf. Fig. 13). In fact, the methodology determined the following sentences for this cluster:

"There is considerable variation in follow-up approaches and few data to support them. In stages I-II melanoma, the intent is to detect loco-regional recurrence early so that the frequency of follow-up examination is usually every 3 months for the first 5 years, whereas for the 6–10th year period investigations every 6 months seem to be adequate."

The second sentence represents a constraint similar to the one depicted by the subprocess in Fig. 13 which leads to the conclusion that it is possible to extract constraints (respectively requirements) using the presented approach. Due to lack of space the third case study based on a selection

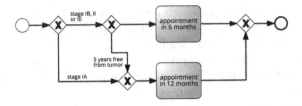

Fig. 13. BPMN model: sonography aftercare subprocess (cf. Fig. 4 in [3])

of privacy documents is not in the paper but can be downloaded (see Footnote 1). Based on the methodology it was possible to group the fragmented documents into clusters having similar topics. For example one cluster treats the legal aspects of privacy and their impact on the society and companies. For this cluster the terms `rights`, `fundamental`, `article`, `impact`, `human`, `protection`, `sources`, `executive`, `fundamental rights`, `impact assessment`, `european union` were derived using word clouds and histograms. Based on these frequent terms, sentences were found that contain implementation instructions.

5 Limitations and Application Scenarios

The evaluation section has demonstrated the applicability of the methodology on several security and one medical document. Many more use cases encountering other domains and document sets are imaginable. But how can organizations and institutions benefit from these results, what limitations might emerge from the methodology and what are application scenarios?

By now, the main focus for visualizing frequent terms is on word clouds and histograms, but for a more in-depth understanding of the documents further techniques might be useful. Dendrograms, i.e., hierarchical clustering of terms are only one possible technique. Another one is to perform association analysis for selected frequent terms and to display the results by, e.g., network maps.

Even though the methodology delivered a reasonable grouping of fragmented documents as well as significant sentences it is necessary to evaluate the completeness of these sentence collections per cluster. It is likely that not every important term is represented by a frequent unigram or bigram. For evaluating the completeness and correctness domain experts should additionally be consulted.

Therefore, in combination with domain knowledge the methodology can extract main characteristics of a series of documents related to planning and implementing of regulatory documents and best practices in existing standards and guidelines. This makes it easier for organizations to accelerate the installation and maintenance of compliance guidelines and related processes, shorten the consulting procedure, and reduce the overall implementation costs. Nevertheless, the methodology does not provide an interpretation of how guidelines should be brought into action in a specific setting.

Another limitation of the approach arises from "hidden" information in pictures and tables. One solution is to use tools that are able to parse such information to make it available for an analysis in R.

Most requirements elicitation approaches demand additional knowledge (cf. Sect. 6). Here, additional knowledge consists of, e.g., overview documents or glossaries, and might decrease implementation effort and increase the output quality during the clustering stage where document fragments are grouped together. Nevertheless, such additional knowledge is not mandatory for this methodology.

6   Related Work

This work touches different areas, i.e., requirements engineering, deriving process-related information from text, and text mining. Thus related work from these areas will be discussed in the following.

Requirements engineering is a broadly investigated field where requirements elicitation constitutes one of the phases in the engineering process. Out of the multiple frameworks for requirements engineering, some approaches suggest (semi-)automated requirements elicitation/identification techniques. The survey presented in [18] has investigated in how far the requirements identification phase is supported in an automated manner. There are some approaches that support or envision (semi-)automatic requirements identification. The majority of these approaches requires additional knowledge, e.g., an ontology. While this is promising for the presented approach as well, the work presented in this paper does not assume additional knowledge. The only automatic approach that does not require additional knowledge (acc. to [18]) is [1]. In contrast to the methodology presented in the paper at hand, [1] assume short forum posts as input. [21] conducted a systematic review on security and privacy requirements elicitation approaches and concluded that there is only little support for automated requirements elicitation. Overall, the approach presented in this work can be seen as support for requirements elicitation and can be employed with any of the requirements engineering frameworks. After extracting requirements, contextualization, i.e., the interpretation of requirements is often useful. An approach using predefined templates is presented in [14].

Implementing requirements contained in regulatory documents often correlates with determining process models and constraints that restrict the implementation. Process model discovery from textual sources is envisaged by several existing approaches (cf. e.g., [2,7,9,10,19,20]), but how to derive these from natural language documents is still an open question (cf. [15]). The reason is that the mentioned approaches have limitations and partly strict requirements on the textual information. The approach presented by [7], for example, requires "the description to be sequential and to contain no questions and little process-irrelevant information". This, in turn, restricts the outcome, i.e., the process models, as well, as real-world processes are often not purely sequential, but involve further patterns such as decisions. For requirements elicitation, in addition, information such as on frequently executed activities can be useful as well (cf. medical case study in Sect. 4.2). Opposed to existing approaches, this work aims at neither imposing restrictions on the text of interest nor on the outcome.

Feinerer et al. [5] outline the prototypical text mining process in R including common corpus transformations as mentioned in Sect. 2. Employed weighting schemes for term-document matrix construction follow best practices as discussed in [17] and [23]. Chunking of long documents in shorter fragments has been proposed by [4] in the context of stylometry of texts. However, their approach focused on better visualization (by having more data points) but ignored the semantic fragmentation we highlight in this paper (where connected fragments and paragraphs capture coherent semantic concepts). Clustering text documents in R was discussed in [12] but focused on clustering techniques and distance measures. We instead use clustering as an explanatory guiding technique that needs specific treatment; e.g., for finding a good number of clusters we used an elbow plot [24] which is useful as an alternative to silhouette plots [22]. The combination of quantitative (classical text mining techniques) and qualitative (expert interview) methods provides hereby a unique contribution for the specific domains considered in this paper.

7   Conclusion and Future Work

This paper outlined a methodology for (semi-) automatically characterizing compliance and regulatory documents by applying well-known text mining and clustering methods like resolving frequent terms or k-means. The evaluation has demonstrated how and to what extent a user can be supported in implementing requirements based on these types of documents.

Future work will encounter conducting user studies in order to further evaluate the usefulness of the methdology for domain experts. The inclusion of topic models in order to improve the results of the presented methodology will be another aspect as well as to try POS tagging for resolving process elements like actions or roles since this assists in implementing requirements more precisely.

References

1. Castro-Herrera, C., Duan, C., Cleland-Huang, J., Mobasher, B.: A recommender system for requirements elicitation in large-scale software projects. In: Symposium on Applied Computing, pp. 1419–1426 (2009)
2. Deeptimahanti, D.K., Babar, M.A.: An automated tool for generating UML models from natural language requirements. In: Proceedings of the 2009 IEEE/ACM International Conference on Automated Software Engineering, ASE 2009, pp. 680–682. IEEE Computer Society, Washington, DC (2009). http://dx.doi.org/10.1109/ASE.2009.48
3. Dunkl, R., Fröschl, K.A., Grossmann, W., Rinderle-Ma, S.: Assessing medical treatment compliance based on formal process modeling. In: Holzinger, A., Simonic, K.-M. (eds.) USAB 2011. LNCS, vol. 7058, pp. 533–546. Springer, Heidelberg (2011). doi:10.1007/978-3-642-25364-5_37

4. Feinerer, I.: An introduction to text mining in R. R News **8**(2), 19–22 (2008). http://CRAN.R-project.org/doc/Rnews/
5. Feinerer, I., Hornik, K., Meyer, D.: Text mining infrastructure in R. J. Stat. Softw. **25**(5), 1–54 (2008)
6. Feldman, R., Sanger, J.: The Text Mining Handbook: Advanced Approaches in Analyzing Unstructured Data. Cambridge University Press, Cambridge (2007)
7. Friedrich, F., Mendling, J., Puhlmann, F.: Process model generation from natural language text. In: Mouratidis, H., Rolland, C. (eds.) CAiSE 2011. LNCS, vol. 6741, pp. 482–496. Springer, Heidelberg (2011). doi:10.1007/978-3-642-21640-4_36
8. Garbe, C., Peris, K., Hauschild, A., Saiag, P., Middleton, M., Spatz, A., Grob, J.J., Malvehy, J., Newton-Bishop, J., Stratigos, A., et al.: Diagnosis and treatment of melanoma: European consensus-based interdisciplinary guideline. Eur. J. Cancer **46**(2), 270–283 (2010)
9. Ghose, A., Koliadis, G., Chueng, A.: Rapid business process discovery (*R*-BPD). In: Parent, C., Schewe, K.-D., Storey, V.C., Thalheim, B. (eds.) ER 2007. LNCS, vol. 4801, pp. 391–406. Springer, Heidelberg (2007). doi:10.1007/978-3-540-75563-0_27
10. Gomez, F., Segami, C., Delaune, C.: A system for the semiautomatic generation of E-R models from natural language specifications. Data Knowl. Eng. **29**(1), 57–81 (1999). http://www.sciencedirect.com/science/article/pii/S0169023X98000329
11. Hill, T., Lewicki, P.: Statistics: Methods and Applications: A Comprehensive Reference for Science, Industry, and Data Mining. StatSoft, Inc., Tulsa (2006)
12. Hornik, K., Feinerer, I., Kober, M., Buchta, C.: Spherical k-means clustering. Journal of Statistical Software **50**(10), 1–22 (2012). http://www.jstatsoft.org/v50/i10
13. Bank for International Settlements: Basel 3: International framework for liquidity risk measurement, standards and monitoring (2010)
14. Koliadis, G., Desai, N.V., Narendra, N.C., Ghose, A.K.: Analyst-mediated contextualization of regulatory policies. In: 2010 IEEE International Conference on Services Computing (SCC), pp. 281–288. IEEE (2010)
15. Leopold, H.: Natural Language in Business Process Models. Springer, Heidelberg (2013)
16. IT Governance Ltd.: ISO 27001 Global Report (2016). http://pribatua.org/wp-content/uploads/2016/08/ISO27001-Global-Report-2016.pdf
17. Manning, C.D., Raghavan, P., Schütze, H.: Introduction to Information Retrieval. Cambridge University Press, New York (2008)
18. Meth, H., Brhel, M., Maedche, A.: The state of the art in automated requirements elicitation. Inf. Softw. Technol. **55**(10), 1695–1709 (2013). https://doi.org/10.1016/j.infsof.2013.03.008
19. More, P., Phalnikar, R.: Generating UML diagrams from natural language specifications. Int. J. Appl. Inf. Syst. **1**(8), 19–23 (2012)
20. Omar, N., Hassan, R., Arshad, H., Sahran, S.: Automation of database design through semantic analysis. In: Proceedings of the 7th WSEAS International Conference on Computational Intelligence, Man-Machine Systems and Cybernetics, CIMMACS, vol. 8, pp. 71–76 (2008)
21. Rinderle-Ma, S., Ma, Z., Madlmayr, B.: Using content analysis for privacy requirement extraction and policy formalization. In: Enterprise Modelling and Information Systems Architectures, pp. 93–107 (2015)
22. Rousseeuw, P.J.: Silhouettes: a graphical aid to the interpretation and validation of cluster analysis. J. Comput. Appl. Math. **20**, 53–65 (1987)
23. Salton, G., Buckley, C.: Term-weighting approaches in automatic text retrieval. Inf. Process. Manag. **24**(5), 513–523 (1988)
24. Thorndike, R.L.: Who belongs in the family? Psychometrika **18**(4), 267–276 (1953)

A Scalable Smart Meter Data Generator Using Spark

Nadeem Iftikhar[1]([✉]), Xiufeng Liu[2], Sergiu Danalachi[1],
Finn Ebertsen Nordbjerg[1], and Jens Henrik Vollesen[1]

[1] University College of Northern Denmark, Aalborg, Denmark
{naif,1028752,fen,hnv}@ucn.dk
[2] Technical University of Denmark, Kongens Lyngby, Denmark
xiuli@dtu.dk

Abstract. Today, smart meters are being used worldwide. As a matter of fact smart meters produce large volumes of data. Thus, it is important for smart meter data management and analytics systems to process petabytes of data. Benchmarking and testing of these systems require scalable data, however, it can be challenging to get large data sets due to privacy and/or data protection regulations. This paper presents a scalable smart meter data generator using Spark that can generate realistic data sets. The proposed data generator is based on a supervised machine learning method that can generate data of any size by using small data sets as seed. Moreover, the generator can preserve the characteristics of data with respect to consumption patterns and user groups. This paper evaluates the proposed data generator in a cluster based environment in order to validate its effectiveness and scalability.

Keywords: Smart meter · Scalable · Synthetic data generator · Time series

1 Introduction

Nowadays, with the popularity of Internet of Things (IoT) and cloud computing, the size of data grows exponentially, posing new challenges to data analysis and management systems, such as the ability to handle petabytes of data. Traditionally, simple benchmarks have been largely used for evaluating the systems in order to prevent unnecessary complexity. On the other hand, we believe that benchmarking should meet a certain diversity and workload requirement for obtaining meaningful results. In addition, it is preferable to use realistic data, however, it is quite challenging to obtain a considerable size of domain dependent data for benchmarking and experimentation purposes. For example, limited public data sets are available in the energy sector. Often, it is difficult to obtain a truthful data source, primarily due to data privacy laws or high data storage cost. Storing petabytes of data is still fairly expensive, although it is much cheaper than before. For example, one TB standard hard drive costs about $80,

© Springer International Publishing AG 2017
H. Panetto et al. (Eds.): OTM 2017 Conferences, Part I, LNCS 10573, pp. 21–36, 2017.
https://doi.org/10.1007/978-3-319-69462-7_2

approximately $0.08 per GB. Similarly, the price for one PB of disk space approximately costs about $80,000. Hence, it is meaningless to store petabyte data only for testing purposes. In addition to data storage, it is also costly to transport large amounts of data over the network, which may consume bandwidth and time. For that reason, scalable data should be produced and used as needed.

In the energy sector, smart meter data management and analysis have received considerable effort in recent years, due to the widespread deployment of smart meters. A smart meter reads energy consumption at a regular time interval, typically every 15 min and sends readings back to an energy data management system for monitoring and billing purposes [1]. Thus, it is essential to evaluate the performance, robustness of energy data management systems and to investigate suitable technologies and algorithms for smart meter data analytics [2–4]. In order to test these systems, it is feasible to generate scalable data sets that should reflect the characteristics of real-world energy consumption patterns. For example, residential energy consumption usually follows a regular pattern based on the consumption habits of a household. Figure 1, illustrates a typical weekly electricity consumption time series from Irish open data [5]. It can be observed that this household have roughly a fixed consumption pattern. The time series has a morning peak roughly at 7–8 o'clock during the workdays. Further, the morning peak delays to around 10 o'clock during the weekend. In the evening, there is a considerable evening peak between 18:00 and 23:00, when all the family members are home and the electric appliances might be turned on, such as dish washer, cooking range, washing machine, television and so on.

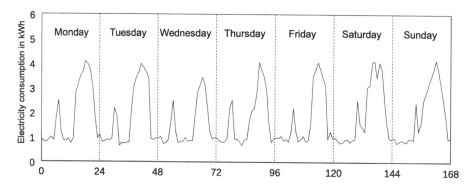

Fig. 1. Weekly consumption pattern of a typical private household

In this paper, we present a scalable data generator that can generate huge volume of realistic synthetic data. The data generator takes as input a real-world energy consumption time series as seed and generates synthetic time series based on historical consumption patterns. In doing so, the generator first creates an adjusted time series using a moving average time series model. The moving average reduces the periodic variations from the actual time series by smoothing the peak periods. Then, it uses autoregressive time series model to predict meter readings. In the end, the periodic variations are added back to the newly predicted meter readings to reflect the pattern and variance of the real-world energy consumption. The data generator is implemented by using the memory-based distributed computing framework, Spark, which can generate scalable data sets on a cluster based environment.

This paper is a significant extension of the previous work [6]. In the previous work, the concept of prediction-based smart meter data generation was introduced, however, it remains to prove that the single machine based data processing platform introduced in [6] also works for cluster-based platform. A scalable data generator is the next step. In this paper, the single machine based technique is extended by introducing the cluster-based technique.

Our main contributions in this paper are as follows:

- We propose a scalable smart meter data generator using Spark.
- We propose a novel method of generating realistic data sets that can preserve the characteristics of real-world energy consumption time series, including patterns and user-groups.
- We evaluate the data generator in terms of effectiveness and scalability of generating scalable data sets, with relatively small data as seed.

The paper is structured as follows. Section 2 describes the methodology used by the proposed data generator. Section 3 describes the implementation on Spark. Section 4 evaluates the generator. Section 5 presents the related work. Section 6 concludes the paper and points out the future research directions.

2   Methodology

2.1   Overview

We now describe the rationale of the proposed data generation solution. The solution uses a *quantitative model*, expressed in mathematical notation. The quantitative model is further divided into a causal model and a time-series model, where the latter is chosen for modeling the consumption time series. The time series model produces predictions according to historical consumption patterns. The time series of residential energy consumption normally comprises the following patterns: *trend*, *cyclic* and *seasonal/periodic*. The periodic pattern is usually resulted from the periodical factors such as the days, which have a fixed and known period [7], e.g., 24-hour. Therefore, it is possible to generate consumption time series with these pattern characteristics.

Further, Fig. 2 gives an overview of the data generation process. The data generation is seeded by a small real-world data set. First, the seed data is deseasonalized in order to flatten the periodic variations. Next, a regression model is trained using the flattened time series. This model is then used to predict new consumption values. In the end, the generated time series is reseasonalized, in other words, the periodic variations are added back. The rationale of using the adjusted periodic variations is that the data that does not have or has reduced

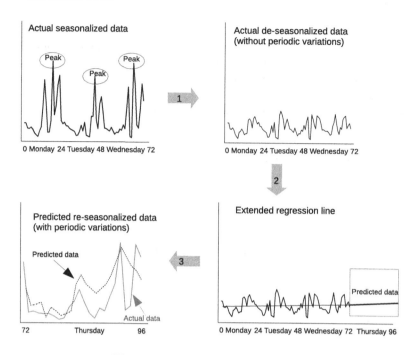

Fig. 2. Data generation overview [6]

periodic variations can lead to more accurate predictions than with variations [8]. The time series with reduced periodic variations also allows us to determine the best regression model for the prediction.

Furthermore, there are two ways of representing energy consumption. First, a smart energy meter measures a cumulative consumption, i.e., the consumption always increases. Second, a smart meter measures consumption in a given (fixed) interval. i.e., an aggregated value in a time window, e.g., 30 min. The generator proposed in this paper is based on the second approach.

2.2   Algorithm Description

We now describe the data generation process and the algorithms used. The data generation process comprises of two methods: training process and generation process. The training process includes flattening of time series fluctuations, deseasonalization and generation of data models, while the generation process includes generating data using the model and reseasonalization. Both of the processes are described in the following subsections.

Training Process. For the proposed data generator, we consider generating data based on daily consumption profiles. During the training process (see Algorithm 1), each time series from the seed data set will be transformed into a key-value pair, of which *meterID* is the key, and the list of meter readings is the

Algorithm 1. Training process

1. Transform a time series into a *key-value* pair.
2. Process the *key-value* pair:
   - (i) Flatten fluctuations by *centered moving averaging* method.
   - (ii) Deseasonalize time series by *periodic indexing* method.
   - (iii) Train *autoregressive (AR)* model for predictions using the deseasonalized *time series*.
   - (iv) Write the output: *meterIDs*, *periodic-indices*, *AR-coefficients* and *flatten-time-series*.

value. The readings in the list are sorted in an ascending order according to the timestamps.

Next, the key-value pair is processed through the following four steps (Algorithm 1) that include flattening of periodic fluctuations, deseasonalization, autoregression and writing the output:

(i) Flatten Periodic Fluctuations: We use the *centered moving averaging (CMA)* method to reduce the impact of periodic fluctuations [9]. CMA replaces the original time series with a new flatten time series where each point is centered at the middle of the data values being averaged.

For the daily profile (24-hour), the CMA of an even period is defined as:

$$A(i) = \frac{1}{2} \left(\frac{y_{i-12} + .. + y_i + .. + y_{i+11}}{24} \right) + \frac{1}{2} \left(\frac{y_{i-11} + .. + y_i + .. + y_{i+12}}{24} \right) \tag{1}$$

where y_i is the i-th observation in a time series of the seed data set.

(ii) Deseasonalization: To deseasonalize a time series, we first need to compute the *raw-index* or *Ratio-to-Moving-Average*, which is computed as below:

$$R(i) = \frac{y_i}{A(i)} \tag{2}$$

We then compute the *periodic indices* by using the resulting raw index values (see Eq. 3). For each hour of the day, a corresponding periodic index is computed, which is the mean value of all the raw index values at that particular hour. For example, $P(0)$ represents the mean of all R values at 0 o'clock in all days for a given time series. Therefore, the total number of resulting periodic indices will be 24.

$$P(h) = \frac{1}{n} \sum_{i=0}^{n-1} R(h + 24i) \tag{3}$$

where, n represents the total number of days for each meter in the time series, and h is the hour of the day, i.e., 0–23. Since there are some chances to encounter data precision problems, e.g., due to the floating point, we need to adjust the computed $\mathcal{P}$ value [10]. Equation 4 normalizes the periodic indices, which ensures that the sum of the adjusted $\mathcal{P}'$ values is 1.0.

$$\mathcal{P}'(h) = \frac{24 * \mathcal{P}(h)}{\sum_{h=0}^{23} \mathcal{P}(h)} \tag{4}$$

In the end, we use this adjusted periodic indice to deseasonalize a time series, which simply divides each data point of the time series (see Eq. 5).

$$y_i' = \frac{y_i}{\mathcal{P}'(h)} \tag{5}$$

where $h = i \bmod 24$ and $\mathcal{P}'$ is the normalized periodic indices.

(iii) Training Autogressive Model and (iv) Writing Output. In the end, we use the flatten (deseasonlized) time series to train an autoregressive model and this model will be used to generate new values by prediction. The resulting coefficients of AR model, the periodic indices and the flatten-time-series, $\{y_i' | i = 0, ..., n-1\}$, will be written to the Hadoop distributed file system (HDFS). The results are stored into two separate files, with the formats of *(meterID, periodic-indices)* and *(meterID, (AR-coefficients, flatten-time-series))*. The reason to save the results for the same meterID into two separate files is to make the data generation model flexible enough to generate synthetic time series with different variances. In this case, the periodic indices could be from a separate time series within the same cluster. In the data generation process, these two files will served as input.

Generation Process. Algorithm 2, describes the data generation process. The data generator uses the files (generated from the preprocessing process) as input. The data from the two files are read as two Resilient Distributed Datasets (RDDs), $\mathcal{PI}$ and $\mathcal{AR}$ in Spark. The *theta join* [11] will apply on the two tables (RDDs) at the condition that the meterIDs are not equal. For each record of the join results, we apply the following three steps to generate a new time series:

Algorithm 2. Data generation

---

1. Read the data from the two input files and create Spark tables *(RDDs)*: $\mathcal{PI}$=*(meterID, peridoc-indices)* and $\mathcal{AR}$ = *(meterID, AR-coefficients, flatten-time-series)*.
2. Perform the *Theta join* on $\mathcal{PI}$ and $\mathcal{AR}$ where $\mathcal{PI}.meterID \neq \mathcal{AR}.meterID$.
3. For-each query results:
    (i) Predict a new reading by using the AR model and the flatten time series values.
    (ii) Reseasonalize the new reading.
    (iii) Add *base load* and *white noise* to the reseasonalised reading in order to simulate reality.

(i) Generate New Reading: We use the AR model and the values from flatten time series (with the order of p) to generate a new value, which is expressed in the following equation:

$$y_i'' = c + \sum_{\lambda=1}^{p} \alpha_i y_{i-\lambda}' \tag{6}$$

where c is the intercept with the y-axis (a constant), α is the AR coefficient and y_i' are the last values from the flatten time series of (with p consecutive values before i).

(ii) Reseasonalization and (iii) Add Base Load and White Noise: The final resulting time series is expressed in Eq. 7.

$$y_i''' = y_i'' * \mathcal{P}'(h) + baseLoad + \epsilon_i \tag{7}$$

where $h = i \bmod 24$ and $i = 0, ..., n$.

The reseasonalization is simply multiplying the adjusted periodic index. In the generated time series, we add a base load, which is a constant value greater or equal to zero. A base load typically represents the energy consumed by the appliance that is always on, e.g., refrigerator. And, we add a Gaussian white noise, $\epsilon \sim \mathcal{N}(0, 1.0)$, to simulate slight variations.

2.3 Optimization

We now optimize our data generator in order to better simulate the real-world data. As mentioned in Sect. 1, energy consumption data follows a certain pattern, due to the daily routine of a household, e.g., having a daily pattern with morning and evening peaks. Moreover, the time series of different households may have similar patterns, which can be identified by grouping/clustering. This technique is often used by utilities to segment the customers in order to offer personalized energy-efficiency services. In order not to lose this information, we optimize data generation by adding the pre-processing process (see Fig. 3). The pre-processing will first cluster the seed, then uses the clustered data for training the models. Recall that in the data generation process, we use the theta join on the resulting models to create data generators. If the models were not generated by the clustered seed, the resulting synthetic data may lose the clustering information.

Moreover, clustering the seed time series according to daily patterns is a two step process: First, we find the typical daily load pattern for each time series, which is done by averaging the consumption of each hour for all days. This results the following averaging load of daily profile for the i-th time series:

$$\mathcal{TS}_i = \{r_{i,0}, r_{i,1}, .., r_{i,23}\} \tag{8}$$

where r represents the average consumption of a meter at each hour of the day, h.

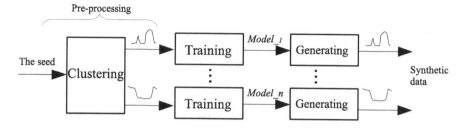

Fig. 3. Optimize data generation with the pre-processing of the seed

Second, we cluster the daily load patterns of all time series using k-means clustering algorithm [12]. In general, k-means clustering algorithm uses Euclidean distance, e.g., [13,14], which is defined as follow. Suppose there are two daily load profiles of $\mathcal{TS}_i$ and $\mathcal{TS}_j$, the distance is

$$euclDist\left(\mathcal{TS}_i, \mathcal{TS}_j\right) = \sqrt{\sum_{h=0}^{23}\left(r_{i,h} - r_{j,h}\right)^2} \tag{9}$$

However, using the Euclidean distance may still not the best to reflect similarity of two load patterns. For example, Fig. 4(a) and (b) both have the Euclidean distance of $\sqrt{3}$, however, the patterns in Fig. 4(b) are totally different.

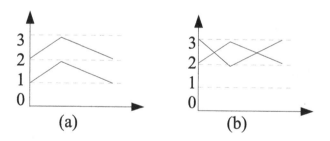

Fig. 4. The two patterns with the same Euclidean distance of $\sqrt{3}$

To further optimize, we adopt the Pearson correlation distance [15], which measures the distance based on the correlation between two patterns. The correlation is defined as follow:

$$corr\left(\mathcal{TS}_i, \mathcal{TS}_j\right) = \frac{\sum_{h=0}^{23}\left(r_{i,h} - \mu_i\right)\left(r_{j,h} - \mu_j\right)}{\sqrt{\sum_{h=0}^{23}\left(r_{i,h} - \mu_i\right)^2}\sqrt{\sum_{h=0}^{23}\left(r_{j,h} - \mu_j\right)^2}} \tag{10}$$

where μ represents the daily average consumption for each meter.

The correlation distance is defined as:

$$corrDist\left(\mathcal{TS}_i, \mathcal{TS}_j\right) = 1 - corr\left(\mathcal{TS}_i, \mathcal{TS}_j\right) \tag{11}$$

The distance of zero represents perfectly correlated (correlation $= 1$) time series. The distance of less than approximately 0.5 indicates that there is a good similarity between two patterns, while the distance of 2 (correlation $= -1$) indicates having an opposite pattern.

3 Implementation on Spark

The proposed data generator is implemented into two modules, training module and data generation module, which are both implemented using *Spark* for generating scalable data. The implementations are described as follows.

The seed data have been processed by grouping/clustering. The training process will take a clustered seed data as the input to create the models. Listing 1.1 shows the code snippet of training process, which takes the parameters of *inputPath*, *outputPath* and *frequency* (line 1). The input path locates a clustered seed data that comprise a set of time series with similar daily consumption patterns. The output path denotes the location of saving the resulting models in HDFS and the frequency indicates the number of occurrences of a meter reading per unit time. For example, frequency $= 48$ represents the reading frequency per day, since the meter is read every 30 min. The input files are the CSV files with the format of *(meterID, timestamp, reading)*, where meterID is taken as the *key* and (timestamp, reading) is taken as the *value*. The function (line 3) will sort and group the readings based on meter id and time as well as cache the data in memory for iterative processing. Second, the periodic indices are computed for each time series, and the seed is deseasonalized (lines 6–7). Third, the AR model is trained (by using the spark-timeseries library) using the deseasonalized time series (line 8). Fourth, three deseasonalized lagged (past period) readings are extracted (with order $= 3$), which will be used for forecasting the new value in the data generation process (line 9). Fifth, the results are mapped as periodic indices, coefficients and lagged readings (line 10). Sixth, the results with undefined coefficients are filtered out (line 12). Last, the results are stored to HDFS directly (lines 14–15).

The training process is run only once for each clustered data set from the seed. The two resulting files have the following format: <*meter identifier, periodic indices*> and <*meter identifier, AR-coefficients, flatten-time-series*>. An example of the rows are <*1460, 1.619, 1.353, 1.208, 0.982,..., 1.776*> and <*1460, 0.224, 0.584, −0.111, 0.095, 0.180, 0.184, 0.195*>. The first row represents that a meter (with meterID $= 1460$) has 48 periodic indices (as the number of occurrences of a meter reading is per half-hour). The second row represents that the meter (with meterID $= 1406$) has an intercept, three AR coefficients (with order $= 3$), and last three lagged readings of the deseasonalized seed data set.

```
1  train(inputPath, outputPath, frequency)
2  {
3    seed = getReadingsForEachMeterID(inputPath).cache()
4
5    output = seed.mapValues(readings => {
6        PI = getPeriodicIndices(readings, frequency)
7        DS = getDeseasonalizedSeed(readings, PI)
8        coefficients = ARIMA.fitModel(3, 0, 0, DS, true).coefficients
9        lagged = Vectors.dense(DS.takeRight(3))
10       (PI, coefficients, lagged)})
11
12   .filter(!coefficients(0).isNaN)
13
14   output.map(tuple => (meterID, PI)).save(outputPath+"/PI")
15   output.map(tuple => (meterID, (coefficients, lagged))).save(
         outputPath+"/AR")
16 }
```

Listing 1.1. The code snippet of training

The implementation of data generation is shown in Listing 1.2, which takes the resulting models as the input (indicated by inputPath) as well as other parameters including the outputPath, the frequency, the number of time series to generate, the number of days and base load. The program first reads the period indices (PI) and Autogressive models (AR) from the input files into the memory (line 3–4). Then, it does the theta join and returns the desired number of rows (equal to the number of generated time series) (line 6). Third, it does the forecasting using the AR model (line 8–9) and the resulting predicted value is reseasonlized. In addition, the base load and the white noise is also added in the predicted value to simulate reality (line 11). Last, the generated data is written to HDFS (line 15).

The synthetic data has the format of <*meter identifier, timestamp, reading*> and an example of the rows is <*100, 201706041900,0.389*>, representing that a meter (with meterID = 100) has used 0.3 kWh electricity in the previous half an hour.

```
1  generate(inputPath, outputPath, frequency, nTimeSeries, nDays,
         baseLoad)
2  {
3    PI = readPI(inputPath+"/PI")
4    AR = readAR(inputPath+"/AR")
5
6    results = thetaJoin(PI, AR).get(nTimeSeries)
7      .map((meterId, (coefficients, lagged, PI)) => {
8        newValues = new ARIMAModel(3, 0, 0, coefficients, true)
9          .forecast(lagged, frequency * nDays)
10       .map(x => {
11         reading = x * PI(hour) + baseLoad + Random.nextGaussian()
12         reading})
13       (meterId, newValues)})
14
15   results.save(outputPath)
16 }
```

Listing 1.2. The code snippet of data generation

4 Evaluation

In this section, we evaluate the data generator in terms of effectiveness and scalability. The effectiveness will be evaluated by comparing the patterns between the real-world and synthetic data. The scalability will be evaluated by measuring the execution performance. The Irish electricity consumption will be used as the seed for training the models.

The experiments are conducted on a 4-node cluster: all the nodes act as slave, and one of them also acts as master. All the machines have the same settings: Intel(R) Xeon(R) CPU E5-2650 (3.40 GHz, 4 Cores, hyper-threading is enabled, two hyper-threads per core), 8 GB RAM, and a Seagate Hard driver (1 TB, 6 GB/s, 32 MB Cache and 7200 RPM), running 64 bit-Ubuntu 12.04 LTS with Linux 3.19.0 kernel.

4.1   Effectiveness

We now evaluate the effectiveness of the proposed smart meter data generator. As mentioned in Sect. 2.3, the data generator first, uses clustered data as the seed to generate the models, then it generates time series. We use the correlation distance metric for the clustering in the pre-processing of the seed. Before validating the generated time series, we would like to further explain by demonstrating a real example.

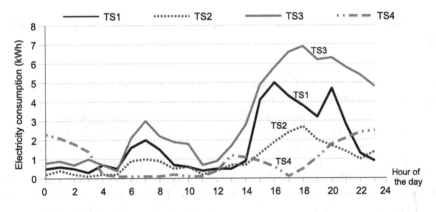

Fig. 5. Daily activity load profile time series

Figure 5, demonstrates four daily load profiles from different households. TS_1 represents a medium energy use household; TS_2 represents a low energy use household, whereas, TS_3 represents a high energy use household. Visually, we could observe that TS_1, TS_2 and TS_3 have a similar pattern, e.g., with morning and evening peaks almost at the same range of the time, although they are within different consumption categories. In contrast, TS_4 is showing a quite different pattern, without morning peak. Hence, according to the consumption patterns, TS_1, TS_2 and TS_3 should be assigned to the same group regardless of their consumption amount, while TS_4 should belong to a different group.

In order to assign the time series to the desired cluster based on the similarity, we compute the distance function. Euclidean function is commonly used as a distance function when performing the clustering. In Sect. 2.3, we have

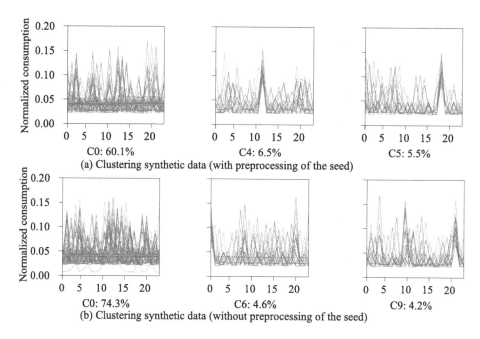

Fig. 6. Comparison of the pattern preservation with and without reprocessing of the seed

mentioned that Euclidean function may not give accurate results and we have recommended to use correlation based distance function instead.

Table 1, shows the comparison between the two distance functions. If we observe the distances, the correlation distances between (TS_1, TS_2) and (TS_1, TS_3) are smaller than the distance between (TS_1, TS_4). The reason that TS_1, TS_2 and TS_3 have smaller correlation distances is due to the fact that they have similar patterns, whereas, TS_4 has a larger distance for the reason that it has a different pattern with respect to TS_1, TS_2 and TS_3 (note that the distance of zero means perfectly correlated). In contrast, the Euclidean distance between (TS_2, TS_4) is the smallest, which may result in wrongly assigning TS_4 to the same group as TS_2. Thus, it is more preferable to choose the correlation distance.

We now demonstrate the importance of preprocessing the seed in order to preserve the information of customer segmentation. We compare the clustering information of the resulting synthetic data sets when we use the seed with and without being preprocessed. We cluster the daily patterns into 20 clusters for

Table 1. Comparison of the two distance metrics

	(TS_1, TS_2)	(TS_1, TS_3)	(TS_1, TS_4)	(TS_2, TS_3)	(TS_2, TS_4)	(TS_3, TS_4)
euclDist	6.13	9.12	9.64	11.5	4.73	12.4
corrDist	0.12	0.13	1.06	0.12	0.76	1.10

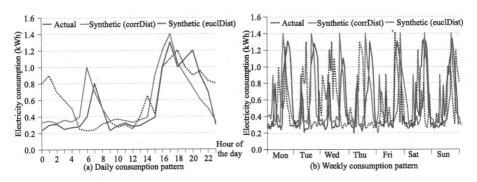

Fig. 7. Comparison of consumption patterns

the two data sets using the adaptive clustering method [16] and compare the top three clusters shown in Fig. 6(a) and (b). According to the top three clusters, we could observe that the patterns are more visible in Fig. 6(a) (where the seed is preprocessed) as compared to Fig. 6(b) (where seed is not preprocessed). Based on these observation, we can conclude that the data generator trained with preprocessed seed can achieve better pattern preservation.

Further, we evaluate the effectiveness by comparing the patterns of the real-world and synthetic data. Figure 7(a) and (b), show the daily and weekly patterns generated from a typical household, respectively. We compare the patterns of the actual and synthetic data. The synthetic data is generated by the data generators trained by clustered seed using corrDist and euclDist. The actual pattern in Fig. 7(a) shows that there is a morning peak (6–9) and a evening peak (16–21) in the pattern. The pattern of synthetic (corrDist) indicates a good matching to the actual pattern, with slight drift. In contrast, the synthetic (euclDist) does not show a perfect fit, for example, having a peak at 1–2 o'clock but there is no peak for the actual pattern. Figure 7(b) shows the weekly patterns, where synthetic (corrDist) also shows better than the synthetic (euclDist) to fit the actual data pattern.

4.2   Scalability

In this section, we evaluate the scalability of the proposed data generator. Note that this study will not measure the execution time of the preprocessing and the training process for the reason that they are performed only once, thus their results can be reused during the data generation process. Figure 8, shows the execution time of generating the data scaled from 50 to 300 GB using all nodes (a total of 16 cores). The results show that the execution time increases almost linearly with the size of the data generated.

Figure 9, shows the speedup of generating a fixed size of data set (100 GB) by varying the number of cores. The speedup is calculated as follow: $speedup = t_4/t_n$, where t_4 is the execution time with 4 parallel cores, and t_n is the execution time with n parallel cores (n with the values of 4, 8, 12 and 16). According to the results, the data generator can achieve a good speedup, when the number of cores increased to 16.

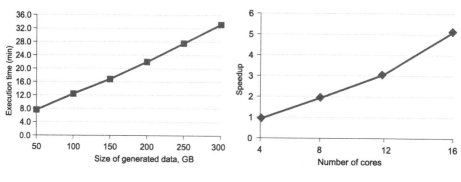

Fig. 8. Scale-up

Fig. 9. Speedup of generating 100 GB data

To summarize, the proposed data generator has the ability to generate realistic time series data with a good performance and the generated data has comparable characteristics with the actual data, in terms of patterns and groups/clusters.

5   Related Work

Synthetic data generation has been studied extensively across several disciplines. *DBGEN* is a well-known data generation tool that can generate up to 10 TB of data for the TPC-H/R database schema [17]. Similarly, synthetic weather data generation has also been extensively studied by [18–22]. The weather generators typically use stochastic models to simulate synthetic weather data. Furthermore, a vehicle crash data generator uses actual vehicle crash data as seed to produce new realistic data using Fourier transformation [23]. The generated data contains different acceleration peaks to test and verify crash management components in a car without running actual crash tests. Time series forecasting has also attracted much research attention in recent years. A hybrid time series forecasting model based on autoregressive integrated moving average (ARIMA) and neural networks is proposed by [24]. Likewise, a periodic autoregressive moving average model (PARMA) for time series forecasting is also suggested by [25]. PARMA model can explicitly describe seasonal/periodic fluctuations in terms of mean, standard deviation and autocorrelation. Based on that, PARMA derives more realistic time series forecasting models and simulations. In addition, a template-based time series generation tool (loom) that utilizes ARIMA as the underlying forecasting model is presented by [26]. Additionally, a survey is conducted on the forecasting models by [27]. It has reported that ARIMA and neural networks are heavily used in time series forecasting. Based on all these works, it can be concluded that models such as stochastic, ARIMA, PARMA, neural networks play a crucial role in time series forecasting. In resemblance with these works, the foundation of the proposed data generator is based on autoregressive centered moving average (ARCMA) model.

Smart metering, as an emerging technology has gained widespread attention recently. A lot of work has been reported in the area of smart meter data analytics, however, to the best of our knowledge, the smart meter synthetic data generation still needs to be extensively studied. Some literature has been found with respect to smart meter synthetic data generation by [2, 4, 28]. The work by [28], uses Markov chain model, while [2, 4] use periodic auto-regression (PAR) to generate synthetic time series in order to benchmark Internet of Things (IoT) and smart meter analytics systems. In contrast to all these works, the focus of the current work is to generate time series based on energy consumption patterns, in a distributed data processing environment.

6   Conclusions and Future Work

Smart meter data management and analytics systems require a large amount of data for benchmarking and testing purposes. In this paper, we have presented a scalable smart meter data generator using the Spark framework. We have used the supervised machine learning method to create the models for generating synthetic data. In addition, we have introduced an optimization method that preserves user-groups/clusters information, i.e., using clustered seeds. We have comprehensively evaluated the data generator by comparing its effectiveness and scalability. The results have demonstrated that the data generator can generate scalable smart meter data that can simulates well to the reality.

For the future work, we could consider to add more features to the data generation models, for example, seasonality (winter, spring, summer and autumn). In addition, the current generator could be extended or modified to generate other types of meter data, such as water, gas and heating.

Acknowledgement. This research is supported by UCN-FOU funding (Project-6/2016-17) and the CITIES project by Danish Innovation Fund (1035-00027B).

References

1. Smart Meter From Wikipedia. https://en.wikipedia.org/wiki/Smart_meter
2. Liu, X., Golab, L., Golab, W., Ilyas, I.F.: Benchmarking smart meter data analytics. In: Proceedings of the 18th International Conference on Extending Database Technology, pp. 385–396 (2015)
3. Liu, X., Golab, L., Golab, W., Ilyas, I.F., Jin, S.: Smart meter data analytics: systems, algorithms, and benchmarking. In: ACM Transactions on Database Systems (TODS), **42**(1), Article no. 2. ACM Press, New York (2017)
4. Liu, X., Golab, L., Ilyas, I.F.: SMAS: a smart meter data analysis system (Demo). In: Proceedings of the 31st International Conference on Data Engineering, pp. 147–1479 (2015)
5. ISSDA. www.ucd.ie/issda/data/commissionforenergyregulationcer
6. Iftikhar, N., Liu, X., Nordbjerg, F.E., Danalachi, S.: A prediction-based smart meter data generator. In: 19th International Conference on Network-Based Information Systems, pp. 173–180. IEEE (2016)

7. Time Series Components. www.otexts.org/fpp/6/1
8. Zhang, G.P., Qi, M.: Neural network forecasting for seasonal and trend time series. Eur. J. Oper. Res. **160**(2), 501–514 (2005)
9. Weiers, R.: Introduction to Business Statistics. Cengage Learning, Boston (2010)
10. Lawrence, K.D., Klimberg, R.K., Lawrence, S.M.: Fundamentals of Forecasting using Excel. Industrial Press Inc., Norwalk (2009)
11. Okcan, A., Riedewald, M.: Processing theta-joins using MapReduce. In: Proceedings of SIGMOD, pp. 949–960 (2011)
12. Wu, J.: Advances in K-means Clustering: A Data Mining Thinking. Springer Science & Business Media, Heidelberg (2012)
13. Parsian, M.: Data Algorithms: Recipes for Scaling Up with Hadoop and Spark. O'Reilly Media Inc., Sebastopol (2015)
14. Liao, T.W.: Clustering of time series data—a survey. Pattern Recogn. **38**(11), 1857–1874 (2005)
15. Black, K.: Business Statistics: For Contemporary Decision Making. Wiley, Hoboken (2011)
16. Peng, B., Wan, C., Dong, S., Lin, J., Song, Y., Zhang, Y., Xiong, J.: A two-stage pattern recognition method for electric customer classification in smart grid. In: Smart Grid Communications (SmartGridComm), pp. 758–763 (2016)
17. Poess, M., Floyd, C.: New TPC benchmarks for decision support and web commerce. ACM Sigmod Rec. **29**(4), 64–71 (2000)
18. Breinl, K., Turkington, T., Stowasser, M.: Simulating daily precipitation and temperature: a weather generation framework for assessing hydrometeorological hazards. Meteorol. Appl. **22**(3), 334–347 (2014)
19. Li, Z., Brissette, F., Chen, J.: Finding the most appropriate precipitation probability distribution for stochastic weather generation and hydrological modeling in nordic watersheds. Hydrol. Process. **27**(25), 3718–3729 (2013)
20. Breinl, K., Turkington, T., Stowasser, M.: A weather generator for hydrometeorological hazard applications EGU general assembly conference. In: EGU General Assembly Conference Abstracts, vol. 16, p. 10522 (2014)
21. van Paassen, A.H., Luo, Q.X.: Weather data generator to study climate change on buildings. Build. Serv. Eng. Res. Technol. **23**(4), 251–258 (2002)
22. Shamshad, A., Bawadi, M.A., Hussin, W.W., Majid, T.A., Sanusi, S.A.M.: First and second order markov chain models for synthetic generation of wind speed time series. Energy **30**(5), 693–708 (2005)
23. Cuddihy, M.A., Drummond Jr., J.B., Bourquin, D.J.: Ford motor company, vehicle crash data generator. U.S. Patent No. 5,608,629 (1997)
24. Zhang, G.P.: Time series forecasting using a hybrid ARIMA and neural network model. Neurocomputing **50**, 159–175 (2003)
25. Anderson, P.L., Meerschaert, M.M., Zhang, K.: Forecasting with prediction intervals for periodic autoregressive moving average models. J. Time Ser. Anal. **34**(2), 187–193 (2013)
26. Kegel, L., Hahmann, M., Lehner, W.: Template-based time series generation with loom. In: EDBT/ICDT Workshops, vol. 1558 (2016)
27. De Gooijer, J.G., Hyndman, R.J.: 25 years of time series forecasting. Int. J. Forecast. **22**(3), 443–473 (2006)
28. Arlitt, M., Marwah, M., Bellala, G., Shah, A., Healey, J., Vandiver, B.: IoTA bench: an internet of things analytics benchmark. In: 6th ACM/SPEC International Conference on Performance Engineering, pp. 133–144. ACM Press, New York (2015)

Network-Aware Stochastic Virtual Machine Placement in Geo-Distributed Data Centers
(Short Paper)

Hana Teyeb[1,2]([⊠]), Nejib Ben Hadj-Alouane[3], and Samir Tata[4]

[1] Faculty of Sciences of Tunis, University of Tunis El Manar, OASIS, Tunis, Tunisia
hana.teyeb@gmail.com
[2] SAMOVAR, Telecom SudParis, CNRS, University of Paris-Saclay,
9 Rue Charles Fourier, 91011 Evry, France
[3] National Engineering School of Tunis, OASIS, Tunis, Tunisia
nejib_bha@yahoo.com
[4] IBM Research - Almaden, 650 Harry Rd, San Jose, CA 95120, USA
stata@us.ibm.com

Abstract. In this work, we focus on the stochastic network-aware virtual machine placement (VM) problem in geodistributed data centers (DCs). We consider the uncertainty of the inter-VMs traffic while making placement and migration decisions. First, we propose a stochastic program with the objective of minimizing inter-DCs traffic. Then, we propose an equivalent optimization model using sampling methods and we present a two-step approach to solve the problem. Experiments show the effectiveness of the proposed approach.

Keywords: Cloud computing · IaaS · VM placement · VM migration · Stochastic integer programming

1 Introduction

In order to achieve reliability and serve world wide users, large-scale cloud providers are relying on a geo-distributed infrastructure where data centers (DCs) are built in different locations and interconnected within a backbone network. In such a context, network congestion is a crucial issue. The increasing traffic exchange of the applications hosted in the VMs may cause bottlenecks in the network resulting in performance degradation of the cloud system. Thus, it is important to minimize the traffic circulating within the backbone traffic. Recent studies [1–3] have shown that the workload of VMs is highly dynamic which may cause the existent placement and migration schemes to be inefficient. Most of the existent works [4–6] make migration decisions based on deterministic demand estimation and workload characterization without considering stochastic properties. There are several optimization techniques to solve the VM placement problem. Among these techniques, we cite deterministic and Stochastic Integer

© Springer International Publishing AG 2017
H. Panetto et al. (Eds.): OTM 2017 Conferences, Part I, LNCS 10573, pp. 37–44, 2017.
https://doi.org/10.1007/978-3-319-69462-7_3

Programming (SIP) [7]. In contrast to deterministic approach, the SIP technique considers uncertain parameters. Many existing works have used SIP to deal with resource provision, load balancing and capacity planning problems [1,8,9]. Others have studied VM consolidation problem with stochastic demand [10,11]. In [12], the authors have proposed a joint approach that combines VM placement and bandwidth allocation. The main limitation of the aforementioned works is the fact that they did not consider inter-VMs communication while making the placement decisions. In this paper, we propose stochastic integer programming (SIP) formulation that aims to solve the network-aware VM placement problem in geo-distributed DCs while minimizing the inter-DCs traffic. The remainder of this paper is organized as follows. In Sect. 2, it presents the problem formulations as well as the proposed heuristic for solving the network-aware VM placement problem in geo-distributed DCs. As for Sect. 3, presents and discusses the experiment results. Finally, we conclude in Sect. 4.

2   Problem Formulation

Table 1 presents the notations used in the presented formulations. DC edge routers are responsible for connecting the DCs to Wide Area Network (WAN) [13]. In the following, we use the decision variables defined below.

- φ_i^h defines the amount of traffic originated from the VM $i \in V$ and sent from the DC $h \in D$ (i.e. the traffic sent to the backbone network).
- x_i^h is equal to 1 if the VM $i \in V$ is placed in the DC $h \in D$, 0 otherwise.
- z_i^h is equal to 1 if the VM i is migrated from the DC $h \in D$.
- f_{hk}^i which denotes the amount of traffic originated from the VM $i \in V$ and circulating between the DCs $h \in D$ and $k \in D$.

Table 1. Notations.

Symbol	Description
D	The set of data centers
V	The set of virtual machines
R	The set of hardware resources (CPU, RAM, storage)
M_i	The migration cost of the VM i ($i \in V$). It is the amount of data transferred when the VM i is migrated
a_i^k	Takes 1 if the VM i can be placed in the DC k, 0 otherwise ($i \in V$, $k \in D$)
c_{hr}	The capacity of the DC h in terms of resource r ($h \in D$, $r \in R$)
u_{ir}	The amount of resource r consumed by the VM i ($r \in R$, $i \in V$)
E_h	The bandwidth capacity of the edge router of the DC $h \in D$
x_{0i}^h	is equal to 1 if the VM i is already placed in the DC h. ($i \in V$, $h \in D$)

2.1 Stochastic Optimization Model

In this formulation, we consider a random variable $\widetilde{d_{ij}}$ that describes the amount of traffic exchanged between each pair of VMs (i.e. inter-VMs bandwidth demand). The variable follows a probability distribution that can be estimated from runtime measurement. We assume that the distribution can be obtained using statistical process to analyze historical data. Many studies [1,10,11] have shown that the resource demand of VMs follows a *Normal* distribution $\mathcal{N}(\mu, \sigma^2)$. Thus, we consider that the inter-VMs bandwidth demand follows also the *Normal distribution*. Our aim is to minimize the amount of traffic circulating between the different DCs.

$$\min \sum_{i \in V} \sum_{h \in D} M_i.z_i^h + \varphi_i^h \tag{1}$$

Subject to the following constraints:

$$\varphi_i^h \geqslant \sum_{j \in V} \widetilde{d_{ij}}.x_i^h - \sum_{j \in V} x_j^h.\widetilde{d_{ij}} \qquad \forall i \in V, \forall h \in D \tag{2}$$

$$\sum_{k \in D} f_{hk}^i - \sum_{k \in D} f_{kh}^i = \sum_{j \in V} \widetilde{d_{ij}} x_i^h - \sum_{j \in V} \widetilde{d_{ij}}.x_j^h \qquad \forall i \in V, \forall h \in D \tag{3}$$

$$Pr(\sum_{i \in V} M_i.z_i^h + \sum_{i \in V} \varphi_i^h \leqslant E_h) \geqslant 1 - \epsilon \qquad \forall h \in D \tag{4}$$

$$\sum_{h \in D} x_i^h = 1 \qquad \forall i \in V \tag{5}$$

$$x_i^h \leqslant a_i^h \qquad \forall i \in V, \forall h \in D \tag{6}$$

$$\sum_{i \in V} u_{ir}.x_i^h \leqslant c_{hr} \qquad \forall r \in R, \forall h \in D \tag{7}$$

$$z_i^h \geqslant x_{0i}^h - x_i^h \qquad \forall i \in V, h \in D \tag{8}$$

$$z_i^h \in \{0.1\} \qquad \forall i \in V, h, k \in D$$

$$x_i^k \in \{0.1\} \qquad \forall i \in V, k \in D$$

$$\varphi_i^h \geqslant 0 \qquad \forall i \in V, h, k \in D$$

$$f_{hk}^i \geqslant 0 \qquad \forall i \in V, h, k \in D$$

The constraints (2) and (3) are both flow conservation constraints. They are both stochastic due to the random variable $\widetilde{d_{ij}}$ which refers to the inter-VMs communication traffic. The constraint (4) ensures that for each DC edge router, the total traffic, which includes the inter-VMs communication traffic and the migration traffic, does not exceed the bandwidth capacity of the edge router with a high probability $(1 - \epsilon)$. The constraint (4) ensures the service quality guarantee with a predefined threshold ε. The constraint (5) is a demand satisfaction constraint. As for the constraint (6) it restricts the placement of VMs in a particular number of DCs. The constraint (7) represents the capacity constraint on the DCs. It ensures that the amount of resources consumed by different VMs placed in a given DC does not exceed the resource capacities of the DC. The constraint (8), ensures that only already existing VMs can be considered as candidates for the migration. Suppose that it has finite support, hence, we can enumerate the set of all different scenarios. We can then, formulate an equivalent deterministic optimization problem that can be solved as a Mixed Integer Linear Program (MILP). However, the size of the problem space can grow very large as the number of scenarios increases. Therefore, in the next section, we propose an alternative solution using sampling-based methods.

2.2   Equivalent Optimization Formulation

In order to solve the stochastic problem, we propose an equivalent formulation using sampling methods. Let us consider the function $g(x) = \sum_{j \in V} \widetilde{d_{ij}}.x_i^h - \sum_{j \in V} x_j^h.\widetilde{d_{ij}}, \quad \forall i \in V, \forall h \in D$ It is clearly impossible to enumerate all the possible outcomes. Hence, sampling techniques are a commonly used tool. In order to discretize the stochastic function $g(.)$, we apply *Sample Average Approximation* (SAA) method [14]. Sampling-based methods often approximate well, with a small number of samples, problems that have a very large number of scenarios [15]. In this work, we use Monte Carlo methods to generate samples of $N = \{1, \ldots, n\}$ replications of the random variable $\widetilde{d_{ij}}$ using the *Normal distribution* $\mathcal{N}(\mu_{ij}, \sigma_{ij}^2)$, where μ_{ij} is the mean and σ_{ij}^2 is the variance. Let us consider the function $g_n(.)$, as the discretization of the stochastic function $g(.)$ by applying SAA methods, $g_n(x) = \frac{1}{n} \sum_{i=1}^{n} g(x, \xi_i)$ Where ξ_i is a random element such that: $\widetilde{d_{ij}} = \frac{1}{n} \sum_{k=1}^{n} \xi_{ij}^k$ and n is the number of iterations. Hence, the equivalent deterministic constraint of (2) is obtained by replacing $g(x)$ by $g_n(x)$.

$$g_n(x) = \sum_{j \in V} \left(\frac{1}{n} \sum_{k=1}^{n} \xi_{ij}^k\right).x_i^h - \sum_{j \in V} \left(\frac{1}{n} \sum_{k=1}^{n} \xi_{ij}^k\right).x_j^h \quad \forall i \in V, \forall h \in D \quad (9)$$

Thus, the constraint (2) becomes $\varphi_i^h \geqslant g_n(x) \quad \forall i \in V, \forall h \in D$. As mentioned above, we consider that the inter-VMs bandwidth demand $\widetilde{d_{ij}}$ follows the *Normal distribution* $\mathcal{N}(\mu_{ij}, \sigma_{ij}^2)$. Hence, we can estimate $\widetilde{d_{ij}}$ by its mean μ_{ij} ($\widetilde{d_{ij}} \simeq \mu_{ij}$). Let us consider the following equation, $\varphi_i^h = \sum_{k \in D} f_{hk}^i, \quad \forall h \in D, \forall i \in V$. The value of f_{hk}^i can be obtained from the flow conservation constraint (3) $\sum_{k \in D} f_{hk}^i = \sum_{k \in D} f_{kh}^i + \sum_{j \in V} \widetilde{d_{ij}} x_i^h - \sum_{j \in V} \widetilde{d_{ij}}.x_j^h, \quad \forall i \in V, \forall h \in D$. If we replace (Sect. 2.2) in (4), we obtain:

$$Pr\left(\sum_{i \in V} M_i.z_i^h + \sum_{i \in V}\left(\sum_{k \in D} f_{kh}^i + \sum_{j \in V} \widetilde{d_{ij}} x_i^h - \sum_{j \in V} \widetilde{d_{ij}}.x_j^h\right) \leqslant E_h\right) \geqslant 1 - \epsilon \quad \forall h \in D$$

$$(10)$$

Since, $\widetilde{d_{ij}}$ follows the Normal distribution $\mathcal{N}(\mu_{ij}, \sigma_{ij}^2)$, then, because we assume that the traffic of each pair of VM $(i, j) \in V$, is independent if $i \neq j$, the aggregate traffic demand $\sum_{i \in V} \sum_{j \in V} \widetilde{d_{ij}}$ follows the Normal distribution

$\mathcal{N}(\sum_{i \in V} \sum_{j \in V} \mu_{ij}, \sum_{i \in V} \sum_{j \in V} \sigma_{ij}^2)$ according to the property of normal distribution and Central Limit Theorem (CLT). Note that, the term $\sum_{i \in V} M_i . z_i^h$ is deterministic, thus, it does not follow a probability distribution. Let us denote by $\alpha^h = \sum_{i \in V} \sum_{j \in V} \widetilde{d_{ij}} x_i^h - \sum_{j \in V} \widetilde{d_{ij}} . x_j^h$. We need to estimate the *Normal* distribution parameters μ_{α^h} and $\sigma_{\alpha^h}^2$. Since, $x_i^h \in \{0,1\}, \forall i \in V, h \in D$, and because we assume that the traffic of each pair of VM $(i,j) \in V$, is independent if $i \neq j$, then, by applying SAA methods, we can estimate μ_{α^h} and $\sigma_{\alpha^h}^2$ as follows $\mu_{\alpha^h} = \sum_{i \in V} \sum_{j \in V} \mu_{ij} x_i^h - \sum_{i \in V} \sum_{j \in V} \mu_{ij} . x_j^h$, $\quad \forall h \in D$, $\sigma_{\alpha^h}^2 = \sum_{i \in V} \sum_{j \in V} \sigma_{ij}^2 x_i^h + \sum_{i \in V} \sum_{j \in V} \sigma_{ij}^2 . x_j^h$, $\quad \forall h \in D$. Hence, it easy to show that the constraint (10) is equal to the overloading probability constraint presented in (11), where $\phi^{-1}(.)$ is the inverse of the cumulative distribution function of the *Standard Normal* distribution. In this work, we consider that $\epsilon \leqslant 0.5$ and $\phi^{-1}(1 - \epsilon) \geqslant 0$.

$$\frac{E_h - \sum_{i \in V} \sum_{k \in D} f_{kh}^i + \sum_{i \in V} M_i . z_i^h + \sum_{i \in V} \sum_{j \in V} \mu_{ij} x_i^h - \sum_{i \in V} \sum_{j \in V} \mu_{ij} . x_j^h}{\sqrt{\sum_{i \in V} \sum_{j \in V} \sigma_{ij}^2 x_i^h + \sum_{i \in V} \sum_{j \in V} \sigma_{ij}^2 . x_j^h}} \geqslant \phi^{-1}(1 - \epsilon)$$

$$(11)$$

The constraint can be written as follows:

$$E_h \geqslant \sum_{i \in V} M_i . z_i^h + \sum_{i \in V} \sum_{k \in D} f_{kh}^i + \mu_{\alpha^h} + \phi^{-1}(1 - \epsilon) . \sqrt{\sigma_{\alpha^h}^2} \quad \forall h \in D \qquad (12)$$

The objective function as well as the rest of constraints are the same. We replace $\widetilde{d_{ij}}$ by μ_{ij}. One of the challenges of this formulation is the non-linearity of the constraint (12). The linearization of this constraint leads to a very large number of variable which will enlarge the research space. To cope with this problem, we propose, in the next section, an iterative two-step approach.

2.3 Network-Aware Stochastic VM Placement Algorithm

Let us denote by $\beta^h = \phi^{-1}(1 - \epsilon) . \sqrt{\sigma_{\alpha^h}^2}$ and $\gamma^h = \sum_{i \in V} M_i . z_i^h + \sum_{i \in V} \sum_{k \in D} f_{kh}^i + \mu_{\alpha^h}$. We denote by $SIP_{\gamma + \beta}$, the optimization program presented in Sect. 2.2. We define SIP_{γ}, the optimization program where the constraint (12) is replaced by $E_h \geqslant \gamma^h \quad \forall h \in D$. Let us define the list $PrefList$ which contains the different values of ε. $PrefList$ is sorted by increasing value of ε. Smaller ε means that the risk of overloading must be very low. The algorithm tries to solve the stochastic optimization problem within two iterations. In the first iteration, it solves the SIP_{γ} model without considering the non-linear term β^h. In fact, the optimization program SIP_{γ} provides a deterministic VM placement scheme as it does not consider the variance of inter-VMs communication traffic in the future. In addition, the term γ ensures that the overall traffic sent via the DC edge router does not exceed its capacity. Afterword, the algorithm evaluates the term β^h by considering the solution provided by SIP_{γ}. In the second iteration, it tries to solve the $SIP_{\gamma + \beta}$ model by adding the term β^h to the overloading probability constraint. If the $SIP_{\gamma + \beta}$ model is feasible, then the new placement scheme is the solution provided by $SIP_{\gamma + \beta}$, otherwise, we relax

---

Algorithm 1. STOCHASTIC VM PLACEMENT ALGORITHM.

Input: Initial Placement scheme, stochastic traffic matrix, $PrefList$
Output: New VMs Placement Plan
1 $\varepsilon \leftarrow PrefList[0]$
2 Solve SIP_γ
3 Record solutions
4 Calculate β^h with the solutions provided by SIP_γ
5 Solve $SIP_{\gamma+\beta}$
6 **if** $SIP_{\gamma+\beta}$ *is feasible* **then**
7 $\quad|\quad$ New Placement Plan $\leftarrow$ Solutions of $SIP_{\gamma+\beta}$
8 **else**
9 $\quad|\quad$ $i \leftarrow i+1$
10 **end**
11 **while** $SIP_{\gamma+\beta}$ *is not feasible* **and** $i \leqslant PrefList.size()$ **do**
12 $\quad|\quad$ $\varepsilon \leftarrow PrefList[i]$
13 $\quad|\quad$ Check feasibility of $SIP_{\gamma+\beta}$
14 $\quad|\quad$ $i \leftarrow i+1$
15 **end**
16 **if** $SIP_{\gamma+\beta}$ *is feasible* **then**
17 $\quad|\quad$ New Placement Plan $\leftarrow$ Solutions of $SIP_{\gamma+\beta}$
18 **else**
19 $\quad|\quad$ New Placement Plan $\leftarrow$ Solutions of SIP_γ
20 $\quad|\quad$ Add SLA violation
21 **end**
22 **return** New Placement Plan

---

the value of ε and try to solve it again. The new placement scheme is either the solution provided by $SIP_{\gamma+\beta}$, if it exists, or the solution provided by SIP_γ with an SLA violation due to the high risk of network overload.

3 Performance Evaluation

The different experiments were carried out on a machine that has an Intel Xeon 3; 3 GHz CPU and 8 GB of RAM. We have used the commercial solver CPLEX 12.5 [16] to solve the different formulations. In all experiments, we have fixed the number of DCs to six. We generated for each pair of VMs traffic a sample of 10000 replications according to Normal distribution. Then, we applied SAA to approximate the values of the traffic matrix. We randomly generated groups of (mean, variance range) for the inter-VM traffic and set each pair of VMs traffic to a value generated by a randomly chosen group. We assume that client's demands are independent. We have implemented the deterministic equivalent model in Java with the above listed parameters. We studied first the performance of the proposed algorithm in terms of computational time. The results depicted in Fig. 1 show that the value of ε has no considerable impact on the execution time of the model. We note that the execution time does not exceed 10 s for a total

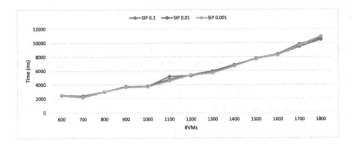

Fig. 1. Execution Time versus the total number of VMs.

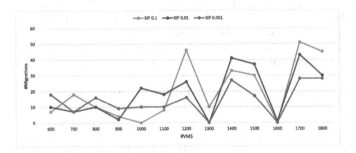

Fig. 2. Variantion of the number of migrations for $\varepsilon \in \{0.1, 0.01, 0.001\}$.

number of VM $|V| = 1800$. The value of the parameter ε affects the bandwidth utilization. Smaller ε requires the system to reserve more bandwidth in order to accommodate the possible variance of inter-VM traffic. To ensure the non violation of the overloading probability with smaller ε, some VMs may have to be migrated to another DC. We have varied the value of ϵ and we plotted the number of migrations performed for each value. We have considered that the bandwidth capacity of the DC edge router are large enough to satisfy the bandwidth demand for all values of ε. The results depicted in Fig. 2 show that smaller ε produces smaller number of migration. This can be explained by the fact that smaller ε means that the risk of network overloading is very small. Since, migration produces additional traffic, $SIP_{\gamma+\beta}$ tries to minimize the number of migration in order to prevent from extensive migrations.

4   Conclusion

In this work, we proposed a SIP formulation that aims to solve the network-aware VM placement problem in geo-distributed DCs. We considered the uncertainty of inter-VMs traffic. Our objective was to minimize the overall traffic circulating in the backbone network in order to prevent from congestion problems. In order to solve the problem, we proposed an equivalent optimization model based on sampling methods as well as an efficient algorithm to cope with non-linearity problem.

References

1. Yu, L., Chen, L., Cai, Z., Shen, H., Liang, Y., Pan, Y.: Stochastic load balancing for virtual resource management in datacenters. IEEE Trans. Cloud Comput. **PP**(99), 1 (2016)
2. Benson, T., Akella, A., Maltz, D.A.: Network traffic characteristics of data centers in the wild. In: Proceedings of the 10th ACM SIGCOMM Conference on Internet Measurement, pp. 267–280 (2010)
3. Kandula, S., Sengupta, S., Greenberg, A., Patel, P., Chaiken, R.: The nature of data center traffic: measurements & analysis. In: Proceedings of the 9th ACM SIGCOMM Conference on Internet Measurement Conference, pp. 202–208 (2009)
4. Beloglazov, A., Buyya, R.: Managing overloaded hosts for dynamic consolidation of virtual machines in cloud data centers under quality of service constraints. IEEE Trans. Parallel Distrib. Syst. **24**(7), 1366–1379 (2013)
5. Xiao, Z., Song, W., Chen, Q.: Dynamic resource allocation using virtual machines for cloud computing environment. IEEE Trans. Parallel Distrib. Syst. **24**(6), 1107–1117 (2013)
6. Gong, Z., Gu, X., Wilkes, J.: Press: Predictive elastic resource scaling for cloud systems. In: International Conference on Network and Service Management, pp. 9–16 (2010)
7. Shapiro, A., Dentcheva, D., Ruszczynski, A.: Lectures on Stochastic Programming - Modeling and Theory, vol. 16, 2nd edn. SIAM, Philadelphia (2014)
8. Maguluri, S.T., Srikant, R., Ying, L.: Stochastic models of load balancing and scheduling in cloud computing clusters. In: INFOCOM Proceedings IEEE, pp. 702–710 (2012)
9. Ghosh, R., Longo, F., Xia, R., Naik, V.K., Trivedi, K.S.: Stochastic model driven capacity planning for an infrastructure-as-a-service cloud. IEEE Trans. Serv. Comput. **7**(4), 667–680 (2014)
10. Wang, M., Meng, X., Zhang, L.: Consolidating virtual machines with dynamic bandwidth demand in data centers. In: INFOCOM Proceedings IEEE, pp. 71–75 (2011)
11. Jin, H., Pan, D., Xu, J., Pissinou, N.: Efficient VM placement with multiple deterministic and stochastic resources in data centers. In: 2012 IEEE Global Communications Conference, pp. 2505–2510 (2012)
12. Chase, J., Niyato, D.: Joint optimization of resource provisioning in cloud computing. IEEE Trans. Services Comput. **PP**(99), 1 (2015)
13. Corporate Headquarters: Data center networking: enterprise distributed data centers solutions reference nework design. In: Solutions Reference Network Design, Cisco Systems Inc (2003)
14. Kim, S., Pasupathy, R., Henderson, S.G.: A Guide to Sample Average Approximation. Handbook of Simulation Optimization. International Series in Operations Research & Management Science, pp. 207–243. Springer, New York (2015). doi:10.1007/978-1-4939-1384-8_8
15. Homem-de Mello, T., Bayraksan, G.: Monte carlo sampling-based methods for stochastic optimization. Surveys Oper. Res. Manag. Sci. **19**(1), 56–85 (2014)
16. IBM Corporation ILOG CPLEX: http://www.ilog.com/products/cplex/. Accessed 04 Feb 2013

Finding Process Variants in Event Logs
(Short Paper)

Alfredo Bolt[(✉)], Wil M.P. van der Aalst, and Massimiliano de Leoni

Eindhoven University of Technology, Eindhoven, The Netherlands
{a.bolt,w.m.p.v.d.aalst,m.d.leoni}@tue.nl

Abstract. The analysis of event data is particularly challenging when there is a lot of variability. Existing approaches can detect variants in very specific settings (e.g., changes of control-flow over time), or do not use statistical testing to decide whether a variant is relevant or not. In this paper, we introduce an unsupervised and generic technique to detect significant variants in event logs by applying existing, well-proven data mining techniques for recursive partitioning driven by conditional inference over event attributes. The approach has been fully implemented and is freely available as a ProM plugin. Finally, we validated our approach by applying it to a real-life event log obtained from a multinational Spanish telecommunications and broadband company, obtaining valuable insights directly from the event data.

Keywords: Process variant detection · Process mining · Event data

1 Introduction

Organizations can record the execution of business processes supported by process aware information systems into event logs [1]. Process mining is a relatively young research discipline that is concerned with discovering, monitoring, and improving real processes by extracting knowledge from event logs [1]. Processes are affected by variability that is not only related to the *control-flow* perspective (e.g., a process may skip risk assessment steps for gold customers), but can also be related to other perspectives, such as *performance*. For example, if two branches of a company execute their processes in the same way (i.e., same control-flow) there could still be performance differences between the branches.

In this paper, we briefly discuss a novel technique to detect relevant *process variants* (i.e., groups of process executions) in an event log using the control-flow, performance and context attributes of events in an interactive and exploratory way, where only relevant results are presented. The full version of this paper containing extended discussions, formalizations and results is presented in [2].

It is important to note that the type of analysis performed with our approach can also be achieved by combining other approaches and standard data mining techniques. However, such techniques require extensive and manual ad-hoc parametrization and configuration to achieve the same results that our approach can

© Springer International Publishing AG 2017
H. Panetto et al. (Eds.): OTM 2017 Conferences, Part I, LNCS 10573, pp. 45–52, 2017.
https://doi.org/10.1007/978-3-319-69462-7_4

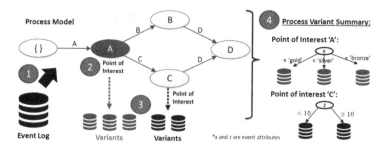

Fig. 1. Overview and steps of our approach: (1) Given an event log, a process model is created. (2) Points of interest are identified in the process model. (3) For each point of interest, the set of cases that reach it is partitioned into process variants. (4) A summary of process variants is produced, where the splitting criteria and the resulting variants are shown for each point of interest.

obtain in a much easier way. We achieve this by leveraging on process models to identify points of interest in the process (e.g., a given *state* in the process). Then, the same variability analysis is automatically performed in each point of interest and the summarized results for the whole process are presented to the user as result. Figure 1 illustrates the overview and steps of our approach. Note that our technique provides, for each point of interest, a clear partitioning criteria that allows one to easily identify and characterize process variants. The resulting process variants can be analyzed individually, but can also be compared using process comparison techniques such as [3]. Our approach has been implemented and evaluated in a real case study.

2   Preliminaries

Let $\mathcal{E}$ be the universe of events, $\mathcal{N}$ be the universe of attribute names and $\mathcal{V}$ be the universe of possible attribute values. Events can have values for given attributes through the function $\# : \mathcal{N} \to (\mathcal{E} \nrightarrow \mathcal{V})$. For an attribute $a \in \mathcal{N}$, the partial function $\#(a) : \mathcal{E} \nrightarrow \mathcal{V}$, denoted as $\#_a$, can relate events to values of the attribute a.

Let $\sigma \in \mathcal{E}^*$ be a trace. A trace records the execution of an *instance* of a process and is a finite sequence of events. Let $L \subseteq \mathcal{E}^*$ be an event log, i.e., a set of traces. Each event is unique and appears only once in one trace within the event log, i.e., for any event $e \in \mathcal{E} : \left| \{(\sigma, i) \mid \sigma \in L \wedge i \in \{1, \dots, |\sigma|\} \wedge \sigma(i) = e\} \right| \leq 1$.

A process variant $V \subseteq L$ is defined as a *set of traces*. The traces in a process variant also contain *similarities* in other event attributes. Process variants also should have *differences* with respect to other process variants. The traces in such process variants should be similar to traces in the same variant, but are different to traces in other process variants.

In this paper, we leverage on the same log augmentation techniques defined in [4] (i.e., trace manipulation operations) to extend events with obtain additional attributes, such as the *elapsed time* of an event within its case, or the *next activity* to be executed in a case.

The **first step** in our approach is to create a process model from the event log. Transition systems are very simple process models that are composed of *states* and of *transitions* between them. A transition is defined by an activity being executed, triggering the current state to move from a *source* to a *target* state. Prefixes of traces can be mapped to states and transitions using representation functions that define how these prefixes are interpreted.

Definition 1 (Transition System). Let $L \in \mathcal{E}^*$ be an event log, P_L the set of all the prefixes of traces of L, E_L the set of all the events of L, r^s a state representation function and r^a an activity representation function. A *transition system* $TS^{(r^s,r^a,L)}$ is defined as a triplet (S, A, T) where $S = \{s \in R^s \mid \exists_{\sigma \in P_L} \, s = r^s(\sigma)\}$ is the set of states, $A = \{a \in R^a \mid \exists_{e \in E_L} \, a = r^a(e)\}$ is the set of activities and $T = \{(s_1, a, s_2) \in S \times A \times S \mid \exists_{\sigma \in P_L \setminus \{\langle\rangle\}} \, s_1 = r^s(pref^{|\sigma|-1}(\sigma)) \wedge a = r^a(\sigma(|\sigma|)) \wedge s_2 = r^s(\sigma)\}$ is the set of valid transitions between states.

3   Finding Process Variants in Event Logs

Defining Points of Interest in a Transition System (Step 2): Given a transition system $TS^{(r^s,r^a,L)} = (S, A, T)$, we define $P \subseteq S \cup T$ as the set of *points of interest*. Given an event log L and a transition system $TS^{(r^s,r^a,L)} = (S, A, T)$, every point of interest $p \in S \cup T$ can be related to a set of traces through the function $tr : (S \cup T) \to \mathcal{P}(L)$.

Finding Variants in a Point of Interest (Step 3): We find process variants in the points of interest defined above by using Recursive Partitioning by Conditional Inference (RPCI) techniques [5] over event attributes. This technique is able to split a set of *instances* based on dependent and independent attributes (i.e., features, variables).

A trace cannot correspond directly to an instance because it may have several different values for the same attribute. For example, an `elapsed time` attribute can have different values for each event in the trace. For this purpose, we choose the attribute values of a single event of a trace to represent it as an *instance*. The choice of which event should be used is related to the definition of points of interest discussed before. Since we know that for a given point of interest p, any trace $\sigma \in tr(p)$ reaches it at some point, we could simply choose the last event of the smallest prefix of σ that reaches p.

Given a point of interest p, the set of events that represent the traces in $tr(p)$ is defined as $E_p = \biguplus_{\sigma \in tr(p)} E(p)(\sigma)$, where every event $e \in E_p$ corresponds to an *instance*. For each point of interest p, we aim to find the relevant partitions of its corresponding set of events (i.e., instances) E_p (denoted as simply E) based on their event attributes.

Let E be a set of events, and $A(E) = \{a \in \mathcal{N} \mid dom(\#_a) \cap E \neq \emptyset\}$ the set of event attributes associated with the events in E. For each attribute $a \in A(E)$,

$n_a(E) = \{\#_a(e)|e \in dom(\#_a) \cap E\}$ defines the set of values of the attribute a over the set of events E.

We choose one of the event attributes $d \in A(E)$ as our *dependent attribute* (chosen by the user), for which we will reduce the variability by partitioning any combination of the other $A(E) \setminus \{d\}$ event attributes, namely *independent attributes*.

Our approach leverages on the Recursive Partitioning by Conditional Inference (RPCI) approach [5] to partition the set of events E. RPCI provides a unbiased selection and binary splitting mechanism by means of statistical tests of independence between the splitting attributes and the dependent attribute. The details of how RPCI works are out of the scope of this paper, and the reader is referred to [5] for the specific mechanisms that RPCI uses to deal with different types of distributions and combinations of attributes.

In a nutshell, RCPI is described for a set of events E by the following steps:

1. Given a dependent attribute $d \in A(E)$, find the independent attribute $i \in A(E) \setminus \{d\}$ with the strongest significant correlation with d.
2. If such independent attribute i does not exist (i.e., no correlation is significant), stop the recursion. If it does exist, an optimal binary partition of the dependent attribute d is obtained, such that E is split into $E_1 \subset E$ and $E_2 = E \setminus E_1$.
3. Repeat step 1 and 2 for E_1 and E_2 recursively.

As a result of RPCI, a set of events E can be partitioned into a set of subsets $S_E = \{\lambda_1, \ldots, \lambda_n\}$. Given the recursive nature of this approach, the exact total number of partitions to be evaluated depends on the characteristics and distributions of the attributes. Every subset $\lambda \in S_E$ corresponds to a set of events. RPCI provides, for each $\lambda \in S_E$ a set of conditions that define it.

Given the way that E was built and the nature of events being unique, every event in λ is related to a different trace. Therefore, $S_E = \{\lambda_1, \ldots, \lambda_n\}$ can be transformed into a set of process variants $V = \{v_1, \ldots, v_n\}$ of the same size where given an point of interest p, a variant v is defined as $v_i = \{\sigma \in tr(p)|\exists_{e \in \lambda_i} : e \in \sigma\}$ for any $i \in \{1, \ldots, n\}$. Therefore, the variants are guaranteed to be disjoint.

The approach discussed in this section is repeated for the sets of events related to each point of interest in the transition system defined by the user.

A Summary of Process Variants (Step 4): According to RPCI, the traces related to a point of interest can be split into process variants or not, depending on the significance of the correlation between dependent and independent attributes. We present a summary of only the points of interest where process variants were found. For each point of interest, the splitting criteria obtained from RPCI is clearly presented, and the process variants are available to the user for other types of analysis. A concrete visual representation of the summary is presented in [2].

4   Implementation and Case Study

We have implemented our approach as a ProM [6] plugin named "Process Variant Finder" included in the *VariantFinder* package.[1]

In this paper, we report on the results of a case study obtained by applying our approach to an event log provided by a Spanish broadband and telecommunications company. The provided event log refers to a *claim handling* process related to three services that this company provides, codenamed: Globalsim, SM2M and Jasper. In total, the event log contains 8296 cases (i.e., claims) processed between January 2015 and December 2016. Each claim has, on average 5 activities. Claims correspond to traces of this process and can have four severities: slight, minor, major and critical. In total, there are 40965 events in the event log.

Customers of the company *create* a claim which is *activated* by an employee of the company when he/she starts working on it. Claims with missing information can be *delayed*. If the service was interrupted, the first step is to work on the *restoration* of the service. If there was no interruption, or the service has been restored, resources work on *solving* the problem that caused the claim. Once a problem has been solved, it is informed to the customer, which can *close* the claim. Customers can also *cancel* claims at any moment.

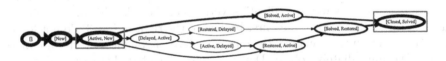

Fig. 2. Transition system representing the claim handling process. States are defined by the last two activities executed in a trace prefix. Thickness represents frequency. Points of interest (i.e., states and transitions) with a frequency of 5% of claims or less were filtered out. States "Active, New" and "Closed, Solved" are highlighted. (Color figure online)

Figure 2 illustrates this process as a transition system, in which a state is defined by the last two activities executed in a prefix of a trace. We used our approach to discover process variants in all states and transitions of the transition system shown in Fig. 2. In every state and transition, we searched for process variants. Because of space limitations, the remainder of this section discusses only a few process variants detected in the "Active, New" and the "Closed, Solved" states of the transition system presented in Fig. 2 (highlighted in red and blue respectively).

The variants detected in the "Active, New" state (shown in Fig. 3) was obtained by selecting the `next activity` attribute (described in Sect. 2) as the dependent attribute, and using all the other attributes as independent attributes.

---

[1] The reader can get this package via the ProM Package Manager.

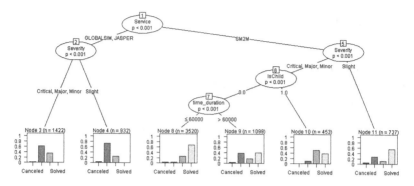

Fig. 3. Partition detected in the "Active, New" state, defining six *Control-flow* variants with differences in the "next activity" to be executed. The labels in each bar chart are (from left to right): Canceled, Delayed, Restored, Solved. The dependent attribute is the next activity to be executed. All other attributes are considered as independent attributes.

Therefore, the resulting variants of this partition can be considered as *control-flow variants*. On the one hand, we can observe that the claims related to the Globalsim and Jasper services (i.e., first branch to the left in Fig. 3) have a higher tendency to get delayed than claims related to the SM2M service. This is accentuated in claims with a "Slight" severity. On the other hand, claims associated to the SM2M service (i.e., first branch to the right in Fig. 3) do not follow this pattern. From these claims, the ones that have a "Slight" severity are more likely to be immediately solved. Domain experts related this to the fact that slight severity claims usually do not involve an interruption of the service (thus, no restoration) and can be immediately solved.

More severe claims are divided whether they belong to a "parent claim" (1) or not (0). This is indicated by the "isChild" attribute (a claim can be subdivided into smaller claims). Claims that belong to a parent claim are more likely to become "restored". This make sense because bigger or more complex claims are more likely to have child claims, and are also more likely to have a service interruption. Claims that do not belong to any parent claim can be split into two main variants: the ones that take one minute or less to be activated and those that take more than one minute. We can observe that the faster claims are more likely to be solved, but the slower ones get delayed more often. This could be related to "easier" claims being processed first.

Figure 4 shows *performance* process variants detected in the "Close, Solved" state (i.e., when a claim has been solved and then closed) where the splitting attributes and criteria are represented in a tree-fashion. We can observe that claims related to the Globalsim service have the longest throughput time (i.e., the time between a claim is created until it is closed), followed by claims related to the Jasper service. Note that claims related to the S2M2 service are the fastest to be closed in average, but the time distribution is more spread than claims related to the Jasper service. This can be observed on the position of quartiles

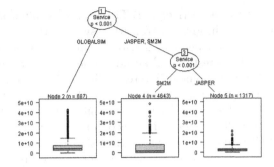

Fig. 4. Performance variants detected in the "Closed, Solved" state. Elapsed time is measured in milliseconds and is presented as box plots for each variant. The dependent attribute is Elapsed Time. All other attributes are considered as independent attributes.

in the box plots shown in Fig. 4. Domain experts explained the fact that, in average, Globalsim claims took longer to be closed by the fact that there was a change in the management of this service in May 2016, which resulted, among other consequences, in the massive closeup of claims. Most of such claims were declared as "Solved" several months before, but were never officially closed. It is important to note that the company is only responsible for claims until they are solved, since the closing of a claim depends on the customer, hence it is not included in the company's SLAs.

Naturally, the obtained process variants can be compared. We refer the reader to [2] for an comparative analysis of obtained process variants.

5   Related Work

We grouped existing process variant detection techniques into four categories: Concept drift detection, Trace clustering, Performance analysis, and Attribute Correlation. In this paper we only discuss Attribute Correlation approaches since it is the category in which our paper falls into. For a more detailed discussion of related work, we refer the reader to the full version of this paper [2].

Attribute Correlation techniques aim to group cases depending on any event attributes. The only other approach that belongs in this category (besides our approach) is [4], that focuses on classifying specific selections of events by building decision or regression trees with the attributes of such events that are later used to classify traces into process variants. Our approach is closely related to [4].

The similarities between these approaches are: (1) Behavioral features are annotated into events as extra attributes via trace manipulation functions. (2) Selection of events are partitioned into subgroups using such event attributes.

The differences between these approaches are that, in our approach: (1) Process variants are always guaranteed to be disjoint (see Sect. 3). This is only guaranteed in [4] for the event filters EF_2 and EF_3, which select either the first or the last event of a trace respectively. (2) The required configuration is simpler than in [4]: In our approach, given a transition system, the user only needs to select the dependent and independent attributes, and the same analysis is performed for all points of interest. In [4] an ad-hoc analysis use case needs to be manually designed for each point of interest. Therefore our approach presents a summary of process variants in many points of the process. In [4], the result is a single decision tree describing variants in a single point of the process. (3) Events are split using RPCI instead of Decision or Regression trees.

Arguably, if RPCI would be used in [4], then they could replicate the results provided by our approach in processes without loops (see first difference), but it would require to manually configure several analysis use cases (see second differences).

6   Conclusions

The problem of detecting process variants in event logs has been tackled by several authors in recent years. Many authors have successfully solved specific scenarios where the focus in on specific attributes, such as time. Some have even provided general solutions, but they fail to filter out irrelevant splits. This paper presents an approach that is able to detect relevant process variants in any available event attribute by automatically splitting any other (combination of) event attributes in many points of the process. The approach has been implemented and is publicly available. We were able to successfully identify points of process variability inside in a real-life event log and we were able to detect process variants without the use of domain knowledge, confirming such variability using process comparison techniques. Therefore, our approach provides a viable solution to process variant detection, even when no domain knowledge is available.

References

1. van der Aalst, W.M.P.: Process Mining: Data Science in Action, 2nd edn. Springer, Heidelberg (2016)
2. Bolt, A., van der Aalst, W.M.P., de Leoni, M.: Finding process variants in event logs. Research Report BPM-17-04. BPMCenter.org (2017)
3. Bolt, A., de Leoni, M., van der Aalst, W.M.P.: A visual approach to spot statistically-significant differences in event logs based on process metrics. In: Nurcan, S., Soffer, P., Bajec, M., Eder, J. (eds.) CAiSE 2016. LNCS, vol. 9694, pp. 151–166. Springer, Cham (2016). doi:10.1007/978-3-319-39696-5_10
4. de Leoni, M., van der Aalst, W.M., Dees, M.: A general process mining framework for correlating, predicting and clustering dynamic behavior based on event logs. Inf. Syst. **56**, 235–257 (2016)
5. Hothorn, T., Hornik, K., Zeileis, A.: Unbiased recursive partitioning: a conditional inference framework. J. Comput. Graph. Stat. **15**(3), 651–674 (2006)
6. van Dongen, B.F., de Medeiros, A.K.A., Verbeek, H.M.W., Weijters, A.J.M.M., van der Aalst, W.M.P.: The ProM framework: a new era in process mining tool support. In: Ciardo, G., Darondeau, P. (eds.) ICATPN 2005. LNCS, vol. 3536, pp. 444–454. Springer, Heidelberg (2005). doi:10.1007/11494744_25

Interactive and Incremental Business Process Model Repair

Abel Armas Cervantes[1(✉)], Nick R.T.P. van Beest[3], Marcello La Rosa[1],
Marlon Dumas[2], and Luciano García-Bañuelos[2]

[1] Queensland University of Technology, Brisbane, Australia
{abel.armascervantes,m.larosa}@qut.edu.au
[2] University of Tartu, Tartu, Estonia
{marlon.dumas,luciano.garcia}@ut.ee
[3] Data61, CSIRO, Brisbane, Australia
nick.vanbeest@data61.csiro.au

Abstract. It is common for the observed behavior of a business process
to differ from the behavior captured in its corresponding model, as workers devise workarounds to handle special circumstances, which over time
become part of the norm. Process model repair methods help modelers to
realign their models with the observed behavior as recorded in an event
log. Given a process model and an event log, these methods produce a
new process model that more closely matches the log, while resembling
the original model as close as possible. Existing repair methods identify points in the process where the log deviates from the model, and
fix these deviations by adding behavior to the model locally. In their
quest for automation, these methods often add too much behavior to
the model, resulting in models that over-generalize the behavior in the
log. This paper advocates for an interactive and incremental approach to
process model repair, where differences between the model and the log
are visually displayed to the user, and the user repairs each difference
manually based on the provided visual guidance. An empirical evaluation
shows that the proposed method leads to repaired models that avoid the
over-generalization pitfall of state-of-the-art automated repair methods.

Keywords: Process model repair · Conformance checking · Visual analytics · Process mining

1 Introduction

Modern information systems maintain detailed data about the execution of the
business processes they support. These data can generally be extracted in the
form of *event logs* consisting of sets of *traces*, i.e. sequences of *events* produced
during the execution of a process case, where each event records the occurrence
of a process activity.

Process mining is a family of methods for analyzing business processes based
on event logs [24]. Among others, process mining methods allow analysts to

© Springer International Publishing AG 2017
H. Panetto et al. (Eds.): OTM 2017 Conferences, Part I, LNCS 10573, pp. 53–74, 2017.
https://doi.org/10.1007/978-3-319-69462-7_5

compare the actual execution of a process against its expected execution captured in a process model. This model-to-log comparison operation is known as *conformance checking*. A conformance checking method takes as input a process model and an event log, and identifies a set of discrepancies between the behavior observed in the log and that allowed by the model. Once an analyst has identified relevant discrepancies between a process model and an event log via conformance checking, they may wish to modify the process model in order to better reflect reality. This operation is known as *process model repair*. Process model repair methods take as input a process model, an event log and a set of discrepancies between the model and the log, and produce a model that resembles the original model as much as possible, but does not have the designated discrepancies.

The quality of a process model repair method can be captured via three metrics: *structural similarity* (how much the produced model structurally resembles the original model), *fitness* (how much behavior observed in the event log is captured by the repaired process model), and *precision* (how much behavior is allowed by the repaired model but never observed in the log). A repaired model should be as structurally similar as possible to the original model, it should have a higher fitness than the original model (since the designated discrepancies are fixed) and it should not degrade precision (i.e. it should not add behavior that is not observed in the log).

Existing process model repair methods [10,21] identify points in the process where the log deviates from the model, and determine the model change operations required to reconcile such deviations. While the aim of [10] is to automatically generate repaired models with higher fitness w.r.t. the original model, [21] aims solely at finding the change operations to be performed. The latter method can associate different costs to the change operations and control the model change via a budget. Further, [21] shows that using only two change operations (insert and skip a task), which can be automatically applied over the model, it is possible to obtain a repaired model with higher fitness w.r.t. the original model. The authors in [21] acknowledge the importance of other metrics in addition to fitness, though the identification of techniques to improve on such metrics during repair is left to future work. In their quest for automation, these methods often add too much behavior to the original model, and thus the repaired model tends to grossly over-generalize the event log. In other words, these methods focus on maximizing fitness at the expense of precision.

This paper advocates for an *interactive* and *incremental* approach to process model repair for models in the BPMN language, where differences between the control-flow of the model and the log are visually displayed to the user and the user repairs each difference manually, at their discretion, based on visual guidance. In fact, some of these differences may underpin positive deviations, e.g. workarounds introduced to improve process performance, and as such, the user may wish to repair the model accordingly, so that these workarounds can become standard practices. On the other hand, differences that point to negative deviations, e.g. the violation of some compliance rule, should not be incorporated into

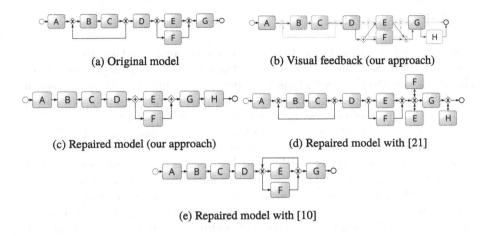

(a) Original model   (b) Visual feedback (our approach)

(c) Repaired model (our approach)   (d) Repaired model with [21]

(e) Repaired model with [10]

Fig. 1. Examples of process model repairs using different approaches

the model. The paper purports that this approach allows users to strike a better tradeoff between the above metrics by giving them a more ample range of choices at each step of the repair process. This hypothesis is validated via an empirical evaluation on a battery of synthetic model-log pairs capturing recurrent change patterns as well as a real-life model-log pair.

To illustrate the benefits of our approach w.r.t. existing ones we consider the process model in Fig. 1a and the log $\{\langle A, B, C, D, E, F, G, H \rangle, \langle A, B, C, D, F, E, G, H \rangle\}$. Our approach provides the visual difference feedback to the user shown in Fig. 1b. Assuming this difference relates to a positive deviation, the user may repair the model in the way indicated in Fig. 1c. This model is similar to the original one, fixes the discrepancy shown in the visual feedback and does not add behavior w.r.t. the original model. Figures 1d and e are the repaired versions generated by the automatic repair methods in [10,21], respectively.[1] Even though the three models Fig. 1c–e can fully replay the log (perfect fitness), the models Fig. 1d and e have lower precision. For instance, tasks B, C, E, F, H can be repeated any number of times in Fig. 1d.

The paper is organized as follows. Section 2 discusses the limitations of existing process model repair methods. Section 3 introduces the conformance checking method presented in [12], which we use as a starting point. Next, Sect. 4 presents the proposed model repair approach, while Sect. 5 discusses the results of the empirical evaluation. Finally Sect. 6 summarizes the contribution and outlines future work directions.

---

[1] Both methods produce Petri nets but for simplicity we present the repaired models in BPMN.

2   Related Work

Process model repair methods take as a starting point a set of discrepancies identified via a conformance checking method such as *trace alignment* [2,17]. This method computes a set of *optimal trace alignments* between each trace of a log and the closest corresponding trace of the model. An alignment is a pair of traces, which, in addition to symbols representing tasks, may also contain *silent moves*. A silent move represents a deviation between the trace of the log and the trace of the model. It may be a *move on log* (a task is observed in the log at a point where it is not allowed in the model) or conversely, a *move on model*. A trace alignment is optimal if it requires a minimum amount of moves.

In [10], the authors present a process model repair method based on alignments. This method starts by computing the optimal alignments between the log and the model, then identifies the non-conforming parts between them and, finally, adds (i) loops, (ii) subprocesses, and (iii) skips of tasks. This approach guarantees a repaired model that fits the log perfectly. Another method, presented in [21], is also based on alignments. However, it only has two types of repair operations: skip a task and insert a new task loop. Unlike [10], this method seeks to maximize fitness while controlling the amount of changes by assigning a cost to each repair operation and setting a maximum budget.

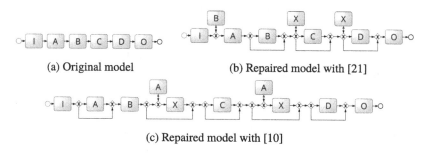

(a) Original model   (b) Repaired model with [21]

(c) Repaired model with [10]

Fig. 2. Examples of automatic process model repairs that over-fit (low precision)

The methods in [10,21] are based on change operations that add behavior to the model. While the former adds subprocesses, loops and skips, the latter adds skips and self-looping tasks. Although these changes have a positive impact on fitness, they negatively affect precision. Consider for example the model in Fig. 2a and log $\{\langle I, A, B, X, C, O\rangle, \langle I, A, B, X, D, O\rangle, \langle I, B, A, X, C, O\rangle, \langle I, B, A, X, D, O\rangle\}$. The repaired models produced by [10,21] are shown in Figs. 2b and c, respectively. Even though both models perfectly fit the log, they allow additional behavior that is neither allowed in the original model nor observed in or implied by the log. For example, in Fig. 2b there are two tasks with label X that can be repeated before and after C, while in Fig. 2c, the same applies to task A. Furthermore, in both repaired models, tasks C and D can co-occur, even though this never happens in the log.

The work in [7] proposes a genetic algorithm that given a reference model and a log, discovers a new model that is similar to the reference model and more closely fits the log. This method optimizes the result along five dimensions: fitness, precision, simplicity, generalization and structural similarity to the reference model. The method generates candidate models (represented as process trees), which are evolved until one of them is found to be optimal w.r.t. a given threshold on the allowed changes. Unlike process model repair methods, this method does not update the original model but discovers a new (possibly very different) one.

The authors of [22] consider the problem of model repair in a context where the log is not necessarily reliable (i.e. there is missing or incorrect data). Hence the analyst needs to indicate to what extent they trust the model and to what extent they trust the log. If the analyst trusts the model but not the log, the log is fully re-generated so that it matches the original model. If the analyst trusts the log but not the model, the model is re-discovered from the event log using an automated process model discovery method, creating a model that can potentially be very distant from the original one. If the user partially trusts the model and the log, a new (repaired) model-log pair is generated, such that the repaired model and log match, and their respective behavior is within a certain distance of the original model and log.

Finally, process model repair has also been approached in the context of unsoundness. For example, [11] uses a multi-objective optimization technique to automatically turn an unsound Petri net into a corresponding sound model in a minimal number of change operations, in an attempt to keep a high structural similarity to the original model. In this case, however, the input to the technique is only an (unsound) model.

3   Behavioral Alignment

Behavioral alignment [12] is a conformance checking method that identifies events or behavioral relations between tasks occurrences that are observed at a given point in the log but not allowed in the corresponding state of the model, and vice-versa. This method operates in four steps. First, it compresses the event log into a graph of behavioral relations between task occurrences known as an *event structure* [19]. Second, it expands the process model into another event structure. Third, it computes a (partially) synchronized product between these two event structures. Finally, it extracts a set of difference statements from the product.

A Prime Event Structure (PES) [19] is a graph of events representing occurrences of tasks. Events are linked via three relations: (i) *Causality* $(e < e')$ indicates that event e is a prerequisite for e'; (ii) *Conflict* $(e \# e')$ implies that e precludes the execution of e'; and (iii) *Concurrency* $(e \parallel e')$ indicates that e and e' co-occur in any order. A PES represents the computations of a system by means of *configurations*, sets of events that can occur together. Figure 3b shows the PES extracted from the log shown in Fig. 3a. A label e:A represents

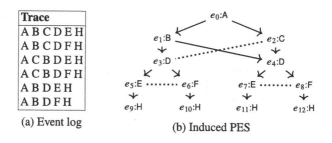

(a) Event log   (b) Induced PES

Fig. 3. Example of an event log and its corresponding PES

an event e signifying an occurrence of task A. For example, e_9:H, e_{10}:H, e_{11}:H and e_{12}:H are events representing different occurrences of task H. A directed black arc between events represents causality, while a dotted edge represents conflict. Concurrency is denoted by the lack of any (direct or transitive) link between two events, e.g. events e_1:B and e_2:C are concurrent. The PES of a log represents every trace in the log as a configuration. E.g., the first trace in Fig. 3a is represented by the configuration $\{e_0$:A, e_1:B, e_2:C, e_4:D, e_7:E, e_{11}:H$\}$.

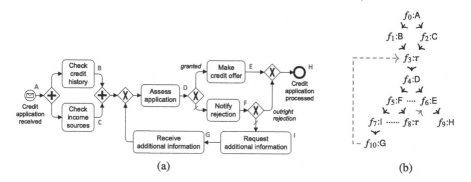

(a)   (b)

Fig. 4. Loan application as BPMN model (a) and PES prefix (b) (Color figure online)

Any BPMN process model can be expanded into a PES using a technique known as unfolding. In this alternative representation of the behavior of the model, each event is associated to a single task, thus it is always possible to determine which task originated a given event. Consider the loan application process model shown in Fig. 4a, which is unfolded into a corresponding PES in Fig. 4b. For conciseness, the PES prefix uses short labels A, B, ..., I (shown next to each task in the BPMN model) instead of the full task labels. The PES of a model can also contain *cutoff-corresponding event relations*, graphically represented as directed red dotted arrows (see Fig. 4b), to denote "jumps" in the continuation of the process execution (e.g., in the case of looping behavior).

The event structure mirrors the BPMN model: the first event is f_0:A and is followed by concurrent events f_1:B and f_2:C. A silent event (f_3:τ) captures their synchronization. Subsequently, $f_4:D$ is followed by two events in conflict ($f_5:F$ and $f_6:E$). Event $f_6:E$ can only be followed by $f_9:H$, while $f_5:F$ can be followed either by event $f_7:I$ (itself followed by $f_{10}:G$) or by a silent event $f_8:\tau$ which is in conflict with $f_7:I$. The cutoff-corresponding relation ($f_{10}:G, f_3:\tau$) tells us that the process can jump back to the point just before D is executed, while ($f_8:\tau, f_6:E$) captures the jump from F to H.

The event structure of the log and that of the model are compared by computing a *partially synchronized product* (PSP) [4,12]. A PSP is a state machine where each state is a triplet $\langle C^l, C^r, \xi \rangle$, where C^l is a configuration of the PES of the log, C^r is a configuration of the PES of the model, and ξ is a partial mapping between them. The construction of the PSP starts with the empty configurations. At each step, a pair of events from each PES is matched (added to ξ) if and only if their labels are the same and their causal relations with the other matched events coincide. When an event cannot be matched, it is "hidden" to allow the comparison to proceed. A "hide" edge captures behavior observed in the log but not allowed in the model (*lhide*), or behavior allowed in the model but not observed in the log (*rhide*). The method relies on an A^* heuristic to build a PSP that finds, for each configuration in the log, the most similar configuration in the model (a.k.a. *optimal*), i.e., the one with the minimum number of hides.

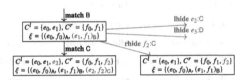

Fig. 5. Fragment of the PSP for PES from Figs. 3b and 4b

Figure 5 presents an excerpt of the PSP of the events structures in Figs. 3b and 4b. The topmost box is the state where configurations $C^l = \{e_0, e_1\}$ (log PES) and $C^r = \{f_0, f_1\}$ (model PES) have been processed, resulting in the mapping $\{(e_0, f_0)_A, (e_1, f_1)_B\}$. Given the above state, the events $\{e_2:C, e_3:D\}$ from log PES would be enabled, and so is $f_2:C$ from the other PES. Under such conditions, four moves are possible in the PSP: (i) the matching of events e_2 and f_2, both carrying the label C, (ii) the (left) hiding of $f_2:C$, and (iii) the (right) hiding of $e_2:C$ and $e_3:D$. Figure 5 presents only the states reached after operations "match C" and "rhide $f_2:C$". However, the full PSP will contain optimal matchings for all runs in the PES of the event log.

The dissimilarities between a model and a log can be of two types, mismatching behavior and behavior only observed in the model (not observed in the log); while the former is captured by means of hide operations in the PSP, the latter is the model behavior not observed in the log. The mismatching behavior patterns defined in [12] are shown in Fig. 6, while their verbalizations are displayed in Table 1. In the verbalizations, the capital letters are placeholders for

Table 1. Verbalization of mismatch patterns

Pattern	Statement
TaskReloc	*In the log, B occurs after [A] instead of [A, C, D]*
ConcConf	*In the model, after [A], B and C are concurrent, while in the log they are mutually exclusive*
CausConc	*In the model, after [A], B occurs before C, while in the log they are concurrent*
CausConf	*In the model, after [A], B occurs before C, while in the log they are mutually exclusive*
TaskSub	*In the log, after [A], B is substituted by X*
TaskAbs/Ins	*In the log, C occurs after [A, B] instead of [A, C]*
UnmRepetition	*In the log, A is repeated after [B]*
TaskSkip	*In the log, after [A], B is optional*
UnobsAcyclicInter	*In the log, tasks [A, B, . . .] do(es) not occur after tasks [D, E, F]*
UnobsCyclicInter	*In the log, the cycle involving tasks [A, B, . . .] does not occur after tasks [D, E, F]*

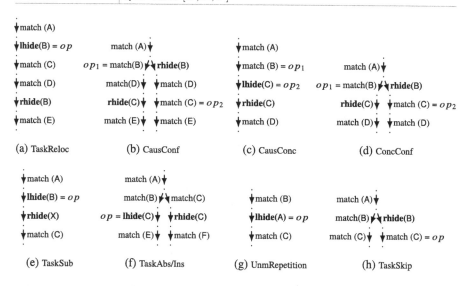

Fig. 6. Mismatch patterns in the PSP

the activities involved in the differences. Note that UnobsAcyclicInter and UnobsCyclicInter in Table 1 refer to the behavior solely contained in the model.

Roughly speaking, given a PSP both the mismatching behavior patterns and the additional behavior in the model are identified, verbalized and reported. For instance, the fragment of the PSP shown in Fig. 5 captures the pattern in Fig. 8c that is verbalized as *"In the event log, task C can be skipped, while in the model it cannot"*.

4   Extending Conformance Checking to Process Model Repair

In this section, we first describe two improvements over the behavioral alignment method introduced above: four of the existing patterns are redefined to consider sequences of tasks (also referred to as *intervals*) instead of individual tasks, and an order for the detection of differences is established in a way that more specific patterns are detected before more general ones. Next, we describe a method for interactive process model repair based on the visualization of the differences detected by the revised behavioral alignment method.

4.1   Extension, Order and Impact of Differences

The patterns in Table 1 define combinations of hide and match operations to describe predefined templates expressing behavioral differences. Those patterns capture differences involving single tasks (TaskSkip, TaskSub, UnmRepetition, TaskReloc and TaskAbs), pairs of tasks (CausConc, ConcConf and CausConf) or even sequences of tasks (UnobsAcyclicInter and UnobsCyclicInter). Patterns involving sequences of tasks can offer condensed dissimilarities, which otherwise should be spelled out one by one. Thus, as a first contribution, we redefined four patterns, those displayed in Fig. 8, to consider sequences of tasks instead of single tasks. In addition, to obtain a more condensed feedback, the new definitions fix some of the issues of the original definitions. For example, even though the PSP in Fig. 7 shows the case when tasks B and C are optional (i.e., they occur in both the model and the log, or only in the model), the original definition considers only the optionality of a single task.

An interval (Int) is a sequence of tasks that can occur one after the other either in the model or in the log. For instance, the occurrences of tasks B and C in Fig. 7 represent an interval, because they occur consecutively in the model after A. For the sake of brevity, the $rhide, lhide$ or $match$ of an interval Int is denoted as $rhide(Int), lhide(Int)$ or $match(Int)$, respectively. Thus, given an interval $Int = [B, C], rhide(Int)$ denotes $rhide(B)$ followed by $rhide(C)$, and analogously for $lhide$ and $match$. The modified patterns to consider intervals instead of single tasks are depicted in Fig. 8. Observe that in the case of Fig. 8b, Int-1 and Int-2 are two different intervals.

The detection of differences consists of traversing the PSP and once a pattern is found, all the involved hide operations are discarded. As such, a hide operation cannot be reported as part of two differences. However, a single hide operation can be explained by various patterns. For example, given the PSP in Fig. 7, it is possible to identify two TaskAbs patterns (i.e., *"In the model, B occurs after [A] and before [D]"* and *"In the model, C occurs after [A] and before [D]"*) or a single TaskSkip pattern (i.e., *"In the model, after [A], the interval [B, C] is optional"*). The latter offers a deeper insight into the difference, in a more compact manner, and is thus preferred. Therefore, as a second contribution, we define a specific order on the detection of the differences, such that those involving intervals are identified first, then differences involving pairs of tasks are identified, and finally

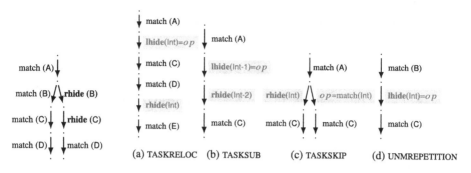

Fig. 7. Interval skip   Fig. 8. Patterns extended to consider intervals

those involving only single tasks. The order for each of the patterns is shown in the first column of Tables 2, 3 and 4.

The proposed repair approach starts from the premise that not all existing differences shall be reconciled but only those pointing to positive deviations, and among them, some may have higher priority than others. For example, differences involving critical tasks, or differences that refer to a particular type of mismatch pattern (e.g., TaskAbs), or affecting a larger number of traces in the log, may be given higher priority. In this regard, as a third contribution we propose a notion of *impact* for each mismatch pattern. We define this notion based on the frequency of the events in the log, such that differences with higher impact shall be reconciled first.[2] The construction of the PES enriched with information about the frequency of the events, can be found in [23]. Consider the patterns in Fig. 6 and a log with X amount of traces. The impact of a given pattern is defined as Y/X, where Y is (i) the frequency of the event (or the minimum frequency of the events in an interval) involved in the operation op for the pattern **TaskReloc, TaskSub, TaskAbs/Ins, UnmRepetition** and **TaskSkip**; (ii) the minimum frequency of the event in op_1 and in op_2 for the patterns **CausConf** and **CausConc**; and (iii) the frequency of the event in op_1 plus the frequency of the event in op_2 for the pattern **ConcConf**. Intuitively, this notion represents the proportion of traces involving the events in a given difference. This simple impact measure can be replaced with a more sophisticated notion, e.g. one that depends on factors that are exogenous to the log such as the cost of rectifying the model according to a given difference.

4.2   Visualization of Differences

The cornerstone of our interactive and incremental model repair approach is the visualization of the differences. It exploits the fact that diagrams are powerful tools for presenting information in a more concise and precise manner than text [18]. Thus, the visualization of differences can be easier to understand than

---

[2] The same rationale of reconciling changes with higher impact first is proposed in [21].

the textual description generated by the conformance checker. Intuitively, every mismatch pattern is translated into a graphical representation, which can be overlaid on the model, and suggests the change to be done for reconciling a given difference. This alternative representation uses standard BPMN notation, so no new symbols are added, and uses a color code to represent the suggested changes. Variations in color are easily distinguishable, more than changes in shapes [18], and can help coping with potential model complexity. This idea incarnates the principle of *graphical highlighting*, which have been shown in [15] to lead to more understandable process models.

Some techniques that have approached the problem of representing differences between graphs and/or models can be broadly categorized into two groups: those that use color-coding of differences in a merged graph for the visualization of differences (see e.g. [6,13,14,20]), and those that overlay the two compared models, such that both models are visible in the same picture [3]. In the visualization proposed in this paper, however, the differences are not directly represented as such to the user. Rather, the required *change* is shown, indicating what needs to be changed in order to repair the model such that it matches the behavior observed in the log.

The color code used for the representation of the differences is as follows. An element in the BPMN model – task, sequence flow (i.e. arc) or gateway – involved in a difference can either be grey (element to be removed) or red with thicker lines (element to be inserted or task affected by the difference); whereas the elements that are not involved in the difference are left unchanged. The proposed visualizations for all mismatch patterns are displayed in Tables 2, 3 and 4. For instance, consider the fourth pattern in Table 2, where task b has to be relocated after o. As a result, task b in the model is grayed out (along with its incoming and outgoing arcs), while a new task b is inserted after o. The background of the new task is colored white, so that it is easier to distinguish new tasks from those already present in the model (those with a yellow background).

The differences are classified with respect to two criteria, *scope* and *type* of change. The scope can be local or cross-context. Local changes occur in a single part of the model, e.g., **TaskAbs** denotes the case when a task does not occur in the log, and thus it has to be removed from the model; whereas, cross-context changes involve two different parts in the model, e.g., **TaskReloc** represents the case when an interval of tasks has to be relocated in the model. The second criterion is the type of change: tasks modification, sequence flows modification, or gateways modification. In our context, a modification implies either the removal, or the removal and insertion of elements.

A local change can be further subdivided into two classes: *interval* and *binary*. A local-interval change can be formally defined as a triplet $\langle I_1, I_2, C \rangle$, such that I_1 is an interval of tasks in the model, I_2 is an interval of tasks to be inserted and C is a configuration. A local-binary change is a triplet $\langle e_1, e_2, C \rangle$, where e_1 and e_2 are tasks in the model, and C is a configuration. Finally, a *cross context* change is defined as $\langle I, C_1, C_2 \rangle$, where I is an interval, and C_1 and C_2 are configurations. C_1 is the source of the difference and C_2 is the target of the

Table 2. Tasks modification patterns

Order	Type	Input	Suggested repair	Sentence
16	TaskAbs	$I_1 = \{b\}$ $I_2 = \emptyset$ $C = \{i, a\}$		In the model, Task b occurs after $[i, a]$ and before $[o]$
6	UnobsAcyclic	$I_1 = \{a, b\}$ $I_2 = \emptyset$ $C = \{i\}$		In the log, the interval $[a, b]$ does not occur after $[i]$
8	TaskSub	$I_1 = \{a\}$ $I_2 = \{c\}$ $C = \{i\}$		In the log, after i, the interval $[a]$ is substituted by Task c
1	TaskReloc	$I = \{b\}$ $C_1 = \{i, a\}$ $C_2 = \{i, a, b, o\}$		In the log, the interval $[b]$ occurs after $[i, a, o]$ instead of $[i, a]$
2	TaskReloc	$I = \{b\}$ $C_1 = \{i, a, o\}$ $C_2 = \{i, a\}$		In the model, the interval $[b]$ occurs after $[i, a, o]$ instead of $[i, a]$

modification. We refer to the start (resp. end) of a difference as the element in the model that precedes (resp. follows) the tasks in $I_1, I_2, \{e_1, e_2\}$ and I, depending on the scope of the change. In order to build the visualization of a difference, we take as input the triplets generated by the conformance checker, and obtain the elements in the model involved in such differences. Subsequently, we generate a triplet $\langle Y, H, A \rangle$, such that Y contains the elements to be grayed out, H contains the elements to be highlighted and A contains the elements to be added. Each set of patterns, grouped by the type of change in the model, is presented below.

Tasks Modification. This set of differences covers three cases: tasks need to be removed from the model, tasks need to be relocated and tasks need to be substituted. Table 2 presents the patterns in this category and shows an example of both their visualization and their verbalization. The graphical representation of these differences consists in graying out the tasks (and their arcs) in the interval I_1 for the local-interval changes, and the tasks in the interval I for the cross-context changes. New arcs, and tasks in the case of task substitution, are inserted to connect the start with the end of each difference (i.e., the tasks around the grayed out elements). Finally, for the last two cross-context changes, new tasks and arcs are inserted representing the relocation of the intervals to the target of the modification. For instance, the fourth pattern – **TaskReloc** in Table 2 suggests to relocate task b after $\{i, a, o\}$. Thus, task b and its incoming/outgoing arcs are grayed out, a new arc is inserted to connect a with o, and a new task b is added after o.

Sequence Flows Modification. The differences in this category cover the cases when existing arcs need to be removed, and new gateways and/or arcs need to be inserted. Thus, no task needs to be grayed out or highlighted. The patterns in this category are presented in Table 3. The arcs to be grayed out are the incoming and outgoing arcs of the tasks in I for cross-context changes,

Table 3. Sequence flows modification patterns

Order	Type	Input	Suggested repair	Sentence
15	TaskAbs	$I = \{b\}$ $C_1 = \{i, a\}$ $C_2 = \{d, c\}$		In the log, Task b occurs after [d, c] and before [e]
3	TaskSkip	$I_1 = \{b\}$ $I_2 = \emptyset$ $C = \{i, a\}$		In the log, after a, the interval [b] is optional
7	UnmRepetition	$I_1 = \{b\}$ $I_2 = \emptyset$ $C = \{i, a\}$		In the log, the interval [b] is repeated after a
9	CausConc	$e_1 = a$ $e_2 = b$ $C = \{i\}$		In the model, after i, Task a occurs before Task b, while in the log they are concurrent
14	CausConf	$e_1 = a$ $e_2 = b$ $C = \{i\}$		In the model, after i, Task a occurs before Task b, while in the log they are mutually exclusivethe log they are concurrent

in I_1 for local-interval changes, and in $\{e_1, e_2\}$ for local-binary changes. Finally, depending on the pattern, new gateways have to be inserted with corresponding arcs for connecting the elements involved by the difference. For instance, the second pattern (**TaskSkip**) in Table 3 suggests that task b should be optional. The incoming and outgoing arcs of b are grayed out, and new XOR gateways are added to allow the skip of b after a.

Gateways Modification. The last set of differences are changes that affect the gateways present in the model, i.e. a gateway needs to be deleted or replaced by another gateway (e.g. an XOR gateway is replaced by an AND gateway). The patterns in this category are shown in Table 4. The elements to be highlighted are the tasks in the interval I and I_1 in the case of **TaskSkip** and **UnobsCyclic**, respectively, and tasks e_1 and e_2 for the rest of the differences. The elements grayed out are the relevant gateways (AND gateways for the first two patterns and XOR gateways for the last four), and their outgoing and incoming arcs. Finally, depending on the pattern, new gateways and arcs are inserted to connect the elements involved in the pattern. For example, the first and third patterns in Table 4 suggest to remove the gateways and to define a causality order between a and b. In these two patterns, a (the task occurring first) is connected to i, and b (occurring last) is connected to o. In the case of the second and the fourth patterns in Table 4, the existing gateways have to be substituted by another type, thus changing the parallel behavior between tasks a and b to exclusive, or vice versa. The substitution is as follows: existing gateways and their incoming and outgoing arcs are grayed out, and new gateways and arcs are inserted. Specifically, in both **ConcConf** patterns, the fork gateway is connected to i, a and b, emulating the connections of the fork gateway to substitute, while the join gateway is connected to a, b and o.

Table 4. Gateways modification patterns

Order	Type	Input	Suggested repair	Sentence
10	CausConc	$e_1 = a$ $e_2 = b$ $C = \{i\}$		In the log, after i, Task a occurs before Task b, while in the model they are concurrent
12	ConcConf	$e_1 = a$ $e_2 = b$ $C = \{i\}$		In the model, after i, Task a and Task b are concurrent, while in the log they are mutually exclusive
13	CausConf	$e_1 = a$ $e_2 = b$ $C = \{i\}$		In the log, after i, Task a occurs before task Task b, while in the model they are mutually exclusive
11	ConcConf	$e_1 = a$ $e_2 = b$ $C = \{i\}$		In the log, after i, Task a and Task b are concurrent, while in the model they are mutually exclusive
4	TaskSkip	$I = \{b\}$ $C_1 = \{i, a\}$ $C_2 = \{i, a, o\}$		In the model, after i, the interval $[b]$ is optional
5	UnobsCyclic	$I_1 = \{b\}$ $I_2 = \emptyset$ $C = \{1, a\}$		In the log, the cycle involving $[b]$ does not occur after $[i]$

5   Evaluation

We implemented our approach as part of the OSGi plugin *Compare* [5] for the Apromore online process analytics platform.[3] This plugin takes as input a BPMN process model and a log in MXML or XES format. Its output is a set of textual differences according to the mismatch patterns described in this paper. When selected, a difference is also represented graphically on top of the input process model. The user can apply a difference at a time to repair the model accordingly. At each application, the differences between the model and the log are recomputed. Once a difference has been selected, it can be automatically applied by the tool. Users can apply differences until the model and the log capture the same behavior, or until desired.[4] An example of the graphical representation of a difference, i.e., **TaskSkip** pattern, over a model is depicted in Fig. 9.

Using this tool, we conducted a two-pronged evaluation to compare our approach to the two existing baseline approaches in [21] (hereafter called Base1) and [10] (Base2). First, we applied each approach on a battery of synthetic event logs generated from a real-like model of a loan origination process, to assess how each of them performs in identifying and repairing elementary changes. Next, we applied each approach on a real-life model-log pair for a road traffic fines management process.

---

[3] Available at http://www.apromore.org.
[4] A screencast of the tool can be found at https://youtu.be/3d00pORc9X8.

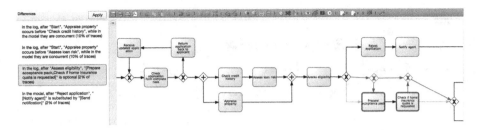

Fig. 9. Example of the visualization of a difference in *Compare*

We compared the quality of the repaired models produced by the three approaches in terms of their fitness, precision and F-Score (the harmonic mean of fitness and precision) w.r.t. the event log, as well as in terms of their structural similarity w.r.t. the original model. We used fitness and precision based on trace alignment [1,2], as these metrics can be computed reasonably quickly, and because the two baseline methods were designed to optimize the fitness measure based on trace alignments. Trace alignment-based fitness [2] measures the degree to which every trace in the log can be aligned with a trace produced by the model, while trace alignment-based precision [1] measures how often the model escapes these aligned traces by adding extra behavior not recorded in the log. We computed model similarity as one minus the graph-edit distance between the two models. The graph-edit distance measures the number of node and edge insertions, removals and substitutions to transform one graph into the other. We used the measure in [8] (with a greedy matching strategy), as it has been shown to provide a good compromise between matching accuracy and performance.

5.1   Experiment with Synthetic Datasets

In the first experiment, we used a battery of 17 synthetic model-log pairs. Starting from a textbook example of a process for assessing loan applications [9] (see Fig. 10), we generated 17 altered versions of this base model by applying different change operations. Next, we used the BIMP simulator[5] to generate an event log from each altered process model. By pairing the base model with these logs, we obtained 17 model-log pairs.

To avoid bias towards any of the evaluated approaches, we selected the change operations to apply from an independent taxonomy of *simple change patterns* [25], which constitute solutions for realizing commonly occurring control-flow changes in information systems. As such, this taxonomy of changes is different from the mismatch patterns presented in this paper, which are based on the difference as observed in the PSP. However, each pattern in one taxonomy can be expressed by one or more patterns in the other taxonomy.

The simple change patterns from [25], summarized in Table 5, capture elementary ways of modifying a process model, such as adding/removing a frag-

---

[5] http://bimp.cs.ut.ee.

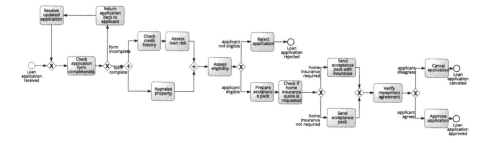

Fig. 10. BPMN model of a loan origination process (source: [9])

ment, putting a fragment in a loop, swapping two fragments, or parallelizing two sequential fragments. Non-applicable patterns such as changing branching frequency or inlining a subprocess were excluded, resulting in eleven change patterns. These patterns can be grouped into three categories based on their type: Insertion ("I"), Resequentialization ("R") and Optionalization ("O"). From these categories, following the same method as in [16,23], we constructed six *composite change patterns* by subsequently nesting simple change patterns from each category within each other: "IOR", "IRO", "OIR", "ORI", "RIO", and "ROI". For example, the composite pattern "IRO" can be obtained by adding a fragment ("I"), putting it in parallel with an existing fragment ("R"), and skipping the latter ("O"). As a result, we obtained a total of 17 change patterns.

Table 5. Simple control-flow change patterns from [25]

Simple pattern	Explanation	Category
Add/remove	Add/remove fragment	I
Cond./Seq.	Make two fragments conditional/sequential	R
Conc./Seq.	Make two fragments concurrent/sequential	R
Loop	Make fragment loopable/non-loopable	O
Skip	Make fragment skippable/non-skippable	O
Cond. move	Move fragment into/out of conditional branch	I
Conc. move	Move fragment into/out of concurrent branch	I
Synchronize	Synchronize two parallel fragments	R
Duplicate	Duplicate fragment	I
Replace	Substitute fragment	I
Swap	Swap two fragments	I

In order to use our approach without user input, we automatically selected the difference retrieved from our tool that has the highest impact in terms of involved log traces, applied that to the original model to obtain a repaired model, recomputed the differences and picked again the most impactful difference, until no more differences existed or five differences had been selected. This *mechanized* version of our approach only removes the interaction with the user but preserves its incremental nature. However, given the limit to maximum five differences, there is no guarantee that all the discrepancies between model and log would be repaired.

Table 6 reports the results of the first experiment, where for each bidirectional pattern (e.g. Add/Remove), we applied the pattern in both directions and reported the average measurements. Our approach always achieves the highest structural similarity w.r.t. the original model (on average 0.92, with values ranging from 0.86 to 0.97). This is substantially higher than the similarity obtained by Base1 (avg $= 0.72$, min $= 0.60$, max $= 0.88$) and by Base2 (avg $= 0.79$, min $= 0.71$, max $= 0.88$). Despite the higher similarity, the models repaired by our approach have the highest F-Score in all but one case. In fact, while both baselines aim to maximize fitness, obtaining a perfect fitness in most cases, our approach keeps the fitness high while improving precision, often substantially (avg $= 0.97$ against 0.68 for Base1 and 0.94 for Base2), hence striking a better balance between the two accuracy measures. The only exception is the synchronization pattern, where our F-Score is 0.95 against 0.97 for the two baselines. The repair introduced in this case was more specific than the behavioral difference reported by our approach, resulting in a lower fitness compared to the two baselines (0.94 instead of 1.00), despite having a slightly higher precision (0.97 instead of 0.95).

In summary, despite the relative simplicity of the introduced changes, the two baseline approaches generate models that are much more distant from the original model, yet less accurate in capturing the log behavior, than the models produced by our approach.

5.2 Experiment with Real-Life Dataset

In the second experiment, we used a real-life model-log pair of a process for managing road traffic fines in Italy. The normative process model, available as a Petri net (see Fig. 11), is obtained from a textual description of this process [17]. The log[6] is extracted from the information system of a municipality. It contains 150,370 traces of which 231 are distinct and a total of 561,470 events. This log contains a number of anomalies, presumably due to noise and other factors. Examples are traces ⟨*Create fine* → *End*⟩ and ⟨*Create Fine* → *Send Fine* → *End*⟩, which cannot be replayed in the model.

Covering all log behavior will naturally increase fitness, but at the same time will result in a highly complex and over-fitting model. From the results reported in Table 7, we can see that both baselines cover *all* log behavior (perfect fitness), but result in a very low precision and hence F-Score, and a repaired model that is very different from the original one. In this table, we also report model size as the sum of the number of places and transitions in the Petri net.

---

[6] http://dx.doi.org/10.4121.

Table 6. Evaluation results on the synthetic datasets

		Add/Remove	Cond./Seq.	Conc./Seq.	Loop	Skip	Cond. move	Conc. move	Synchronize	Duplicate	Replace	Swap	IOR	IRO	OIR	ORI	RIO	ROI
Base1 [21]	Similarity	0.74	0.68	0.88	0.57	0.82	0.72	0.70	0.88	0.67	0.73	0.63	0.60	0.70	0.62	0.82	0.80	0.72
	Fitness	1.00	1.00	1.00	1.00	1.00	1.00	1.00	1.00	1.00	1.00	1.00	1.00	1.00	1.00	1.00	1.00	1.00
	Precision	0.68	0.94	0.92	0.36	0.87	0.70	0.72	0.95	0.52	0.64	0.55	0.59	0.45	0.46	0.92	0.84	0.50
	F-Score	0.81	0.97	0.96	0.53	0.93	0.82	0.84	0.97	0.69	0.78	0.71	0.74	0.62	0.63	0.96	0.91	0.67
Base2 [10]	Similarity	0.79	0.79	0.88	0.78	0.77	0.75	0.74	0.88	0.82	0.79	0.71	0.82	0.88	0.71	0.79	0.83	0.74
	Fitness	1.00	1.00	1.00	1.00	1.00	1.00	1.00	1.00	1.00	0.95	1.00	0.95	0.81	0.76	1.00	1.00	0.8
	Precision	0.97	0.98	0.92	0.95	0.85	0.93	0.93	0.95	0.98	0.97	0.93	0.97	0.97	0.88	0.95	0.94	0.83
	F-Score	0.99	0.99	0.96	0.98	0.92	0.97	0.96	0.97	0.99	0.96	0.97	0.96	0.88	0.82	0.98	0.97	0.86
Ours	Similarity	0.86	0.92	0.97	0.95	0.93	0.90	0.88	0.92	0.94	0.93	0.95	0.86	0.90	0.87	0.91	0.97	0.91
	Fitness	1.00	1.00	1.00	1.00	1.00	1.00	1.00	0.94	1.00	1.00	1.00	1.00	1.00	1.00	1.00	1.00	1.00
	Precision	0.97	0.98	0.92	0.96	0.97	0.98	0.98	0.97	0.98	0.97	0.97	0.95	0.97	0.96	0.95	0.98	0.95
	F-Score	0.99	0.99	0.96	0.98	0.99	0.99	0.99	0.95	0.99	0.99	0.99	0.98	0.98	0.98	0.98	0.99	0.98

These results exacerbate the differences between the three approaches already exposed in Table 6, clearly demonstrating the advantages of our approach over the two baselines. Our approach reduces the fitness of the original model by 0.07, but dramatically increases the precision and, thus the F-score. In addition, it produces a much more readable model (the size is almost half of that of the baselines) which is very close to the original model (the similarity is 0.90 vs. 0.46 for Base1 and 0.55 for Base2). Figure 12 shows the repaired model obtained by our approach, while Fig. 13 shows the model obtained by Base2 (for the sake of comparison, we show the model in Petri nets).

Table 7. Evaluation results on the real-life dataset

	Similarity	Fitness	Precision	F-score	Size
Original model	–	0.99	0.77	0.87	28
Base1 [21]	0.46	**1.00**	0.45	0.62	46
Base2 [10]	0.55	**1.00**	0.49	0.65	50
Ours	**0.90**	0.92	**0.90**	**0.91**	29

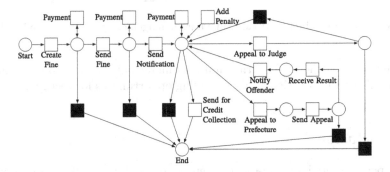

Fig. 11. Normative model of the road traffic fines management process (source: [17])

In this second experiment, we only applied the top two differences in terms of number of affected traces, as identified by our tool. Next, we tried a subsequent repair iteration: this increased the precision to 0.93 and the F-Score to 0.92 at the cost of reducing the similarity to 0.86 and increasing the size to 31 nodes. Given that in this dataset the original model is a normative specification, it is up to the user to select which model-log discrepancies to repair based on domain knowledge. In fact, unfitting behavior could be the result of non-compliance, and as such related discrepancies should not be applied to the model, but rather provide opportunities to rectify current practices. Alternatively, they may expose practical workarounds to improve performance, which could in principle be imported

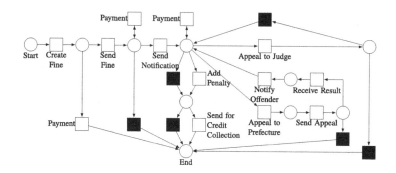

Fig. 12. Our repaired model of the road traffic fines management process

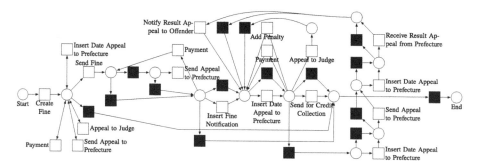

Fig. 13. Process model for the road traffic fines management process repaired by Base2

into the normative model. In turn, additional model behavior may point to norms that are ignored in practice, again providing opportunities for rectifying current practices.

6 Conclusion

This paper presented a process model repair approach that differs from previous proposals in that it does not seek to fix each discrepancy automatically by adding behavior, but rather, it overlays each discrepancy on top of the model, and lets the user decide incrementally which discrepancies to fix and how, based on the provided visual guidance. Further, the fixes suggested through visual guidance are more accurate (in the sense that they do not add behavior w.r.t. the log), than those provided by state-of-the-art model repair methods. This characteristic is confirmed by the empirical evaluation, which showed that our approach leads to repaired models with a higher F-Score and higher structural similarity relative to two state-of-the-art process model repair methods. The empirical evaluation is limited to a collection of synthetically modified model-log pairs, and one real-life model-log pair. This restricted dataset is a potential threat to the validity of the findings. Conducting a more comprehensive evaluation with further real-life model-log pairs is thus an avenue for future work.

While interactivity is a strength of our approach, it is also a potential limitation insofar as the effort required to repair model-log pairs with many discrepancies may be prohibitive. Another avenue for future work is to automatically identify sets of compatible discrepancies that affect the same model fragment, which can be repaired together in a way that leads to a very similar model with higher F-Score.

Acknowledgments. We thank Artem Polyvyanyy and Raffaele Conforti for their feedback on earlier versions of this work. This research is funded by the Australian Research Council (grant DP150103356) and the Estonian Research Council (grant IUT20-55).

References

1. Adriansyah, A., Muñoz-Gama, J., Carmona, J., van Dongen, B., van der Aalst, W.: Measuring precision of modeled behavior. ISeB **13**(1), 37–67 (2015)
2. Adriansyah, A., van Dongen, B.F., van der Aalst, W.M.P.: Conformance checking using cost-based fitness analysis. In: Proceedings of the EDOC. IEEE Computer Society (2011)
3. Andrews, K., Wohlfahrt, M., Wurzinger, G.: Visual graph comparison. In: 2009 13th International Conference on Information Visualisation, pp. 62–67. IEEE (2009)
4. Armas-Cervantes, A., Baldan, P., Dumas, M., García-Bañuelos, L.: Diagnosing behavioral differences between business process models: an approach based on event structures. Inf. Syst. **56**, 304–325 (2016)
5. Armas-Cervantes, A., van Beest, N.R.T.P., La Rosa, M., Dumas, M., Raboczi, S.: Incremental and interactive business process model repair in apromore. In: Proceedings of the BPM Demos. CRC Press (2017, to appear)
6. van den Brand, M., Protić, Z., Verhoeff, T.: Generic tool for visualization of model differences. In: Proceedings of the 1st International Workshop on Model Comparison in Practice, pp. 66–75. ACM (2010)
7. Buijs, J.C.A.M., La Rosa, M., Reijers, H.A., van Dongen, B.F., van der Aalst, W.M.P.: Improving business process models using observed behavior. In: Cudre-Mauroux, P., Ceravolo, P., Gašević, D. (eds.) SIMPDA 2012. LNBIP, vol. 162, pp. 44–59. Springer, Heidelberg (2013). doi:10.1007/978-3-642-40919-6_3
8. Dijkman, R., Dumas, M., García-Bañuelos, L.: Graph matching algorithms for business process model similarity search. In: Dayal, U., Eder, J., Koehler, J., Reijers, H.A. (eds.) BPM 2009. LNCS, vol. 5701, pp. 48–63. Springer, Heidelberg (2009). doi:10.1007/978-3-642-03848-8_5
9. Dumas, M., La Rosa, M., Mendling, J., Reijers, H.: Fundamentals of Business Process Management. Springer, Heidelberg (2013). doi:10.1007/978-3-642-33143-5
10. Fahland, D., van der Aalst, W.M.P.: Model repair - aligning process models to reality. Inf. Syst. **47**, 220–243 (2015)
11. Gambini, M., La Rosa, M., Migliorini, S., Ter Hofstede, A.H.M.: Automated error correction of business process models. In: Rinderle-Ma, S., Toumani, F., Wolf, K. (eds.) BPM 2011. LNCS, vol. 6896, pp. 148–165. Springer, Heidelberg (2011). doi:10.1007/978-3-642-23059-2_14

12. García-Bañuelos, L., van Beest, N.R., Dumas, M., La Rosa, M.: Complete and interpretable conformance checking of business processes. IEEE Trans. Softw. Eng. (2017, to appear)
13. Geyer, M., Kaufmann, M., Krug, R.: Visualizing differences between two large graphs. In: Brandes, U., Cornelsen, S. (eds.) GD 2010. LNCS, vol. 6502, pp. 393–394. Springer, Heidelberg (2011). doi:10.1007/978-3-642-18469-7_38
14. Kriglstein, S., Wallner, G., Rinderle-Ma, S.: A visualization approach for difference analysis of process models and instance traffic. In: Daniel, F., Wang, J., Weber, B. (eds.) BPM 2013. LNCS, vol. 8094, pp. 219–226. Springer, Heidelberg (2013). doi:10.1007/978-3-642-40176-3_18
15. La Rosa, M., ter Hofstede, A.H.M., Wohed, P., Reijers, H.A., Mendling, J., van der Aalst, W.M.P.: Managing process model complexity via concrete syntax modifications. IEEE Trans. Ind. Inform. 7(2), 255–265 (2011)
16. Maaradji, A., Dumas, M., La Rosa, M., Ostovar, A.: Fast and accurate business process drift detection. In: Motahari-Nezhad, H.R., Recker, J., Weidlich, M. (eds.) BPM 2015. LNCS, vol. 9253, pp. 406–422. Springer, Cham (2015). doi:10.1007/978-3-319-23063-4_27
17. Mannhardt, F., de Leoni, M., Reijers, H.A., van der Aalst, W.M.P.: Balanced multi-perspective checking of process conformance. Computing 98(4), 407–437 (2016)
18. Moody, D.: The "physics" of notations: toward a scientific basis for constructing visual notations in software engineering. IEEE Trans. Softw. Eng. 35(6), 756–779 (2009)
19. Nielsen, M., Plotkin, G.D., Winskel, G.: Petri nets, event structures and domains, part I. Theoret. Comput. Sci. 13, 85–108 (1981)
20. Ohst, D., Welle, M., Kelter, U.: Differences between versions of UML diagrams. In: ACM SIGSOFT Software Engineering Notes, vol. 28, pp. 227–236. ACM (2003)
21. Polyvyanyy, A., van der Aalst, W.M.P., ter Hofstede, A.H.M., Wynn, M.T.: Impact-driven process model repair. ACM Trans. Softw. Eng. Methodol. 25(4), 28:1–28:60 (2016)
22. Rogge-Solti, A., Senderovich, A., Weidlich, M., Mendling, J., Gal, A.: In log and model we trust? A generalized conformance checking framework. In: La Rosa, M., Loos, P., Pastor, O. (eds.) BPM 2016. LNCS, vol. 9850, pp. 179–196. Springer, Cham (2016). doi:10.1007/978-3-319-45348-4_11
23. van Beest, N.R.T.P., Dumas, M., García-Bañuelos, L., La Rosa, M.: Log delta analysis: interpretable differencing of business process event logs. In: Motahari-Nezhad, H.R., Recker, J., Weidlich, M. (eds.) BPM 2015. LNCS, vol. 9253, pp. 386–405. Springer, Cham (2015). doi:10.1007/978-3-319-23063-4_26
24. van der Aalst, W.M.P.: Process Mining: Data Science in Action. Springer, Heidelberg (2016). doi:10.1007/978-3-662-49851-4
25. Weber, B., Reichert, M., Rinderle-Ma, S.: Change patterns and change support features: enhancing flexibility in process-aware information systems. Data Knowl. Eng. 66(3), 438–466 (2008)

Control Flow Structure Preservation During Process Fragment Anonymization

(Short Paper)

Kristof Böhmer[(✉)] and Stefanie Rinderle-Ma

Faculty of Computer Science, University of Vienna, Vienna, Austria
{kristof.boehmer,stefanie.rinderle-ma}@univie.ac.at

Abstract. Existing process anonymization work does not yet support control flow anonymization. However, as the control flow is one core component of business processes and represents confidential information, this can result in an information leakage. Hence, a control flow anonymization approach is proposed. It merges multiple control flows from the same process collection to minimize the similarity between the original and the anonymized one. At the same time, the original control flow structure is preserved to foster the representativity of the anonymized flows. The approach is prototypically implemented and evaluated with 10,987 business process models from multiple process collections.

Keywords: Business process · Fragment · Sub-graph · Anonymization

1 Introduction

Business processes represent *confidential* business behavior so that organizations are typically unwilling to share them [18]. Literature tackles this challenge based on *anonymization* approaches, cf. [18]. Generally, anonymization aims at abstracting confidential information while preserving certain properties cf. [18,20]. Hence, this paper aims at abstracting confidential process details (Objective 1) while preserving the original control flow structure (Objective 2), such that, for example, benchmarks conducted with the anonymized and original non-anonymized processes reveal the same bottlenecks. To address these objectives we propose to extract Representative Process Fragments (RPF). RPFs, that solely consist of the control flow structure, are already sufficient for various application areas, such as but not limited to, process benchmarking cf. [19,20].

We define RPFs as anonymized process fragments which enable to reconcile two apparently contradictory aims. This is, that on one hand the structure of the fragment generation processes is *preserved* (addressing Objective 2), while on the other hand the original control flow can hardly be deduced from RPFs to increase the control flow confidentiality (addressing Objective 1). The first aspect is a necessity to ensure the representativity of the fragments so that they can replace their original counterpart during analysis and reasoning [20]. If the anonymized

H. Panetto et al. (Eds.): OTM 2017 Conferences, Part I, LNCS 10573, pp. 75–83, 2017.
https://doi.org/10.1007/978-3-319-69462-7_6

fragments are not representative to the original ones then misleading results could follow. For example, a process execution engine could be optimized (based on benchmarks conducted with non representative anonymized fragments) for a control flow structure that hardly ever occurs in an organization's processes. The second aspect focuses on hardening the extraction of the original control flow (i.e., deanonymization) from given RPFs. *Deanonymization* would require to determine which parts of the anonymized control flow origin from the original one or were added/abstracted by the proposed anonymization approach.

Structure preservation (see Objective 1) refers to the possible paths that can be followed (i.e., the amount – *complexity* – and order – *paths* from nodes without ingoing edges to nodes without outgoing edges – of control flow nodes) in an anonymized fragment (e.g., a RPF) compared to its original, non-anonymized, fragment counterpart. It is assumed that the more similar these two properties (i.e., complexity and paths) are between the original and the anonymized fragment the more representative the anonymized fragments are when being, for example, applied for process benchmarking purposes. In this work we opted to focus on the process structure (i.e., the control flow perspective) as it is sufficient for a wide range of application areas but also abstracts from some confidential information, such as, activity labels, cf. [18,20]. Because RPFs can be generated with a configurable anonymization strength (i.e., measured based on the *similarity* of the original and the anonymized fragment, e.g., a low similarity indicates a better anonymization) they can be customized for specific use cases.

*RPF*s preserve, in comparison, e.g., to abstracted synthetic original fragment representations, more of the original fragment's control flow structure. To achieve this the proposed approach conducts an anonymization that enriches the original non-anonymized fragment structure with details extracted from the structure found in a given set of processes R to anonymize it (see Objective 2). Through this all anonymization driven fragment control flow structure modifications are also representative for the control flow structure in R.

This paper is organized as follows: The proposed anonymization approach is presented in Sect. 2. Section 3 holds the evaluation. Section 4 discusses related work. Conclusions and future work is given in Sect. 5.

2 Control Flow Anonymization

This paper proposes a multi stage process subgraph anonymization approach. It exploits the graph like nature of business processes and combines approaches from graph and process theory. For this, each process P is represented by a directed control flow graph, i.e., $P := (N, E)$. Sets N and $E \subseteq N \times N$ hold P's **N**odes and **E**dges. Function t maps each node onto its type, i.e., $t : N \rightarrow \{activity, gateway, \cdots, start, end\}$, i.e., this work focuses on the BPMN 2.0 core [23]. Functions $n\bullet := \{n' \in N \mid (n, n') \in E\}$ and $\bullet n := \{n' \in N \mid (n', n) \in E\}$ determine the preceding or succeeding nodes of a node n. Such a brief definition is sufficient because: *(a)* this work focuses on the control flow as we assume that all processes P were already *pre-anonymized* by removing node labels [18]; and *(b)* BPMN 2.0 core is sufficient to represent most BPMN process models, cf. [23].

Figure 1 provides an overview on the proposed business process subgraph anonymization heuristic. Initially, see ①, a set of processes R and a process $P \in R$ to extract anonymized RPFs from P is given. The first idea is to identify similar control flow structures (i.e., subgraphs) in P and all $P' \in R \setminus \{P\}$, see ② and ③.

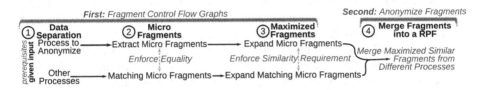

Fig. 1. RPF generation/fragment anonymization approach – overview

The first step (②) fragments all processes in R into subgraphs of size $sms \in \mathbb{N}$. Imagine that step ② is applied on following control flow graph: $P := (\{A1, A2, A3\}, \{(A1; A2), (A2; A3)\})$. When assuming $sms := 2$ then subgraphs P_1/P_2 are generated: $P_1 := (\{A1, A2\}, \{(A1; A2)\})$ and $P_2 := (\{A2, A3\}, \{(A2; A3)\})$.

Typically this step will use a relatively small value for sms as the subgraphs generated in step ② will be expanded in the following steps while being utilized as starting points for RPF generation. Hence, the initially generated subgraphs will be relatively small – so we refer to them as micro subgraphs or *micro fragments*. The term "fragment" is driven by the process modeling domain. It refers to connected control flow graphs with relaxed completeness and consistency criteria compared to executable process graphs, cf. [16].

We slightly extend the previous definition by specifying that a, micro, fragment (or RPF resp.) is not necessarily related to a process model. Here this means that it can also represent a control flow graph that is not a subgraph of any $P \in R$. So, a fragment f is defined as $f := (N, E, P, N_0)$, i.e., a set of **N**odes and **E**dges. Moreover, N_0 is the set of nodes that was initially used to construct f. If f refers to a process model P, f is a subgraph of P.

The second step, see ③, identifies similar fragments (i.e., subgraphs) with maximum size in P and $R \setminus \{P\}$ respectively. This requires to tackle four challenges: *(1)* Each micro fragment f must be able to "grow" by expanding it with additional nodes and edges held by $f.P$; *(2)* the similarity between different fragments must be measured; *(3)* it must be decided when a fragment has reached its maximum size; *(4)* for each maximized fragment $f := (N, E, P, N_0)$, a list of similar maximized fragments from processes in $R \setminus \{P\}$ must be found.

The first challenge is addressed by an *expansion function* ϵ. Its key idea is to expand a fragment $f := (N, E, P, N_0)$ by adding nodes and edges to N and E which are in a direct neighborhood relationship to N based on P. By applying ϵ multiple times a fragment can grow stepwise, e.g., till it covers all nodes in P.

Assume a fragment $f := (\{A2\}, \{\}, P, \{A2\})$ where $P := (\{A1, A2, A3, A4\}, \{(A1; A2), (A2; A3), (A3; A4)\})$. Then $\epsilon(f) := f'$ so that $f'.N = \{A1, A2, A3\}$,

$f'.E := \{(A1; A2), (A2; A3)\}$, and $f'.N_0 = \{A2\}$. Note, $f'.N$ does not hold $A4$ because it is not covered by the expansion in 1 step. N_0 is preserved as the proposed similarity calculation and anonymization approaches exploit it.

The second challenge is met by presenting a control flow graph *similarity* notion. Here, the similarity between control flows is measured by a novel greedy heuristic that combines two strategies. First, to measure the similarity of individual nodes based on their node type equality. Secondly, to measure the similarity of fragments by dividing them into micro-fragments and iteratively (up to user chosen $c \in \mathbb{N}_{>0}$ steps) expanding and aligning the most similar ones. Hereby, the neighborhood of each node is incorporated into the similarity calculation. Fragments are referred to as similar if their similarity is above $ms \in [0,1]$.

The third challenge: each micro fragment f that was extracted from P is expanded until the number of processes in $R \setminus \{P\}$ that contain a fragment which is similar to f falls below a certain percentage $mk \in [0,1]$ of $|R \setminus \{P\}|$.

Finally, challenge four must be met. For a given maximized fragment f and for all processes X in $R \setminus \{P\}$ the set of maximized similar fragments that are similar to f must be determined (i.e., determine $SF_f := \{(X, RF) \mid X \in R \setminus \{P\} \wedge RF := \{f' = (N', E', X, N_0') \mid f' \text{ similar to } f\}\}$).

Then the proposed anonymization technique is applied, see ④. Each fragment f and the set of similar fragments in SF_f are merged using an adjacency matrix based merging heuristic. For this, artificial labels are generated to represent both fragment control flows as adjacency matrices. Subsequently, those matrices are superimposed and transformed into a merged control flow graph. The space of possible label assignments is heuristically explored. While doing so the quality of an identified merge result is assumed as being inversely proportional to the increase in complexity of the original fragment compared to the merged one.

The merged fragments, i.e., RPF's, are only similar (i.e., structure is preserved), but not equal (e.g., they stem from multiple processes). So an attacker will struggle to decide if a node/edge in a RPF belongs to the original f or if it was introduced due to the proposed anonymization approach.

3   Evaluation

The test data which was used for the evaluation was extracted from $114,577$ real world process models. The models stem from two process collections created by the Cloud Process Execution Engine (CPEE), cf. http://cpee.org/, and the BPM Academic Initiative (BPMAI), cf. http://bpmai.org/. All available models were filtered based on *(a)* minimal node count, i.e., 12; and *(b)* minimal amount of gateways, i.e., 4. Overall a set of $10,987$ models remained.

Metrics and Evaluation. The evaluation analyzes the feasibility of the presented anonymization approach, i.e., if anonymized RPFs *(a)* are dissimilar to their original non-anonymized fragment counterpart *of* to harden their deanonymization; and *(b)* have a similar complexity as *of*; and *(c)* are representative (i.e., preserve the structure of *of* without adding too much new paths and nodes) so that a *rpf/of* show similar analysis (e.g., benchmark) results;

and *(d)* can be generated with different anonymization strengths to meet differ-ent requirements.

Hence, the proposed approach processes each $P \in R$ to extract all possible anonymized RPFs. Four key values are determined for each RPF (rpf respec-tively), its non-anonymized counterpart of, and a synthetic fragment sf that was generated from of to represent existing graph anonymization approaches, e.g., applied in differential graph privacy, cf. [9]. First, the *similarity* between rpf/sf and of is calculated to quantify the achieved anonymization. It is assumed that a low similarity hardens it to deduce which parts in rpf's or sf's have an equal counterpart in of. Secondly, the complexity of rpf/sf is compared with of because a similar complexity fosters the representativity of the anonymized fragments so that rpf/sf and of yield similar analysis, e.g., benchmark, results.

Thirdly, *path preservation* measures if the paths in of are preserved in rpf and sf. For this, all possible paths are extracted from of and rpf/sf and their differences (e.g., if a gateway is expected based on of but an activity was found) are counted as the path preservation. So, a lower path preservation value for rpf and sf indicates that the anonymized fragments are more representative. Fourthly, *path pollution* represents the number of possible paths in of in com-parison to rpf/sf. Through this path pollution indicates how many paths were added or lost in rpf/sf relatively to of. Overall, the path pollution should be close to one so that the path count in of and rpf/sf is as equal as possible. In the following rpf's are generated with two different anonymization strengths (i.e., low/high anonymization) to show the configurability of the proposed approach.

The *synthetic fragment* (sf) generation is motivated by differential privacy graph anonymization, cf. [9]. For this a sf is randomly generated from a statisti-cal representation of of (e.g., the number of activities). An *optimal* anonymiza-tion of of has a low similarity, similar complexity, path preservation of zero (i.e., no difference between the paths), and path pollution of one (i.e., equal path count). As these aspects are diametrical (i.e., structure preservation ↔ low similarity) the proposed approach aims at identifying a compromise between them. The given results are an average of 100 evaluation runs to even out the randomness of the applied approaches (e.g., the random generation of sf).

Results. The results were generated by applying a Java 1.8 based proof-of-concept implementation of the presented approach on the CPEE/BPMAI process models. On average it only took seconds to anonymize a fragment and minutes to anonymize all fragments extracted from a process model on a 2.6 Ghz Intel Q6300 CPU and 32 GB RAM. This suggests an applicability on larger processes (the average process node count was 32.5) and process repositories.

Primary tests were applied to identify appropriate configuration values for the presented approach. This is the size of the generated micro fragments, $sms := 2$. Moreover, when expanding the micro fragments, the minimal fragment similarity was $ms := 0.5$ (high anonymization) and $ms := 0.6$ (low anonymization). The minimal amount of processes in R' that must contain similar fragments was set to $mk := 0.4$ (high anon.) and $mk := 0.5$ (low anon.). Finally, the neighborhood size for the neighborhood similarity calculating was defined as $c := 6$.

It was found that ms and mk have a significant impact on the achieved anonymization strength. They enable to control the *balance* between *anonymization* and *structure preservation*. For example, lower ms/mk values were found to result in a stronger anonymization (i.e., a lower similarity between of/rpf) but a worse preservation of the control flow structure (i.e., the rpf had a higher complexity and more paths than of). In comparison varying the sms and c values mainly had an effect on the required computing time, e.g., larger values for sms resulted in lower computation times as less micro fragments are generated. In comparison larger values for c increased the required computation time as the similarity calculation incorporates more information when c is increased.

Both compared anonymization approaches provided a similar anonymization strength (i.e., similar similarity of of and the anonymized fragments). In addition the synthetically generated fragments (i.e., sf) were found to have a complexity which is close to the complexity of the original fragment (i.e., 1.02 and 1.05 on average). This was expected as synthetic fragments are generated in a way that ensures that their amount of edges/nodes is close to the original fragment. In comparison the proposed approach merges multiple similar fragments, which not only achieves the anonymization but also increases the amount of nodes/edges, i.e., its complexity is 1.29 to 1.55 compared to the original fragment of.

In regards to the representativity of the anonymized fragments (i.e., the preservation of the control flow structure) a different picture emerges. The path preservation value that indicates the number of differences between the paths in the anonymized/non-anonymized fragments is between 1.64 and 11.69 for the proposed approach and 3.96 to 15.34 for the comparison approach. Hence, the proposed approach achieves a better preservation of the structure. This increases the likelihood that the analysis results generated with RPF fragments are comparable to the ones generated with the original fragment (e.g., during benchmarks). This observation is also supported by the measured path pollution. While this value was found to be greater than 1 for the RPF based approach (i.e., more paths are represented in the anonymized fragment than in the original one) it was lower than 1 for the fragments generated by the comparison approach. Hence, less paths were present in sf than in the original fragment of. This indicates that control flow structure was lost during the anonymization.

But more importantly, the generated RPFs are assumed by us as being more representative than the fragments generated by the compared anonymization approach as they preserve more of the control flow structure defined in the original fragments. Through this the RPFs, likely, result in analysis and application results which are more comparable to the original fragments (of) than the results generated based on synthetic fragments sf (i.e., the compared approach) (Table 1).

Table 1. Aggregated evaluation results for all extracted fragments and collections

rpf is left of |, sf is right of |	Similarity	Complexity	Path preservation	Path pollution
BPMAI low anonymization	0.54 | 0.52	1.55 | 1.05	1.64 | 3.96	2.10 | 0.85
BPMAI high anonymization	0.40 | 0.43	1.42 | 1.03	3.67 | 5.23	1.46 | 0.68
CPEE low anonymization	0.54 | 0.44	1.29 | 1.02	10.58 | 13.71	1.04 | 0.52
CPEE high anonymization	0.42 | 0.37	1.49 | 1.04	11.69 | 15.34	1.11 | 0.29

4   Related Work

To the extent of our knowledge a process anonymization approach that focuses on BPMN 2.0 fragment anonymization was not yet proposed. Comparable work, e.g., [18] anonymizes only BPEL models (e.g., by removing labels) and does not provide structure anonymization. Generic graph based approaches, such as, *differential privacy* [9], lack the support for BPMN 2.0 and also do not provide structure preservation. In comparison multiple non-anonymization focused fragmentation approaches are available, e.g., [5, 7, 15–17, 20]. Unfortunately these works only extract exact clones, do not support cyclic graphs (loops), do "only" identify all recurring fragments or require node labels, e.g. [15], which are assumed as being not available for this work. A similar situation becomes obvious when analyzing graph focused fragmentation approaches, such as, [3, 13, 21].

The work in [6, 8, 12] identifies approximate clones in process model collections, i.e., these techniques could also be applied to identify similar fragments which are required by the proposed approach. But [6, 8, 12] is not applicable here because they only support Single Entry Single Exit (SESE) fragments while the proposed approach utilizes Multiple Entries Multiple Exists (MEME) fragments that provide more flexibly and through this aid the achieved anonymization strength. In general related process similarity calculation work [1, 4, 14, 22] applies a less rigid focus on the control flow than this work. These approaches frequently rely on labels as a core component of their similarity calculation approaches, cf. [16]. APROMORE [10] also provides similarity calculation techniques but only focuses on calculating the similarity between two process models or to identify selected patterns in process model collections.

This work is also related to process merging. Existing approaches, such as, [2, 7, 11], often focus on the merging of complete process model definitions, apply SESE based approaches, or are not capable of merging fragments which only hold structural information (e.g., because labels are mandatory and exploited to identify similar process nodes).

5   Conclusion and Discussion

To apply the proposed approach the anonymized process collections are required to hold processes with a minimum structural diversity. Imagine, for example, that all process models in a process collection only consist of solely sequential control flows. Then, the proposed merge based anonymization would likely not find enough control flow diversity to achieve the shown anonymization performance. However, we assume such a situation as being rather unlikely.

In future work, we will evaluate if the presented approach can be extended to preserve additional model characteristics (e.g., the data flow) during the anonymization. Moreover we will apply RPFs to generate anonymized business process models and process model collections.

References

1. Becker, M., Laue, R.: A comparative survey of business process similarity measures. Comput. Ind. **63**(2), 148–167 (2012)
2. Böhmer, K., Rinderle-Ma, S.: Difference-preserving process merge. In: Demey, Y.T., Panetto, H. (eds.) OTM 2013. LNCS, vol. 8186, pp. 718–721. Springer, Heidelberg (2013). doi:10.1007/978-3-642-41033-8_92
3. Cordella, L.P., et al.: A (sub) graph isomorphism algorithm for matching large graphs. Pattern Anal. Mach. Intell. **26**(10), 1367–1372 (2004)
4. Dijkman, R., et al.: Similarity of business process models: metrics and evaluation. Inf. Syst. **36**(2), 498–516 (2011)
5. Dumas, M., et al.: Fast detection of exact clones in business process model repositories. Inf. Syst. **38**(4), 619–633 (2013)
6. Ekanayake, C.C., Dumas, M., García-Bañuelos, L., La Rosa, M., ter Hofstede, A.H.M.: Approximate clone detection in repositories of business process models. In: Barros, A., Gal, A., Kindler, E. (eds.) BPM 2012. LNCS, vol. 7481, pp. 302–318. Springer, Heidelberg (2012). doi:10.1007/978-3-642-32885-5_24
7. Gao, X., et al.: Process model fragmentization, clustering and merging: an empirical study. In: Lohmann, N., Song, M., Wohed, P. (eds.) BPM 2013. LNBIP, vol. 171, pp. 405–416. Springer, Cham (2014). doi:10.1007/978-3-319-06257-0_32
8. Ivanov, S., Kalenkova, A., van der Aalst, W.M.: Bpmndiffviz: a tool for bpmn models comparison. In: BPM (Demos), pp. 35–39 (2015)
9. Kasiviswanathan, S.P., Nissim, K., Raskhodnikova, S., Smith, A.: Analyzing graphs with node differential privacy. In: Sahai, A. (ed.) TCC 2013. LNCS, vol. 7785, pp. 457–476. Springer, Heidelberg (2013). doi:10.1007/978-3-642-36594-2_26
10. La Rosa, M., Reijers, H.A., Van Der Aalst, W.M., Dijkman, R.M., Mendling, J., Dumas, M., GarcíA-BañUelos, L.: Apromore: an advanced process model repository. Expert Syst. Appl. **38**(6), 7029–7040 (2011)
11. La Rosa, M., et al.: Business process model merging: an approach to business process consolidation. TOSEM **22**(2), 11: 1–11: 42 (2013)
12. La Rosa, M., et al.: Detecting approximate clones in business process model repositories. Inf. Syst. **49**, 102–125 (2015)
13. Martínez, C., Valiente, G.: An algorithm for graph pattern-matching. Workshop String Process. **8**, 180–197 (1997)
14. Pietsch, P., Wenzel, S.: Comparison of BPMN2 diagrams. In: Mendling, J., Weidlich, M. (eds.) BPMN 2012. LNBIP, vol. 125, pp. 83–97. Springer, Heidelberg (2012). doi:10.1007/978-3-642-33155-8_7
15. Pittke, F., Leopold, H., Mendling, J., Tamm, G.: Enabling reuse of process models through the detection of similar process parts. In: La Rosa, M., et al. (eds.) BPM 2012. LNBIP, vol. 132, pp. 586–597. Springer, Heidelberg (2013). doi:10.1007/978-3-642-36285-9_59

16. Skouradaki, M., Göerlach, K., Hahn, M., Leymann, F.: Application of sub-graph isomorphism to extract reoccurring structures from BPMN 2.0 process models. In: Service-Oriented System Engineering, pp. 11–20 (2015)

17. Skouradaki, M., Leymann, F., Nikolaou, C., Leymann, F.: Detecting frequently recurring structures in BPMN 2.0. In: SummerSOC 2015, pp. 102–116 (2015)

18. Skouradaki, M., Roller, D., Pautasso, C., Leymann, F.: "BPELanon": anonymizing BPEL processes. In: Services and their Composition (2014)

19. Skouradaki, M., et al.: On the road to benchmarking BPMN 2.0 workflow engines. In: Performance Engineering, pp. 301–304 (2015)

20. Skouradaki, M., Andrikopoulos, V., Kopp, O., Leymann, F.: RoSE: reoccurring structures detection in BPMN 2.0 process model collections. In: Debruyne, C., et al. (eds.) COOPIS 2016. LNCS, pp. 263–281. Springer, Cham (2016). doi:10.1007/978-3-319-48472-3_15

21. Ullmann, J.R.: An algorithm for subgraph isomorphism. JACM **23**(1), 31–42 (1976)

22. Van Dongen, B., Dijkman, R., Mendling, J.: Measuring similarity between business process models. In: Information Systems Engineering, pp. 405–419. Springer (2013)

23. Muehlen, M., Recker, J.: How much language is enough? theoretical and practical use of the business process modeling notation. In: Bellahsène, Z., Léonard, M. (eds.) CAiSE 2008. LNCS, vol. 5074, pp. 465–479. Springer, Heidelberg (2008). doi:10.1007/978-3-540-69534-9_35

Diversity-Aware Continuous Top-k Queries in Social Networks

(Short Paper)

Abdulhafiz Alkhouli[✉] and Dan Vodislav

ETIS, ENSEA, Univ. of Cergy-Pontoise, CNRS, Cergy-Pontoise, France
{abdulhafiz.alkhouli,dan.vodislav}@ensea.fr

Abstract. We consider here the problem of adding diversity requirements for the results of continuous top-k queries in a large scale social
network, while preserving an efficient, continuous processing. We propose the DA-SANTA algorithm, which smoothly adds content diversity
to the continuous processing of top-k queries at the social network scale.
The experimental study demonstrates the very good properties in terms
of effectiveness and efficiency of this algorithm.

Keywords: Information streams · Social networks · Diversity · Continuous top-k query processing · Publish/subscribe systems

1 Introduction and Related Work

We consider here the context of *top-k continuous queries* over text information
streams produced in social networks. Efficient processing of such queries at the
social network scale requires *continuous processing techniques* that incrementally maintain the top-k list of each user in reaction to social network events
(new message, interaction with message). However, existing methods have difficulties to handle complex scoring functions, including social network criteria,
and usually focus on content-based relevance and time-based factors favoring
more recent messages. In previous work [1], we proposed an efficient method for
continuous processing of top-k queries over information streams in a large-scale
social network, using a relevance model with content-based, time-based and rich
social network factors (user- and interaction-based). We extend this work here,
to introduce *results diversification* into the continuous processing method.

Result diversification [4] aims at avoiding redundancy and too homogeneous
results to often imprecise user queries. For instance, a query about "olympic
games" could get only items about the 2024 games abundantly discussed recently,
while the user interest may be different. The *content diversity* of a results set is
generally measured either by the average or the minimum distance between all
the results. The general approach to add diversity to top-k querying is to use a
bi-objective scoring function that combines relevance and diversity.

© Springer International Publishing AG 2017
H. Panetto et al. (Eds.): OTM 2017 Conferences, Part I, LNCS 10573, pp. 84–92, 2017.
https://doi.org/10.1007/978-3-319-69462-7_7

Adding results diversification to continuous processing of top-k queries over information streams at a social network scale is a very challenging task. First, the general diversification problem is NP-hard [7]: given a query over a set O of n objects, find $S_k^* \subseteq O$ of size $k \leq n$, which maximizes a bi-objective function that combines relevance and diversity. Specific approximate methods are necessary, especially in the case of continuously arriving data. For instance, [3,5] limit O to a sliding window of the most recent items, while [9] proposes an incremental approach to maintain an approximate diversified top-k set, where O is limited to the new item plus the current top-k. We adopt the latter definition for O, but propose heuristics to select a victim in the current top-k. Next, the constraint of continuous processing at the social network scale requires very efficient algorithms. The methods above are not adapted to large social networks, since they evaluate each new item with all the queries. The only work to date that proposes a diversity-aware method adapted to a large number of queries is [2]. Like us, they index subscription queries to be able to prune queries not affected by a new message, but focus on grouping queries into blocks for efficiency reasons. They also use a victim selection heuristics that considers the oldest message in the top-k. But their relevance function, favoring query grouping, is very specific and they do not consider social network relevance factors.

Finally, when trying to extend an existing relevance-only approach to add diversity, one must face the model mismatch between top-k computation for relevance, at the element level, and diversity, at the set level. Efficient continuous processing of top-k queries at a large scale is based on *index structures for user queries*, with the specificity that they must include information about μ_q, the k-th current score of each query, the limit to overstep to enter the top-k. Most of them use inverted lists, one per query term. In [8], the index list for term t contains subscription queries q containing t, sorted by w_{tq}/μ_q, where w_{tq} is the weight of t in q. In [10] lists are ordered by a threshold value based on the current top-k of each query. Some methods use different structures, for instance [13] employs an original two-dimensional inverted query indexing scheme combining w_{tq} and μ_q. [12] use a double index per query term: an inverted list ordered by query id and a tree structure organized by μ_q. Methods that focus on *grouping strategies* to handle groups of queries instead of individual ones propose specific index structures, for instance [11] uses a graph to index covering relationships between subscription queries. To extend these methods with results diversification, the indexing technique must be flexible enough to support it. This is not the case for most of them, which also explains the fact that they do not consider more complex scoring components, such as social network factors.

In this context, our main contributions are (i) *a model* that smoothly integrates content-based diversity into the continuous top-k processing model presented in [1], including heuristics for approximate diversification and a query indexing structure for efficient processing of diversity-aware top-k queries at the social network scale, and (ii) *an algorithm*, DA-SANTA (Diversity-Aware Social and Action Network Threshold Algorithm), based on this model, whose effectiveness and efficiency are demonstrated through a set of experiments.

2   Data and Processing Models

Social Network Information Streams. We consider the social network model from [1], with asymmetric relation graphs, where each user produces a single information stream of text messages and issues a single *implicit* subscription query, expressed by the *user profile*. Like messages, user profiles are described by a set of weighted terms expressing the user's points of interest. The importance of the content of a message m for user u is measured by a similarity function sim (e.g. cosine similarity) between m and u's profile $p(u)$.

The model also considers an asymmetric *importance function* f, where $f(u_1, u_2) \in [0, 1]$ is the importance of user u_2 for u_1 in the social network. Note that, even if f is defined for any couple of users in the network, in practice each user has a limited number of users of interest (with $f > 0$), which results into reasonable effort to manage this information.

Relevance Scoring Function. We consider the relevance scoring model proposed in [1], combining content-based, time-based and social network factors.

$$tscore(m, u) = score(m, u) \cdot TB(t^m - t_o) \tag{1}$$

$$score(m, u) = a\ sim(m, p(u)) + b\ f(u, u^m) + c\ G(m) \tag{2}$$

Equation (1) gives the time-based relevance of message m for user u. We use a *time bonus function* $TB : \mathbb{R}_+ \to [1, \infty)$, monotonically increasing, with $TB(0) = 1$, where $t^m \in TS$ is the publishing time for message m and $t_o \in TS$ a fixed origin moment. Time-independent relevance $score(m, u)$ expresses the initial importance of m for u at moment t^m. It combines three elements: (i) *content-based similarity* ($sim(m, p(u))$) between the message and the user profile, (ii) *user-dependent importance of the message* in the social network ($f(u, u^m)$), measured by the importance of the message emitter u^m for user u, and (iii) *user-independent importance of the message* in the social network ($G(m)$), measured at publishing time by the global importance of the emitter in the social network.

Diversity Model. We adopt the commonly used *max-sum diversification* bicriteria objective function [7] to combine relevance and diversity into a single scoring function. If we note $u.TL_k = \{m_1, ..., m_k\}$ the top-k result set for user u, its diversity $D(u.TL_k)$ is given by the sum of distances between the set elements, where $dist(m_i, m_j) = 1 - sim(m_i, m_j)$. The combined relevance-diversity score DR is a linear combination between relevance and diversity.

$$DR(u.TL_k) = \nu\ f_R(u.TL_k) + (1 - \nu)\ f_D(u.TL_k) \tag{3}$$

Here, $f_R(u.TL_k) = \sum_{m \in u.TL_k} rel(m, u)$ expresses the relevance of the top-k list (the *rel* scoring function may be (1) or (2)), while $f_D(u.TL_k) = \frac{2}{k-1} D(u.TL_k)$ measures the diversity score, where the homogeneity factor $2/(k-1)$ compensates the fact that f_R sums k values, while for f_D we have $k(k-1)/2$ values.

Processing Model. We adopt the commonly used approach [2,9] in top-k diversification on streams, to limit the set of objects to the new message plus the current top-k. Hence, for a given user u having the top-k result list $u.TL_k$, when a new message m_{new} arrives, the updated top-k list $u.TL'_k$ will be the subset of size k of $u.TL_k \cup \{m_{new}\}$ that maximizes the relevance-diversity score DR defined in (3). Then the condition for the top-k to be updated is:

$$DR(u.TL'_k) > DR(u.TL_k) \tag{4}$$

The basic algorithm for updating the top-k lists would *repeat the above processing for all the users in the social network*, which raises an important efficiency issue. Also, for each user *the top-k update method is expensive*, since it requires k computations of the DR function.

We propose the DA-SANTA algorithm that provides solutions for both these efficiency problems. DA-SANTA proposes a pruning approach to avoid evaluating all users and employs heuristic methods to choose a single message (victim) to be replaced with the new message.

3   The DA-SANTA Algorithm

DA-SANTA Scoring. For a new published message m_{new} and a given user u, an heuristic function designates $m_{vic} \in u.TL_k$ as potential victim. As shown in Sect. 2, the condition for m_{new} to replace m_{vic} in $u.TL_k$ is $DR(u.TL'_k) > DR(u.TL_k)$, where $u.TL'_k = u.TL_k \cup \{m_{new}\} - \{m_{vic}\}$. We note $u.F_k = u.TL_k - \{m_{vic}\} = u.TL'_k - \{m_{new}\}$ the subset of $k-1$ results for u that do not change when m_{new} replaces m_{vic}. By developing (4) and simplifying the common part that corresponds to $u.F_k$, the update condition becomes:

$$dr_u(m_{new}, u.F_k) > dr_u(m_{vic}, u.F_k) \tag{5}$$

We note $dr_u(m, X) = \nu\, rel(m, u) + (1 - \nu)\, \frac{2}{k-1} D_m(X)$ the *simplified relevance-diversity scoring function*, where $D_m(X) = \sum_{x \in X} dist(m, x)$ can be interpreted as the *diversity of the set X relative to message m. $dr_u(m, X)$ com-bines the relevance of m for u with the diversity of X relative to m. Note that evaluating condition (5) is significantly faster than for the equivalent condition on DR.

Victim Selection Heuristics. We explore two heuristics for choosing the victim message in $u.TL_k$: (1) *Minimum relevance* (MR), which selects the message with the smallest relevance to u: $m_{vic} = argmin_{m \in u.TL_k} rel(m, u)$. (2) *Minimum relevance-diversity* (MRD), which introduces a part of diversity into the heuristics, by selecting the message with the smallest simplified relevance-diversity dr_u: $m_{vic} = argmin_{m \in u.TL_k} dr_u(m, u.TL_k - \{m\})$.

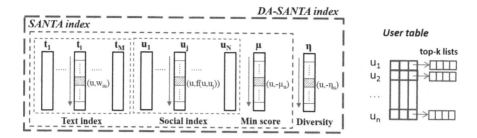

Fig. 1. SANTA and DA-SANTA index and data structures

The DA-SANTA Index. Figure 1 presents the DA-SANTA index structure, as an extension of the SANTA index, composed of sorted lists of users by profile term t_i (text index), by user importance for u_j (social index) and by relevance score limit μ. DA-SANTA adds an extra list η to handle diversity, as follows.

Like for SANTA, we consider a monotonic objective function F_{DA} for the threshold-based strategy, issued from the update condition (5). Here $F_{DA}(m_{new}, u) = dr_u(m_{new}, u.F_k) - dr_u(m_{vic}, u.F_k)$, so (5) is equivalent to $F_{DA}(m_{new}, u) > 0$. By developing dr_u, the update condition becomes:

$$\nu\, rel(m_{new}, u) + (1 - \nu)\, \frac{2}{k-1} D_{m_{new}}(u.F_k) - \nu\, \mu_u - (1 - \nu)\, \frac{2}{k-1}\eta_u > 0 \quad (6)$$

Here $\mu_u = rel(m_{vic}, u)$ is the relevance of the victim for u and $\eta_u = D_{m_{vic}}(u.F_k)$ the diversity of $u.F_k$ relative to m_{vic}. As the choice of m_{vic} is independent of m_{new}, μ_u and η_u are independent from m_{new}, can be computed in advance and maintained after each top-k update.

Like for SANTA, the term in $rel(m_{new}, u)$, when using scoring functions such as (2) or (1) with cosine similarity, is indexed by the text and social indexes. We also have the term in μ_u, indexed by the min-score index, with the difference that the indexed value is here $-rel(m_{vic}, u)$. For the term in η_u, we add a new list η to the index (diversity index), organized like μ but storing the values of $-\eta_u$ in descending order. However, the term in $D_{m_{new}}(u.F_k)$ in (6) cannot be indexed in a similar way. Therefore, we consider an upperbound of $F_{DA}(m_{new}, u)$, by replacing $D_{m_{new}}(u.F_k)$ with $k - 1$, given that $D_{m_{new}}(u.F_k)$ sums $k - 1$ distances $\in [0, 1]$. We note this upperbound F_{DA}^+. With relevance function (2) using cosine similarity, $score(m, u) = a \sum_{t_i \in m} w_{im} w_{iu} + b\, f(u, u^m) + c\, G(m)$, we obtain the following objective function, monotonic in the (underlined) index dimensions:

$$F_{DA}^+(m_{new}, u) = \nu\, (a \sum_{t_i \in m_{new}} w_{im_{new}} \underline{w_{iu}} + b\, \underline{f(u, u^{m_{new}})} + c\, G(m_{new})) +$$
$$2(1 - \nu) - \nu\, \underline{\mu_u} - (1 - \nu)\, \frac{2}{k-1}\underline{\eta_u}$$

DA-SANTA also manages a *user table* to keep for each user in the social network the current $u.TL_k$ and information for score computation.

The Algorithm. Figure 2 presents DA-SANTA. On publication of a new message m_{new}, the *getCandidates* method returns only users that have a chance to integrate m_{new} in their top-k. Each returned candidate is a couple (user, upperbound) - we take advantage here of the capability of the index traversal method to also estimate an upperbound for $dr_u(m_{new}, u.F_k)$ (here u is *c.user*).

For each candidate, its entry *ue* in the user table is necessary to compute the real value of $dr_u(m_{new}, u.F_k)$. To avoid as much as possible this costly operation, we filter out cases when the upperbound is not greater than $dr_u(m_{vic}, u.F_k)$ (stored in $ue.dr_{vic}$). After computing the real score with the *compute-dr* function, if the update condition (5) is fulfilled, we update the top-k list $u.TL_k$, select the new victim by using heuristics MR or MRD, and update $dr_u(m_{vic}, u.F_k)$.

Finally, we update the index lists μ and η, by moving only entries for u, following the new value of $-rel(m_{vic}, u)$, respectively $-D_{m_{vic}}(u.F_k)$.

```
DA-SANTA algorithm                              getCandidates method
Input: message m_new, index I, user table U     Input: message m_new, index I
  On m_new publication                            initTraversal(I, m_new)
    for all c ∈ getCandidates(I, m_new) do        result ← ∅

      ue ← getUserEntry(U, c.user)                threshold ← F⁺_DA(m_new)
      if c.upperbound > ue.dr_vic then            while threshold > 0 do
        s ← compute-dr(ue, m_new)                   u ← nextIndexUser(I)
        if s > ue.dr_vic then                       result ← result ∪ {(u, dr(m_new))}

          ue.TL_k ← ue.TL_k ∪ {m_new} − {ue.m_vic}  threshold ← F⁺_DA(m_new)
          ue.m_vic ← heuristics(ue.TL_k) //MR or MRD end while
          ue.dr_vic ← compute-dr(ue, m_vic)       return result
          Update I.μ, I.η
        end if
      end if
    end for
```

Fig. 2. The DA-SANTA algorithm

The *getCandidates* method traverses the index to prune candidates. Given m_{new}, *initTraversal* selects the related lists from the index and computes the coefficients of the objective function $F_{DA}^+(m_{new}, u)$. The index lists traversal may follow any threshold algorithm strategy (e.g. TA [6]) through the call to *nextIndexUser*, which returns the next user (in some of the lists) not yet seen in the index (new candidate).

The threshold is the maximal value that the objective function F_{DA}^+ may have, and is evaluated by $\overline{F_{DA}^+}(m_{new})$ as being $F_{DA}^+(m_{new}, u)$ applied to the last visited value in each index list. The monotony of F_{DA}^+ and of the index lists implies that for a new candidate u, $F_{DA}^+(m_{new}, u) \leq \overline{F_{DA}^+}(m_{new})$. For the same reasons, we obtain an upperbound for $dr_u(m_{new}, u.F_k)$ through $\overline{dr}(m_{new})$, computed like $\overline{F_{DA}^+}(m_{new})$ but only on the part that corresponds to $dr_u(m_{new}, u.F_k)$. Each new candidate and its upperbound for dr_u are appended to the results list.

Index traversal stops when the decreasing threshold becomes ≤ 0.

4   Experimental Evaluation

Experimental Setting. Our settings are similar to those applied in [1]. The social network is extracted from Twitter, with about 104 000 users and 18 million direct links between them. Computation of f uses the existence of a direct link (u_1, u_2) and the number of actions of u_1 on the messages of u_2. We use about 500 000 tweets extracted from the last 200 tweets for each user. Messages contain 3–4 terms in average. A dictionary of about 187 000 terms was built with message terms employed by at least 5 users. For each user, the profile contains all the dictionary terms that occur in his messages - the average profile size is 125. Note that user profiles and f are not continuously recomputed.

The relevance scoring function (1) and (2) uses the default coefficients $a = 0.5$, $b = 0.375$ and $c = 0.125$, while $G(m)$ uses the Klout score. Time bonus uses a linear function $TB(t^m - t_o) = 1 + (t^m - t_o)/T_b$, where T_b is the period of time after which an extra bonus equal to the initial $score(m, u)$ is earned, with a default value of 15 days. We consider four combinations of factors in the relevance scoring function: *Text-Social-Time* corresponds to the complete function (1), *Text-Social* to (2), *Text-Time* ignores the social components considering $b = c = 0$, and *Text* only keeps the text relevance.

The other default values in the experiments are $k = 10$ and $\nu = 0.75$.

We compare DA-SANTA with two other algorithms. *Baseline* corresponds to the *basic algorithm* (Sect. 2), *Incremental* [9] optimizes the computation of the relevance-diversity scores by using condition (5) with dr_u instead of (4). All the algorithms have an *initialization phase* that processes the first 300 000 messages, followed by *the measure phase* on the remaining 200 000 messages.

Effectiveness. We measure the quality of the results in terms of relevance $(f_R(u.TL_k))$ and diversity $(f_D(u.TL_k))$, while varying the balance parameter ν. The values of f_R and f_D are normalized to [0,1] by division by k. Figure 3 represents this variation for MR and for each type of relevance scoring. Values for $\nu = 1$ correspond to the case without diversity, while $\nu = 0$ is the other extreme case, where only diversity counts. Figure 3a shows a monotonic decrease of relevance in all the cases when ν decreases, to very low values when $\nu = 0$. However, when social criteria are included into the relevance scoring, the decrease is much smoother. This can be explained by a better natural content-diversity of messages when the relevance is not only based on content. Note also that relevance scores are not comparable among the various scoring types.

Figure 3b shows that diversity grows when ν decreases, with a stabilization to high values around $\nu = 0.6$. We notice that relevance functions including more criteria provide increased content diversity. Also, the social network criteria appear to have a good influence on diversity, better than the time bonus.

When using the MRD, measures are very close to those for MR, with noticeable better diversity, although the difference with MR is small.

In conclusion, a small contribution of diversity to the balance with relevance brings a very good diversity to the results, without loosing much of the relevance.

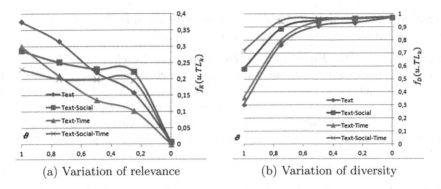

(a) Variation of relevance   (b) Variation of diversity

Fig. 3. Variation with ν of the achieved relevance and diversity, with MR heuristics

Efficiency. We measure the execution time per message, for both MR and MRD. Since time-dependent scoring has a significant impact on the execution time (because of the increased probability of new messages to be relevant and to enter the top-k), we compare two scoring cases, without (*Text-Social*), and with (*Text-Social-Time*) time bonus.

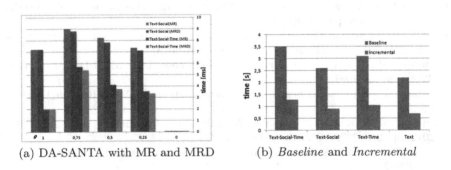

(a) DA-SANTA with MR and MRD   (b) *Baseline* and *Incremental*

Fig. 4. Execution time for DA-SANTA, *Baseline* and *Incremental*

Figure 4a presents the variation of the execution time with ν. In all the cases, the execution time first increases when ν decreases from 1 to around 0.75, then it decreases when ν continues to decrease. The initial increase is explained by the increasing role of the diversity in the global score, provoking more and more updates to the top-k and to the index. Around ν=0.75 the diversity becomes high enough and cannot increase too much anymore; the diversity part in the objective function F_{DA}^+ becomes important enough to produce a quicker termination of the index traversal. In all the cases, MR is slightly faster than the MRD. Time-dependent scoring has a much higher impact on the execution time, which is 1.5 to 2.5 times longer with *Text-Social-Time* than with *Text-Social*.

In conclusion, the execution time of DA-SANTA (ms per message) is adapted to continuous top-k processing. Time-dependent scoring has a real impact on the execution time, but do not change the order of magnitude. The victim selection heuristics has less impact than the other efficiency factors.

Comparison with *Baseline*and *Incremental*. *Baseline* and *Incremental* produce both the same relevance and diversity, since they test each time all the possible victims in the top-k. Comparing DA-SANTA with them evaluates the loss of relevance and diversity by applying a victim selection heuristics. Measures (not shown here for space reasons) indicate a negligible loss of relevance and diversity, which proves the very good quality of results produced by DA-SANTA.

Figure 4b compares the execution time of *Baseline* and *Incremental* with all the DA-SANTA scoring cases. *Incremental* is about 3 times faster than *Baseline*, but unlike DA-SANTA, its execution time (about 1 second/message) is not appropriate for continuous processing of top-k queries at a social network scale.

In conclusion, DA-SANTA delivers a similar quality of results 2–3 orders of magnitude faster than *Baseline* and *Incremental*, with execution times compatible with the continuous processing of top-k queries in large social networks.

References

1. Alkhouli, A., Vodislav, D., Borzic, B.: Continuous top-k queries in social networks. In: CoopIS 2016, pp. 24–42 (2016)
2. Chen, L., Cong, G.: Diversity-aware top-k publish/subscribe for text stream. In: SIGMOD 2015, pp. 347–362. ACM, New York (2015)
3. Drosou, M., Pitoura, E.: Diversity over continuous data. IEEE Data Eng. Bull. **32**(4), 49–56 (2009)
4. Drosou, M., Pitoura, E.: Search result diversification. SIGMOD Rec. **39**(1), 41–47 (2010)
5. Drosou, M., Pitoura, E.: Dynamic diversification of continuous data. In: EDBT 2012, pp. 216–227. ACM, New York (2012)
6. Fagin, R.: Combining fuzzy information: an overview. SIGMOD Rec. **31**(2), 109–118 (2002)
7. Gollapudi, S., Sharma, A.: An axiomatic approach for result diversification. In: WWW 2009, pp. 381–390. ACM, New York (2009)
8. Haghani, P., Michel, S., Aberer, K.: The gist of everything new: personalized top-k processing over web 2.0 streams. In: CIKM 2010, pp. 489–498 (2010)
9. Minack, E., Siberski, W., Nejdl, W.: Incremental diversification for very large sets: a streaming-based approach. In: SIGIR 2011, pp. 585–594 (2011)
10. Mouratidis, K., Pang, H.: Efficient evaluation of continuous text search queries. IEEE Trans. Knowl. Data Eng. **23**(10), 1469–1482 (2011)
11. Rao, W., Chen, L., Chen, S., Tarkoma, S.: Evaluating continuous top-k queries over document streams. World Wide Web **17**(1), 59–83 (2014)
12. Shraer, A., Gurevich, M., Fontoura, M., Josifovski, V.: Top-k publish-subscribe for social annotation of news. Proc. VLDB Endow. **6**(6), 385–396 (2013)
13. Vouzoukidou, N., Amann, B., Christophides, V.: Processing continuous text queries featuring non-homogeneous scoring functions. In: CIKM 2012, pp. 1065–1074 (2012)

Modeling and Discovering Cancelation Behavior

Maikel Leemans$^{(\boxtimes)}$ and Wil M.P. van der Aalst

Eindhoven University of Technology, 5600 MB Eindhoven, The Netherlands
m.leemans@tue.nl

Abstract. This paper presents a novel extension to the process tree model to support cancelation behavior, and proposes a novel process discovery technique to discover sound, fitting models with cancelation features. The proposed discovery technique relies on a generic error oracle function, and allows us to discover complex combinations of multiple, possibly nested cancelation regions based on observed behavior. An implementation of the proposed approach is available as a ProM plugin. Experimental results based on real-life event logs demonstrate the feasibility and usefulness of the approach.

Keywords: Process mining · Process discovery · Cancelation discovery · Cancelation modeling · Process trees · Event logs · Reset nets

1 Introduction

Process mining provides a powerful way to discover and analyze operational processes based on recorded event data stored in *event logs*. These event logs can be found everywhere: in enterprise information systems and business transaction logs, in web servers, in high-tech systems such as X-ray machines, in warehousing systems, etc. [17]. The majority of such real-life event logs contain some form of cancelation or error-handling behavior. A bank loan request may be canceled or declined, a webserver needs to handle a connection error, an X-ray machine may detect a sensor problem, etc. These cancelations can easily be expressed in workflow languages (BPMN, YAWL) and formal models such as reset workflow nets (RWF-nets). When formal process descriptions are not available, outdated or otherwise inaccurate, we turn to *process discovery* techniques. Process discovery aims to learn a process model from example behavior in event logs. Many discovery techniques have been proposed in literature, but few take into account cancelation features.

To illustrate the need for cancelation features, consider a bank loan request example, as modeled in Fig. 1. After a request is registered (a), in parallel the client's credit is checked (d), the request is processed (b, c, f), and a fraud check is performed (g). Once all parallel branches succeed, the loan is granted (h). If the credit check failed, the loan is declined (e), and there is no need to wait for the other activities. Since the credit check can fail at any stage during the request processing and fraud checking, there are $3 \cdot 2 = 6$ scenarios where a decline loan has to be modeled (see the six black transitions in Fig. 1(a)).

© Springer International Publishing AG 2017
H. Panetto et al. (Eds.): OTM 2017 Conferences, Part I, LNCS 10573, pp. 93–113, 2017.
https://doi.org/10.1007/978-3-319-69462-7_8

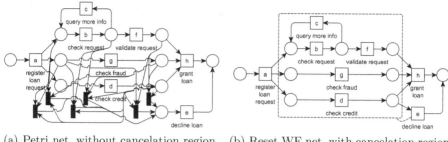

(a) Petri net, without cancelation region   (b) Reset WF net, with cancelation region

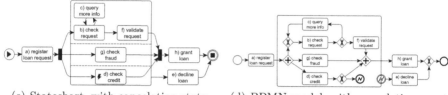

(c) Statechart, with cancelation state   (d) BPMN model, with cancelation event

Fig. 1. Small loan application example with cancelation behavior, modeled in different languages, illustrating the advantage of using cancelation features. As we will show in this paper, these models can be compactly represented using the proposed cancelation process tree (see Definition 4): $\rightarrow(a, \overset{*}{\rightarrow}(\rightarrow(\wedge(\rightarrow(\circlearrowleft(b,c),f),g,\star_d^{\{e\}}),h),e))$

The model in Fig. 1(a) is already rather complex. However, we assumed that the activities are atomic. This is not realistic, because the check credit, request processing and check fraud happen in parallel, and if the check fails, any processing tasks need to be withdrawn. Assuming we also model the start and end for each activity (thus modeling the running state of an activity), the number of scenarios/black transitions increases to $6 \cdot 3 = 18$. Using the cancelation features available in the various languages, we get a much simpler and more precise model, as shown in Figs. 1(b), (c) and (d).

Clearly, there is a need to model cancelation features explicitly, and process discovery should be able to discover these cancelation features. Furthermore, it is important that the discovered model meets certain quality criteria. Obviously, the model should be sound, i.e., all process steps can be executed and an end state is always reachable. Moreover, it is desirable to discover models that can replay all the behavior in the event log, i.e., to discover fitting models that relate to the event log. Recent work proposed a framework for discovery algorithms which find sound, fitting models in finite time, based on process trees [14]. The process tree notation is tailored towards process discovery and is a compact way to represent block-structured models that can easily be represented in terms of, for example: workflow nets, statecharts and BPMN models. However, in its current form, process trees cannot effectively capture cancelation features.

In this paper, *we propose a novel extension to the process tree model to support cancelation behavior, and propose a novel process discovery technique to discover sound, fitting models with cancelation features.* The proposed discovery technique relies on a generic error oracle function, and allows us to discover complex combinations of multiple, possibly nested cancelation regions based on observed behavior.

The approach is outlined in Fig. 2. An implementation of the proposed algorithm is tested and made available via the *Statechart* plugin for the ProM framework [12].

The remainder of this paper is organized as follows. Section 2 positions the work in existing literature. Section 3 introduces formal definitions and the proposed cancelation model. In Sect. 4, we discuss several heuristics for our error oracle. The novel cancelation process discovery technique is explained in Sect. 5. The approach is evaluated in Sect. 6. Section 7 concludes the paper.

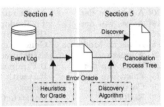

Fig. 2. Approach outline.

2   Related Work

To relate our work to existing approaches in process mining, we provide a systematic comparison of discovery approaches based on four criteria (Sect. 2.1). That is, the approach should provide expressive and sound models, and allow for a trade-off between fitness and simplicity. Next, we discuss and compare the related work (Sect. 2.2). Table 1 summarizes the comparison.

2.1   Criteria for Comparison

For comparing the related work, we define several comparison criteria.

As a basis, any discovery algorithm should yield *sound* process models. That is, all process steps can be executed and an end state is always reachable. In addition, it is desirable to be able to discover models that can replay all the behavior in the event log, i.e., to discover *fitting* models that relate to the event log [17].

Real-life event logs are often messy and challenging for discovery algorithms. In these cases, it may be necessary to trade off fitness for simplicity by filtering out *infrequent* behavior. In addition, some behavior can only be captured by *non-local* constructs such as long-distance dependencies or, to a degree, cancelation features.

For the related work techniques that do discover cancelation features, we look into *how the corresponding cancelation regions are discovered* from the event log. In addition, we will also compare on the complexity of the discovered features. That is, if *multiple regions* can be discovered, possibly in a *nested* configuration.

2.2   Discussion of the Related Work

We divided the related work discussion into several groups based on their comparison characteristics, see also Table 1.

Table 1. Comparison of related techniques from Sect. 2, according to the comparison criteria detailed in Sect. 2.

Author	Algorithm	Formalism	Sound	Fitting	Infrequent	Non-local	Cancel disc.	#Regions	Nested
[20] Aalst, van der	Alpha miner	Petri net	-	-	-	-	-	-	-
[19] Aalst, van der	TS Regions	Petri net	-	-	-	✓	-	-	-
[24] Weijters	(Flexible) Heuristics miner	Heuristics net	-	-	✓	✓	-	-	-
[18] Alves de Medeiros	Genetic miner	Heuristics net	-	-	✓	✓	-	-	-
[2] Augusto	Structured miner	BPMN	-	-	✓	✓	-	-	-
[21] Werf, van der	ILP miner	Petri net	-	✓	-	✓	-	-	-
[23] Zelst, van	ILP with filtering	Petri net	-	✓	✓	✓	-	-	-
[5] Carmona	Genet	Petri net	-	✓	✓	✓	-	-	-
[15] Redlich	Constructs Competition miner	BP(MN)	✓	-	✓	-	-	-	-
[4] Buijs	ETMd miner	Process tree	✓	-	✓	-	-	-	-
[14] Leemans, S.J.J.	Inductive miner (IM)	Process tree	✓	✓	✓	-	-	-	-
[10] Fluxicon / Günther	Disco	Fuzzy Model[†]	n/a	n/a	✓	-	-	-	-
[6] Celonis GmbH	Celonis Process Mining	Fuzzy Model[†]	n/a	n/a	✓	-	-	-	-
[9] Gradient ECM	Minit	Fuzzy Model[†]	n/a	n/a	✓	-	-	-	-
[11] Kalenkova	TS Cancel	RWF-net	-	-	-	±	✓	1	-
[16] Aalst, van der	Generic post-processing	RWF-net	-	-	-	±	±	1	-
[7] Conforti	BPMN miner	BPMN	✓	-	✓	±	±	n	✓
This paper	Cancelation Discovery	C. process tree[*]	✓	✓	✓	±	✓	n	✓

[†] Fuzzy models have no executable semantics.　　　[*] Cancelation process tree, see Definition 4.

There is a wide variety of Petri net based algorithms that can discover non-local constructs [2,5,18,19,21,23,24]. Even though these algorithms do not support cancelation features, they can discover and model complex, real-life behavior. The major downside is that none of these techniques guarantee a sound model.

A small group of algorithms focus on the process tree representation [4,14]. The use of process trees guarantees the discovery of sound workflow nets. However, by design, this limits the discovery search space to models with structured, local features only. Less structured behavior, like cancelation features, cannot be modeled or discovered.

In recent years, several commercial process mining tools emerged on the market [6,9,10]. Compared to academic tools, these commercial tools are easier to use, but provide less functionality. In particular, the commercial models provide no executable semantics, and do not support concurrency [17].

There have been a few attempts at supporting the discovery of cancelation features. In the work of [11], cancelation discovery is based on the behavior found in the event log. By analyzing a transition system (TS) abstraction of the event log, [11] searches for a single cancelation region. In contrast, the technique outlined in [16] uses a post-processing strategy based on conformance techniques.

That is, given a discovered model (using an existing algorithm), it tries to determine where a cancelation region should have been, based on unsuccessful event log replays (i.e., remaining tokens in the Petri net). In the work of [7], another post-processing heuristic is proposed. It assumes the underlying discovery algorithm (the BPMN miner) can correctly identify subprocesses (based on the relations in additional data attributes), and then checks which of these subprocesses should be "upgraded" to a cancelation region.

3   Event Logs and Process Trees

Before we explain the proposed discovery technique, we first introduce some definitions and our novel extension to process trees. We start with some preliminaries in Subsect. 3.1. In Subsect. 3.2 we introduce event logs (our input). Finally, in Subsects. 3.3 and 3.4, we will discuss the process tree model and our novel extension: *cancelation process trees*.

3.1   Preliminaries

We denote the powerset over some set A as $\mathcal{P}(A)$. We denote the set of all multisets over some set A as $\mathcal{B}(A)$. Note that the ordering of elements in a set or multiset is irrelevant.

Given a set X, a sequence over X of length n is denoted as $t = \langle a_1, \ldots, a_n \rangle \in X^*$. The empty sequence is denoted as ε, and we define $head(t) = a_1$, $end(t) = a_n$. We define $a \in t$ iff there is at least one a_i such that $a_i = a$. We write $\cdot$ to denote sequence concatenation, for example: $\langle a \rangle \cdot \langle b, c \rangle = \langle a, b, c \rangle$, and $\langle a \rangle \cdot \varepsilon = \langle a \rangle$. We write $\diamond$ to denote sequence interleaving (shuffle). For example: $\langle a, b \rangle \diamond \langle c, d \rangle = \{ \langle a, b, c, d \rangle, \langle a, c, b, d \rangle, \langle a, c, d, b \rangle, \langle c, a, b, d \rangle, \langle c, a, d, b \rangle, \langle c, d, a, b \rangle \}$.

We write $f : X \mapsto Y$ for a function with domain $dom(f) = X$ and range $rng(f) = \{ f(x) \mid x \in X \} \subseteq Y$.

3.2   Event Logs

The starting point for any process mining technique is an *event log*, a set of *events* grouped into *traces*, describing what happened when. Each trace corresponds to an execution of a process. Events may be characterized by various *attributes*, e.g., an event may have a timestamp, correspond to an activity, denote a start or end, is executed by a particular resource, etc.

For the sake of clarity, we will ignore most event attributes, and use sequences of activities directly, as defined below.

Definition 1 (Event Log). Let $\mathbb{A}$ be a set of activities. Let $L \in \mathcal{B}(\mathbb{A}^*)$ be an event log, a multiset of traces. A trace $t \in L$, with $t \in \mathbb{A}^*$, is a sequence of activities.

Given a set $\Sigma \subseteq \mathbb{A}$ and trace $t \in L$, we write $\Sigma(t) = \Sigma \cap \{a \in t\}$ to denote the set of activities in the intersection of Σ and t.

3.3   Process Trees

In this subsection, we introduce *process trees* as a notation to compactly represent *block-structured models*. An important property of block-structured models is that they are *sound by construction*; they do not suffer from deadlocks, livelocks and other anomalies. In addition, process trees are tailored towards process discovery, and have been used previously to discover block-structured workflow nets [14]. A process tree describes a language; an operator describes how the languages of its subtrees are to be combined.

Definition 2 (Process Tree). We formally define *process trees* recursively. We assume a finite alphabet $\mathbb{A}$ of activities and a set $\bigotimes$ of operators to be given. Symbol $\tau \notin \mathbb{A}$ denotes the silent activity.

- a with $a \in (\mathbb{A} \cup \{\tau\})$ is a process tree;
- Let $P_1, \ldots, P_n$ with $n > 0$ be process trees and let $\otimes \in \bigotimes$ be a process tree operator, then $\otimes(P_1, \ldots, P_n)$ is a process tree. We consider the following operators for process trees:
    - $\rightarrow$ denotes the *sequential execution* of all subtrees;
    - $\times$ denotes the *exclusive choice* between one of the subtrees;
    - $\circlearrowright$ denotes the *structured loop* of loop body P_1 and alternative loop back paths $P_2, \ldots, P_n$ (with $n \geq 2$);
    - $\wedge$ denotes the *parallel (interleaved) execution* of all subtrees.

Definition 3 (Process Tree Semantics). To describe the semantics of process trees, the language of a process tree P is defined using a recursive monotonic function $\mathcal{L}(P)$, where each operator $\otimes \in \bigotimes$ has a language join function $\otimes^l : (\mathcal{P}(\mathbb{A}^*) \times \ldots \times \mathcal{P}(\mathbb{A}^*)) \mapsto \mathcal{P}(\mathbb{A}^*)$:

$$\mathcal{L}(a) = \{\langle a \rangle\} \text{ for } a \in \mathbb{A} \qquad \mathcal{L}(\otimes(P_1, \ldots, P_n))$$
$$\mathcal{L}(\tau) = \{\varepsilon\} \qquad\qquad\qquad = \otimes^l(\mathcal{L}(P_1), \ldots, \mathcal{L}(P_n))$$

Each operator has its own language join function $\otimes^l$. The language join functions below are borrowed from [14,17], with $L_i \subseteq \mathbb{A}^*$:

$$\rightarrow^l(L_1, \ldots, L_n) = \{t_1 \cdot \ldots \cdot t_n \mid \forall 1 \leq i \leq n : t_i \in L_i\}$$
$$\times^l(L_1, \ldots, L_n) = \bigcup_{1 \leq i \leq n} L_i$$
$$\circlearrowright^l(L_1, \ldots, L_n) = \{t_1 \cdot t_1' \cdot t_2 \cdot t_2' \cdot \ldots \cdot t_{m-1} \cdot t_{m-1}' \cdot t_m \mid \forall i : t_i \in L_1, t_i' \in \bigcup_{2 \leq j \leq n} L_j\}$$
$$\wedge^l(L_1, \ldots, L_n) = \{t' \in (t_1 \diamond \ldots \diamond t_n) \mid \forall 1 \leq i \leq n : t_i \in L_i\}$$

Example models and their languages:

$$\mathcal{L}(\wedge(a,b)) = \{\langle a,b\rangle, \langle b,a\rangle\}$$
$$\mathcal{L}(\circlearrowleft(a,b)) = \{\langle a\rangle, \langle a,b,a\rangle,$$
$$\langle a,b,a,b,a\rangle, \dots\}$$

$$\mathcal{L}(\rightarrow(a,\times(b,c))) = \{\langle a,b\rangle, \langle a,c\rangle\}$$
$$\mathcal{L}(\wedge(a,\rightarrow(b,c))) = \{\langle a,b,c\rangle, \langle b,a,c\rangle,$$
$$\langle b,c,a\rangle\}$$

3.4   Cancelation Process Trees

We extend the process tree representation to support cancelation behavior. We add two new tree operators to represent a cancelation region, and a new tree leaf to denote a cancelation trigger.

Definition 4 (Cancelation Process Tree). We formally define *cancelation process trees* recursively. We assume a finite alphabet $\mathbb{A}$ of activities to be given.

- Any process tree is also a cancelation process tree;
- Let $P_1, \dots, P_n$ with $n \geq 2$ be cancelation process trees, then:
  - $\overset{*}{\rightarrow}(P_1, \dots, P_n)$ denotes the sequence-cancel of cancelation body P_1 and mutually exclusive error alternative paths $P_2, \dots, P_n$;
  - $\overset{*}{\circlearrowleft}(P_1, \dots, P_n)$ denotes the loop-cancel of cancelation body P_1 and mutually exclusive error loop back paths $P_2, \dots, P_n$;
- $\star_a^E$ with $a \in \mathbb{A}, E \subseteq \mathbb{A}$ denotes the cancelation trigger. Combined with a cancelation operator $\overset{*}{\rightarrow}, \overset{*}{\circlearrowleft}$, this leaf denotes the point where we execute activity a, and have the option to trigger an error $e \in E$, firing a corresponding cancelation region. This concept is explained and defined in detail below. ⌙

The intuition behind the cancelation operators is described below. We reference to Table 2 for a concrete example tree with a step by step construction of its language. Observe that the new operators enable the modeling of semi-block structured behavior (see the semi-structured loop modeling activity r).

Assume a tree $\overset{*}{\otimes}(P_1, \dots, P_n)$, $\overset{*}{\otimes} \in \{\overset{*}{\rightarrow}, \overset{*}{\circlearrowleft}\}$ with a leaf $\star_a^E$ somewhere in the subtree P_1. When we want to "execute" this tree (i.e., generate a trace in its language), we start with the subtree P_1. At any $\star_a^E$ point, we do activity a as normal (happy flow), and have the option to trigger any error $e \in E$. For example, in Table 2, the leaf $\star_c^{\{e,r\}}$ can trigger either error e or r.

In case an error $e \in E$ is triggered, we need to find a matching cancelation region. A cancelation region $\overset{*}{\otimes}(P_1, \dots, P_n)$ matches an error $e \in E$ iff e is the start activity for a trace in $t \in \mathcal{L}(P_2, \dots, P_n)$. When we trigger the error $e \in E$ at $\star_a^E$, we perform the activity a, but ignore the rest of the subtree P_1. I.e., we take the prefix up to and including a, and fire the cancelation region. We follow

Table 2. Example *cancelation process tree* (left) with its language (right) and corresponding Reset WF net (bottom). Shown are the traces in the language and the corresponding errors that are triggered to generate the trace. The grey arrows in the cancelation process tree indicate the possible error trigger "jumps".

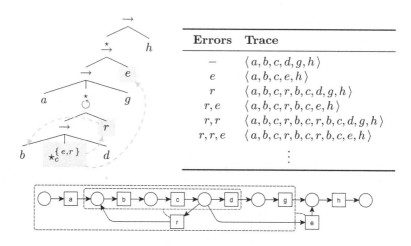

Errors	Trace
$-$	$\langle a,b,c,d,g,h \rangle$
e	$\langle a,b,c,e,h \rangle$
r	$\langle a,b,c,r,b,c,d,g,h \rangle$
r,e	$\langle a,b,c,r,b,c,e,h \rangle$
r,r	$\langle a,b,c,r,b,c,r,b,c,d,g,h \rangle$
r,r,e	$\langle a,b,c,r,b,c,r,b,c,e,h \rangle$
	$\vdots$

with a trace $t \in \mathcal{L}(P_2, \ldots, P_n)$ such that e is the start activity of t. I.e., we make a "jump" and execute a matching trace from one of the non-first subtrees. In Table 2, the node $\overset{\star}{\to}$ matches with error e, and the node $\overset{\star}{\circlearrowright}$ matches with error r.

The difference between $\overset{\star}{\to}$ and $\overset{\star}{\circlearrowright}$ is as follows: In case of $\overset{\star}{\to}$, we have sequential behavior, i.e., after a happy flow or error path, we continue with the rest of the process tree. In case of $\overset{\star}{\circlearrowright}$, we have looping behavior, i.e., after an error path, we loop back and try executing P_1 again. For instance, in the example of Table 2 at the leaf $\star_c^{\{e,r\}}$, we can either continue as normal (happy flow), or jump to r (repetitively) or e (only once).

Definition 5 (Cancelation Process Tree Semantics). We define the semantics of cancelation process trees in multiple steps, and provide an adaptation for the existing process tree semantics.

First, we define the language of the cancelation trigger leaf $\star_a^E$. At this leaf, we can either execute activity a as normal, or execute it and trigger an error $e \in E$.

$$\mathcal{L}(\star_a^E) = \{ \langle \star_a^E \rangle, \langle a \rangle \} \text{ for } a \in \mathbb{A}, E \subseteq \mathbb{A}$$

Next, we define the language for the cancelation operators $\overset{\star}{\to}, \overset{\star}{\circlearrowright}$. There are two common cases for these operators: (1) no error is triggered, and (2) an error is not caught by this operator. We define $\overset{\star l}{\otimes}$ to represent these common cases.

$$\overset{*\,l}{\otimes}(L_1,\ldots,L_n) = \{\, t_1 \mid t_1 \in L_1,\ end(t_1) \neq \star_a^{*\,E} \,\} \cup \{\, t_1 \cdot \left\langle \star_a^{E\backslash S} \right\rangle \mid t_1 \cdot \left\langle \star_a^{E} \right\rangle \in L_1,$$

$$S = \{\, head(t) \mid t \in \bigcup_{2 \leq j \leq n} L_j \,\},\ E \backslash S \neq \emptyset \,\}$$

For the sequence-cancel operator $\overset{*}{\to}$, we extend upon the language of $\overset{*\,l}{\otimes}$ by allowing a matching error path, after which we continue with the rest of the process tree.

$$\overset{*\,l}{\to}(L_1,\ldots,L_n) = \{\, t_1 \cdot \langle\, a\,\rangle \cdot t_e \mid t_1 \cdot \left\langle \star_a^{E} \right\rangle \in L_1,\ head(t_e) \in E,\ t_e \in \bigcup_{2 \leq j \leq n} L_j \,\}$$

$$\cup \overset{*\,l}{\otimes}(L_1,\ldots,L_n)$$

For the loop-cancel operator $\overset{*}{\circlearrowleft}$, we extend upon the language of $\overset{*\,l}{\otimes}$ by allowing a matching error path, after which we loop back and try executing P_1 again.

$$\overset{*\,l}{\circlearrowleft}(L_1,\ldots,L_n) = \{\, t_1 \cdot \langle\, a_1\,\rangle \cdot t_1' \cdot t_2 \cdot \langle\, a_2\,\rangle \cdot t_2' \cdot \ldots \cdot t_{m-1} \cdot \langle\, a_{m-1}\,\rangle \cdot t_{m-1}' \cdot t_m$$

$$\mid t_m \in \overset{*\,l}{\otimes}(L_1,\ldots,L_n),\ \forall i < m : t_i \cdot \left\langle \star_{a_i}^{E_i} \right\rangle \in L_1,\ head(t_i') \in E_i,$$

$$t_i' \in \bigcup_{2 \leq j \leq n} L_j \} \cup \overset{*\,l}{\otimes}(L_1,\ldots,L_n)$$

The existing process tree semantics can easily be adapted for these cancelation semantics by applying a prefix function ϕ to all traces in $\otimes^l$ (for $\otimes^l \in \{\,\to^l, \times^l, \circlearrowleft^l, \wedge^l\,\}$). The idea is to remove any activity after a $\star_a^E$ symbol. For instance $\phi(\langle\, a, b, c\,\rangle) = \langle\, a, b, c\,\rangle$, but $\phi(\langle\, a, \star_b^E, c\,\rangle) = \langle\, a, \star_b^E\,\rangle$. $\lrcorner$

Below are some simple models and their corresponding language:

$$\mathcal{L}(\to(a, \star_b^{\{e\}}, c)) = \{\, \langle\, a, b, c\,\rangle, \langle\, a, \star_b^{\{e\}}\,\rangle \,\} \qquad \mathcal{L}(\overset{*}{\circlearrowleft}(\to(a, \star_b^{\{r\}}, c), r)$$

$$\mathcal{L}(\overset{*}{\to}(\to(a, \star_b^{\{e\}}, c), e)) = \{\, \langle\, a, b, c\,\rangle, \langle\, a, b, e\,\rangle \,\} \qquad = \{\, \langle\, a, b, c\,\rangle, \langle\, a, b, r, a, b, c\,\rangle,$$

$$\langle\, a, b, r, a, b, r, a, b, c\,\rangle, \ldots \,\}$$

4 Heuristics for Error Oracle

We rely on explicitly modeling cancelation triggers and error activities (see Definition 4). For the algorithm in Sect. 5, we assume that the error activities are also explicit in the input. However, for any given event log, this is usually not the case (see Definition 1). To make error activities explicit in the input, we will assume a so-called error oracle function as an additional input.

Definition 6 (Error Oracle). Let $\mathbb{A}$ be a set of activities. Let $isError : \mathbb{A} \mapsto \{\,true, false\,\}$ be an error oracle function, yielding $true$ iff an activity $a \in \mathbb{A}$ is an error activity.   ⌟

There are numerous ways to instantiate such an error oracle function. A simple heuristics is to rely on domain knowledge or keywords in the activity names. For example, a negative activity name like "Cancelled" or "Declined" is often a good candidate. In addition, one can also check activities that break the normal flow: timed triggers, (external) events or asynchronous activities. Alternatively, exception or error data attributes may prove useful.

When none of these heuristics are an option, one can always fall back to an optimization strategy. The intuition is that, if cancelation behavior is present, modeling this behavior with cancelation operators will yield a more fitting and possibly more precise process tree. This is easy to see considering the fact that process trees traditionally capture only block-structured behavior, and cancelation behavior breaks this block-structuredness. Such an optimization strategy would feed several candidate error oracle functions to the discovery algorithm, compute the fitness and precision of the resulting model, and return the best scoring candidate.

Future work should look into more behavioral oriented error oracle heuristics.

5   Model Discovery

In this section, we will detail our discovery approach. We start by introducing the directly follows abstraction over an event log in Subsect. 5.1. Next, we briefly cover the framework our technique is based on in Subsect. 5.2. After that, we detail our proposed approach in Subsect. 5.3.

5.1   Directly Follows Graph and Cuts

The *directly follows relation* describes when two activities directly follow each other in a process. This relation can be expressed in the *directly follows graph* of a log L, written $G(L)$. Nodes in $G(L)$ are the activities of L. An edge (a, b) is present in $G(L)$ iff some trace $\langle \ldots, a, b, \ldots \rangle \in L$. We define the start and end nodes of G, $Start(G)$ and $End(G)$ respectively, based on the start and end activities in L. An *n-ary cut* of $G(L)$ is a partition of the nodes of the graph into disjoint sets $\Sigma_1, \ldots, \Sigma_n$.

Consider a directly follows graph $G(L)$ and error oracle function $isError : \mathbb{A} \mapsto \{\,true, false\,\}$. If an edge (a, b) has $isError(b)$, then we call (a, b) an *error edge*. In any subgraph $G' \subseteq G$, a node a can only be an end node of G' iff it would be an end node without error edges.

5.2   Discovery Framework

Our technique is based on the Inductive Miner (IM) framework for discovering process tree models, as described in [14]. Given a set $\bigotimes$ of process tree operators, [14] defines a framework to discover models using a divide and conquer approach. Given a log L, the framework searches for possible splits of L into sublogs $L_1, \ldots, L_n$, such that these logs combined with an operator $\otimes \in \bigotimes$ can (at least) reproduce L again. The split search is based on finding cuts in the directly follows graph $G(L)$ of the log L. For each operator, a different cut is characterized based on the edges between the nodes in $G(L)$. The framework then recurses on the corresponding sublogs and returns the discovered submodels. Logs with empty traces or traces with a single activity form the base cases for this framework. Note that, by design, each activity only appears once in the produced process tree, and this tree can be a generalization of the original event log.

We use this framework as a basis because of its extensibility (it works independently of the chosen process tree operators), as well as the following properties: the log fits the resulting model, and there exists an implementation with a polynomial run time complexity [14].

5.3   Cancelation Discovery

We generalize the above approach to also support the discovery of cancelation behavior by including the error oracle function *isError* as an additional input, and tracking *cancelation triggers* during discovery. Note that we maintain the rediscoverability and fitness guarantees of the original framework [14]. In Algorithm 1, an overview of the discovery approach is given. In Table 3, an example run is given.

We will first discuss *cancelation triggers* in more detail. The three extension points labelled in Algorithm 1 are discussed next: (1) the cancelation trigger base case (line 5), (2) the cut finding extensions (line 10), and (3) the log splitting for cancelation (line 11).

Cancelation Triggers. A key observation is that we can track cancelation triggers during discovery. Whenever we observe an error activity e with $isError(e)$ in the log, it had to be triggered by the last activity before e. This naturally follows from the prefix cancelation semantics and the fact that each activity only appears once in the produced process tree. We keep track of these predecessors via the *triggers* mapping defined below.

Definition 7 (Cancelation Triggers). Let $triggers : \mathbb{A} \mapsto \mathcal{P}(\mathbb{A})$ be a cancelation triggers mapping, mapping an activity to the set of error activities such that $\forall E \in rng(triggers), e \in E : isError(e)$. ⌟

Algorithm 1: Cancelation Discovery

Input: An event log L and error oracle function $isError : \mathbb{A} \mapsto \{\ true,\ false\ \}$
Output: A cancelation process tree P such that L fits P
Description: Extended framework that takes into account cancelation operators.

DISCOVER(L, $isError$)

```
(1)        if ∀σ ∈ L : σ = ε
(2)            // the log is empty or only contains empty traces
(3)            return τ
(4)        else if ∃x ∈ 𝔸 : ∀σ ∈ L : σ = ⟨x⟩
(5)            // the log only has a single activity  - - - - - - - - - - - - - - - - - - - - (1)
(6)            if triggers(x) ≠ ∅ then return ⋆ₓ^{triggers(x)}   // cancelation trigger case
(7)                          else return x          // normal base case
(8)        else
(9)            // the normal framework cases
(10)           (⊗, (Σ₁, . . . , Σₙ)) = findCut'(G(L), isError)  - - - - - - - - - - - - - (2)
(11)           ((L₁, . . . , Lₙ), triggers) = splitLog'(L, (Σ₁, . . . , Σₙ), isError, triggers)  - (3)
(12)           return ⊗(M₁, . . . , Mₙ)  where  Mᵢ = DISCOVER(Lᵢ, isError)
```

(1) Base Case. In the case the sublog consists of only a single activity, we have two options. To discover a cancelation trigger $\star_a^E$ for an activity a, we simply check the *triggers* mapping. If this is mapping empty, we have a normal activity leaf a, else we have a cancelation trigger leaf with $E = triggers(a)$.

(2) Finding Cuts. We include support for our cancelation tree operators by adding new cuts, and only slightly adapting existing cut definitions from [14]. In Fig. 3, all the graph cuts are depicted informally.

In our cancelation discovery, any non-cancelation cut *cannot* have an error edge between two partitions. In contrast, a cancelation cut is characterized by having error edges from its first partition to all non-first partitions. That is, in a cancelation cut, the first partition is the normal (happy flow) behavior inside the cancelation region. The non-first partitions are the mutually exclusive error paths after triggering the cancelation. The sequence and loop cancelation cuts are formally defined below.

Definition 8 (Sequence Cancel Cut). A sequence cancel ($\overset{\star}{\to}$) cut is a partially ordered cut $\Sigma_1, \ldots, \Sigma_n$, with $n \geq 2$, of a directly-follows graph G such that:

1. All start activities are in the body Σ_1:
$$Start(G) \subseteq \Sigma_1$$

2. Every partition Σ_i has some end activities:
$$\forall i \geq 1 : End(G) \cap \Sigma_i \neq \emptyset$$

3. There are only error edges from Σ_1 to $\Sigma_{i>1}$:
$$\forall i > 1, a_i \in \Sigma_i, a_1 \in \Sigma_1 :$$
$$(a_1, a_i) \in G \Rightarrow isError(a_i)$$

4. There are no edges from $\Sigma_{i>1}$ to $\Sigma_{j \geq 1}$:
$$\forall i > 1, j \geq 1, i \neq j, a_i \in \Sigma_i, a_j \in \Sigma_j :$$
$$(a_i, a_j) \notin G$$

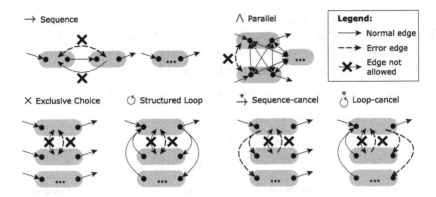

Fig. 3. Cuts of the directly-follows graph for all operators. The grey areas indicate partitions; the arrows indicate required and disallowed edges characterizing the cut.

Definition 9 (Loop Cancel Cut). A loop cancel ($\overset{*}{\circlearrowleft}$) cut is a partially ordered cut $\Sigma_1, \ldots, \Sigma_n$, with $n \geq 2$, of a directly-follows graph G such that:

1. All start and end activities are in the body Σ_1:

    $Start(G) \cup End(G) \subseteq \Sigma_1$

2. There are only error edges from Σ_1 to $\Sigma_{i>1}$:

    $\forall i > 1, a_i \in \Sigma_i, a_1 \in \Sigma_1$:
    $(a_1, a_i) \in G \Rightarrow isError(a_i)$

3. There are only edges from Σ_i to start nodes in Σ_1:

    $\forall i > 1, a_i \in \Sigma_i, a_1 \in \Sigma_1$:
    $(a_i, a_1) \in G \Rightarrow a_1 \in Start(G)$

4. There are no edges from $\Sigma_{i>1}$ to $\Sigma_{j>1}$:

    $\forall i > 1, j > 1, i \neq j, a_i \in \Sigma_i, a_j \in \Sigma_j : (a_i, a_j) \notin G$

5. If Σ_i has an edge to Σ_1, it connects to all start activities:

    $\forall i > 1, a_i \in \Sigma_i, a_1 \in Start(G)$:
    $(\exists a_1' \in \Sigma_1 : (a_i, a_1') \in G) \Leftrightarrow (a_i, a_1) \in G$ ⌟

(3) Splitting Logs. Once a cut $\Sigma_1, \ldots, \Sigma_n$ has been found for an operator $\otimes$, we need to split the log L into sublogs $L_1, \ldots, L_n$, such that these logs combined with operator $\otimes$ can (at least) reproduce L again. For the new cancelation operators, we define the log splits and cancelation trigger update below.

Definition 10 (Sequence Cancel Split). Given a sequence cancelation cut $\Sigma_1, \ldots, \Sigma_n$:

1. Sublog L_1 consists of all maximal prefix subtraces with activities in Σ_1:

$$L_1 = \{\, t_1 \mid t_1 \cdot t_2 \in L, \Sigma(t_1) \subseteq \Sigma_1, (t_2 = \varepsilon \vee (t_2 = \langle e, \ldots \rangle \wedge e \notin \Sigma_1)) \,\}$$

2. Sublog $L_{i>1}$ consists of all maximal postfix subtraces with activities in Σ_i:

$$L_{i>1} = \{\, t_2 \mid t_1 \cdot t_2 \in L, \Sigma(t_2) \subseteq \Sigma_i, (t_1 = \varepsilon \vee (t_1 = \langle \ldots, a_1 \rangle \wedge a_1 \in \Sigma_1)) \,\}$$

3. Update the triggers mapping such that any activity $a \in \mathbb{A}$ ending a trace in L_1 is mapped to all error activities following it in L:

$$triggers(a) = triggers(a) \cup \{\, e \mid t_1 \cdot t_2 \in L, \Sigma(t_1) \subseteq \Sigma_1, e \notin \Sigma_1,$$
$$t_1 = \langle \ldots, a \rangle, t_2 = \langle e, \ldots \rangle \,\}$$ ⌟

For example, consider a log $L = [\langle b, c, d \rangle, \langle c, b, d \rangle, \langle c, e, f \rangle]$ (taken from step 2 in Table 3) and sequence cancelation cut $\Sigma_1 = \{ b, c, d \}$, $\Sigma_2 = \{ e, f \}$. The resulting log splits are $L_1 = [\langle b, c, d \rangle, \langle c, b, d \rangle, \langle c \rangle]$, $L_2 = [\langle e, f \rangle]$, and the resulting triggers mapping is $triggers = \{ c \mapsto \{ e \} \}$.

Definition 11 (Loop Cancel Split). Given a loop cancelation cut $\Sigma_1, \ldots, \Sigma_n$:

1. Sublog L_i consists of all maximal subtraces with activities in Σ_i:

$$L_i = \{ t_2 \mid t_1 \cdot t_2 \cdot t_3 \in L, \Sigma(t_2) \subseteq \Sigma_i, (t_1 = \varepsilon \vee (t_1 = \langle \ldots, a_1 \rangle \wedge a_1 \notin \Sigma_i)),$$
$$(t_3 = \varepsilon \vee (t_3 = \langle a_3, \ldots \rangle \wedge a_3 \notin \Sigma_i)) \}$$

2. Update the triggers mapping such that any activity $a \in \mathbb{A}$ ending a trace in L_1 is mapped to all error activities following it in L:

$$triggers(a) = triggers(a) \cup \{ e \mid t_1 \cdot t_2 \cdot t_3 \in L, \Sigma(t_2) \subseteq \Sigma_1, e \notin \Sigma_1,$$
$$t_2 = \langle \ldots, a \rangle, t_3 = \langle e, \ldots \rangle \}$$

For example, consider a log $L = [\langle b, c, r, b, c, d \rangle]$ (a small snippet from Table 2) and loop cancel cuts $\Sigma_1 = \{ b, c, d \}$, $\Sigma_2 = \{ r \}$. The resulting log splits are $L_1 = [\langle b, c \rangle, \langle b, c, d \rangle]$, $L_2 = [\langle r \rangle]$, and the resulting triggers mapping is $triggers = \{ c \mapsto \{ r \} \}$.

6   Evaluation

In this section, we compare our technique against related, implemented techniques. The proposed algorithm is implemented in the *Statechart* plugin for the process mining framework ProM [12]. In the remainder of this section, we will refer to Algorithm 1 as *cancelation*. We end the evaluation by showing example results obtained using our tool.

6.1   Input and Methodology for Comparative Evaluation

In this comparative evaluation, we focus on the quantitative aspects. That is, the models discovered are precise and fit the actual system. We compare a number of techniques and input event logs on: (1) the *running time* of the technique, (2) the *model quality* (fitness and precision), and (3) the *model simplicity*.

For the running time, we measure the average running time and associated 95% confidence interval over 30 micro-benchmark executions, after 10 warmup rounds for the Java JVM. Each technique is allowed at most 30 seconds for completing a single model discovery. Fitness and precision are calculated using the technique described in [1]. In short, *fitness* expresses the part of the log that is represented by the model; *precision* expresses the behavior in the model that

Table 3. Example Cancelation Discovery on the log $[\langle a,b,c,d,g \rangle, \langle a,c,b,d,g \rangle,$ $\langle a,c,e,f,g \rangle]$ and error oracle *isError* with *isError*$(e) =$ *true* and *false* otherwise. The rows illustrate how the discovery progresses. The highlights indicate the parts of the log and directly follows graph used, and relate them to the corresponding partial model that is discovered. The dashed arrow is an error edge, and the dashed lines indicate the cuts. The resulting Reset WF net is shown at the bottom.

Step	Discovered Model	Event Log	Directly Follows Graph
1			
2			
3			
4/5			

is present in the log. For these experiments we used a laptop with an i7-4700MQ CPU @ 2.40 GHz, Windows 8.1 and Java SE 1.7.0 (64 bit) with 8 GB of RAM.

We selected several real-life event logs as experiment input, covering a range of input problem sizes and complexities. The input problem size is typically measured in terms of four metrics: number of traces, number of events, number of activities (size of the alphabet), and average trace length. The event logs and their sizes are shown in Table 4.

Table 4. The event logs used in the evaluation, with input sizes and applied filters

Event Log	# Traces	# Events	# Acts	Avg. |T|	Filter
[13] NASA CEV	2	48	17	24.00	Only test cases 1 and 10
[3] WABO	1,434	8,577	27	5.98	–
[22] BPIC12, A	13,087	60,849	10	4.64	"A_" subprocess only
[8] Road fine, a	150,370	561,470	11	3.73	–
[8] Road fine, f	150,370	404,009	9	2.68	No "no payment" and "add penalty"

Table 5. The error oracles used for the event logs in the evaluation.

Event Log	Error Oracle Activities
[13] NASA CEV	"cev.ErrorLog.last()"
[3] WABO	"T15 Print document X request unlicensed", "T16 Report reasons to hold request"
[22] BPIC12, A	"A_CANCELLED", "A_DECLINED"
[8] Road fine, a	"Send for Credit Collection"
[8] Road fine, f	"Send for Credit Collection"

The *NASA CEV* [13] event log describes two executions of a software process with errors. This small log was obtained from two existing NASA CEV software tests. The *WABO* [3] event log describes the receipt phase of an environmental permit application process ('WABO') at a Dutch municipality. The *BPIC12* [22] event log is a BPI challenge log that describes three subprocesses of a loan application process. In this evaluation, we only focus on the "A_" subprocess. The *Road fine* [8] event log was obtained from an information system managing road traffic fines. We use two variants of this large event log. The *Road fine, a* variant is the largest, most complex event log in our experimental setup. In variant *Road fine, f*, we filtered out two asynchronous activities to decrease the (directly-follows) complexity.

We compare our discovery algorithm against most of the techniques mentioned in Sect. 2. Unfortunately, we could not compare against the work of [5–7,9,10,15,16] due to invalid input assumptions, absence of semantics, or the lack of a reference implementation. The Inductive Miner (IM) [14] is our baseline comparison algorithm, since our approach builds upon the IM framework. For the Inductive Miner and our derived techniques, we also consider the *paths* setting. This is the frequency cutoff for discovering an 80/20 model: 1.0 means all behavior, 0.8 means 80% of the behavior. In Table 5, we have listed the error oracles we used for our cancelation discovery techniques.

6.2 Comparative Evaluation Results and Discussion

Runtime Analysis. In Table 6, the results for the runtime benchmark are given.

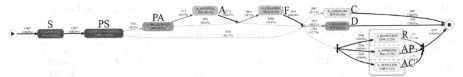

(a) IM (baseline) model result, no cancelation. Skips obfuscate the happy flow.

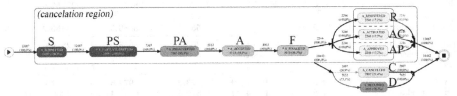

(b) Cancel model result, with cancelation region. Happy flow in cancel reagion.

Fig. 4. Models mined from the BPIC12 log with the Path filter at 0.8, both produced by our ProM plugin [12], and visualized in the Statechart language. Legend: (S) A_Submitted, (PS) A_PartlySubmitted, (PA) A_PreAccepted, (A) A_Accepted, (F) A_Finalized, (C) A_Cancelled, (D) A_Declined, (R) A_Registered, (AP) A_Approved, (AC) A_Activated

The first thing we notice is that, in contrast to the TS Cancel technique, our Cancelation algorithm always discovers a model within the allotted time. When compared to the baseline Inductive Miner, there seems to be a small overhead in running time. There are two explanations for this small overhead. One being the fact that more tree operator cuts have to be checked at each recursive call of the algorithm. But more importantly, the new cancelation operators potentially uncover more structures in the directly follows graph. In cases where the original Inductive Miner might give up and falls back to loops with skips and/or flower models, we can find a cancelation pattern, and recurse on a more structured subproblem. The end result is that we have more recursive calls to uncover all the structures/tree operators, and hence have a larger running time. Nevertheless, our technique successfully scales to larger logs and consistently yields results within seconds.

Model Quality Analysis. In Table 7, the results of the model quality measurements are given. Note that in order to compute model quality scores, the model should be sound.

Observe that, compared to the original Inductive Miner, our Cancelation algorithm always yields an equal or more fitting model. Moreover, we preserve the perfect fitness guarantee of the original Inductive Miner (for *path* 1.0). In addition, in most cases, the resulting model is also more precise.

Table 6. Runtime for the different algorithms, paths filter settings, and event logs.

Algorithm	Path	NASA CEV	WABO	BPIC12, A	Road fine, a	Road fine, f
[20] Alpha miner	-	0.3	8.6	18.7	467.8	260.6
[24] Heuristics	-	2.3	40.5	162.8	1641.7	1047.7
[21] ILP	-	371.0	560.3	501.5	3426.1	2517.8
[23] LP, filtering	-	381.9	673.2	497.2	3395.1	2583.1
[18] Genetic miner	-	3498.1	26033.8	3402.9	$-^T$	25592.0
[4] ETMd miner	-	27836.5	27516.7	27130.5	28722.1	27557.6
[11] TS Cancel	-	$-^T$	$-^T$	139.6	$-^T$	$-^T$
[19] TS Regions	-	18.9	$-^T$	555.1	5089.5	3455.2
[14] IM (baseline)	1.0	1.4	120.1	308.0	5668.9	2859.0
[14] IM (baseline)	0.8	0.9	138.1	300.7	4049.5	2540.2
[14] IM (baseline)	0.5	0.9	135.4	301.4	4132.9	2537.1
Ours Cancelation	1.0	2.0	176.7	373.2	5972.5	3047.9
Ours Cancelation	0.8	1.5	156.4	379.8	6145.4	3289.3
Ours Cancelation	0.5	1.7	154.1	377.7	6180.8	3558.8
		$10^0 \, 10^2 \, 10^4$	$10^2 \, 10^4$	$10^2 \, 10^3 \, 10^4$	$10^3 \, 10^4$	$10^3 \, 10^4$

Avg. runtime (in milliseconds, with log scale plot), over 30 runs, with 95% confidence interval
^{T} Time limit exceeded (30 sec.)

Table 7. Fitness (Fit.) and Precision (Prec.) scores for the different algorithms, paths filter settings, and event logs. Scores range from 0.0 to 1.0, higher is better.

Algorithm	Path	NASE CEV Fit.	NASE CEV Prec.	WABO Fit.	WABO Prec.	BPIC12, A Fit.	BPIC12, A Prec.	Road fine, a Fit.	Road fine, a Prec.	Road fine, f Fit.	Road fine, f Prec.
[20] Alpha miner	-	0.89	0.08	$-^U$	$-^U$	$-^U$	$-^U$	$-^U$	$-^U$	$-^U$	$-^U$
[24] Heuristics	-	$-^U$	$-^U$	0.61	0.98	$-^U$	$-^U$	$-^U$	$-^U$	0.74	1.00
[21] ILP	-	1.00	0.33	1.00	0.12	1.00	0.22	1.00	0.50	1.00	0.53
[23] ILP, filtering	-	1.00	0.33	1.00	0.35	1.00	0.28	0.78	1.00	0.81	1.00
[18] Genetic miner	-	$-^U$	$-^U$	$-^U$	$-^U$	$-^U$	$-^U$	$-^N$	$-^N$	$-^U$	$-^U$
[4] ETMd miner	-	0.74	1.00	0.83	1.00	1.00	0.86	0.79	1.00	0.75	1.00
[11] TS Cancel	-	$-^N$	$-^N$	$-^N$	$-^N$	0.91	0.78	$-^N$	$-^N$	$-^N$	$-^N$
[19] TS Regions	-	0.27	0.61	$-^N$	$-^N$	0.93	0.88	0.86	0.76	0.76	0.82
[14] IM (baseline)	1.0	1.00	0.69	1.00	0.43	1.00	0.89	1.00	0.69	1.00	0.83
[14] IM (baseline)	0.8	0.74	0.73	0.94	0.64	1.00	0.92	0.99	0.48	1.00	0.82
[14] IM (baseline)	0.5	0.62	0.75	0.94	0.63	0.82	1.00	0.76	0.48	0.74	0.77
Ours Cancelation	1.0	1.00	0.70	1.00	0.62	1.00	1.00	1.00	0.66	1.00	0.76
Ours Cancelation	0.8	0.76	0.58	0.94	0.67	1.00	1.00	1.00	0.35	1.00	0.68
Ours Cancelation	0.5	0.64	0.67	0.94	0.66	1.00	1.00	0.90	0.39	0.81	0.73

^{U} Unsound model ^{N} No model (see Table 6)

In all cases, we can see that we outperform the ILP algorithms on precision, and we outperform the ETMd miner and TS based miners on fitness. Overall, we can conclude that the added expressiveness of modeling the cancelation region have a positive impact on the model quality.

On the Simplicity of Models. We compared the discovered models both using a simplicity metric [4] and manually. In most cases, the discovered models are comparably complex, with the cancelation models usually being slightly simpler.

In Fig. 4, two discovered models for the BPIC12 log at paths 0.8 are shown. Note that in the Cancelation model (Fig. 4(b)), we see that the main, happy flow behavior is neatly discovered inside the cancelation region, and the "negative" behavior is modeled separately after triggering the cancelation region. In the IM model (Fig. 4(a)), skips obfuscate the normal happy flow behavior.

Overall, we can conclude that the added expressiveness of modeling the cancelation region has, in most cases, a positive impact on the model simplicity.

7   Conclusion

In this paper, we presented a novel extension to the process tree model to support cancelation behavior, and proposed a novel process discovery technique to discover sound, fitting models with cancelation features. The proposed discovery technique relies on a generic error oracle function, and allows us to discover complex combinations of multiple, possibly nested cancelation regions based on observed behavior. An implementation of the proposed algorithm has been tested and made available via the *Statechart* plugin for the ProM framework [12]. Our experimental results, based on real-life event logs, demonstrate the feasibility and usefulness of the approach.

Future work aims to further aid the user in selecting an error oracle, and (partially) automate the error oracle instantiation. In addition, we aim to support reliability analysis around cancelation features, using additional event log data. Lastly, enabling the proposed techniques in a streaming context could provide valuable real-time insights into (business) processes in their natural environment. Techniques able to operate in a streaming context need less memory and are therefore also valuable for other types of analysis.

References

1. Adriansyah, A.: Aligning observed and modeled behavior. Ph.D. thesis, Eindhoven University of Technology (2014)
2. Augusto, A., Conforti, R., Dumas, M., La Rosa, M., Bruno, G.: Automated discovery of structured process models: discover structured vs. discover and structure. In: Comyn-Wattiau, I., Tanaka, K., Song, I.-Y., Yamamoto, S., Saeki, M. (eds.) ER 2016. LNCS, vol. 9974, pp. 313–329. Springer, Cham (2016). doi:10. 1007/978-3-319-46397-1_25
3. Buijs, J.C.A.M.: Receipt phase of an environmental permit application process ('WABO'), CoSeLoG project (2014). https://doi.org/10.4121/uuid: a07386a5-7be3-4367-9535-70bc9e77dbe6
4. Buijs, J.C.A.M., van Dongen, B.F., van der Aalst, W.M.P.: On the role of fitness, precision, generalization and simplicity in process discovery. In: Meersman, R., Panetto, H., Dillon, T., Rinderle-Ma, S., Dadam, P., Zhou, X., Pearson, S., Ferscha, A., Bergamaschi, S., Cruz, I.F. (eds.) OTM 2012. LNCS, vol. 7565, pp. 305–322. Springer, Heidelberg (2012). doi:10.1007/978-3-642-33606-5_19

5. Carmona, J., Cortadella, J., Kishinevsky, M.: A region-based algorithm for discovering Petri nets from event logs. In: Dumas, M., Reichert, M., Shan, M.-C. (eds.) BPM 2008. LNCS, vol. 5240, pp. 358–373. Springer, Heidelberg (2008). doi:10.1007/978-3-540-85758-7_26

6. Celonis GmbH: Celonis Process Mining. https://www.celonis.com. Accessed 06 July 2017

7. Conforti, R., Dumas, M., García-Banuelos, L., La Rosa, M.: BPMN miner: automated discovery of BPMN process models with hierarchical structure. Inf. Syst. **56**, 284–303 (2016)

8. De Leoni, M., Mannhardt, F.: Road traffic fine management process (2015). https://doi.org/10.4121/uuid:270fd440-1057-4fb9-89a9-b699b47990f5

9. Gradient ECM: Minit. https://www.minit.io/. Accessed 06 July 2017

10. Günther, C.W., Rozinat, A.: Disco: discover your processes. In: Lohmann, N., Moser, S. (eds.) Proceedings of the Demonstration Track of the 10th International Conference on Business Process Management (BPM 2012), vol. 940, pp. 40–44. CEUR Workshop Proceedings (2012)

11. Kalenkova, A.A., Lomazova, I.A.: Discovery of cancellation regions within process mining techniques. Fundamenta Informaticae **133**(2–3), 197–209 (2014)

12. Leemans, M.: Statechart plugin for ProM 6. https://svn.win.tue.nl/repos/prom/Packages/Statechart/. Accessed 24 May 2017

13. Leemans, M.: NASA Crew Exploration Vehicle (CEV) software event log (2017). http://doi.org/10.4121/uuid:60383406-ffcd-441f-aa5e-4ec763426b76

14. Leemans, S.J.J.: Robust process mining with guarantees. Ph.D. thesis, Eindhoven University of Technology, May 2017

15. Redlich, D., Molka, T., Gilani, W., Blair, G., Rashid, A.: Constructs competition miner: process control-flow discovery of BP-domain constructs. In: Sadiq, S., Soffer, P., Völzer, H. (eds.) BPM 2014. LNCS, vol. 8659, pp. 134–150. Springer, Cham (2014). doi:10.1007/978-3-319-10172-9_9

16. van der Aalst, W.M.P.: Discovery, verification and conformance of workflows with cancellation. In: Ehrig, H., Heckel, R., Rozenberg, G., Taentzer, G. (eds.) ICGT 2008. LNCS, vol. 5214, pp. 18–37. Springer, Heidelberg (2008). doi:10.1007/978-3-540-87405-8_2

17. van der Aalst, W.M.P.: Process Mining: Data Science in Action. Springer, Heidelberg (2016)

18. van der Aalst, W.M.P., de Medeiros, A.K.A., Weijters, A.J.M.M.: Genetic process mining. In: Ciardo, G., Darondeau, P. (eds.) ICATPN 2005. LNCS, vol. 3536, pp. 48–69. Springer, Heidelberg (2005). doi:10.1007/11494744_5

19. van der Aalst, W.M.P., Rubin, V., Verbeek, H.M.W., van Dongen, B.F., Kindler, E., Günther, C.W.: Process mining: a two-step approach to balance between underfitting and overfitting. Soft. Syst. Model. **9**(1), 87–111 (2010)

20. van der Aalst, W.M.P., Weijters, A.J.M.M., Maruster, L.: Workflow mining: discovering process models from event logs. IEEE Trans. Knowl. Data Eng. **16**(9), 1128–1142 (2004)

21. van der Werf, J.M.E.M., van Dongen, B.F., Hurkens, C.A.J., Serebrenik, A.: Process discovery using integer linear programming. In: van Hee, K.M., Valk, R. (eds.) PETRI NETS 2008. LNCS, vol. 5062, pp. 368–387. Springer, Heidelberg (2008). doi:10.1007/978-3-540-68746-7_24

22. van Dongen, B.F.: BPI Challenge 2012 (2012). http://dx.doi.org/10.4121/uuid:3926db30-f712-4394-aebc-75976070e91f

23. van Zelst, S.J., van Dongen, B.F., van der Aalst, W.M.P.: Avoiding over-fitting in ILP-based process discovery. In: Motahari-Nezhad, H.R., Recker, J., Weidlich, M. (eds.) BPM 2015. LNCS, vol. 9253, pp. 163–171. Springer, Cham (2015). doi:10. 1007/978-3-319-23063-4_10

24. Weijters, A.J.M.M., Ribeiro, J.T.S.: Flexible Heuristics Miner (FHM). In: 2011 IEEE Symposium on Computational Intelligence and Data Mining (CIDM), pp. 310–317, April 2011

Introducing Collaboration for Locating Features in Models: Approach and Industrial Evaluation

Francisca Pérez[✉], Ana C. Marcén, Raúl Lapeña, and Carlos Cetina

SVIT Research Group, Universidad San Jorge,
Autovía A-23 Zaragoza-Huesca Km. 299, 50830 Zaragoza, Spain
{mfperez,acmarcen,rlapena,ccetina}@usj.es

Abstract. Feature Location (FL) is one of the most important tasks in software maintenance and evolution. However, current works on FL neglected the collaboration of different domain experts. This collaboration is especially important in long-living industrial domains where a single domain expert may lack the required knowledge to fully locate a feature, so the collaboration among different domain experts could alleviate this lack of knowledge. In this work, we address collaboration among different domain experts by automatically reformulating their feature descriptions. With our approach, we extend existing FL approaches based on Information Retrieval and Linguistic rules to locate features in models. We evaluate our approach in a real-world case study from our industrial partner, which is a worldwide leader in train manufacturing. We analyze the impact of our approach in terms of recall, precision, and F-Measure. Moreover, we perform a statistical analysis to show that the impact of the results is significant. Our results show that our approach for collaboration boosts the quality of the results of FL.

Keywords: Collaborative information retrieval · Feature location · Query expansion · Model driven engineering

1 Introduction

Nowadays, work environments are characterized by an emphasis on collaborative team work [9]. Many empirical studies identified collaborative information seeking and retrieval as everyday work patterns in order to solve a shared information need and to benefit from the diverse expertise and experience of the team members [13].

Despite the importance of collaboration, Feature Location (FL) approaches neglected collaboration among different domain experts to find the set of software artifacts (e.g., code or models) that realize a specific feature. Even though collaboration is a useful and often necessary component of complex projects in industrial contexts when the task at hand is difficult or cannot be carried out by one individual [36].

To cope with this lack, the contribution of this paper is the introduction of collaboration for locating a target feature in models from different domain

© Springer International Publishing AG 2017
H. Panetto et al. (Eds.): OTM 2017 Conferences, Part I, LNCS 10573, pp. 114–131, 2017.
https://doi.org/10.1007/978-3-319-69462-7_9

experts. First, each domain expert provides both a feature description and an estimation of confidence level. Then, our approach uses the confidence level to identify relevant feature descriptions. After, our approach automatically reformulates the relevant feature descriptions in a single query using a technique that is based on Rocchio's method [34]. The resulting query is used to find the model fragment that realizes the feature being located using two different FL cores (Information Retrieval (IR) or Linguistic rules) since they obtain the best results in the literature [21,29,30,38].

We analyze the impact of collaboration in a real-world industrial case study from the railway domain. Our industrial partner, Construcciones y Auxiliar de Ferrocarriles (CAF)[1], is a worldwide leader in train manufacturing. CAF provided us with both the models of software that control and manage the trains and the oracle (the realization of features validated by our industrial partner). Then, we involve 19 domain experts from our industrial partner to obtain feature descriptions and confidence levels as the input of our approach. We compare the model fragment that realizes each of the target features that our approach obtains as a result with the oracle (which is considered to be the ground truth) in terms of recall, precision, and F-measure. Finally, we perform a statistical analysis in order to provide quantitative evidence of the impact of the results.

The results of this paper show that introducing collaboration boost the quality of the results in the existing FL approaches: IR obtains an improvement of 37.19% in F-measure, and Linguistic rules obtain an improvement of 29.31% in F-measure. We hope that these results promote the introduction of collaborative mechanisms in FL.

The rest of the paper is structured as follows: Sect. 2 provides the required background. Section 3 presents our approach to introduce collaboration, and the cores of IR and Linguistic rules. Section 4 describes the evaluation carried out. Section 5 describes the threats to validity. Section 6 reviews the related work. Finally, Sect. 7 concludes the paper.

2   Background

The Domain Specific Language (DSL) of our industrial partner has the expressiveness required to describe both the interaction between the main pieces of equipment installed in a train unit and the non-functional aspects that are related to regulation. We present an equipment-focused simplified subset of the DSL for the sake of understandability and legibility and due to intellectual property rights concerns. This subset of the DSL will be used to present a running example throughout the rest of the paper.

Figure 1 shows an example of a product model from a real-world train. It shows two separate pantographs (High Voltage Equipment) that collect energy from the overhead wires and send it to their respective circuit breakers (Contactors), which, in turn, send it to their independent Voltage Converters.

---

[1] www.caf.net/en.

The converters then power their assigned Consumer Equipment: the HVAC on the left (the train's air conditioning system), and the PA (public address system) and CCTV (television system) on the right.

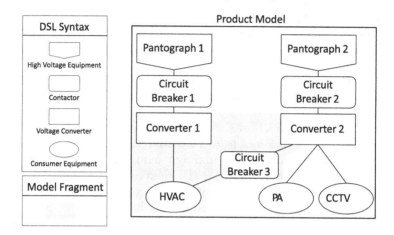

Fig. 1. Example of product model and model fragment (Color figure online)

An example of model fragment is also shown in Fig. 1. The elements of the model fragment are highlighted in green, which are the realization of the feature: HVAC Assistance. This feature allows the passing of current from one converter to the HVAC that is assigned to its peer for coverage in case of overload or failure of the first converter.

3   Our Approach for Introducing Collaboration in FL

Figure 2 presents our approach to introduce collaboration from different domain experts for locating features in models. First, each domain expert involved provides both a feature description and a self-rated confidence level for the feature name as input. Second, one of the feature descriptions is automatically reformulated to include relevant terms from other feature descriptions. To do this, feature descriptions are ordered from the highest to the lowest confidence level. The first feature description is set as base query, and the k subsequent feature descriptions are set as relevant documents. Next, the base query is automatically reformulated to expand it with the most representative terms found in the relevant documents. Finally, both the product model and the reformulated query are taken as input to locate the relevant model fragments. To locate the relevant model fragments, we use two different FL cores (IR or Linguistic rules) since they obtain the best results in the literature [21,29,30,38]. The result after the core is executed is a model fragment to the input reformulated query.

In the next three subsections, we describe how the automatic query reformulation is performed, and how FL can be performed using one of the FL cores: IR or the Linguistic rules.

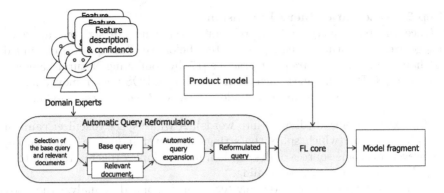

Fig. 2. Our approach for introducing collaboration in feature location

3.1 Automatic Query Reformulation

In order to introduce collaboration from different domain experts' feature descriptions, our approach starts with the selection of one feature description as the base query (Step 1). Afterwards, the base query is automatically reformulated to expand it with the most representative terms found in other domain experts' feature descriptions set as relevant documents (Step 2). These two steps are performed as follows:

(Step 1) Selection of the base query and relevant documents

Our approach sorts the feature descriptions provided by the domain experts from the highest to the lowest self-rated confidence level in order to select the feature description in the first position (i.e., the highest self-rated confidence) as the base query. The self-rated confidence level is supplied for each feature description using a Likert scale ranging from 7 (the highest self-rated confidence) to 1 (the lowest self-rated confidence). Then, our approach selects k feature descriptions sorted by confidence level, where k is the number of domain experts who collaborate to reformulate the base query. Each of the selected feature descriptions is set as a relevant document.

For example, the feature description *"Passing of current from one converter to the HVAC assigned to its peer for coverage in case of overload or failure of the first converter"* provided by Domain expert A is selected as the base query since it has the highest self-rated confidence level (6). Next, two subsequent feature descriptions with the highest self-rated confidence level are set as relevant documents since two exerts are established to collaborate in the reformulation of the query (k = 2). These feature descriptions are: *"The circuit breaker changes to another converter in case of failure in the HVAC converter"* (from Domain expert B since the self-rated confidence level is 4); and *"In case of failure or overload in the converter that provides energy to the air conditioning unit, the circuit breaker provides energy from its converter"* (from Domain expert C since the self-rated confidence level is 3).

(Step 2) Automatic Query Expansion

Once the base query and the relevant documents are set, our approach homogenizes the Natural Language (NL) text before the base query is expanded. Text homogenization is a frequent practice [17] by combining Natural Language Processing (NLP) techniques, such as the analysis of POS tags, removal of stop-words, and stemming. Our approach adopts the NLP techniques as follows:

- The text is tokenized (divided into words). A white space tokenizer can usually be applied (which splits the strings whenever it finds a white space); however, for some sources of description, more complex tokenizers need to be applied such as CamelCase naming.
- The Parts-of-Speech (POS) tagging technique is applied to analyze the words grammatically and to infer the role of each word in the text provided. As a result, each word is tagged, which allows the removal of some categories that do not provide relevant information. For instance, conjunctions (e.g., *or*), articles (e.g., *a*), or prepositions (e.g., *at*) are words that are commonly used and do not contribute relevant information to describe the feature, so they are removed.
- Stemming techniques are applied to unify the language that is used in the text. This technique consists of reducing each word to its root, which allows different words that refer to similar concepts to be grouped together. For instance, plurals are turned into singulars (*doors* to *door*) or verb tenses are unified (*using* and *used* are turned into *use*).
- The Domain Term Extraction and Stopword Removal techniques are applied. In order to carry out these techniques, domain experts provide two separate lists of terms: one list of both single-word and multiple-word terms that belong to the domain and must be kept for analysis, and a list of irrelevant words that have no analysis value. Both kinds of terms can be automatically filtered in or out of the final query.

For example, the terms of the feature description that is set as the base query (*Passing of current from one converter to the HVAC assigned to its peer for coverage in case of overload or failure of the first converter*) are homogenized as follows: *current, convert, hvac, coverag, overload, failur, convert,* and *assign*.

Once the NL text is homogenized, our approach automatically reformulates the base query to expand it with terms of the relevant documents using a technique that is based on Rocchio's method [34], which is perhaps the most commonly used method for query reformulation [37]. Rocchio's method orders the terms in the top K relevant documents based on the sum of the importance of each term of the K documents using the following equation:

$$Rocchio = \sum_{d \in R} TfIdf(t, d) \tag{1}$$

where R is the set of top K relevant documents in the list of retrieved results, d is a document in R, and t is a term in d. The first component of the measure is the Term Frequency (Tf), which is the number of times the term appears in a document; it is an indicator of the importance of the term in the document compared to the rest of the terms in that document. The second component is the Inverse Document Frequency (*Idf*), which is the inverse of the number of documents that contain that term; it indicates the specificity of that term for a document that contains it. Once the terms of the relevant documents are ordered, we consider the first 10 term suggestions to expand the base query, as is recommended in the domain literature [5].

For example, the first 10 terms from the relevant documents set in the previous step (*convert, energi, provid, overload, circuit, breaker, failur, hvac, air, condit*) are used to reformulate the base query by adding these terms. Therefore, the reformulated query is made up of the following terms: *current, convert, hvac, converag, overload, failur, assign, energi, provid, circuit, breaker, air*, and *condit*.

3.2   IR FL CORE

IR [12,23,35] is a sub-field of computer science that deals with the automated storage and retrieval of documents. There are many IR techniques, but most of the efforts show better results when applying LSI [21,29,30]. Hence, we use LSI to recover the model fragment that realizes a feature description as one of the FL cores of our approach.

LSI [19] is an automatic mathematical/statistical technique that analyzes relationships between *queries* and *documents* (bodies of text). It constructs vector representations of both a user *query* and a corpus of text *documents* by encoding them as a *term-by-document co-occurrence matrix*, and analyzes the relationships between those vectors to get a similarity ranking between the reformulated *query* and the *documents*. Each row in the co-occurrence matrix (*term*) stands for each of the words that compose the reformulated query and NL representation of the input model, extracted through the technique presented in [28]. In Fig. 3, it is possible to appreciate an example of matrix in which the rows are a set of representative terms in the domain such as 'pantograph' or 'door'. Each column in the matrix (*document*) stands for one model element from one input model, taken from our real world case study. In Fig. 3, it is possible to appreciate identifiers in the columns such as 'ME1' or 'ME2', which stand for the *documents* of those particular model elements. The final column stands for the *reformulated query*. Each cell in the matrix contains the frequency with which the *term* of its row appears in the *document* denoted by its column. For instance, in Fig. 3, the *term* 'pantograph' appears twice in the 'ME2' *document* and once in the *reformulated query*.

Afterwards, vector representations of the *documents* and the *reformulated query* are obtained by normalizing and decomposing the *term-by-document co-occurrence matrix* using a matrix factorization technique called *Singular Value Decomposition* (SVD) [19]. SVD is a form of factor analysis, or more properly the mathematical generalization of which factor analysis is a special case. In SVD, a rectangular matrix is decomposed into the product of three other matrices. One component matrix describes the original row entities as vectors of derived

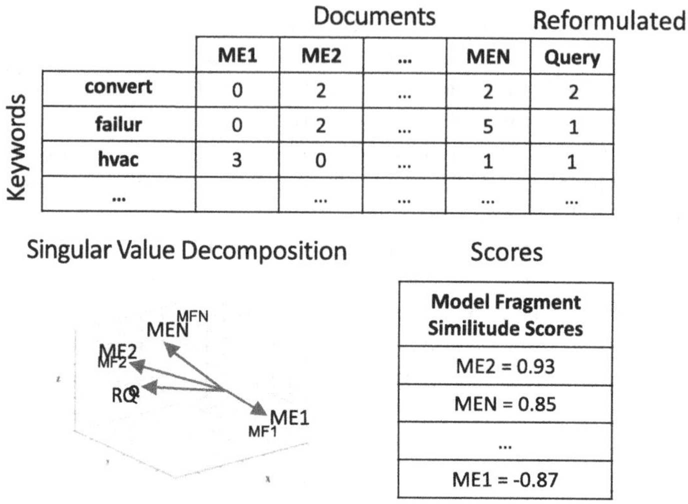

		ME1	ME2	...	MEN	Query
	convert	0	2	...	2	2
	failur	0	2	...	5	1
	hvac	3	0	...	1	1
	...		...	...	...	...

Model Fragment Similitude Scores
ME2 = 0.93
MEN = 0.85
...
ME1 = -0.87

Fig. 3. Latent Semantic Indexing Example using the reformulated query

orthogonal factor values, another describes the original column entities in the same way, and the third is a diagonal matrix containing scaling values such that when the three components are matrix-multiplied, the original matrix is reconstructed.

The relevancy ranking (which can be seen in Fig. 3) is produced according to the calculated similarity degrees. In this example, LSI retrieves 'ME2' and 'MEN' in the first and second position of the relevancy ranking due to *query-documents* cosines being '0.9343' and '0.8524', implying a high similarity degree between the model elements and the reformulated query. On the opposite, the 'M1' model element is returned in a latter position of the ranking due to its *query-document* cosine being '−0.8736', implying a lower similarity degree.

From the ranking of all the model elements, only those model elements that have a similarity measure greater than x must be taken into account. A good heuristic that is widely used is $x = 0.7$. This value corresponds to a 45° angle between the corresponding vectors. Even though the selection of the threshold is an issue under study, the heuristic chosen for this work has yielded good results in other similar works [25, 33].

Following this principle, the elements with a similarity measure equal or superior to $x = 0.7$ are taken to conform a model fragment, candidate for realizing the feature. Through the example provided in Fig. 3, ME2 and MEN are model elements that conform part of the model fragment obtained by this core for the reformulated query, due to their cosine values being superior to the 0.7 threshold.

3.3   Linguistic Rules FL CORE

This core is based on an approach presented by Spanoudakis et al. [38], which is a linguistic rule-based to support the automatic generation of Traceability Links between feature descriptions and models. Specifically, the Traceability Links are generated following two stages:

1. First, a Parts-of-Speech (POS) tagging technique [20] is applied on the feature descriptions that are defined using natural language.
2. Second, the Traceability Links between the feature descriptions and the models are generated through the *description-to-object-model* rules.

The *description-to-object-model* (DTOM) rules are specified by investigating grammatical patterns in feature descriptions. Moreover, the DTOM rules are based on two kinds of relations between feature descriptions and models. On the one hand, *Overlap* relations are understood as the relation between a sequence of terms in a feature description and a class, attribute, association or association end in model. On the other hand, *Requires_Execution_Of* relations are understood as the relation between a sequence of terms in a feature description and an operation in model.

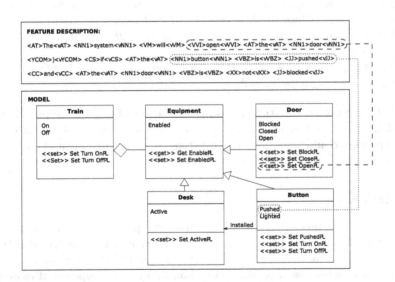

Fig. 4. Example of Traceability Links generation based on DTOM rules

In Fig. 4 the sequence of terms <NN1>button</NN1> <VBZ>is</VBZ> <JJ>pushed</JJ> in the feature description and the attribute Pushed of the class Button satisfy the conditions of a rule and, as a consequence, an *Overlap* relation would be created between them. In Fig. 4 the sequence of terms <VVI>open</VVI> <AT>the</AT> <NN1>door</NN1> in the feature description and the operation Set Open of the class Door satisfy the conditions of a rule and, as a consequence, a *Requires_Execution_Of* relation would be created between them.

In [38], the authors propose rules that we apply between a reformulated feature description and a model. Hence, a set of elements of the model are related to the reformulated feature description. These elements compose the model fragment as traceability result.

4   Evaluation

This section presents the research questions that our work tackles, the data set of our real-world case study, the implementation details, the planning and execution, the results and the statistical analysis.

4.1   Definition

We aim to answer the following research questions:

RQ$_1$: *Does the collaboration produce an improvement in the existing FL approaches that obtain the best results (IR and Linguistic rules)?*
RQ$_2$: *If a positive answer in RQ$_1$, how much is the quality of the solution improved on IR?*
RQ$_3$: *If a positive answer in RQ$_1$, how much is the quality of the solution improved on Linguistic rules?*

The first research question investigates the results of our approach using IR and Linguistic rules, the results of IR, and the results of Linguistic rules. While, the second and third question investigates the improvement of introducing collaboration in IR and Linguistic rules, respectively.

4.2   Data Set

Our industrial partner, CAF, is an international provider of railway solutions all over the world that can be seen in different types of trains (regular trains, subway, light rail, monorail, etc.). The data set that CAF provided us is made up of 23 trains where each product model on average is composed of more than 1200 elements. They are built from 121 different features that can be part of a specific product model.

Furthermore, CAF provided us with the model fragments of 43 features from different trains. Nineteen domain experts from CAF were involved in obtaining different descriptions for each feature. Also, CAF provided us with lists of domain terms and stopwords to process the NL. The domain terms list has around 300 domain terms, and the stopwords list has around 60 words. Both lists were created by CAF domain experts who are associated with the provided products.

4.3   Implementation Details

We have used the Eclipse Modeling Framework to manipulate the models and CVL [14] to manage the model fragments. The techniques used to process the NL have been implemented using OpenNLP [1] for the POS-Tagger and the English (Porter2) stemming algorithm [3] for the stemming algorithm (originally created using snowball and then compiled to Java). The LSI has been implemented using the Efficient Java Matrix Library (EJML [2]).

4.4   Planning and Execution

Figure 5 shows an overview of the process that was planned to answer the research questions taking as input both the documentation from our industrial partner, and the 43 feature descriptions and their self-rated confidence for each of the 19 domain experts from our industrial partner, who were involved.

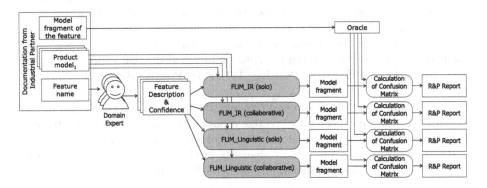

Fig. 5. Evaluation process

First, we execute two variants (solo and collaborative) for each FL core (depicted as shaded boxes in Fig. 5):

- **FLiM_IR (solo):** this variant uses the IR FL core to locate the model fragment that realizes the feature description provided by each of the 19 domain experts for each of the 43 features, i.e., 43 (features) × 19 domain experts' feature descriptions = 817 independent runs.
- **FLiM_IR (collaborative):** this variant enables that different domain experts collaborate to locate the model fragment that realizes each of the 43 target features (as described in Sect. 3.1) using the IR FL core. Specifically, we set $k = 5$ (i.e., one domain expert's feature description is set as base query and five domain experts' feature descriptions are set as relevant documents for the location of the target features). This decision was made based on recommendations found in the literature [5].

- **FLiM_Linguistic (solo):** this variant uses the Linguistic rules FL core to locate the model fragment that realizes the feature description provided by each of the 19 domain experts for each of the 43 features, i.e., 43 (features) × 19 domain experts' feature descriptions = 817 independent runs.
- **FLiM_Linguistic (collaborative):** this variant not only uses the Linguistic rules FL core but also, it enables that different domain experts collaborate to locate the model fragment that realizes each of the 43 target features (as described in Sect. 3.1) by setting k = 5.

When a variant is executed, we obtain as a result a model fragment that realizes a target feature. Then, we compare the model fragment with an oracle as Fig. 5 shows. The oracle is prepared using the model fragments that realize each target feature provided by our industrial partner. The oracle will be considered the ground truth and will be used to calculate a confusion matrix.

The confusion matrix is a table that is often used to describe the performance of a classification model on a set of test data (the solutions) for which the true values are known (from the oracle). In our case, each solution obtained is a model fragment that is composed of a subset of the model elements that are part of the product model. Since the granularity is at the level of model elements, each model element presence or absence is considered as a classification. The confusion matrix distinguishes between the predicted values and the real values classifying them into four categories: True Positive (TP), False Positive (FP), True Negative (TN) and False Negative (FN).

Finally, some performance measurements are derived from the values in the confusion matrix. Specifically, we create a report for the confusion matrix including three performance measurements: recall, precision, and F-measure.

Recall measures the number of elements of the solution that are correctly retrieved by the proposed solution and is defined as follows:

$$Recall = \frac{TP}{TP + FN}$$

Precision measures the number of elements from the solution that are correct according to the ground truth (the oracle) and is defined as follows:

$$Precision = \frac{TP}{TP + FP}$$

F-measure corresponds to the harmonic mean of precision and recall and is defined as follows:

$$F - measure = 2 * \frac{Precision * Recall}{Precision + Recall}$$

Recall values can range from 0% (which means that no single model element obtained from the oracle is present in the solution) to 100% (which means that all the model elements from the oracle are present in the solution). Precision values can range from 0% (which means that no single model fragment from the solution is present in the oracle) to 100% (which means that all the model fragments from the solution are present in the oracle). A value of 100% precision and 100% recall implies that both the solution and the oracle are the same.

4.5   Results

Table 1 shows the mean values of recall, precision, and the F-measure. In terms of recall, FLiM_IR (solo) obtains the best result, providing a precision value of 70.28%. The second best results are obtained by the collaborative approaches (an average value of 59.19% in FLiM_IR (collaborative) and 57.13% in FLiM_Linguistic (collaborative)). The worst result is obtained by FLiM_Linguistic (solo), which obtains an average value of 24.64%. In terms of precision, the collaborative approaches obtain the best results. FLiM_IR (collaborative) obtains the best results in precision, providing an average value of 68.24%, whereas FLiM_Linguistic (collaborative) obtains an average value of 54.49%. FLiM_Linguistic (solo) provides an average value of 33.95%, whereas the worst results are obtained by FLiM_IR (solo) (16.51%). In terms of the F-measure, the collaboration among domain experts improves the results in IR and Linguistic (up to 37.19 and 29.31, respectively).

Table 1. Mean values and standard deviations for Precision, Recall, and F-measure in the industrial case study

	Recall $\pm$ (σ)	Precision $\pm$ (σ)	F-measure $\pm$ (σ)
FLiM_IR (solo)	70.28 $\pm$ 15.93	16.51 $\pm$ 10.86	24.81 $\pm$ 14.15
FLiM_IR (collaborative)	59.19 $\pm$ 13.99	68.24 $\pm$ 14.32	**62.00 $\pm$ 11.35**
FLiM_Linguistic (solo)	24.64 $\pm$ 13.61	33.95 $\pm$ 15.59	25.00 $\pm$ 12.42
FLiM_Linguistic (collaborative)	57.13 $\pm$ 13.37	54.49 $\pm$ 12.89	**54.31 $\pm$ 9.90**

4.6   Statistical Analysis

To answer whether the collaboration produce an improvement (RQ_1), we compare the solo variant with the collaborative variant for the two cores (IR and Linguistic rules). To properly compare the variants, all of the data was analyzed using statistical methods to provide formal and quantitative evidence (statistical significance). The statistical tests provide a probability value, $p - value$. The $p - value$ obtains values between 0 and 1. It is accepted by the research community that a $p - value$ under 0.05 is statistically significant.

To compare the variants, we carry out a Holm's post hoc analysis, which performs a pair-wise comparison among the results of each variant. The $p - Values$ of Holm's post hoc analysis are smaller than the corresponding significance threshold value (0.05) in the comparison between the solo and collaborative variants for each of the two cores.

RQ_1 answer. From the results, we can conclude that introducing collaboration when locating features in models produces an improvement in terms of solution quality using both the IR and Linguistic rules core.

To answer how much is the quality of the solution improved using IR (RQ_2), and how much is the quality of the solution improved using Linguistic rules

(RQ$_3$), we perform statistical analysis of the results since statistically significant differences can be obtained even if they are so small as to be of no practical value. Hence, it is important to assess if the results of our approach are statistically better than another and to assess the magnitude of the improvement. *Effect size* measures are needed to analyze this. For a non-parametric effect size measure, we use Vargha and Delaney's $\hat{A}_{12}$ [39]. $\hat{A}_{12}$ measures the probability that running one approach yields higher values than running another approach. If the two approaches are equivalent, then $\hat{A}_{12}$ will be 0.5.

Table 2 shows the values of the effect size statistics for our approach in IR and Linguistic rules. The second row of the table shows the comparison that entails IR. FLiM_IR (collaborative) would obtain better results than FLiM_IR (solo) in 44.94% of the runs for recall and 99.68% of the runs for precision.

RQ$_2$ answer. From the results, we conclude how much the solution of the quality is improved introducing collaborative in IR. Although the value for recall is near to be statistically equivalent, the value for precision shows a pronounced superiority to perform collaborative feature location in models (99.68% of the runs would obtain better results for precision introducing collaboration).

The third row of Table 2 shows the comparison that entails Linguistic rules, which obtains the largest differences. FLiM_Linguistic (collaborative) would obtain better results than FLiM_Linguistic (solo) in 97.99% of the runs for recall and 92.37% of the runs for precision.

RQ$_3$ answer. The results confirm that introducing collaboration to locate features in models using Linguistic rules has a pronounced superiority since more than 92% of the runs would obtain better results.

Table 2. $\hat{A}_{12}$ statistic for each core vs. its collaborative variant

	Recall	Precision
FLiM_IR (collaborative) vs. FLiM_IR (solo)	0.4494	0.9968
FLiM_Linguistic (collaborative) vs. FLiM_Linguistic (solo)	0.9799	0.9237

5   Threats to Validity

We use the classification of threats of validity of [32, 42], which distinguishes four aspects of validity to acknowledge the limitations of our evaluation.

Construct Validity: To minimize this risk, our evaluation is performed using three measures: precision, recall and F-measure. These measures are widely accepted in the software engineering research community [35].

Internal Validity: We used an oracle (obtained from our industrial partner and considering the ground truth) to evaluate our approach using feature descriptions or reformulated feature descriptions as queries where the expected solution was known beforehand. By doing so, we were able to compute the recall, precision and F-measure. With regard to the number of relevant documents and terms used to expand the query, we used the values of 5 and 10, respectively as recommended in the literature [5]. However, we do not know at this stage how using different values would impact the results.

External Validity: In order to mitigate this threat, our approach has been designed to be applied not only to the domain of our industrial partner but also, to different domains. The requisites to apply our approach are that the set of models where features have to be located conform to MOF (the OMG metalanguage for defining modeling languages), and the query must be provided as a textual description.

Furthermore, query reformulation techniques can only work if the original query is reasonably strong to retrieve at least some of the relevant documents [37]. As occurs in other works [15,37], results depend on the quality of the queries. Poor queries assign high rank to irrelevant model fragments. It is also worth noting that the language used for the textual elements of the models and the feature descriptions in the query provided must be the same. This language is particular for each domain.

Hence, despite our approach can be applied to locate features on MOF-based models from different domains, our approach should be applied to other domains before assuring its generalization.

Reliability: To reduce this threat, the feature descriptions and the product family are provided by our industrial partner, who is not involved in this research.

6 Related Work

Several approaches have been proposed to reformulate queries in a semi-automatic or automatic way by expanding the query of a user [36] based on relevant documents such as source code and Internet sites. For example, Yang and Tan [43] reformulate the query by extracting synonyms, antonyms, abbreviations, and related words from the source code. Rivas et al. [31] add relevant terms from a scientific documental database to a query to improve the documents initially retrieved. Hill et al. [15] also obtain possible query expansion terms from the code. Lu et al. [22] improve code search by expanding the query with synonyms. Marcus et al. [25] expand the query using LSI in order to determine the terms from the source code that are most similar to the query. Other approaches expand the query by adding information from external sources of information such as public repositories [8].

Table 3 compares the above query expansion works with our work. As the table shows, the base query that is going to be expanded is obtained from a human, who can play different roles (developer, user, analyst, and domain expert). The relevant documents used to find the terms to expand the query are usually source code, online documentation, or text. In contrast, to support collaboration in our work, we use other domain experts' feature descriptions as

Table 3. Comparison with query expansion works

Author	Base query	Relevant documents	Industrial domain	Artifact
Yang and Tan [43]	Developer	Source code	No	Code
Rivas et al. [31]	User	Biomedical articles	No	Text
Hill et al. [15]	Developer	Source code	No	Code
Lu et al. [22]	Developer	Internet site	No	Code
Marcus et al. [25]	User	Source code	No	Code
Dimitru et al. [8]	User	Internet sites	No	Product specifications
Our work	Domain expert	Domain experts	Yes	Models

relevant documents in order to enrich the base query feature description with the knowledge of other domain experts. Moreover, in contrast to the above works, our work aims to apply query expansion techniques for introducing collaboration in industry since the context is not the same as in academia [4].

Also, there are many feature location approaches that have been proposed to find features in code by taking textual information as input [7] such as [6,18,40]. Other works such as [11,16,26,27,41,44,45] focus on the location of features in models by comparing the models with each other to formalize the variability among them, whereas Font et al. [10] use an evolutive algorithm to locate features among a family of models. In contrast to these feature location approaches, our work introduces collaboration among different domain experts to locate a target feature in models.

7   Concluding Remarks

Although collaboration is a useful and a necessary component in industrial contexts to take advantage of the experience of different domain experts, it is neglected in existing FL approaches. In this paper, we propose an approach that introduces collaboration in two existing FL approaches (IR and Linguistic rules) to locate features in models. To introduce collaboration, the relevant feature descriptions provided by the domain experts are identified using an estimation of confidence level. After, our approach automatically reformulates the relevant feature descriptions in a single query.

The results show that introducing collaboration for locating features in models boosts the quality of the results of existing FL approaches (IR and Linguistic rules). The statistical analysis of the results assesses the magnitude of the improvement of introducing collaboration. Moreover, our results show that our approach can be applied in real world environments.

As future work, we plan to evaluate the influence in the quality of the solution whether the number of domain experts who collaborate changes. In addition, we plan to evaluate the quality of the solution with new approaches based on Machine Learning [24].

Acknowledgements. This work has been partially supported by the Ministry of Economy and Competitiveness (MINECO) through the Spanish National R+D+i Plan and ERDF funds under the project Model-Driven Variability Extraction for Software Product Line Adoption (TIN2015-64397-R).

References

1. Apache OpenNLP: Toolkit for the processing of natural language text (2017). https://opennlp.apache.org/
2. Efficient java matrix library (2017). http://ejml.org/
3. English (porter2) stemming algorithm (2017). http://snowball.tartarus.org/algorithms/english/stemmer.htm
4. Ambreen, T., Ikram, N., Usman, M., Niazi, M.: Empirical research in requirements engineering: trends and opportunities. Requirements Eng., 1–33 (2016)
5. Carpineto, C., Romano, G.: A survey of automatic query expansion in information retrieval. ACM Comput. Surv. **44**(1), 1:1–1:50 (2012)
6. Cavalcanti, Y.a.C., Machado, I.d.C., Neto, P.A.d.M.S., de Almeida, E.S., Meira, S.R.d.L.: Combining rule-based and information retrieval techniques to assign software change requests. In: Proceedings of the 29th ACM/IEEE International Conference on Automated Software Engineering, ASE 2014, pp. 325–330 (2014)
7. Dit, B., Revelle, M., Gethers, M., Poshyvanyk, D.: Feature location in source code: a taxonomy and survey. J. Softw. Evol. Process **25**(1), 53–95 (2013)
8. Dumitru, H., Gibiec, M., Hariri, N., Cleland-Huang, J., Mobasher, B., Castro-Herrera, C., Mirakhorli, M.: On-demand feature recommendations derived from mining public product descriptions. In: Proceedings of the 33rd International Conference on Software Engineering, ICSE 2011, pp. 181–190 (2011)
9. Fidel, R., Pejtersen, A.M., Cleal, B., Bruce, H.: A multidimensional approach to the study of human-information interaction: a case study of collaborative information retrieval. J. Am. Soc. Inf. Sci. Technol. **55**(11), 939–953 (2004)
10. Font, J., Arcega, L., Haugen, Ø., Cetina, C.: Feature location in model-based software product lines through a genetic algorithm. In: Kapitsaki, G.M., Santana de Almeida, E. (eds.) ICSR 2016. LNCS, vol. 9679, pp. 39–54. Springer, Cham (2016). doi:10.1007/978-3-319-35122-3_3
11. Font, J., Ballarín, M., Haugen, Ø., Cetina, C.: Automating the variability formalization of a model family by means of common variability language. In: Proceedings of the 19th International Conference on Software Product Line (SPLC), pp. 411–418 (2015)
12. Frakes, W.B., Baeza-Yates, R.: Information Retrieval: Data Structures and Algorithms. Prentice-Hall, Inc., Upper Saddle River (1992)
13. Hansen, P., Shah, C., Klas, C.P.: Collaborative Information Seeking: Best Practices, New Domains and New Thoughts, 1st edn. Springer Publishing Company, Incorporated, Berlin (2015)
14. Haugen, Ø., Moller-Pedersen, B., Oldevik, J., Olsen, G., Svendsen, A.: Adding standardized variability to domain specific languages. In: 12th International on Software Product Line Conference, SPLC 2008, pp. 139–148, September 2008
15. Hill, E., Pollock, L., Vijay-Shanker, K.: Automatically capturing source code context of NL-queries for software maintenance and reuse. In: Proceedings of the 31st International Conference on Software Engineering, ICSE 2009, pp. 232–242. IEEE Computer Society, Washington, DC (2009)

16. Holthusen, S., Wille, D., Legat, C., Beddig, S., Schaefer, I., Vogel-Heuser, B.: Family model mining for function block diagrams in automation software. In: Proceedings of the 18th International Software Product Line Conference, vol. 2. pp. 36–43 (2014)

17. Hulth, A.: Improved automatic keyword extraction given more linguistic knowledge. In: Proceedings of the 2003 Conference on Empirical Methods in Natural Language Processing, pp. 216–223 (2003)

18. Kimmig, M., Monperrus, M., Mezini, M.: Querying source code with natural language. In: Proceedings of the 2011 26th IEEE/ACM International Conference on Automated Software Engineering, ASE 2011, pp. 376–379 (2011)

19. Landauer, T.K., Foltz, P.W., Laham, D.: An introduction to latent semantic analysis. Discourse Process. **25**(2–3), 259–284 (1998)

20. Leech, G., Garside, R., Bryant, M.: Claws4: the tagging of the British National Corpus. In: Proceedings of the 15th Conference on Computational Linguistics, vol. 1, pp. 622–628. Association for Computational Linguistics (1994)

21. Liu, D., Marcus, A., Poshyvanyk, D., Rajlich, V.: Feature location via information retrieval based filtering of a single scenario execution trace. In: Proceedings of the Twenty-Second IEEE/ACM International Conference on Automated Software Engineering, ASE 2007, pp. 234–243. ACM, New York (2007)

22. Lu, M., Sun, X., Wang, S., Lo, D., Duan, Y.: Query expansion via wordnet for effective code search. In: 2015 IEEE 22nd International Conference on Software Analysis, Evolution, and Reengineering (SANER), pp. 545–549, March 2015

23. Manning, C.D., Raghavan, P., Schütze, H., et al.: Introduction to Information Retrieval, vol. 1. Cambridge University Press, Cambridge (2008)

24. Marcén, A.C., Pérez, F., Cetina, C.: Ontological evolutionary encoding to bridge machine learning and conceptual models: approach and industrial evaluation. In: Proceedings of the 36th International Conference on Conceptual Modeling (2017)

25. Marcus, A., Sergeyev, A., Rajlich, V., Maletic, J.I.: An information retrieval approach to concept location in source code. In: Proceedings of the 11th Working Conference on Reverse Engineering, WCRE 2004, pp. 214–223 (2004)

26. Martinez, J., Ziadi, T., Bissyand, T.F., Klein, J., le Traon, Y.: Automating the extraction of model-based software product lines from model variants (T). In: 2015 30th IEEE/ACM International Conference on Automated Software Engineering (ASE), pp. 396–406, November 2015

27. Martinez, J., Ziadi, T., Bissyandé, T.F., Klein, J., Traon, Y.L.: Bottom-up adoption of software product lines: a generic and extensible approach. In: Proceedings of the 19th International Conference on Software Product Line, pp. 101–110 (2015)

28. Meziane, F., Athanasakis, N., Ananiadou, S.: Generating natural language specifications from UML class diagrams. Requirements Eng. **13**(1), 1–18

29. Poshyvanyk, D., Gueheneuc, Y.G., Marcus, A., Antoniol, G., Rajlich, V.: Feature location using probabilistic ranking of methods based on execution scenarios and information retrieval. IEEE Trans. Softw. Eng. **33**(6), 420–432 (2007)

30. Revelle, M., Dit, B., Poshyvanyk, D.: Using data fusion and web mining to support feature location in software. In: IEEE 18th International Conference on Program Comprehension (ICPC), pp. 14–23, June 2010

31. Rivas, A., Iglesias, E., Borrajo, L.: Study of query expansion techniques and their application in the biomedical information retrieval. Sci. World J. (2014)

32. Runeson, P., Höst, M.: Guidelines for conducting and reporting case study research in software engineering. Empirical Softw. Eng. **14**(2), 131–164 (2009)

33. Salman, H.E., Seriai, A., Dony, C.: Feature location in a collection of product variants: combining information retrieval and hierarchical clustering. In: The 26th International Conference on Software Engineering and Knowledge Engineering, pp. 426–430 (2013)
34. Salton, G.: The SMART Retrieval System-Experiments in Automatic Document Processing. Prentice-Hall Inc., Upper Saddle River (1971)
35. Salton, G., McGill, M.J.: Introduction to Modern Information Retrieval. McGraw-Hill, Inc., New York (1986)
36. Shah, C.: Collaborative information seeking: a literature review. Exploring the Digital Frontier Advances in Librarianship, vol. 32 (2010)
37. Sisman, B., Kak, A.C.: Assisting code search with automatic query reformulation for bug localization. In: Proceedings of the 10th Working Conference on Mining Software Repositories, MSR 2013, pp. 309–318 (2013)
38. Spanoudakis, G., Zisman, A., Pérez-Minana, E., Krause, P.: Rule-based generation of requirements traceability relations. J. Syst. Softw. **72**(2), 105–127 (2004)
39. Vargha, A., Delaney, H.D.: A critique and improvement of the CL common language effect size statistics of McGraw and Wong. J. Educ. Behav. Stat. **25**(2), 101–132 (2000)
40. Wang, S., Lo, D., Jiang, L.: Active code search: incorporating user feedback to improve code search relevance. In: Proceedings of the 29th ACM/IEEE International Conference on Automated Software Engineering, ASE 2014, pp. 677–682 (2014)
41. Wille, D., Holthusen, S., Schulze, S., Schaefer, I.: Interface variability in family model mining. In: Proceedings of the 17th International Software Product Line Conference: Co-located Workshops, pp. 44–51 (2013)
42. Wohlin, C., Runeson, P., Höst, M., Ohlsson, M.C., Regnell, B., Wesslén, A.: Experimentation in Software Engineering. Springer, Heidelberg (2012)
43. Yang, J., Tan, L.: Inferring semantically related words from software context. In: Mining Software Repositories (MSR), pp. 161–170 (2012)
44. Zhang, X., Haugen, Ø., Møller-Pedersen, B.: Augmenting product lines. In: Software Engineering Conference (APSEC), vol. 1, pp. 766–771 (2012)
45. Zhang, X., Haugen, Ø., Moller-Pedersen, B.: Model comparison to synthesize a model-driven software product line. In: Proceedings of the 2011 15th International Software Product Line Conference (SPLC), pp. 90–99 (2011)

Context-Aware Access Control with Imprecise Context Characterization Through a Combined Fuzzy Logic and Ontology-Based Approach

A.S.M. Kayes[1(✉)], Wenny Rahayu[1], Tharam Dillon[1], Elizabeth Chang[2], and Jun Han[3]

[1] La Trobe University, Melbourne, Australia
{a.kayes,w.rahayu,t.dillon}@latrobe.edu.au
[2] University of New South Wales, Canberra, Australia
elizabeth.chang@adfa.edu.au
[3] Swinburne University of Technology, Melbourne, Australia
jhan@swin.edu.au

Abstract. Context information plays a crucial role in dynamically changing environments and the different types of contextual conditions bring new challenges to access control. This information mostly can be derived from the crisp sets. For example, we can utilize a crisp set to derive a patient and nurse are co-located in the general ward of the hospital or not. Some of the context information characterizations cannot be made using crisp sets, however, they are equally important in order to make access control decisions. For example, a patient's current health status is *"critical"* or *"high critical"* which are imprecise fuzzy facts, whereas *"95% level of maximum blood pressure allowed"* is precise. Thus, there is a growing need for integrating these kinds of fuzzy and other conditions to appropriately control context-specific access to information resources at different granularity levels. Towards this goal, this paper introduces an approach to *Context-Aware Access Control using Fuzzy logic (FCAAC)* for information resources. It includes a *formal context model* to represent the fuzzy and other contextual conditions. It also includes a *formal policy model* to specify the policies by utilizing these conditions. Using our formal approach, we combine the fuzzy model with an ontology-based approach that captures such contextual conditions and incorporates them into the policies, utilizing the ontology languages and the fuzzy logic-based reasoning. We justify the feasibility of our approach by demonstrating the *practicality* through a prototype implementation and a healthcare case study, and also evaluating the *performance* in terms of response time.

Keywords: Context-aware access control · Fuzzy facts · Contextual conditions · Context model · Fuzzy reasoning model · Policy model

1 Introduction

Over the years, access control mechanisms have shifted from a fixed desktop environment to dynamic environments (e.g., pervasive, cloud and mobile

© Springer International Publishing AG 2017
H. Panetto et al. (Eds.): OTM 2017 Conferences, Part I, LNCS 10573, pp. 132–153, 2017.
https://doi.org/10.1007/978-3-319-69462-7_10

computing environments) [1]. Due to this paradigm shift, the role of dynamically changing *context information* has gained great importance for *context-specific decision making*, where users need seamless access to information resources and services from anywhere and at anytime fashion, even when they are on the move. In terms of *context-aware access control* systems [2,3], context means information about the state of a relevant entity or the state of a relevant relationship between entities, where an entity can be a user, resource or their environments.

The gathering of relevant context information as the major underlying mechanism in today's dynamic world is crucial and thus demanding for further studies on many aspects of access control to information resources and services. Among the significant factors, an access controller needs to be *context-aware* by incorporating the different types of dynamic context information. In particular, there is a need for an even seamless integration of *precise fuzzy conditions* and *other relevant contextual conditions* subsequently with access control policies, in order to manage an access to information resources at different granularity levels. Consider a healthcare scenario where a doctor Jane is needed to access the medical records of a patient Bob, who is currently admitted to a hospital due to a severe heart attack. In general, only the emergency doctors have access to all of the medical records for patients who are admitted for emergency treatment, including their medical history and personal health records. However, Jane, while not being an emergency doctor, can play the *emergency doctor role* from the *emergency ward* of the hospital when Bob's health status is *"high critical"* and consequently can access *all of his medical records* to save his life. Therefore, an access controller needs to consider such kinds of fuzzy facts/conditions when making access control decisions. In particular, there is a need to quantify the fuzzy conditions more precisely (e.g., Bob's health status is *"high critical"* with *"criticality level 95%"*). Context-specific access control to information resources together with such conditions can provide an extra level of safety for patients in such emergency medical situations. In order to achieve *context-awareness* and integrate the different types of fuzzy and other contextual conditions into the access control processes, the following research issues need to be addressed.

(R1) How to derive precise contextual conditions from imprecise fuzzy facts for context-specific decision making?
(R2) How to integrate these derived fuzzy conditions and other relevant contextual conditions with access control policies to facilitate context-specific access to information resources at different granularity levels?

Context-aware access control is a mechanism to determine whether a user's request to limit the access permissions to information resources based on the dynamically changing contextual conditions (e.g., the interpersonal relationship between patient and nurse is "assigned nurse", the patient's health status is "66% normal" with "criticality level 34%", etc.). In the literature, there has been a significant amount of research work in developing context-aware access control approaches. A number of such access control approaches consider the *spatial information* (e.g., [4]), the *temporal information* (e.g., [5]), the *event-driven information* such as surgery in progress (e.g., [6]), and other *environment context information* such as the range of IP addresses (e.g., [7–9]), as contextual conditions when making access control decisions. In this context, our group has a successful track record in developing context-aware access control systems by considering a wide variety of contextual conditions: the *general context information* about the state of the users, resources and their environments [2,10], the *relationship context information* utilizing the process of inferring implicit knowledge [11], and the *purpose-oriented situation information* based on the currently available context information [3,12]. We also propose a context-aware access control policy model in our earlier research [13], incorporating these relevant contextual conditions into the access control policies. These contextual conditions usually derive from the crisp sets (e.g., the doctor is located in the "emergency ward" of the hospital or "not"), and these traditional approaches are not adequate to deal with imprecise context characterization. However, there are other types of contextual conditions which only can be derived from the fuzzy sets by utilizing the low-level fuzzy facts, and they are equally important in order to make access control decisions at different granularity levels.

Other than the above-mentioned traditional context-aware access control approaches, several research works consider the use of fuzzy conditions (e.g., computing resource owners' trust degrees [14], quantifying risks [15], measuring trust levels [16], calculating user-permission strengths [17]) for making access control decisions. However, these approaches are not context-aware and robust enough to integrate both the fuzzy conditions and other dynamic contextual conditions with access control policies for context-specific decision making. Using successful experience from our group's earlier research on fuzzy linguistic representations for capturing the semantics of warehoused data [18], we develop our fuzzy model that is used in this paper to deal with imprecise context characterization.

The above-identified gap in the literature suggests that there is still a need for a new form of dynamic access control approach that can further limit the applicability of the available access permissions to information resources, integrating both the fuzzy facts and other contextual conditions together with access control policies for context-specific decision making. Our paper makes the following contributions towards achieving this goal.

(C1) **Formal Access Control Approach:** We introduce a new form of access control approach, Context-Aware Access Control using Fuzzy logic (FCAAC), specifically addressing the following aspects:

    (i) **Context Representation and Reasoning Model:** We present a formal analysis of the fuzziness of (imprecise) context information. We introduce a formal context model to represent the fuzzy and other contextual conditions from the low-level information.

    (ii) **Policy Model:** We present a formal analysis of the context-specific access control decision making by taking into account the relevant fuzzy and other contextual conditions.

(C2) **Ontology-based FCAAC Approach:** Using our formal context and policy models, we introduce an ontology-based approach to model and reason about the relevant fuzzy and other contextual conditions, and consequently model the context-specific access control policies, incorporating the relevant conditions into the access control processes.

(C3) **Evaluation:** Other than the above two main contributions, we justify the feasibility of our approach by demonstrating the following factors:

(i) **Practicality:** We develop a prototype of the FCAAC approach that assists software practitioners in rapid prototyping. Using this prototype, a case study from the healthcare domain is presented which demonstrates the practicality of the proposed approach.

(ii) **Performance:** We conduct two sets of experiment in a healthcare environment and evaluate the applicability of our access control approach by means of response time.

The rest of this paper is organized as follows. We first present an application scenario in Sect. 2 to motivate our work. Section 3 introduces our formal access control approach, including the context representation and reasoning model and its associated policy model. Using the formal context and policy models, Sect. 4 introduces an ontology-based access control approach. Section 5 demonstrates the practicality of our approach against a healthcare case study and the performance in terms of response time. Section 6 briefly presents the related work. Finally, Sect. 7 concludes the paper and outlines future work.

2 Significance of Our Research and General Requirements

This section presents an extended application scenario from our earlier work [2]. In addition, we identify the general requirements of developing a new access control approach by integrating both the fuzzy conditions and other contextual conditions together with access control policies.

2.1 Application Scenario

Let us consider our extended healthcare scenario where *a patient Bob who is currently admitted in the emergency department of the hospital due to a severe heart attack. Jane, who is a hospital doctor, is required to access the necessary medical records of Bob to treat him and save his life from such life-threatening situation. After getting emergency treatment, Bob is shifted to the general ward of the hospital and assigned a registered nurse Mary to monitor his health status.*
In general, the emergency doctors, including a patient's treating physician, can access all the necessary health records of patients, such as the medical records, past medical history and private medical records. However, Jane, while not being an emergency doctor, is able to access the necessary medical records by playing the emergency doctor role from the emergency ward of the hospital when Bob's health status is *"high or 95% critical"*. When the context changes (e.g., Bob's health status becomes *"66% normal"*), a decision on a further access

request by Jane to Bob's emergency medical records may need to change accordingly (e.g., an access permission should be *denied*). That is, Jane is only authorized to play the hospital doctor role, and consequently can access Bob's normal medical records when his health condition is *"66% normal"*.

Normally, a registered nurse, who is assigned to look after a patient (or a group of patients), is able to access the daily medical records during her ward duty time and when she is present in the general ward where the patient is located. However, in the mentioned emergency scenario, Mary is able to access Bob's medical records when she is co-located with Jane, who is currently treating Bob by playing the emergency doctor role, and only when his health status is *"high critical"*. When the context changes (e.g., Mary *leaves the emergency department* or *outside of duty time*), a decision on a further access request by Mary to Bob's medical records may need to change accordingly (e.g., an access permission should be *denied*). That is, Mary, by playing the assigned nurse role, is only able to access Bob's daily medical records during her ward duty time and only when they both are co-located in the general ward of the hospital.

The different types of conditions are involved in this scenario, e.g., the location and request time of a nurse, the health status of a patient, etc. Therefore, an access controller needs to exploit such conditions directly or indirectly when making access control decisions. The normal conditions such as the location and request time can be obtained directly from the context sources. The health status is not able to obtain directly but can be derived from the available low-level data such as the body temperature and pulse rate. As such, it is necessary to further process the retrieved low-level imprecise data or fuzzy facts automatically to precisely obtain the relevant results (e.g., the health status is *"66% normal"* with *"criticality level 34%"*). In order to limit the access permissions to resources exploiting such fuzzy and other conditions is both a strength and a challenge.

2.2 General Requirements

The general requirements of developing the context-specific access control with imprecise fuzzy characterization are as follows:

(Req. 1) There is a need for a new form of access control approach to capture the low-level imprecise fuzzy facts and consequently derive the precise fuzzy conditions from them. In this respect, we introduce a *context representation and reasoning approach* to represent the raw facts from the context sources and infer the relevant conditions from them.

(Req. 2) Also, an access controller needs to take into account both the fuzzy conditions and other relevant contextual conditions for context-specific decision making. As such, we introduce a *policy model and a software prototype* to incorporate these conditions into the access control policies.

3 Our Formal FCAAC Approach

In this section, we introduce an approach to Context-Aware Access Control using Fuzzy logic (FCAAC), including context and policy models.

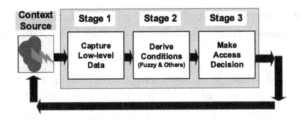

Fig. 1. Our FCAAC approach

Figure 1 presents the conceptual FCAAC approach, which includes 3 basic steps: capture low-level data, derive relevant information (fuzzy and other contextual conditions) and make access control decision. Stage 1 is the process of gathering low-level data from the context sources. Stage 2 is the process of inferring relevant fuzzy and other contextual conditions from the low-level data. Finally, stage 3 is the process of making access control decision based on the relevant conditions. In the following, we present a formal analysis of the approach.

3.1   Context Model

The development of a relevant Context-Aware Access Control (CAAC) approach is a complex task because of the need to accommodate for a wide variety of contextual conditions. The first step in achieving this is to define these conditions.

Representation of Fuzzy and Normal Contextual Conditions: In the literature, many researchers have defined the context information. The most well accepted definition is given by Dey [19], *context is any information about the situation of an entity, where an entity can be a person, place or object.* In general, it is a broad and generalized vision of what the context means for context-aware applications. However, based on our application scenario, we need to represent the different types of contextual conditions as some conditions which only can be derived by utilizing fuzzy sets and fuzzy logic-based reasoning.

Definition 1 *(Fuzziness of Context Information). According to the degree of fuzziness of context information, we classify contextual conditions into fuzzy conditions and normal conditions, i.e., contextual conditions (CC) is the set of all fuzzy conditions (FC) and all normal conditions (NC).*

$$CC \ = \ FC \ \cup \ NC \tag{1}$$

Definition 2 *(Fuzzy Contextual Condition). A fuzzy contextual condition is an implicit context information and it can be derived from a fuzzy set by means of a concept (i.e., contextual condition) with its values. On the basis of the fuzzy set theory [18], a decimal point or truth value ranging from 0 to 1 is generally used to characterize the degree of membership of the values to a concept.*

The elements (fuzzy contextual conditions) of a fuzzy set have the truth values (tValues) ranging from 0 for non-membership to 1 for full-membership.

$$\mu_{fc(v)} \in [0,1] \tag{2}$$

In the above expression, 'fc' denotes a fuzzy condition ($fc \in FC$) and '$\mu_{fc(v)}$' denotes a membership degree of a concept 'fc' for a certain value 'v'.

Example 1. *A patient's current health status (PCHState) is 95% critical, which is a fuzzy contextual condition. The degree of membership is represented in the following expression.*

$$\mu_{PCHState(critical)} = 0.95, \; i.e.,$$
$$PCHState = "critical", \; where \; tValue = 0.95 \tag{3}$$

Definition 3 *(Normal Contextual Condition). A normal contextual condition is an implicit context information and it can be derived from a classical crisp set by means of a concept with its values. On the basis of the classical crisp set theory, a truth value 0 or 1 is generally used to characterize the degree of membership of the values to a concept.*

The elements (normal conditions) of a crisp set have the truth values either 0 for non-membership or 1 for full-membership. The degree of membership of a concept 'nc' ($nc \in NC$) to its value 'v' is represented in the following expression.

$$\mu_{nc(v)} \in \{0,1\} \tag{4}$$

Example 2. *In our application scenario, the interpersonal relationship (inter-Relationship) between Bob and Mary is assigned nurse, which is a normal contextual condition. The degree of membership is represented in the following expression.*

$$\mu_{interRelationship(assignedNurse)} = 1, \; i.e.,$$
$$interRelationship = "assignedNurse" \tag{5}$$

Example 3. *In the same application scenario, the relationship between Bob and Jane is non-treating physician. The degree of membership is represented in the following expression.*

$$\mu_{interRelationship(treatingPhysician)} = 0, \; i.e.,$$
$$interRelationship = "non-treatingPhysician" \tag{6}$$

Reasoning About Fuzzy and Normal Contextual Conditions: The context reasoning part includes two types of inference rules to derive fuzzy and normal contextual conditions. The first set of rules are used to infer the fuzzy contextual conditions for the precise linguistic labels and the crisp boundary values (e.g., a patient's current health status is *"66% normal"* with *"criticality level 34%"*) from the low-level fuzzy facts through fuzzy-logic based reasoning. The second set of rules are used to infer the normal contextual conditions from the low-level context information through normal rule-based reasoning.

Further details of the reasoning about these conditions using fuzzy logic-based and ontology-based inference rules are discussed in Sect. 4.2.

3.2 Policy Model

Role-Based Access Control [20] is an emerging model of access control and is well recognized for its many advantages in large-scale authorization management [21]. It provides the core concepts of user-role and role-permission assignments in which a user can exercise organizational functions that are associated with the roles. Our core CAAC policy model [2] extends the traditional RBAC model to support context-oriented access control according to normal contextual conditions. This section introduces a formal FCAAC policy model, which extends our core CAAC policy model to a further coverage of fuzzy contextual conditions.

Definition 4 *(FCAAC Policy Model).* *A Fuzzy logic-based Context-Aware Access Control (FCAAC) policy model is denoted by a 4-tuple relation.*

$$FCAAC = \langle U, R, CC, P \rangle \tag{7}$$

In the above relation, 'U' represents a set of system users who are the resource requesters, 'R' represents a set of roles, 'CC' represents a set of contextual conditions, and 'P' represents a set of permissions or rights to perform some operations on resources (read or write) by the users who initiate access requests.

If 'u' represents a user ($u \in U$), 'r' represents a role ($r \in R$), 'cc' represents a contextual condition ($cc \in CC$, $CC = FC \cup NC$) and 'p' represents a permission ($p \in P$), then, together the elements '$Users$' ($U = \{u_1, u_2, ..., u_m\}$), '$Roles$' ($R = \{r_1, r_2, ..., r_i\}$), '$Contextual\ Conditions$' ($CC = \{cc_1, cc_2, ..., cc_j\}$) and '$Permissions$' ($P = \{p_1, p_2, ..., p_n\}$) form the *FCAAC Policy Model*.

Definition 5 *(A FCAAC Policy).* *A FCAAC policy specifies whether a user in an appropriate role is granted a permission associated with that role to access the information resource(s) in order to perform some operations on that resources(s), when the relevant contextual conditions are satisfied. We consider the contextual conditions as the policy constraints and they can be formed by integrating the relevant fuzzy and/or normal contextual conditions.*

Example 4. *Consider the application scenario presented in Sect. 2, where Mary wants to access certain medical records of patient Bob, the FCAAC policy determines whether the access permission is granted or denied. An example FCAAC policy associated with this scene can be read as: "a user by playing a registered nurse (RN) role is permitted to access the daily medical records (DMR) of a patient, during her ward duty time from the location where the patient is located in the general ward, and if she is assigned to monitor his health status, and only when his current health status is within normal ranges". The rule shown in Table 1 expresses the policy, $fcaac_1 = \langle Mary, RN, cc_1, DMR \rangle$.*

In this example, the access control decision is based on the following constraints: *who* the user is (e.g., *Mary*), *what* role the user can play (e.g., *RN*), *what* resource is being requested (e.g., write operation on DMR, *writeDMR*)

Table 1. An example FCAAC policy for the registered nurses

If
$FCAACPolicy(fcaac_1) \wedge User(u_1) \wedge hasUser(fcaac_1, u_1) \wedge equal(u_1, \text{``Mary''})$
$\wedge \ Role(r_1) \wedge hasRole(fcaac_1, r_1) \wedge \ equal(r_1, \text{``RN''}) \wedge Permission(p_1)$
$\wedge \ hasPermission(fcaac_1, p_1) \wedge equal(p_1, \text{``writeDMR''})$
$\wedge \ ContextualCondition(cc_1) \wedge hasCondition(fcaac_1, cc_1)$
$\wedge \ NormalCondition(nc_1) \wedge FuzzyCondition(fc_1) \wedge hasContext(cc_1, nc_1 \vee fc_1)$
Then
$canAccess(u_1, p_1)$

and under *what* contextual conditions (e.g., cc_1). Looking at our application scenario, the contextual condition 'cc_1' is based on a normal condition 'nc_1' (e.g., Mary's location address is "general ward" and request time is "duty time", and the interpersonal relationship between Mary and Bob is "assigned nurse") and a fuzzy condition 'fc_1' (e.g., Bob's current health status is "66% normal" with "criticality level 34%"), and it can be represented as, $cc_1 = nc_1 \vee fc_1$.

Further details of the FCAAC policy specification using ontology-based languages are discussed in the following section (see Sect. 4.3).

4   Ontology-Based FCAAC Approach

This section introduces an ontology-based approach, to realize the formal models.

We introduce the FCAAC ontology to model the contextual conditions, utilizing user-defined inference rules to derive the relevant conditions from the low-level context information. In the FCAAC ontology, we also model the access control policies, incorporating these contextual conditions. Riboni and Bettini [22] have shown that ontologies are well-suited for representing and modelling dynamic contextual conditions and are very useful semantic technologies for pervasive computing applications. The FCAAC ontology is defined in Web Ontology Language (OWL) [23]. We have chosen OWL rather than other ontology languages, because it is more expressive to specify the contextual conditions and policies in an easy and natural manner, than others [22]. Also, it is a widely used ontology language in semantic Web. In order to infer new knowledge, the expressivity of OWL is extended by incorporating the SWRL (Semantic Web Rule Language) rules [24] to the FCAAC ontology.

The FCAAC ontology, as depicted in Fig. 2, has the core concepts *User*, *Role*, *ContextualCondition*, *Permission*, *Resource*, *Operation* and *AccessDecision*, which are organized into a *FCAACPolicy* hierarchy. It is divided into three layers. The top layer, which extends our core CAAC policy ontology [2] to a further coverage of fuzzy contextual conditions and includes the concepts for modelling the FCAAC policies. The middle layer includes the core concepts for modelling the fuzzy and normal contextual conditions. The bottom layer includes the core concepts for modelling the context information.

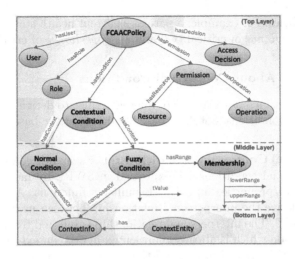

Fig. 2. The FCAAC ontology

The detailed representation of a wide range of context information is out of the scope of this paper, which can be found in our earlier research [2,3,11].

4.1   Modelling Contextual Conditions

The middle layer in Fig. 2 has the concepts *NormalCondition*, *FuzzyCondition* and *Membership*, which are organized into a *ContextualCondition* hierarchy. The relationships between these concepts are represented by object and data type properties. The links between a concept and its attributes are achieved via data type properties, and the links between two concepts are achieved by means of object properties (built-in and user-defined) with '*rdfs:domain*' and '*rdfs:range*'.

A contextual condition consists of the relevant fuzzy and normal conditions. Thus, the *ContextualCondition* class has an object property named *hasContext*, which is used to link the *ContextualCondition* class and the union of *NormalCondition* and *FuzzyCondition* classes. The normal and fuzzy contextual conditions are composed of the relevant context information specific to access control, using an object property named *composedOf*. The *NormalCondition* and *FuzzyCondition* classes use the concepts (*ContextInfo*) from the core context ontology, which is already introduced in our earlier work [2,10]. The object property *hasRange* is used to link the classes *FuzzyCondition* and *Membership*.

The *FuzzyCondition* class contains a '*xsd:float*' type data property named *tValue*, which denotes a membership degree (or truth value) of a concept for a certain value. For example, concerning our application scenario, Bob's current health status is "*66% normal*", which means that the criticality level (*tValue*) is 0.34. The class *Membership* has two '*xsd:float*' type data properties, named *lowerRange* and *upperRange*, which denote the ranges of membership degree for a fuzzy condition. These properties are used to specify the fuzzy conditions in the FCAAC policies. For example, a patient's current health status is *"normal"*, which has a *lowerRange* of criticality 0 and an *upperRange* of criticality 0.50.

4.2 Reasoning About Contextual Conditions

The reasoning part includes two sets of inference rules to derive the normal and fuzzy contextual conditions: ontology-based and fuzzy logic-based rules.

Inferring Normal Contextual Conditions: The semantic rules that are used to derive the normal conditions are expressed in SWRL by means of FCAAC ontology concepts/properties and SWRL built-ins functions. An example reasoning rule to derive the interpersonal relationship between user and patient is specified in Table 2. The interpersonal relationship is inferred from the low-level context information which is already represented in our context ontology [2,10], i.e., from the user's personal profile and the patient's social profile information.

Table 2. A reasoning rule to infer the interpersonal relationship

User(?u) ∧ Role(?role) ∧ hasRole(?u, ?role) ∧ swrlb:equal(?role, "RN") ∧ Owner(?o) ∧ Resource(?r) ∧ isOwnedBy(?r, ?o) ∧ InterpersonalRelationship(?rel) ∧ hasRelationship(?u, ?rel) ∧ hasRelationship(?o, ?rel) ∧ **PersonalProfile**(?pp) ∧ hasProfile(?u, ?pp) ∧ userIdentity(?pp, ?userID) ∧ roleIdentity(?pp, ?roleID) ∧ **SocialProfile**(?sp) ∧ hasProfile(?o, ?sp) ∧ connectedPeopleIdentity(?sp, ?connID) ∧ connectedPeopleRoleIdentity(?sp, ?connRoleID) ∧ **swrlb:equal**(?userID, ?connID) ∧ **swrlb:equal**(?roleID, ?connRoleID) → interRelationship(?rel, **"assignedNurse"**)

Inferring Fuzzy Contextual Conditions: The inference rules that are used to derive the fuzzy conditions are expressed in "if-then statements" by means of the specification of linguistic labels, where the first part (*if*) contains the input conditions and the second part (*then*) contains an action output. An example set of fuzzy logic-based reasoning rules to derive the current health status of the patients is specified in Table 3. The first rule in Table 3 can be read as, if *PAge* is *"Young"* and *PulseR* is *"T4"*, then *PCHState* is *"Normal"*. Further details can be found in prototype implementation section (see Sect. 5.1).

Table 3. A set of reasoning rules to infer the current health status

If	$PAge(Young) \land PulseR(T4)$	***Then***	$PCHState(Normal)$	
If	$PAge(Young) \land PulseR(T5)$	***Then***	$PCHState(Normal)$	
If	$PAge(MiddleAge) \land PulseR(T4)$	***Then***	$PCHState(Normal)$	
If	$PAge(MiddleAge) \land PulseR(T5)$	***Then***	$PCHState(Critical)$	

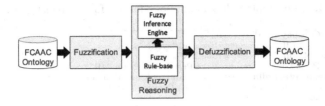

Fig. 3. The fuzzy context information system

One of the main contributions of this research is to derive the fuzzy contextual conditions from the low-level information, utilizing fuzzy-logic-based context reasoning. Towards this goal, Fig. 3 shows our fuzzy context information system, which includes three main steps for mapping between crisp and fuzzy datasets: fuzzification, fuzzy reasoning and defuzzification [18]. Our FCAAC ontology captures the low-level data from the context sources and sends them for fuzzification. Fuzzification is the process of representing these inputs (from the crisp values) into their linguistic labels using membership functions. Fuzzy reasoning is the process of deriving the linguistic outputs from the given linguistic inputs in terms of fuzzy logic. As such, it selects the required reasoning rules from a fuzzy rule-base and executes them using the fuzzy inference engine. Defuzzification is the process of combining all linguistic outputs into a single/composite crisp result. Finally, Our FCAAC ontology stores such inferred result/condition.

4.3   FCAAC Policy

We use the OWL ontology language to represent the FCAAC policy concepts and their relationships (see top layer in Fig. 2). OWL-based reasoning rules are not always sufficient to infer the implicit information from the low-level information. For example, in order to compare the first and second arguments (e.g., they are the 'same', 'less than' or 'greater than'), we use the SWRL language and its built-in functions to represent the fuzzy contextual conditions in our ontology, in terms of their linguistic labels and the ranges of their degree of membership. As such, we codify the FCAAC policies with OWL and SWRL languages.

An Example FCAAC Policy: Let us consider the registered nurses' policy shown in Table 1. In this policy, the access decision is based on the following constraints: *who the requester/user* is (e.g., registered nurse, *RN*), *what resource* is being requested (e.g., daily medical records (DMR) on write operation) and *under what contextual conditions* the user sends the request (current health status, request time, and interpersonal and co-located relationships). The FCAAC policy rule in OWL is shown in the top part in Table 4 (the core policy concepts are specified in *Line #1 to 7*), including the definition of contextual condition (which is defined in *Line #8 to 20*). The bottom part in Table 4 shows the specification of contextual conditions and other policy constraints (e.g., fuzzy

Table 4. An example policy in ontology format for the registered nurses

1	**\<FCAACPolicy** rdf:ID="fcaac$_1$">
2	\<hasUser rdf:resource="#User_canPlay_RN"/>
3	\<hasRole rdf:resource="#Role_RN"/>
4	\<hasPermission rdf:resource="#Permission_writeDMR"/>
5	\<hasCondition rdf:resource="#ContextualCondition_cc$_1$"/>
6	\<hasDecision rdf:resource="#AccessDecision_Granted"/>
7	\</**FCAACPolicy**>
8	\<owl:Class rdf:ID="**ContextualCondition**">
9	\<owl:ObjectProperty rdf:ID="**hasContext**">
10	\<rdfs:domain rdf:resource="#ContextualCondition"/>
11	\<rdfs:range>
12	\<owl:Class>
13	\<**owl:unionOf** rdf:parseType="Collection">
14	\<owl:Class rdf:about="#**NormalCondition**"/>
15	\<owl:Class rdf:about="#**FuzzyCondition**"/>
16	\</**owl:unionOf**>
17	\</owl:Class>
18	\</rdfs:range>
19	\</owl:ObjectProperty>
20	\</owl:Class>
21	FCAACPolicy(?fcaac$_1$) ∧ User(?u) ∧ hasUser(?fcaac$_1$, ?u) ∧ Role(?r) ∧
22	hasRole(?fcaac$_1$, ?r) ∧ canPlay(?u, ?r) ∧ roleIdentity(?r, "**RN**") ∧
23	Permission(?per) ∧ hasPermission(?fcaac$_1$, ?per) ∧ Resource(?res) ∧
24	hasResource(?per, ?res) ∧ resourceIdentity(?res, "**DMR**") ∧
25	Owner(?o) ∧ isOwnedBy(?res, ?o) ∧ Operation(?op) ∧
26	hasOperation(?per, ?op) ∧ action(?op, "**Write**") ∧
27	**ContextualCondition**(?cc$_1$) ∧ hasCondition(?fcaac$_1$, ?cc$_1$) ∧
28	**NormalCondition**(?nc$_1$) ∧ **hasContext**(?cc$_1$, ?nc$_1$) ∧
29	InterpersonalRelationship(?rel) ∧ hasRelationship(?u, ?rel) ∧
30	hasRelationship(?o, ?rel) ∧ interRelationship(?rel, "**assignedNurse**") ∧
31	RequestTime(?rt) ∧ hasRequestTime(?u, ?rt) ∧ requestTime(?rt, "**dutyTime**")
32	∧ Co-locatedRelationship(?col) ∧ hasRelationship(?u, ?col) ∧
33	hasRelationship(?o, ?col) ∧ isColocatedWith(?col, **yes**) ∧
34	**composedOf**(?nc$_1$, ?rel) ∧ **composedOf**(?nc$_1$, ?rt) ∧
35	**composedOf**(?nc$_1$, ?col) ∧
36	**FuzzyCondition**(?fc$_1$) ∧ **hasContext**(?cc$_1$, ?fc$_1$) ∧ **PCHState**(?hs) ∧
37	**composedOf**(?fc$_1$, ?hs) ∧ **swrlb:equal**(?hs, "**normal**") ∧ **tValue**(?fc$_1$, ?tv) ∧
38	Membership(?m) ∧ hasRange(?fc$_1$, ?m) ∧ lowerRange(?m, ?lr) ∧
39	**swrlb:equal**(?lr, **0**) ∧ upperRange(?m, ?ur) ∧ **swrlb:equal**(?ur, **0.50**) ∧
40	**swrlb:greaterThan**(?tv, lr) ∧ **swrlb:lessThan**(?tv, ur) ∧
41	AccessDecision(?dec) ∧ hasDecision(?fcaac$_1$, ?dec) → decision(?dec, "**Granted**")

conditions, role identity) in SWRL (where the main conditions/constraints are represented in bold type). The user and role specifications are shown in *Line #21 to 22*, the permission specification is shown in *Line #23 to 26*, the contextual condition construction is specified in *Line #27*, the normal condition composition is specified in *Line #28 to 35*, the fuzzy condition composition is specified in *Line #36 to 40*, and the access decision is specified in *Line #41*. In the previous section, an example SWRL-based reasoning rule in Table 2 is used to determine the user and patient have a '*assignedNurse*' relationship, and an example set of fuzzy logic-based reasoning rules in Table 3 is used to determine a patient's current health status is '*normal*'. The reasoning rules to derive the *request time* and *co-located relationship* can be found in our earlier work [2].

One of the key features of our FCAAC ontology is its ability to specify the fuzzy contextual conditions at different membership/criticality levels (see the middle layer in Fig. 2). For example, in the above policy, Mary can access

Bob's DMR when his current health status is *"normal"*, which means that the criticality levels of the degree of membership are between 0 (*lowerRange*) to 0.50 (*upperRange*). However, Mary is not granted access to Bob's DMR from the general ward of the hospital, when his current health status is *"high critical"* or *"critical"*, as he needs to admit immediately in the emergency department of the hospital in such a situation. That is, our FCAAC policy model provides access control decisions by taking into account the fuzzy contextual conditions.

5   Prototype and Evaluation

In this section, we first present a prototype architecture to assist application developers in rapid prototyping. Using this prototype, we develop a healthcare application, called *eHealthcare*, to validate the functionalities of our FCAAC approach. In particular, we present a case study from the healthcare domain to demonstrate the practicality of our access control approach. Furthermore, the deployment of *eHealthcare* application is performed for measuring the performance of our approach. The performance results are presented in Sect. 5.2.

5.1   Practicality

Prototype: Figure 4 shows an architecture of the software prototype, which extends our earlier prototype [2], utilizing both the fuzzy logic and ontology-based reasoning capabilities. It includes environment, middleware and application layers. The environment layer includes the sensors or data sources and the middleware layer includes the context provides, FCAAC ontology, context reasoner and access control processor. The context providers receive the raw context facts from the data sources and the FCAAC ontology captures the low-level information from the context providers. The access control policies are also stored in FCAAC ontology. The context reasoner derives the relevant contextual conditions by using the information from the ontology. The access control processor includes the FCAAC PDP (policy decision point), which is implemented in Java to determine the access request is "granted" or "denied", according to the applicable policies and the necessary contextual conditions. The application layer includes the FCAAC PEP (policy enforcement point), which forwards the request to the FCAAC PDP. The detailed implementation of the context providers, PEP and PDP can be found in our earlier prototype [2]. We in this paper mainly discuss the implementation of the context reasoner to derive the contextual conditions.

The FCAAC ontology is defined by using ontology languages OWL [23] and SWRL [24], and the ontology has been generated with the Protégé-OWL graphical tool [25]. We develop an ontology rule base to derive the normal contextual conditions from the low-level information using ontology-based reasoning rules, which have been generated with the Protégé-SWRLTab. We have used a rule engine that is written in Java, named Jess [26] to facilitate reasoning tasks for

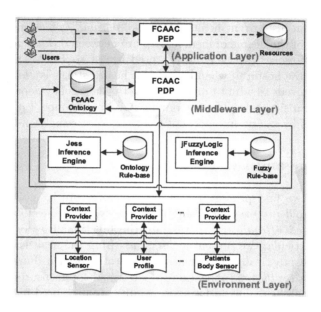

Fig. 4. Overview of the prototype architecture of FCAAC

executing such rules. We develop a fuzzy rule base to derive the fuzzy contextual conditions from the imprecise fuzzy facts using fuzzy reasoning rules, which have been expressed in the form of fuzzy conditional "if-then" statements. For executing such fuzzy rules, we have used the fuzzy inference engine, named jFuzzyLogic [27], which is written in Java. We have already shown the fuzzy reasoning processes in Fig. 3. In order to execute these reasoning rules and consequently derive the implicit information (normal and fuzzy conditions), we have implemented a context reasoner in Java. In particular, we have implemented two Java functions, the first function is used to execute the reasoning rules and infer the implicit information using low-level data from the FCAAC ontology, and the other function is used to transfer the inferred information in the ontology.

Case Study: We evaluate our FCAAC prototype using an *eHealthcare* application scenario described in Sect. 2. The *eHealthcare* application provides the healthcare professionals (e.g., emergency doctors, treating doctors, registered nurses) to access different medical records of patients based on the dynamic context information (normal and fuzzy contextual conditions).

Consider the motivating example where Mary wants to access the daily medical records (DMR) of Bob, an access request is submitted to the *FCAAC PEP* for evaluation. The *FCAAC PEP* forwards the request to the *FCAAC PDP* to determine whether the access request is "granted" or "denied", according to the current contextual conditions in effect and the applicable access control policies. The applicable FCAAC policy is already specified in Table 4, which defines the permission is granted when both of the two Boolean conditions "nc_1" and "fc_1"

are true. The normal contextual condition nc_1 is composed based on the following sub-conditions (context information): the nurse is *assigned* to monitor the patient's health condition and they both are *co-located* in the general ward during her *duty time*. The fuzzy contextual condition fc_1 is composed of the context information: the patient's current health status (*PCHState*). In the following, we further discuss how our proposed approach captures the *PCHState* of Bob.

For simplicity, in our *eHealthcare* application, we consider the pulse rate (*PulseR*) and age of a patient (*PAge*) are the two input fuzzy sets to derive the *PCHState* (an output fuzzy set). We also consider three fuzzy age groups: *VeryYoung*, *Young* and *MiddleAge*, a normal pulse rate that is between 75 to 110 beats per minute (bpm) (which represents seven fuzzy sets, *T1* to *T7*), and a patient's current health status which is represented using three fuzzy sets: *Normal*, *Critical* and *HighCritical*. Based on the experience from our group's earlier research on fuzzy linguistic representations [18], these input and output fuzzy sets are characterized by triangular and trapezoidal membership functions (see Fig. 5) and Mamdani's center of gravity (COG) method in conjunction with max-min inference is used for fuzzy reasoning (see Fig. 6). We have specified 21 linguistic rules to cover all the possible values of *PAge* and *PulseR*.

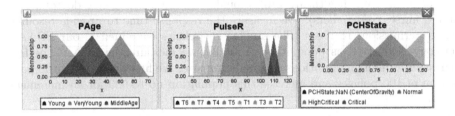

Fig. 5. Inputs and output membership functions

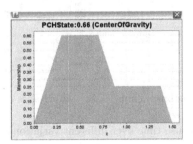

Fig. 6. PCHState

We assume that Bob's age is 35, which belongs to the fuzzy sets *Young* and *MiddleAge* and his pulse rate is captured as 102 bpm, which belongs to the fuzzy sets *T4* and *T5*. These inputs are fired four rules, which are already specified in

Table 3. Finally, Bob's *PCHState* is derived using the COG max-min inference method (see Fig. 6). In this scenario, Mary is assigned to look after Bob and we can observe that she is granted access to Bob's DMR in his normal health condition (i.e., *"66% normal with criticality level 0.34"*).

In FCAAC, we model the criticality ranges of the *normal*, *critical* and *high critical* health status are *[0, 0.50]*, *[0.50, 0.75]* and *[0.75, 1.0]*, respectively. However, Mary is not granted access to Bob's DMR when the context changes (e.g., Bob's health condition is critical or high critical again, i.e., the criticality level is beyond the normal ranges). In summary, the purpose of the case scenario and prototype testing is to provide a walkthrough of the whole FCAAC approach.

5.2 Performance

We conduct two sets of experiments in our simulated healthcare environment with the aim of measuring the response time and scalability of our FCAAC proposal. The conducted tests are carried out in a Windows PC with an Intel Core i7@3.6 GHz Processor and 16 GB of RAM. The results have been obtained by executing the experiments 10 times and computing their arithmetic mean.

In our first set of experiments, we vary the number of FCAAC policies with respect to different healthcare professional roles (e.g., emergency doctors, registered nurses, researchers). We measure the response time to provide resource access permissions to users. The number of policies contained in our FCAAC ontology is referred as population. Actually, we measure the FCAAC performance with different variations of population size. We first define an initial population of 100 policies and increase this population up to 500 for an increment of 100.

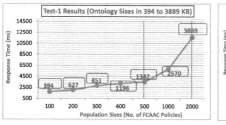
(a) Populations vs Response Time

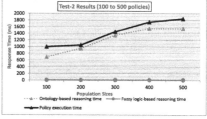
(b) Stages of Response Time

Fig. 7. Average response time over different variations of population size

Figure 7(a) depicts how the response time varies, measured in milliseconds (ms), considering different population sizes associated to the policies. We observe that the response time is linearly increased according to the number of policies up to 500 and it varies from 1.7 to 3.5 s approximately. For all populations, the difference in response time between the sizes of 394 KB and 1342 KB of ontology is around a few seconds. We can say that the performance is acceptable in such a computer setup with limited computing resources.

The FCAAC reasoning model based on the fuzzy and ontology-based inference rules is one of the important parts of our proposed access control approach. In order to check the reasoning time and its scalability, we conduct another set of experiments. Actually, we measure the different breakdowns of the response time, where we observe the following main stages: time taken to (i) derive the fuzzy contextual conditions, (ii) derive the normal contextual conditions, and (iii) execute the access control policies for making decisions.

Figure 7(b) depicts the time, measured in milliseconds (ms), depending on the different stages of response time breakdown. We observe that the fuzzy logic-based reasoning in order to derive the implicit knowledge which does not have a great impact in total reasoning time (fuzzy reasoning and ontology-based reasoning). This is due to the following reasons. In our experiments, the current health status of a patient (i.e., an output fuzzy set) is derived from the pulse rate and age of the patient (i.e., two input fuzzy sets). However, the numbers of input and output fuzzy sets usually appear to be limited according to the inherent nature of context-aware access control (CAAC) applications. We also note that it does not even impact the size of the FCAAC ontology when we increase the number of fuzzy inference rules. Thus, the time taken to derive the fuzzy condition seems a straight line in Fig. 7(b).

In these two sets of experiments, we separate the ontology loading time from the access request processing time and we only consider the access request processing time as the total response time. However, the ontology loading occurs once when the system runs the first time. Regarding the performance of our FCAAC approach, the fuzzy logic-based reasoning time has a very low impact in the overall response time to process a user's request to access the resources (see Fig. 7(b)), as the search space is limited to a small number of fuzzy inference rules. On the other hand, when we linearly increase the number of policies in our FCAAC ontology, the response time also increases linearly. However, the results fluctuate greatly at the point when we specify a large number of policies and they are more stable up to 500 policies (see Fig. 7(a)). This is due to the growing numbers of users, roles, contextual conditions and reasoning rules in the ontology. In this sense, we can conclude that the population size (i.e., the number of policies) mainly affects the overall system performance of our FCAAC approach. Furthermore, the linearity property behind the results allows us to deduce that a better computer system with powerful computing resources would obtain a lower response time. Based on the experience from our previous work on improving system performance [28], we may adopt RDF language to build a new approach as an alternative of using OWL language.

6   Related Work and Discussion

This section provides a short overview of the relevant access control approaches.

Context-Aware Access Control Approaches: Different approaches have been proposed in literature to model role-based access control policies in conjunction with context information. Mostly these policies are based on involving the normal contextual conditions, which can be derived from the crisp sets.

Joshi et al. [5] have proposed a role-based access control (RBAC) approach and incorporated the temporal information into the RBAC policies. Bertino et al. [4] have proposed another RBAC approach, incorporating the spatial information into the policies. However, these temporal and spatial approaches are not context-aware and adequate enough to capture and infer a wide variety of dynamically changing conditions of the environments (e.g., the relationships).

On the other hand, Bonatti et al. [6] have introduced an event-driven extension to the temporal RBAC approach. They provide an implementation of RBAC in which access control is managed by means of context information (e.g., location, time, an event such as "surgery in progress"). Schefer-Wenzl and Strembeck [7] have proposed a context-aware RBAC approach to ubiquitous systems, incorporating the context information such as time and location into the policies. Similar to [7], Hosseinzadeh et al. [8] and Trnka and Cerný [9] have proposed the context-aware RBAC approaches. Using these approaches, users can access the resources by playing the appropriate roles and based on the context information. For example, in the healthcare domain, a doctor is restricted to read the medical history of the patients after the office time or outside the hospital locations. Different from these approaches, our FCAAC approach utilizes fuzzy sets to derive the fuzzy conditions from the low-level fuzzy facts, and incorporates such fuzzy conditions along with normal contextual conditions into the policies. However, these existing context-aware RBAC approaches are not adequate to exploit the relevant contextual conditions together with fuzzy conditions for context-specific decision making at different granularity levels.

We have a successful history of using a wide range of contextual conditions for context-oriented decision making. In [2,10], we have introduced an ontology-based context-aware RBAC approach to information resources, where we consider the context information about the state of the users, resources and their surrounding environments (e.g., patients' profiles, users' locations, users' request times). In [11], we have introduced an ontology-based relationship-aware RBAC approach, incorporating the relationship context information (e.g., the different granularity levels of relationship, the relationship types, the relationship strengths) into the policies. In [3,12], we have introduced an ontology-based situation-aware RBAC approach, where we incorporate the purpose-oriented situation information (e.g., normal/emergency treatment purpose, research purpose) into the policies. Similar to above-mentioned context-aware approaches, however, our earlier approaches do not provide adequate functionalities to derive and incorporate the fuzzy contextual conditions into the access control policies.

Overall, the existing context-aware RBAC approaches are not adequate to deal with imprecise context characterization and consequently derive the fuzzy conditions from the low-level fuzzy facts. For example, concerning our application scenario, Bob's current health status is "66% normal with criticality level 0.34" only can be derived from Bob's pulse rate and body temperature.

Fuzzy Logic-Based Access Control Approaches: Different access control approaches have been proposed in literature to model policies based on involving the fuzzy conditions, which can be derived from the fuzzy sets.

In [14], the authors have proposed a trust-based access control approach based on the trust values [29], allowing only authorized users to access sensitive data (and information resources) that are usually confidential. They also propose a trust model to dynamically derive the trust degrees of high, medium and low. Cheng et al. [15] have proposed a risk-adaptive access control approach for an organization to protect its sensitive information. They quantify risk as the expected value of damage and consider risk to make access control decisions (e.g., the access decision is "denied" because the risk is too high). Takabi et al. [16] have proposed a trust-based RBAC approach to online services based on trustworthiness which is fuzzy in nature. They use fuzzy relations to compute trust values from the relevant attributes (e.g., behavioral, personal). In [17], the authors have proposed a fuzzy RBAC approach to deal with authorization-related imprecise information through fuzzy relations. They consider the various strengths of user-permission assignments as fuzzy relations to deal with such imprecise information and consequently propagate them to make access decisions.

However, these fuzzy logic-based access control approaches are not context-aware and still limited to incorporate a wide variety of access-control specific normal contextual conditions together with fuzzy conditions into the access control policies for context-specific decision making. Different from these fuzzy logic-based approaches, our FCAAC approach provides context-specific access permissions to users exploiting both the fuzzy and normal contextual conditions, and further limits the users' access to information resources accordingly.

Discussion: Following the traditional context-aware RBAC approaches, they are not adequate to derive the fuzzy conditions from the low-level fuzzy facts and incorporate them into the access control policies for decision making. On the other hand, the fuzzy logic-based approaches are not context-aware and robust enough to capture and derive the dynamically changing contextual conditions from the low-level information. In this respect, different from these existing access control approaches, our proposed FCAAC approach exploits the raw imprecise fuzzy facts, derives the fuzzy conditions from them and incorporates such conditions together with other contextual conditions into the access control policies for context-specific decision making at different granularity levels.

7   Conclusion and Future Research

The FCAAC approach described in this paper represents a flexible policy specification solution to the problem of incorporating fuzzy contextual conditions, in the domain of access control to information resources utilizing the benefits of fuzzy sets. Our approach significantly differs from the existing access control approaches in that it integrates the fuzzy conditions together with other relevant contextual conditions into the access control policies for context-specific decision making. We have presented the formal and ontology-based approaches to represent and reason about the fuzzy and other contextual conditions, and specify the access control policies by taking into account these conditions.

Furthermore, we have demonstrated the feasibility of our approach by considering the factors such as practicality and performance. In particular, we have developed a software prototype in order to assist the engineers in rapid prototyping. Using this prototype, software practitioners can build context-specific access control applications to cope with the complexities in the integration of fuzzy and other contextual conditions. Using this prototype, we have demonstrated the practicality of our approach by showing a case-based proof of the applicability of the FCAAC concepts against a healthcare case study. In addition, we have conducted two sets of experiment with our prototype and measured the response time and scalability of our proposal. Both the prototype implementation and the performance analysis results show that the new approach to access control using fuzzy logic is efficient and can be used in practice.

In this paper, we have defined the membership functions using the necessary information from the existing literature (e.g., the criticality ranges of the degree of membership for a "normal" health status are specified from 0 to 0.50). However, it may require special modelling to define the membership functions, which are domain dependent, and thus, further investigation to effectively represent them using the crisp boundary conditions is required in the future.

References

1. Weiser, M.: Some computer science issues in ubiquitous computing. Commun. ACM **36**(7), 75–84 (1993)
2. Kayes, A.S.M., Han, J., Colman, A.: OntCAAC: an ontology-based approach to context-aware access control for software services. Comput. J. **58**(11), 3000–3034 (2015)
3. Kayes, A.S.M., Han, J., Colman, A.W.: An ontological framework for situation-aware access control of software services. Inf. Syst. **53**, 253–277 (2015)
4. Bertino, E., Catania, B., Damiani, M.L., Perlasca, P.: GEO-RBAC: a spatially aware RBAC. In: SACMAT, pp. 29–37 (2005)
5. Joshi, J., Bertino, E., Latif, U., Ghafoor, A.: A generalized temporal role-based access control model. IEEE Trans. Knowl. Data Eng. **17**(1), 4–23 (2005)
6. Bonatti, P., Galdi, C., Torres, D.: Event-driven RBAC. J. Comput. Secur. **23**(6), 709–757 (2015)
7. Schefer-Wenzl, S., Strembeck, M.: Modelling context-aware RBAC models for mobile business processes. IJWMC **6**(5), 448–462 (2013)

8. Hosseinzadeh, S., Virtanen, S., Rodríguez, N.D., Lilius, J.: A semantic security framework and context-aware role-based access control ontology for smart spaces. In: SBD@SIGMOD, pp. 1–6 (2016)
9. Trnka, M., Cerný, T.: On security level usage in context-aware role-based access control. In: SAC, pp. 1192–1195 (2016)
10. Kayes, A.S.M., Han, J., Colman, A.: An ontology-based approach to context-aware access control for software services. In: Lin, X., Manolopoulos, Y., Srivastava, D., Huang, G. (eds.) WISE 2013. LNCS, vol. 8180, pp. 410–420. Springer, Heidelberg (2013). doi:10.1007/978-3-642-41230-1_34
11. Kayes, A.S.M., Han, J., Colman, A., Islam, M.S.: RelBOSS: a relationship-aware access control framework for software services. In: CoopIS, pp. 258–276 (2014)
12. Kayes, A.S.M., Han, J., Colman, A.: PO-SAAC: a purpose-oriented situation-aware access control framework for software services. In: Jarke, M., Mylopoulos, J., Quix, C., Rolland, C., Manolopoulos, Y., Mouratidis, H., Horkoff, J. (eds.) CAiSE 2014. LNCS, vol. 8484, pp. 58–74. Springer, Cham (2014). doi:10.1007/978-3-319-07881-6_5
13. Kayes, A.S.M., Han, J., Colman, A.: A semantic policy framework for context-aware access control applications. In: TrustCom, pp. 753–762 (2013)
14. Almenárez, F., Marín, A., Campo, C., García R., C.: TrustAC: Trust-based Access Control for pervasive devices. In: Hutter, D., Ullmann, M. (eds.) SPC 2005. LNCS, vol. 3450, pp. 225–238. Springer, Heidelberg (2005). doi:10.1007/11414360_22
15. Cheng, P.C., Rohatgi, P., Keser, C., Karger, P.A., Wagner, G.M., Reninger, A.S.: Fuzzy multi-level security: an experiment on quantified risk-adaptive access control. In: IEEE Symposium on Security and Privacy, pp. 222–230. IEEE (2007)
16. Takabi, H., Amini, M., Jalili, R.: Trust-based user-role assignment in role-based access control. In: AICCSA, pp. 807–814. IEEE (2007)
17. Martínez-García, C., Navarro-Arribas, G., Borrell, J.: Fuzzy role-based access control. Inf. Process. Lett. **111**(10), 483–487 (2011)
18. Feng, L., Dillon, T.S.: Using fuzzy linguistic representations to provide explanatory semantics for data warehouses. IEEE Trans. Knowl. Data Eng. **15**(1), 86–102 (2003)
19. Dey, A.K.: Understanding and using context. Pers. Ubiquitous Comput. **5**(1), 4–7 (2001)
20. Sandhu, R.S., Coyne, E.J., Feinstein, H.L., Youman, C.E.: Role-based access control models. IEEE Comput. **29**, 38–47 (1996)
21. Ferraiolo, D.F., Sandhu, R., Gavrila, S., Kuhn, D.R., Chandramouli, R.: Proposed NIST standard for role-based access control. ACM TISSEC **4**(3), 224–274 (2001)
22. Riboni, D., Bettini, C.: OWL 2 modeling and reasoning with complex human activities. Pervasive Mob. Comput. **7**, 379–395 (2011)
23. OWL: Web ontology language (2017). http://www.w3.org/2007/owl/
24. SWRL: Semantic web rule language (2017). http://www.w3.org/submission/swrl/
25. Protégé: Protégé-OWL API (2017). http://protege.stanford.edu/
26. Jess: Jess rule engine (2017). http://herzberg.ca.sandia.gov/
27. jFuzzyLogic: Fuzzy concepts and fuzzy control system in Java (2017). http://sourceforge.net/projects/jfuzzylogic
28. Wong, A.K.Y., Wong, J.H.K., Lin, W.W.K., Dillon, T.S., Chang, E.J.: Semantically Based Clinical TCM Telemedicine Systems. SCI, vol. 587. Springer, Heidelberg (2015). doi:10.1007/978-3-662-46024-5
29. Chang, E., Hussain, F., Dillon, T.: Trust and Reputation for Service-Oriented Environments: Technologies for Building Business Intelligence and Consumer Confidence. Wiley, London (2006)

Semi-supervised Log Pattern Detection and Exploration Using Event Concurrence and Contextual Information

Xixi Lu[1]([✉]), Dirk Fahland[1], Robert Andrews[2], Suriadi Suriadi[2],
Moe T. Wynn[2], Arthur H.M. ter Hofstede[2], and Wil M.P. van der Aalst[1]

[1] Eindhoven University of Technology, Eindhoven, The Netherlands
{x.lu,d.fahland,w.m.p.v.d.aalst}@tue.nl
[2] Queensland University of Technology, Brisbane, Australia
{r.andrews,s.suriadi,m.wynn,a.terhofstede}@qut.edu.au

Abstract. Process mining offers a variety of techniques for analyzing process execution event logs. Although process discovery algorithms construct end-to-end process models, they often have difficulties dealing with the complexity of real-life event logs. Discovered models may contain either complex or over-generalized fragments, the interpretation of which is difficult, and can result in misleading insights. Detecting and visualizing behavioral patterns instead of creating model structures can reduce complexity and give more accurate insights into recorded behaviors. Unsupervised detection techniques, based on statistical properties of the log only, generate a multitude of patterns and lack domain context. Supervised pattern detection requires a domain expert to specify patterns manually and lacks the event log context. In this paper, we reconcile supervised and unsupervised pattern detection. We visualize the log and help users *extract* patterns of interest from the log or obtain patterns through unsupervised learning automatically. Pattern matches are visualized in the context of the event log (also showing concurrency and additional contextual information). Earlier patterns can be extended or modified based on the insights. This enables an interactive and iterative approach to identify complex and concrete behavioral patterns in event logs. We implemented our approach in the ProM framework and evaluated the tool using both the BPI Challenge 2012 log of a loan application process and an insurance claims log from a major Australian insurance company.

Keywords: Pattern detection · Log pattern · Semi-supervised learning

1 Introduction

Process discovery offers automated construction of an end-to-end process model from an event log, recorded when executing a real-life business process, to gain useful insights into process executions. However, real-life logs often contain complex behavior which discovery algorithms struggle to handle. Consequently, discovered models may contain either very complex or over-generalized fragments,

© Springer International Publishing AG 2017
H. Panetto et al. (Eds.): OTM 2017 Conferences, Part I, LNCS 10573, pp. 154–174, 2017.
https://doi.org/10.1007/978-3-319-69462-7_11

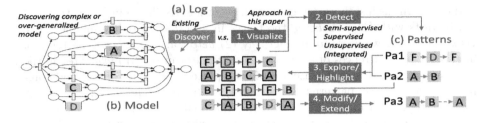

Fig. 1. An example of a process discovery problem versus the approach in this paper.

the interpretation of which is difficult. This can result in misleading insights. Figure 1(a) exemplifies an event log which contains four traces over five activities; from the log, an existing discovery algorithm (e.g., the Inductive Miner [1]) may return an over-generalized model allowing all five activities to be executed or skipped in any order, as shown in Fig. 1(b).

Discovery algorithms struggle because they aim to represent all variants and contexts of an activity within one single model. Instead, the use of patterns to untangle these complex behaviors and address them separately, as well as the detection of these patterns on event logs, can uncover more accurate insights into the complex behavior in process executions [2–9]. However, existing approaches to pattern detection are either *unsupervised learning* or *supervised learning*, with each having their own limitations, as depicted by Fig. 2.

Unsupervised learning [3,7] generates patterns based on statistical properties of the log, such as frequency, support and confidence, and suffers from a problem known as "pattern explosion". Patterns that are less frequent are more difficult to detect. Moreover, to the best of our knowledge, no existing unsupervised approach in log pattern detection supports retrieving occurrences of patterns. In contrast, most *supervised approaches* [5,6] assume a pattern set will be given, e.g., manually drawn by the experts, which puts the burden on the modeler.

In this paper, we propose a semi-automated approach to detecting *log patterns* from an event log (where events of traces could be totally ordered or *partially ordered*); an overview of the approach is shown in Fig. 1. We define the log patterns as short acyclic process fragments (formally: *partial orders* of activities where direct and indirect succession of activities are specified) and the semantics

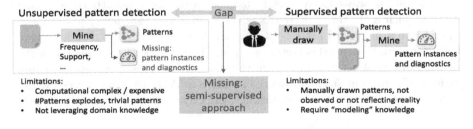

Fig. 2. Problem analysis: missing semi-automated pattern detection for log exploration.

of the patterns as instances detected in an event log. As shown in Fig. 1, the approach starts with (1) *visualizing* the traces in a dotted-chart-like manner. The visualization allows users to interactively (2) *detect* and *extract* log patterns of interest. The occurrences of a log pattern in traces can be (3) *highlighted* in the log visualization, to support users *exploring* patterns in their context and comparing occurrences of patterns. Users can (4) *modify* or *extend* patterns based on the occurrences (by extending them with additional activities, different ordering relations, etc.) or create new patterns based on existing ones. This way we enable the user to explore the patterns in their context (where and how frequently they occur) and modify patterns interactively and iteratively, aiming to help the user to (i) explore an event log with ease, (ii) obtain more concrete and accurate insights into process behavior than is possible from some end-to-end discovered models, and (iii) balance between unsupervised and supervised learning.

We implemented the functionalities for all these four steps in a log visualization tool as a plug-in for the ProM process mining framework[1]. For step (2) detecting log patterns, we proposed three semi-automated detectors and integrated two existing unsupervised approaches (by converting the detected patterns into our log patterns). Moreover, we support users to convert traces into partial orders. The aim is to *complement* and *not* substitute existing pattern detection approaches. We evaluated the approach in two case studies conducted using real-life event data sets. The results show concrete insights into process behavior patterns which existing discovery algorithms cannot reveal.

The remainder of this paper is organized as follows. Section 2 discusses related work. Section 3 discusses the preliminaries and the input for our approach. In Sect. 4, we formally define log patterns and pattern instances. In Sect. 5, we explain our approach to detect patterns and find pattern instances. Section 6 reports results for the two case studies conducted. Section 7 concludes the paper and discusses future work.

2   Related Work

We use Table 1 to discuss and structure related work. Here we discuss related work on log pattern detection on a conceptual level; an evaluation of the unsupervised techniques on a real-life event log is discussed in Sect. 6.

Log Pattern Definition. Patterns in the process mining literature have been defined in various ways. Early work focuses on clustering frequently co-occurring activities; such a cluster of activities is considered as a (low-level) pattern and mapped into a high level activity [4,10]. Later, more specialized definitions are used. Bose et al. [3] defined patterns as repeated sequences. [7,8] defined patterns as partial orders of activities in which the edges only represent eventually-cause relations. The work in [5,11] considered patterns as Petri nets. In our case, we use partial orders, distinguishing concurrence, directly-cause and eventually-cause.

---

[1] http://www.promtools.org/.

Table 1. Comparison of related pattern detection approaches.

	S/U/M/V[a]	Provide unsuper. support	Change/ define patterns?	Explore pattern instances?
Bose et al. Pattern abstraction [3]	U	+ (beh.)[b]	−	−
Günther et al. Fuzzy mining [4]	U	+ (act.)	−	−
Mannhardt et al. Log abstraction [5]	S	− (act.)	+	+/−
Maggi et al. LTL checker [6]	S	− (com.)	+/−	+/−
Leemans et al. Episodes miner [7]	U	+ (beh.)	−	−
Diamantini et al. Pattern discovery [8]	U	+ (beh.)	−	−
Baier et al. Activity maching [12]	M	− (act.)	−	−
Ferreira et al. Label abstraction [10]	U	+ (act.)	−	−
Tax et al. Local models [11]	U	+ (beh.)	−	−
Song et al. Dotted chart [13]	V	− (beh.)	−	−
Shneiderman et al. EventFlow [2]	S.V.	− (beh.)	+	+
Lu et al. Pattern explorer	M.S.V.U.	+ (beh.)	+	+

[a]S for supervised; U for unsupervised; M for Semi-Supervised; V for visualization.
[b] In parentheses, the aim of the technique is abbreviated: beh. for behavior analysis; act. for low level activity abstraction; com. for compliance checking.

Unsupervised Log Pattern Detection. Unsupervised log pattern detection approaches take an event log as input and generate patterns based on predefined measures [3,7,8]. This has some limitations. Firstly, such unsupervised approaches are computationally complex and expensive, generating a massive amount of possible patterns based on their frequencies or other measures. If one sets the values for the measures too high, then only very frequent, trivial patterns are returned, thereby missing many interesting results. By setting the values too low, too many patterns are returned (so called "pattern explosion") [3,7,8]. Secondly, as a result of not leveraging domain knowledge, many of the patterns generated by unsupervised learning are not of interest or are meaningless. Finally, most unsupervised approaches do not return pattern instances or additional contextual information for the detected patterns, thus obstructing the user from exploring and analyzing the patterns [7].

Supervised Log Pattern Detection. Supervised pattern detection approaches in process mining take patterns and logs as input and detect pattern instances as results [5]. Such supervised approaches require the user to model patterns in a formal language such as Petri nets or LTL constraints. This relies on the expertise of the user, who may need to formalize a large set of pattern descriptions. Moreover, the user may miss potentially important patterns through incomplete specification or model idealized patterns that are not observed in reality.

Log Explorer and Visualization. Advanced log visualization analytics have also been proposed as a way to help the user observe patterns. The dotted chart [13] is a simple way of visualizing event logs and helping the user spot and interpret patterns such as batch processing. However, no pattern extraction approaches are supported, nor is it possible to query for all instances of the observed patterns. EventFlow [2] has been proposed as a more advanced tool for visualizing event sequences. It allows for advanced querying, but also requires the user to create patterns (queries) and does not support generating patterns in an unsupervised or semi-unsupervised way. In our case, we allow semi-supervised pattern detection; the detected patterns are suggested to the user and can be used as queries. Moreover, we support partially ordered events and help the user detect and explore causal dependencies between events [14].

3 Preliminaries

In this section, we recall the basic concepts such as *partial orders*, *event logs*, and *partially ordered traces* and discuss the *oracle functions* that convert sequential traces of an event log into partially ordered traces. These are used later in the paper.

Let X be a set of elements and $X' \subseteq X$ a subset of X. A relation $R \subseteq X \times X$ between X is a *partial order* over X if and only if R is irreflexive, anti-symmetric, and transitive. Let $G = (N, <)$ be a *directed acyclic graph (DAG)* with $< \subseteq N \times N$. We use $<^-$ to denote transitive reduction of $<$ and $<^+$ to denote the transitive closure of $<$. The relation $<^+$ will be used to denote reachability and defines a partial order. In this paper, for all nodes $n, n' \in N$, $n \not<^+ n'$ and $n' \not<^+ n$, we say n and n' are *concurrent* and use $n||_< n'$ to denote this. We use $\downarrow$ to denote a *projection function*, i.e., $X \downarrow_{X'} = X \cap X'$ and $R \downarrow_{X'} = R \cap (X' \times X')$. Let $G = (N, <)$ be a DAG and $N' \subseteq N$. We overload the projection function and define the projection for a graph, $G \downarrow_{N'} = (N \downarrow_{N'}, < \downarrow_{N'})$, also known as *induced subgraph*.

An *event log* represents the observed behavior of a process. Each case going through the process results in a trace of events in the event log.

Definition 1 (Event, trace, event log). *Let $\mathcal{E}$ be the universe of unique events, i.e., the set of all possible event identifiers. A trace $\sigma = \langle e_1, \cdots, e_n \rangle \in \mathcal{E}^*$ is a finite sequence of events $e_1, \cdots, e_n \in \mathcal{E}$. An event log $L = \{\sigma_1, \sigma_2, \cdots, \sigma_n\} \subseteq \mathcal{E}^*$ is a set of traces.*

Here we assume that no event appears twice in the same trace nor in the same log. We use E_σ for the set of events in trace σ and E_L for the set of events in log L. We use $\pi_{act}(e)$ to return the activity associated with e; and $\pi_{time}(e)$ to return the timestamp of e. In this paper, we use π_{act} as our default labeling function for events and assume that both π_{act} and π_{time} are universally available for $\mathcal{E}$. Figure 3(a) shows an example of totally ordered trace $\sigma = \langle e_1, e_2, ..., e_5 \rangle$. Event e_1 has activity $\pi_{act}(e_1) = Injury$ and is executed on $\pi_{time}(e_1) = 08/09/2016\text{-}00{:}30{:}00$.

Many recent papers consider partial orders of events [14–16], instead of totally ordered event sequences. One reason for this consideration is that a particular total order of events may be unreliable or unknown. For example, if events a and b are recorded only on day granularity (not seconds), then a totally ordered log may contain the sequence $\langle a, b \rangle$ whereas in reality $\langle b, a \rangle$ occurred. Representing events as a partial order (where a and b can occur "unordered" or "concurrent") alleviates this problem and allows us to represent more accurate contextual information of events [14, 16].

Definition 2 (Partially Ordered Trace). *A partially ordered trace $\varphi = (E, \prec)$ is a Directed Acyclic Graph (DAG), in which $\prec$ denotes the inferred "cause" relations[2] over events E. If $e \prec e'$, we say e caused e'. We use $\prec^-$ to denote directly-cause, $\prec^+$ to denote eventually-cause, and $||_\prec$ to denote the concurrent relation.*

Note that the *eventually-cause* relation $\prec^+$ forms a partial order. Figure 3(b) shows a partially ordered trace $(E, \prec)$, in which $E = \{e_1, e_2, \cdots, e_5\}$ and $\prec = \{(e_1, e_2), (e_1, e_3), (e_2, e_4), (e_3, e_4), (e_4, e_5)\}$. In this particular case, $\prec^- = \prec$ and $\prec^+ = \prec \cup \{(e_1, e_4), (e_1, e_5), (e_2, e_5), (e_3, e_5)\}$. The concurrence relation $||_\prec = \{(e_2, e_3), (e_3, e_2)\}$. Note that $\prec$, $\prec^-$, and $\prec^+$ are irreflexive and acyclic, and $||_\prec$ is irreflexive and symmetric.

Conversion. Partial orders of events may be obtained from totally ordered traces. Some approaches [14, 15, 17] assume to have an oracle that indicates the set of activities that are concurrent or unordered and use this oracle to convert totally ordered events into partial orders. Such an oracle could be obtained by interviewing domain experts or computed from logs [14].

We overload the symbol φ to denote an *conversion oracle function* that, for a trace σ, returns a partially ordered trace $\varphi(\sigma) = (E_\sigma, \prec_\sigma)$. We deploy the oracle that considers the events occurring within a short time to be concurrent [14] as our default oracle. Let $\sigma = \langle e_1, \cdots, e_n \rangle$ be a trace. $\varphi_{time}(\sigma, dt) = (E_\sigma, \prec_\sigma)$ in

---

[2] We use the term *"cause"* *(causality)* only to distinguish the relations in a partially ordered trace from the *follow* relations (i.e., *directly-follow* and *eventually-follow*) in totally ordered traces.

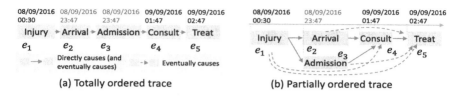

Fig. 3. A sequential trace and its converted partially ordered trace.

which $\prec_\sigma = \{(e_i, e_j) \mid 1 \leq i < j \leq n \wedge \exists_{i \leq k < j} \pi_{time}(e_{k+1}) - \pi_{time}(e_k) \geq dt\}$. In addition, we also overload the φ function to handle a log $L = \{\sigma_1, \cdots, \sigma_n\}$ and return a set of partially ordered traces, one for each trace in L, i.e., $\varphi(L) = \{\varphi(\sigma_1), \cdots, \varphi(\sigma_n)\}$. An example of such a conversion based on the timestamps is shown in Fig. 3, with $dt = 1.0$ s.

4   Patterns and Pattern Instances

Having defined partially ordered traces of an event log, we now present the concept of log patterns. In Sect. 4.1 we motivate and define the log patterns and pattern instances. In Sect. 4.2, we discuss pervasiveness measures for the patterns, such as support, confidence and coverage. In Sect. 5, three approaches to pattern detection are presented.

4.1   Core-Activity, Pattern and Pattern Instance

To support process analysts in expressing and modifying log patterns with ease, the patterns should be simple. Moreover, if such a pattern resembles our input traces, it would be easier for the user to observe and recognize these patterns. We therefore define a pattern as a labeled DAG (similar to a partially ordered trace) and allow a pattern to express causal dependencies that occur in a partially ordered trace, namely *directly-cause* and *eventually-cause*. The *concurrence* relation is then deduced for any two events that are not eventually causing one another. Figure 4(a) and (d) show two examples of patterns.

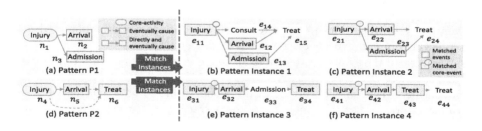

Fig. 4. Two patterns and four highlighted pattern instances, two for each pattern.

In addition to considering patterns as labeled DAGs, we, for two reasons, explicitly include a notion called a *core-activity* in the patterns. Firstly, the *core-activity* notion allows for unambiguous notions of pattern matches and unambiguous computation of measures such as frequency. Without a core-activity, having a pattern P_{ab} that says "a eventually-caused b" and a trace $\sigma_1 = \langle a, b, b, a \rangle$, should this be considered as one instance or two? Given another trace $\sigma_2 = \langle a, b, a, b, c \rangle$, if the number of combinations is considered, such an approach results in three pattern instances in σ_2 while there are only two a's and two b's, causing confusion. In our case, the core-activity anchors the pattern in a particular perspective. We simply count the number of distinct events that match the core-activity and satisfy the pattern and call these events *core-events*; we ignore whether the events that match to the other nodes in a pattern occur several times. Take the same example, if a is the core-activity of P_{ab}, then we have one a for σ_1 and two a's for σ_2. Similarly, if the pattern P_{ab} has b as the core-activity, then we have two b's for σ_1 and also two b's for σ_2. In addition, having a core-activity also allows to unambiguously compute the events (*anti-instances*) that do *not* satisfy a pattern (not just the traces). For example, having the same pattern P_{ab} with a as the core-activity, the second a in σ_1 does not satisfy the pattern; without the core-activity, σ_1 may be consider as compliant or not, depending on the interpretation of the pattern.

Definition 3 (Log Pattern). *A log pattern $P = (N, \mapsto, \rightsquigarrow, \alpha, c)$ is a directed acyclic graph, where:*

- *N is a set of nodes,*
- *$\mapsto$ is a set of edges among N and denotes the directly-cause relation (see footnote 2),*
- *$\rightsquigarrow$ is a set of edges among N and denotes the eventually-cause relation (see footnote 2),*
- *$\alpha : N \rightarrow A$ is a partial function that assigns a label $\alpha(n)$ to any node $n \in N$,*
- *$c \in N$ is the core-activity of the pattern*

and satisfies the following constraints:

1. *$\mapsto \subseteq \rightsquigarrow$, i.e., the directly-cause relation is a subset of the eventually-cause relation;*
2. *$(N, \rightsquigarrow)$ is a partial order from which the concurrence $||_{\rightsquigarrow}$ can be deduced;*
3. *for all $n, n' \in N$, if there is $n'' \in N$ such that $n \rightsquigarrow n'' \rightsquigarrow n'$, then there is no $n \mapsto n'$.*

We also say c has context P and call $N \backslash \{c\}$ the context-nodes of c.

Note that Definition 3 only defines the syntax of a log pattern. The semantics of a log pattern is defined by its instances in Definition 4. A pattern may be regarded as a core-activity (activity) that occurred in a particular context. Accordingly, an instance of the pattern is an occurrence of the core-activity in the log in the same context.

Definition 4 (Pattern Instance and Pattern Trace). *Let* $P = (N, \mapsto, \rightsquigarrow, \alpha, c)$ *be a log pattern,* $\varphi = (E_\varphi, \prec)$ *a partially ordered trace,* $E' \subseteq E_\varphi$ *a subset of events, and* $e_c \in E'$ *an event.* (e_c, E') *is an instance of pattern* P *if and only if there is a bijective function* $I : E' \rightarrow N$ *such that*

- e_c *(called the* core-event *of* (e_c, E')) *is mapped to the core-activity, i.e.,* $I(e_c) = c$;
- *for each event* $e \in E'$, *event* e *and the corresponding node* $I(e)$ *have the same label, i.e.,* $\pi_{act}(e) = \alpha(I(e))$;
- *for all events* $e, e' \in E'$, *the relations between* e *and* e' *satisfy the relations between* $I(e)$ *and* $I(e')$, *i.e.,* (1) $I(e) \mapsto I(e') \Rightarrow e \prec^- e'$, (2) $I(e) \rightsquigarrow I(e') \Rightarrow e \prec^+ e'$, *and* (3) $I(e')||_{\rightsquigarrow}I(e) \Rightarrow e'||_\prec e$

If (e_c, E') *is an instance of* P, *we also say* e_c *satisfies* P *and say* φ *is a trace of* P.

The behavior specified by a pattern is preserved in the instances of the pattern. Figure 4(b), (c), (e) and (f) exemplify four pattern instances highlighted in their partially ordered traces: highlighted e_{11} and e_{21} satisfy P1; e_{31} and e_{41} satisfy P2. It is important to note that changing the core-activity of a pattern does not change the behavioral relations of the pattern, but does change the instances that match the pattern.

Definition 5 (A Maximal Set of Pattern Instances). *Let* $P = (N, \mapsto, \rightsquigarrow, \alpha, c)$ *be a pattern,* L *an event log,* φ *the conversion oracle. A maximal set of pattern instances* $PI(P, \varphi(\sigma))$ *of trace* $\sigma \in L$ *is defined as a maximal set of all instances of* P *in* $\varphi(\sigma)$ *that differ in their core-event, i.e., for any instance* (e', E') *of* P, *if* $(e', E') \notin PI(P, \varphi(\sigma))$ *then there exist* $(e', E'') \in PI(P, \varphi(\sigma))$. *We write* $PI(P, L, \varphi) = \bigcup_{\sigma \in L} PI(P, \varphi(\sigma))$ *for the union of a maximal set of pattern instances of all traces in log* L. *We write* $PIC(P, L, \varphi)$ *to denote the set of all core-events that satisfy* P, *i.e.,* $PIC(P, L, \varphi) = \{e \mid (e, E') \in PI(P, L, \varphi)\}$.

The set of *anti-pattern instances* $AntiPIC(P, L, \varphi)$ is defined as the set of events that do not satisfy P, i.e., $AntiPIC(P, L, \varphi) = \{e \in E_L \mid \pi_{act}(e) = \alpha(c)\} \backslash PIC(P, L, \varphi)$. Note that independent of which maximal set of pattern instances is returned, the set of all core-events and the set of anti-pattern instances of a pattern remain the same. Consider for example pattern P2 in Fig. 4(d) and the four partially ordered traces shown on the right-hand side to be the set of all partially ordered traces in $\varphi(L)$. A maximal set of pattern instances of P2 in $\varphi(L)$ always contains four pattern instances with e_{11}, e_{21}, e_{31} and e_{41} as the core-events that satisfy P2. As e_{41} (*Injury*) in Fig. 4(f) already satisfies P2 with context-event e_{43} (*Treat*), e_{44} (*Treat*) is not considered as a context event. However, if n_6 (*Treat*) was considered as core-activity of P2, we would obtain five instances with the core-events $e_{15}, e_{24}, e_{34}, e_{43}$, and e_{44}. Furthermore, e_{31} and e_{41} are anti-pattern instances of pattern P1.

4.2   Pattern Support, Confidence and Coverage

To help the user assess the pervasiveness of a pattern, we define the following five measures of a pattern based on the set of all pattern instances. Let $P = (N, \mapsto, \leadsto, \alpha, c)$ be a pattern. Given a log L and the partially ordered traces $\varphi(L)$, we have the set of all core-events $PIC(P, L, \varphi) = \{e_1, e_2, \cdots, e_n\}$ that satisfy P.

- *Pattern support* indicates how many distinct events satisfy P. i.e., $P\text{-}supp(P, L, \varphi) = |\ PIC(P, L, \varphi)\ |$.
- *Pattern confidence* is the number of events that satisfy P divided by the total number of events that have the same label as the core-activity; this measure indicates how often is the occurrence of the core-activity a predictor for the occurrence of the entire pattern. i.e., $P\text{-}conf(P, L, \varphi) = \frac{P\text{-}supp(P, L, \varphi)}{|\{e \in E_L | \pi_{act}(e) = \alpha(c)\}|}$.
- *Case support* is the number of traces that have at least one pattern instance satisfying P, i.e., $C\text{-}supp(P, L, \varphi) = |\ \{\sigma \in L\ |\ \exists e \in PIC(P, L, \varphi), e \in E_\sigma\}\ |$.
- *Case confidence* is the case support of P divided by the number of cases that have an event with the same label as c, i.e., $C\text{-}conf(P, L, \varphi) = \frac{C\text{-}supp(P, L, \varphi)}{|\{\sigma \in L | \exists e \in \sigma, \pi_{act}(e) = \alpha(c)\}|}$.
- *Case coverage* is the case support of P divided by the number of cases in the log, i.e., $C\text{-}cover(P, L, \varphi) = \frac{C\text{-}supp(P, L, \varphi)}{|L|}$.

We note that the desirability of pervasiveness measures (highly pervasiveness or otherwise) is context/application dependent. For example, for patterns representing non-complaint behavior or data quality issues, we would prefer to see low pervasiveness.

5   Pattern Detection and Pattern Instance Matching

We now introduce operations to *extract* and *extend* patterns in the context of a log in an explorative approach. To help the reader envision this explorative approach, we first briefly introduce the *Log Pattern Explorer (LPE)* tool, for which a screenshot is shown in Fig. 5. The right-hand side panel shows the log patterns, manually or automatically extracted. The user may create or modify a pattern. On the left-hand side, each log trace is visualized as a partial order in its own panel: each event is visualized as a square tile; tiles can be colored based on event attributes; concurrent events are stacked on top of each other; the labels and arcs are omitted for the sake of simplicity. When a user selects one or more patterns in the right-hand panel, all pattern instances are highlighted on the left (by a color-coded frame around the tiles of the satisfying events).

In Sect. 5.1, we discuss various ways to extract, detect, and modify patterns using supervised, semi-supervised or unsupervised approaches. We then discuss an approach to compute a maximal set of pattern instances of a pattern (Definition 5) in Sect. 5.2.

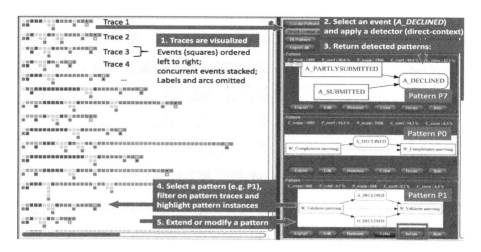

Fig. 5. Snippet of the tool after it found 9 patterns using event *A_DECLINED* and filtered on the pattern traces of pattern P1.

5.1 Pattern Detection Approaches - Partially Ordered Traces to Patterns

Definition 3 and the tool allow the user to create any pattern of interest. Nevertheless, as shown in Fig. 1, we would like to support the user in detecting these patterns from a log with ease. For example, an expert glances through the partially ordered traces visualized in Fig. 5 and may observe some events (patterns) reoccur in many traces, e.g., through the color coding. The user can then mark all events (tiles) in a trace that make up the pattern. An automated operation is defined in the following for extracting the marked events and their relations from the trace to build a complete pattern definition.

Pattern Extraction from Partially Ordered Traces. In essence, the marked events in a partially ordered trace φ are used to extract a subgraph of φ as the pattern. Formally, let $\varphi = (E, \prec)$ be a partially ordered trace, $E' \subseteq E$ a set of events marked by the user and $e_c \in E'$ a chosen core-event. The function $extractPattern(\varphi, e_c, E') = P$ returns the log pattern P of E' induced on φ, i.e., $extractPattern(\varphi, e_c, E') = P = (N, \mapsto, \leadsto, \alpha, c)$ where $N := E'$, $\mapsto := (\prec^-)\downarrow_{E'}$, $\leadsto := (\prec^+)\downarrow_{E'}$, $\alpha := (\pi_{act})\downarrow_{E'}$, and $c := e_c$.

Figure 6 (Step 1) shows extraction of a pattern from the partially ordered trace by marking events $e_1(Injury)$, $e_2(Arrival)$ and $e_3(Admission)$ and considering e_1 as the core-event. The extracted pattern $P1 = (N, \mapsto, \leadsto, \alpha, c)$ with $N = \{n_1, n_2, n_3\}$, $n_1 \mapsto n_2, n_1 \mapsto n_3$, $n_1 \leadsto n_2$, and $n_1 \leadsto n_3$; n_2 and n_3 are concurrent; for α, the labels of nodes remain unchanged; the core-activity c is n_1 (*Injury*), abstracted from the core-event e_1. In the tool shown in Fig. 5, the user can mark a set of tiles (events) on the left and apply the pattern extraction function, the first tile is abstracted as the core-activity; the extracted pattern will appear in the panel on the right.

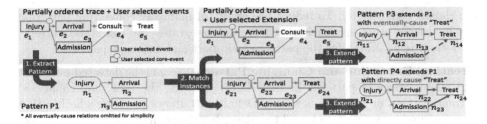

Fig. 6. Iteratively extracting and extending patterns from partially ordered traces.

Semi-supervised Pattern Detection. To help the user detect patterns of interest, we also present three detectors that identify patterns for a user-chosen activity: (1) *concurrence detector*, (2) *direct-predecessor/successor detector*, and (3) *direct-context detector*. (1) The *concurrence detector* obtains a set of distinct patterns that describe different sets of activities which occurred concurrently to the user-chosen activity. Formally, let $\mathcal{P}$ be the set of patterns we have detected so far. Let L be a log, φ a conversion oracle, and c a core-activity of interest. Let $E_c = \{e \in E_L \mid \pi_{act}(e) = \alpha(c)\}$ be candidates for core-events. For event $e \in E_c$, let φ be the trace containing e, and let C_e be the events that are concurrent with e in φ. If $(e, C_e \cup \{e\})$ is not an instance of any pattern $P \in \mathcal{P}$, then we obtain a new pattern $P' = extractPattern(\varphi, e, C_e \cup \{e\})$ and add P' to $\mathcal{P}$. For (2) the *direct-predecessor (successor) detector*, C_e is the set of events that are directly-causing (that directly-caused by) core-event e. For (3) the *direct-context detector*, C_e contains directly preceding, succeeding and all concurrent events of e. The relevance of an extracted pattern can be assessed using the measures of Sect. 4.2.

Integrating Unsupervised Pattern Detection. To also leverage on existing unsupervised pattern detection techniques, their output has to be converted to our pattern notion (Definition 3). We discuss this conversion for two techniques [3,7]. For the technique in [3], the output patterns are subsequences of activities. Any such pattern also satisfies our pattern definition (Definition 3). However, events that originally were totally ordered may now be independent (concurrent) due to the usage of partially ordered traces and may therefore no longer satisfy the converted pattern. In such cases, the user may find low *P-supp* and *P-conf*, explore anti-pattern instances and modify the pattern accordingly. The output patterns in [7] are partial orders of events in which the relations represent eventually-follow and no relations represent co-occurrence. We retain all eventually-follow relations and choose to consider co-occurrence as concurrent. Regarding choosing the core-activities, the user may specify an activity (label) of interest to be automatically selected; otherwise, a random node is selected. The user can run such an unsupervised detection in the tool shown in Fig. 5. The returned and converted patterns are shown in the right panel. The user can explore the pattern instances in the left panel. Note that the pervasiveness of a pattern (such as *P-supp* and *P-conf*) are recomputed in our case, depending on the chosen core-activity. To reduce the efforts of manually converting patterns obtained from other approaches that are not integrated, we can standardize a file format (e.g., XML) for the patterns to import and exchange patterns.

Creating, Extending and Changing Patterns. Seeing the detected patterns and all instances of patterns in their larger context, the user may change a current pattern or create a new one. Definition 3 allows to check whether the new pattern is syntax-correct, and the algorithm that computes the instances of the pattern updates the semantics of the pattern, discussed in Sect. 5.2. Step 3 (extend pattern) in Fig. 6 exemplifies two different extensions of P1 that differ in how the added node *Treat* is relates to its predecessors.

5.2   Computing a Maximal Set of Pattern Instances

Having patterns obtained using one of the aforementioned methods, the semantics and the relevance of a pattern are defined by its pattern instances, as discussed in Sect. 4.1. To compute a *maximal set of instances* of a pattern (Definition 5), we present the following method, divided into three phases. First, all events that can be matched to core-activity c of pattern P are computed; we call these events the candidates of c and use E_c to denote this set of events. In the second phase, for each candidate $e \in E_c$, we try to find a pattern instance for P with e as core-event. This is done through an incremental construction of the mapping I (Definition 4) with backtracking and pruning. If we can complete the construction of I mapping to events E' in the trace, then (e, E') is a pattern instance and added to the maximal set. Else, e is an anti-pattern instance. The formal algorithm for computing the pattern instances is listed in [18]

The running-time complexity is exponential w.r.t. the size of the pattern (i.e., $|N|$), but polynomial in the size of E_c. In the best case, we try one combination and it already is an instance, then the algorithm runs in linear time. In the worst case, one may have to try every combination, then the algorithm runs in exponential time. However, we can incrementally check the validity of the chosen candidates so far through the projection function and efficiently prune the search space. Moreover, note that by only searching for a maximal set of pattern instances instead of all pattern instances, we evade exploring exponentially many matches for the same core-activity.

6   Evaluation and Discussion

The *Log Pattern Explorer (LPE)* is implemented as a log visualizer in the *LogPatternExplorer* package in the ProM framework (see footnote 1). We conducted two case studies. As discussed in Sect. 1, one objective is to evaluate whether (i) the semi-supervised approach supports the user in exploring log behavior and detecting complex patterns of interest. Another objective is to evaluate that (ii) the exploration of the occurrences of patterns in the log context helps to gain important and concrete insights into process behavior, especially on the parts where discovery algorithms have difficulties with. Furthermore, using some representative patterns detected in the first case study, the semi-supervised approach is compared to the existing unsupervised approaches to show that (iii) the proposed approach complements the existing unsupervised approaches by detecting complex, infrequent patterns and anti-pattern instances.

Guidelines for Exploring Log Patterns. The results in the case studies were obtained according to the following guidelines, using the tool described in Sect. 5 for steps 3–6.

1. A-priori, a process model was discovered using an existing discovery algorithm (e.g., the Inductive Miner [1]).
2. In the discovered model, unstructured parts may be detected (e.g., flower loops or activities with many or no connections). To understand these complex subprocesses and obtain more insights, the involved activities are used as a starting point for further analysis.
3. For each activity of interest, a first set of patterns (where the core-activity is labeled with the activity) is detected automatically, for instance, using the direct-context detector, to reveal different contexts (in the log) in which the activity occurs.
4. With each pattern detected, the corresponding pattern instances and pattern traces may be explored. A patterns is *interesting*, for example, if the pattern or the traces where the pattern occurred show behavior different from the discovered model.
5. During exploration, some detected patterns and their pattern traces may immediately show distinctive behaviors compared to the discovered model; some patterns may reveal similar behaviors and be grouped together; other patterns and their traces may be compared and reveal significant difference in their behavior.
6. During exploration, the same pattern may occur in distinct contexts (other events "around" the pattern). In such a case, the pattern is extended or modified to new patterns. The pattern traces of the new patterns are compared to gain more insights.

Evaluation using BPI Challenge 2012 Log. The BPI Challenge 2012 event log[3] was recorded for a loan application process in a Dutch financial institute. There are 13,087 cases in the log having in total 262,200 events and 24 activities and 4,366 distinct process instances.

Applying the Inductive Miner [1] with default settings (noise threshold: 0.2) on the BPI challenge log, we obtained a model shown in Fig. 7. This model indicates that, after the two activities, *A_SUBMITTED (A_SU)* and *A_PARTLYSUBMITTED (A_PA)*, all other activities are in a loop and all of them can be skipped (may be executed repetitively in any order) as highlighted in Fig. 7. The allowed behavior of this model is therefore rather general and does

---

[3] 10.4121/uuid:3926db30-f712-4394-aebc-75976070e91f.

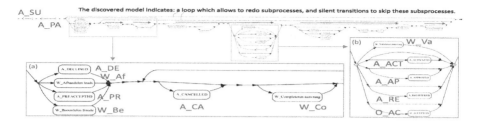

Fig. 7. The process model discovered on the BPI challenge 2012 log using IM; the model allows a large subprocesses to be executed in any order and any number.

not help to understand the specific behavior of the process exhibited by the log, even though the model appears to be structured.

Starting from the discovered model, we select activities from the two submodels (a) and (b) highlighted in Fig. 7 and apply the LPE. We start with the four activities A_DECLINED (A_DE), W_Afhandelen leads (W_Af), A_PREACCEPTED (A_PR), and W_Beoordelen fraude (W_Be) which according to the model may directly occur after the A_PA and can be followed by any other activity. We used the default oracle $\varphi_{time}(L, dt)$, with $dt = 1.0$ s (e.g., events are concurrent if they happened within a second), to obtain partially ordered traces. We discuss our main findings[4].

Results Regarding A_DECLINED (A_DE). The direct-context detector returned 9 patterns P0, ..., P8 for the core-activity A_DE, indicating that A_DE occurs in 9 distinct contexts which we could group into three variants as follows. (1) The pattern traces of pattern P5 (and P7) show that when A_PA directly-causes A_DE, the process ends immediately after A_DE (in 3429 cases, 44.9% of all cases), exemplified in Fig. 8; no exceptional case was found. Thus, A_DE directly-preceded by A_PA does not occur repeatedly in a large subprocess (as suggested by the model of Fig. 7) but always at the end of a trace (as seen by the patterns in Fig. 8). (2) The pattern traces of P0, P2 and P3 show that A_DE

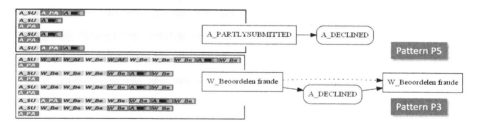

Fig. 8. Patterns P5 and P3 and their *pattern traces* show simple behavior, different from the discovered model in Fig. 7 and different from each other.

---

[4] The resuls and detected patterns can be downloaded at prom-svn://Documentation/ LogPatternExplorer.

Fig. 9. Pattern P8 and its pattern traces which show different and more complex behavior when compared to the pattern traces of P3 and P5 as shown in Fig. 8.

occurs between repeated occurrences of the same activities (W_Co, W_Af, W_Be; Fig. 5 shows P3); all traces where P0, P2, P3 occur show the same more general pattern: the cases start with the same sequence of activities, only 8 out of the 24 process activities occurred, and each case is declined with P0, P2, or P3. (3) P1, P6, P8 show A_DE occurs concurrently to $O\_DECLINED$ with various predecessors/successors; traces containing these patterns are more complex and contain activities (including "acceptance" activities) that do not occur in the cases discussed in (1) and (2); see Fig. 9. Altogether, we could identify 3 distinct variants. Moreover, where the model suggests complex behavior, such as the large sub-process after A_DE that allows complex behavior and can be repeated, the detected patterns show that all traces of these 9 patterns immediately ended after A_DE (plus one more activity).

Results Regarding W_Af and A_PR. Following the same procedure, we investigated activities W_Af, and A_PR. We found 40 patterns (direct-contexts) for W_Af (of which 12 patterns have C-cover higher than 1%) and 9 patterns for A_PR. Exploring the pattern traces, we found two identical contexts for W_Af and A_PR: (P9) W_Af preceded by A_PA and (P10) A_PR preceded by A_PA. Visual inspection of the occurrences of P9 and P10 on the event log shows that A_DE occurs more frequently when P9 occurs than when P10 occurs. To confirm this, we extended P9 and P10 with "... eventually causes A_DE" to patterns P9a and P10a, respectively, as described in Sect. 5.1 (see Fig. 10). The relative share of the extended patterns $\frac{C-supp(P9a)}{C-supp(P9)} = 67.6\%$ and $\frac{C-supp(P10a)}{C-supp(P10)} = 19.4\%$ confirms our observation. Similarly, we observe that P9 "...eventually causes $O\_ACCEPTED\ (O\_AC)$" in 12% of the cases whereas this holds for P10 in 34.4% of the cases. These long-term probabilistic dependencies cannot be observed in the model of Fig. 7 (or be discovered with any existing process discovery algorithm); however they showed up distinctly in our approach.

Results Regarding $W\_Beoordelen\ Fraude\ (W\_Be)$. Activity W_Be concerns the analysis of fraud in an application. For W_Be as core-activity, we discovered 34 patterns using the direct-context detector. Among those, pattern P11 (W_Be preceded by A_PA) stands out. If W_Be occurs in the context of P11, then only in 1.5% of the cases, the application was accepted (O_AC) and rejected in all other cases. If W_Be occurs in any other context, then 70.7% of the cases are accepted. In case of P11, W_Be is executed by the system whereas it is executed manually in all other cases. This suggests that the system-based

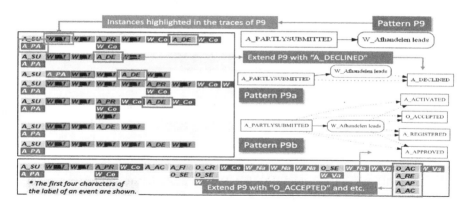

Fig. 10. The traces where pattern P9 occurs show a significant number of
A_DECLINED (A_DE); P9 is extended with eventually-cause *A_DE*.

fraud detection leads to rejection of possible fraud cases more rigorously than
when fraud is decided manually.

Results Regarding *O_AC*. The discovered model in Fig. 7 suggests that the
four activities in part (b) can be executed in any order and also be skipped. We
checked this hypothesis and extracted patterns using the *concurrent detector*.
For *O_AC* as core-activity, we find that *A_ACT*, *A_AP*, *A_RE* indeed occur
concurrently in all 2243 cases (P-conf is 1.0), whereas the model allows skipping
any of these. For *A_ACT* as core activity (pattern P12 in the following), *A_AP*
and *A_RE* occur concurrently with confidence 1.0, but *O_AC* is *not* in 3 of
2246 cases. In these 3 anti-pattern instances of P12, *O_AC* is not executed
at all, which may have severe financial impact according to the data owner:
"*From a business point of view, this* [the cases where *O_AC* was skipped while
A_ACT was executed] *implies that the customer never accepted an offer on a
loan application, but the money was transferred nonetheless. In total, for 63,000
euro in these three cases.*"

Evaluation Using an Insurance Log. For this evaluation we used a real-life
insurance claims log which included 10,228 cases, 195,872 events spread over 18
event classes. A process model was obtained (see Fig. 11) using the Inductive
Miner [1] with default settings (noise threshold: 0.2). The model shows a flower-
like behavior. Thus, the allowed behaviors of the model are general and hide
specific behaviors.

Next we applied one of the built-in pattern detector (the Episode Detector)
which returned 22 patterns, each of which involved the following activities: *conduct file review*, *incoming correspondence* and *follow up requested*. Twelve of the
patterns had *incoming correspondence* as the core-activity and 9 of the patterns
had *conduct file review* as the core-activity. Each of the three activities appeared
as predecessor(s), successor(s) and concurrent with the core-activity across the
detected pattern set.

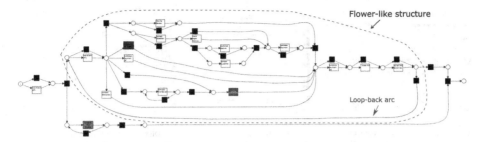

Fig. 11. Original process model discovered using Inductive Miner. The flower-like structure is outlined in blue dashed line, the arc indicating that the activities in the flower may be executed any number of times is shown in red. Corresponding colors have been applied to highlight the activities identified by the Episode Detector. Silent transitions are indicated by black boxes. (Color figure online)

By coloring these three activities in the Log Pattern Explorer visualization, we can see that these activities can happen almost in any order and at any point in time in a process (see Fig. 12(a)). We also note that, collectively, these three activities make up 68.5% of the total activities in the log. The relative positions of these three activities are shown in the discovered model. It is interesting to note that the activity *conduct file review* is not part of the flower structure in the discovered model but is part of a separate loop structure.

The LPE visualization provides an unfolded view of traces which clearly shows the interspersion of these three activities across all cases. This facet of behavior will be hidden in a process model (or at least, it will not be obvious to those who are not trained to interpret process models carefully). Regardless, the magnitude of the interspersion will not be revealed by the simple process model visualization.

The visualization of the spread of these activities strongly suggests that these activities do not have any strong temporal relationships with any specific activities in the process. In other words, these activities are likely to be 'auxiliary activities' that support the main process but are not temporally dependent on any other core claim process activities. In order to discover the main insurance process flow, we decided to remove these three auxiliary activities from the log to obtain the core process activities (though they may be relevant, for instance when analyzing performance).

Removal of Auxiliary Activities. After filtering the activities *conduct file review*, *incoming correspondence*, and *follow up requested*, we sought to find behaviors representative of core insurance processes. We did so by (i) sorting the activities by frequency, and (ii) coloring the (top 10) most frequent activities accordingly. Furthermore, we also color the two *New Claim (IPI)* and *New Claim (CP)* activities as they seem to be referring to two types of claims and often occurred at the beginning of each case. This coloring exercise provides an interesting view: the distribution of activities between those cases that were (generally) started with *New Claim (IPI)* and *New Claim (CP)* as shown in

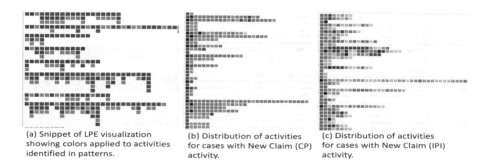

(a) Snippet of LPE visualization showing colors applied to activities identified in patterns.

(b) Distribution of activities for cases with New Claim (CP) activity.

(c) Distribution of activities for cases with New Claim (IPI) activity.

Fig. 12. Three snippets of LPE visualization showing different log patterns.

Fig. 12(b) and (c) is quite distinct. Armed with this newly-gained insight, we split the log into *Variant 1* (those with activity *New Claim (IPI)*), *Variant 2* (those with activity *New Claim (CP)*), and *Variant 3* (the remaining cases).

Results. The final process models that we obtained for each of the three variants exhibit significantly-reduced flower behaviors. Using the Projected Recall and Precision 3-activities metric [19]: the precision measure of the process model has increased from 0.47 (for the process model obtained after removing auxiliary events) to >0.7 (*Variant 1*), >0.9 (*Variant 2*), and >0.5 (*Variant 3*); the precision measure of the original model shown in Fig. 11 is 0.46.

Comparison to Unsupervised Approaches. Next, we compared our pattern detection approach to existing detection techniques regarding their ability to discover particular kinds of patterns. From the patterns discussed in the first case study, we tried to discover the 7 patterns listed in Table 2 using the 4 different unsupervised detection techniques [3,7,8,11] discussed in Sect. 2; The frequent pattern P0 (sequence of activities) in Fig. 5 could be identified by all techniques; the infrequent P1 (concurrent activities) in Fig. 5 can be detected by unsupervised techniques only when lowering threshold parameters (which usually leads to very large result sets); the extremely infrequent pattern P3 (C-cover < 0.005) in Fig. 8 can not be detected due to for example implementation limitations (e.g., the tool of [7] requires frequency (C-cover) ≥ 0.05; the tool of [11] computes 500 patterns maximally). P9a (combines directly- and eventually-follows relations) in Fig. 10 cannot be detected by [3] (only directly-follows relations); the other techniques may detect it but do not distinguish directly and eventually-follows (only eventually-follows). P9b (combines directly-follows, eventually-follows and concurrency) in Fig. 10 cannot be detected by existing techniques as these would have to enumerate the possible interleavings of the pattern activities with all other process activities to find the pattern. For P12anti, no existing unsupervised approach supports detecting anti-pattern instances. Overall, existing approaches have difficulties with detecting patterns that are infrequent or contain a large set of concurrent (independent) events. The other patterns detected by these approaches can be converted into the *log patterns* as discussed in Sect. 5 and be further explored in the LPE tool.

Table 2. Comparison of related pattern detection approaches: $(+)$ can detect; $(+/-)$ may detect with suitable parameters; $(-)$ cannot detect

	P0	P1	P3	P9a	P9b	P12	P12 (anti)
Pattern abstraction [3]	+	+	−	−	−	+/−	−
Episodes miner [7]	+	+/−	−	+/−	−	+/−	−
Pattern discovery [8]	+	+/−	−	+/−	−	+/−	−
Local models [11]	+	+	−	+/−	−	+/−	−
Log pattern explorer	+	+	+	+	+	+	+

7 Conclusion and Future Work

In this paper, we proposed a semi-supervised approach for log pattern detection. We defined our patterns as partial orders and distinguished a core-activity to help the user detect patterns of interest and count instances of patterns. We use concurrency and contextual information of the core-events and support the user in extracting, modifying, and extending patterns. The two case studies show that the proposed approach helps to (1) detect complex patterns and infrequent patterns of interest, (2) modify/explore patterns (detected by existing unsupervised approaches or by our approach) and (3) gain insights into the parts of discovered models that are unstructured or over-generalized. However, in contrast to fully automated detection, some manual efforts are needed in our approach. Moreover, in the log visualization, an overview of the overall behavior of the log is missing. A more extensive user study is needed to evaluate and improve the effectiveness of the approach. Future work aims at empirically evaluating the approach and the tool with more involvement of users and domain experts. Moreover, we would like to integrate log cleaning operations, such as event abstraction, event relabeling, event filtering etc., and recommend such operations for the patterns detected.

References

1. Leemans, S.J.J., Fahland, D., van der Aalst, W.M.P.: Discovering block-structured process models from event logs - a constructive approach. In: 2013 Proceedings of Application and Theory of Petri Nets and Concurrency, pp. 311–329 (2013)
2. Monroe, M., Lan, R., Lee, H., Plaisant, C., Shneiderman, B.: Temporal event sequence simplification. IEEE Trans. Vis. Comput. Graph. **19**(12), 2227–2236 (2013)
3. Jagadeesh Chandra Bose, R.P., van der Aalst, W.M.P.: Abstractions in process mining: a taxonomy of patterns. In: Dayal, U., Eder, J., Koehler, J., Reijers, H.A. (eds.) BPM 2009. LNCS, vol. 5701, pp. 159–175. Springer, Heidelberg (2009). doi:10.1007/978-3-642-03848-8_12
4. Günther, C.W., Rozinat, A., van der Aalst, W.M.P.: Activity mining by global trace segmentation. In: Rinderle-Ma, S., Sadiq, S., Leymann, F. (eds.) BPM 2009. LNBIP, vol. 43, pp. 128–139. Springer, Heidelberg (2010). doi:10.1007/978-3-642-12186-9_13

5. Mannhardt, F., de Leoni, M., Reijers, H.A., Aalst, W.M.P., van der Toussaint, P.J.: From low-level events to activities - a pattern-based approach. In: La Rosa, M., Loos, P., Pastor, O. (eds.) BPM 2016. LNCS, vol. 9850, pp. 125–141. Springer, Cham (2016). doi:10.1007/978-3-319-45348-4_8

6. Maggi, F.M., Montali, M., Westergaard, M., van der Aalst, W.M.P.: Monitoring business constraints with linear temporal logic: an approach based on colored automata. In: Rinderle-Ma, S., Toumani, F., Wolf, K. (eds.) BPM 2011. LNCS, vol. 6896, pp. 132–147. Springer, Heidelberg (2011). doi:10.1007/978-3-642-23059-2_13

7. Leemans, M., van der Aalst, W.M.P.: Discovery of frequent episodes in event logs. In: Ceravolo, P., Russo, B., Accorsi, R. (eds.) SIMPDA 2014. LNBIP, vol. 237, pp. 1–31. Springer, Cham (2015). doi:10.1007/978-3-319-27243-6_1

8. Diamantini, C., Genga, L., Potena, D.: Behavioral process mining for unstructured processes. J. Intell. Inf. Syst. 47(1), 5–32 (2016)

9. Suriadi, S., Andrews, R., ter Hofstede, A.H., Wynn, M.T.: Event log imperfection patterns for process mining: towards a systematic approach to cleaning event logs. Inf. Syst. 64, 132–150 (2017)

10. Ferreira, D.R., Szimanski, F., Ralha, C.G.: Improving process models by mining mappings of low-level events to high-level activities. J. Intell. Inf. Syst. 43(2), 379–407 (2014)

11. Tax, N., Sidorova, N., Haakma, R., van der Aalst, W.M.P.: Mining local process models. J. Innovation Digital Ecosyst. 3(2), 183–196 (2016)

12. Baier, T., Rogge-Solti, A., Mendling, J., Weske, M.: Matching of events and activities: an approach based on behavioral constraint satisfaction. In: SAC, pp. 1225–1230. ACM (2015)

13. Song, M., van der Aalst, W.M.P.: Supporting process mining by showing events at a glance. In: Proceedings of WITS, pp. 139–145 (2007)

14. Lu, X., Fahland, D., van der Aalst, W.M.P.: Conformance checking based on partially ordered event data. In: Fournier, F., Mendling, J. (eds.) BPM 2014. LNBIP, vol. 202, pp. 75–88. Springer, Cham (2015). doi:10.1007/978-3-319-15895-2_7

15. Ponce-de-León, H., Rodríguez, C., Carmona, J., Heljanko, K., Haar, S.: Unfolding-based process discovery. In: Finkbeiner, B., Pu, G., Zhang, L. (eds.) ATVA 2015. LNCS, vol. 9364, pp. 31–47. Springer, Cham (2015). doi:10.1007/978-3-319-24953-7_4

16. Mokhov, A., Carmona, J., Beaumont, J.: Mining conditional partial order graphs from event logs. In: Koutny, M., Desel, J., Kleijn, J. (eds.) Transactions on Petri Nets and Other Models of Concurrency XI. LNCS, vol. 9930, pp. 114–136. Springer, Heidelberg (2016). doi:10.1007/978-3-662-53401-4_6

17. Diamantini, C., Genga, L., Potena, D., van der Aalst, W.M.P.: Building instance graphs for highly variable processes. Exp. Syst. Appl. 59, 101–118 (2016)

18. Lu, X., et. al.: Semi-supervised log pattern detection and exploration using event concurrence and contextual information (extended version). BPM Center report BPM-17-01 (2017)

19. Leemans, S.J.J., Fahland, D., van der Aalst, W.M.P.: Scalable process discovery and conformance checking. Soft. Syst. Model. (2016)

Cloud of Things Modeling for Efficient and Coordinated Resources Provisioning

Elie Rachkidi[1,2]([✉]), Djamel Belaïd[1], Nazim Agoulmine[2], and Nada Chendeb[3]

[1] SAMOVAR, Télécom SudParis, CNRS, Université Paris Saclay,
9 Rue Charles Fourier, 91011 Evry Cedex, France
`djamel.belaid@telecom-sudparis.eu`
[2] COSMO, Université d'Evry Val d'Essonne, Université Paris Saclay,
23 Boulevard de France, 91000 Evry, France
`elie.rachkidi@ibisc.univ-evry.fr, nazim.agoulmine@ufrst.univ-evry.fr`
[3] Faculty of Engineering, Lebanese University, Beirut, Lebanon
`nchendeb@ul.edu.lb`

Abstract. The shift towards the Cloud of Things (CoT) requires a seamless integration of Cloud Computing and the Internet of Things (IoT). Such transition promotes a holistic approach for managing and orchestrating cloud and IoT infrastructures. However, existing platforms manage each domain individually. They provision cloud and IoT resources separately without a global vision on the underlying infrastructure and network state. Moreover, resource models provided by these platforms do not cope with the CoT requirements. In this paper, we study existing resource models for cloud and the IoT. Afterward, we select the Open Cloud Computing Interface (OCCI) specifications to represent the CoT environment on the Infrastructure as a Service (IaaS) and the Platform as a Service (PaaS) levels. We extend the chosen standard to model the CoT for IaaS and PaaS. We show through two example scenarios the benefit of our model in representing the CoT and performing an efficient global provisioning of incoming requests. Furthermore, we rely on the OCCI CoT extension to formulate an analytical model and show through a simulation the benefit of such global orchestration.

Keywords: Cloud Computing · Internet of Things · Cloud of Things · Modeling · Mapping · Provisioning · Optimization · OCCI

1 Introduction

The integration of Cloud Computing and the IoT can not be achieved without filling the gap between these two worlds. As today, the orchestration process of IoT applications on the cloud undergoes two separate stages: (a) cloud-related services provisioning, and (b) IoT devices selection and integration to the cloud. These two processes are executed in an arbitrary order. Cloud platforms manage the step (a) while cloud-based IoT platforms control the step (b). As a result, the orchestration process of IoT applications is not executed in a holistic

© Springer International Publishing AG 2017
H. Panetto et al. (Eds.): OTM 2017 Conferences, Part I, LNCS 10573, pp. 175–193, 2017.
https://doi.org/10.1007/978-3-319-69462-7_12

manner and the two stages are optimized individually without a global vision. We can then state that existing cloud-based IoT platforms do not achieve a complete integration of Cloud Computing and the IoT. There is a need to leverage them to build the Cloud of Things (CoT) vision. The CoT promotes a holistic management of both domains.

CoT platforms might manage in a homogeneous way resources hosted in cloud data centers as well as connected objects. They might intervene at all layers of the Cloud Computing namely: Infrastructure as a Service (IaaS), Platform as a Service (PaaS), and Software as a Service (SaaS). Moreover, CoT platforms provide IoT related service models such as the Sensing as a Service (S^2aaS) [19] as part of the PaaS and the SaaS. They should be able to orchestrate in one stage requests involving both Cloud Computing and IoT resources. Such integrated orchestration mechanism will permit a global optimization of the underlying infrastructure resources and eventually the introduction of new types of services. However, existing Cloud Computing and IoT management platforms are not able to describe such an integrated CoT resources model [4]. Also, they prevent one stage provisioning mechanism for CoT requests.

Therefore, there is a clear need to leverage existing resource models to design a new one that is able to support the CoT vision and its service provisioning models for the infrastructure (IaaS), the development environment (PaaS), and the CoT services (SaaS). It is worth noting that the SaaS and PaaS models are similar. Their difference will only be in the scope of the control each layer provides. This work will then focus on the proposition of a new model for CoT IaaS and PaaS. Also, it will provide a provisioning mechanism based on the proposed model to optimize in one stage the orchestration of a CoT request onto a CoT substrate. The CoT request consists of a description of interconnected cloud services and IoT data sources to be deployed. However, a CoT substrate is the graph model of the CoT infrastructure.

The paper is organized as follows. First, we present and discuss in Sect. 2 existing standards for modeling Cloud Computing and IoT resources. We select a suitable design pattern for the foundation of our CoT model. Secondly, we cite related works which contribute to the CoT modeling in Sect. 3. Then in Sects. 4 and 5, we study the CoT infrastructure and platform requirements and extend the core model[1] accordingly. In Sect. 6, we provide two example scenarios to show the capability of our proposed model to describe a CoT environment and perform one stage resources provisioning. Afterward, we present an analytical formulation of the placement problem based on the defined model in Sect. 7 and discuss simulations results in Sect. 8. Finally, we conclude the paper with an overview outlining key aspects of our contribution and discuss future works in Sect. 9.

---

[1] We use the Sparx Enterprise Architect v12 to Create the UML Diagrams.

2   Existing Models and Standards

To the best of our knowledge, there exist no resource-oriented standards for the CoT. Current works dealing with the convergence of Cloud Computing and the IoT are significantly different from one to another [4]. Moreover, proposed management platforms in each domain have distinct resource models and Application Programming Interfaces (APIs), even though they offer sometimes similar functionality. This heterogeneity of approaches that exist already in each domain has motivated several resource-oriented standardization initiatives to be launched. These models indeed do have part of the information that is required to build the CoT model since several concepts are similar. Therefore, the objective of this work was to analyze, reuse, extend, and adapt current standards to leverage them to the level that is required in the CoT.

2.1   Internet of Things Environment

IoT related specifications could be divided into two categories: sensor web and sensor semantic web. These groups of solutions are complementary and can work alongside one another.

Sensor Web Initiatives. The Sensor Web Enablement (SWE) [2] suite of specifications define models to describe sensor objects and their data. It is developed by the Open Geospatial Consortium (OGC). The main adopted SWE standards are the Sensor Model Language (SensorML), the Observations and Measurements (O&M) model, and the Sensor Observations Service (SOS) interface. The SensorML describes sensor systems capabilities, properties, measurements, and processes. The O&M offers an XML schema for encoding observed and measured data. The SOS interface describes methods for managing, discovering, requesting, filtering, and retrieving sensors descriptions and their generated data. The OGC derive also the SensorThings APIs which are lightweight and designed specifically for the IoT. It is based on existing SWE standards.

Similarly, the Devices Profile for Web Services (DPWS) [15] brings Web services to constrained devices. It is based on the W3C Web services standards. However, the DPWS uses a minimal set of Web services specifications to enable messaging, discovery, description, and eventing for low-powered devices. Moreover, it adopts the SOAP-over-UDP binding to minimize the connection overhead and to use multicast addressing for discovery mechanisms.

Sensor Semantic Web Initiatives. The W3C produced the Semantic Sensor Network (SSN) ontology [6]. It describes sensors (e.g. location, type), their properties (e.g. precision, resolution, unit), and their observed measurements (e.g. values). Furthermore, the SSN enables context awareness by linking sensors observations to features of interest and events. Hence, data evolve from a sensor measurement (e.g. temperature, humidity) to a part of a broader context (e.g. soil condition) enabling a higher level of knowledge representation.

However, the SSN does not capture the full extent of the IoT which stimulated several initiatives for creating an IoT ontology. Although many ontologies were designed for the IoT, not all of them apply good practices as discussed in [18]. In this perspective, Authors in [18] present the Internet of Things Ontology (IoT-O) which respects ontology design rules and defines some missing concepts relevant to the IoT such as `Thing`, `Actuator`, and `Actuation`. Although the IoT-O is not an approved standard, it aligns with existing ontologies. Similarly, the W3C is integrating in an ongoing work the SSN and the Sensor, Observation, Sample, and Actuator (SOSA)[2] ontologies for a better representation of the IoT.

2.2 Cloud Infrastructure Management Initiatives

Cloud Computing standards deal with management interfaces (e.g. CIMI[3]), applications portability (e.g. TOSCA[4]), virtual appliances packaging (e.g. OVF[5]), and many others. In our work, we focus on standards providing a comprehensive modeling of Cloud Computing resources. Therefore, we study current cloud resources management interfaces. We identify the Cloud Infrastructure Management Interface (CIMI), the Open Cloud Computing Interface (OCCI)[6], and the Cloud Application Management for Platforms (CAMP) [8].

Cloud Infrastructure Management Interface. The CIMI standard is developed by the Distributed Management Task Force (DMTF) and describe the IaaS model in Cloud Computing. It defines a data model and a RESTful communication protocol for cloud platforms. The CIMI represents the cloud infrastructure as a set of resources based on key entities managed at the IaaS layer (e.g. machines, storage volumes). Each resource has three forms: (a) template resource, (b) configuration resource, and (c) resource instance. Firstly, the template resources are the operator's predefined catalog of offerings. Secondly, a configuration resource is the client's modified version of an existing template. Finally, a resource instance is a deployed and running resource in the cloud infrastructure. The CIMI defines a root endpoint (i.e. `CloudEntityPoint`) which describes and locates the resources available in the cloud infrastructure.

Open Cloud Computing Interface. The OCCI is a set of specifications delivered by the Open Grid Forum (OGF) and lead by community contributions. The OCCI defines a RESTful protocol and API for resources management frameworks. An extensible and domain independent core model is at the heart of the OCCI specifications. It can be extended to describe various resources, relations, and possible actions on both. At first, the OCCI was used to create

---

[2] http://www.w3.org/ns/sosa/.

[3] https://www.dmtf.org/standards/cmwg.

[4] https://www.oasis-open.org/committees/tosca/.

[5] https://www.dmtf.org/standards/ovf.

[6] http://occi-wg.org/.

an interface for managing IaaS [13] resources for cloud platforms. Such initiative enabled IaaS level operations such as the deployment of virtual machines via standard APIs. Then, the community extended existing models to manage additional aspects of the cloud (e.g. PaaS [21], IaaS monitoring and automatic scaling [14]).

Cloud Applications Management Initiatives. The Organization for the Advancement of Structured Information Standards (OASIS) advances the CAMP specification. It describes APIs, models, mechanisms, and protocols for packaging and deploying applications in the cloud. It enables a cloud provider independent interface to perform PaaS level activities such as provisioning, monitoring, and control on applications. Therefore, the CAMP eliminates vendor lock-in by providing a standardized application description model. CAMP promotes interoperability among PaaS clouds by specifying artifacts and defining APIs which are necessary to manage the building, running, administration, monitoring and patching of applications in any PaaS cloud. Accordingly, it defines platform level resources such as `assembly`, `service`, and `component`.

2.3 Synthesis of Existing Works on Cloud and IoT Models and Standards

Sensor and semantic web models focus on describing connected objects capabilities and represent their data in a unified manner. They model these resources on the application level (i.e. equivalent to PaaS and SaaS in the cloud). However, they achieve their objectives with different technologies. Sensor web standards use syntactic approaches to enforce interoperability while semantic web specifications rely on technology standards defined by the W3C (e.g. Resource Description Framework (RDF), Ontology Web Language (OWL)) to design and describe data on the web. Syntactic approaches are rigid and difficult to extend. In particular, sensor web approaches are tightly coupled and focus on modeling data types rather than device networks. They are not flexible nor easily extensible. On the contrary, ontologies are more flexible and able to describe complex systems. Moreover, they rely on the W3C standards to encode defined classes, properties, and relations. However, extending ontologies requires some effort to validate new classes and properties alignment with existing definitions.

Separately, the CIMI focuses on the IaaS level, the CAMP targets the PaaS, while OCCI specifications cover both. Furthermore, the OCCI IaaS specification is adopted more than the CIMI model [1] by cloud management frameworks (i.e. OpenStack) and European projects (e.g. OCCIware[7], EGI[8]). The OCCI provides an extensible core which enables the representation of loosely coupled entity-relationship models for different domains. On the contrary, CIMI and CAMP standards provide rigid with tightly coupled resources and links definitions. Furthermore, they define complete data model structures in contrast to the OCCI

---

[7] http://www.occiware.org/.

[8] https://www.egi.eu/.

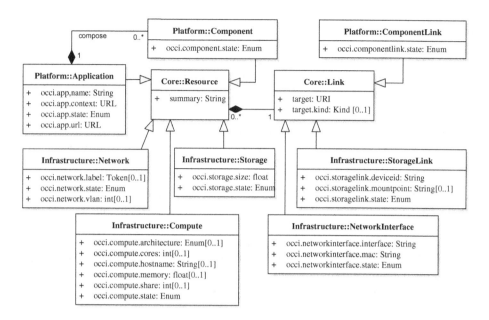

Fig. 1. OCCI infrastructure and platform specifications.

which introduces the `Mixin`. The `Mixin` class enables augmenting a given data model with additional capabilities without affecting its initial structure.

The OCCI provides a high extensibility in contrast to other cloud specifications and IoT related standards. Also, it is the only analyzed standards which describes PaaS and IaaS service models simultaneously. Therefore, we select the OCCI specifications to model the CoT environment. However, we reuse cloud and IoT specifications when necessary to stay aligned with predefined concepts and vocabulary. Figure 1 illustrates current OCCI IaaS and PaaS specifications.

3   Related Works

Ciuffoletti et al. [5] define an `Aggregator` and `RealWorldObject` `Resources` with a `Sensor` `Link`. They consider that one or multiple sensors observe a real world object and report measured information to aggregators. They perform an orchestration scenario to show how their model can be used to deploy a CoT request in a converged cloud and IoT environments. However, their approach do not consider actuators nor represent the IaaS service model in the CoT environment.

Authors in [12] define an extension of the OCCI model to fill the gap between the cloud and the robotics world. The work focuses on developing an OCCI enabled gateway to hide underlying heterogeneous and mobile robots. Unlike our approach, they consider solely a gateway integration pattern (see Sect. 4) and define a domain specific extension.

4   Cloud of Things Infrastructure Model

IoT devices and cloud data centers have different hardware properties. The CoT should capture these differences and enable a unified description of the infrastructure. In this section, we describe the OCCI CoT infrastructure model.

4.1   Network Graph Model

The provisioning process in the CoT is similar to the Virtual Network Embedding (VNE) problem described in [9]. The set of requested virtual machines, sensors, actuators, and the relations between them can be considered as the Virtual Network (VN). Furthermore, the set of cloud data centers, IoT devices, and network links can be mapped as the Substrate Network (SN). Hence, the OCCI CoT should enable a network graph representation of its resources.

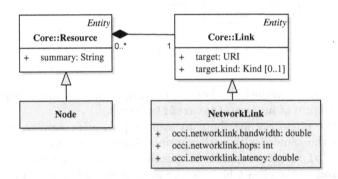

Fig. 2. OCCI extensions (colored boxes) to enable a CoT graph representation.

A graph is a set of interconnected nodes. In the CoT, nodes represent cloud data centers and IoT devices. Moreover, edges are network links connecting these nodes. Hence, we define Node and NetworkLink entities as shown in Fig. 2. A Node instance uses Link items to identify its available and provided resources (e.g. Compute, Component). A NetworkLink instance connects two Node objects and provides their network connection attributes such as latency (NetworkLink.latency) and bandwidth (NetworkLink.bandwidth).

4.2   Sensor and Actuator Resources

Sensors and actuators are the main functional blocks of the IoT. Sensors observe a physical or logical object property and provide therefore sensing resources. However, actuators act on a property of an object. Hence, they offer actuating resources. We represent such connected objects capabilities by the Sensor and Actuator resources as illustrated in Fig. 3. The sensor and actuator entities are presented in IoT standards under different classes but with similar attributes.

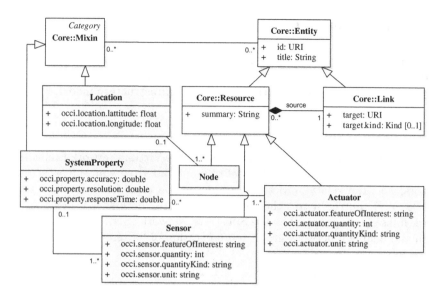

Fig. 3. Extensions (colored boxes) of the OCCI infrastructure for the Cloud of Things.

These entities are identified by their type (`quantityKind`), measurement unit (`unit`), and feature of interest (`featureOfInterest`).

We add the `quantity` attribute to avoid representing similar sensors and actuators on the same device with multiple `Sensor` and `Actuator` instances. Furthermore, IoT resources operate in specific geographical areas. Therefore, we use a `Location Mixin` class which can be associated with sensors or actuators. A location point has `Location.longitude` and `Location.latitude` attributes.

Previously described properties represent functional requirements. However, several non-functional requirements such as accuracy, response time, and resolution can be associated with sensors and actuators. The SSN ontology considers these requirements as subclasses of the `SystemProperty` class. We did not model all possible non-functional requirements, however, we illustrate how additional attributes are added with a `SystemProperty Mixin`.

4.3   Things Virtualization

IoT devices have different capabilities regarding virtualization resulting in distinct characteristics, thus the need to specify how to model this aspect in the CoT. It is possible that an IoT device does not retain any virtualization capabilities incurring no pool of available compute and storage resources. However, it still provides sensing and actuating resources through an API. Another type of possible virtualization is the network-level virtualization [11]. This kind of virtualization aims to create sub-networks by deploying an overlay network on top of connected objects or by physically clustering a group of IoT devices. The latter virtualization type is performed during the deployment process and does not affect the resources provisioning step thus not considered in the model.

The last kind is the node-level virtualization. It depends on the virtualization technology supported by the node's hardware. It consists of: (a) hypervisor-based, (b) OS-based containers, (c) application-based containers, and (d) threading virtualization. On the one hand, hypervisor-based and OS-based containers virtualization are on the infrastructure level. Nodes capable of such virtualization are able to provide computing, storage, networking, sensing, and actuating resources. On the other hand, the application-based containers and threading virtualization are on the platform service level. Therefore, related nodes do not have virtualized infrastructure resources such as compute and storage. However, they can host concurrent software components (e.g. containers, threads). Hence, they are linked to one or more `Component` resources.

4.4   Things Integration Patterns

Integration patterns represent different ways IoT devices connect to the Internet and get accessed by third party applications. There exist three integration patterns: (a) direct connectivity, (b) gateway-based connectivity, and (c) cloud-based connectivity. A direct connectivity means that the IoT device has an Internet Protocol (IP) address and is able to communicate directly over the Internet. The device might be a high-powered connected object with computing, storage, sensing and actuating resources (e.g. Raspberry PI). However, it might be also a constrained device offering sensing and actuating resources via a RESTful or WS interface with no virtualized resources.

Gateway-based connectivity represents connected objects in a non-IP network (e.g. IEEE 802.15.4). These IoT devices lie behind a gateway which performs protocol translation between the Internet and its internal network. Attached IoT devices might be uniquely identified and accessible from the Internet or hidden behind the gateway. In the first case, we represent them individually in the infrastructure (i.e. each one is a `Node` instance). In the second case, there is no need to model them independently. Therefore, we represent the gateway as the association of attached connected objects. It describes all the sensing and actuating resources provided by abstracted IoT devices. Such aggregation of available resources reduces the number of nodes within the infrastructure. Likewise, the cloud-based connectivity consists of a virtual gateway in the cloud. Its representation is identical to gateway-based connectivity in both cases.

5   Cloud of Things Platform Model

A PaaS level request in the cloud consists of interconnected atomic components. However, a CoT PaaS application requires data streams from connected objects towards some of its components. Therefore, additional aspects such as data delivery methods, IoT devices selection, and many others should be considered during the CoT PaaS provisioning process. In this section, we describe the OCCI CoT platform model which enables such operations.

5.1 Cloud of Things Deployment Options

Consuming IoT devices requires deploying device-specific services of `DeviceComponent` to access IoT resources. These components might be deployed locally (i.e. on-device) or on remote hardware such as gateways to access IoT devices (e.g. Wireless Sensor Network (WSN)). However, these services operate on low-powered devices with specific hardware architecture and expose a proprietary interface. Therefore, IoT middleware offered by `CollectorComponents` abstract the latter device-specific services and provide their functionalities through standard APIs. IoT middleware are hosted on powerful hardware (e.g. high-powered devices, cloud servers), therefore, they can perform additional processing such as aggregation, filtering, access control, etc. They also aim to manage non-functional aspects for device-specific services. Several IoT middleware were introduced in the literature [16]. The `DeviceComponent` and `CollectorComponent` extensions are defined as `Mixins` for the OCCI `Component` as shown in Fig. 4.

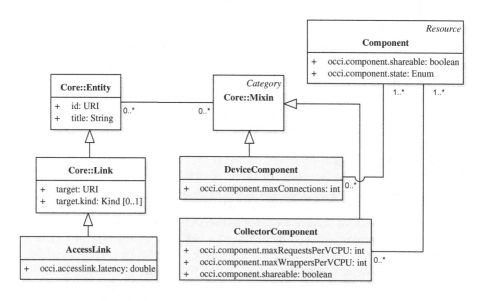

Fig. 4. Extensions (colored boxes) of the OCCI platform for the Cloud of Things.

It's worth mentioning that since the device-specific services have proprietary APIs, establishing a connection between these services and IoT middleware can be challenging. However, multiple researchers have already addressed this issue and provided valuable solutions. For example, the GSN middleware [3,17] generates device-specific wrappers relying on devices defined description files. These wrappers are then used to connect IoT devices to the GSN middleware. Furthermore, recent specifications for constrained devices such as DPWS enable more standardized and defined interfacing with constrained devices [10]. In our model, we use the `ComponentLink` to indicate whether a `DeviceComponent` connects to a particular `CollectorComponent` or not.

As a result, provisioning data streams for a CoT PaaS request consists of orchestrating `CollectorComponents` and `DeviceComponents` on available Cloud Computing and IoT resources respectively. Hence, such provisioning enables a one-stage mapping for all resources types. Each `Node` is linked to the `CollectorComponent` and `DeviceComponent` components it can host.

5.2   Data Components Sharing

In CoT, data streams might be reused to maximize resources utilization and minimize energy consumption. Several works such as [7] show that sharing sensing and actuating resources among various applications increases usage efficiency. Hence, `CollectorComponents` instances which already exist in the CoT environment might be shared among multiple applications. Existing `CollectorComponents` result from previous deployed requests or from instances provided by third party data providers. These instances can be partially or completely reused by incoming requests based on their sensing/actuating needs. Furthermore, reusing existing components speed up the provisioning and deployment process.

As a result, deployed `CollectorComponent` and `DeviceComponent` entities should be considered during the provisioning process. The `Component.state` attribute specifies whether components are instantiated or not. Also, these components should be shareable among applications. However, a newly requested application might require IoT resources which are not entirely managed by existing `CollectorComponents`. In such case, it is possible to utilize already deployed services and add missing resources to them. Therefore, the provisioning process should be aware of existing services and their maximal capacity.

We add the `CollectorComponent` attributes to extend the capabilities of the existing `Component` class. The `CollectorComponent` includes the `shareable` attribute to indicate whether an instance is shareable or not. Furthermore, the capacities of the component are given by `maxRequestsPerVCPU` and `maxWrappersPerVCPU` attributes. They represent the maximal number of data streams the component can collect from devices or dispatch to applications based on its allocated infrastructure resources. In some cases, a connected object might be linked to two different `CollectorComponents`. Therefore, we need to know how much connections it can accept. If it only accepts one connection, it means the connected object cannot be shared among IoT middleware. We represented the `DeviceComponent` `Mixin` to describe IoT devices related `Components`.

In addition, we attach deployed `CollectorComponent`s to their managed sensing and actuating resources via `AccessLink` instances. Therefore, the provisioning process verifies directly available resources at a collector component without scanning the entire CoT graph for its allocated IoT resources.

6   Example Scenarios

The extended OCCI model enables representing CoT IaaS and PaaS requests/substrates as network graphs. Also, it facilitates one stage mapping of incoming requests on available resources. We illustrate two example scenarios to show the ability of the proposed model to describe the CoT.

6.1   Resources Description

We define only a substrate graph for the demonstration, however, creating a CoT request graph is similar. The main difference is that a request graph represents needed resources, while a substrate graph describes available resources. Figure 5 illustrates the needed OCCI instances and their relations for representing a CoT substrate.

The described infrastructure graph corresponds to two physical nodes `infra_n1:Node` and `infra_n2:Node`. The first one is a Raspberry PI with four cores and 1 GB of RAM. The other one is a cloud server. The Raspberry PI

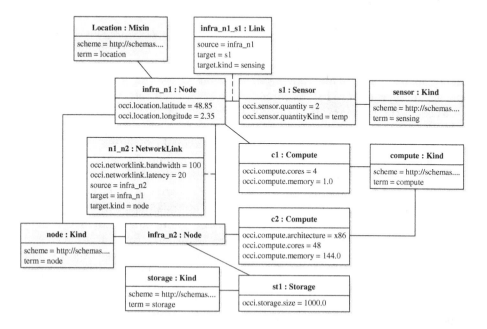

Fig. 5. CoT substrate graph description with the proposed OCCI CoT model.

is a directly accessed IoT device with 2 connected temperature sensors, while the cloud server has a 1000 GB storage unit. We can associate nodes with the Location Mixin to extend their attributes. Furthermore, both physical nodes are connected with a 100 Gbps network link having a latency of 20 ms. In this case, the Raspberry PI can also be associated with an on-device Component (e.g. RaspTemp). The latter Component might be augmented with a DeviceComponent Mixin instance to indicate that the RaspTemp is able to manage up to 10 concurrent connections (i.e. occi.component.maxConnections = 10).

6.2   Mapping Process

Moreover, the proposed OCCI CoT model enables mapping requests onto a CoT substrate. We represent an IaaS level mapping. However, a PaaS level modeling is similar but requires the additional Component resource. In order to simplify the mapping example in Fig. 6, we represented graphically the OCCI CoT instances and their relations. We give a brief description for nodes as well. It includes their Node.id attribute and attached infrastructure resources. Moreover, NetworkLink instances are presented as inter-nodes connectors.

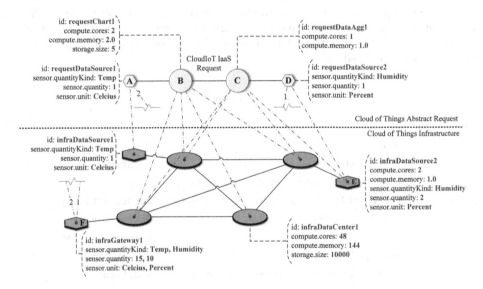

Fig. 6. Mapping Process Example Scenario for CoT IaaS request onto CoT infrastructure.

In the mapping process, candidate hosts are identified for each requested node based on infrastructure resources (e.g. and components availability for PaaS). We noticed that nodes requiring sensing and actuating resources are mapped towards IoT devices and gateways. However, nodes demanding more computational power are mapped to cloud servers. Moreover, some nodes are mapped to several hosts types. For example, the request node C (i.e. `requestDataAgg1`) has an additional resource constrained connected object (i.e. `infraDataSource2`) candidate. Consequently, Cloud Computing and IoT requests are combined, and they can be mapped to available hosts in both domains infrastructures simultaneously. Hence, we achieved a one-stage mapping.

Furthermore, the `infraGateway1` is exposing temperature and humidity sensors collectively, thus, minimizing the problem size. Otherwise, request nodes A and B would have been mapped to additional candidate hosts. In large-scale IoT infrastructures, such aggregation of available resources reduces considerably the problem size and the solution computation time. Hereafter, an efficient selection process is needed to pick best candidates for hosting the CoT request.

7   Cloud of Things Placement Problem Formulation

Let's consider a CoT substrate graph (N^s, E^s) where N^s is the set of substrate nodes and E^s is the set of edges interconnecting these nodes. Moreover, the CoT request graph is represented as (N^v, E^v). The set N^s has a cardinality m, while N^v has p requested nodes. Table 1 summarizes the notations used in this section and shows its equivalence in the proposed OCCI CoT model. Notations having the s superscript portray elements in the SN and those with the v superscript represent VN variables. For example, a compute resource c_i can be offered by an substrate node n_i^s and therefore represented as c_i^s, or requested by a virtual node n_i^v and annotated as c_i^v.

Table 1. Notation table

Parameter	Definition	OCCI CoT equivalence
n_i	Node i	`Node`
g_i	Location of node n_i	`Location`
c_i	Amount of compute resources attached to n_i	`Compute`
$o_{i,j}$	Amount of IoT resource i attached to node n_j	`Sensor/Actuator`
$e_{i,j}$	Edge between nodes n_i and n_j	`NetworkLink`
$l_{i,j}$	Latency of edge $e_{i,j}$	`networklink.latency`
$b_{i,j}$	Bandwidth of edge $e_{i,j}$	`networklink.bandwidth`
$s_{i,j}^s$	j^{th} service available at substrate node n_i^s	`Component`
s_i^v	Service requested by node n_i^v	`Component`

A substrate node n_i^s might be a cloud data center or an IoT device with a location g_i^s and an amount of available compute (c_i^s) and IoT resources. We represent available sensing and actuating resources of type i at substrate node n_j^s as $o_{i,j}^s$. We consider a total amount u of available connected objects types in the infrastructure. Furthermore, an edge $e_{i,j}^s$ connects two nodes n_i^s and n_j^s. An edge $e_{i,j}^s$ has an available bandwidth $b_{i,j}^s$ and a network latency $l_{i,j}^s$.

We formulate the problem as a Linear Programming (LP) model. The variable $\alpha_{i,j}$ represents the mapping of a node n_i^v to the host n_j^s. We use a numeric variable to enable mapping sensing or actuating resources to different gateways and devices at the same time. We introduce the $y_{i,j,v,w}$ to linearize product operations similarly to [20]. Furthermore, we introduce prices for computing (ϕ_i^c), networking ($\phi_{i,j}^e$), sensing, and actuating ($\phi_{i,j}^o$) resources. ϕ_i^c is the price of a compute unit at data center n_i^s. $\phi_{i,j}^o$ represents the price of the sensing/actuating unit o_i^s at data center n_j^s. $\phi_{i,j}^e$ corresponds to the price of a data unit on edge $e_{i,j}$. Our objective is to map a CoT request graph (N^v, E^v) onto a CoT substrate graph (N^s, E^s) while minimizing the resource cost. Therefore, we define the LP model as follows:

$$\min_{\alpha} F(\alpha) = \beta F^c(\alpha) + \gamma F^o(\alpha) + \sigma F^n(\alpha) \quad \textbf{s.t.} \ \beta + \gamma + \sigma = 1 \qquad (1)$$

With:

$$F^c(\alpha) = \sum_i \sum_j (\alpha_{i,j} c_i^v \phi_j^c + s_i^v \phi_j^s) \quad \forall i : 1 \to p, \quad \forall j : 1 \to m \qquad (1a)$$

$$F^o(\alpha) = \sum_i \sum_j \sum_k \alpha_{i,j} o_{i,k}^v \phi_{j,k}^o \forall i : 1 \to p, \quad \forall j : 1 \to m, \quad \forall k : 1 \to u \qquad (1b)$$

$$F^n(\alpha) = \sum_i \sum_j \sum_v \sum_w y_{i,j,v,w} b_{i,v}^v \phi_{j,w}^n \forall i, v : 1 \to p, \quad \forall j, w : 1 \to m \qquad (1c)$$

Subjected to :

$$\sum_i \sum_v y_{i,j,v,w} b_{i,v}^v \leq b_{j,w}^s \quad \forall j, w : 1 \to m \qquad (1d)$$

$$\sum_j \sum_w y_{i,j,v,w} l_{j,w}^s \leq l_{i,v}^v \quad \forall i, v : 1 \to p \qquad (1e)$$

$$\sum_i \alpha_{i,j} (c_i^v - c_j^s) \leq 0 \quad \forall j : 1 \to m \qquad (1f)$$

$$\sum_i \alpha_{i,j} o_{i,k}^v \leq o_{j,k}^s \quad \forall j : 1 \to m, \quad \forall k : 1 \to u \qquad (1g)$$

$$\alpha_{i,j} (g_j^s - g_i^v) = 0 \quad \forall i : 1 \to p, \quad \forall j : 1 \to m \qquad (1h)$$

$$\alpha_{i,j} (s_{i,j}^s - s_i^v) = 0 \quad \forall i : 1 \to p, \quad \forall j : 1 \to m \qquad (1i)$$

$$\sum_j \alpha_{i,j} = 1 \quad \forall j : 1 \to m \qquad (1j)$$

$$\alpha_{i,j} \geq 0 \quad \forall i : 1 \to m, \forall j : 1 \to p \qquad (1k)$$

$$\sum_j y_{i,j,v,w} = \alpha_{v,w} \quad \forall i, v : 1 \to m, \quad \forall j : 1 \to p \qquad (1l)$$

$$y_{i,j,v,w} = y_{v,w,i,j} \quad \forall i, v : 1 \to m, \quad \forall j, w : 1 \to p \qquad (1m)$$

F^c (1a), F^o (1b), and F^n (1c) correspond to the compute, IoT, and network resources costs respectively. Firstly, constraints (1d), (1f), and (1g) make sure that requested nodes are mapped to substrate nodes with sufficient resources. Secondly, constraint (1e) preserves the Quality of Service (QoS) requested by the CoT request regarding latency. Also, constraints (1h) and (1i) enforce each virtual node requested geographical location and verify its component availability. Finally, constraints (1j) and (1k) verify that all requested resources are mapped and satisfied while (1l) and (1m) define the relation between $\alpha_{i,j}$ and $y_{i,j,v,w}$.

8   Simulation and Results

We use IBM CPLEX to implement the proposed model in Java. The simulation compares between a global one stage mapping in the CoT and a two stages mapping of Cloud Computing and IoT resources. We generate CoT request graphs with $p = 1 \rightarrow 15$ requesting a total number $a = 20 \rightarrow 300$ of connected objects. The created CoT graphs are then divided into separate cloud and IoT graphs which are also mapped with the proposed model. This is possible since the model takes into consideration cloud and IoT resources simultaneously. This way, we simulate a two stages mapping. We measure the gain of both orchestration approaches. Results are shown in Fig. 7.

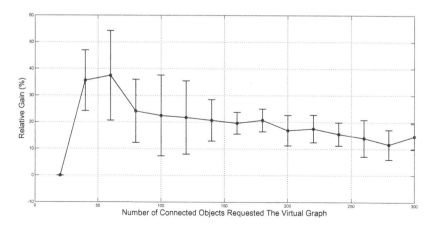

Fig. 7. The gain of a one stage mapping over a two stage mapping in the CoT.

We notice the benefit of mapping the CoT graph with a global representation of the infrastructure. This difference is mainly due to the cost of data units exchanged between cloud and IoT nodes. During a two stages mapping, such information cannot be accounted for. Either nodes in both graphs are mapped separately, or one of the two mappings uses the results of the other to coordinate its nodes placement. However, in the latter cases, at least one of the mappings will aim for less expensive hosts while ignoring the network cost. Hence, when the number of connected objects increase, the transmitted data between the cloud data centers and IoT devices increases, leading to additional costs. Our model captures the network state between cloud and IoT resources and therefore is able to adapt the orchestration of resources accordingly to minimize the overall cost. When the number of nodes increases, the gain decreases because the one stage mapping process selects more costly nodes when the network links bandwidth is not sufficient. However, the two stages mapping do not consider these constraints because it is not aware of the bandwidth consumption between IoT and cloud resources. Hence, it always maps to the less costly substrate nodes.

9 Conclusion and Perspectives

Combining Cloud Computing and the Internet of Things evolved over the years. Nowadays, IoT components were migrated to the cloud to benefit from its characteristics (e.g. elasticity). However, sensing and actuating resources were still managed separately from cloud resources. In fact, provisioning mechanisms were divided into two stages: (a) the cloud services orchestration, and (b) the IoT resources selection and integration. Such partitioning prevented a global optimization of both domains infrastructures. Therefore, the next logical evolution of such convergence is the unified management of Cloud Computing and IoT resources. The seamless integration of both domains is referred to as the Cloud of Things. In this paper, we modeled the CoT resources on the IaaS and PaaS levels. Our objective is to enable a joint management and provisioning of underlying resources. Hence, we targeted one stage provisioning for achieving an overall optimization of CoT resources. In this perspective, we studied existing standards for the cloud and the IoT to identify the most suitable one for modeling the CoT. We selected the OCCI standard due to its extensibility and comprehensive modeling of the cloud. We extended the existing OCCI model with respect to the CoT characteristics for IaaS and PaaS. Also, we showed through two example scenarios how the proposed model enables the representation of CoT graphs and one stage mappings. Furthermore, we used the proposed model to formulate an analytical model for the CoT request placement problem. A holistic provisioning process proves to be 10%–20% more efficient than two separate orchestration processes for cloud and IoT resources respectively. In the future, we intend to optimize further the analytical model to enable large scale deployments of IoT applications. Furthermore, we will provide an implementation for the proposed information model and demonstrate an end-to-end deployment of IoT applications in a CoT substrate.

References

1. Boutaba, R., da Fonseca, N.L.S.: Cloud Services, Networking, and Management, vol. 53. Wiley, Hoboken (2015)
2. Broring, A., Echterhoff, J., Jirka, S., Simonis, I., Everding, T., Stasch, C., Liang, S., Lemmens, R.: New generation sensor web enablement. Sensors 11(12), 2652–2699 (2011)

3. Calbimonte, J.p., Sarni, S., Eberle, J., Aberer, K.: XGSN: an open-source semantic sensing middleware for the web of things. In: 7th International Workshop on Semantic Sensor Networks (2014)
4. Cavalcante, E., Pereira, J., Alves, M.P., Maia, P., Moura, R., Batista, T., Delicato, F.C., Pires, P.F.: On the interplay of Internet of Things and cloud computing: a systematic mapping study. Comput. Commun. **89–90**, 17–33 (2016)
5. Ciuffoletti, A.: OCCI-IOT: an API to deploy and operate an IoT infrastructure. IEEE IoT J. **4662**(c), 1 (2017)
6. Compton, M., Barnaghi, P., Bermudez, L., Garcia-Castro, R., Corcho, O., Cox, S., Graybeal, J., Hauswirth, M., Henson, C., Herzog, A., Huang, V., Janowicz, K., Kelsey, W.D., Le Phuoc, D., Lefort, L., Leggieri, M., Neuhaus, H., Nikolov, A., Page, K., Passant, A., Sheth, A., Taylor, K.: The SSN ontology of the W3C semantic sensor network incubator group. Web Semant. Sci. Serv. Agents World Wide Web **17**, 25–32 (2012)
7. Dinh, T., Kim, Y.: An efficient interactive model for on-demand sensing-as-a-services of sensor-cloud. Sensors **16**(7), 992 (2016)
8. Durand, J., Otto, A., Pilz, G., Rutt, T.: Cloud application management for platforms version 1.2. Technical report, OASIS Committee Specification Draft 01 (2017)
9. Fischer, A., Botero, J.F., Beck, M.T., De Meer, H., Hesselbach, X.: Virtual network embedding: a survey. IEEE Commun. Surv. Tutor. **15**(4), 1888–1906 (2013)
10. Han, S.N., Khan, I., Lee, G.M., Crespi, N., Glitho, R.H.: Service composition for IP smart object using realtime web protocols: concept and research challenges. Comput. Stan. Interfaces **43**, 79–90 (2016)
11. Khan, I., Belqasmi, F., Glitho, R., Crespi, N., Morrow, M., Polakos, P.: Wireless sensor network virtualization: a survey. IEEE Commun. Surv. Tutor. **18**(1), 553–576 (2016)
12. Merle, P., Gourdin, C., Mitton, N.: Mobile cloud robotics as a service with OCCI-ware. In: Proceedings of the 2nd IEEE International Congress on Internet of Things, IEEE ICIOT 2017, vol. 1 (2017)
13. Metsch, T., Edmonds, A., Parák, B.: Open cloud computing interface - infrastructure. Technical report, Open Grid Forum (2016). http://ogf.org/documents/GFD.224.pdf
14. Mohamed, M., Belaïd, D., Tata, S.: Extending OCCI for autonomic management in the cloud. J. Syst. Softw. **122**, 416–429 (2016)
15. Moritz, G., Zeeb, E., Prüter, S., Golatowski, F., Timmermann, D., Stoll, R.: Devices profile for web services and the REST. In: IEEE International Conference on Industrial Informatics (INDIN), pp. 584–591 (2010)
16. Ngu, A.H.H., Gutierrez, M., Metsis, V., Nepal, S., Sheng, M.Z.: IoT middleware: a survey on issues and enabling technologies. IEEE IoT J. **4**(1), 1 (2016)
17. Perera, C., Zaslavsky, A., Christen, P., Salehi, A., Georgakopoulos, D.: Connecting mobile things to global sensor network middleware using system-generated wrappers. In: Proceedings of the 11th ACM International Workshop on Data Engineering for Wireless and Mobile Access, MobiDE 2012, p. 23. ACM, New York (2012)
18. Seydoux, N., Drira, K., Hernandez, N., Monteil, T.: IoT-O, a core-domain iot ontology to represent connected devices networks. In: Blomqvist, E., Ciancarini, P., Poggi, F., Vitali, F. (eds.) EKAW 2016. LNCS, vol. 10024, pp. 561–576. Springer, Cham (2016). doi:10.1007/978-3-319-49004-5_36
19. Sheng, X., Tang, J., Xiao, X., Xue, G.: Sensing as a service: challenges, solutions and future directions. IEEE Sens. J. **13**(10), 3733–3741 (2013)

20. Stefanello, F., Aggarwal, V., Buriol, L.S., Resende, M.G.C.: Hybrid algorithms for placement of virtual machines across geo-separated data centers. Eur. J. Oper. Res. (2016)
21. Yangui, S., Tata, S.: An OCCI compliant model for paas resources description and provisioning. Comput. J. **59**(3), 308–324 (2016)

A Framework for Integrating Real-World Events and Business Processes in an IoT Environment

Sankalita Mandal[(✉)], Marcin Hewelt, and Mathias Weske

Business Process Technology Group, Hasso Plattner Institute,
University of Potsdam, Potsdam, Germany
{sankalita.mandal,marcin.hewelt,mathias.weske}@hpi.de

Abstract. Business process management is essential for companies to document, execute, monitor, and optimize their business processes. These processes are often influenced by external events occurring in the process context, especially when considering Internet of Things (IoT) scenarios. Modeling constructs for different types of events are part of the Business Process Model and Notation (BPMN) standard. However, when the integration of external events needs to be supported by process-oriented information systems, the gap between conceptual process model and its implementation needs to be bridged. We elicited the requirements for this integration using an use case from the IoT domain. Based on them, we propose a framework that outsources the management of events to an event processing platform that the process engine subscribes to. The BPMN process model is extended with annotations to specify the type of expected events. Further, we implement a system that realizes the proposed integration..

Keywords: Process execution · Event processing · BPMN

1 Introduction

Business processes are omnipresent in companies. In today's digital age, huge amount of data is being produced every moment and processes try to take advantage of those streams of dynamic event data. In an Internet of Things (IoT) scenario, sources like smart devices and sensors generate tons of events, which can be filtered, combined, and aggregated to trigger and drive business processes. Proper aggregation and analysis of the events makes the processes more flexible, robust, and efficient. BPMN 2.0 (Business Process Model and Notation) offers a rich variety of constructs to model different types of events, e.g. start, intermediate, and end events that can be further differentiated into throwing and catching events

To support a business process with IT, e.g. by enacting it in a process engine, the gap between the conceptual level of the model and the detailed, technical level required for running the process needs to be bridged. For catching events, which represent that a process instance waits and reacts to some environmental

© Springer International Publishing AG 2017
H. Panetto et al. (Eds.): OTM 2017 Conferences, Part I, LNCS 10573, pp. 194–212, 2017.
https://doi.org/10.1007/978-3-319-69462-7_13

occurring, this means to specify how this event can be detected by the process engine, how its information is extracted and mapped to process variables, and how it is correlated to the correct process instance. Throwing events are produced by the process instance, hence they do not need to be detected and correlated. However, process variables have to be packaged into the produced event. The gap between process model and its implementation hinders the fast deployment and subsequent optimization of business processes in a company.

Our contribution aims at bridging this gap by (a) providing a conceptual framework for the integration of events and processes and (b) implementing the proposed framework in the process engine Chimera [8]. We gather the requirements to come up with the framework with reference to a use case from IoT scenario. Namely, we address the following major aspects in our framework:

- Separation of concerns between process behavior and event processing
- Aggregation of events and representation of event hierarchies
- Execution of event integration into business processes

Complex event processing and business process management are individually well explored fields. But the integration of these two worlds is still in its early stage. Our framework establishes the required steps for enabling the communication between events and processes. The prototypical implementation offers an end-to-end solution that encompasses (a) the modeling of processes, data, and event types, (b) the deployment of process models into the process engine and of event types and event annotation into the event platform, and (c) the execution of process models integrated with external events.

We suggest to outsource detection and correlation of catching events to a dedicated event processing platform, in our case Unicorn [19]. This supports separation of concerns and hides the complexity of dealing with external events, especially event adapters and aggregation, from the process engine. High-level events [6] aggregated from primitive events encapsulate complexity and serve as interface for the process model. When modeling event nodes, modelers can refer to the expected high-level event, e.g. a positive market trend, instead of having to deal with hundreds of individual stock tick events, because the event platform takes care of aggregating those into a higher-level event. This eases process modeling and keeps the annotations required for process execution simple. Also, the separation of process control and event processing logic improves maintainability, in case there is a change in event aggregation rules.

The paper is structured as following. Section 2 illustrates the fundamentals of business process management and complex event processing as the ground of our framework. Section 3 elicits the requirements using a motivational use case. The related works in both the fields relevant to our requirements are described in Sect. 4. Section 5 presents the conceptual framework for using events in processes and describes how we address the elicited requirements in the proposed framework. The prototypical implementation is detailed in Sect. 6. Finally, Sect. 7 concludes the paper and mentions the future research possibilities.

2   Foundations

In this section, we present the fundamental knowledge required to build our framework. There are two main domains addressed in the work, namely, business process management and complex event processing. The basic concepts from those two fields are discussed below.

2.1   Business Process Management

A business process is a sequence of activities performed in an organizational context. These activities collectively achieve a business goal [21]. The activities and their orchestration are represented with business process models. BPMN 2.0 is the de facto standard for modeling business processes. A business process model can be considered as a blueprint for multiple process instances. Similarly, an activity model can be instantiated for a set of activity instances.

An activity instance goes through several state transitions as shown in Fig. 1. Each activity in the process is initialized and is in state *init* as soon as a process instance is started. When the incoming flow of an activity is triggered, the instance is in state *ready*. The state changes to *running* once the activity starts execution. Finally, the activity ends and goes to *terminated* state. If the activity instance is *not started* but before it starts the process instance follows a different path, then it directly goes to *skipped* state. The occurrence of an attached boundary event can change the state of a running activity instance to *canceled*.

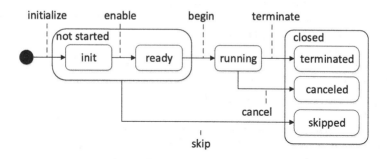

Fig. 1. Activity instance life cycle

The process flow can be enriched by information about the occurrences in the environment represented as events. BPMN describes different usage of events based on the position of the event in the process, namely, *start*, *intermediate* or *end* events. Start events are used to trigger a process instance. Intermediate events are produced or consumed by the process to use the information for further execution. If the event is received in the process, then it is called a *catching event*. The event produced by a process is named as *throwing event* in contrast. We will focus on the catching events as they are generated in the environment and used in the process.

Intermediate events such as boundary events or event-based gateways can be used to determine the process flow. Boundary events are associated with an activity and they can be interrupting or non-interrupting. If the event occurs after the activity is started and before it finishes, then an exceptional path is triggered. In case of interrupting boundary event, the ongoing activity is canceled. An event-based gateway is a decision gateway that depends on event occurring instead of data. The first occurrence among the events after the gateway causes that branch to be executed further.

2.2 Complex Event Processing

Using BPMN, one can model processes with catching events that are needed in the course of process execution. But BPMN does not talk about the source of these events, the information carried by the events or how to receive these events. On the other hand, such concepts are well explored in the event processing field [15]. Events are the environmental occurrences that can be relevant for a process. Event objects represent these occurrences in a computing system [12]. While an event can be a road accident or a sudden temperature fall, the representative event objects can be a traffic update from traffic API or a weather update from a sensor, respectively. The terms *event* and *event object* are often used interchangeably. Atomic events do not take time to occur, e.g., events generated by sensors. Atomic events can be aggregated to generate higher-level or complex events. A set of temporally totally ordered associated events form an event stream. Operations can be performed on single events (Simple Event Processing), multiple events in a single event stream (Event Stream Processing) or multiple events in multiple event streams (Complex Event Processing).

It is important to distinguish among three different kinds of events, which are denoted by the same term, although they are largely different. First, the term event can refer to the modeling construct of BPMN that is used to model catching or throwing events. We will use the term *BPMN event* for disambiguation. Second, it can refer to state transitions in lifecycles, e.g. the event `begin` that puts an activity instance into state *running* (see Fig. 1). In the remainder, we will use the term *lifecycle transition* to denote this meaning. Finally, event can mean an external event object that is present in the event processing platform and thus represents some real-world happening. Hence, these events are already abstractions of real-world happenings represented in an IT system. We will be using the term *external event* to refer to this kind of events.

3 Requirements Analysis

The goal of this contribution is to find an end-to-end solution for integrating real-world events into business process execution. Now, to come up with the framework it was necessary to elicit the requirements for the integration. Therefore, we explored the state-of-the-art of standard process engines as well as complex event processing techniques. The BPMN specification [17] built the foundations

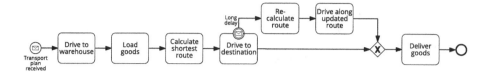

Fig. 2. Use case from logistics domain

for usage of activities and events in a process model. On the other hand, the literature survey discussed in Sect. 4 gave us insight about several concepts needed to be considered while integrating external events into processes. The project partners and domain experts from both academia and industry contributed vastly to extract use cases in IoT environment.

The process model in Fig. 2 represents one of those use cases from logistics domain. The process starts when the truck driver receives the transport plan from the logistics company. Then she drives to the warehouse to load goods. After the goods are loaded, the driver follows the shortest route to the destination. While the driving activity is ongoing, the driver gets notified if there is a long delay caused by an accident or traffic congestion. If the notification for long delay is received then the driver stops following the same route. Rather, she calculates alternate routes which might be faster and follows the best of those. Once the destination is reached the goods are delivered and the process ends. Based on the above scenario, the requirements for using events in processes are identified and described in the rest of this section.

Requirement 1: Separation of Concerns

Using external events in business processes is essentially connecting the two fields of business process management and complex event processing. As seen in the use case, event information can improve the process execution with respect to flexibility, monitoring and efficiency by reacting on occurrences in the environment in a timely manner. Process engines could directly connect to event sources by querying their interfaces, listening to event queues, or issuing subscriptions. However, from a software design perspective this design decision would dramatically increase the complexity of the engine and violate established principles like single responsibility and modularity. Therefore, we consider separation of concerns between process behavior and event processing as major requirement.

Different event sources produce events in different formats, e.g., XML, CSV, JSON, plain text, and over different channels, e.g. REST, web service, or a messaging system. In the example scenario, the probable event sources are the logistics company, the GPS sensor in the vehicle, the traffic API and each of them might have their own format of producing events. If the process engine were directly connected with event sources, it would need to be extended with adapters for each of the sources to parse the events. On the other hand, certain events can be interesting for more than one consumer. For example, the long delay event might be relevant not only for the specific truck driver, but also for other cars following the same route. CEP platforms are able to connect to different event sources as well as they can perform further operations on event streams [13]. Single event streams can be filtered based on certain time window, specific number of event occurrences or attribute values of the events. Also, multiple events from multiple event streams can be aggregated based on predefined transformation rules to create complex events relevant for a process.

To include all these functionalities in a process engine will increase the complexity and redundancy of the engine to a great extent. Instead, it is more efficient to use a separate event platform for complex event processing. The event consumers can then subscribe to the event platform for being notified of the relevant events. This separation of event processing logic is also efficient from the maintenance perspective. If there is a need to change the event source or the aggregation logic, then the process model does not need to be touched.

Requirement 2: Representation of Event Hierarchies

Simple event streams generated from multiple event sources can be aggregated to create complex or higher-level events. One could argue to use BPMN parallel multiple events to represent the event hierarchy, at least to show the connection among simple and complex events. However, using that approach one cannot express the different dimensions of event aggregation such as sequence, time period, count of events or the attribute values. Different patterns of event sequences are thoroughly discussed in [15] whereas a structured classification for composite events can be found in [3]. Moreover, this would complicate the process model and defeat their purpose to give an overview of business processes for business users. As an user of BPM, one would be interested to see the higher-level event that influences the process, rather than the source or the structure of the event. For example, the driver is only interested to know if there is a long delay that might impact her journey, but she does not care what caused the delay.

Using event hierarchies, the process model includes only the high-level business events relevant for the process and easily understandable by business users. The model is not burdened with details of event sources and aggregations, which instead are dealt with by event hierarchies in the event platform. Event hierarchies also improve maintainability, because the process model need to be adapted whenever event sources or the format of events changes. Therefore, we consider event hierarchies represented through event processing techniques as requirement for successful integration.

Requirement 3: Execution of Event Integration

Incorporating the above two requirements, the logical distribution is made from the architectural point of view as well as the representations of event processing and process execution. But the technical requirements from the implementation aspects are still remaining. We define following three technical requirements to realize the integration of events and processes.

R3.1: Binding Events. The higher-level events modeled in the process model needs to be mapped with the event hierarchy defined in the CEP platform to make sure that the correct event information is fed to the process. E.g., the driver should be informed only about the delay in the route she is following.

R3.2: Receiving Events. The process engine should listen to specific event occurrences relevant for the process execution. In other words, the driver must subscribe for the `Long delay` event to get notification, as modeled in Fig. 2.

R3.3: Reacting on Events. The driver needs to decide if the alternate route is faster than the current one and for that she needs to know the duration of the delay. Therefore, information carried by the events should be stored for later use in the process.

4   Related Work

Over the last decade, the BPM community adopted concepts from the field of complex event processing (CEP) and event-driven architectures (EDA). Several approaches were presented aiming to extend BPMN with modeling constructs for concepts of CEP [1,3,11]. Some of those approaches provide execution support in the form of an engine. Decker and Mendling [9] present a conceptual framework for process instantiation based on external events. Other approaches use external events to monitor running business processes [2,5,13], predict deviations [6], check compliance to the process model [20], and calculate KPIs [11]. In this section we describe these related approaches and check them against the requirements.

Barros et al. [3] discuss events in business processes and touch on many of the topics we consider in our contribution, like event subscription, occurrence, matching, and unsubscription. The authors present a catalog of event patterns used in real-world business processes and find that most of those patterns are neither supported by BPEL nor BPMN. Similar to *Requirement 2*, they identify support for event hierarchies, i.e. aggregation of low-level events into high-level business events, as important but yet unsupported. They suggest to integrate descriptions of event patterns into process modeling languages and consequently extend engines to handle such patterns.

This contradicts *Requirement 1* to separate concerns between process execution and event processing, but can be understood in light of the limited types of events supported by BPEL (only message receive and timer events). Related concepts to *Requirement 3* are discussed in this work. The causal ordering among event subscription, event occurrence and event consumption proposed by them is shown in Fig. 3. According to it, an event can be consumed only if there exists a subscription and the event has already occurred. Though we have followed the same ordering in our implementation, the authors in [3] did not implement the suggested engine.

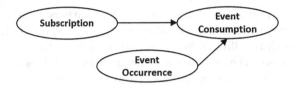

Fig. 3. Causal ordering between event subscription, occurrence and consumption

Estruch and lvaro [11] propose an IT solution architecture for the manufacturing domain that integrates concepts of SOA, EDA, business activity monitoring (BAM), and CEP. They suggest to embed event processing and KPI calculation logic directly into process models and execute them in an extended BPMN engine. This complicates the understanding of process models and contradicts *Requirement 1*. The authors sketch such an engine for executing their extended process models, but refrain from giving technical details, like handling of subscriptions or event format, thus failing to meet *Requirement 3*. However, they suggest that some processes collect simple events, evaluate and transform them, and provide high-level events for use in other process instances, realizing an event hierarchy (*Requirement 2*).

Processes from the logistics domain, which contain long-running activities, e.g. a task 'shipment by truck', needs continuous monitoring. In these scenarios external events, e.g. GPS locations sent by a tracking device inside the truck, can provide insight into when the shipment task will be completed. Appel et al. [1] integrate complex event processing into process models by means of event stream processing tasks that can consume and produce event streams. These are used to monitor the progress of shipments and terminate either explicitly via a signal event or when a condition is fulfilled, e.g. the shipment reached the target address. While these tasks are active, they can trigger additional flows if the event stream contains some specified patterns. The authors provide an implementation by mapping the process model to BPEL and connecting the execution to a component called eventlet manager that takes care of event processing. Thus *Requirement 1 and 3* are fulfilled, however, *Requirement 2* is not supported. Authors in [16] facilitate an integrated architecture for using events to monitor and predict process execution. The butterfly architecture analyzes the need of external event input in the process and generates CEP rules from historical data. But they do not talk about the conceptual and technical challenges of the integration.

For processes, in which some tasks are not handled by the POIS, monitoring of events can be used to determine the state of these tasks, e.g. to detect that an user task terminated. When a process is not supported by a POIS at all, monitoring can still capture and display the state of the process by means of events. For example, Herzberg et al. [13] introduce Process Event Monitoring Points (PEMPs), which map external events, e.g. a change in the database, to expected state changes in the process model, e.g. termination of a task. Whenever the specified event occurs, it is assumed that the task terminated, thus allowing to monitor the current state of the process. The authors separate the process model from the event processing and allow the monitored events to be complex, high-level events thus fulfilling *Requirements 1 and 2*. The approach has been implemented, however the event data is not used by and does not influence the process activities. Rather the engine uses them to determine the current state of the process instance. Therefore, we consider *Requirement 3.3* to be unfulfilled.

A framework for predictive monitoring of such continuous tasks in processes is presented in the work by Cabanillas et al. [6]. The framework defines monitoring points and expected behavior for a task before enactment. Then event information from multiple event streams are captured and aggregated to have a meaningful interpretation. These aggregated events are then used to train the classifier and later the classifier can analyze the event stream during execution of the task to specify whether the task is following a safe path or not. [20] derive event queries from the control flow of a process model, deploy them to a event engine and use them to find violations of the control flow. A similar derivation of event queries from the process model is done by [2]. These work have in common that they use external event processing (*Requirement 1*) and are implemented. However, just like [13], the events are not used to drive the process instance, but rather to find out something about it.

On the other hand, Decker and Mendling [9] conceptually analyze how processes are instantiated by events. They propose a framework named CASU which specifies when to create new process instances (C), which control threads are activated due to this instantiation (A), which are the remaining start events that the process instance should still subscribe to (S), and when should the process instance unsubscribe from these events (U). Because of its conceptual nature, *Requirement 3* is not fulfilled and we cannot judge *Requirement 1*, as the paper does not mention an architecture. However, the CASU framework satisfies *Requirement 2*, although partially. They focus on process instantiation and therefore, concentrates only on single or composite start events. On the other hand, we consider not only the start events, but also the intermediate or boundary events as well as the event based gateway.

Finally, we consider the state-of-the-art for implementing a process engine that supports event integration. Namely, we look into the popular open source process engine Camunda [7]. Using events for executing processes is also an area which has gained a lot of interest in past few years. The standard process engines like Camunda support BPMN events for starting a process instance or to choose between alternative paths following a gateway. However, Camunda does not care about the receiving part of the message event. The engine has interfaces that can be connected to a JMS queue or a REST interface but the reception of messages is not implemented. Also, there is no existing process engine that supports complex event processing. On the other hand, the event processing platforms do not have any engine to implement the generated events. Moreover, the mapping between external events and BPMN events is not there.

5 Conceptual Framework

This section presents the conceptual framework for integrating events into processes. Keeping in mind the requirements specified in Sect. 3, we discuss the aspects to be considered. Also, the proposed solutions that we came up with for each aspect are mentioned in the context of the use case presented before.

5.1 Event Generation and Aggregation

In our use case, we need two events for the process execution, a catching start event and a catching interrupting boundary event. The start event is created based on the input from the logistics company. The transport plan contains the location of warehouse to load goods, the destination for delivery and the deadline for delivery. This is an example of simple event which might be sent to the truck driver via email or even as a text message directly from the logistics company. The boundary event, on the other hand is definitely a higher-level event. Considering *Requirement 1: Separation of Concerns* this complex event is created in the event processing platform. Since we did not have access to real "truck positions", we used the sensor unit Bosch XDK developer kit[1], a package with multiple integrated sensors for prototyping of IoT applications. The unit sends measurement values over wireless network to a gateway. The gateway then parses the proprietary format of the received data and forwards it to Unicorn using the REST API. The traffic updates was received from *Tomtom Traffic Information*[2]. If there is a delay above a threshold and the location of the source of delay is ahead of the current GPS location of the truck, then a LongDelay event is produced.

In Unicorn, event aggregation rules are written accordingly to generate the high-level event LongDelay. Since Unicorn has the Esper engine at its core, we used Esper Event Processing Language (EPL) [10] for writing event aggregation rules. The event types can be registered in Unicorn as following:

```
CREATE schema Disruption
(latitude float, longitude float,
reason string, delay double);
CREATE schema CurrentLocation
(latitude float, longitude float, destination string);
CREATE schema LongDelay
(reason string, delay double, destination string);
```

---

[1] see http://xdk.bosch-connectivity.com.

[2] see https://www.tomtom.com/en_gb/sat-nav/tomtom-traffic/.

The aggregation rule for creating `LongDelay` may look like the following. The function `distance()` is not defined in EPL though. We implemented it to find out if the disruption is ahead of the truck or not.

```
INSERT INTO LongDelay
SELECT d.reason as reason,
d.delay as delay,
l.destination as destination
FROM pattern[every d=Disruption-> l=CurrentLocation
WHERE distance(d.latitude, d.longitude, destination)
< distance(l.latitude, l.longitude, destination)];
```

5.2   Event Binding Points

Event binding points are those elements of process model where events with different properties (see Sect. 2) are mapped to each other. For example, the external event `LongDelay` is needed to be mapped to the BPMN event `Long delay` that has been modeled in the process. To enable that, process models have to be extended by *event annotations* that are used as event binding points. These event binding points specify which events to listen to at what point in the model. In BPMN, external events are usually modeled with the help of catching message events (see Fig. 2). To receive these events, we need to make sure that subscriptions for the events are made. Therefore, a subscription query for each event is added to the process model in design time. Receiving the subscribed events allows to create new process instances, similarly to process instantiation in [9], and to react on the intermediate events from external event sources. To simplify the annotation language, simple queries are added in the model to subscribe to the aggregated high-level events. More complex event queries to produce these high-level events are generated by aggregation rules inside the event processing platform, as per *Requirement 2: Representation of Event Hierarchies*. For example, the annotation for the start event looks like `SELECT * FROM LongDelay` which abstracts from the complexity of event queries dealt in CEP platform. Figure 4 shows an example of modeling event queries represented as event annotation.

On the other hand, lifecycle transitions, like `terminate` in the activity lifecycle, also provide event binding points [21]. These event binding points are required to automatically change the state of an activity instance, thus enabling their monitoring [5]. The start and end of each activity like `Load goods` or `Deliver goods` are required to monitor the status of the shipment. Whereas, the cancellation of the activity `Drive to destination` suggests that the previously calculated delivery time might be postponed due to delay on the way.

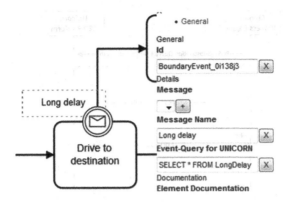

Fig. 4. Modeling of event queries

5.3   Event Subscription and Correlation

Before they can be executed, process models have to be deployed from the modeling component to the process engine. During the deployment, event annotations, like the other information in the process model, are parsed and stored in a database. Once the queries are registered to the CEP platform, whenever a matching event occurs, the platform notifies all the subscribers.

When it comes to registering event queries the *lifespan* of event annotation plays a central role. By lifespan we mean the time between registering an event query at the event processing platform and removing the subscription. A process model works as a blueprint for several process instances [21] and subscription can be done by a process model or a process instance. The annotation of the start event binding point needs to be registered right after deployment, as it is needed for process instantiation. So, subscription for `Transport plan received` is done at process deployment. In our case, the truck driver might register to the mailing list of the logistics company to receive transport plans. Other annotations, e.g., annotation for event binding point of `Long delay` is registered later for each process instance separately.

Often, event queries have a limited lifespan during a process execution. For example, we no longer need to listen to a boundary event that might occur after the activity it was attached to has terminated and can unregister the query. In our use case, the driver stops listening to `Long delay` once she changes the route or reaches the destination. On the other hand, process trigger queries can only be unregistered when the process model is undeployed from the engine.

To handle the subscription, we extend the *execute*-method of the event nodes. When the process execution flow reaches this event node, the subscription query and a notification path is sent to Unicorn. Unicorn sends an UUID in response which is then stored in Chimera as a correlation key. When an event occurs, Unicorn checks if there is an existing subscription for this event. If a matching query is found, then Unicorn sends the notification to the provided path along with the UUID. Chimera matches this UUID to the one stored before and correlates

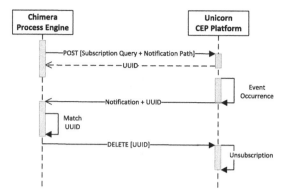

Fig. 5. Event subscription and correlation using Chimera and Unicorn

the event to the process instance. Once the event is consumed, the *leave*-method of the event node performs a DELETE operation for unsubscription. The above sequence is depicted in Fig. 5.

The attribute values of events can be used to filter out events irrelevant for the current process instance. For example, only events that occur in the same route as the truck is following might be interesting for that particular transport. Therefore, the annotated event queries may contain expressions referring to event attribute values. These expressions follow the dot-notation known from object oriented programming, e.g. Disruption.route. In some cases, however, the correlation is less direct and the filter criterion is either not available or not restricted inside the scope of the process instance. In such cases we assume that the event processing platform provides a correlation key, as described above.

5.4   Reaction on Events

In many cases, receiving an event notification from the event processing platform simply causes an BPMN event to occur. The further reaction follows the BPMN execution semantics [17]. The notification of a start event can start a new instance of a process. For example, each customer complaint can start one handling process for a manufacturer. Again, a notification of a boundary error event causes the abortion of the associated running activity and enables error handling, as discussed in our example process.

While in these cases only the fact that the event occurred is relevant, in other cases the content of the event is also of interest. E.g., the Long delay event certainly abides by the BPMN behavior for boundary events, but the driver might look into the information carried by the event to know how much delay has been caused and whether the alternate route will be faster or not. Therefore, notifications need to have a defined structure that allows to access the contained data for further use in the process.

To use event data in the further execution, we suggest to map the data contained in the notification to the attributes of a data object. This data object

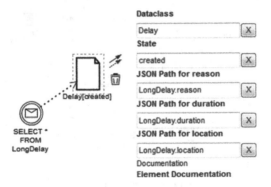

Fig. 6. Event data is written into newly created data object

might already exist at the time the event notification is received, or it can be created anew. The mapping is specified in the outgoing data object, which for each of its attributes has an expression specifying how to derive the attribute value from the notification. This is depicted in Fig. 6 which shows the boundary event and the property editor for the data object. Technically, this mapping is achieved by representing event notifications in the JSON notation[3] and giving a path expression for each attribute of the target data object that defines how the value can be derived from the notification.

An alternative would have been to directly use the event object to reuse data later on in the process, instead of mapping it to a data object. We decided against this option, because events are singular occurrences, whereas data objects have a lifecycle and can be changed again. For example, each temperature measurement of the sensor is an unique event that might cause the sending of an event notification when a matching query is registered. However, for the duration of a process instance execution we would like to have one data object that holds the current temperature, which of course changes, when new notification arrives for sensor events.

The third kind of reaction that the engine can perform upon receiving an event notification is to conduct a lifecycle transition. In context of our use case, the GPS location of the truck is checked against the coordinates of the destination. When the locations match, a higher-level event is generated to notify the process engine that the driver has reached the destination. The engine then changes the state of the activity `Drive to destination` from *running* to *terminated*.

6   Implementation

This section briefly describes the implemented systems used to realize the integration of processes and events, and their interplay. A coarse architecture is

---

[3] see http://goessner.net/articles/JsonPath/.

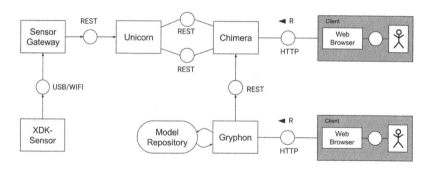

Fig. 7. Architecture

depicted in Fig. 7 with a sensor gateway as one specific example of an event source (left side).

6.1 Unicorn Event Processing Platform

Unicorn, first described in [13], is the event processing platform of choice for the implementation of our approach. It is build around Esper[4] and manages event types, event queries, and notifications both via a web-based UI and a REST API. Notifications are delivered via email, the Java Message Service (JMS), by calling back registered REST endpoints, or viewed in the UI.

There are different ways to connect Unicorn to external event sources. Event sources can simply use Unicorn's REST API to send events or publish them to a specific JMS channel to which Unicorn listens. This is feasible if the code of the event source can be changed or an intermediary gateway is used, which collects events, for example from sensors, and forwards them to Unicorn.

Unicorn also supports active pulling of events by means of adapters, which periodically call webservices. Adapters have to be configured programmatically for each event source that should be accessed. However, Unicorn offers a framework to easily extend existing adapters.

Historic events available as comma-separated values (csv) or spreadsheets (xls) can also be parsed and imported as event stream into Unicorn. Replaying such events keeps their order and time-lag, which allows to test pattern detecting queries and aggregation rules. Finally, Unicorn has a built-in event generator that uses value ranges and distributions to generate realistic events that can be used for testing event-driven applications.

Creation of high-level events is handled by *aggregation rules*, i.e. event queries that transform a pattern of events into another, higher-level event. These aggregation rules are defined by domain experts for each business scenario.

---

[4] see http://www.espertech.com/products/esper.php.

6.2   Gryphon Case Modeler

The second component is Gryphon, a web-based modeler for process models. Additionally, it allows to create a data model, i.e. a specification of data classes and attributes used in process model. Each data class defines possible states and valid state transitions for their instances, i.e. data objects, at runtime, called object life cycle. Process models can be directly deployed to a running Chimera instance. Gryphon builds on a node[5] stack and uses bpmn.io[6] which is an open-source BPMN modeler implemented in Javascript, while the other components are developed by our team.

For this contribution we extended Gryphon with the functionalities to annotate process elements with event annotations and model event types. The data model editor distinguishes between event types and data classes. While both are named sets of typed attributes, the former need to be registered with the event processing platform Unicorn when the model is deployed. Event annotations can be attached to certain elements in process fragments models, corresponding to the event binding points defined in Sect. 5.2. We decided to reuse the symbol for catching message events to model waiting for external events, because event notifications can be considered messages. For lifecycle binding points, the annotations are stored as property of the transitions in the lifecycle diagram, and for model-level binding points they can be specified in the model overview. As we use the event processing platform Unicorn that builds atop of Esper, event annotations have to use the EPL query language.

6.3   Chimera Case Engine

The final component is Chimera, a process engine that can also execute case models according to the fragment-based case management approach (fCM) [14]. It supports user activities with forms for data entry, as well as automatically executed email and web-service tasks. Attributes of data objects can be used as variables in email text, web-service calls, and gateway conditions. Variables are substituted by attribute values when sending email, calling web-service, or checking which sequence flow to enable.

The front-end displays all available process models and allows users to start new cases or work on running cases. Chimera follows the common worklist approach, displaying enabled activities to knowledge workers who can select and start them. Enablement of activities depends on sequence flow, i.e. preceding activities need to have terminated. Also, the data flow i.e. required data objects need to be available in the state as specified by the data input set of an activity. When terminating a running activity, knowledge workers can enter data stored in data objects. However, the resulting state of the data object needs to conform to the data output set specified in the model.

Discussion. As the architecture was developed as an academic prototype highlighting research challenges, we abstained from implementing well-understood

---

[5] see http://node.js.

[6] see http://bpmn.io.

features required for business use, like user and role management, or database
accessors.

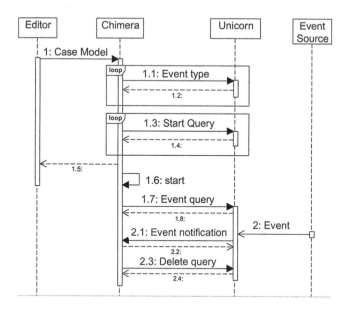

Fig. 8. Interaction sequence of components

The interaction sequence among the above described components is depicted
in Fig. 8. We used the implemented architecture and depicted communication
sequence between events and processes to realize several use cases from different application domains to evaluate the practicality of our framework. As
already mentioned, Chimera supports case management along with process execution. Therefore, the catching start event functionality has been tested to trigger process instances as well as cases and fragments[7]. Different use cases had
different usage of events such as boundary events or event-based gateways. Our
architecture was able to handle all of them efficiently. To connect the event
processing platform to different event sources we used sensors as well as Web
API or live RSS Feeds[8]. The aggregation rules were specified according to the
input extracted from the corresponding domain experts for each use case.

7   Conclusion and Future Work

The work presented here addresses a relevant situation of the current business
world, as web services and IoT increase the amount of external events relevant

---

[7] see screencast: https://bpt.hpi.uni-potsdam.de/Chimera.

[8] e.g., http://www.eurotunnelfreight.com/uk/contact-us/travel-information/.

for business processes. Our work focuses on integrating such external events into business processes and making use of them for process execution. To bridge the gap between the conceptual level of the process model and the technical details necessary to execute it, certain aspects need to be considered as shown by our use case driven requirement elicitation. Based on those requirements, we present a conceptual framework for the integration of external events, defining event binding points, event annotations, as well as subscription and correlation mechanisms. These concepts are implemented into a system consisting of a modeling component (Gryphon), a process engine (Chimera), and an event platform (Unicorn), thus enabling the integration of real-world events into business processes.

Although our solution architecture handles the basic BPMN event constructs such as message or timer events, boundary events and event-based gateways, other event constructs like signal or error events have not been considered and are left to be implemented. Along with the events, decisions also play a big role in business processes [4]. The explicit use of decisions in processes becomes more popular with the recently released standard Decision Model and Notation (DMN) [18]. As discussed in the motivating example, the truck driver can check the duration of the delay caused by the disruption and decide whether to take an alternative route or not. Therefore, the next logical extension of our framework is to integrate decision management.

References

1. Appel, S., Frischbier, S., Freudenreich, T., Buchmann, A.: Event stream processing units in business processes. In: Daniel, F., Wang, J., Weber, B. (eds.) BPM 2013. LNCS, vol. 8094, pp. 187–202. Springer, Heidelberg (2013). doi:10.1007/978-3-642-40176-3_15
2. Backmann, M., Baumgrass, A., Herzberg, N., Meyer, A., Weske, M.: Model-driven event query generation for business process monitoring. In: Lomuscio, A.R., Nepal, S., Patrizi, F., Benatallah, B., Brandić, I. (eds.) ICSOC 2013. LNCS, vol. 8377, pp. 406–418. Springer, Cham (2014). doi:10.1007/978-3-319-06859-6_36
3. Barros, A., Decker, G., Grosskopf, A.: Complex events in business processes. In: Abramowicz, W. (ed.) BIS 2007. LNCS, vol. 4439, pp. 29–40. Springer, Heidelberg (2007). doi:10.1007/978-3-540-72035-5_3
4. Batoulis, K., Meyer, A., Bazhenova, E., Decker, G., Weske, M.: Extracting decision logic from process models. In: Advanced Information Systems Engineering - Proceedings of 27th International Conference, CAiSE 2015, Stockholm, Sweden, 8–12 June 2015, pp. 349–366 (2015)
5. Baumgrass, A., Herzberg, N., Meyer, A., Weske, M.: BPMN extension for business process monitoring. In: EMISA, pp. 85–98 (2014)
6. Cabanillas, C., Di Ciccio, C., Mendling, J., Baumgrass, A.: Predictive task monitoring for business processes. In: Sadiq, S., Soffer, P., Völzer, H. (eds.) BPM 2014. LNCS, vol. 8659, pp. 424–432. Springer, Cham (2014). doi:10.1007/978-3-319-10172-9_31
7. Camunda: Camunda BPM platform. https://www.camunda.org/
8. Chimera: Case engine. https://bpt.hpi.uni-potsdam.de/Chimera
9. Decker, G., Mendling, J.: Process instantiation. Data Knowl. Eng. **68**(9), 777–792 (2009). http://dx.doi.org/10.1016/j.datak.2009.02.013

10. EsperTech: Esper Event Processing Language EPL. http://www.espertech.com/esper/release-5.4.0/esper-reference/html/
11. Estruch, A., Heredia Álvaro, J.A.: Event-driven manufacturing process management approach. In: Barros, A., Gal, A., Kindler, E. (eds.) BPM 2012. LNCS, vol. 7481, pp. 120–133. Springer, Heidelberg (2012). doi:10.1007/978-3-642-32885-5_9
12. Etzion, O., Niblett, P.: Event Processing in Action. Manning Publications Co., Greenwich (2010)
13. Herzberg, N., Meyer, A., Weske, M.: An event processing platform for business process management. In: EDOC. IEEE (2013)
14. Hewelt, M., Weske, M.: A hybrid approach for flexible case modeling and execution. In: La Rosa, M., Loos, P., Pastor, O. (eds.) BPM 2016. LNBIP, vol. 260, pp. 38–54. Springer, Cham (2016). doi:10.1007/978-3-319-45468-9_3
15. Luckham, D.C.: The Power of Events: An Introduction to Complex Event Processing in Distributed Enterprise Systems. Addison-Wesley, Boston (2010)
16. Mousheimish, R., Taher, Y., Zeitouni, K.: The butterfly: An intelligent framework for violation prediction within business processes. In: Proceedings of the 20th International Database Engineering & Applications Symposium, IDEAS 2016, pp. 302–307. ACM, New York (2016). http://doi.acm.org/10.1145/2938503.2938541
17. OMG: Business Process Model and Notation (BPMN), Version 2.0., January 2011
18. OMG: Decision Model and Notation (DMN), Version 1.1., June 2016
19. UNICORN: Complex event processing platform. https://bpt.hpi.uni-potsdam.de/UNICORN/WebHome
20. Weidlich, M., Ziekow, H., Mendling, J., Günther, O., Weske, M., Desai, N.: Event-based monitoring of process execution violations. In: Rinderle-Ma, S., Toumani, F., Wolf, K. (eds.) BPM 2011. LNCS, vol. 6896, pp. 182–198. Springer, Heidelberg (2011). doi:10.1007/978-3-642-23059-2_16
21. Weske, M.: Business Process Management: Concepts, Languages, Architectures, 2nd edn. Springer, Heidelberg (2012). doi:10.1007/978-3-642-28616-2

Methodological Support for Coordinating Tasks in Global Product Software Engineering

Carolus B. Widiyatmoko[1]([✉])[iD], Sietse J. Overbeek[2][iD],
and Sjaak Brinkkemper[2][iD]

[1] PT Telkomunikasi Indonesia, Tbk., Bandung, Indonesia
Carolusbw@telkom.co.id
[2] Department of Information and Computing Sciences,
Utrecht University, Utrecht, The Netherlands
{S.J.Overbeek,S.Brinkkemper}@uu.nl

Abstract. Distributing software processes by software producing organizations (SPOs) is emerging increasingly due to benefits that global software engineering (GSE) brings in terms of cost reduction, leveraging competencies, and market expansion. However, these organizations are facing communication and project control issues that can slow down the overall organization performance. Therefore, SPOs should be able to manage inter-dependencies among the tasks distributed to the globally dispersed teams. We analyze existing works and product software companies' best practices in coordinating tasks in GSE. This paper specifically focuses on constructing methodological support for task coordination that can be influenced by the situational factors at the companies. The support comprises a framework and a method developed by using a method engineering approach. We introduce the framework that depicts the aspects that should be examined by companies and the method that elaborates the practices to guide companies to coordinate tasks in GSE projects. The validation results show that the framework and the method are accepted by experts regarding completeness and applicability to help SPOs in managing coordination among globally distributed teams.

Keywords: Design science · Global software engineering · Method engineering · Software producing organization · Task coordination

1 Introduction

Software producing organizations (SPOs) are companies that focus on developing software as a product to be delivered to a targeted market [31]. The extensive client considerations and technical factors such as platform variability of prospective clients make these companies have more software engineering complexity than software companies who perform on software development projects for specific customers [37].

Some countries (India, China, and parts of eastern European countries to name a few) offer a large number of human resources such as software developers with lower salaries than developed countries such as in Europe and the

© Springer International Publishing AG 2017
H. Panetto et al. (Eds.): OTM 2017 Conferences, Part I, LNCS 10573, pp. 213–231, 2017.
https://doi.org/10.1007/978-3-319-69462-7_14

US [1]. SPOs can get the economic value by utilizing these qualified human resources from other parts of the world. These companies apply what is called global software engineering (GSE) where parts of their engineering processes are conducted collaboratively in remote facilities located in other countries. By doing GSE, SPOs can also gain other benefits such as opportunities for market expansion and focus on product development competencies by leveraging some of the business functions to other organizational units [1,14].

In GSE, teams interact with each other working on tasks in the software engineering cycle and create an internationally distributed collaborative network. Apparently, there are differences in geographical location, time-zone, socio-cultural, and organization among these teams [15,21]. Problems then arise, when the teams begin to have difficulties in organizing tasks to manage dependencies on resources or processes caused by the distribution. To that end, SPOs need to be able to coordinate tasks to build better communication, synchronize work, and balance knowledge among these distributed teams [30].

The literature on GSE is rich as existing frameworks, guidelines, and GSE best practices can be identified [13,24,25]. However, most of the studies are not critically discussing coordination practices in globally distributed teams. It is believed that an approach cannot be easily implemented in the same way on every organization because each organization has different situational factors [4]. To this end, this paper surveys the state of the art of task coordination approaches and proposes methodological support that presents the abstraction of coordination methods for SPOs that perform software engineering globally.

In this paper, our goal is to identify the coordination studies and practices and to present the elements related to task coordination that SPOs must understand and be aware of in the context of GSE. This paper is organized as follows: Sect. 2 illustrates our research methods where the execution will be elaborated in subsequent sections. Section 3 describes the literature review to build the foundation of knowledge in this topic. Next to that, Sect. 4 reports how the research artifacts are constructed and then assessed to ensure that the method meets the expectation of those who need this method. After that, Sect. 5 recounts the research process and provides a critical reflection on the benefits and limitations of the method. Finally, Sect. 6 concludes the paper by discussing our findings and limitations.

2   Research Method: The Design Cycle

This research is conducted by adopting an iterative problem-solving design science method [36]. Design science cycle is a subset of the engineering cycle which is a continuing investigation includes design processes to solve a problem by creating an artifact. The design science approach comprises three main stages which are: Problem investigation, solution development, and solution validation. The problem investigation phase and the solution development phase are executed by using method association technique, and the solution validation is accomplished through expert opinion. The stages are elaborated in the following section.

2.1   The Method Association Approach

For the analysis of the existing frameworks, techniques, and methods in global software engineering, we use the method engineering approach [4]. Where software engineering pays attention to all aspects pertains to software production, method engineering focuses on the construction of method that fall into the software engineering domain. Method engineering is defined as "the engineering discipline to design, construct and adapt methods, techniques, and tools" [4]. Hence, this stage will represent the investigation and the solution development of the design cycle that we follow.

In this section, we present our approach to constructing a reference method for task coordination by adopting the method association approach (MAA) [19]. The MAA is used to create meta-models for the processes and data perspectives constructed from the established methods gathered by studying the state of the art of task coordination from literature and best-practices obtained from companies' experiences. Hence, we choose process-delivery diagram (PDD) [35] as the meta-modeling technique to present the methodological support in a uniform and formal representation. The MAA approach adopted from [19] in this research consists of eight steps as depicted in Fig. 1.

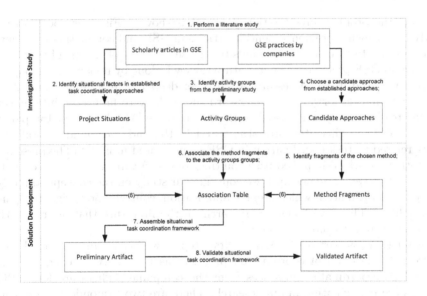

Fig. 1. The MAA design cycle, adapted from [19]

2.2   Method Evaluation Approach

The validation sessions were performed by involving researchers and practitioners to get feedback from a broader perspective. The proposed artifact will be assessed by these experts who reflect how such an artifact will interact with problem contexts and then predict what effects that they think the artifact would have by using expert opinion [36]. Validation of the proposed method by expert opinion will work if the experts understand the artifacts which enables them to imagine problem contexts and predict the effects of the artifacts in the contexts.

The validation strategy is following the Framework for Evaluating in Design Science [32]. We selected the Human Risk and Effectiveness strategy as the proposed artifacts are user oriented that should be evaluated with real users in their real context. Formative assessment starts the evaluation and progressively engages more summative assessment focusing on the applicability of the artifacts. Therefore, we use two-steps validation by involving experts that have scientific backgrounds and experts from business practitioners. The scientific experts will do the criteria-based assessment on the designed method based on meta-modeling criteria which are completeness, consistency, efficiency, reliability, and applicability [5]. Business practitioners will concentrate on usefulness and ease of use of the designed method [16,26].

3 Problem Investigation

3.1 Literature Study

To start the problem investigation phase as is show in Fig. 1, we performed a study by reviewing previous scientific articles in GSE and conducting interviews with several SPOs in the Netherlands to expose the state of the art of task coordination in GSE practices. The literature review is done by reviewing the paper's abstraction, and if it is necessary, we also get deeper by examining contents of the article and do forward and backward snowballing to find more information related to the concepts discussed in the main article [34]. A total of 155 papers were involved in this literature study, and in the end, we found several concepts related to task coordination in GSE as presented in Fig. 2 (The presented semantic network is simplified based on the global software engineering context). Through this semantic network, we continue our study on the concepts directly-related to coordination which are: communication, control, knowledge, tool, and stakeholders. Then, we focus our research on the literature that addresses the practices related to those concepts.

After that, as part of the MAA's first step as well as to understand the situational background of coordination practices as the MAA's second step, we studied the coordination practices from the companies. There are four SPOs which were participating in our research. There are two respondents from each of the companies from various job positions who have experiences in global software engineering. The company's names are replaced with AlphaSoft, BetaSoft, GammaSoft, and DeltaSoft for the reason of confidentiality. The interviews were performed from December 2016 until February 2017. During the interviews, the concepts found in the literature study and the company's practices in coordinating interdependencies during performing GSE were discussed. To this end, the interviewer posed several questions such as: *"What kind of challenges does the*

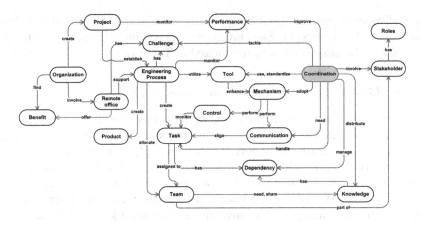

Fig. 2. Semantic net of task coordination at SPO in GSE environment

organization have by doing GSE?" and *"What kind of instruments do you use to support you in managing tasks?"*

The interviews captured several topics: company background, job roles and functions, partners or remote offices profiles, product profiles, company's vision in

Table 1. Task coordination practices comparison

Aspects	AlphaSoft	BetaSoft	GammaSoft	DeltaSoft
Remote office (RO) location	Belgium, Romania, India (susp.)	Malaysia	Mainly in Poland and India	Romania
RO ages	>6 years (Romania)	17 years	>2 years	±2.5 years
RO functions	Development, testing	System design, development, testing	System design, development, testing	Development
Team size	±40	>100	>100	6
Engineering method	Scrum of Scrum	SAFE[a]	Traditional	Scrum[a]
Main market	Dutch companies	Global	Global and internal	Global and internal
Challenges	Communication, trust, timezone difference	Communication tools quality	Communication, expertise, time-zone, org. silos, culture	Lack of explicit knowledge
Communication	Direct	Direct (technical area), indirect (enterprise)	Indirect	Direct
Control	Proactive, mutual adjustment	Reactive, mutual adjustment	Reactive, direct supervision	Proactive, standardization
Knowledge sharing	Document sharing, site visit	Formal training, document sharing	Mentoring (socialization), documentation	Socialization
Tool	Vicon, site visit, skype, TFS	Webex, Skype, Sharepoint	Webex, Skype, OneVision, Sharepoint	Regular site visit, Slack, Skype
Important roles	Scrum master, unit manager	Product manager, feature owner, dev manager	Service coordinator	Team leader

[a] self-build, customized, or similar approach

GSE, challenges in performing GSE, approaches in GSE practices, and stakeholders involved in GSE projects. Each interview was conducted between 45–60 min. The results of the interviews are summarized in Table 1.

3.2 The Global Task Coordination Framework

Before starting to assembly the method, we introduce the Global Task Coordination framework showing that coordinating tasks in GSE comprises three main blocks that build practices to achieve coordinated output: the task coordination mechanisms, the coordination mechanisms supports, and the organization's situational factors that can influence the appropriate coordination mechanisms of the SPOs (Fig. 3). We define coordinated output as a situation where an organization is able to manage task ownerships, synchronize jobs and hand-over to achieve an integrated results among its distributed teams.

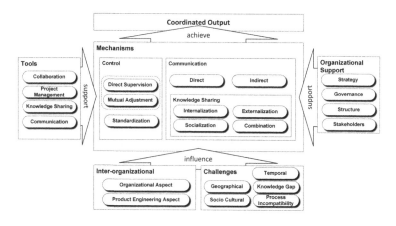

Fig. 3. The global task coordination framework

Based on this framework, SPOs should consider three main elements of task coordination in GSE as follows:

Coordination Mechanism. Managers or leaders in SPOs should be aware of two coordination mechanisms which are the control mechanism and the communication mechanism. Stakeholders who hold the management function have a common function: to manage the dependencies by synchronizing the activities that bring all the team members together at the same time and place for some pre-arranged purpose [9,20]. There are different types of mechanisms perceived from coordination mechanisms by Mintzberg [20]. Distributed teams that can organize dependencies by themselves, managers may perform a mutual adjustment mechanism by helping the teams in managing the work without getting involved directly in the decision making in problem-solving and task distribution. Meanwhile, in some organizations, managers still should supervise directly and take the lead in controlling the task and problem management. Beyond that two practices, the standardization of work and deliverable are the notable practices to ensure the teams have a standard guideline to perform their jobs and the task transition can be achieved smoothly.

Next, regarding the communication mechanism, we suggest SPOs to encourage direct communication among the distributed teams as much as possible by adjusting the working time or providing communication supports (e.g. tools, protocols) to build strong teamness. However, because of the GSE challenges, direct communication will never be enough. Indirect communication in a large company can be in a hierarchical form following the information flow mechanism within the organization top to bottom and vice versa. It is necessary to break the network into smaller groups to facilitate communication [8].

Communication in software engineering is seen as a knowledge-intensive activity [3]. Without an effective knowledge sharing, the project can suffer due to the failure of coordination problems that encourage collaboration [14]. Companies need to recognize the cognitive level of team members to know what kind of knowledge is needed by the team members. Our study identified several mechanisms in disseminating or distributing knowledge based on the knowledge transformation categorization, which are: Socialization, externalization, combination, and internalization [22]. SPOs can create a mapping between the available knowledge and the knowledge required to provide a knowledge gap analysis to determine the need of tools, technologies, or methods can help the transfer of knowledge [18].

Coordination Support. From the literature review and interviews, we found that SPOs aware the importance of coordination in dispersed software engineering teams. Therefore, SPOs facilitate the coordination practices by providing organizational infrastructure and tools. There are organizational supports that can be recognized such as strategy [28,29], governance [2,12], organization structure [29], and clear roles and job functions [28]. To promote task coordination, SPOs should support the practices with tools. The purposes of the tools can vary, such as providing collaboration space, enabling direct communication, amplifying the distribution of knowledge, and enhancing project control [6,27].

Tools can be utilized to provide an ongoing project activities overview at different levels of detail, to support communication and knowledge transfer, and to bring improvements to shared spaces in an integrated development environment [7]. The divergence in tools, the inadequateness of the supporting system, and the imbalance level of expertise in using the tools can limit collaboration and impede communication which ultimately delays the project [15].

Coordination Situational Background. We identified that approaching coordination can be affected by the internal organizational factors and the challenges faced by SPOs. There are several internal factors of the organization that can affect to how the organization prepares and manage task interdependencies. Organization distribution specifies how large the dispersed teams are, how legal relationships between scattered organizations are, and how organizations divide the engineering work. Process distribution describes the relationship among the tasks, the proportion of overlapped tasks, and the process chain between the distributed teams. Dependency shows how the artifacts are shared or transferred among the distributed teams.

Meanwhile, GSE challenges also provide variability in determining appropriate coordination practices. Problems emerge from the incompatibility of processes, tools, and issues related to collaboration bottlenecks because the teams that do not stay in one place are expected to impact job settings and dependencies. The geographical distance shows how teams are distributed in different locations spatially that restrict the organization to have direct communication. Temporal challenges and socio-cultural problems frequently become the communication barriers that inhibit the achievement of mutual understanding.

3.3   Identifying Activity Groups

The third step in the MAA is determining activity groups. An activity group can be seen as a class of similar tasks that represent particular functions or requirements. A chain of activity groups will describe the flow of the method, integrate the involved concepts, and elaborate the detail steps in each of the activity. By using the literature that was used during the preliminary study, we identified all the activities and concepts from the articles on the literature study and practices by the participating companies in the interviews. The study resulted in 46 unique activities and 33 concepts related to the activities. Thereafter, by analyzing the logic and categorizing the associated activities, we circumscribed the activities into 5 activity groups which are: (1) Business Analysis, (2) Situational Factor Analysis, (3) Support Analysis, (4) Task Coordination Mechanisms, and (5) Finalization and Improvement. These activity groups will be used to build the association table to map the candidate activities and concepts as the fragments of the proposed method.

3.4   Identifying Method Fragments

To facilitate the development of the association tables, we reduce the number of articles involved by selecting six articles to represent other articles (The MAA Step 4). These selected articles are chosen subjectively based on several criteria such as the number of citations and depth of discussion related to coordination practices. In the end, we tied our investigation on six papers representing the other articles: [11,17,23,28,29]. The combination of these papers covers the concepts found in the study literature as illustrated in Fig. 2.

The following step of the MAA (Step 5), we identified all the fragments found in the literature and interviews before mapping the fragments to the activity groups [19]. To avoid ambiguity, manage fragments' granularity, and improve clarity, we standardized new terms for the fragment names. For example, "Assign a liaison officer" and "Assign a service coordinator" have two different concepts namely "liaison officer" and "service coordinator." However, these concepts can be understood as a single concept: "On-site Coordinator." Another activity might consist of two activities, such as "Collaboratively develop, communicate, and distribute work plan" should be split into "Develop work plan" and "Distribute work plan."

4 Solution Design: Construction of a Global Task Coordination Method

4.1 Method Association

To start the solution development phase, the sixth step of the MAA requires that each activity and concept found in the preliminary studies are mapped to the activity groups in the association tables as depicted in Fig. 4.

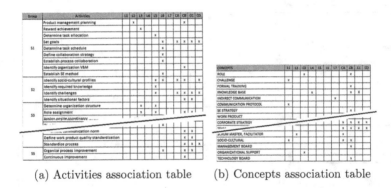

(a) Activities association table   (b) Concepts association table

Fig. 4. Association tables for the Global Task Coordination Method

There are some codes used in Fig. 4. The codes are described as follows:

- L1..L6: Literature selected to represent other literature, which are [11, 17, 23, 25, 28, 29];
- C1..C4: The codification for the participating companies, C1: AlphaSoft, C2: Betasoft; C3: GammaSoft; C4: DeltaSoft; and
- S1..S5: The five activity groups.

4.2 Method Assembly

To assemble the designed method (step 7), we use PDD language to present the method. That supports assembly-based method engineering approach for constructing situational analysis and design methods [35]. The designed method is derived based on the activity groups as listed in Sect. 3.3 and consists of the steps digested from the association tables. In the end, we present the Global Task Coordination Method. Due to page limitations, we will only present the final validated versions as depicted in Fig. 5. Additional activities are also added based on our subjectivity to maintain the logical order and flow of the activities within the method.

As the method needs to elaborate the coordination practices in a more detail, the method contains three open activities, which are *"Perform routine activities"*, *"Determine appropriate control mechanism"*, and *"Determine appropriate communication mechanism"*. These open activities are presented in the Appendix A.

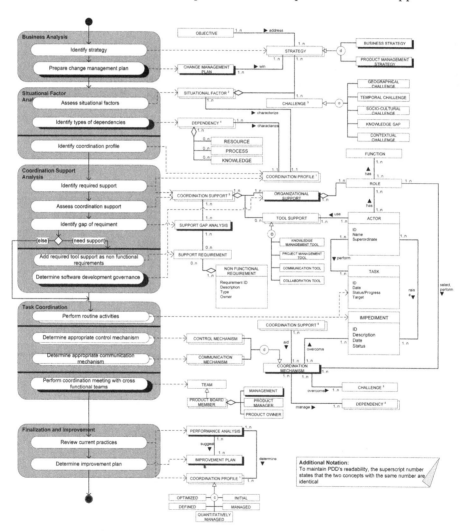

Fig. 5. The Global Task Coordination Method: Main level

4.3 Method Validation

The participating experts consist of scientific experts and business practitioners. The rationale for inviting the researcher is to obtain his feedback and critics from a person who has a broader perspective in global software engineering domain from the scientific standpoint. Other experts would be expected to provide their feedback and critics from their daily practices to assess the usability of the artifacts. The first evaluation adopts the criteria-based approach. We consider evaluating the model based on a set of criteria in assessing a method designed by the method assembly approach, which are: Completeness, consistency, efficiency, reliability, and applicability [5].

For the remaining evaluations, we involve real users to assess our designed artifacts with a natural setting that offers more critical face validity and also assures more rigorous assessment of the acceptance of the artifact [32]. We adopt two constructs from Technology Acceptance Model (TAM) which are Perceived Usefulness and Perceived Ease of Use [10]. TAM usually is used to test the behavioural acceptance or intention of using information technology such an application framework [26], software process engineering tools [33], and a newly designed method in software engineering [16]. Perceived usefulness is defined as "the degree to which a person believes that using a particular system would enhance his or her job performance." Meanwhile, perceived ease of use refers to "the degree to which a person believes that using a particular system would be free of effort" [10, p. 320]. As method engineering is used in the engineering of methods and tools in information system and technology domain [4], we assume that the adoption of TAM will be useful to evaluate the designed artifacts. The results of the validation are summmarized as presented in Table 2.

Table 2. Evaluation session summary

Criteria	Session					Remarks
	#1	#2	#3	#4	#5	
Completeness	±	+	+	+	+	Some details should be added (e.g. socio-cultural level, stakeholders, change management, governance)
Consistency	−	n/a	+	−	+	Overlapped concepts
Efficiency	−	±	+	−	+	Difficult for non-technical users. The method application does not need huge effort
Reliability	−	−	+	−	+	Several unclear terminologies
Applicability	n/a	+	+	−	+	Easy to be followed
Usefulness	n/a	+	+	n/a	+	Covers both theoretical and practical, broad aspects
Ease of use	n/a	+	+	n/a	+	The high-level PDD is useful for higher-level users
Intention to use	n/a	+	+	n/a	+	Practitioners intent to use in different ways

+: satisfied; −: unsatisfied; ±: partly satisfied; n/a: no feedback related to the criteria

Criteria-Based Evaluation. This sub section describes the results of the criteria-based evaluation that consists of five criteria of method engineering [5], which are listed below:

Completeness. The participants were satisfied with the framework and the method. The practitioners indicated that the method covers the real practices and describes the roles who are responsible for the specific activities in managing the distributed work and team members. At the same time, the framework provides a holistic overview of the theoretical perspective of a task coordination approach for PSOs to assess their situational background and the required support, as expressed by the Service Delivery Manager from GammaSoft, "*I like the overview that you have that really helps me. It's more than just theoretical. I've learned a lot.*"

Consistency. The attempt to provide a guideline at more detailed levels threaten the coherence of the developed method. The first-round evaluation directly criticized the consistency issue related to the relationship between communication and knowledge sharing in the domain of software engineering. In the subsequent rounds of assessment, participants found that the concepts and the activities are autonomous and mutually consistent.

Efficiency. The scientific experts argued that the method will not be easy to be followed by non-technical users due to the complexity and granularity. Indeed, as noticed by the practitioners, the artifacts cover all task coordination aspects in global software engineering because the artifacts attempt to cover broad topics. It is a challenge to provide a solution that comprises broad issues, while on the other hand the solution should present a clear explanation and applicative guideline.

Reliability. During the evaluation sessions, some disagreements and suggestions of the terminologies were conveyed by the participants. In the first session, the expert suggests using more specific and general terminologies to avoid misperception and uncertainty, while in the second session the expert suggested that the control mechanism should be elaborated. Then, we find it difficult to keep the method compact. After the fourth session, based on the suggestion from the expert we modified the model and optimized the documentation to make the method more concise. It is easier to maintain the reliability and consistency of concepts and activities presented in the method.

Applicability. The practitioners are satisfied with the designed artifacts and indicated that both the method and the framework could be applied as a reference guideline in their daily practices. The following sub section discusses the applicability from the perspective of behavioral intention to use by discussing the perceived usefulness and the perceived ease of use of the artifacts.

Perceived Usefulness and Perceived Ease of Use Evaluation. The practitioners as the participants of the expert validation sessions indicated to have an intention to use the GSE task coordination method. The participants from

BetaSoft were enthusiastic and considered the usefulness and ease of use of the method even though they have been doing global software engineering for more than ten years. The Product Manager from BetaSoft conveyed to augment the Technology Director comments by reflecting their past experiences, "*The model is useful and we recognize a lot of things... There's part of the method that can help us in different ways (of coordination). We also think that it's easy to use because we already used to it. We still can use the guidelines.*" Meanwhile, the participant from AlphaSoft indicated that the framework could be useful for those who have higher management roles and the detailed guidelines will be helpful for line managers and team leaders. We noticed that the experts preferred to see the method as a set of best practices guidelines where they can come back anytime, assess their current situation to detect the coordination deficiencies while enhancing their coordination practices. The practitioners could see the benefits of the method. They notified that they are very pleased with the method and desire to use the method in their daily practices.

5 Discussion

Our investigative study was completed by a literature study and interviews with SPOs so that we got a full picture from a scientific side as well as a practical side. Similarly, by performing iterative validation in which each update of each validation session becomes the input for the next session, we can assure that the method presented after the last validation is the complete and applicable artifact. The benefits and drawbacks of the method were obtained by evaluating the method based on the FEDS approach. We compare the results with the meta-modeling criteria and the intention to use test.

We incorporated experts' feedback in newer method versions. We observed a number of benefits by the assessment of the method. First, the GTC framework offers the abstraction of coordination elements that can be used as a reference that can help managers in PSOs to analyze the extent to which the organization has been able to prepare for coordination. Second, the method can be used to increase awareness of stakeholders to find out who are involved and when these roles perform particular coordination activities. Third, the method is perceived to improve the effectiveness of communication and project management by showing the variety of problems, best practices that can be emulated, and the resources needed to solve the problem.

This research aims to present the coordinating elements in the GSE comprehensively as well as its best practices to be a reference to all levels of stakeholders in SPOs. The attempt to fulfill both the conceptual as well as the detail presentation becomes an obstacle in maintaining the level of granularity that becomes the main drawback of this method which is suspected can lead to rejection of the application of this method. Although we conduct investigations and validations with rich GSE experienced business-practitioners, there is no evidence to present the applicability of this method in the real situation. However, with a positive response from experts, we perceive that this method is acceptable and has a great opportunity to be applied as a reference in coordinating on the GSE by SPOs.

In order to judge the quality of this research, we used multiple data sources to guarantee the construct validity by using multiple data sources through literature study and investigative interviews. However, expert opinion focuses more on the desire to use the method. This research could not provide an evidence where the method can effectively improve the performance of SPOs in quantitative results. The interviews involve different stakeholders with different perspectives from different companies with different characteristics. We also performed the validation phase by involving both scientific experts and business experts to ensure that the method is built comprehensively examined and gained objective judgments not only from a single point of view to maintain the external validity. Nonetheless, it may be possible that another investigation phase and validation phase at another organization outside the Netherlands yields different results. Last but not least, the limitation regarding the reliability is that the results are heavily dependent on the experience of the experts, which possibly will raise a threat to the reliability of this research.

6   Conclusions and Future Research

Software producing companies involve complex factors in their software engineering processes. As the complexity increases in situations where engineering processes are carried out in a globally distributed environment, the need to coordinate tasks will be influenced by the differentiating factors that make coordination practices unique for each organization. Therefore, we present the Global Task Coordination Method as a guideline for SPOs to determine the appropriate coordination practices based on their situational backgrounds. The method provides a reference for a better understanding of the existing aspects related to task coordination among distributed teams and to suggest adequate proposals to identify and analyze the various alternatives in managing interdependencies distributed teams. During the process, our research develops a comprehensive understanding of existing knowledge base of task coordination methods by elaborating and connecting methods which have been studied and approaches by SPOs on how tasks are allocated in GSE projects. In addition, the methodological support enhance the theoretical base in the software engineering domain by adding sources of knowledge in task coordination regarding project planning and execution management through the MAA as a method engineering approach.

Our proposal is developed based on industry inputs which are headquartered in Netherlands. We tried to maintain the external validity by selecting companies with a different characteristic of global distribution. Nonetheless, it may be possible that another investigation phase and validation phase at another organization outside the Netherlands yields different results.

To further help practitioners, we intend to encourage researchers to conduct a longitudinal multi-case study on the practices of companies. To that way, we will able to provide more objective evidence on related challenges and successful practices when coordinating tasks among distributed teams globally as well as measuring the application of the method quantitatively.

Appendix A   The Global Task Coordination Method

See Figs. 6, 7 and 8.

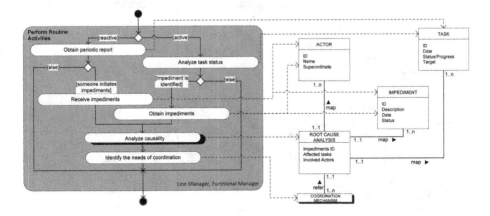

Fig. 6. The Global Task Coordination Method: Daily routines

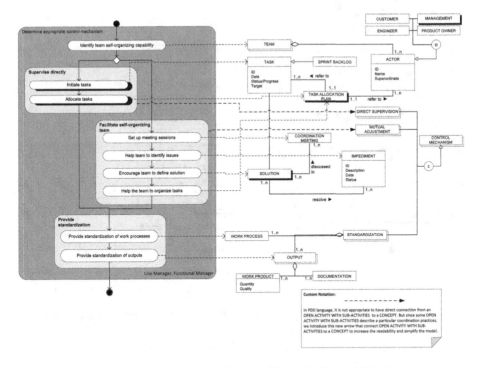

Fig. 7. The Global Task Coordination Method: Control Mechanism

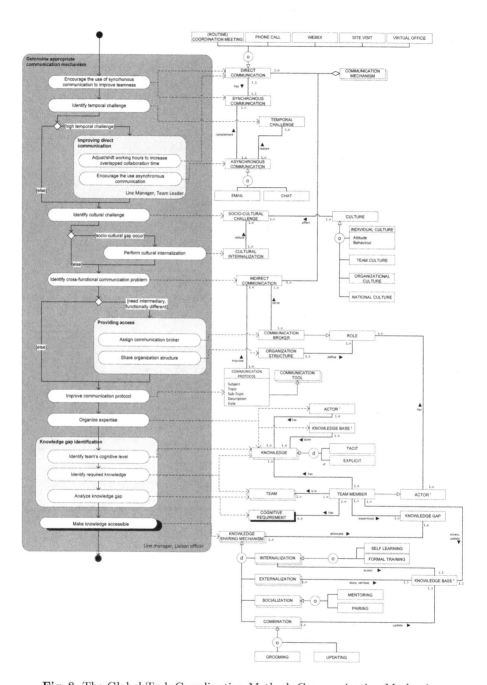

Fig. 8. The Global Task Coordination Method: Communication Mechanism

References

1. Ågerfalk, P.J., Fitzgerald, B., Olsson, H.H., Conchúir, E.O.: Benefits of global software development: the known and unknown. In: Wang, Q., Pfahl, D., Raffo, D.M. (eds.) ICSP 2008. LNCS, vol. 5007, pp. 1–9. Springer, Heidelberg (2008). doi:10.1007/978-3-540-79588-9_1
2. Bannerman, P.L.: Software development governance: a meta-management perspective. In: Proceedings of the 2009 ICSE Workshop on Software Development Governance, pp. 3–8. IEEE Computer Society (2009)
3. Bjørnson, F.O., Dingsøyr, T.: Knowledge management in software engineering: a systematic review of studied concepts, findings and research methods used. Inf. Softw. Technol. **50**(11), 1055–1068 (2008)
4. Brinkkemper, S.: Method engineering: engineering of information systems development methods and tools. Inf. Softw. Technol. **38**(4), 275–280 (1996)
5. Brinkkemper, S., Saeki, M., Harmsen, F.: Meta-modelling based assembly techniques for situational method engineering. Inf. Syst. **24**(3), 209–228 (1999)
6. Cataldo, M., Wagstrom, P.A., Herbsleb, J.D., Carley, K.M.: Identification of coordination requirements. In: Proceedings of the 2006 20th Anniversary Conference on Computer Supported Cooperative Work - CSCW 2006, p. 353. ACM Press, New York (2006)
7. Chadli, S.Y., Idri, A., Ros, J.N., Fernández-Alemán, J.L., de Gea, J.M.C., Toval, A.: Software project management tools in global software development: a systematic mapping study. SpringerPlus **5**(1), 2006 (2016)
8. Chiu, M.L.: An organizational view of design communication in design collaboration. Des. Stud. **23**(2), 187–210 (2002)
9. Colomo-Palacios, R., Casado-Lumbreras, C., Soto-Acosta, P., Garcia-Penalvo, F.J., Tovar, E.: Project managers in global software development teams: a study of the effects on productivity and performance. Softw. Qual. J. **22**(1), 3–19 (2014)
10. Davis, F.D.: Perceived usefulness, perceived ease of use, and user acceptance of information technology. MIS Q. **13**(3), 319 (1989)
11. Deshpande, S., Beecham, S., Richardson, I.: Global software development coordination strategies - a vendor perspective. In: Kotlarsky, J., Willcocks, L.P., Oshri, I. (eds.) Global Sourcing 2011. LNBIP, vol. 91, pp. 153–174. Springer, Heidelberg (2011). doi:10.1007/978-3-642-24815-3_9
12. Ebert, C.: Managing global software projects. In: Ruhe, G., Wohlin, C. (eds.) Software Project Management in a Changing World. LNCS, pp. 223–246. Springer, Heidelberg (2014). doi:10.1007/978-3-642-55035-5_9
13. Espinosa, J.A., Lerch, J., Kraut, R.: Explicit vs. implicit coordination mechanisms and task dependencies: one size does not fit all. Technical report, Carnegie Mellon University, Pennsylvania, USA (2002)
14. Herbsleb, J., Moitra, D.: Global software development. IEEE Softw. **18**(2), 16–20 (2001)
15. Jain, R., Suman, U.: A systematic literature review on global software development life cycle. ACM SIGSOFT SEN **40**(2), 1–14 (2015)
16. Koc, H., Timm, F., Espana, S., Gonzalez, T., Sandkuhl, K.: A Method for Context Modelling in Capability Management. Research Papers 43, January 2016
17. Kotlarsky, J., van Fenema, P.C., Willcocks, L.P.: Developing a knowledge-based perspective on coordination: the case of global software projects. Inf. Manage. **45**(2), 96–108 (2008)

18. Kristjánsson, B., Helms, R., Brinkkemper, S.: Integration by communication: knowledge exchange in global outsourcing of product software development. Expert Syst. **31**(3), 267–281 (2014)
19. Luinenburg, L., Jansen, S., Souer, J., van de Weerd, I., Brinkkemper, S.: Designing web content management systems using the method association approach. In: Proceedings of the 4th International Workshop on Model-Driven Web Engineering, MDWE 2008, pp. 106–120 (2008)
20. Mintzberg, H.: The Structuring of Organizations. Pearson, Upper Saddle River (1979)
21. Niazi, M., Mahmood, S., Alshayeb, M., Riaz, M.R., Faisal, K., Cerpa, N., Khan, S.U., Richardson, I.: Challenges of project management in global software development: a client-vendor analysis. Inf. Softw. Technol. **80**, 1–19 (2016)
22. Nonaka, I., Takeuchi, H.: The knowledge-Creating Company: How Japanese Companies Create the Dynamics of Innovation. Oxford University Press, New York (1995)
23. Olsson, H.H., Conchúir, E.Ó., Ågerfalk, P.J., Fitzgerald, B.: Global software development challenges: a case study on temporal, geographical and socio-cultural distance. In: Proceedings of the IEEE International Conference on Global Software Engineering, ICGSE 2006, pp. 3–11. IEEE, Florianopolis, October 2006
24. Olsson, H.H., Fitzgerald, B., Ågerfalk, P.J., Conchúir, E.Ó.: Agile practices reduce distance in global software development. Inf. Syst. Manage. **23**(3), 7–18 (2006)
25. Paasivaara, M., Lassenius, C.: Could global software development benefit from agile methods? ICGSE 2006, pp. 109–113. IEEE, Florianopolis (2006)
26. Polančič, G., Heričko, M., Rozman, I.: An empirical examination of application frameworks success based on technology acceptance model. J. Syst. Softw. **83**(4), 574–584 (2010)
27. Portillo-Rodríguez, J., Vizcaíno, A., Piattini, M., Beecham, S.: Tools used in global software engineering: a systematic mapping review. Inf. Softw. Technol. **54**(7), 663–685 (2012)
28. Richardson, I., Casey, V., Burton, J., McCaffery, F.: Global software engineering: a software process approach. In: Mistrík, I., Grundy, J., Hoek, A., Whitehead, J. (eds.) Collaborative Software Engineering, pp. 35–56. Springer, Heidelberg (2010)
29. Smirnova, I., Münch, J., Stupperich, M.: A canvas for establishing global software development collaborations. In: Dregvaite, G., Damasevicius, R. (eds.) ICIST 2014. CCIS, vol. 465, pp. 73–93. Springer, Cham (2014). doi:10.1007/978-3-319-11958-8_7
30. Smite, D.: Global software development improvement. Ph.D. thesis, University of Latvia (2007)
31. Vähäniitty, J.: Do small software companies need portfolio management? Ph.D. thesis, Helsinki University of Technology (2006)
32. Venable, J., Pries-Heje, J., Baskerville, R.: FEDS: A Framework for Evaluation in Design Science Research. Eur. J. Inf. Syst. **25**(1), 77–89 (2016)
33. Wagenaar, G., Overbeek, S., Helms, R.: Describing criteria for selecting a scrum tool using the technology acceptance model. In: Nguyen, N.T., Tojo, S., Nguyen, L.M., Trawiński, B. (eds.) ACIIDS 2017. LNCS (LNAI), vol. 10192, pp. 811–821. Springer, Cham (2017). doi:10.1007/978-3-319-54430-4_77
34. Webster, J., Watson, R.T.: Analyzing the past to prepare for the future writing a literature review. MIS Q. **26**(2), xiii–xxiii (2002)

35. van de Weerd, I., Brinkkemper, S.: Meta-modeling for situational analysis and design methods. In: Handbook of Research on Modern Systems Analysis and Design Technologies and Applications, pp. 38–58 (2009)
36. Wieringa, R.J.: Design Science Methodology: For Information Systems and Software Engineering. Springer, Heidelberg (2014)
37. Xu, L., Brinkkemper, S.: Concepts of product software. Eur. J. Inf. Syst. **16**(5), 531–541 (2007)

Enhancing Process Models to Improve Business Performance: A Methodology and Case Studies

Marcus Dees[1,2]($\boxtimes$), Massimiliano de Leoni[2], and Felix Mannhardt[2]

[1] Uitvoeringsinstituut Werknemersverzekeringen (UWV), Amsterdam,
The Netherlands
marcus.dees@uwv.nl
[2] Eindhoven University of Technology, Eindhoven, The Netherlands
{m.d.leoni,f.mannhardt}@tue.nl

Abstract. Process mining is not only about discovery and conformance checking of business processes. It is also focused on enhancing processes to improve the business performance. While from a business perspective this third main stream is definitely as important as the others if not even more, little research work has been conducted. The existing body of work on process enhancement mainly focuses on ensuring that the process model is adapted to incorporate behavior that is observed in reality. It is less focused on improving the performance of the process. This paper reports on a methodology that creates an enhanced model with an improved performance level. The enhancements of the model limit incorporated behavior to only those parts that do not violate any business rules. Finally, the enhanced model is kept as close to the original model as possible. The practical relevance and feasibility of the methodology is assessed through two case studies. The result shows that the process models improved through our methodology, in comparison with state-of the art techniques, have improved KPI levels while still adhering to the desired prescriptive model.

1 Introduction

One of the main targets of every company is to continuously improve its business processes to lower the costs, increase the revenue and guarantee higher customer satisfaction. The range of possibilities of analysing how processes are executed and of finding bottlenecks, pitfalls, etc. has been greatly enlarged by the growth of available data.

Process enhancement/improvement belongs to the realm of process mining [1], which aims to extract business knowledge from logging data that record the events linked to executions of business processes. The lion's share of attention of Process Mining has typically been about discovering models representing the actual executions of business processes as well as about checking the compliance/conformance of process executions against predefined normative models. However, less attention has been paid to process enhancement.

H. Panetto et al. (Eds.): OTM 2017 Conferences, Part I, LNCS 10573, pp. 232–251, 2017.
https://doi.org/10.1007/978-3-319-69462-7_15

We start from the belief that process improvement can happen through improving process models. Different process models can be used to discuss alternative ways to execute business processes. An improved process model describes an improved way to execute business processes. Process models can not only have a descriptive nature, but also a prescriptive purpose: they can be used to automate the process executions and enforce how processes are executed. An *improved* process model can be used to enforce a *better way* to execute business processes.

In this paper, improving a process model corresponds to repairing it to reflect reality. The existing body of work focuses on ensuring that the repaired model allows for all behavior observed in a reference event log [2]. However, this is often too extreme. First, process models are initially designed to comply with laws and regulations; therefore, new behavior can be incorporated into the model only if the new parts comply with laws and regulations. Second, one should only include extra behavior that is linked to an improvement in the performance: Adding behavior that is not linked to better performance would just make the model unnecessarily complex without any benefit.

This paper provides a methodology to enhance a process model so as to incorporate observed behavior that is not allowed in the original model. This behavior is only incorporated if it is not in violation with laws and regulations and provides an improvement of performance. Our methodology prevents behavior not related to good performance levels from being incorporated into the model. Thus we obtain models that are simple but, yet, guarantee better performance levels.

The starting point is an existing process model and an event log. The event log consists of multiple traces, each of which records the events referring to one execution of the process. A Key Performance Indicator (KPI) is attached to each log trace and is typically associated with characteristics of the traces, such as its duration (e.g. the difference between the timestamp of the last and the first event), the value of certain attributes or the number of events (in total or related to a certain activity). Our methodology builds on top of the model-repair techniques proposed by Fahland et al. [2]. In order to determine which deviations to incorporate in the improved model, the model-repair technique is combined with classification-tree learning techniques. The classification tree determines if and what kind of correlation exists between KPI levels and the observed behavior. Then, the model is enhanced by incorporating behavior that is not-allowed in the original model and that is correlated with better KPI levels while it is not violating laws and company regulations.

The benefits and the practical feasibility of the methodology are assessed through two case studies. The first case study is performed with UWV, a company which provides unemployment benefits for Dutch residents. In particular, this case study illustrates how certain deviations are worth including in the enhanced process model whereas others must not be incorporated as they are either leading to poor KPIs or are not compatible with Dutch laws. To illustrate a more generally applicability of our method beyond the single case of interest, we also report on improvement for a SAP procurement process example.

Section 2 introduces the existing body of work on which our methodology builds. Section 3 illustrates the different steps of the methodology. Section 4 shows the application of the methodology. Section 5 analyses the state of the art; finally, Sect. 6 concludes the paper.

2 Preliminaries

The research reported in this paper builds on a body of existing research in process mining. We assume that models are represented as Petri nets (see Sect. 2.1). Our methodology uses alignments of Petri nets and Event Logs to pinpoint deviations (see Sect. 2.2).

2.1 Petri Nets and Event Logs

Process models describe how activities in the process must be performed. Our approach is applicable to any modelling language (e.g. BPMN, EPC, etc.). Here, we opt for the Petri net modelling language because it has simple and clear semantics.

Definition 1 (Petri net). *Let P be a set of places, T be a set of transitions and $F \subseteq (P \times T) \cup (T \times P)$ be a flow relation between places and transitions (and between transitions and places). A **Petri net** N is a tuple $N = (P, T, F)$.*

Figure 1(a) shows an example of a Petri net. In a Petri net, transitions represent process activities. The only exceptions are the *invisible* transitions, which do not represent pieces of the process' work but are necessary to properly model some types of routing of the process. These transitions are not recorded by IT systems. For further information on Petri nets, readers are referred to [1,3]. For the Petri net in Fig. 1(a), transition names are depicted inside the transitions.

Places contain tokens; while the structure of the Petri net never changes, tokens are created and consumed. A transition is enabled if at least one token

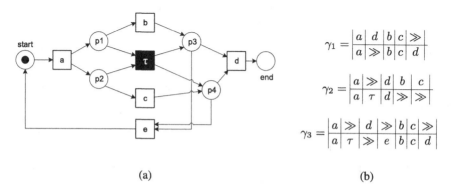

(a)   (b)

Fig. 1. A Petri net (a) and three examples of alignments of trace $\langle a, d, b, c \rangle$ with it (b)

exists in each input place of the transition. By executing (i.e., firing) a transition, a token is consumed from each input place and a token is produced for each of its output places. The state of a Petri net is determined by the distribution of tokens over places, which is denoted as *the marking*. Business processes have a precise initial and final state. Hence, when Petri nets represent business processes, they have an initial and a final marking. For the Petri net in Fig. 1(a), the markings with respectively one token in place *start* or in place *end* are the initial and final marking (and no tokens in any other place). A complete firing sequence is a sequence of transitions leading from the initial marking to the final marking. It indicates a complete execution of a process instance. The set of all complete firing sequences of a Petri net N is denoted by Ψ_N.

Real executions of processes are recorded by information systems in event logs:

Definition 2 (Event, Trace, Log). *Let $N = (P, T, F)$ be a Petri net. Each execution of a transition $e \in T$ is an **event**. A **trace** $\sigma \in T^*$ is a sequence of events. An **event log** $\mathcal{L}$ consists of a multi set of traces, i.e. $\mathcal{L} \in \mathbb{B}(T^*)$.*

2.2   Aligning Petri Nets and Event Logs and Repairing Event Logs

Not all traces in an event log may be reproduced by a Petri net, i.e. not all traces may correspond to a complete firing sequence. Conformance checking aims to verify whether the observed behavior recorded in an event log matches the intended behavior represented as a process model. The notion of **alignments** [4] provides a robust approach to conformance checking, which makes it possible to pinpoint the deviations causing nonconformity. For this aim, the events in the event log need to be related to transitions in the model, and vice versa. Building this alignment is far from trivial, since the log may deviate from the model at an arbitrary number of places. We need to relate "moves" in the log to "moves" in the model in order to establish an alignment between a process model and an event log. However, it may be that some of the moves in the log cannot be mimicked by the model and vice versa. We explicitly denote such "no moves" by $\gg$.

Definition 3 (Alignment Moves). *Let $N = (P, T, F)$ be a Petri net and $\mathcal{L}$ be an event log. A legal alignment move for N and $\mathcal{L}$ is represented by a pair $(s_L, s_M) \in (T \cup \{\gg\} \times T \cup \{\gg\}) \setminus \{(\gg, \gg)\}$ such that:*

- *(s_L, s_M) is a move in log if $s_L \neq \gg$ and $s_M = \gg$,*
- *(s_L, s_M) is a move in model if $s_L = \gg$ and $s_M \in T$,*
- *(s_L, s_M) is a synchronous move if $s_L = s_M$.*

An alignment is a sequence of alignment moves:

Definition 4 (Alignment). *Let $N = (P, T, F)$ be a Petri net with an initial marking and final marking denoted with m_i and m_f Let $\mathcal{L}$ be an event log. Let LA_N be the universe of all alignment moves for N and $\mathcal{L}$. Let $\sigma_{\mathcal{L}} \in \mathcal{L}$ be a log* *trace. Sequence $\gamma \in LA_N^*$ is an* alignment *of N and $\sigma_{\mathcal{L}}$ if, ignoring all occurrences of $\gg$, the projection on the first element yields $\sigma_{\mathcal{L}}$ (i.e. also denoted as process projection) and the projection on the second yields a sequence $\sigma'' \in T^*$ such that $m_i \xrightarrow{\sigma''} m_f$.*

A move in log for a transition t indicates that t occurred when not allowed; a move in model for a visible transition t indicates that t did not occur, when, conversely, expected. Many alignments are possible for the same trace. For example, Fig. 1(b) shows three possible alignments for a trace $\sigma_1 = \langle a, d, b, c \rangle$. Note how moves are represented vertically. For example, as shown in Fig. 1(b), the first move of γ_1 is (a, a), i.e., a synchronous move of a, while the the second and fifth move of γ_1 are a move in log and model, respectively. We aim at finding a complete alignment of $\sigma_{\mathcal{L}}$ and N with a minimal number of deviations for visible transitions, also known as an **optimal alignment** [4]. Aligning an event log $\mathcal{L}$ means to compute an optimal alignment for each trace $\sigma_{\mathcal{L}} \in \mathcal{L}$. Clearly, different traces in $\mathcal{L}$ generally have different alignments, because they contain different events and deviations. With reference to the alignments in Fig. 1(b), γ_1 and γ_2 contain two moves in model and/or log for visible transitions (γ_1 has one move in model for an invisible transition but it does not count for computing optimal alignments). Conversely, γ_3 has three moves for visible transitions, specifically one log move for d (3rd move) and two model moves for e (4th move) and d (last move). Since no alignment exists with only one move in model and/or log for visible transitions, both γ_1 and γ_2 are optimal alignments and either of these can be returned with an equal probability. For the sake of space, we assume here that all deviations (i.e., moves in model for visible transitions and moves in log) have the same severity. In [4], Aalst et al. show how this assumption can be removed. In the remainder, given a trace $\sigma_{\mathcal{L}}$ and a Petri net N, function *computeAlignment*$(N, \sigma_{\mathcal{L}})$ non-deterministically returns one of the optimal alignments for $\sigma_{\mathcal{L}}$ and N.

Alignments can be used to **repair a process model**. In Fahland et al. [2] techniques are presented to repair a model so it can replay all behavior of the event log. Alignments can also be used to **repair an event log**. Repairing an event log consists of repairing every trace of the log. A trace can be repaired by taking the process projection of the computed optimal alignment after removing all moves on model for invisible transitions. For example, trace σ_1 can be repaired as $\langle a, b, c, d \rangle$ if we consider alignment γ_1 in Fig. 1(b) as optimal alignment. Please note that several optimal alignments are possible and, hence, the trace can be repaired in multiple ways. The choice does not influence the applicability of the proposed methodology. Repairing a trace based on a model move means adding the missing event to the original trace. A log move is repaired by removing the event from the original trace. This means that, after repairing the event log, the number of traces does not vary. However, traces can become shorter or longer, depending on whether the alignment respectively contained moves on log or moves on model. When repairing an event log wrt. an alignment, it is also possible to not repair some of the deviations. For log or model moves, this respectively means that the corresponding event is kept or not added. For trace

Algorithm 1. Repair Event Log

Input: Event Log $\mathcal{L} \in \mathbb{B}(T^*)$, a Petri Net Model N, a set of legal moves that should not be repaired $D \subseteq \text{LA}_N$.
Result: Repaired Event Log

$\mathcal{L}' = \{\}$
foreach $\sigma_{\mathcal{L}} \in \mathcal{L}$ **do**
    $\gamma = computeAlignment(N, \sigma_{\mathcal{L}})$
    $\sigma \leftarrow \langle \rangle$ // the repaired trace
    for $i \leftarrow 1$ **to** $length(\gamma)$ **do**
        $(l, m) \leftarrow \gamma(i)$
        if $(l, m) \in D \wedge m => \gg \wedge l \neq \gg$ **then** // if log move not to be repaired
        | $\sigma \leftarrow \sigma \oplus \langle l \rangle$ // keep the event l
        else if $(l, m) \notin D \wedge m \neq \gg \wedge l = \gg$ **then** // if model move not to be repaired
        | $\sigma \leftarrow \sigma \oplus \langle m \rangle$ // add the missing event m
        else if $l \neq \gg \wedge m \neq \gg$ **then** // if synchronous move
        | $\sigma \leftarrow \sigma \oplus \langle l \rangle$ // keep the event l
    end
    $\mathcal{L}' \leftarrow \mathcal{L}' \cup \{\sigma\}$ // add the repaired trace to the repaired log
end
return $(\mathcal{L}')$

σ_1, not repairing deviations for d, would lead to the following repaired trace: $\langle a, d, b, c \rangle$, i.e. event d is not added for the corresponding move on model at the end of σ_1, nor is event d removed in the second position of the trace. Algorithm 1 describes how we repair an event log based on an alignment.[1] The input consists of an event log to be repaired, a Petri net model and a set of legal moves, i.e. deviations, that should not be repaired as inputs. The result is a repaired event log where all deviations, except the ones used as input, are fixed.

3   The Methodology

Figure 2 shows the steps of our methodology to enhance a process model. The basic inputs are an event log and a Petri net. The Petri net can be resulting from applying process discovery techniques or may have been designed by process owners according to how the process is prescribed or expected to be executed.

Step 1. Deviation Analysis. Deviations are detected and a set of rules is discovered that correlate deviations to a selected KPI, which is assumed to be present in the event log as an attribute. The event log is clustered according to the rules: for each rule, the corresponding cluster contains all traces that comply with the rule.

---

[1] In the algorithm, symbol $\oplus$ identifies the concatenation of two sequences. Given an alignment γ, $\gamma(i)$ denotes the i-th element of the alignment.

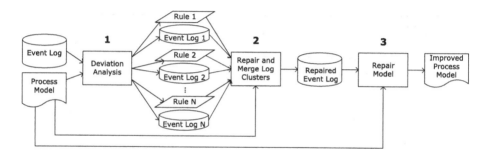

Fig. 2. Our approach to enhance a process model

Step 2. Repair and Merge Log Clusters. Traces in the different clusters are repaired to only retain those deviations that have a positive impact on the value of the KPI. All log clusters are then merged to obtain a single repaired event log.

Step 3. Repair Model. Finally the repaired log is used to repair the model: the process model is modified in such a way that it can replay all the behavior of the repaired event log. In the repaired event log we have repaired all deviations corresponding to behavior that does not improve the KPI. In this way the repair-model technique will only modify the model to make the desired deviating behavior possible.

Sections 3.1, 3.2 and 3.3 provide further details of each of the three steps of the methodology.

3.1 Deviation Analysis

The aim of the deviation analysis is to determine which deviations have a positive effect on the process performance. To measure process performance, we introduce the concept of Key Performance Indicators (KPI).

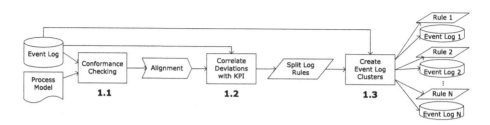

Fig. 3. Details of the deviation analysis.

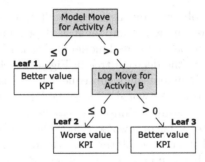

Fig. 4. Example of a decision tree built only upon model moves and log moves. A name is assigned to leaves for easy reference.

Definition 5 (Key Performance Indicator). *Let $\mathcal{L}$ be an event log. Let $\mathcal{U}$ be the universe of possible values for a key performance indicator. A key performance indicator is a pair (κ, K) consisting of a function $\kappa : \mathcal{L} \to \mathcal{U}$ that assigns a KPI value $\kappa(\sigma)$ to each trace σ and of a set $K \subset \mathcal{U}$ that contains the KPI values that are satisfactory from a business viewpoint.*

Typically, the KPI value of a trace corresponds to or is a function of the attributes present in the event log. However, in this paper we remain general on how the KPI values of process executions (i.e., traces) are computed.

Step 1.1. The first step is checking conformance of the event log and the process model. This is done to determine all deviations that are observed between the log and the model. The result of conformance checking is an alignment as discussed in Sect. 2.2.

Step 1.2. The number of model moves and log moves for the recorded activities is correlated with the chosen KPI. We want to ensure that, when the model is improved, it remains compliant with rules and regulations. This step requires analysts to provide us with a set of *disallowed activities* (i.e., transitions) $G_D \subseteq T$ that should never become part of the process model as well as with a set of *mandatory activities* be $G_M \subseteq T$, which should never become optional or be removed from the model.

During this step, we build a set of so-called *observation instances*, which are used to train a classification tree. We build one observation instance for each trace $\sigma_{\mathcal{L}} \in \mathcal{L}$. Given a trace $\sigma_{\mathcal{L}} \in \mathcal{L}$ and a key performance indicator (κ, K), an observation instance is built with the following features:

- The number of model moves in the optimal alignment of T for each allowed activity $a \in T \backslash G_D$.
- The number of log moves in the optimal alignment of T for each non-mandatory activity $a \in T \backslash G_M$.
- The KPI value for σ, namely $\kappa(\sigma)$.

Based on the set of observation instances, built from all traces, we learn a classification tree. We use the KPI component as dependent variable, namely to be predicted, and the number of log and model moves as independent variables.[2]

Consider, for example, the decision tree in Fig. 4, which is constructed after computing the optimal alignments and, subsequently, the classification tree. Each leaf is associated with a better or worse value of the KPI. In this example, to keep the explanation simple, we assume to only have two values for the KPI (i.e., $\mathcal{U}$ contains two values): a better value or a worse value. The leaves that are related to a better value are selected and can be interpreted as follows: *Activity A should never be skipped* (Leaf 1) or when skipped, *Activity B should be executed* (Leaf 3).

In the remainder, we represent the decision trees that are computed in this set as classification rules.

Definition 6 (Classification Rule). *Let* $A_L = \{a \in LA_N | a = (l, \gg)\}$ *and* $A_M = \{a \in LA_N | a = (\gg, m)\}$ *be the set of all log and model moves, respectively. Let* (κ, K) *be a KPI defined over a universe* $\mathcal{U}$ *of possible values. A* classification rule *is a tuple* $(f, v) \in ((A_L \cup A_M) \nrightarrow 2^{\mathbb{N}_0}) \times \mathcal{U}$.

For the example of Fig. 4 the leaves of the decision tree can be described as the following classification rules:[3]

- $Rule_1 = (\{(\gg, A) \rightarrowtail \{0\}\}, BetterKPI)$
- $Rule_2 = (\{(\gg, A) \rightarrowtail \mathbb{N}, (B, \gg) \rightarrowtail \{0\}\}, WorseKPI)$
- $Rule_3 = (\{(\gg, A) \rightarrowtail \mathbb{N}, (B, \gg) \rightarrowtail \mathbb{N}\}, BetterKPI)$

The determination whether a given resulting decision tree is satisfactory, e.g. in terms of f-score support, is here left to the analyst, who can use domain knowledge to assess the quality of the tree.

Step 1.3. The classification tree can be seen as a clustering of the traces of an event log. Each leaf is a different cluster and the path from the root to the leaf provides a clustering rule. For reliability, for each cluster, we discard the event log traces that are wrongly classified, namely which are classified to have KPI values that are actually different from the actual observed values. Wrongly-classified traces might potentially affect the repair-model phase by allowing behavior in the model that would not be actually linked to good KPI values. The next definition specifies how these log-trace clusters are created.

Definition 7 (Log Trace Clustering). *Let* (κ, K) *be a KPI. Let* $\{(f_1, v_1),$ $..., (f_n, v_n)\}$ *be the set of decision tree rules for a tree with n leaves. For each* (f_i, v_i), *a log trace cluster* $\mathcal{L}_i$ *is created and contains all traces* $\sigma \in \mathcal{L}$ *such that* $\kappa(\sigma) = v_n$ *and, for each move type* $a \in A_L \cup A_M$, *the number of moves of type* a *in the optimal alignment* $\gamma = computeAlignment(N, \sigma)$ *is* $\#a \in f_i(a)$.

---

[2] Here we talk in term of classification tree, which can be a decision or regression tree. Decision trees are used when the KPI values are discrete; otherwise, we use regression tree.

[3] Notation $\{d_1 \rightarrowtail c_1, \dots, d_n \rightarrowtail c_n\}$ indicates a function f with domain $\{d_1, \dots, d_n\}$ in which $f(d_1) = c_1, \dots, f(d_n) = c_n$.

3.2 Repair and Merge Log Clusters

The log clusters that have been created in Step 1.3 need to be repaired, based on the repair rules that are derived in Step 1.2. After that, the log clusters need to be merged to present a single repaired log for usage in Step 3. Figure 5 shows the details of this phase of the approach.

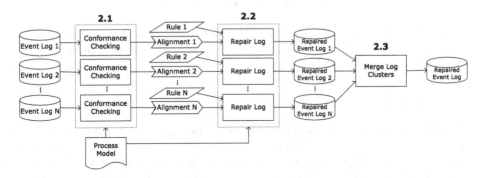

Fig. 5. Details of repair and merge log clusters.

Step 2.1. Conformance Checking is done with the original process model and each log cluster. The alignment reproduces the deviations that were part of the Conformance Checking result of Step 1.1.[4]

Step 2.2. This step is repeated for each cluster $\mathcal{L}_i$, associated with a classification rule (f_i, v_i). If $\mathcal{L}_i$ is linked with a satisfactory KPI value (i.e., $v_i \in K$), we use Algorithm 1 to repair all deviations in each trace of $\mathcal{L}_i$, except for those being part of the repair rule associated with the log cluster. For each cluster, we repair all deviations linked to behavior that is not correlated to improved KPI values and/or that would be violating laws and regulations. This guarantees that each repaired log cluster only retains those deviations that, on the one hand, are worth incorporating in the enhanced model and, on the other hand, would create a model that still adheres to the applicable laws and regulations. The other deviations are repaired and, hence, not eligible for repair in the next methodology steps. For instance, consider Fig. 4. For the event log cluster associated with Leaf 3 we repair all deviations except model moves for *Activity A* and log moves for *Activity B*, as the combination of model moves for *Activity A* and log moves for *Activity B* leads to a better KPI value. For the event log cluster associated with Leaf 2, we repair all deviations because the KPI value is poor. For the event log cluster associated with Leaf 1, we also repair all deviations except for the model move of *Activity A*. However, the clustering rule specifies that all cluster traces have no model moves for *Activity A*; hence, in fact, all deviations are also repaired.

---

[4] Step 2.1 is a conceptual step. In practice, one does not need to recompute the alignments for the cluster logs as one can simply reuse the alignments obtained as result of Step 1.1.

Definition 8 (Repaired Log Cluster). *Let $\mathcal{L}_i$ be an log trace cluster created on the basis of a classification rule (f_i, v_i). Let N be a Petri net model and (κ, K) a KPI. The corresponding* Repaired-Log *cluster is*

$$\mathcal{L}'_i = \begin{cases} repairLog(\mathcal{L}_i, N, dom(f_i)) & \text{if } v_i \in K \\ repairLog(\mathcal{L}_i, N, \emptyset) & \text{otherwise.} \end{cases}$$

Step 2.3. We merge all repaired log clusters into a single event log. The repaired log $\mathcal{L}_R$ is equal to $\bigcup \mathcal{L}'_i$. This is a requirement to apply the next step, namely repairing the process model.

3.3   Repair Model

This phase repairs the model. The input is the original model and the repaired log $\mathcal{L}_R$, obtained as result of Step 2.3. The process model is modified in such a way that it can play out all the behavior of the repaired event log. Consider again the example in Fig. 4: the model is modified so that *Activity A* may be skipped and *Activity B* is now allowed at certain points of the process (Leaf 3 in the figure). These modifications can be incorporated into the model because they have shown to achieve better KPI values.

Unfortunately, we cannot always manually modify the model to allow for extra behavior because it is not always clear which part of the model should be modified. For instance, for the example referring to Fig. 4, *Activity B* should become allowed at certain points of the process. But, at which points exactly? As another example, the same activity - say X - may be prescribed to happen at different points of the process. From the classification tree, we can hypothetically observe that activity X can be skipped while the KPI becomes better. However, at which point can activity X be skipped? In the light of above, we leverage on the repairing technique by Fahland et al. [2] to determine how to modify the model. This technique will modify the model so as to obtain a new model that can replay all the behavior observed in the repaired event log. The repairing technique supports a frequency threshold parameter to apply a filtering such that the repaired model only incorporates behavior with a frequency higher than this threshold. For our methodology, this option is very interesting: if a certain deviating behavior has shown to guarantee good KPI levels, it should only be kept if it is occurring in a sufficient enough number of traces.

4   Implementation and Evaluation

The entire methodology is supported by plug-ins of the process mining framework ProM 6.6[5]. In particular, Steps 1.1 and 2.1 can be carried out with the plug-in *Replay a Log on Petri Net for Conformance Analysis*, which operationalizes the alignment technique discussed in [4]. Steps 1.2 and 1.3 are implemented

---

[5] http://www.promtools.org/doku.php?id=prom66.

as plug-in *Perform Predictions of Business Process Features* (see [5]). Step 2.2 is implemented as plug-in *Repair Log With Respect to Alignment*. Step 2.3 is implemented as plug-in *Concatenate/Union of Event Logs*. Finally, Step 3 is implemented as plug-in *Repair Model*, which operationalizes the technique by Fahland et al. [2].

We evaluated our methodology through two case studies. Section 4.1 reports on the first case study in collaboration with UWV (Uitvoeringsinstituut Werknemersverzekeringen) and refers to the provisioning of unemployment benefits for the Netherlands' residents. The second case study is presented in Sect. 4.2 and refers to a procurement process as implemented in a SAP system.

4.1   UWV Case Study

UWV is the social security institute of the Netherlands and responsible for the execution of a number of employee related insurances. The case study focuses on the unemployment benefits process of UWV. When employees become unemployed, they may be entitled to these benefits. Employees have to file a claim at UWV. UWV then decides whether they are entitled to benefits. When claims are accepted employees receive benefits with a regular frequency until they find a new job or the maximum period for their entitlements is reached. UWV refers to employees who are making use of their services as customers, therefore we use the term customer in the remainder of the paper.

UWV translates the legal text of the law into a process design that is executable. Within the boundaries of the law there is some flexibility in the way the process can be designed and executed. UWV can for example choose how to communicate with its customers. Communication can happen through several channels like internet, a letter or a telephone call. On the other hand, some parts of the process need to adhere to the Dutch laws. Figure 6 shows a prescriptive process model that encodes the relevant Dutch law and UWV's protocols. The process model was designed in collaboration with a process specialist of UWV.

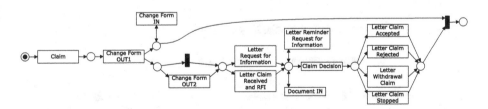

Fig. 6. The original process model of the unemployment benefits claim handling process at UWV.

In particular, UWV aims to improve this process by reducing the claim's throughput time. This would likely reduce the costs and improve the satisfaction of customers, who receive a faster answer to claims. Therefore, **the throughput**

time is the KPI that we aim to minimize by employing our methodology. Regarding the definition of KPI (κ, K), $\kappa(\sigma)$ is defined as the timestamp of the last event of σ minus the timestamp of the first event in σ (i.e. trace duration), and K contains any duration smaller than 8 days.

Next to a process model, our methodology requires an event log. The event log consists of 25476 traces and 161365 events, which refer to 21 different activities. The event log recorded executions of several activities that are not part of the original model, for example: *Call Claim Received and RFI, Letter Information to Customer, Call Information to Customer* and *Call OUT*. These events were not foreseen in the reference model, but, in fact, they were executed by UWV employees. During the application of our methodology, we will hence investigate whether some of these activities ought to be added to the process model. In particular, we aim to incorporate any of those activities in the right positions if their execution has shown to lead to good KPI levels. From a business perspective, only the following model activities can be made optional or made possible at a different point of the process execution: *Letter Claim Received and RFI, Letter Request for Information* and *Letter Reminder Request for Information*. All other activities in the model are mandatory and may not be moved to other positions or removed. Similarly, only the following log activities may be added to model: *Call Claim Received and RFI, Call Information to Customer, Call OUT* and *Letter Information to Customer*.

For the purpose of the validation, the event log was randomly split in two groups: 80% of the traces are used to improve the model using our methodology and 20% of the traces are used later to test the quality of the improved model. This is, in fact, performed in line with the validation techniques employed, e.g., in data mining.

The first step of the approach is the deviation analysis. It consists of building alignments between the model in Fig. 6 and the extracted event log (Step 1.1). The alignments are used along with the event log to correlate the deviations with the KPI values (Step 1.2 in Fig. 3). The resulting tree is shown in Fig. 7. We used a regression tree because the dependent variable, throughput time of

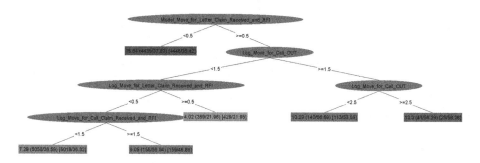

Fig. 7. Regression tree for the case study that correlates the deviations with the throughput time. Two leaves with the lowest throughput time are selected as repair rules. They are highlighted with a dashed box.

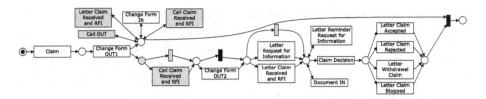

Fig. 8. Improved model using our methodology. The gray transitions are those which are added to the model.

the claim process, is an attribute that has a numeric data type. The tree has six leaves. There are two leaves with an average throughput time below the desired value of 8 days, i.e., the attribute used as KPI. For those two leaves, we generated the corresponding log clusters and we repaired the deviations that are not present in the regression tree, in accordance with the methodology step defined in Definition 8. Also, for the other leaves, we repair every deviation. Only correctly classified traces for each leaf are selected to be exported into a log cluster. The error threshold is set to 15% (Step 1.3). In this way, we retain the traces referring to process executions with a KPI value relatively close to the expected value of the log cluster. The other traces are considered to be noise, which can negatively affect the final result of Step 3: the application of the model-repair technique to obtain a model that incorporates behavior that has not been associated with better KPI values. The log clusters are merged into a new log (Step 2.3) that is compliant with the model, except for the deviations that are correlated with a better KPI value.

Finally the repaired event log is used together with the original model to improve the model (Step 3). Only behavior that has occurred at least 50 times is taken into account. Figure 8 shows the improved model. Activities *Call Claim Received and RFI*, *Call OUT* and *Letter Claim Received and RFI* are allowed to be executed multiple times after the first *Change Form* has been sent. Before *Change Form OUT2* now *Call Claim Received and RFI* is optionally allowed to be executed. Finally, the activities *Letter Claim Received and RFI* and *Letter Request for Information* are now allowed to be skipped.

In essence, the repaired model represents the fact that calling the customer for a confirmation of a claim reception along with, when necessary, asking for extra information, is faster than sending a letter. This is also understandable from a domain perspective: waiting for a customer to respond to a letter would take time. When the customer ultimately does respond, the employee handling the case, has to recall the details of the claim. By calling the customers, the employee can, most of the time, handle the claim in one go without having to wait. We assessed the usefulness of our methodology by comparing this model with the model obtained by only employing the model-repair technique of Fahland et al. [2]. For a fair comparison, the same parameters where used, namely only considering traces that occur at least 50 times. The model obtained without our methodology is shown in Fig. 9. This model contains several activities that can be executed an arbitrary number of times in any order; e.g. see the part of the model that is within the dashed box. Clearly, this model is underfitting the event log because it allows for a lot of behavior that is not observed in reality. The model also allows behavior that violates Dutch laws as well as UWV's internal regulations and protocols. This clearly manifests itself in the fact that sending one of the mandatory claim-outcome letters can be skipped.

As a final validation, we measured the KPI level of the traces that record executions compliant with the original model and we compared it with the KPI level of the executions compliant with the enhanced model, both with our methodology and with the technique by Fahland et al. [2]. As mentioned before, for this evaluation we employ the 20% log traces that were left aside, hereafter named test log.

If we consider the traces of the test log that are compliant with the original model in Fig. 6, the average throughput time is 15.44 days. If we consider traces of the test log that are compliant with the model enhanced with our methodology in Fig. 8, the average throughput time is 10.88 days, leading to an improvement of 29.5%. When using the model enhanced only with the technique of Fahland et al. [2] (Fig. 9), the average is 11.18 days, leading to a decrease of throughput time of around 27.6% compared with the original model. It is clear that the overall improvement with our methodology is marginal compared with the technique by Fahland et al., if we look at the improvement of the KPI, only. However, first, our model is compliant with regulations, whereas the other model could not be used by UWV since it violates the regulations. Moreover, our model is simpler and, very importantly, it is not underfitting, i.e. allowing for a lot of behavior that should be disallowed. If that behavior is incorporated, the model becomes less insightful on which behavior is really linked to better KPI levels. In other words, the extra allowed behavior is not all guaranteed to be connected to better KPI levels.

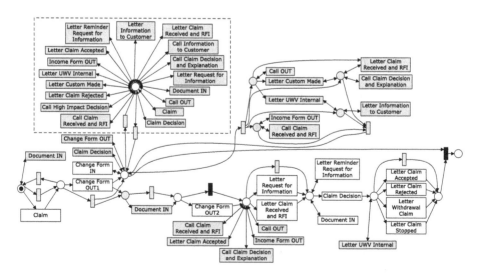

Fig. 9. Improved model based on the Fahland technique [2]. The gray transitions are those which are added to the model.

4.2 SAP Procurement Case Study

The procurement process we used, is part of the SAP Material Management (SAP MM) module. This module supports the procurement and inventory functions occurring in day-to-day business operations. Figure 10 shows the process model we used in this case study. During this process changes may occur to the price of the goods, the quantity and the vendor where the goods are ordered. Changes incur higher costs. Therefore, the KPI that we want to **minimize** is **the number of purchase orders with a change.**

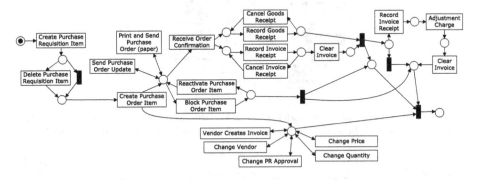

Fig. 10. The original process model of the SAP procurement process.

All activities that are shown in the model in Fig. 10 are present in the event log. Additionally there are some activities that are not part of the model, but exist in the event log. Examples of these are *Delete Purchase Order Item* and *Send Overdue Notice*. The event log is created consisting of 105708 traces. In the same way as for the case study with UWV discussed in Sect. 4.1, we split the event log in a training log with 80% of the remaining traces and in a test log that accounts for 20% of the traces. For the sake of space, we cannot discuss all the intermediate steps and we limit ourselves to discuss the final models. Figure 11 shows the model obtained by using the technique of Fahland et al. whereas Fig. 12 illustrates the model obtained when applying our methodology. This model is clearly simpler and easier to understand since it is a lot more similar to the original model.

If we employ the test log and use it together with the original model, then 52% of the fitting cases have a change. When the same test log is used together with the model in Fig. 11 obtained by the technique of Fahland et al., 38% of the cases have a change. Finally, when it is used together with the model obtained through our approach in Fig. 12, 44% of the cases have a change. While our methodology guarantees an improvement of the KPI in comparison with the original model, the model enhanced with the technique of Fahland et al. has an even better performance improvement. In terms of simplicity and not being underfitting, while also showing all the positive considerations drawn for the UWV case study at the end of Sect. 4.1, it is again clear that our model is more

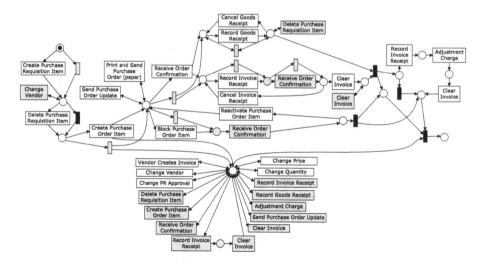

Fig. 11. The repaired process model of the SAP procurement process obtained by the Fahland technique. The gray transitions are those which are added to the model.

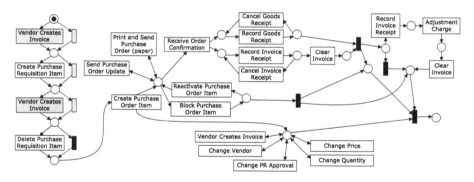

Fig. 12. The repaired model for the SAP procurement process using our approach. The gray transitions are those which are added to the model.

valuable. It is also worth highlighting that we still obtain an improvement of 8% in the number of changes occurring in purchase orders with only a few changes of the original model.

5   Related Work

In Sect. 1 we already discussed one limitation of the model-repair technique of Fahland et al. [2]: This technique considers all deviations to be eligible to be incorporated in the model, independently whether or not a certain deviation typically causes poor performance levels or violates laws and regulations. The evaluation showed that a lot of behavior is actually not necessary to obtain a model with good performance levels. The same limitation is also observed in works [6–8]; in particular, in [8] the decision whether or not to incorporate certain deviations into the model is driven by many quality dimensions that, however, does not include KPI fulfilment and adherence to legislations.

Schunselaar et al. [9] try to find a best configured process model given a set of configurable process models and a set of KPIs. Using a Pareto front, they select those process models that have the best performance on the set of KPIs. While this approach needs a set of configurable process models to capture variation in the KPIs values, our approach only needs a single process model.

While our paper reports on the use of the techniques by Fahland et al. for repairing the process model, our methodology is not bounded to that repair technique only. We could alternatively use other techniques, such as that by Buijs et al. [8], which mediates several model quality criteria. Similarly to [2], in [10] authors propose to repair the model while improving the correspondence, i.e. fitness, between model and log as much as possible. Costs are associated with repair actions and an optimal strategy is determined to minimize the costs of the repairs, but it is not possible to improve on a KPI. In [11], authors propose to repair a process model to make it similar to a collection of process models. Differently from our approach, they do not use an event log but a collection of models and try to minimise the edit distance to all models in the collection.

Repairing a process model can also be regarded as ensuring that the model is sound and, hence, does not contain deadlocks. Gambini et al. [12] and Lohmann [13] propose techniques along these lines. Repairing a process model can also be seen as simplifying the model while allowing for the same behavior, such as what is proposed by Fahland et al. [14]. Clearly, ensuring soundness and simplifying models pursue a different goal than what we aim at in our work.

The methodology proposed in this paper is also in line with the DMAIC methodology of Six Sigma [15], which comprises five steps: define the goals, measure key aspects, analyse data, improve and control the improvement. In fact, the first step of our methodology (Deviation Analysis) corresponds to the analyse data step of DMAIC and the second and third step corresponds to improving (the model of) the process. The DMAIC's step of measuring key aspects corresponds to having the KPI values as an event log attribute. The step of controlling the improvement is beyond the scope of this paper but corresponds to the natural follow-up once the model is improved. Namely one would like to verify whether the improved model actually leads to better KPI values.

6 Conclusion

Process mining is not just about discovering the control-flow or diagnosing non compliance. Enhancing business processes to improve the performance is perceived equally important [1] and this can be viewed from many perspectives. Here, we aim to repair the model thus incorporating some of the behavior that is observed in reality but disallowed by the model. However, the model should be extended so as to only allow additional behavior that *(i)* does not violate company rules and national regulations and that *(ii)* has shown to lead to better KPI levels. As discussed in Sect. 5, the existing body of work in the Process Mining context does not look at these aspects.

This paper has proposed a methodology that considers the two aspects above: the adherence to rules/laws and KPI improvement. Our proposed methodology unifies an existing approach for model repair proposed by Fahland et al. [2] with other existing works in the field of conformance checking and deviation-to-KPI correlation.

A detailed evaluation has been carried out through two case studies. One case study is based on the SAP procurement process and the other is conducted with UWV, a Dutch financial institute that provides social security benefits to the residents in the Netherlands. The results clearly show the practical usefulness of the methodology: we improve the SAP model so less changes will occur during the process, leading to lower costs. The UWV model is improved to allow all the behavior observed in the event log which is correlated with better KPI levels, i.e. a lower throughput time, while preventing violations of the Dutch unemployment-benefit laws.

As future work, we aim to extend our methodology so that the improved model only retains behavior allowed by the original model that yields better KPI levels. Currently, our methodology only allows one to extend the model to incorporate behavior observed in reality that led to KPI improvements. Our methodology should be extended so that the new model forbids certain behavior allowed by the original model if it has shown to yield worse KPI levels. Also, it is worth working further on our methodology to consider concept drift [16]. For instance, it would be relevant to investigate which drifts contribute to improve KPIs and to only incorporate those into the model. At the moment we can only improve one KPI at the time. Thus, improving multiple KPIs at the same time, is also a direction for future work.

We acknowledge that one of the limitations of our framework is that we only consider mandatory and disallowed activities to ensure compliance. However, generally speaking, there might be other criteria (e.g. precedency between activities) to inspect to ensure compliance. The consideration of the latter is a possible avenue of future work.

Whereas we show the relevancy of the methodology, one of the drawbacks is that all steps of the methodology need to be manually performed. This is clearly tedious and error-prone. However, this can be automated through scientific workflows. Scientific Workflow Management systems help users to design, compose, execute, archive, and share workflows that represent some type of analysis or experiment [17]. As future work, we will implement this methodology as an scientific workflow: This is far from being difficult as every step can be easily automated as an activity of a scientific workflow.

References

1. van der Aalst, W.M.P.: Process Mining: Data Science in Action, 2nd edn. Springer, Heidelberg (2016)
2. Fahland, D., van der Aalst, W.M.P.: Model repair - aligning process models to reality. Inform. Syst. **47**, 220–243 (2015)

3. Murata, T.: Petri nets: properties, analysis and applications. Proc. IEEE **77**(4), 541–580 (1989)
4. van der Aalst, W.M.P., Adriansyah, A., van Dongen, B.F.: Replaying history on process models for conformance checking and performance analysis. Wiley Interdisc. Rev Data Min. Knowl. Discov. **2**(2), 182–192 (2012)
5. de Leoni, M., van der Aalst, W.M.P., Dees, M.: A general process mining framework for correlating, predicting and clustering dynamic behavior based on event logs. Inform. Syst. **56**, 235–257 (2016)
6. Kalsing, A.C., do Nascimento, G.S., Iochpe, C., Thom, L.H.: An incremental process mining approach to extract knowledge from legacy systems. In: Proceedings of the 14th IEEE International Enterprise Distributed Object Computing Conference, pp. 79–88. IEEE (2010)
7. Sun, W., Li, T., Peng, W., Sun, T.: Incremental workflow mining with optional patterns and its application to production printing process. Int. J. Intell. Contr. Syst. **12**(1), 44–55 (2007)
8. Buijs, J.C.A.M., La Rosa, M., Reijers, H.A., van Dongen, B.F., van der Aalst, W.M.P.: Data-Driven Process Discovery and Analysis. Proceedings of the Second International Symposium Data-Driven Process Discovery and Analysis (SIMPDA 2012). LNBIP, vol. 162, pp. 44–59. Springer, Heidelberg (2013)
9. Schunselaar, D.M.: Configurable process trees: elicitation, analysis, and enactment. Ph.D. thesis, Eindhoven University of Technology, Eindhoven (2016)
10. Polyvyanyy, A., Van der Aalst, W.M.P., Ter Hofstede, A., Wynn, M.: Impact-driven process model repair. ACM Trans. Softw. Eng. Methodol. (TOSEM) **25**(4), 28 (2016)
11. Li, C., Reichert, M., Wombacher, A.: Discovering reference models by mining process variants using a heuristic approach. In: Dayal, U., Eder, J., Koehler, J., Reijers, H.A. (eds.) BPM 2009. LNCS, vol. 5701, pp. 344–362. Springer, Heidelberg (2009). doi:10.1007/978-3-642-03848-8_23
12. Gambini, M., La Rosa, M., Migliorini, S., Hofstede, A.H.M.: Automated error correction of business process models. In: Rinderle-Ma, S., Toumani, F., Wolf, K. (eds.) BPM 2011. LNCS, vol. 6896, pp. 148–165. Springer, Heidelberg (2011). doi:10.1007/978-3-642-23059-2_14
13. Lohmann, N.: Correcting deadlocking service choreographies using a simulation-based graph edit distance. In: Dumas, M., Reichert, M., Shan, M.-C. (eds.) BPM 2008. LNCS, vol. 5240, pp. 132–147. Springer, Heidelberg (2008). doi:10.1007/978-3-540-85758-7_12
14. Fahland, D., van der Aalst, W.M.P.: Simplifying discovered process models in a controlled manner. Inform. Syst. **38**(4), 585–605 (2013)
15. International Organization for Standardization: ISO 13053:2011 quantitative methods in process improvement - Six Sigma - part 1: DMAIC methodology, September 2011
16. Bose, R.P.J.C., van der Aalst, W.M.P., Žliobaitė, I., Pechenizkiy, M.: Handling concept drift in process mining. In: Mouratidis, H., Rolland, C. (eds.) CAiSE 2011. LNCS, vol. 6741, pp. 391–405. Springer, Heidelberg (2011). doi:10.1007/978-3-642-21640-4_30
17. Bolt, A., de Leoni, M., van der Aalst, W.M.P.: Scientific workflows for process mining: building blocks, scenarios, and implementation. STTT **18**(6), 607–628 (2016)

TraDE - A Transparent Data Exchange Middleware for Service Choreographies

Michael Hahn[(✉)], Uwe Breitenbücher, Frank Leymann, and Andreas Weiß

Institute of Architecture of Application Systems (IAAS),
University of Stuttgart, Stuttgart, Germany
{hahnml,breitenbuecher,leymann,weissas}@iaas.uni-stuttgart.de

Abstract. Due to recent advances in data science the importance of data is increasing also in the domain of business process management. To reflect the paradigm shift towards data-awareness in service compositions, in previous work, we introduced the notion of data-aware choreographies through cross-partner data objects and cross-partner data flows as means to increase run time flexibility while reducing the complexity of modeling data flows in service choreographies. In this paper, we focus on the required run time environment to execute such data-aware choreographies through a new Transparent Data Exchange (TraDE) Middleware. The contributions of this paper are a choreography language-independent metamodel and an architecture for such a middleware. Furthermore, we evaluated our concepts and TraDE Middleware prototype by conducting a performance evaluation that compares our approach for cross-partner data flows with the classical exchange of data within service choreographies through messages. The evaluation results already show some valuable performance improvements when applying our TraDE concepts.

Keywords: Service choreographies · Data-awareness · Cross-partner data flow · Transparent data exchange · BPM

1 Introduction

Service-oriented architectures (SOA) have seen wide spread adoption. The concept of composing self-contained units of functionality as services over the network has found application in many research areas and application domains [18]. For example, in Business Process Management (BPM), Cloud Computing, the Internet of Things, or eScience. The composition of services can be specified through a broad variety of modeling languages which can be grouped into two categories: service orchestrations and service choreographies.

While service orchestrations, also known as *processes*, are specified from the viewpoint of one party that acts as a central coordinator, service choreographies provide a global view on the potentially complex conversations between multiple interacting services without relying on a central coordinator [3]. Therefore, the notion of service choreographies focuses on services taking part in a

© Springer International Publishing AG 2017
H. Panetto et al. (Eds.): OTM 2017 Conferences, Part I, LNCS 10573, pp. 252–270, 2017.
https://doi.org/10.1007/978-3-319-69462-7_16

collaboration, as so-called *participants*, and their interplay with other services by specifying corresponding conversations through message exchanges between them [3]. Prominent modeling languages for service orchestrations are the Business Process Management Notation (BPMN) [13] and the Business Process Execution Language (BPEL) [12]. Service choreographies can be modeled, for example, with modeling languages such as BPMN or BPEL4Chor [9].

With recent advances in data science the importance of data is increasing also in the domain of business process management [11,15]. For improving the level of data-awareness of service choreographies, we introduced an extended management life cycle [8] for data-aware choreographies and proposed an approach for enabling transparent data exchange (TraDE) in choreographies motivated on shortcomings of current choreography modeling languages [7]. The overall goal is to support data capabilities already on the level of the choreography to reduce modeling complexity while increasing run time flexibility.

In this work, we focus on the required middleware and its integration with process engines to support the run time perspective of data-aware choreographies as briefly outlined in Hahn et al. [7]. More specifically, the contributions of this paper can be summarized as follows: (i) we introduce and discuss an internal metamodel for a TraDE Middleware, (ii) present an architecture of such a middleware together with a prototypical implementation, and (iii) evaluated the prototype and the underlying TraDE concepts. The rest of this paper is structured as follows: Sect. 2 provides an overview of our previous work on the concepts of transparent data exchange in choreographies and the role of a corresponding middleware. Based on that, we introduce a modeling language-independent metamodel and a generic TraDE Middleware architecture and discuss how the middleware can be integrated with corresponding process engines in Sect. 3. In Sect. 4, we present a prototypical implementation of the architecture and its integration with a process engine solution. Section 5 presents an evaluation of the performance alteration when applying our TraDE concepts to a choreography execution by integrating our TraDE Middleware prototype. Finally, the paper discusses related work in Sect. 6, and concludes with our findings together with an outlook on future work in Sect. 7.

2 Transparent Data Exchange Approach

As background, in the following, we shortly outline our previously presented concepts for modeling and execution of data-aware choreographies through a transparent data exchange as discussed in detail in Hahn et al. [7]. Therefore, we briefly explain our modeling extensions, namely *cross-partner data objects* and *cross-partner data flows*, and how these extensions can be supported during run time by our new TraDE Middleware which is in the focus of this paper.

2.1 Cross-Partner Data Objects and Cross-Partner Data Flows

Figure 1 shows a motivation example of a data-aware choreography model with three interacting participants with our applied modeling extensions. We use the

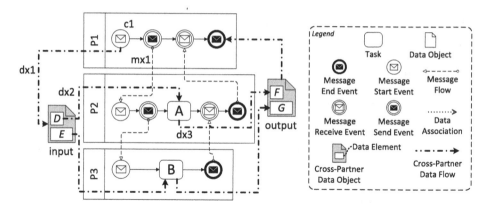

Fig. 1. Example choreography illustrated as BPMN collaboration model with applied *cross-partner data objects* and *cross-partner data flows*.

Business Process Management Notation (BPMN) [13] as a basis to illustrate our modeling extensions. However, the underlying concepts and the middleware presented in this work are not bound to BPMN and can therefore be applied or integrated with any other choreography modeling language.

The conversations between the participants shown in Fig. 1 are modeled by BPMN message intermediate events and message flows, e. g., *mx1* in Fig. 1. Each of the participants is instantiated through a corresponding BPMN message start event, e. g., *c1* in P1, which consumes an incoming request message and extracts the contained data for processing it within the choreography. Choreography data is modeled by our cross-partner data objects, e. g., *input* in Fig. 1, and the reading and writing of the cross-partner data objects from tasks and events is specified through cross-partner data flows, e. g., *dx1* or *dx3* in Fig. 1. To avoid confusion between BPMN data objects as language-specific constructs and cross-partner data objects as a general concept, we use the generic term *data container* when we talk about modeling constructs that allow the specification of data on the level of a specific modeling language, e. g., BPMN data objects or BPEL variables. While data between participants is normally transferred through the exchange of messages within conversations, the notion of cross-partner data objects and cross-partner data flows allows us to decouple the exchange of data from the exchange of messages. For example, instead of forwarding the data of the initial request from participant P1 to participant P2 through the message flow mx1, we can directly specify a cross-partner data flow to task *A* of participant P2 where the data are actually processed. The same applies for the result data of task A, instead of forwarding it to other participants through the exchange of messages, it can be directly stored in cross-partner data object *output* by the specified cross-partner data flow *dx3* as shown in Fig. 1.

The notion of cross-partner data objects allows us to specify all data relevant for a choreography by specifying a set of cross-partner data objects. Such a set expresses the commonly agreed data of a choreography shared by and accessible from all participants and can therefore be seen as a *choreography data model*. This enables modelers to specify required data and their structures in a self-contained and consolidated manner within a choreography model. A *cross-partner data object* has a unique identifier and contains one or more *data elements*. For example, the cross-partner data object *input* contains the two data elements D and E as shown in Fig. 1. A data element has a name and contains a reference to a definition of its structure, e. g.,using a build-in type system or an XML Schema Definition [16]. The actual data values during run time are held by these data elements. Therefore, they are comparable to the classical data containers of standardized choreography and orchestration modeling languages.

By introducing cross-partner data flows we support modelers so that they are able to intuitively specify data flows within and across participants in a choreography. While in classical choreography modeling languages, such as BPMN, data can only be passed across participants through message flows, cross-partner data flows allow to decouple the exchange of data from the exchange of messages. This means that instead of introducing additional modeling constructs for passing the value of a data container from one participant to another through a message flow (e. g.,mx1 in Fig. 1), cross-partner data flows allow to model the exchange of data between participants and globally shared cross-partner data objects, e. g.,dx2 or dx3 in Fig. 1. Since a lot of choreography modeling languages do not allow to specify directly executable models, an established approach is to transform the choreography models into a collection of private process models [4]. The resulting private process models can then be manually refined by adding corresponding internal logic for each participant. We extended this transformation step for data-aware choreographies in our previous work [7] by translating all cross-partner data objects into standard data containers on the level of the private process models again, e. g.,using data objects in BPMN or variables in BPEL. The reason for this translation is that it allows modelers to refine the private processes using both locally and globally shared data containers in the same manner.

2.2 Towards a TraDE Middleware

After we have introduced our modeling extensions for transparent data exchange, the open research question is: "How can we support these modeling extensions during run time in order to actually execute a modeled data-aware choreography?".

Towards this goal in Hahn et al. [7] we outlined an overall system architecture for a modeling and run time environment that enables the execution of data-aware choreographies by supporting cross-partner data objects and cross-partner data flows. Therefore, we discussed our concept of introducing links between the data containers in the private process models of a choreography and the cross-partner data objects and data elements they represent. To provide and manage the cross-partner data objects of a data-aware choreography model we sketched a new middleware layer, the TraDE Middleware. This middleware acts as a data hub between the choreography participants and therefore supports the process engines that execute the private process models with the realization of the modeled cross-partner data flows. In the following, we want to shortly recap our vision presented in Hahn et al. [7] by describing how such a TraDE Middleware is used to execute the example choreography shown in Fig. 1. This will provide us the basis for the introduction and detailed discussion of an underlying architecture and a prototypical implementation of a corresponding TraDE Middleware solution within the context of this paper.

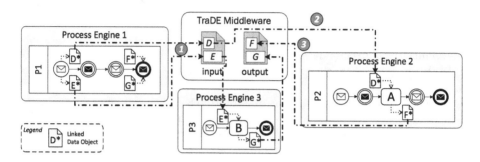

Fig. 2. Execution of cross-partner data flows of example choreography shown in Fig. 1.

Figure 2 shows the private process models of all three participants from our motivation scenario depicted in Fig. 1. For the sake of simplicity we omitted the message flows between the participants. Furthermore, we only describe the execution of the first three cross-partner data flows, namely dx1, dx2 and dx3, depicted in Fig. 1 as an example. Each task and event that was connected through a data flow with a cross-partner data object in Fig. 1 has a corresponding BPMN data object associated. As indicated by the *, these BPMN data objects are linked with the respective data elements of cross-partner data objects provided by the TraDE Middleware as depicted in Fig. 2.

In the following, we use the term *choreography instance* to refer to the collection of interconnected instances of the private process models implementing the choreography. Thus, in this example, a new choreography instance is created whenever participant P1 receives a new request message which is modeled by the BPMN message start event. The request message contains values for both data containers D and E of private process model P1. Process Engine 1 extracts these values from the request message to store them in the associated data containers. Since the data containers D and E are linked with the respective data elements D and E of cross-partner data object *input*, the process engine detects this linking and instead of storing the data internally, it directly uploads the data to the corresponding data elements in the TraDE Middleware (step 1 in Fig. 2). Subsequently, participant P1 invokes participant P2 through a message exchange (cf. *mx1* in Fig. 1). As soon as participant P2 reaches task A, Process Engine 2 reads the value of its data container D. Again the process engine detects that this data container is linked to a cross-partner data object in the middleware and, therefore, retrieves the value directly from data element E of cross-partner data object input at the middleware (step 2 in Fig. 2). After task A is completed, the process engine stores the tasks' result data in data container F. Based on the linking, the data is directly uploaded to data element F of cross-partner data object *output* from where it can be retrieved by participant P1.

3   The TraDE Middleware

While in Hahn et al. [7] we outlined the overall vision and concepts of data-aware choreographies focusing on their modeling, in this paper, our main focus is on the TraDE Middleware. In the following we introduce a choreography language-independent metamodel and a detailed architecture for such a middleware as well as describe how it can be integrated with process engines.

3.1   Metamodel

The middleware and its underlying concepts should not be bound to any specific choreography and process modeling languages or related run time environments. Therefore, the TraDE Middleware has its own internal metamodel shown in Fig. 3.

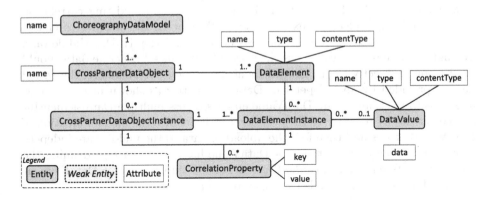

Fig. 3. Metamodel of the TraDE Middleware

The collection of cross-partner data objects of a choreography is represented by a ChoreographyDataModel entity within the TraDE Middleware. Therefore, a ChoreographyDataModel has a qualified *name* and contains one or more Cross-PartnerDataObject entities which represent the cross-partner data objects of a choreography. A CrossPartnerDataObject has a unique *name*, a reference to its ChoreographyDataModel and contains one or more DataElement entities. A DataElement has a *name*, a *type* and a *contentType* definition. While the type allows to specify a concrete data structure and its syntax, e. g.,in XML or JSON, the contentType enables to specify the kind of data and its semantics. Therefore, the middleware can handle binary data without its interpretation while still being aware of its content and how to represent it. This allows us to support various types of data, e. g.,structured and unstructured data, videos or pictures, as well as data formats and representations, e. g.,XML, plain text, MPEG, or PNG. Such content type information can be specified, for example, using *Media Types*[1].

Since we have to represent and manage the data of multiple instances of a choreography model referring to the same CrossPartnerDataObject and DataElement entities during run time, we apply the well-known concept of *model instances* from BPM to our metamodel. Therefore, we introduce corresponding CrossPartnerDataObjectInstance and DataElementInstance entities which allow us to represent concrete instances of cross-partner data objects and their data elements for one specific instance of a choreography model. In order to correlate the data managed by the TraDE Middleware, i. e.,CrossPartnerDataObjectInstance and DataElementInstance entities, with a choreography model instance, corresponding correlation information have to be supported and provided by the metamodel. Therefore, we associate a set of CorrelationProperty entities to the DataObjectInstance and DataElementInstance entities. These CorrelationProperty entities allow to uniquely identify a choreography instance on the level of a process engine as well as to identify the data that belongs to this instance on the level of the TraDE Middleware. Since the concept of property-based correlation is well known in the domain of BPM, we therefore reuse existing correlation mechanisms as provided, for example, by BPMN or BPEL in order to enable the instance correlation between process engines and the TraDE Middleware. To enable the reuse of data across choreography instances, concrete data should not be bound directly to one DataElementInstance entity. Therefore, the actual data is provided by an independent DataValue entity as shown in Fig. 3. This allows us to reuse and share DataValue entities across multiple DataElementInstance entities by referencing them. Moreover, it enables the manual creation of DataValue entities and therefore the upload of data to the middleware independent of a choreography instance. A DataValue has a *name*, a *type* defining the structure of its data and a *contentType* definition similarly as for DataElement entities. Furthermore, it holds the concrete *data*.

3.2 Architecture

Figure 4 presents the TraDE Middleware architecture. The design follows the three layer architecture pattern [5], which we describe in a top-down manner.

The *Presentation* layer provides a *Web User Interface* (UI) and a (set of) *REST API(s)* which enable the interaction with the TraDE Middleware, and its integration with other systems. By following the REST architectural style, each entity of our internal TraDE metamodel shown in Fig. 3 is represented as a

---

[1] Internet Assigned Numbers Authority (IANA), Media Types: https://www.iana.org/assignments/media-types/media-types.xhtml.

resource and can, therefore, be easily accessed, referenced, and shared through a Uniform Resource Locator (URL). For example, process engines can use the REST API to integrate with the middleware in order to upload or retrieve data. Modeling tools can use the REST API to support modelers with deploying their specified collections of cross-partner data objects, i. e.,choreography data models, to the middleware, so that they are available to the process engines during choreography run time.

The *Business Logic* layer contains the core functionality of the middleware which is grouped into the following components. The *TraDE Instance Models* component contains instances of the metamodel shown in Fig. 3 to represent concrete cross-partner data objects and their instances within the middleware. All functionality related to data management is provided by the *Data Management* component which is the core component of the middleware. It supports the access and inspection of data associated to corresponding cross-partner data objects through the REST API. Furthermore, it provides the functionality to upload and retrieve data for a corresponding DataValue or DataElementInstance entity. Related to that, it also handles the correlation of the TraDE Instance Models with choreography instances in order to enable the process engines to access and retrieve the correct TraDE Instance Models and their associated data. Moreover, it is responsible for the management of the life cycle of the TraDE Instance Models. Therefore, it provides and implements corresponding life cycle operations such as create, instantiate, archive and delete. The *TraDE Node & Network Management* component is responsible for enabling a decentralized deployment of multiple middleware nodes and their connection into networks to allow more efficient data placement and staging as well as further optimizing the data exchange between the choreography participants. In order to decouple the life time of the data from its choreography instance the *Persistence* component stores both the internal metamodels of the middleware as well as the managed data in an underlying data source in order to guarantee its availability

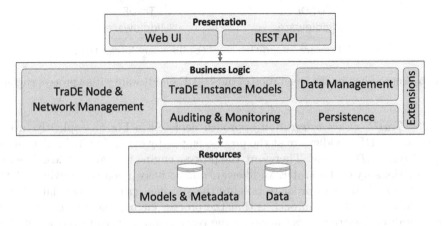

Fig. 4. Architecture of the TraDE Middleware

for later (re)use. The *Auditing & Monitoring* component provides an associated life cycle for each of the entities in the above introduced metamodel. This enables the auditing and monitoring of all entities by emitting corresponding events whenever the life cycle of an entity changes. Furthermore, these internal events can be consumed by any interested component within the middleware, for example, allowing the Data Management component to trigger corresponding actions on state changes in order to realize the life cycle management of the TraDE Instance Models. The whole middleware is designed to be extensible in order to integrate new or adapt existing components. Therefore, the *Extensions* component provides corresponding functionality and mechanisms to plug-in new functionality as well as extensions or variants of existing components. For example, the default persistence component can be replaced by a new implementation that uses a different technology stack by adding it as an extension to the middleware.

The *Resources* layer contains all required resources used within the Business Logic layer. This comprises data sources for the persistence of TraDE Instance Models and related metadata about nodes and networks (*Models & Metadata* data source in Fig. 4) as wells as a data source for the actual data managed by the TraDE Middleware (*Data* data source in Fig. 4).

3.3 Integration with Process Engines

In the following we want to briefly discuss two approaches how the TraDE Middleware can be integrated with a process engine as shown in Fig. 5.

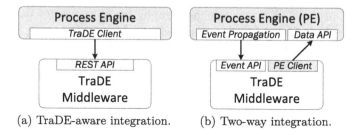

(a) TraDE-aware integration.   (b) Two-way integration.

Fig. 5. Approaches for the integration of the TraDE Middleware with a process engine.

The *TraDE-aware integration* approach depicted in Fig. 5a explicitly introduces the TraDE Middleware at the process engine by extending its implementation with a *TraDE Client*. Therefore, the process engine is aware of the existence and functionality of the TraDE Middleware and actively uploads or retrieves data from the middleware in order to execute the specified cross-partner data flows. The advantage of this approach is that the process engine remains in control of the overall choreography execution or the corresponding private processes it is responsible for, respectively. The main disadvantage is that the process engine implementation has to be extended in order to integrate the client and introduce required functionality to identify and handle the linking of data containers to cross-partner data objects. Especially, if the collaborating partners use different process engine solutions this integration approach requires too much effort.

In contrary, the *two-way integration* approach depicted in Fig. 5b integrates the process engine and the TraDE Middleware in a loosely coupled manner through corresponding APIs. The basic idea of this integration approach is to extract any data-related knowledge from the data-aware choreography models, e. g.,which participant requires or produces which cross-partner data objects, to move the control of executing specified cross-partner data flows to the TraDE Middleware. Therefore, the process engines have to expose the execution state of their private process model instances through a corresponding event propagation mechanism, e. g.,using messaging. The emitted state change events can then consumed by an *Event API* at the TraDE Middleware, so that it is always aware of the current execution state of the overall choreography instances for which it executes the cross-partner data flows. Furthermore, the process engine implementations have to expose an *Data API* which allows external parties, such as the TraDE Middleware, to retrieve and write data from and to data containers of the private process model instances executed by a process engine. The advantage of this integration approach is that the process engine implementation is not directly coupled with the TraDE Middleware. Instead it has to be only extended with generic event propagation functionality and expose its data management capabilities through an API. Some process engine implementations potentially already provide such capabilities and if not, the required extensions are not only bound and therefore usable for the integration with the TraDE Middleware. The main disadvantage of this approach is that the TraDE Middleware has to keep track of the execution state of all choreography instances to fully take over control of the execution of the cross-partner data flows which increases the complexity of the TraDE Middleware implementation and requires (to a certain extend) control over and access to the process engines.

4 Validation

To validate the practical feasibility of our concepts, we prototypically implemented the TraDE Middleware architecture shown in Fig. 4 and integrated it with a process engine following the TraDE-aware integration approach shown in Fig. 5a. For the implementation of our TraDE concepts, we extended the choreography modeling language BPEL4Chor and the process modeling language BPEL. The extension of BPEL4Chor allows us to specify cross-partner data objects and cross-partner data flows and the extension of BPEL enables us to link BPEL variables with cross-partner data objects in the resulting private process models constituting the overall data-aware choreography. An extended version of the open source BPEL engine Apache *Orchestration Director Engine* (ODE)[2] is used as process engine solution. In order to support the reading and

---

[2] The Apache Software Foundation, Apache ODE: http://ode.apache.org.

writing of cross-partner data objects, we extended the underlying implementation of Apache ODE and integrated it with our TraDE Middleware following the TraDE-aware integration approach discussed in Sect. 3.3. Therefore, the communication between the process engine and the TraDE Middleware is realized by integrating a REST API client into Apache ODE. On top of this client, we introduced a new *TraDE Manager* component that encapsulates logic for the retrieval and upload of data to the TraDE Middleware. For example, this comprises the creation and resolution of DataElementInstance and DataValue entities for a private process instance in order to upload data to the TraDE Middleware.

The TraDE Middleware itself is realized as a Java-based web server which exposes its functionality through a REST API. As underlying web server we are using Eclipse *Jetty*[3] in embedded mode. The REST API is specified and documented in form of a *Swagger Specification*[4] and implemented based on the *Jersey* RESTful Web Services framework[5]. For the implementation of the REST API, we are following an API-first approach which means that we developed against the API specification. Therefore, we use the related Swagger tooling support to generate client code as well as server code skeletons directly from the API specification. This approach has two major advantages. First, only the relevant business logic of the REST API has to be implemented and provided and second, changes in the API specification are directly reflected on the level of the code keeping the API and its implementation in sync. For the persistence of the TraDE instance models and the associated (business) data, we support MongoDB as a document-oriented database and the local file system as persistence layer for the middleware at the moment. Which persistence layer to use and a lot of other configuration options for the middleware can be specified in configuration files. At the moment, we only support a single-node deployment of the TraDE Middleware, but for future work, we are aiming at supporting also multi-node deployments by leveraging the capabilities of corresponding distributed data grid frameworks as provided, for example, by the *Hazelcast In-Memory Data Grid*[6]. The complete open source code of the middleware is available on GitHub[7].

5   Evaluation

In the following, we introduce a performance evaluation comparing cross-partner data flows with the classical exchange of data through messages within service choreographies. Therefore, we first present the underlying evaluation methodology we apply, followed by a description of the experimental setup and finally a discussion of the evaluation results.

---

[3] The Eclipse Foundation, Eclipse Jetty: https://www.eclipse.org/jetty/.
[4] TraDE Swagger Specification: https://github.com/traDE4chor/trade-core/blob/master/server/swagger.json.
[5] Oracle Corporation, Jersey: https://jersey.java.net/.
[6] Hazelcast, Inc., Hazelcast IMDG: https://hazelcast.org/.
[7] TraDE Middleware: https://github.com/traDE4chor/trade-core.

5.1   Evaluation Methodology and Experimental Setup

The focus of the evaluation is at empirically analyzing the performance variation when introducing our TraDE concepts. We therefore use the example choreography depicted in Fig. 1 to measure variations of the response time perceived by a user invoking the choreography. As a baseline, we use the same choreography model without our concepts applied, i. e.,the data is passed within messages through the modeled message flows between the participants of the choreography. For example, the data contained in the initial request sent to participant P1 for data elements D and E is not uploaded to the TraDE Middleware, instead the data is stored in corresponding local data containers at the process engine and then passed within a message exchange to participant P2. The same applies for all other data exchanges depicted through cross-partner data flows in Fig. 1. For the sake of completeness, the standards-based choreography model used as baseline is shown in Fig. 6.

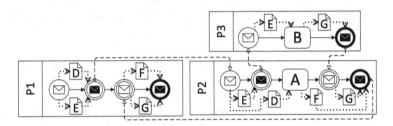

Fig. 6. Standards-based version of the example choreography shown in Fig. 1.

The two choreography models are implemented using BPEL4Chor [4] as choreography modeling language and are transformed [14] to three BPEL process models, one for each participant, which are manually refined in order to make them executable. The tasks A and B of participant P2 and P3 are implemented through BPEL assign activities so that they duplicate the random data contained in data containers D and E by its concatenation and store the result in data containers F and G. To guarantee that for both, the baseline scenarios and the TraDE scenarios, data has to be actually transferred through messages or the TraDE Middleware, respectively, we deploy each of the resulting executable private process models to a separate process engine instance using an extended version of Apache ODE with its default configuration. The BPEL process models for the TraDE scenarios are furthermore extended with corresponding TraDE annotations so that the process engine is aware of the linking of the BPEL variables with the cross-partner data objects managed by the TraDE Middleware.

In order to also measure a potential impact of the size of the data being processed, we introduce three scenarios with increasing data size for each of the input data elements (data object *input*, data elements D and E): 1 KB, 128 KB and 256 KB. While for the baseline scenarios all data is stored locally in corresponding data containers at the process engines, in the TraDE evaluation scenarios the data are uploaded once to the corresponding cross-partner data objects in the middleware and retrieved directly from there only when required. For each of the six scenarios summarized in Table 1, a workload consisting of randomly generated request messages with the above mentioned data element sizes is created. The workload is distributed among a *warm-up phase* ($w(t_0)$) with 10 messages followed by an *experimental phase* comprising a set of 310 requests sent in five load bursts according to the following function over time:

$$m(t_i) = w(t_o) + \sum_{i=1}^{5} 2^{i-1} \cdot k \mid k = 10, w(t_0) = 10$$

The experimental environment is set up in an on-premise private cloud infrastructure on two virtual machines (VM). The evaluation VM is configured with 8 virtual CPUs, Intel® Xeon® CPU E5-2690 v2 3.00 GHz, 32 GB RAM, 120 GB disk space, and is running an Ubuntu 14.04.4 64bit server distribution. We use Docker[8] within this VM to deploy the required three separate instances of Apache ODE and in addition one TraDE Middleware instance in the TraDE scenarios. The idea behind this level of nesting and using Docker for the deployment of the evaluation environment is that we want to have a clean and therefore identical setup for each of the conducted experiments towards creating reproducible evaluation results.

In order to setup the evaluation environment and to conduct the workload for each of the defined evaluation scenarios, we use Apache JMeter 3.2[9] as load driver which is deployed in a separate VM, with the following configuration: 2 virtual CPUs, Intel® Xeon® CPU E5-2690 v2 3.00 GHz, 4 GB RAM, 40 GB disk space, running an Ubuntu 14.04.2 64 bit desktop distribution. We created a JMeter test plan for each of the defined six scenarios, i. e.,three baseline scenarios and three TraDE scenarios with data sizes of 1 KB, 128 KB and 256 KB each, which concurrently sends the above defined workload for five concurrent users to the endpoint of the BPEL process model implementing participant P1. To alleviate the effect of outliers in the experimental results, we execute ten rounds of each scenario and calculate the average response time for each load burst while excluding the samples which are timed-out at the process engine.

5.2   Experimental Results

Figure 7a shows the experimental results comparing the user-perceived performance (average response time) of the load bursts of all scenarios. If we compare the baseline (B1-B3) with the TraDE (T1-T3) scenarios, there exists an overall beneficial impact to the user-perceived performance when introducing cross-partner data flows. However, this impact greatly varies on the size of the data being exchanged as well as on the workload applied throughout the load bursts. Comparing the scenarios with 256 KB data size (B3 vs. T3) shown in Fig. 7a

---

[8] Docker Community Edition: https://www.docker.com/community-edition.

[9] Apache JMeter: http://jmeter.apache.org/.

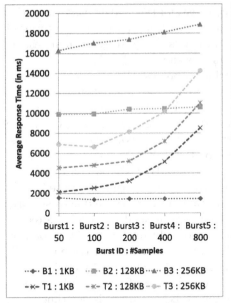 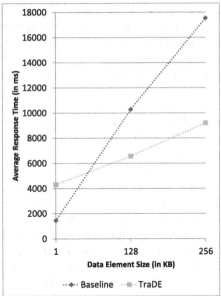

(a) Average response time in milliseconds (ms) for the load bursts of all scenarios.

(b) Average response time (in ms) of scenarios based on data element size.

Fig. 7. Evaluation results for the defined six scenarios.

the performance is improved by approximately 90% in total. When we have a look at the different load bursts in detail, this improvement decreases from approximately 136% in burst 1 to approximately 32% in the last load burst. Therefore, we can assume that when increasing the load on the middleware, the improvement will further degrade and actually convert to an overall performance deterioration.

This is also underpinned when comparing the 128 KB scenarios (B2 vs. T2) shown in Fig. 7a where the performance is still improved by approximately 56% in total. However, again the performance alters from an improvement of approximately 117% in burst 1 to a small performance degradation of approximately 0.04% in the last load burst. Comparing the scenarios with 1 KB data size (B3 vs. T3) shown in Fig. 7a, the performance is degraded by approximately 66% in total when introducing the middleware and cross-partner data flows. There the overhead of introducing additional communication between the process engines and the TraDE Middleware to conduct the cross-partner data flows is higher than the improvements gained by reducing the amount of data to be exchanged.

Furthermore, Fig. 7a shows that for the baseline scenarios with message-based data exchange, the performance maintains quite stable in terms of increasing the workload across the load bursts but decreases significantly when increasing data element sizes. This fact is also underpinned when comparing the overall average response time among all load bursts of the six scenarios based on the processed data element sizes as shown in Fig. 7b. In contrast, for the TraDE-based scenarios with cross-partner data flows, the performance maintains quite stable in terms of data element sizes as shown in Fig. 7b, but decreases significantly when increasing the workload throughout the five load bursts as shown in Fig. 7a.

Both types of scenarios are not fully able to process the complete workload in all load bursts without having a set of samples that timeout at the process engine (by default after 120 s for Apache ODE). For the baseline scenarios this is especially the case in scenario B3 with a data element size of 256 KB, where about 30% of the samples in load burst 5 (after approximately 556 success-ful samples) result in timeouts. A reason for this behavior might be the large amount of data (1550 instances $*$ 3840 KB $\approx$ 5.9 GB) ODE is not capable of handling in its default configuration at a certain point in time. For the TraDE scenarios such samples causing timeouts are randomly distributed across nearly all scenarios and load bursts, but also with a peak in load burst 5. The reason for such an unpredictable behavior is most probably related to the resolution of required data from the TraDE Middleware through the process engine. To correlate process instances and data object instances and to finally retrieve data element values, the process engines poll the middlewares' REST API by sending repeated requests every second as long as the process instance is not timed-out. These requests are queued up at the TraDE Middleware while throttling its per-formance for a certain amount of time which again results in timeouts at the process engines. The average amount of timed-out samples as well as a summary of the scenarios and their average response times is shown in Table 1.

Table 1. Summary of the experimental evaluation scenarios and their results.

Scenario	ID	Data element Size (in KB)	Total data size (in KB/instance)	Timed-out Req. (in %)	Avg. Resp. Time (in ms)
Baseline	B1	1	15	0.04	1451.58
	B2	128	1920	0.32	10272.86
	B3	256	3840	6.49	17540.36
TraDE	T1	1	6	0.20	4313.86
	T2	128	768	0.47	6560.86
	T3	256	1536	0.61	9214.08

In summary the evaluation results show that introducing a TraDE Middle-ware layer and applying our concept for cross-partner data flows in service choreographies provide valuable performance improvements already for rela-tively small data sizes above 128 KB. To alleviate the performance degradation when increasing the load at the middleware, in future work, we will improve our prototypical implementation and its integration with Apache ODE so that its performance maintains stable when increasing the workload. Therefore, future experiments will aim at investigating current capacity limitations of the TraDE

Middleware when increasing the data sizes as well as the number of concurrent users and requests. The complete evaluation result data and any related material, e. g.,BPEL process models and JMeter test plans, are available on GitHub[10].

6 Related Work

In this section, we will compare our TraDE approach and the introduced middleware with related work on cross-partner data flows in service choreographies. Since our focus is on improving and extending the role of data in classical control flow driven choreography and process modeling languages such as BPMN, BPEL4Chor, and BPEL, we focus on related work following the same paradigm.

Meyer et al. [10] introduce a model-driven approach towards improving the modeling and enactment of data exchange in choreographies through messages. Through an extension of the BPMN modeling language with annotations on BPMN data objects they enable the specification and enactment of message extraction from and message storage to local databases. This allows them to completely automate the exchange of data across participants and also to enrich model transformations with data-related aspects. However, our approach introduces the TraDE Middleware as an abstraction layer and data hub, instead of directly binding data containers to databases on the level of the models. This allows us to decouple the exchange of data from the exchange of messages towards increasing run time flexibility while reducing modeling complexity of choreographies.

Barker et al. [1] introduce MAP as new language for executable service choreographies. By introducing the concept of so-called *peers* they provide a mechanism to apply extra functionality that enables web services to participate in a choreography without requiring to adapt the underlying service implementations. In contrast to their approach, we are building on top of standardized languages and tools in order to support cross-partner data flows in choreographies.

Furthermore, Barker et al. [2] introduce the *Circulate approach* which combines the advantages of the orchestration and choreography paradigm. While the control flow remains orchestration-based the data flow is realized in a choreography-based manner. Similar to the above mentioned peers, *proxies* are introduced to enable the transfer of data across services. Based on that, the process engine communicates with the proxies in order to invoke services and exchange data between them. In general, our approach is similar since we introduce the TraDE Middleware as an intermediary to realize the data exchange. However, instead of explicitly invoking proxies to conduct the data exchange, we propose to introduce cross-partner data flows which are transparently conducted through the TraDE Middleware and its integration with the process engines. With our approach, models are enriched instead of changed to preserve their portability.

---

[10] TraDE Evaluation: https://github.com/traDE4chor/trade-core-evaluation/tree/master/initial-evaluation.

Habich et al. [6] provide an approach for cross-partner data flows similar to ours but with focus on the level of process models and BPEL in particular. They try to solve the problem of BPEL's *by value* semantics for data exchange resulting from the centralized and implicitly specified data flows in BPEL through variables and assign activities. Therefore, they extend BPEL with the notion of *BPEL data transitions* and apply their concept of Data-Grey-Box Web Services. These web services provide enhanced interfaces specifying related data aspects and therefore allow to define which parameters are passed by value or by reference. With the help of the introduced data transitions, explicit data flows between the composed Data-Grey-Box Web Services can be specified in BPEL process models. The combination of both concepts further allows to integrate specialized data propagation tools and logic, e. g.,Extract Transform Load (ETL) tools, to implement the modeled data transitions which act as mediators to provide and resolve data by reference between the composed Data-Grey-Box Web Services during run time. In contrast to introducing explicit data flows between interacting services on the level of BPEL, or process models in general, we argue that cross-partner data flows can be specified much easier and more intuitively on the level of choreography models since choreographies provide a global view on the interactions and conversations between the services. Our overall goal is to hide cross-partner data flows on the level of the process models by transparently providing the required logic and functionality through the TraDE Middleware.

7   Conclusions and Outlook

To support the notion of data-aware service choreographies, we previously introduced our concepts for transparent data exchange through cross-partner data objects and cross-partner data flows in choreographies. In this work, we focused on the execution of data-aware choreographies with the help of a new middleware layer: the TraDE Middleware. Therefore, we introduced an architecture and an internal metamodel for the TraDE Middleware and discussed its integration with a process engine. To evaluate the feasibility and applicability of our approach and the middleware, we conducted a performance evaluation comparing our approach for cross-partner data flows with the classical exchange of data through messages within service choreographies. The evaluation results already show interesting performance improvements for relatively small data sizes when applying our TraDE concepts and integrating the TraDE Middleware with process engines.

In future work, we will provide a Web UI for our TraDE Middleware prototype in order to enable its use by human users and also ease manual data access and inspection. Furthermore, we plan to tackle the performance weaknesses identified within the evaluation towards improving the overall performance and robustness of the middleware as well as to improve its integration with Apache ODE, e. g.,introducing a callback mechanism instead of periodically polling resource changes. Moreover, we are planning to integrate our middleware with the *ChorSystem* middleware introduced in Weiß et al. [17]. The goal is to leverage the capabilities of the ChorSystem middleware to ease and improve the deployment and management of data-aware choreographies in the future.

Acknowledgment. This research was supported by SimTech (EXC 310/2) and SmartOrchestra (01MD16001F).

References

1. Barker, A., Walton, C., Robertson, D.: Choreographing web services. IEEE Trans. Serv. Comput. **2**(2), 152–166 (2009)
2. Barker, A., Weissman, J.B., Van Hemert, J., et al.: Reducing data transfer in service-oriented architectures: the circulate approach. IEEE Trans. Serv. Comput. **5**(3), 437–449 (2012)
3. Decker, G., Kopp, O., Barros, A.: An introduction to service choreographies. Inf. Technol. **50**(2), 122–127 (2008)
4. Decker, G., Kopp, O., Leymann, F., Weske, M.: Interacting services: from specification to execution. Data Knowl. Eng. **68**(10), 946–972 (2009)
5. Fowler, M.: Patterns of Enterprise Application Architecture. Addison-Wesley Professional, Boston (2002)
6. Habich, D., Richly, S., Preissler, S., Grasselt, M., Lehner, W., Maier, A.: BPEL^{DT}-Data-Aware Extension for Data-Intensive Service Applications. In: Gschwind, T., Pautasso, C. (eds.) Emerging Web Services Technology, Volume II. Whitestein Series in Software Agent Technologies and Autonomic Computing, vol. II, pp. 111–128. Birkhäuser, Basel (2008). doi:10.1007/978-3-7643-8864-5_8
7. Hahn, M., Breitenbücher, U., Kopp, O., Leymann, F.: Modeling and Execution of Data-aware Choreographies: An Overview. Springer CSRD (2017)
8. Hahn, M., Karastoyanova, D., Leymann, F.: A management life cycle for data-aware service choreographies. In: Proceedings of ICWS 2016. IEEE (2016)
9. Kopp, O., Leymann, F., Wagner, S.: Modeling choreographies: BPMN 2.0 versus BPEL-based approaches. In: Proceedings of EMISA 2011. LNI, GI (2011)
10. Meyer, A., Pufahl, L., Batoulis, K., Kruse, S., Lindhauer, T., Stoff, T., Fahland, D., Weske, M.: Automating data exchange in process choreographies. In: Jarke, M., Mylopoulos, J., Quix, C., Rolland, C., Manolopoulos, Y., Mouratidis, H., Horkoff, J. (eds.) CAiSE 2014. LNCS, vol. 8484, pp. 316–331. Springer, Cham (2014). doi:10.1007/978-3-319-07881-6_22
11. Meyer, S., Sperner, K., Magerkurth, C., Pasquier, J.: Towards modeling real-world aware business processes. In: Proceedings of WoT 2011, pp. 81–86. ACM (2011)
12. OASIS: Web Services Business Process Execution Language Version 2.0 (2007). http://docs.oasis-open.org/wsbpel/2.0/OS/wsbpel-v2.0-OS.html
13. OMG: Business Process Model And Notation (BPMN) Version 2.0 (2011). http://www.omg.org/spec/BPMN/2.0/
14. Reimann, P., Kopp, O., Decker, G., Leymann, F.: Generating WS-BPEL 2.0 Processes from a Grounded BPEL4Chor Choreography. Technical report 2008/07, University of Stuttgart (2008)
15. Schmidt, R., Möhring, M., Maier, S., Pietsch, J., Härting, R.-C.: Big data as strategic enabler - insights from central European enterprises. In: Abramowicz, W., Kokkinaki, A. (eds.) BIS 2014. LNBIP, vol. 176, pp. 50–60. Springer, Cham (2014). doi:10.1007/978-3-319-06695-0_5

16. W3C: XML Schema Definition Language (XSD) 1.1 Part 1: Structures (2012). http://www.w3.org/TR/xmlschema11-1/

17. Weiß, A., Andrikopoulos, V., Sáez, S.G., Hahn, M., Karastoyanova, D.: Chorsystem: a message-based system for the life cycle management of choreographies. In: Debruyne, C., et al. (eds.) OTM 2016. LNCS, vol. 10033, pp. 503–521. Springer, Cham (2016). doi:10.1007/978-3-319-48472-3_30

18. Zimmermann, O.: Microservices tenets. Springer CSRD, pp. 1–10 (2016)

Building the Most Creative and Innovative Collaborative Groups Using Bayes Classifiers

Gabriela Moise, Monica Vladoiu[✉], and Zoran Constantinescu

The Department of Computer Science and Information Technology,
Mathematics and Physics, CerTIMF Research Center,
Petroleum-Gas University of Ploiesti, Ploieşti, Romania
{gmoise, mvladoiu, zoran}@upg-ploiesti.ro

Abstract. Building "the best" creative and innovative groups that have common goals and tasks to perform, efficiently and effectively, is difficult. The complexity of this undertaking is significantly increased by the necessity to first understand and then measure what "the best" goal means for the individuals in the groups, but also for each group as a whole. We present here our Bayes classifiers-based technique for building "the best" groups of students to work together in collaborative learning situations. In our case, "the best" goal means *the most creative and innovative teams* possible in a given learning situation based on some particular attributes: *individual creativity, motivation, domain knowledge,* and *inter-personal affinities.* However, both the proposed model and method are general and they may be used for building collaborative groups in any situation, with the appropriate "the best" goal and attributes. A case study on using this method with our Computer Science students is also included.

Keywords: Creative and innovative collaborative learning · Collaborative work · Bayes classification

1 Introduction

Construction of groups that have common goals and tasks to perform in collaborative situations, e.g. in working or learning scenarios, is very usual nowadays. During cooperative activities, individuals work together to accomplish shared objectives and to obtain results beneficial to both themselves and other group members. However, coming up with a technique to build "the best" groups of people to optimally collaborate, both efficiently and effectively, while pursuing the achievement of common aims is not straightforward. The complexity of this undertaking is significantly increased by the necessity to first understand and then measure what "the best" means for the individuals in the groups, but also for each group as a whole. Therefore, in any given collaborative context, clustering the most suitable individuals to form the highest performance groups aiming at reaching some specific outcomes is very challenging.

Cooperative learning consists of the instructional use of small groups of students that work together to maximize their own and each other's learning. Thus, after receiving instruction from their instructor, class members are usually split into small groups that work through the assignment until all the group members have successfully understood

H. Panetto et al. (Eds.): OTM 2017 Conferences, Part I, LNCS 10573, pp. 271–283, 2017.
https://doi.org/10.1007/978-3-319-69462-7_17

and completed it [1]. But ad hoc grouping of students does not necessarily mean that cooperative groups have been built, given that the purpose of cooperation is to maximize both achievement of the proposed goals and learning. Based on the performance in this respect, several categories of learning groups exist, namely pseudo, traditional classroom, cooperative, and high-performance cooperative learning groups [2]. For example, in cooperative learning groups students work together to accomplish common aims, while seeking outcomes beneficial to all group members. Students embark together on the learning journey, helping each other throughout the way. The main benefit is that the group becomes more than the sum of its parts and that virtually all students perform higher than they would if they worked alone. The high-performance ones outperform all the expectations for cooperative learning groups, the group members being highly committed to each other and to the group's success, so that the group becomes a team. The benefits of cooperative learning in educational processes are plentiful [1–4]. Thus, students learn more by doing and by involving actively and cooperatively in learning activities. Less motivated students can be appealed to continue working on difficult aspects by their teammates. Even strong students can benefit from clarifying the approached issues to their colleagues by improving their own understanding and mastering. Moreover, timely delivery of assignments is much more frequent [3].

We present here our work towards developing a technique, based on Bayes classifiers, for building "the best" groups of students to work together in collaborative learning situations. In our case, "the best" goal means *the most creative and innovative teams* possible, in a given learning situation, based on some particular attributes: *individual creativity, motivation, domain knowledge,* and *inter-personal affinities.* Our focus is dual: first, we introduce both a model and a method for grouping individuals in creative groups in collaborative situations (using Bayesian Networks-based classification), and, second, we instantiate and apply them in learning contexts. However, *both the proposed model and method are general* and they may be used for building collaborative groups in any situation, with the appropriate goal and attributes for that context. In the long run, we aim at the development of an intelligent system able to support organizing individuals in the most creative and innovative groups, in any given collaborative situation, and at offering it as an open project [5]. A case study on using this method with our Computer Science students is also included.

The structure of the paper is as follows: the next section presents the related work; the third one introduces both our model and method for building collaborative groups with instantiation for learning situations. Section 4 presents the results of our experimenting with the resulted technique, while in Sect. 5 they are evaluated and discussed. The last section includes some conclusions and future work ideas.

2   Related Work on Collaborative Creativity

The approach of creativity in the literature has shifted from focusing on gifted individuals to acknowledging that each person can be creative, and, finally, to recognizing that social structures have a strong influence on individual and group creativity [6–10]. Moreover, research shows that in order to understand how an individual contributes to group creativity a large variety of factors needs to be considered (individual characteristics, organizational environment, social relationships, etc.) as such and also combined. In [6], three research directions are proposed: *group creativity in context, group-level creative synergy, and strategies for developing group creativity*. In [7], the authors approach the factors that influence team creativity and innovation based on the triad *Input–Process–Output*. The Input shows the team composition based on the members' individual characteristics. The Process includes the activities undertaken by the team members to carry out some tasks or to solve some problems, while the Output consists of the creativity and innovation of the team (team effectiveness).

One of the most representative models for collaborative creativity is introduced in [8]. The input variables for this model are *Group Member Variables, Group Structure, Group Climate, and External Demands*. Three categories of processes are taken into account: cognitive, motivational, and social. The output consists of team creativity and innovation. Using this model, the authors show how *the group member attributes* (personality, task relevant knowledge, skills, and abilities, intrinsic motivation, cognitive flexibility, creative self-efficacy, etc.), *the group structure* (diversity, size, communication mode, cohesiveness, leadership style etc.), *the group climate* (commitment to task, conflict, trust, norms of participation/risk-taking/innovation etc.), and *the external demands* (creative mentors and models, rewards and penalties, freedom/autonomy/self-management, support for creativity, intergroup and intra-group competition, task structure, performance feedback etc.) influence the cognitive, social, and motivational processes that collaborative creativity relies on.

Despite the interest on increasing group creativity, a few experiments of grouping people in the most creative groups exist, in general, and, in learning, in particular. Moreover, most of them do not use data mining techniques, machine learning, nor intelligent data analysis. Thus, in [4], it is shown that in case of one wanting to teach her course effectively, ability heterogeneity should be her primary criterion. Also, if the groups need to meet outside class, forming teams of students who have common blocks of unscheduled time could be suitable. This work also points out some of the downsides of groups composed exclusively of strong students, who are likely to distribute the work rather than *engaging in the group discussions and informal tutoring sessions* that lead to many of the confirmed instructional benefits of cooperative learning. In [11], the authors have analyzed the cause-effect relationships between 6 factors: *team creativity, exploitation, exploration, organizational learning culture, knowledge sharing, and expertise heterogeneity*. They have also built a General Bayesian Network, which has as a target node the team creativity and that shows the dependencies between these factors. The main question addressed in this work was dual, i.e. *(1) how do the processes of creative revelation—exploitation and exploration—engaged in by team members contribute to building team creativity, and (2) how do environmental factors—organizational learning culture, knowledge sharing, and expertise heterogeneity—affect team creativity*. The results obtained using scenario-based simulations show that a direct relationship exists between team creativity and exploitation, exploration, organizational learning culture, knowledge sharing, and expertise heterogeneity. Our approach differs from the one in [11], which establish dependencies between team creativity and some specific factors. Thus, we use Bayes classifiers to build the most creative and innovative groups based on particular values of some individual characteristics (related to creativity) of the group members.

3 A Bayes Classifier-Based Model and Method for Building Optimally Creative and Innovative Groups

3.1 Bayesian Network Classifiers

Classification is very important in many domains, for example in applications for object recognition (forms, human faces, characters, etc.), detecting spam e-mails or intruders in computer networks, and so on. The concept of a classifier is seen often as a correspondence between a data set (values of attributes or features of an object) and a class (category) to which the object belongs [12, 13].

Formally, a classifier is defined as follows. Given a set of attributes $\{A_1, A_2, ..., A_n\}$ with finite domains and C the class variable, also with finite domain, that corresponds to possible classes, a classifier is a correspondence f: $A_1 \times A_2 \times ...\times A_n \rightarrow C = \{c_1, c_2, ..., c_m\}$. An object is described by the values of the considered attributes $(v_1, v_2, ..., v_n)$ and a classifier $f(v_1, v_2, ..., v_n) = c_i$ shows the class to which the considered object belongs.

A Bayesian (Belief) Network (BBN) is a graphical model based on probabilistic directed acyclic graphs that can be used for representing uncertain knowledge and reasoning techniques. Each node of the graph represents a random variable, while the arcs define a probabilistic dependency between variables. These dependencies are quantified by the conditional probabilities between variables.

One of the high performance classifiers with regard to prediction of the class to which a particular object pertain are naïve Bayes classifiers [12, 14]. In case of naïve Bayes classifiers, each attribute has only one parent, namely the class variable. Naïve Bayes classifiers use Naïve Bayes Structures and Bayes' rule to predict the most probable class to which an object pertain based on the training data set. Bayes' rule is used to calculate the probability that some object pertain to a class as it is shown further on. Given an object "o", a naïve Bayes classifier estimates the probability that the object belongs to each class c_k $P(c_k|v_1, v_2,...,v_n)$ to find the maximum value, according to Bayes's rule:

$$P(C = c_k|v_1, v_2, ..., v_n) = P(C = c_k) * P(v_1, v_2, ..., v_n|C = c_k) / P(v_1, v_2, ..., v_n) \tag{1}$$

Using the chain rule $P(v_1, v_2, ...,v_n| C = c_k)$ can be written as:

$$P(v_1, v_2, ..., v_n|C = c_k) = P(v_1|v_2, ..., v_n, C = c_k) * P(v_2|v_3, ..., v_n, C = c_k)$$
$$* ... * P(v_{n-1}|v_n, C = c_k) * P(v_n|C = c_k) \tag{2}$$

The naïve assumption is that the attributes are independent given the class: an attributes v_i is independent of attribute v_j for $i <> j$ given de class c_k. Thus, the following relations are true:

$$P(v_i|v_{i+1}, \ldots, v_n, c_k) = P(v_i|c_k), P(v_1, v_2, \ldots, v_n|c_k) = \Pi_i P(v_i|c_k) \qquad (3)$$

Given the input, $P(v_1, v_2, \ldots, v_n)$ is constant. So, the classification rule for a new object is described by $(v_1, v_2, \ldots, v_n)$ is as follows:

$$C^{new} < - \text{argmax}_{ck} P(C = c_k) * \Pi_i P(v_i|c_k), \text{ where } C^{new} \text{ is the estimated class.} \qquad (4)$$

A detailed description of the classification techniques based on Bayesian Networks can be found in [12, 13, 15, 16].

3.2 General Model for Building "the Best" Collaborative Groups

In this sub-section, we introduce our model for building "the best" groups of people from a cohort of individuals. The best could mean *the most creative, the most innovative, the most effective, the most proficient*, etc. The main idea is to take into account a group characteristic that is the most relevant for the proposed *"the best"* goal and to maximize it by grouping and re-grouping people based on the values of some particular individual characteristics. Our model includes three stages and is shown in Fig. 1.

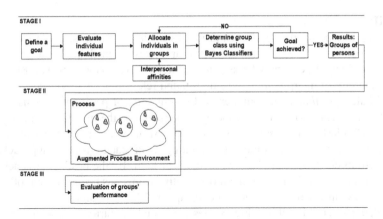

Fig. 1. Model for building "the best" collaborative groups.

Stage I

The algorithm that distributes the individuals in increasingly better groups with regard to the relevant characteristic considered includes the following steps:

- Allocate each learner to a group (ad-hoc or clique based);
- Determine the class of each of the resulted groups using Bayes classifiers;
- *If* the obtained group classification satisfies the proposed goal (for example, given 5 groups, 2 of them are in class High, 2 are in class Medium, and 1 is in class Low) then the distribution stops;

- *Else* another trial is undertaken by various combinations such as: first between the group members of the class L, second between the members of the classes L and M, and, in the end, a total re-distribution of all people.

Stage II

Within the second phase, a particular process takes place (i.e. working, learning, competing, etc.). The environment in which the process takes place can provide for achieving "the best" goal. For instance, in case of aiming at building increasingly creative learning groups, the creative contextual learning environment can be augmented with creativity triggering activities such as: promoting the importance of creativity - learners have to be aware of creativity's role in education and in everyday life, including motivation tasks and advising tasks, using different instructional strategies (mainly the ones focused on problem-based learning and project-based learning), providing for development of social and collaborative skills, developing various teaching and learning scenarios using critical thinking models, allowing questions sessions, not over-structuring the lessons or lectures, keeping a balance between the learner control and the machine control regarding the management of the learning process, designing multicultural and multidisciplinary tasks, and including information aggregation tasks [17].

Stage III

In this stage, the performance of each group, as a whole, is assessed. For example, to evaluate group creativity, in general, the four scales defined in Torrance Tests of Creative Thinking (TTCT) may be used to measure the following aspects [18]: *fluency*: the total quantity of interpretable, meaningful, and relevant ideas produced in response to the stimulus; *flexibility*: the number of different sorts of pertinent responses; *originality*: the statistical rarity of the responses; *elaboration*: the quantity of detail in the responses. Similarly, in learning processes, an instructor evaluates the learning outcomes (ideas, products, solutions, etc.), along with each group's approach. The obtained group creativity may be low, medium or high. If "the best" goal is achieved, then the objective of constructing the most creative teams has been fulfilled, otherwise, during the next instructional session, the groups will be re-organized.

The data gathered in all the three stages are stored for further use as training data.

3.3   Model Instantiation for Building the Most Creative Learning Groups

To instantiate the model presented in the previous section, we evaluate first a set of learners' characteristics that are known to have an impact of group creativity, such as *individual level of creativity, personal motivation, domain expertise*, or any other factors that influence creativity and that we can measure. The Naïve Bayesian structure for these three attributes is shown in Fig. 2.

The level of creativity may be established using various tests such as tests of divergent thinking, creative personality, etc. A well-known and easy to use test is Gough's Creative Personality Scale [19], which output range is between −12 and 18. Domain expertise is determined by assessing specific knowledge and skills (it ranges between 1 and 10). The intrinsic motivation level is evaluated using a specific

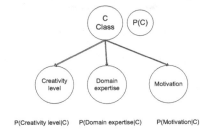

Fig. 2. The Naïve Bayes structure for groups classification.

questionnaire, which may result in 0 (low motivation), 1 (medium motivation), and 2 (high motivation). The Bayesian Classifier makes predictions starting with a training data set, which is obtained by performing several experiments during educational processes over long periods of time. Learners are grouped and re-grouped repeatedly until the objective with regard to obtaining the most creative groups is achieved. From the training data set, the classifier determines the conditional probability for each attribute of each individual pertaining to a certain class. Then, by applying the Bayes' rule, the probability of falling within one class for a given set of attributes is computed. The class with the highest posterior probability is the predicted class.

4 Case Study – A Real-World Scenario

We used the method and the algorithm presented in the previous section to perform a real-world grouping of students in "the best" creative and innovative teams possible with 20 of our third year Computer Science students enrolled in the Software Engineering course. The data presented here have been collected and processed during a period of five months, however, we have grouped students this way for several years now (of course, only the ones willing to participate in this educational scenario, while the others have grouped themselves either on cliques or ad-hoc).

Our "the best" goal was to obtain at least three groups with creativity class medium or higher. The learning achievements are assessed by the grade obtained in the Software Engineering course, which is based on several criteria related to domain expertise, to soft skills achieved, to the creativeness and innovativeness of the solutions, etc. The final grade measures both how well they have achieved the course requirements with respect to the domain knowledge and how well they work together in small developers' teams that aim to complete a common software development project and to properly present their work and the final product.

First, we performed a Gough-based evaluation of creativity of our students and we obtained a creativity score distribution that is presented in Fig. 3. The creativity score mean is 2.55.

The values obtained for the two other attributes considered in the classification are shown in Table 1. The domain expertise is the grade obtained at the Data Structures and Algorithms class, while they were freshmen. We have chosen this grade because

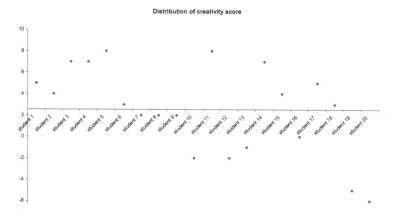

Fig. 3. Distribution of the creativity score of sophomores.

the programming part of the Software Engineering project consists of developing computer applications with fundamental data structures and algorithms in Java. The motivation attribute has been determined using an adapted questionnaire based on MSLQ, which is a multi-item self-report Likert-scaled instrument designed to assess motivation and use of learning strategies by college students [20].

During the experiment, the students have grouped themselves in small cliques based on their inter-personal affinities (they were buddies). Four uneven cliques resulted this way (learners' IDs are presented): (1, 2, 3, 4, 5, 6), (7, 8, 9, 10), (11, 12, 13, 14, 15, 16, 17), and (18, 19, 20). After a learning session, the creativity of each group has been assessed with the following results: no team was in the high creativity class, two teams had medium creativity (coded with value 2), and two teams had low creativity (value 1): groups 1 and 3 had medium creativity and groups 2 and 4 low creativity. As our proposed goal was to have *at least three teams* with creativity class medium or higher, we used the method presented here to re-group the students. For each student, a creativity class has been predicted using a Naïve Bayes classifier (nb1) trained with a predefined data set, as it can be seen in Fig. 4 (a Matlab cropped screenshot). For each student the values obtained, respectively are as follows: 3, 2, 3, 3, 2, 2, 2, 1, 1, 1, 3, 1, 1, 2, 2, 1, 1, 1, 1, 1.

Working our way to reaching the established goal, we re-grouped the students so that at least three of five students were buddies (this having a positive effect on team creativity in our experience and in the literature [21]). If a minimum of three of the five students are buddies, a clique is formed and, consequently, the clique attribute in Fig. 6 is set to 1 (column 6 in the left screenshot). As our objective *to obtain at least three groups with creativity class medium or higher* has not been achieved we continued further the grouping process. To determine each group's creativity a second classifier nb2 has been used. The creativity classes for the obtained groups are presented in Fig. 5 (right screenshot) – group1 = (1, 2, 4, 9, 18), group2 = (11, 12, 14, 15, 19), group3 = (3, 5, 7, 8, 13), and group4 = (6, 10, 16, 17, 20). In the left screenshot, on the first line, the data of team 1 may be read as follows: in the first column, the influence class of the student 1 to the team creativity, resulted from the Bayes-based classification

Table 1. Sophomores' attributes used in classification

Learner ID	Gough score	Domain expertise	Motivation	Learner ID	Gough score	Domain expertise	Motivation
Learner 1	5	8	2	Learner 11	8	10	1
Learner 2	4	8	1	Learner 12	−2	7	1
Learner 3	7	8	2	Learner 13	−1	6	2
Learner 4	7	10	2	Learner 14	7	7	1
Learner 5	8	8	1	Learner 15	4	8	1
Learner 6	3	8	2	Learner 16	0	5	2
Learner 7	2	7	0	Learner 17	5	5	2
Learner 8	2	6	0	Learner 18	3	5	0
Learner 9	2	6	1	Learner 19	−5	6	0
Learner 10	−2	5	0	Learner 20	−6	6	0

```
>> nb1=NaiveBayes.fit(training_nb1, class_nb1);
>> cpre1 = predict(nb1,date_nb1);
>> disp(cpre1);
```

Fig. 4. Naïve Bayes classifier for individual creativity

on the Gough score, the domain expertise, and the motivation; in the following columns, 2 to 5, of line 1 we find the influence classes for the students 2, 4, 9, and 18. The next lines include the data of teams 2, 3, and 4, respectively.

The algorithm stops because "the best" goal has been achieved (i.e. at least 3 teams have creativity medium or higher) (Fig. 6).

At the end of the first stage of our method, the groups obtained fulfill "the best" goal established. During the second stage, the learning process takes place within the augmented environment, while in the third stage the final assessment of each group's creativity takes places. In our case, the performance of the groups is measured by the grade obtained in the Software Engineering course, which is granted based on several criteria that measure the performance of each group as a whole, taking also in consideration each individual contribution. These criteria assess the developed software, the related documentation, the difficulty of the problem, the creative and innovative solutions used during development and for the presentation of the final product, the complexity of the algorithms, the cost-effectiveness of the solution and so on. As it can be seen below in Table 2, the performance of the majority of students is higher after this collaborative learning experience (first average grade is in Data Structures and Algorithms course, while the second is in the Software Engineering one). And this is

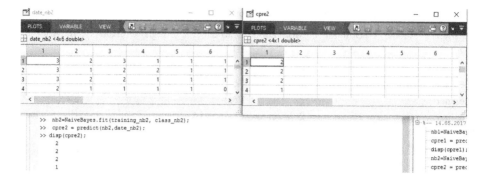

Fig. 5. Left screenshot: the influence classes for each student; Right screenshot: the creativity classes for the groups 1 to 4

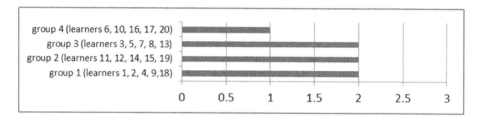

Fig. 6. Creativity classes for the groups

Table 2. Evaluation of learning achievements before and after the grouping

Group	Average grade DSA	Average grade SE
Group 1	7.40	8.20
Group 2	7.60	9.40
Group 3	7.00	8.40
Group 4	5.80	8.00

not just an isolated situation, as we have already performed this grouping, in similar circumstances, for 5 years now, and the results are consistent and show increased learning with respect to both domain expertise and soft skills achieved.

5 Discussion

In order to make our technique easier to use, we explain here further some of the particularities of the method. For the time being, we use as classifying technique the naïve Bayes classifier, but some other similar classifiers may be used (for instance neural network based classifiers, decision trees, and support vector machines).

The main shortcoming is presuming that the characteristics taken into account are independent of each other, which allowed us to use Naïve Bayesian Network based classifiers. This assumption may simplify things, but it can be false sometimes. Even in our case, there can be dependencies between the individual creativity level and the domain expertise or between the motivation and the domain expertise. In the case study presented here, we have modeled the interactions inside groups using cliques. In our testing, a group was been considered to be a clique if 3 out of the 5 members of a group were buddies.

As for the first results, it can be noticed that our students in the experiment fall in the following classes with respect to creativity: 4 are in the high class (value 3 in Fig. 5), 6 are in the medium class (value 2), and 10 in the low class (value 1). This observation lead us to the conclusion that in some particular situations, when the individual creativity scores or the other two attributes are low, some bold goals cannot be achieved whatsoever. This understanding is something to consider for any kind of collaborative activity.

To experiment further, we plan to include in the parameter set some sort of evaluation of the potential for creativity obtained by external observation of both instructional and extra-curricular activities.

6   Conclusions and Future Work

Nowadays, amazing technological progress and tremendous changes in the global economy have both contributed massively to a paradigm shift with regard to collaboration among people. Increasing the efficiency and effectiveness of groups of individuals performing together specific activities to achieve common goals in given contexts is of crucial importance. Nevertheless, building high performance groups is quite challenging regardless the domain they activate it. We presented here our work on clustering individuals in groups so that a global objective with respect to the quality of groups is achieved. The grouping is based on both individual characteristics and inter-personal interactions. The proposed model and method are general and can be used for several collaborative activities such as working, learning, competing, etc. To test them, we instantiated and used them to construct the most creative and innovative collaborative learning groups. After using this method in several learning situations, we have learned that trying to semi-automatically group individuals in high performance teams with regard to some particular objectives is a laborious task that involves knowledge and instruments in various fields such as education sciences, social and personality psychology, computer science, and machine learning.

Further work ideas, besides the ones presented in the previous section, include using several creativity evaluation scales, adding contextual and organizational factors, testing the method in other activities and in other fields, improving the algorithm, performing experiments with control groups, and, eventually, offering the resulted technique as an online open service.

References

1. Johnson, D.W., Johnson, R.T., Holubec, E.J.: The New Circles of Learning: Cooperation in the Classroom and School. Association for Supervision and Curriculum Development, Alexandria (1994)
2. Johnson, D.W., Johnson, R.T.: Making Cooperative Learning Work. Theor. Pract. **38**(2), 67–73 (1999). Building Community through Cooperative Learning
3. Felder, R.M., Brent, R.: Cooperative learning. In: Mabrouk, P.A. (ed.) Active Learning: Models from the Analytical Sciences. ACS Symposium Series, vol. 970, pp. 34–53. American Chemical Society, Washington (2007)
4. Felder, R.M., Brent, R.: Effective strategies for cooperative learning. J. Coop. Collab. Coll. Teach. **10**(2), 69–75 (2001)
5. Moise, G., Vladoiu, M., Constantinescu, Z.: GC-MAS – a multiagent system for building creative groups used in computer supported collaborative learning. In: Jezic, G., Kusek, M., Lovrek, I., Howlett, R.J., Jain, L.C. (eds.) Agent and Multi-Agent Systems: Technologies and Applications. AISC, vol. 296, pp. 313–323. Springer, Cham (2014). doi:10.1007/978-3-319-07650-8_31
6. Zhou, C., Luo, L.: Group creativity in learning context: understanding in a social-cultural framework and methodology. Creat. Educ. **3**(4), 392–399 (2012)
7. Reiter-Palmon, R., Wigert, B., de Vreede, T.: Team creativity and innovation: the effect of group composition, social processes, and cognition. In: Handbook of Organizational Creativity. Elsevier Science & Technology (2011)
8. Paulus, P.B., Dzindolet, M.T.: Social influence, creativity and innovation. Soc. Influ. **3**, 228–247 (2008)
9. Fischer, G., Giaccardi, E., Eden, H., Sugimoto, M., Ye, Y.: Beyond binary choices: integrating individual and social creativity. Int. J. Hum. Comput. Stud. Comput. Support Creat. **63**(4–5), 482–512 (2005)
10. Kenny, A.: 'Collaborative creativity' within a jazz ensemble as a musical and social practice. Think. Skills Creat. **13**, 1–8 (2014)
11. Choi, D.Y., Lee, K.C., Seo, Y.W.: Scenario-based management of team creativity in sensitivity contexts: an approach with a general bayesian network. In: Lee, K.C. (ed.), Digital Creativity Individuals, Groups, and Organizations (2013)
12. Friedman, N., Geiger, D., Goldszmidt, M.: Bayesian network classifiers. Mach. Learn. **29**, 131–163 (1997)
13. Jensen, F.V., Nielsen, T.D.: Bayesian Networks and Decision Graphs. Springer-Verlag, New York (2007)
14. Duda, R.O., Hart, P.E.: Pattern Classification and Scene Analysis. Wiley, New York (1973)
15. Pearl, J., Russel, S., Bayesian networks. In: Arbib, M. (ed.), Handbook of Brain Theory and Neural, Technical Report Networks, MIT Press (2001)
16. Störr, H.P.: A compact fuzzy extension of the Naive Bayesian classification algorithm. In: Phuong, N.H., Nguyen, H.T., Ho, N.C., Santiprabhob, P. (eds.), Proceedings InTech/VJFuzzy 2002, pp. 172–177 (2002)
17. Moise, G.: Fuzzy enhancement of creativity in collaborative online learning. In: Chiu, D.K. W., Wang, M., Popescu, E., Li, Q., Lau, R. (eds.) ICWL 2012. LNCS, vol. 7697, pp. 290–299. Springer, Heidelberg (2014). doi:10.1007/978-3-662-43454-3_30
18. Torrance, E.P.: Torrance tests of creative thinking. Scholastic Testing Service, Bensenville (1966)
19. Gough, H.G.: A creative personality scale for the adjective check list. J. Pers. Soc. Psychol. **37**, 1398–1405 (1979)

20. Pintrich, P.R., Smith, D.A.F., Garcia, T., McKeachie, W.J.: Reliability and predictive validity of the motivated strategies for learning questionnaire (MSLQ). Educ. Psychol. Meas. **53**(3), 801–813 (1993)
21. Mueller, J., Cronin, M.A.: How relational processes support team creativity. In: Mannix, E. A., Goncalo, J.A., Neale, M.A. (eds.) Creativity in Groups, Research on Managing Groups and Teams Series, vol. 12, pp. 291–310. Emerald Group Publishing Ltd, Bingley (2009)

A New Collaborative Paradigm of Computer Science Student Contests: An Experience

Monica Vladoiu$^{(\boxtimes)}$, Zoran Constantinescu, and Gabriela Moise

The Department of Computer Science and Information Technology,
Mathematics and Physics, CerTIMF Research Center,
Petroleum-Gas University of Ploiesti, Ploiesti, Romania
{mvladoiu,zoran,gmoise}@upg-ploiesti.ro

Abstract. For more than 30 years now, the hallmark of competitions in Computer Science has been the *time-to-complete programming contest*. Though its success is indubitable for some students and some objectives, lately new paradigms have appeared, aiming at reaching more students, more Computer Science areas, and more learning objectives. This paper reports on and reflects about our academic experience with our Computer Science competition, spanning on more than three decades. The focus is on its evolution and continuous adaptation to current learning needs, goals, and contexts. The last paradigm has a free format, is fun- and skill-oriented, and aims at boosting students' creativity and learning. The competition is based on the *three-words-from-a-hat* challenge and starts with the teams choosing three words from a hat, these being the keywords to be covered by the developed software application. We explain here the motivation of shifting from the traditional way, the new paradigm's core ideas, while highlighting the lessons learnt. Having assessed it for two editions already, we can provide some conclusions on its impact on team-based competitions. Our main finding is that it inspires students to work with new ideas, to build up on their strengths and to work on their weaknesses, whilst competing for both fun and success, along with their peers. The demographic of the contest has also changed, with more girls entering competition each time. In our common view, both students and academics, the main benefit of this new paradigm is increased learning of both domain knowledge and soft skills.

Keywords: Computer Science competition · Teamwork · Creativity · Learning · Education

1 Introduction

No common point of view regarding the benefits of competitions in learning exists among education scholars and practitioners. Some claim that humans are born to compete and, therefore, evolution is strongly related to competition, while others are convinced that humans are ought to cooperate and that competition is only culturally induced. First recorded human competitions are in sport and, only later, in area of military exercises. The first notable science contest was established in 1894 in Hungary in mathematics; later, competitions in other areas have been launched as well, in domains such as physics, chemistry, informatics, philosophy, biology, astronomy, and

© Springer International Publishing AG 2017
H. Panetto et al. (Eds.): OTM 2017 Conferences, Part I, LNCS 10573, pp. 284–297, 2017.
https://doi.org/10.1007/978-3-319-69462-7_18

geography [1]. In the early years, finding solutions for important mathematical and scientific problems has been the main reason of holding prize competitions, mainly within academia. Thus, many educational institutions and various associations, inside and outside of the formal education system, have held regional, national, and international contests, Olympiads, and other forms of competitions, in various scientific areas and, lately, also in interdisciplinary domains.

In Computer Science, for more than three decades now, the prominent paradigm of prize competitions has been the *time-to-complete programming contest,* which is proven successful for some students and some objectives. However, currently, new approaches seem more suitable for reaching more students, for covering more Computer Science areas, and for achieving more learning objectives.

This paper reports on and reflects about the academic experience we have had in the last 32 years with our Computer Science competition, focusing on its evolution and continuous adaptation to current learning needs, goals, and contexts. We started with the time-to-complete paradigm as well and we have been keeping that format for more than 25 years. In the last two editions, we have come up with a new approach that has a free format, is fun- and skill-oriented, and aims at boosting students' creativity and learning. The competition is based on the *three-words-from-a-hat challenge* and starts with the enrolled teams choosing three words from a hat. These three picks are the keywords to be covered by the developed software application, which must fulfill a set of quality criteria, such as being new and useful.

We explain also here the motivation of shifting from the traditional way, what are the new paradigm's core ideas, while emphasizing the lessons learnt. Having assessed it for two editions already, we can provide some conclusions on its impact on team-based competitions. Our main finding is that it inspires students to work with new ideas, to build up on their strengths and to work on their weaknesses, whilst competing for both fun and success, along with their peers. Another positive fact is that the demographic of the contest has also changed, with more girls entering competition each time. In our common view, both students and academics, the main benefit of this new paradigm is increased learning of both domain knowledge and soft skills, both of these being really sought-after in the labor-market.

In the next section, we present a short history of Computer Science competitions. In the third section, we report on more than 30 years of experience in our university with respect to CS competitions (paradigms, evolution, and feedback). We emphasize also the motivation to continuously change the approach to fit better the learning needs and goals of current generations of students, and also their learning and working contexts. The latest experiences with the most recent paradigm are also presented here. The forth section includes a brief discussion of the benefits of competition in education, in general, and reflections on our experience, in particular. The fifth section is dedicated to both the conclusions and some future development ideas.

2   Computer Science Contests

The ACM International Collegiate Programming Contest (ICPC), first hosted in 1977 in Atlanta, and the International Olympiad in Informatics (IOI), first hosted by Bulgaria in 1989, are the oldest and most prestigious international Computer Science competitions

[3, 4]. Both of them and many other similar competitions have been organized, at least in their beginnings, as programming contests.

2.1 Computer Programming Contests

The ACM ICPC is a multitier, team-based programming competition, which takes place under the auspices of ACM and provides for creativity, teamwork, and innovation in building new software programs, while enabling students to test their capacity to perform under pressure [3]. The primary goal of the IOI, initiated by UNESCO, is dual, i.e. to stimulate interest in Informatics (Computer Science) and Information Technology and to bring together outstandingly talented students from several countries and to have them share scientific and cultural experiences [4].

During these programming contests, the participants (either individuals or teams) have to solve a set of programming problems, in a limited time *(time-to-complete* or *pressure cooker model)*, having access to only one computer. Problems with various degrees of complexity are typically proposed in such a session. The very difficult ones often combine multiple advanced concepts in algorithms or mathematics. The references [1, 3–6] are valuable resources on such problems, while reports on deploying such contests can be found in [5, 7, 8]. The contestants may get support as coaching by faculty members prior to the contest and, during the competition, in form of test files for the problems to be solved, access to reference materials, and feedback regarding the correctness of submissions. Contest strategies have been addressed in [7, 9, 10] and consulting them constitutes a valuable aid for training either individuals or teams. The number of problems solved in a limited time, the least total time, the earliest time of submittal of the last accepted *run* (a run is a solution to problems that has been submitted for judging), and, in some cases, the penalties for incorrect submissions are taken into account for scoring in such competitions [7]. The runs are tested with the well-known *black-box testing*, where only the output of a program is measured when running on some particular test data.

As some research shows, competitions among teams provide for higher learning performances than competition among individuals [11]. Some teams are formed by a coder, an algorithm designer, and a debugger and all of them try to solve all the problems [6], while in others the tasks are distributed among team members based on either the level of difficulty or their expertise. Anyway, the team dynamic proves to be of chief importance in a contest [11, 12].

Benefits for the participants are primarily pedagogical. Tangible rewards are not the most important benefits, as most of the participants confirm [5, 7, 8, 12]. Deeper understanding of core concepts in Computer Science (not always covered in typical curricula), as well as building some cross-cutting skills (as teamwork, efficient working under pressure, leadership, communication, creativity and innovation, and so on) and task enjoyment are the main benefits reported by most of the contestants [7, 12].

2.2 New Paradigms

In the last 10–15 years, other contest paradigms have appeared aiming at approaching other aspects of Computer Science and Information Technology, which cannot be reduced to merely programming. As a result, more and more non-programmer students have enrolled in CS competitions.

Project-based competitions have long lasting impact for students because both the experiences themselves and the developed skills are great assets for their future careers [13]. Such contests help the participants learn to apply theoretical and practical knowledge, to collaborate as close-knit teams, to use various tools, to choose the best strategies to find solutions for given problems, to face occurring challenges, to test and evaluate the results, and to present them in a well suited manner, all of these contributing to acquiring further knowledge, but also additional project-management, time-scheduling, and leadership skills [2]. Learning based on excitement is triggered by the new paradigm of competition, like Formula Student or RoboCup, where teams of students implement their ideas and test them along with products of other teams during the contest [2]. The major goal of such competitions is "to foster science and technologies for evolving industries and for advancing societies" [14].

In other paradigms, the time-to-complete approach is replaced with *quality of process and competing for fun and success*, to make participants feel confident in their success and to best promote their ideas [9, 15]. Sherrell and McCauley [16] designed a programming paradigm where the participant teams had to design and implement within a month a documented adventure game to promote their ideas and increase their responsibility. Speed programming is also replaced with speed thinking. Combining *art show and science fair*, where creativity in using technologies to solve real-world problems is furthered, is reported in [17]. Similarly, Intel ISEF (Intel International Science and Engineering Fair) and Siemens Competition in Math, Science and Technology encourage students to undertake individual or team *research projects*, aiming at improving students' understanding of the usefulness of scientific study as such, but also for solving real-world problems [18, 19].

As Schank [20] and Kolb [21] say, significant implicit learning can be achieved through games, experience, and other hands-on activities that increase students' involvement (*learning by doing*). Thus, project-based contests provide for an implicit learning, mostly learning by doing [20], and, surely, they increase the participants' engagement and responsibility. Moreover, this highly contributes to their readiness for a job in the industry [2].

3 Our Experience with Computer Science Contests

In this section, we present our experience with organizing Computer Science contests at academic level for more than 30 years now. Various paradigms have been tried during this time, the most significant being illustrated further on.

3.1 Evolution of Our Contest

In our university, the best Computer Science students are traditionally trained to participate to regional and national competitions, including the internal Computer Science contest that takes place almost annually, since 1985 [12]. Up to 2011, the contest (named "The best programmer") put in practice the classical time-to-complete paradigm of computer programming contest. Since the beginning, the competition has been sponsored by local and national companies and organizations (IT&C or not). During the entire 32 years of competition, the contest management team has changed, but the know-how has been continuously carried on. Thus, one of the authors of this paper has been part of the organizing (and jury) team since the early years of the competition, in 1991, and the others in its last five editions.

The individual contestants had to solve on site, with no Internet access nor other resources, 2 to 4 problems of different difficulty, in a limited time (2 to 3 h). The majority of the problems included advanced mathematical issues, such as analytical geometry, combinatorial optimization, or matrix theory. Only individual undergraduates were allowed to participate in the programming contest. The subjects to be covered (data structures, programming languages, programming techniques, searching, sorting, object oriented programming, and artificial intelligence) and a reference list of fundamental readings were made available early on.

The juries, formed by internal faculty, performed *white-box testing*. They made on site both *objective evaluations* (regarding the correctness of the solutions, the number of code lines, invalid or special input data handling, etc.) and *semi-objective evaluations* (regarding readability and organization of the code, its documentation, readability of all submitted data, innovation, easiness of debugging, elegance of the solution, the time and space complexity etc.). The quality of all artifacts was carefully assessed. Since 2008, given the increasing number of participants, a qualification stage has become necessary, which has been usually held a day or two before the finals. In 2011, the internal contest was opened also to participants from another university in our region and that year the first prizes went to students in the two universities.

For the next two years (2012 and 2013), the competition shifted to a more flexible manifestation, called "Computer Science: A Successful Career", having three sections: "The best programmer", "The best CS project" and the "The best CS scientific paper". A local newspaper reported the contest results, thus better promoting of CS university programs to high school students. In 2013, a foreign student in Applied Informatics in Chemistry won the 3rd prize of the programming section.

The projects' and scientific papers' sections were opened also to graduate students and mixed teams of undergraduates and graduates. However, only individual participants submitted their work. In our experience, the reason for this behavior seemed to be students' habit to work alone, especially when in competition, in spite of having experience on working together on common projects (that have been even graded as such) on some curricular activities. The project themes were freely chosen by the participants and the projects had to be original, relevant, of medium or high complexity, and finalized in three weeks. The academic jury thoroughly verified every project and every paper for plagiarism and no such case was found.

The best project in 2012 was a 3D graphic rendering application for the university campus, while in 2013 it was a complex graphic simulator showing how intelligent agents imitating ants' behavior can solve combinatorial problems; the author, student in the 3rd bachelor year, used the Ant Colony Optimization technique. His very intuitive simulator that uses many application scenarios is used since then in the Metaheuristics class for graduates. For the scientific papers section, the contestants reported results of their research on various topics of Computer Science. The best papers in 2012 and

2013 focused on innovations in cryptography and belonged to graduates. One of them handled PGP, while the other dealt with biometric cryptography and quantum cryptography. The latter emphasized the novelty and benefits of (1) quantum key distribution, which detects eavesdropping since the reading data encoded in a quantum state changes that state and (2) using elliptical curves mathematics to factorize the products of big prime numbers.

3.2   A New Paradigm: Three-Words-from-a-Hat

A relatively small number of students submitted papers in 2012 and 2013 and the participation to the programming section decreased constantly for more than 5 years in a row. In contrast, an increasing number of participants attended the projects' section year after year. After debating these trends with our students, the conclusion has been that nowadays good students prefer to be challenged to creatively apply their knowledge to obtain operational hardware and software, to choose the newest appropriate technologies for building innovative solutions, to be independent learners who can research new topics and solve problems with little or no guidance. Hence, we have realized that a *new paradigm* has become necessary and that it should focus only on *project-based competition*. The core idea of this new paradigm had to do with encouraging participants' creativity in problem solving, with reinforcing good project development practices, and, especially, with team work. This new paradigm based on *the-three-words-from-a-hat* challenge is presented further on.

The competition (called 3ITC) starts with teams of 2 or 3 participants choosing *three words from a hat* (for example, key, mobile, bit, three, calendar, sound, art, career, optimization, student, network, intrusion, autonomous, planet, season, etc.). Each team decides a theme for their project that is required to cover all three words. In two days, a short description of each project's theme has to be submitted to the software platform, after authentication, which is not public until the finals. For the next four weeks, each team works on its idea, design a solution, then implement it and test it till an operational software product is obtained. In the finals, each team has the opportunity to present its work and results (for approximately 15 min) in front of both the jury (formed by academics and industry representatives) and their colleagues. Each member of the jury gets a file that contains the detailed description for each project, which is uploaded by the participants, and may ask questions.

The results are evaluated against a set of *judging criteria*, which are multifaceted and regard various quality factors of software systems such as applicability, utility, creativity, originality, complexity, programming effort, user interface, description, conclusions, presentation, demo, efficiency, robustness, and maintenance. After all the contestants had presented their work, the jury retires to deliberate. The final scoring board is in accordance with the cumulated score for all jury members. Moreover, during the jury debate, other useful ideas may arise. For example, in 2016, some new criteria have been proposed for future competitions. The necessity to have two development sections has also become obvious for most of the jury members, i.e. one for participants using frameworks and tools and the other for those who develop their solutions from scratch. All the participants receive some form of reward such as electronic devices, internships, etc. and/or certificates of recognition.

3.3   Experiencing the New Paradigm

In the last two editions, in 2014 and 2016, the new paradigm was put into practice and tested. In addition, in 2016 a digital art section was added, but very few students pursued it. In 2014, eight in twelve enrolled teams finalized their projects and presented their results in the finals. In 2016, eight in fourteen enrolled teams finalized their projects and presented them in the finals. The competition took place each time at the middle of the first semester, as in our educational system that is not a very busy period neither for students nor staff.

Administration of the contest. Organizing a good competition is a major challenge and a labor-intensive task, starting from the contest's promotion to handing out the prizes to the top participants. To cope with the significant amount of work, one of the co-authors of this paper had developed a dedicated software platform to allow contestants to upload their files, to facilitate communication of the organizing team with the participants in each stage of the contest, and to publicly present the final results. A day before the public finals, each team is ought to upload several artifacts as follows: (1) a reference documentation for the project, so that the most knowledgeable (to specific subjects) three academic judges have the time to analyze the idea, its development, and its originality; (2) a template-based poster synthetically presenting the project, and (3) a preliminary version of the software (improvements are possible until the start of the presentations). Just before the finals begins, the participants must upload the source code, a demo (or a video recording of the application run), and any other files necessary to present their project in front of the jury. The posters and all the other relevant information may be found on the contest's webpage [22]. In the final day of the competition, the posters are displayed in the university hall. Moreover, they are to be used further in following events as a token of our students' academic achievements.

In 2014 the jury was formed by 8 faculty members. In 2016 the jury was formed by 6 academics, an independent artist (given the digital art section), and 6 industry representatives from four companies (two local and two international); one of the latter is our Computer Science MSc program's graduate and also the winner of the best programmer contest 16 years ago; he is now with an international company. Given the cumulated expertise of the jury, various aspects of contestants' work and results have been considered. The most frequent question asked by the jury was which were the biggest challenges encountered and what were the solutions found by each team.

The prizes in 2014 consisted in various IT&C devices and, in 2016, in internships to two IT&C companies. The winners of 2016 said that they preferred the internships to the devices. The jury may offer special awards for excellence or for the most original project.

Examples of submitted projects, along with the keywords to be covered and a short description, are as follows:

- *Student, education, creative* - educational platform able to virtually join students worldwide in programming contests, in a game-like manner, funnier, and more efficient. Many difficulty levels allow deep learning for the beginners also;
- *University, energy, sound* - multiplayer competitive simulator of sound-based bat intelligence, necessary to find the exit from a labyrinth in a university;

- *Graphic, student, time* - educational platform to manage classes, courses, tasks, resources, questionnaires, virtual communication by real time or offline messages, post-class reports regarding the subject understanding, private consultations, quizzes and so on;
- *Optic, adaptive, optimization* - interactive game to test a prototype of optic weapon using a rendering engine where the number of particles used is limited for performance optimization;
- *Optimization, student, mobile* - task scheduling Android application for students, working with events such as exams, projects, meetings, etc.;
- *Learning, carrier, calendar* - a mobile application that helps users to become aware of significant days worldwide and to discover specific facts and traditions from different countries and cultures for such days (for example, national holidays or specific folk traditional customs);
- *Cell, photography, key* - project to early diagnose mammal cancer, detecting cancer cells in photographic mammograms;
- *Intelligent, architecture, remote control* - a remote desktop control application for user support in troubleshooting computers.

Other ideas concretized in remote control applications, land management applications in agriculture, one-player or multiplayer cooperative or competitive games, interactive applications on the Web, and so on.

4   Is the New Paradigm Beneficial for Computer Science Education?

4.1   Is Competition Beneficial to Education?

Many works in the literature sustain the beneficial role of competition in evolution, pointing out the competing nature of humans, which results in having both individuals and societies advancing through competition [2]. It is a fact that some children desire to compete within groups of common interests and the pleaders of these theories claim that "it is necessary to incorporate competition into education to help children get used to it in later life" [1]. Either formal or informal, sound competitions may bring advantages to science evolution and human development by crossing knowledge boundaries and finding novel ideas and solutions to problems. However, the drawbacks of ill competitions must not be ignored. A testimonial on sacrificing deep learning for the sake of winning, especially if the competition is mandatory or grades depends on its results, is given in [2]. In an academic setting, *competition-based learning* based on Game Theory has been tested and the results obtained suggest that combining game theory with friendly competitions provides a strong motivation for students, helping them to increase their learning performance [23].

Our experience shows that competition is pedagogically beneficial mainly to really competitive students. Their motivation to evolve is certainly related to good or poor results in comparison with others. Also, some non-competitive students can make use of success or failure in some forms of competitions. However, for a significant number of students competing anxiety is not motivating at all, just crimping their creativity and self-expression, and, therefore, additionally increasing their insecurity. Moreover, solutions to problems found by cooperation are often more tenable, more rapidly detected, more efficient and, therefore, more valuable.

As a result of such contradictory ideas, a harmonious way to combine advantages of both cooperation and competition in education is, in our view, organizing contests outside the curricula, where teams compete with each other and their work is evaluated against various judging criteria that assess multiple facets of their knowledge, skills, effort, and results.

4.2   Reflecting on the New Paradigm

To organize such a competition is not an easy task and, therefore, some issues had to be approached before deciding what is to be done in the future. Thus, after each competition had taken place, common discussions involving both the organizing team and the attending students (both active participants and audience) had taken place. We asked ourselves and each other many intriguing questions such as *Why do we organize year after year this contest? Is it really worthy to have a paradigm shift? Is this beneficial within the new (lifelong and life-wide) learning contexts?*

To respond to these questions, the philosophy and objectives of this new paradigm have to be taken into account, as they are illustrated further on. First of all, *the contest is a learning framework*. The students are challenged in many ways: to come up with a useful idea based on three random concepts, to build a consistent design aiming at putting that idea into practice, to implement it in an operational software product during a development process, to use the adequate languages and software development environments, to synthetically present the project work in both a presentation and a poster, and to convince the jury that their results fully meet the required criteria. Even the students who are not contestants may benefit from this experience, likewise the faculty and industry representatives, as it is shown later here.

The contest is viewed also as *a marketing tool*, as it promotes Computer Science university programs. High-school students and teachers are also invited to attend the finals, so that future prospective students make contact with real projects of elder colleagues based on some advanced theoretic concepts in their K-12 Computer Science curriculum. The marketing function is also acquired by making public the competition and its results, each step of the way, on department's website and also via local media channels.

The contest is *a motivation factor for students* in various directions. For example, they are stimulated to approach novelty in Computer Science areas. Over the years, we have acknowledged their accentuated desire to draw out from the curricula routine. To work on projects in formal classroom environments may attract certain students, but not all of them. Good students want to have the opportunity to work on more complex projects, even interdisciplinary ones. Obtaining good results or even just participation in Computer Science contests are viewed as key achievements in their resumes after graduation, taking into account that employers tend to hire graduates with some working experience, even at junior level. The developers retain the copyright on their work that being another motivation factor. The contest prizes are additional strong motivations for some of them.

The competition stimulates the participants *to make responsible decisions about their work*, while being aware of their consequences for their projects' success. They also acknowledge the importance of all the aspects involved in professionally working together to complete a job. They form affinity-based teams, regardless of the year of study or graduation field and choose original high-impact team names; the teams are heterogeneous and can include, for example, a programmer, a philology student, and a Web designer. They may either use existent software application development environments or other software development tools, or create them from scratch. They use a variety of ways to present their results: as a team with all the members taking turns or there may be one presenter who gets help from his/her colleagues (or not) to present a demo or to run the software. They also include artistic effects in their presentations (both in the real presentation and computer-based one), spiced with ad hoc jokes and witty comments.

In our view, the results of the two last editions of the internal Computer Science competition have proved that the new paradigm is tightly adapted to students' learning motivations, except for the digital art section. The most important conclusion is that *project-based competitions are well suited in Computer Science at academic level*. In such contests, the participants may make better use of their skills and, therefore, can develop competencies related to analytical and algorithmic thinking, modeling, design, and implementation, but also to public speaking about personal work. Our experience shows, for example, that they prefer to use new software tools, frameworks, or programming languages that are not covered by the standard curricula, to test themselves similarly to a job in the real-world, where they will have to adapt and to learn new things continuously in order to evolve professionally.

Contestants' motivations are both extrinsic and intrinsic. From the received feedback after interviewing a significant amount of participants and after our common round tables, we have learnt that students' interest regarding the involvement in the contest is dual. First, they want to have the opportunity to work on some interesting idea, along with their buddies, to put to good use their knowledge, skills, and effort, to prove themselves that they are able to finish successfully a complex software project. But, they are equally interested in their weaknesses and how they can be addressed. They are also very keen to learn from each other, both domain knowledge and soft skills. For example, the public presentations of the projects have generally been well organized, concise, clear, with focus on personal contributions, strong points, and future developments; in addition, issues needing improvements have been discovered, and all the students (including the ones from the audience) seemed to have learned from that experience. Regularly, some students attend to consecutive editions and increasing improvements are visible. On the other hand, participants boosted their self-confidence and improved their professional approach either in class, at the semester exams and even at the graduation exam. Very few participants have proved the expectancy theory [24], which states the dependence of motivation on prediction of reward, on importance of the reward, and on the expectation of achieving the reward. In our experience, only a small percentage of students are interested mainly in wining. After the competition, the participants acknowledge additional advantages, such as deep learning, exposing to new areas of Computer Science, innovating, gaining exposure to potential employers, reinforcing and developing new knowledge, skills, and values, improving their capabilities to deep and reflective learning, etc.

With regard to the art section, despite having about 10–15% of students each year who are interested in expressing themselves in artistic ways, some of them being really talented, the participation was unimpressive. Following up this issue with our students, we have learnt that the main reason they have not either enrolled or present their final project has been their fear of failure. However, the projects that have been successfully completed and presented in front of the jury were in trends with the latest development in digital art and digital culture, in general, and have been highly appreciated by both the audience and the jury.

What is worth to be mentioned regards the demographic of the participants, in the sense that the number of girls enrolling in the competition has increased constantly. Thus, when the traditional programming contest was taking place, the participation of girls has been rather thin, being generally less than 10% of the total number of participants. However, starting with 2012, when the contest extended to the three sections dedicated to programming, projects and scientific papers, the number of girls entering the competition has increased (2012 – 15%, 2013 – 16%). The last two editions that comply with the new paradigm, the increase has been even more significant with 30% girls in 2014 and 25% girls in 2016. The main reason of this increase seems to be the valuable contribution that good communicators bring to software development teams, this being a fact very well documented in the literature and sustained by the practice.

The participants' enthusiasm has combined seamlessly with the one of the organizing team and faculty management (the dean was present at the finals and delivered a motivational and supporting message accented with some interesting examples of immersive digital art).

By inviting industry representatives and potential employers to the event, *a better interrelation between academia and industry* is cultivated. From another point of view, the industry takes a note on the potential of the upcoming professionals and may guide them towards the actual trends in the market. The students who outperform in such contests are generally recruited to work on research and development projects and may be preferentially recommended by the faculty to employers.

Despite the number of completed projects being relatively small, *the diversity of ideas was striking*, especially given that only around 150 students are enrolled in our CS programs. The applying domains varied from education, networking, image rendering, and scheduling to gaming, agronomy, and medicine. Their colleagues, either at university or high-school level, appreciated also the wide variety of applications created by their peers. Some participants appreciated that freshmen are generally discouraged to enroll because the contest is opened to students in every year of study.

Such competitions may have *beneficiaries* other than the persons directly involved, such as contestants, jury, management, etc. For example, the invited K-12 students become accustomed with academia and are initiated on what a Computer Science career means, the K-12 teachers make note of the skills needed to prospective Computer Science students, the IT&C industry and sponsors have the opportunity to briefly promote their identity, services, or products in front of the participants, the other students gain confidence in their own possibilities to concretize their knowledge and skills, and, finally, all the participants, being contestants or not, have a deep meaningful experience to reflect on.

5 Conclusions

Despite the ambivalent attitude towards competition that hovers in the arena of education, extracurricular competitions for students are generally very well received and have good pedagogical results. A balanced solution seems to be having team competitions with scoring criteria that consider multiple facets of the contestants' knowledge, skills, effort, and results.

In Computer Science, the classical programming-core competition framework is more and more replaced by other paradigms that either relax this highly constrained model or extend it to promote creativity, innovation, achievement of soft skills, and so on. In our experience, working under pressure proved to be a factor that lowers creativity in solving problems. On the other hand, interesting and challenging projects developed by teams seem to be necessary ingredients for pedagogically valuable Computer Science competitions. When no negative stress is put on the participants, they can concentrate on what is of chief importance to them, such as putting to good use their knowledge, skills, and effort to complete successfully complex software projects. Because reassurance with regard to their opportunities for getting a good job after graduation is what matters the most for the majority of them, their success and learning during the contest is a promising premise for reaching that major goal.

Our academic department's experience of 26 years of classical programming competitions has lead us to a new paradigm of Computer Science contest, which is presented and discussed in detail this work. Thus, our approach has shifted considerably to better adapt to students' needs of creatively express themselves, while improving their knowledge and skills that have an important contribution to securing a good job after graduation. Under the new paradigm, breaking the routine of curricular activities and providing for multiple ways of creative expression is already a fact, as at the core of this new paradigm is boosting creativity and innovation in solving (real-world) problems. In our case, project-based competitions between teams of students have already proved to provide for this goal. Mainly because the participants get to decide crucial aspects of their work and also to benefit from the unfolded opportunities, such as good use of theoretical and practical knowledge, skills, and values that are necessary to successfully complete the work, accomplishment of learning and professional objectives, working on their weaknesses, and so on. All this experience is much more similar with real-world work than traditional programming contests.

After already experimenting with the new paradigm for two editions, future development ideas occurred, such as having separate sections, one for teams which build their software (or tools) from scratch and one for teams that use existing software, taking into account the number of team members when scoring, including a discussion session immediately after the finals where the students share the lessons learned with each other and reflect together on the (learning) experience, supplementing the event with recreational activities, including more ways to get sponsors, and better promoting of the contest in local media.

References

1. Verhoeff, T.: The role of competitions in education. Technische Universiteit Eindhoven, Eindhoven, Netherlands (1997). http://olympiads.win.tue.nl/ioi/ioi97/ffutwrld/competit.pdf
2. Kristensen, F., Troeng, O., Safavi, M., Narayanan, P.: Competition in higher education – good or bad? Report, Department of Electrical and Information Technology, Lund University, Lund, Sweden (2015). http://portal.research.lu.se/portal/files/5680982/8519800.pdf
3. The ACM-ICPC International Collegiate Programming Contest. https://icpc.baylor.edu/
4. IOI - International Olympiad in Informatics. http://www.ioinformatics.org/
5. Deimel, L.: ACM international scholastic programming contest. SIGCSE Bull. **16**(3), 7–12 (1984)
6. Skiena, S., Revilla, M.: Programming Challenges: The Programming Contest Training Manual. Texts in Computer Science. Springer, New York (2003). doi:10.1007/b97559
7. Bloomfield, A., Sotomayor, B.: A programming contest strategy guide. In: Proceedings of the 47th ACM Technical Symposium on Computing Science Education (SIGCSE 2016), pp. 609–614. ACM, New York (2016)
8. Comer, J.R., Wier, R.R., Rinewalt, J.R.: Programming contests. In: Proceedings of the 14th SIGCSE Technical Symposium on Computer Science Education, SIGCSE 1983, pp. 241–244. ACM, New York (1983)
9. Khera, V., Astrachan, O., Kotz, D.: The internet programming contest: a report and philosophy. In: Proceedings of the 24th SIGCSE Technical Symposium on Computer Science Education, pp. 48–52. ACM, New York (1993)
10. Trotman, A., Handley, C.: Programming contest strategy. Comput. Educ. **50**(3), 821–837 (2008)
11. Fu, F.L., Wu, Y.L., Ho, H.C.: An investigation of coopetitive pedagogic design for knowledge creation in web-based learning. Comput. Educ. **53**(3), 550–562 (2009)
12. Yair, G.: Key educational experiences and self-discovery in higher education. Teach. Teach. Educ. **24**(1), 92–103 (2008)
13. Constantinescu, Z., Nicoara, S., Vladoiu, M., Moise, G.: Computer science student contests: individuals or teams? In: Proceedings of the 16th RoEduNet International Conference: Networking in Education and Research, Petru-Maior University of Targu Mures, Romania (2017)
14. RoboCup Federation. http://www.robocup.org/
15. Bowring, J.F.: A new paradigm for programming competitions. In: Proceedings of the 39th SIGCSE Technical Symposium on Computer Science Education, pp. 87–91. ACM, New York (2008)
16. Sherrell, L., McCauley, L.: A programming competition for high school students emphasizing process. In: Proceedings of the 2nd Annual Conference on Mid-South College Computing Conference, Mid-South College Computing Conference, Little Rock, Arkansas, USA, pp. 173–182 (2004)
17. Fitzgerald, S., Hines, M.L.: The computer science fair: an alternative to the computer programming contest. SIGCSE Bull. **28**(1), 368–372 (1996). Proceedings the 27th SIGCSE Technical Symposium on Computer Science Education. ACM, New York
18. Intel International Science and Engineering Fair (Intel ISEF). https://student.societyforscience.org/intel-isef
19. Siemens Competition in Math, Science and Technology. https://siemenscompetition.discoveryeducation.com/
20. Schank, R.: Designing World-Class e-Learning: How IBM, GE, Harvard Business School and Columbia University are Succeeding at e-Learning. McGraw-Hill, New York (2002)

21. Kolb, D.A.: Experiential Learning: Experience as the Source of Learning and Development. Prentice Hall, Pearson Education, Upper Saddle River (1984)
22. 3ITC Contest Web Page. https://timf.upg-ploiesti.ro/3ITC/
23. Burguillo, J.C.: Using game theory and competition-based learning to stimulate student motivation and performance. Comput. Educ. **55**(2), 566–575 (2010). https://doi.org/10.1016/j.compedu.2010.02.018
24. Vroom, V.H.: Work and Motivation. McGraw Hill, New York (1964)

Ranking-Based Evaluation of Process Model Matching
(Short Paper)

Elena Kuss[1]([✉]), Henrik Leopold[2], Christian Meilicke[1],
and Heiner Stuckenschmidt[1]

[1] Research Group Data and Web Science, University of Mannheim,
68163 Mannheim, Germany
{elena,christian,heiner}@informatik.uni-mannheim.de
[2] Department of Computer Science, Vrije Universiteit Amsterdam,
De Boelelaan 1081, 1081 HV Amsterdam, The Netherlands
h.leopold@vu.nl

Abstract. Process model matching refers to the automatic detection of semantically equivalent or similar activities between two process models. The output of process model matchers is the basis for many advanced process model analysis techniques and, therefore, must be as accurate as possible. Measuring the performance of process model matchers, however, is a difficult task. On the one hand, it is hard to define which correspondences are actually correct. On the other hand, it is challenging to appropriately take the output of matchers into account, because they often produce confidence values between zero and one. In this paper, we propose the first evaluation procedure for process model matchers that addresses both of these challenges. The core idea is to rank both the computed and the desired correspondences based on their confidence values and compare them using the Spearman's rank correlation coefficient. We perform an in-depth evaluation in which we apply the new evaluation procedure and illustrate how it helps gaining interesting insights.

Keywords: Process model matching · Ranking-based evaluation · Non-binary gold standard

1   Introduction

Process models are conceptual models used for a variety of purposes ranging from business process documentation to requirements definition [6]. Process model *matching* is concerned with the automatic identification of semantically equivalent or similar activities between such models. The application scenarios of process model matching are manifold. They include the analysis of model differences [10], harmonization of process model variants [11], and process model search [7]. The challenges associated with the matching task are considerable. Among others, process model matchers must be able to deal with heterogeneous

© Springer International Publishing AG 2017
H. Panetto et al. (Eds.): OTM 2017 Conferences, Part I, LNCS 10573, pp. 298–305, 2017.
https://doi.org/10.1007/978-3-319-69462-7_19

vocabulary, different levels of granularity, and the fact that typically only a few activities from one model have a corresponding counterpart in the other. In recent years, a significant number of process model matchers have been defined to address these problems (cf. [3,8,12,18,19]).

One important question that concerns all these matchers is how to evaluate whether they actually perform well. To measure the performance of process model matchers, their final output is compared to a manually annotated gold standard. A key problem in this context is that it is hard to define which correspondences are actually correct. A recently introduced evaluation procedure for process model matchers addresses this problem by introducing the notion of a non-binary gold standard [9]. The idea of a non-binary gold standard is to associate each activity correspondence with a confidence value instead of defining it as correct or incorrect. However, this evaluation procedure still assumes that the output of the matcher is binary. In fact, many matchers produce confidence values that indicate the reliability of the identified correspondences. The transformation of these confidence values into binary values does not only come with the loss of information, but also results in a less accurate assessment of the performance of the matching technique.

In this paper, we therefore introduce the first evaluation procedure for process model matchers that takes the non-binary output of matchers as input and compares it against a non-binary gold standard. To this end, we rank the correspondences produced by the matcher and the gold standard based on their confidence values and compare them using the Spearman's rank correlation coefficient. We perform an in-depth evaluation where we apply the new evaluation procedure and illustrate how it helps in gaining interesting insights.

2 Problem Statement

The goal of evaluation procedures for process model matching is to assess which of the correspondences identified by a matcher are correct. However, there are several problems associated with this task. To illustrate these problems, consider the example depicted in Fig. 1. It shows two simplified process models from the Process Model Matching Contest 2015 [2]. Possible correspondences are denoted using gray shades.

Upon closer inspection of the correspondences shown in Fig. 1, it becomes clear that some of the correspondences are actually disputable. Consider, for instance, the correspondence between "Receive online application" from University 1 and "Receive application form" in the process of University 2. On the one hand, we can argue in favor of this correspondence because both activities deal with the receipt of an application document. On the other hand, we can argue that these activities do not correspond to each other because the former relates to an online procedure whereas the second refers to a paper-based application. Similar arguments can be brought forward for the correspondence between "Invite for interview" and "Invite for aptitude test". We could argue that both activities represent a means to select a promising candidate. However,

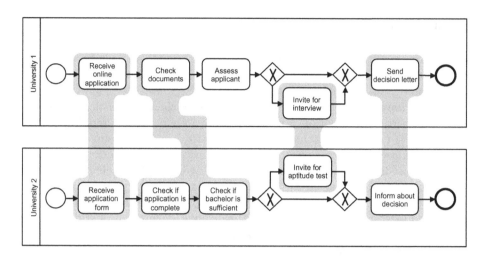

Fig. 1. Two process models and possible correspondences as shown in [9]

we can also argue that an interview is clearly a different assessment instrument than an aptitude test.

These examples illustrate that it may be hard and, in some cases, even impossible to agree on a single *correct* set of correspondences. Traditional process model matching evaluation procedures, however, assume that such a set of correct correspondences, a so-called *gold standard*, is available. Based on this gold standard, they distinguish between correct and incorrect correspondences generated by a matcher and compute the traditional evaluation metrics precision, recall, and F-measure (cf. [2,4,12,18,19]). Recently, Kuss et al. [9] introduced the notion of a non-binary gold standard. Such a non-binary gold standard assigns a confidence value to correspondences instead of defining them as either correct or incorrect. What this approach, however, still does not take into account is that also matchers often generate confidence values between zero and one. The transformation of these confidence values into binary values for the purpose of evaluating the performance of a matcher with precision, recall, and F-measure does not only come with the loss of information, but also does not result in a fair assessment of the performance.

Recognizing these shortcomings, we use this paper to propose a ranking-based evaluation procedure for process model matching. To this end, we replace the question whether a correspondence is correct with the question in how far the confidence estimated by a matcher resembles the confidence values in the gold standard.

3   Probabilistic Evaluation of Process Model Matching

Given two process models $\mathcal{M}_1$ and $\mathcal{M}_2$, let $\mathcal{G}$ be a non-binary alignment between $\mathcal{M}_1$ and $\mathcal{M}_2$ that represents the manually created gold standard and $\mathcal{A}$ be a non-binary alignment between $\mathcal{M}_1$ and $\mathcal{M}_2$ that was generated by a matching techniqe. In the following, we show how to compute and use the Spearman's rank correlation coefficient [17] to measure the quality of $\mathcal{A}$ given the manually created gold standard $\mathcal{G}$. Let n be the number of correspondences with a confidence value higher than zero in $\mathcal{G}$ or $\mathcal{A}$, i.e., $n = |\{(a_1, a_2) \in act(\mathcal{M}_1) \times act(\mathcal{M}_1) \mid \mathcal{A}(a_1, a_2) > 0 \vee \mathcal{G}(a_1, a_2) > 0\}|$. To compute the rank correlation, the following steps need to be performed independently for both $\mathcal{G}$ and $\mathcal{A}$.

Normalized Ranks. The n correspondences in $\mathcal{G}$ and $\mathcal{A}$ have to be ranked according to their confidence values (in increasing order). This leads to a rank of 1 through n for each correspondence. If there are correspondences with the same confidence value, their ranks are normalized. In these cases, which we refer to as ties, the rank of each correspondence with the same confidence value is given by the arithmetic mean of the ranks occupied by these correspondences.

Correction Term for Ties. The number of times each value is observed in the alignment is counted. This is denoted by $t_{\mathcal{A},k}$ with respect to $\mathcal{A}$ and $t_{\mathcal{G},k}$ with respect to $\mathcal{G}$. The index k is used here to refer to the different values (or ranks). As a result of this counting, we obtain $\sum_k t_{\mathcal{A},k} = \sum_k t_{\mathcal{G},k} = n$. In the final formula, we need to use the correction terms $T_{\mathcal{A}} = \sum_k \left((t_{\mathcal{A},k})^3 - t_{\mathcal{A},k} \right)$ and $T_{\mathcal{G}} = \sum_k \left((t_{\mathcal{G},k})^3 - t_{\mathcal{G},k} \right)$.

We can now use the following formula to compute Spearman's rank correlation coefficient, where d_i denotes the difference between the normalized ranks of the i-correspondence from those correspondences that have a positive confidence value in $\mathcal{G}$ or $\mathcal{A}$:

$$\rho = \frac{n^3 - n - \frac{1}{2}T^{\mathcal{G}} - \frac{1}{2}T^{\mathcal{A}} - 6\sum_{i=1}^{n} d_i^2}{\sqrt{(n^3 - n - T^{\mathcal{G}})(n^3 - n - T^{\mathcal{A}})}}.$$

4   Evaluation Experiments

In this section, we demonstrate the capabilities of the proposed evaluation procedure by applying it to a set of seven process model matchers. The goal of the evaluation is to show that our evaluation procedure represents a viable alternative to binary evaluation procedures and provides useful analytical insights.

4.1   Setup

To evaluate the proposed evaluation procedure, we applied it to the output of seven matchers that participated in the Process Model Matching Contest 2015 and the Process Model Matching Track at the OAEI 2016. Note that we had to limit our evaluation to seven matchers because not all of the matchers participating in these events provided confidence values for the correspondences they generated.

The *data set* that was used in these events consists of 36 model pairs derived from nine process models (referred to as *University Admission* data set), which describe the application procedures for accepting graduate students of nine German universities. The models vary in size and consist between 10 to 44 activities. In the context of both events, the task was to match these models pair-wise. For more details about the data set, we refer the reader to [1, 2].

For the creation of the *non-binary gold standard*, eight individuals were asked to independently create a binary gold standard for the University Admission data set. The resulting eight individually created binary gold standards were merged into a non-binary gold standard. Each correspondence in the non-binary gold standard has an associated non-binary confidence value which represents the share of the eight individual gold standards that contain the respective correspondence.

4.2 Results

As a result of applying our evaluation procedure to the output of the considered seven matchers, we obtained a respective rank correlation coefficient for each matcher. Table 1 gives an overview of the results. It shows the evaluation metrics and the rank (R) for three different evaluation procedures:

- *nB-nB*: The non-binary evaluation procedure introduced in this paper. The performance is captured using the rank correlation coefficient (ρ).
- *B-nB*: The probabilistic evaluation procedure from [9], which compares the binary output of a matcher against a non-binary gold standard. The performance is captured using the probabilistic F-measure (ProFM), probabilistic precision (ProP), and probabilistic recall (ProR).
- *B-B*: The classical evaluation procedure comparing the binary output of a matcher against a binary gold standard. The performance is captured using the F-measure (FM), precision (Prec), and recall (Rec).

The results from Table 1 reveal that there is a rather weak correlation between the output of the matchers and the non-binary gold standard. Three matchers

Table 1. Results for the seven considered matchers from the Process Model Matching Contest 2015 and the OAEI 2016 for three evaluation procedures.

Matcher	nB-nB		B-nB				B-B			
	R	ρ	R	ProFM	ProP	ProR	R	FM	Prec	Rec
AML	1	**.245**	1	**.424**	.806	.288	1	**.702**	.719	**.685**
Match-SSS	2	.223	5	.314	**.828**	.194	2	.608	**.807**	.487
LogMap	3	.153	2	.418	.680	.302	5	.481	.449	.517
Know-Match-SSS	4	.120	3	.409	.676	.293	3	.544	.513	.578
TripleS	5	−.008	6	.300	.519	.211	4	.485	.487	.483
AML-PM	6	−.266	4	.407	.411	**.404**	6	.385	.269	.672
pPalm-DS	7	−.295	7	.276	.230	.346	7	.253	.162	.578

even have a negative correlation coefficient. This outcome can be explained by the characteristics of the matchers as well as the characteristics of the gold standard. To understand how the characteristics of the matchers can explain this outcome, consider the metrics from the other two evaluation procedures (i.e. B-nB and B-B). All three matchers with a negative correlation coefficient have a particularly low precision, i.e. smaller than 0.5 for the classical and 0.519 for the probabilistic version. Apparently, a negative correlation coefficient primarily relates to a high number of false positives. A notable characteristic of the non-binary gold standard that contributed to the weak correlation is the high number of correspondences with a low support value. The non-binary gold standard contains a total of 831 correspondences, of which about 20% have the lowest rank, i.e. at most one of the eight annotators has voted for them. It is, thus, not surprising that many matchers miss these correspondences. While the penalty for missing them is rather low, the recall values reveal that they also explain the overall correlation coefficient. Abstracting from the absolute values, we see that the correlation coefficient allows us to rank the matchers according to their performance. What is particularly interesting is that the ranking obtained through the evaluation procedure presented in this paper does not always deviate from the ranking we obtain when using the other evaluation procedures. In fact, the matcher AML is always considered to perform best and the matcher pPalm-DS is always considered to be worst.

All in all, the results highlight a major difference of the presented evaluation procedure to existing ones: The confidence of the matcher is taken into account. If a matcher identifies a correspondences that is not part of the gold standard with high certainty, the penalty is much higher than if the certainty is low. This is an important difference to both the B-nB and B-B evaluation procedures where the output of the matcher is considered as zero or one. This particular feature of our evaluation procedure also explains the different ranking in Table 1. Matchers that identify false positives with high certainty receive a bigger penalty than matchers that identify false positives with low certainty.

5   Related Work

To the present day, the evaluation using precision, recall, and F-measure still represents the standard procedure for assessing the performance of process model matching techniques, see for example the reports of the Process Model Matching Contests [2,4]. In fact, this also applies to the related fields of schema and ontology matching, which also aim at identifying relations between different conceptual models [14,16]. However, these fields use a broader range of evaluation metrics to also address the needs related to specific application scenarios (see e.g. [13]).

One of the first evaluation procedures that builds on a non-binary alignment as input has been proposed by Sagi and Gal [15]. They adapt precision and recall metrics in such a way that they can be directly applied to first-line-matching results with non-binary confidence values. Their approach, however, still requires a binary gold standard as input. In [5], the authors directly compare the confidence values of the matchers to the confidence values of a gold standard. However, the confidence values are not normalized to the same range. As a result, the performance evaluation is quite questionable, since many matchers use largely differing ranges of confidence values. This is a weakness we address with the evaluation procedure proposed in this paper.

6   Conclusion

In this paper, we addressed the problem of how to properly evaluate the quality of process model matchers. Recognizing that binary evaluation procedures based on F-Measure, precision, and recall are insufficient, we introduced the first evaluation procedure for process model matchers that takes the non-binary output of a matcher as input and compares it against a non-binary gold standard. Our evaluation procedure builds on ranking the correspondences produced by the matcher and the gold standard based on their confidence values and comparing them using the Spearman's rank correlation coefficient. The core idea is that the confidence value distribution of the matcher should resemble the confidence value distribution of the gold standard as closely as possible.

To illustrate the usefulness and applicability of our evaluation procedure, we applied it to the output of seven process model matchers that participated in the Process Model Matching Contest 2015 and in the Process Model Matching Track of the OAEI 2016. The results show that our evaluation procedure indeed delivers useful results. While the assessment with respect to the best and the worst performance is congruent with other evaluation procedures, our non-binary procedure also assesses some matchers differently. By considering the confidence values produced by the matchers, it is able to assign a bigger penalty to those matchers that identify false positives with high certainty than to those techniques that identify false positives with little certainty. As a result, the performance of matchers is more accurately assessed.

In future work, we set out to apply the novel evaluation procedure in the context of comparative evaluation experiments. Our goal is to increase the awareness about the necessity to use metrics other than F-Measure, precision, and recall to assess uncertain problems such as process model matching.

References

1. Achichi, M., Cheatham, M., Dragisic, Z., Euzenat, J., Faria, D., Ferrara, A., Flouris, G., Fundulaki, I., Harrow, I., Ivanova, V., Jiménez-Ruiz, E., Kuss, E., Lambrix, P., Leopold, H., Li, H., Meilicke, C., Montanelli, S., Pesquita, C., Saveta, T., Shvaiko, P., Splendiani, A., Stuckenschmidt, H., Todorov, K., Trojahn, C., Zamazal, O.: Results of the ontology alignment evaluation initiative 2016. In: CEUR Workshop Proceedings, vol. 1766, pp. 73-129. RWTH (2016)
2. Antunes, G., et al.: The process model matching contest 2015. In: 6th International Workshop on Enterprise Modelling and Information Systems Architectures (2015)
3. Cayoglu, U., Oberweis, A., Schoknecht, A., Ullrich, M.: Triple-s: A matching approach for Petri nets on syntactic, semantic and structural level. Technical report Karlsruhe Institute of Technology (KIT) (2013)

4. Cayoglu, U., et al.: The process model matching contest 2013. In: 4th International Workshop on Process Model Collections: Management and Reuse (PMC-MR 2013) (2013)
5. Cheatham, M., Hitzler, P.: Conference v2.0: an uncertain version of the OAEI conference benchmark. In: Mika, P., et al. (eds.) ISWC 2014. LNCS, vol. 8797, pp. 33–48. Springer, Cham (2014). doi:10.1007/978-3-319-11915-1_3
6. Dumas, M., Rosa, M., Mendling, J., Reijers, H.: Fundamentals of Business Process Management. Springer, Heidelberg (2013)
7. Jin, T., Wang, J., La Rosa, M., Ter Hofstede, A., Wen, L.: Efficient querying of large process model repositories. Comput. Ind. **64**(1), 41–49 (2013)
8. Klinkmüller, C., Weber, I., Mendling, J., Leopold, H., Ludwig, A.: Increasing recall of process model matching by improved activity label matching. In: Daniel, F., Wang, J., Weber, B. (eds.) BPM 2013. LNCS, vol. 8094, pp. 211–218. Springer, Heidelberg (2013). doi:10.1007/978-3-642-40176-3_17
9. Kuss, E., Leopold, H., van der Aa, H., Stuckenschmidt, H., Reijers, H.A.: Probabilistic evaluation of process model matching techniques. In: Comyn-Wattiau, I., Tanaka, K., Song, I.-Y., Yamamoto, S., Saeki, M. (eds.) ER 2016. LNCS, vol. 9974, pp. 279–292. Springer, Cham (2016). doi:10.1007/978-3-319-46397-1_22
10. Küster, J.M., Gerth, C., Förster, A., Engels, G.: Detecting and resolving process model differences in the absence of a change log. In: Dumas, M., Reichert, M., Shan, M.-C. (eds.) BPM 2008. LNCS, vol. 5240, pp. 244–260. Springer, Heidelberg (2008). doi:10.1007/978-3-540-85758-7_19
11. La Rosa, M., Dumas, M., Uba, R., Dijkman, R.: Business process model merging: an approach to business process consolidation. ACM Trans. Softw. Eng. Methodol. (TOSEM) **22**(2), 11 (2013)
12. Leopold, H., Niepert, M., Weidlich, M., Mendling, J., Dijkman, R., Stuckenschmidt, H.: Probabilistic optimization of semantic process model matching. In: Barros, A., Gal, A., Kindler, E. (eds.) BPM 2012. LNCS, vol. 7481, pp. 319–334. Springer, Heidelberg (2012). doi:10.1007/978-3-642-32885-5_25
13. Mena, E., Kashyap, V., Illarramendi, A., Sheth, A.: Imprecise answers in distributed environments: estimation of information loss for multi-ontology based query processing. Int. J. Coop. Inf. Syst. **9**(04), 403–425 (2000)
14. Rahm, E., Bernstein, P.A.: A survey of approaches to automatic schema matching. VLDB J. **10**(4), 334–350 (2001)
15. Sagi, T., Gal, A.: Non-binary evaluation for schema matching. In: Atzeni, P., Cheung, D., Ram, S. (eds.) ER 2012. LNCS, vol. 7532, pp. 477–486. Springer, Heidelberg (2012). doi:10.1007/978-3-642-34002-4_37
16. Shvaiko, P., Euzenat, J.: Ontology matching: state of the art and future challenges. IEEE Trans. Knowl. Data Eng. **25**(1), 158–176 (2013)
17. Spearman, C.: The proof and measurement of association between two things. Am. J. Psychol. **15**(1), 72–101 (1904)
18. Weidlich, M., Dijkman, R., Mendling, J.: The ICoP framework: identification of correspondences between process models. In: Pernici, B. (ed.) CAiSE 2010. LNCS, vol. 6051, pp. 483–498. Springer, Heidelberg (2010). doi:10.1007/978-3-642-13094-6_37
19. Weidlich, M., Sheetrit, E., Branco, M.C., Gal, A.: Matching business process models using positional passage-based language models. In: Ng, W., Storey, V.C., Trujillo, J.C. (eds.) ER 2013. LNCS, vol. 8217, pp. 130–137. Springer, Heidelberg (2013). doi:10.1007/978-3-642-41924-9_12

Analysis and Re-configuration of Decision Logic in Adaptive and Data-Intensive Processes (Short Paper)

Lina Ochoa[ID] and Oscar González-Rojas[✉][ID]

Systems and Computing Engineering Department, School of Engineering,
Universidad de los Andes, Bogotá D.C., Colombia
{lm.ochoa750,o-gonza1}@uniandes.edu.co

Abstract. Decision logic related to business processes can be specified with decision tables. Current approaches analyze decision tables in isolation without considering the dependencies with other tables, related tasks, and data flows. This paper presents an approach to integrate data flows with DMN and BPMN models in adaptive and data-intensive processes. We use feature modeling to specify interdependencies among decision logic models in one or multiple business process tasks. We created a constraint-based algorithm that encodes the feature-based specification and propagates decisions among decision logic models according to the process data flow. Therefore, an integrated process model can be configured and its data flow integrity can be guaranteed during process execution.

Keywords: Process analysis · Data-intensive processes · Decision logic management · Rules propagation · DMN · Feature modeling

1 Introduction

Decision logic related to the execution of business process tasks can be specified with decision tables. Given the critical business knowledge captured by *Decision Model and Notation* (DMN) decision logic models, there is an increasing need to interrelate them as part of the business process specification. In that way, data flow integrity and decision propagation during process execution is guaranteed.

Current approaches [1,2,7,8] interrelate process and decision logic models with modeling languages (*e.g.* Declare-R-DMN, BPMN, SBVR) to improve expressiveness and control at the specification level. Calvanese et al. [3] represent DMN decision tables as iso-oriented hyper-rectangles in a N-dimensional space to detect overlapping and missing rules. However, there are no techniques to analyze the dependencies of attribute values among decision tables. Decision tables are usually analyzed in isolation without considering the dependencies with other decision logic models, related tasks, and data flows. Moreover, a technique to support automatic decision logic propagation, once stakeholders

© Springer International Publishing AG 2017
H. Panetto et al. (Eds.): OTM 2017 Conferences, Part I, LNCS 10573, pp. 306–313, 2017.
https://doi.org/10.1007/978-3-319-69462-7_20

start the process execution, is missing. We use a loan approval business process to motivate these challenges (see Sect. 2).

This paper presents an approach to specify decision logic interdependencies within a business process and to support automatic decision propagation based on process data inputs (see Sect. 3). On the one hand, we specify the integration of independent DMN, BPMN, and data models of a given business process by using an extended feature model [10]. Feature modeling allows the specification of constraints among non-connected features of the three types of models, defining how attributes in different decision logic models are related. Moreover, decision logic information related to different tasks is maintained along the process execution, while attribute information is kept as a variable input. On the other hand, we implemented a constraint-based algorithm to find, within the integrated specification, a process configuration that can propagate decisions depending on dynamic input data. This allows to control interdependencies among decision logic models; and to instantiate the input attributes of dependent decision tables based on the expected data flow and defined constraints.

We created different data input configurations to validate the correct propagation of decision logic for multiple scenarios within a loan approval process (see Sect. 4). Conclusions and future work are presented in Sect. 5.

2   Challenges for Decision Logic Re-configuration: A Loan Approval Motivation Scenario

DMN is the standard notation for representing decision-making requirements and models in organizational contexts [9]. We use a loan approval process [9] as an example to motivate the challenges related to decision logic representation and consideration during the execution of a business process. Figure 1 illustrates three different perspectives that must be interrelated to propagate decisions.

First, the *process model* is used for decision-making coordination and is represented with the Business Process Management Notation (BPMN). Second, the *decision requirements* specify the needed process decisions and the corresponding related data inputs. Third, *decision logic* details a group of *business knowledge models* that specify business rules, which take a set of *input* values and generate a different set of *output* values. *Decision tables* are a DMN-supported format for representing business knowledge models.

The process starts when *application data is collected* and based on this information, a *routing decision* is taken (*i.e.* accept, decline). If it is accepted, the *loan is offered* to the customer, otherwise the *loan is declined*. Application data is dynamically collected for each process execution. This data is the input of the decision requirements model, which interrelates three decisions (the *routing* decision depends on the *application risk* decision). Two of these decisions are related to the *decide routing* and the *offer loan* tasks of the process model.

Each decision requirement is related to a decision table within the decision logic model (*i.e. risk rating, loan type,* and *routing*). Input and output attributes are shown as column names for which input and output entries scope their set of

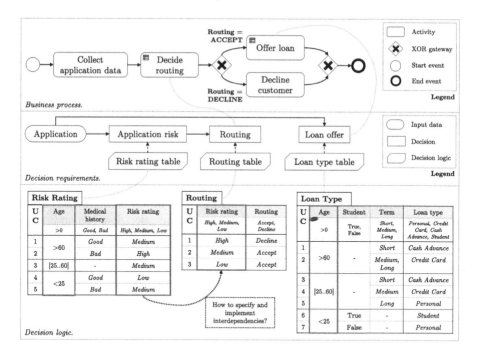

Fig. 1. Integrated models and perspectives of a decision-making process.

accepted values. For example, the *medical history* input attribute within the *Risk Rating* decision table only accepts *good* or *bad* values. The − symbol is known as the *irrelevant* value, which represents any input value. Each row represents a *decision rule* and the first cell in the row represents its priority. For instance, the third rule states that: **If** Age ∈ [25..60] **And** Medical history ∈ {*Good, Bad*} **Then** Risk rating = *Medium*. The *complete value (C)* indicator states that any possible combination of input entries triggers at least one decision rule; and the *unique policy (U)* hit indicator states that only one rule can be triggered, then all rules must be disjoint.

Lastly, decision rules of two decision tables are directly executed from the application data, whereas the input data for the *routing table* depends on the output of the *risk rating table*. This means that the output entry obtained in the *risk rating* context will define the acceptance or decline of a given application in the loan approval process. At this point, two main challenges remain in literature.

C1. Decision logic interdependencies specification. There is a need to specify interdependencies among decision logic models in one or multiple business process tasks. For instance, if the applicant is *65 years old*, she has a *bad* medical history, then the risk rating will be defined as *high* (*cf.* rule 2 in the *risk rating* table), and therefore the loan will be *declined* because of the existing interdependency among the risk rating table and the routing table. Current approaches [1,2,7,8] use different models to increase business expressiveness and control over the process specification. Nevertheless, no modeling technique has been proposed to represent interdependencies among decision logic models, instead they are specified and analyzed in isolation.

C2. Decision logic propagation. Once decision logic models are interrelated, decisions propagation should be validated before process execution. In this way, interdependencies among decision logic models do not only affect the output entry, but also the path followed in the represented process. For instance, if the loan is *declined*, then no *loan type* should be sought. None of the studied approaches contemplates the propagation of decisions during a process analysis stage. This gap must be closed to avoid errors during the process execution.

3   Supporting Decision Propagation in Business Processes

3.1   Feature-Oriented Representation of Decision Logic

We propose to translate decision logic of a given business process to an extended Feature Solution Graph (e-FSG), a structure that interrelates a set of attribute-based feature models through cross-model constraints [4,10]. A *Feature Model* (FM) [6] is a tree-structure that represents a domain's variability, where different configurations or feature selections can be derived. Each node represents a *feature* of the domain, and each edge represents a *tree-constraint* that defines relations among parent and child features. One tree-constraint correspond to the *mandatory* relation, which means that if the parent feature is selected, then the child feature should also be included. Moreover, FMs can be extended with additional metadata known as *feature attributes*, which are usually defined with a name, a data type, and a domain. A *Cross-Tree Constraint* (CTC) forces or prohibits the selection of features in the same FM through logical formulas, whereas a *Cross-Model Constraint* (CMC) constraints this selection in different FMs.

In our approach each decision table is mapped to a FM within a process-specific e-FSG. Figure 2 illustrates two FMs representing two interrelated decision tables (*i.e.* risk rating and *routing*) of the loan approval process. All features are *mandatory* in each FM given the need to represent complete tables. In addition, each root feature has exactly two child features representing the set of *input attributes* and *output attributes*. Each of these branches has a variable number of children depending on the number of input and output columns represented in the decision tables. For example, there are two features in the *input attributes* branch of the *risk rating* table, namely *age* and *medical history*. In the *output attributes* branch there is only one feature, namely *risk rating*.

Each leaf feature is related to one feature attribute. For instance, the *medical history* feature has one feature attribute with a *string* data type, and a domain defined with the enum list { *Good, Bad* }. Feature attributes do not have a static value in the model, they are dynamically configured based on the process data flow. Only child features of *input attribute* features need to be configured since the remaining attributes obtain their value by decision propagation. For example, an input configuration of the e-FSG corresponds to $C = \{Age = 65, MedicalHistory = Bad\}$.

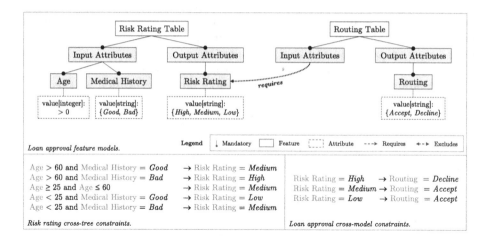

Fig. 2. e-FSG representing decision tables and their interdependencies.

Finally, decision rules are translated to CTCs if all input and output attributes are represented in the same FM, otherwise decision rules are translated to CMCs. For example, all decision rules in the *risk rating table* are represented as CTCs because they do not have any dependency in an input or output attribute of other decision table (*i.e. age, medical history*, and *risk rating*). However, decision rules specified in the *routing table* are represented as CMCs because the involved input and output attributes are specified in different FMs. In this case, we define a *requires* CMC, which means that the *routing table* requires the configuration of the *risk rating* feature.

The e-FSG is required as an intermediate model that can be interpreted by different execution platforms. Without this previous step, communication and business decisions are lost in a low level specification.

3.2   Constraint-Based Analysis of Decision Logic

Once decision logic and its interdependencies are specified as a valid e-FSG, the model is encoded as a *Constraint Programming* (CP) implementation to support automatic decision propagation during process execution. CP searching strategies are used to support business decisions. CP is a paradigm that defines *Constraint Satisfaction Problems* (CSP), where variables are labeled to values in their domain based on a set of constraints. We propose the specification of the following CSP to support decision propagation in multi-perspective scenarios.

Variables and domains. Integer variables are used to represent the input and output attribute values used along the e-FSG. Other data types like *string* or *boolean* should be represented using an integer domain. In addition, an array R of fixed rules is initialized based on the CTCs and CMCs information. Lastly, a *configuration* array is instantiated according to the process data flow, assigning initial values to input attributes that do not depend on previous decisions.

Algorithm 1. Propagate decisions.

Require: *rules* and *configuration* arrays, and *attributes* hash map and variables are initialized
Ensure: Input and output attributes values are returned
1: **for** i = 0 **to** R **do**
2:     Initialize *ifConstraints* and *elseConstraints* arrays
3:     **for** j = 0 **to** I + O **do**
4:         Initialize *constraint*
5:         $constraint = Arithm(attributes.get(rules[i].getAttribute(j)),$
6:         $rules[i].getOperator(j), rules[i].getAttrValue(j))$
7:         **if** j < I **then**
8:             $ifConstraints.add(constraint)$
9:         **else**
10:            $elseConstraints.add(constraint)$
11:        **end if**
12:        Post constraint: $If(And(ifConstraints))$
13:        $Then(And(elseConstraints))$
14:    **end for**
15: **end for**
16: **for** i = 0 **to** I **do**
17:    Post constraint: $Arithm(attributes.get(configuration[i].getAttribute()), =,$
18:    $configuration[i].getAttrValue())$
19: **end for**

Constraints. Algorithm 1 presents the procedure for propagating decisions in an interrelated decision logic environment. First, a loop is defined to iterate over all R rules, which correspond to the CTC and CMC constraints in the e-FSG (*cf.* line 1). Two temporal arrays, *ifConstraints* and *elseConstraints* are instantiated to store arithmetic CP constraints related to the rule input and output attributes respectively (*cf.* line 2). Then, we iterate over both the I input and O output attributes related to the current rule (*cf.* line 3). An arithmetic *constraint* is defined per rule attribute. The *Arithm* constraint has three parameters: i) the input or output attribute which is obtained from an *attributes* hash map based on its name (*cf.* lines 4–5); ii) the operator related to the jth expression of the ith rule (*cf.* line 6); and iii) the related value of the jth expression (*cf.* line 6).

The constraint is added to the *ifConstraints* array if the attribute is an input attribute (*cf.* lines 7-8), otherwise it is added to the *elseConstraints* array (*cf.* lines 9–11). Lastly, an *If-Then* CP rule is post; the *If* and *Else* statements consider the conjunction of all elements in the *ifConstraints* and *elseConstraints* arrays respectively (*cf.* lines 12–14). Finally, we iterate over all non-dependent input attributes configuration (*cf.* line 17). An *Arithm* constraint, which defines an equality relation among an attribute and a value obtained from the business process data flow, is post per loop (*cf.* lines 18–20). This algorithm was implemented by using the Java CP solver Choco version 3.3.1.

4   Evaluation

We aim to evaluate the specification and implementation of decision logic interdependencies for supporting decision propagation. We made a CP-based *Proof of Concept* (PoC), where we consider the business process presented in Fig. 1. We defined five scenarios where decision rules in each decision table varied from one scenario to another. We employed an existing DMN tool [3] to obtain correct and complete decision tables for each scenario. Then, we provided the complete e-FSG information (created from the original decision tables) as input to the CP-based implementation. For each scenario we defined five different input configurations in order to propagate decisions representing the process data flow, and to evaluate the procurement of the expected outputs[1].

The results of this evaluation validated the correctness of the derived process configurations. For all input configurations of the five scenarios, the automated output configuration responds to the decision logic originally defined in decision tables; the values of the *output attributes* of the CP-based approach are equal to the expected outcomes, which were manually computed prior to the evaluation execution. All specified decision rules are fulfilled in all cases, building and resolving the CSP model in around 167 ms and 34 ms, respectively. For instance, Table 1 shows the five decision configurations used for the first scenario (*cf.* Fig. 1). Entries for the *input attributes* columns are obtained from the process data flow, and correspond to the *application* input data. This initial information propagates to *output attributes* variables thanks to the definition of constraints in the CSP. To illustrate this, the first input configuration considers the following data $C_I = \{Age = 17, MedicalHistory = Good, Student = False, LoanTerm = Medium\}$. According to the specified decision logic captured with an e-FSG model, the resulting CP output corresponds to $C_O = \{RiskRating = Low, Routing = Accept, LoanType = Personal\}$.

Table 1. Proof of concept of decisions propagation for Scenario 1.

Id	Input attributes				Output attributes			Build(sec)	Res. (sec)
	Age	Medical h	Student	L. term	Risk r	Routing	L. Type		
1	17	Good	False	Medium	Low	Accept	Personal	0.073	0.015
2	24	Bad	True	Long	Medium	Accept	Student	0.071	0.015
3	43	Good	False	Short	Medium	Accept	Cash Advance	0.073	0.015
4	57	Bad	False	Medium	Medium	Accept	Credit Card	0.074	0.015
5	88	Bad	False	Long	High	Decline	-	0.072	0.015

5   Conclusions and Future Work

This paper presents an approach for specifying interdependencies among BPMN and DMN models, and for supporting decision propagation during process execution. The incorporation of an extended feature model that is used for configuring product lines is a novel approach for integrating process and decision logic models, and the existing interdependencies among these models. This specification is then encoded as a CP-based algorithm that enacts the propagation of decisions, generating an output configuration that responds to the process decision rules and to the input data. Therefore, process analysts avoid expending additional time and effort to execute individual approaches for process analysis on decision consistency, data dependencies, and decisions propagation.

---

[1] Code and results are available at https://github.com/CoCoResearch/DMNTables.

Multiple improvements are considered for further research. First, multiple techniques must be studied to interpret the integrated feature model in terms of performance and scalability. Second, process analysis capabilities can be improved by integrating decision logic analysis with process risk analysis [5] to implement processes that are less error-prone at runtime. Finally, we plan to automatically generate the multi-model representation from baseline BPM, DMN, and data models to validate the approach with real world process scenarios.

References

1. Batoulis, K., Meyer, A., Bazhenova, E., Decker, G., Weske, M.: Extracting decision logic from process models. In: Zdravkovic, J., Kirikova, M., Johannesson, P. (eds.) CAiSE 2015. LNCS, vol. 9097, pp. 349–366. Springer, Cham (2015). doi:10.1007/978-3-319-19069-3_22
2. Biard, T., Le Mauff, A., Bigand, M., Bourey, J.-P.: Separation of decision modeling from business process modeling using new "Decision Model and Notation" (DMN) for automating operational decision-making. In: Camarinha-Matos, L.M., Bénaben, F., Picard, W. (eds.) PRO-VE 2015. IAICT, vol. 463, pp. 489–496. Springer, Cham (2015). doi:10.1007/978-3-319-24141-8_45
3. Calvanese, D., Dumas, M., Laurson, Ü., Maggi, F.M., Montali, M., Teinemaa, I.: Semantics and analysis of DMN decision tables. In: La Rosa, M., Loos, P., Pastor, O. (eds.) BPM 2016. LNCS, vol. 9850, pp. 217–233. Springer, Cham (2016). doi:10.1007/978-3-319-45348-4_13
4. Chavarriaga, J., Noguera, C., Casallas, R., Jonckers, V.: Propagating decisions to detect and explain conflicts in a multi-step configuration process. In: Dingel, J., Schulte, W., Ramos, I., Abrahão, S., Insfran, E. (eds.) MODELS 2014. LNCS, vol. 8767, pp. 337–352. Springer, Cham (2014). doi:10.1007/978-3-319-11653-2_21
5. González-Rojas, O., Lesmes, S.: Value at risk within business processes: an automated IT risk governance approach. In: La Rosa, M., Loos, P., Pastor, O. (eds.) BPM 2016. LNCS, vol. 9850, pp. 365–380. Springer, Cham (2016). doi:10.1007/978-3-319-45348-4_21
6. Kang, K., Cohen, S., Hess, J., Novak, W., Peterson, A.: Feature-Oriented Domain Analysis (FODA) Feasibility Study. Technical report CMU/SEI-90-TR-021, SEI (1990)
7. Kluza, K., Honkisz, K.: From SBVR to BPMN and DMN models. proposal of translation from rules to process and decision models. In: Rutkowski, L., Korytkowski, M., Scherer, R., Tadeusiewicz, R., Zadeh, L.A., Zurada, J.M. (eds.) ICAISC 2016. LNCS, vol. 9693, pp. 453–462. Springer, Cham (2016). doi:10.1007/978-3-319-39384-1_39
8. Mertens, S., Gailly, F., Poels, G.: Enhancing declarative process models with DMN decision logic. In: Gaaloul, K., Schmidt, R., Nurcan, S., Guerreiro, S., Ma, Q. (eds.) CAISE 2015. LNBIP, vol. 214, pp. 151–165. Springer, Cham (2015). doi:10.1007/978-3-319-19237-6_10
9. Object Management Group (OMG): Decision Model and Notation (DMN) 1.1. Standard, OMG (2016)
10. Ochoa, L., González-Rojas, O., Verano, M., Castro, H.: Searching for optimal configurations within large-scale models: a cloud computing domain. In: Link, S., Trujillo, J.C. (eds.) ER 2016. LNCS, vol. 9975, pp. 65–75. Springer, Cham (2016). doi:10.1007/978-3-319-47717-6_6

Contingency Management for Event-Driven Business Processes

John Wondoh[(✉)], Georg Grossmann, and Markus Stumptner

University of South of Australia, Adelaide, Australia
john.wondoh@mymail.unisa.edu.au, {georg.grossmann,mst}@cs.unisa.edu.au

Abstract. In the past two decades, business process research has focused on process flexibility to facilitate the operation of business processes in an open and dynamic environment. This is important to ensure that processes accurately reflect and handle changes occurring in the real-world. While substantial existing work has investigated changes in business processes, the contingency management of running processes did not receive sufficient attention, mainly because events are considered to be immutable. Yet high-level business events have been shown to be subject to changes. To be able to capture such changes, business events have to be considered as bitemporal, where the occurrence (scheduled) time and detection time of events are differentiated. Modifying an event's content may result in a contingency that has to be handled appropriately. For instance, the scheduled time of a planned event in a process may change, which has an impact on subsequent events. In this work, we present an approach to capture bitemporal mutable events in business processes, assess the scope of changes and provide an approach for specifying contingency plans.

Keywords: Contingency plans · Bitemporal mutable events · Business processes

1 Introduction

In today's enterprise, it is crucial for business processes to be capable of operating in an open and dynamic environment. This requirement has resulted in rigorous business process research aimed towards ensuring the flexibility and adaptability of processes during runtime [1,17]. This requires the anticipation and categorisation of possible contingencies that may result from changes in the execution environment.

In *event-driven business process management* (ED-BPM), events are the main citizens in a business process, and are used for monitoring and controlling the business process [3]. Flexibility in ED-BPM is driven mainly by events. The management of contingencies is driven by the occurrence of events associated with each contingency. While substantial existing work has focused on the issue of categorising contingent events to facilitate runtime flexibility, there has been little work towards the handling of modification of business events. This is because

© Springer International Publishing AG 2017
H. Panetto et al. (Eds.): OTM 2017 Conferences, Part I, LNCS 10573, pp. 314–333, 2017.
https://doi.org/10.1007/978-3-319-69462-7_21

traditionally, events are considered to be immutable. Yet high-level business events have been shown to be subject to changes [7,10,20,25]. Since business processes are incapable of detecting such changes in events, there are no existing approaches for handling such changes. Our work focuses on providing techniques for handling contingencies associated with immutability of events in business processes.

Contingency management is an important aspect of business process management (BPM) [21]. It usually follows these steps [9]: (1) identify potential contingency, (2) develop contingency plans, and (3) develop preventive measures. Since contingencies that result from changes in events are not considered in business processes, contingency plans cannot be developed, and preventive plans cannot be implemented. It is crucial to firstly, identify contingencies that may result from changes in events, and secondly, provide plans to handle them. The focus of this paper is to identify contingencies resulting from event modification and synthesising plans to handle them.

Modification of events is a phenomenon that has been investigated in various research areas. For instance, in areas such as web and database monitoring, existing research has investigated the detection and effect of changes in an event [7,10,20]. The previous information of an event, i.e., its old state and the new information of the event, i.e., its new state can still be assessed. To be able to capture changes to events, we need to consider the occurrence and detection times of the event states, i.e., the detection time of the new state will always be greater than the detection time of the previous state. We use the term *bitemporal events* to capture the dual temporal properties of events. Bitemporality of events is necessary to capture the ordering of old and new event information.

This work is aimed at anticipating changes in events, and managing contingency that may result from such changes. The main contributions of our work are as follows:

- We introduce an approach for detecting changes in events, and analysing the scope of the impact of such changes to the business process.
- We provide an approach for designing contingency plans aimed at handling contingencies resulting from event mutation.
- We provide an approach for specifying rules to trigger contingency plans when event contents' are changed.

The remainder of this paper is organised as follows: Sect. 2 provides a discussion of the fundamental concepts used in this work and the running example. Section 3 provides the approach overview. Section 4 provides our approach for determining the scope of bitemporal contingencies in business processes, and Sect. 5 provides our approach for contingency planning. We provide a brief discussion of our overarching project in Sect. 6, followed by a review of related work in Sect. 7. We conclude the paper and discuss future work in Sect. 8.

2   Background

In this section, we introduce the fundamental concepts in this work, and provide an example for better illustration. The central concepts considered in this work are *event modification* and the *bitemporal* nature of events.

2.1   Bitemporal Mutable Events

The properties of an event include an *event name*, an *identification number*, a *timestamp*, and a *payload*. In conventional event processing systems, only a single timestamp of an event is captured to determine when the event occurred. Other temporal dimensions as described in the temporal database literature are not exploited [18]. This work focuses on bitemporal events [7,24]. The latter has associated temporal properties (1) *occurrence (scheduled) time*; the time the event occurs or is supposed to occur, and *detection time*; the time when the event is detected. Bitemporality is necessary for capturing the history of events after modification.

High-level business events such as requests events may be modified during a process lifecycle. Event modification may occur as a result of two factors: (i) new information about an event is obtained [7,20], or (ii) errors are detected within the event's information that requires correction [20]. In both cases, the result is the modification of the event.

Event modification results in a new state of the event. A state of an event is a representation of the attributes of an event at a given point in time. The state of an event changes when the event content is modified. Given the initial state of an event s_0 and its new state s_1, let dt_0 be the detection time of s_0 and let the detection time of s_1 be dt_1. The relations between the detection time is given as follows: $dt_0 < dt_1$. This relationship is important as it facilitates the ordering of event states. The content of s_1 differs from the content of s_0. The occurrence time of an event is used to represent when the event occurred. For planned events, i.e., events expected to occur in the future, the actual occurrence time is in the future. In this work, we use the term *scheduled time* instead of occurrence time to allow for the inclusion of planned events.

We considered two types of modifications in this paper: (1) modifying the content of an event, and (2) modifying the scheduled time of an event. The content of an event is modified by replacing the value of an attribute of the event with a new one, deleting an attribute, or adding new attributes to the event. These modifications are demonstrated in the example below.

2.2   The Homecare Example

Homecare organisations provide home care services to clients with physical disabilities, clients who require carers during their health recovery process, and the elderly. The homecare organisation provides these clients (patients)[1] with carers

---

[1] Client is a preferred term in homecare organisation while patient is the preferred term in hospitals.

who visit them at home and provide necessary services to them. Initially, the homecare organisation is made known of a possible client and the necessary information required to provide the service to the client. A typical example is when a hospital makes a request for home care for a patient planned to be discharged from the hospital. Based on the planned discharge date and all the necessary information provided by the hospital, the homecare organisation plans the care schedule for its clients. Once the patient is discharged, homecare services will be provided to the client according to the planned schedule. Constant changes are likely to occur before, during, or after the planning or service provision stage.

The processes in the homecare organisation consist of events, activities, gateways. Some of these events are high-level events such as message events, while others are primitive events signifying the initiation or termination of activity instances. We selected four important processes in the homecare organisation for illustrating potential event changes in the process. The selected processes are the planner, dispatcher, client, and carers process. These processes, modelled in *business process model and notation* (BPMN), are shown in Fig. 1.

Planner. The planner process in Fig. 1 is responsible for planning home care services for each client. After receiving the home care request from the hospital, the homecare planner process first assesses the needs of the patient. Based on this assessment, the carers with the required skill level are selected. Firstly, an available carer is selected and his/her skill level is determined. If the skill level is not within what is required for the client, a new available carer is selected. Else the carer is contacted to verify his/her availability for the shifts intended to be allocated to him/her. Once the availability of the carer has been verified, it is recorded in the schedule (homecare plan database). If a carer is unavailable to cover his/her shifts, the database is updated with the carer's availability status for later use. This search is repeated until the plan for each client is completed.

The information required to plan the home care for the client includes the client's name, health status, care type requirement, planned discharge date, and care intensity and duration for the client. This information is sent to the homecare organisation as the request event in the planner process in Fig. 1. In the real-world, this event may undergo changes such that the hospital updates the message sent to the homecare organisation. The planned discharge date may be moved to an earlier or later date, the care duration or intensity may change, and/or the care type may change as well. Simply, the scheduled time of the event, as well as its content may be modified. In extreme cases, the entire request may no longer be necessary and the entire message event cancelled. Another event that may change is the response from carer after they have been contacted to determine their availability, i.e., the response event in Fig. 1. The initial response may change from 'no' to 'yes', or from 'yes' to 'no'. Processes should be capable of handling such resulting contingencies.

Dispatcher. The dispatcher process in Fig. 1 is responsible for notifying carers of the next client they need to attend to according to their verified schedule.

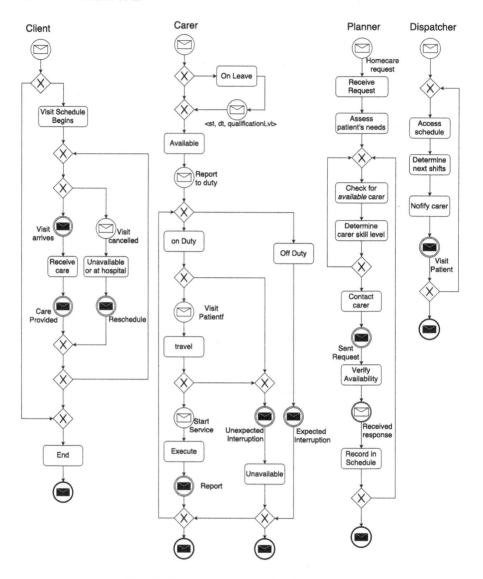

Fig. 1. Homecare organisational processes

It does this by accessing the carer's schedule, determining the next shift and notifying them. Usually, the notification should take into consideration the distance the carer has to travel to get to the home of the client. Changes in the dispatcher process are external and may affect the carer and/or client processes.

Client. The client process in Fig. 1 shows the care services provided to the client after being discharged from the hospital. The client receives care visits from the carers where home care services are provided to him/her. Occasionally, home care services cannot be provided because the client may be re-admitted to the hospital, or become unavailable due to personal reasons. The business event likely to change in the client process in Fig. 1 is the visit cancelled event which represents the cancellation of an initiated visit. This event may be dismissed (cancelled) if new information is obtained regarding the return to the client to his/her home. Processes should be capable of handling event cancellations.

Carer. The carer is responsible for providing all the necessary home care services to the patient. The process for each carer is given in Fig. 1. A carer may be on leave, but upon returning should notify the organisation of any changes in their physical condition that may impede their work output. For instance, a carer with a high level of qualification may be required to do less work if they develop a back injury. Once the availability of a carer is determined, they are required to report to duty. A carer may be off duty if they have completed their work for the day, or on duty for their shifts. Off duty carers will notify the homecare organisations that they have complete all client services for the day. The on duty carer receives a notification from the dispatcher about the next client to visit. The carer travels to the client's home to provide the services. The carer may not be able to reach the client's home due to some *unexpected interruption*. For instance, the carer's car may breakdown. The carer will notify the homecare organisation about how long he/she may not be available.

Some of the changes that can occur in the events in the carer process are the visit patient event may be changed depending on whether the patient is available at home or not. The dispatch process may change the parameters of the visit patient event to delay the visit by say 1 h. The *unexpected interruption* may not be as severe as initially thought and the carer may reduce the number of hours that he/she may be unavailable. These changes, among others, are represented by the changes in events. The business process should be equipped to handle such changes. In the next section, we shall discuss how we handle changes to events in business processes.

3 Approach Overview

The overview of our approach is given in Fig. 2. This consists of two main components, i.e., the event modification component, and the contingency management component. The event modification component detects modification of an event's content or its scheduled time and adds the modified event to the modified event set. A modified event may cause other events within the processing system to be modified as well. This is termed as modification propagation [25]. The provenance component captures the current and previous states of all modified events. This aspect of the work was the focus of our initial work [25]. The output of this component is a set of modified events serving as an input to the contingency management component.

The contingency selector is responsible for selecting the right contingency plan for managing a change in an event. The selection can be done automatically by bitemporal business rule engine or by incorporating the assistance of

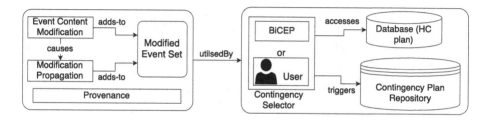

Fig. 2. Approach framework

a domain expert. *Bitemporal Complex Event Processing* (BiCEP) engine was designed by [7] purposely for managing updates in bitemporal web events. Our rule engine is based on BiCEP. It takes as input the modified event set, evaluates a set of conditions, and triggers the contingency plan necessary for correcting the deviation. The contingency plans are modelled and stored in process repository and can be accessed by domain experts or the BiCEP engine.

Process models are designed and instantiated during runtime, where activities are executed according to how they are specified. We term this as the passive behaviour of the process. When we trigger contingency plans, the process instance deviates from its design time specification. It no longer passively executes activities, but exhibits a dynamic behaviour. We term this behaviour the *active behaviour* of the process instance. We actively control the behaviour of the process to successfully manage contingencies.

4   Contingency Scope

Our contingency plans are triggered by bitemporal rules. These rules, inspired by [7] are a variant of the *event-condition-action* (ECA) rules, where modified events can be detected and processed. When we detect a modification of an event's content, evaluate a set of conditions associated with the rule specific to the event, and then perform a set of actions if the conditions are satisfied. The condition part of a bitemporal rule evaluates the following:

- Event Type: Rules are specified for event types and are applicable to their associated event instances. Once an event instance is modified, its associated rule is activated.
- Nature of deviation: The nature of deviation (NoD) resulting from event modification can be put into two main categories:
  - Nature of Scheduled Time Deviation (NoD_{st}): This nature of deviation results from modifying the scheduled time of an event. NoD_{st} is a tuple (δ, α), where δ is the modification type of the scheduled time modification, and α is a the modification scope of impact on the process.
  - Nature of Event Content Deviation (NoD_{ec}): The NoD_{ec} resulting from event modification deals with changes in the content of an event. NoD_{ec} is a tuple (a, a', θ) where a is an attribute of the event in its initial state,

a' is the attribute of the event in its new state, and θ is the significant scope of change between a' and a.

- Additional Conditions: These are additional conditions required for triggering contingency plans and are specified by domain experts. These conditions may be omitted during contingency planning, however once specified, they must be taken into account when evaluating the rule's conditions.

We shall now proceed to discuss how the nature of deviation can be determined in the following sections.

4.1 Modification Type

We ascertain the modification type by comparing the previous and new scheduled times of the modified events. We denote the previous scheduled time as st and the new scheduled time as st'. Corresponding to the scheduled times are the detection times of the events. The detection time of the previous event state is denoted as dt and the detection time for the new event state is denoted as dt'. We denote the current wall clock time as NOW. The latter corresponds to the current real world time.

Based on the temporal properties associated with events, there exist various relationships between st and st'. Such relationships, as provided in [7,10], are illustrated in Fig. 3. The relationships are put into three main categories, namely announcement, modification, and cancellation. The announcement category deals with only one scheduled time; there is no need to draw a distinction. In this category, the scheduled time of the event is not modified. In the second category, the scheduled time of the event has been modified resulting in st'. The cancellation category deals with the cancellation of an event, and consequently its scheduled time.

We have excluded the trivial case of no actual change, i.e., when a modification results in the new scheduled time being the same as the previous scheduled time ($st = st'$). Further, since the first category contains no changes, we only introduce it as the initial state of the scheduled time. The two non-trivial categories are the modification and cancellation categories. These categories may potentially result in changes in the running process.

The event announcement category deals with the detection of events, i.e., when the event becomes known in the process. In this category, the focus is on finding the relationship between the detection time and the scheduled time of an event. In Fig. 3, NOW is the time when the event becomes known in the business process management system (BPMS), and this is compared to the scheduled time of the event. In categories modification and cancellation, the event is already known in the BPMS, however, changes are compared to the current wall clock time that coincides with the time the change occurred.

Example 1. From our home care example (see Sect. 2.2), changes in scheduled time of some events will fall under one of the modification types in Fig. 3. For instance, the planned discharge time of the patient from the hospital is a scheduled time. This time is important as it may affect the homecare schedule of

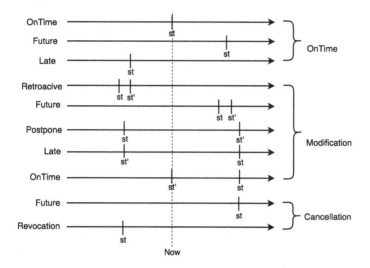

Fig. 3. Qualitative temporal comparisons of scheduled time states [7]

the patient. As discussed in the example, the discharge time may be modified. Assume the current time has exceeded the planned discharge time. If the planned discharge time is modified to a later time than the current time, then the modification type for the planned discharge time is future under modification category. Similarly, if the discharge time is modified to a time that is less than the current time, then the modification type for the discharge time is retroactive under modification category.

The modification type provides a qualitative description of the change in the scheduled time of the event. The quantitative effect of this change on the entire process needs to be determined as well. This is addressed in the next section.

4.2 Modification Scope

The modification scope captures the impact of scheduled time modification to process specification. The quantitative temporal distance between the new and previous scheduled times is used to determine to extent of the change. The temporal distance can be calculated as follows:

$$t_d(\delta) = st' - st$$

Where t_d is the temporal distance, and modification type is replaced by the specific type in the modification category. A negative value for t_d represents a change where the scheduled time is modified to an earlier time, i.e., $st' < st$, while a positive value represents a change where the scheduled time is modified to a later time, i.e., $st' > st$.

The temporal distance created during modification may not have an impact on the running process if the change is not significant. To determine whether a

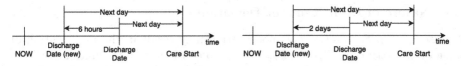

(a) Insignificant Scheduled Time Modification

(b) Significant Scheduled Time Modification

Fig. 4. Illustration of scheduled time modification significance

change actually affects process execution, we need to identify if specified threshold values have been exceeded. A *threshold value* is a value specified by a domain expert such that any change that does not exceed this value is disregarded. The threshold value is set for each modification type such that it is only evaluated if the change matches its modification type. We denote the threshold for a modification type as THRESHOLD_δ.

The temporal distance introduced by a modification type, and the threshold value for the modification type can be used to determine the *effective temporal distance* of the change in schedule time, i.e., the scope α. The effective temporal distance is the temporal distance that actually affects the process. This is the modification scope α and its value can range from $-\text{THRESHOLD}$ to t_d, i.e., $-\text{THRESHOLD}_\delta \leq \alpha \leq t_d$. The value of α can be determined as follows:

$$\alpha = t_d - \text{THRESHOLD}_\delta$$

Example 2. In our home care example, let us assume that care services are supposed to begin for the client a day after they have been discharged from the hospital. In Fig. 4, we show two cases where the planned discharge date for the patient has been modified. The new discharge date is 6 h earlier than the original date in Fig. 4a, while the new discharge date is 2 days earlier in Fig. 4b. In both cases, the modification type is future modification as both the new and original discharge times are in the future (i.e., $>\text{NOW}$). Assume care services are supposed to be provided to the patient a day after they have been discharged from the hospital. In the first case, moving the planned discharge date to 6 h earlier may not affect the start date of 1 day after discharge provided the new discharge date is on the same day as the original discharge date. The change is therefore insignificant. In the second case, the new discharge date is two days earlier which makes it impossible to provide services to day patient on the scheduled day. The threshold in this example is 1 day and t_d is 2 days. The modification scope is therefore 1 day. The homecare organisation will have to accommodate the change by adding schedules for the extra day introduced at the beginning of its planned schedule.

4.3 Nature of Event Content Deviation

As discussed in Sect. 2.1, event modification may result in the modification of some event attributes. The NoD_{ec} takes into consideration the initial attribute a, the new attribute a', and the significance of the change θ. Content modification of an event consists of attribute deletion, attribute addition, and attribute modification. Each of these types of content modification results in a different NoD_{ec} tuple.

- Attribute modification: In attribute modification, the value of a modified such that $value(a) \neq value(a')$.
- Attribute deletion: In this content modification type, an event attribute is deleted from the event content. This is represented by the tuple $(a, -, \theta)$. Attribute a is deleted from the content of the event in the new state and represented by $-$ in the tuple.
- Attribute addition: The addition of an attribute a' to an event content in the new event state is presented by $(-, a', \theta)$. Since there is no previous version of a' in the previous state of the event's content, its non-existence in the previous state is represented by $-$.

If $value(a) = value(a')$, then there was no change in the attribute value and, therefore no significance. This is because not all event attributes in the new state will be modified versions of the old state. In attribute modification, if the modified attribute is not necessary for the execution of the process, then it is not significant. However, if the attribute is necessary for process execution, then the significance of modifying that attribute a is the difference between the a and a'. Similarly, if an attribute is necessary for the execution is deleted, then the deletion of the attribute may result in a significant contingency. If a new attribute is added to the content of an event, this content modification is significant if the new attribute is necessary for the execution of the process. Therefore, for attribute deletion and addition, the significance of NoD_{ec} may be either true or false. For attribute modification, θ is a value representing the difference between a and a'.

Example 3. In the homecare example, the request event has as part of its attribute *care duration* and *care type*. Assume the care duration was originally 36 days but has been increased to 40 days, then θ is 4 days. The initial care type may only be to provide support with *activities of daily living* (ADLs)[2]. Extra requirements such as the provision of emotional support may be added to the care type after request event's content modification. The θ for the care type modification is the extra requirement of emotional supported added to the original attribute value.

---

[2] ADLs are activities that people do daily without requiring assistance such as eating, bathing, dressing, toileting, walking, and continence. Some clients may require assistance with these activities (see http://www.investopedia.com/terms/a/adl.asp).

5   Contingency Planning

Contingency planning is a very important requirement for dealing with the impact of event modification on a business process. Contingency planning for business processes falls under the broad scope of *contingency theory*. The latter holds that business organisations should have the ability to adapt to changes in their contextual environment [6]. Contingency planning has been adopted in BPM for adapting business processes to changes in their execution context [12,14]. It has also been adopted in other areas such as web services composition [4], as well as, operations management research [19].

In this work, contingency plans (CPs) are required to resolve issues that may arise in a business process as a result of scheduled time modification. CPs are modelled during design time, although additional contingency plans can be added later. We now proceed to discuss the design of contingency plans and their application during runtime.

5.1   Design Time

At design time, with the assistance of domain experts, contingencies can be anticipated and CPs designed to handle them when they occur during runtime. Synthesising contingency plans should be intuitive, i.e., it should imitate how an actual worker will do it. Ideally, selecting the right contingency plan should be semi-automated. That is, the contingency plan should either be selected automatically using bitemporal ECA rules or manually by a user (see Fig. 2). The manual approach should involve providing all the necessary information regarding available intervention strategies to a domain expert. The latter is required to make the final decision regarding that strategy to consider. During runtime, if a contingency occurs, a contingency plan is automatically selected. The latter may be changed by the user after it has been selected.

Modelling Contingency Plans. CPs are modelled similarly to the process models. The difference between a CP and the process model is that a CP is only triggered when a particular contingency occurs. The CP is similar to a process fragment. The latter is a term used to describe an atomic task, a sub-process, or a sub-graph [23] which replaces a placeholder in a process model based on execution context [1]. Unlike process fragments, CPs do not necessarily replace a placeholder but can be executed simultaneously to the process model. In addition, a CP model has distinctive start and event events.

CPs are stored in a repository which we term as the contingency plan repository (see Fig. 2). A CP becomes active during runtime only if its pre-specified conditions are satisfied. The effect of a CP on the business process can be put into three main categories:

- Reactive Contingency: A CP can be designed to realign a deviating business process with its goal. A carer may be unable to attend to a patient because

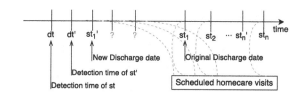

(a) Temporal Representation of Discharge Date Modification

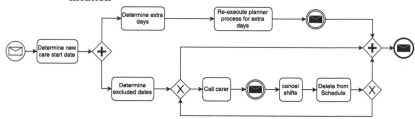

(b) Contingency Plan for Handling Discharge Date Modification

Fig. 5. Discharge date modification and its corresponding contingency plan

his car broke down. A CP aimed at quickly assisting the carer with a new transportation or getting another carer to take his place is a reactive CP.

- Proactive Contingency: Some CPs are executed to ensure the long term success of the running process. For instance, if the care duration increase from 36 days to 40 days, shifts need to be planned for the extra 4 days. The scheduling of these extra shifts may not be urgent but are required to ensure that the shifts are available when required.
- Compensation Contingency: These are important exceptional handling techniques that can be utilised for contingency planning. We can include activities that compensate changes in requirement. Assume the planned discharge date of the patient is modified to a later date. If carers have already been contacted, we will need to cancel shifts that have been cut off due to the change of date. We can include compensating activities call carers and cancel shifts to resolve the issue for the included time period.

For each type of CP, the process designer needs to identify the right conditions for triggering a particular contingency plan, and the set of actions required to mitigate the effect of the contingency on the process.

A CP process model is essentially a 'small', 'specific' process model with activities, events, and gateways. The start and end events of a cp process model are both message events, i.e., the CP process model required a start event to be initiated and it sends out an end event after it has terminated. The following example describes a cp process model.

Example 4. The planner process in Fig. 1 requires a request event with the planned discharge date of the client to initiate planning homecare visits for the client (see Sect. 2.2). Assume the planned discharge date of the patient is modified to a much earlier date as shown in Fig. 5a. We need to consider the extra days included between the initial discharge date and the new discharge date, i.e., the modification scope. If the care duration is not modified, then some shifts will be unnecessary for the client. From Fig. 5a, the original schedule for the patient ends after scheduled visit s_n. Modifying the discharge to an earlier date without modifying the care duration will result in the last scheduled visit being s'_n. The shifts planned between s'_n and s_n are unnecessary and must be terminated. The CP is designed to handle these changes. Once the CP has been initialised, the new discharge date is determined, after which two simultaneous paths are executed. The first path deals with the additional shift requirement by determining the number of extra days (scope) and then executing the planner process from Fig. 1 for those days. The second path deals with cancelling shifts for the excluded days. The carers are informed that their shifts for those days have been cancelled, after which the shifts are deleted from the schedule.

Specifying Bitemporal Contingency Rules. CPs form the action part of bitemporal ECA rules for business processes. When designing CPs, the conditions for activating a plan is known beforehand. That is, we need to identify possible contingency and consequently design plans to handle them. After designing the CP, we need to include the rule for activating the CP in our system. Each cp process model is identified by an identification number which is used in the rule. The format of the bitemporal ECA rules is given as follows:

```
ON:  modified_Event
IF:  (modification_type  = 'specifiedModificationType' AND
      modification_scope = 'specifiedModificationScope' AND
      NoD_of_EventContent = 'specifiedNoD_{ec}' AND
      optional_Additional_Condition)
THEN: cp_Process_ID
```

The ON (trigger) part of the rule is basically the event part of an ECA rule with the exception that the rule is not activated with the occurrence of an event, but with the modification of an event. The condition and action parts of a bitemporal rule have the same semantics as that of an ECA rule. Each part of the rule may have more than one element. These elements can be separated by logical connections such as AND ($\wedge$) and OR ($\vee$). The conditional part of the rule usually has more than one element and these are connected by logical connectors as shown in the rule format above. Logical connectors can also be used when the trigger or action part of a rule has more than one element.

Two important factors to consider about bitemporal rule specification is *rule prioritisation* and *rule composition*. Bitemporal contingency rules may be specified such that one event modification may trigger two or more rules. Since we cannot determine which rule to activate and which ones to ignore, a strategy needs to be put in place to select the appropriate rule. We can use rule prioritisation to indicate the order in which rules should be prioritised. Prioritisation is done using numbers. A higher number signified a lower priority and a lower

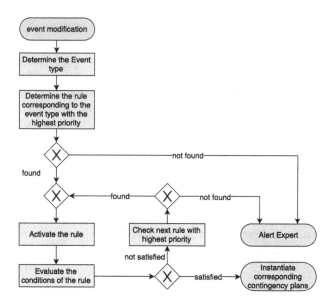

Fig. 6. Runtime contingency plan selection

number signifies a higher priority. When numbers are not used, rules are prioritised from top to bottom. When a rule with a higher priority is evaluated and its
conditions are not satisfied, the next rule with the highest priority is evaluated.

Rule composition is the combination of two or more rules into one rule provided that each individual rule has the same trigger. Constructs for capturing
alternative paths are used in rule composition. We use *else if* construct to facilitate rule composition. The structure of composite bitemporal contingency planning rules is similar to that of nested conditional statements in programming
languages. Prioritisation is inherently captured in rule composition. The individual rules are nested in the main rule in order of priority, that is, the condition
of the rule with the highest priority is added to the first 'if' condition and the
next 'else if' conditions are ordered in a similar manner. Rule composition is
an alternative for rule prioritisation with the following benefits: it reduces the
number of rules and omits the need to use numbers to prioritise rules. The drawback of rule composition is that it becomes too complex when the composition
becomes too large. In such cases, it becomes difficult to read and debug rules.
Rule prioritisation can be used alongside rule composition to enhance readability.

5.2   Runtime

Ideally, the process model should be sufficient to achieve the organisation's goals.
The execution of the process model follows the specification provided during
design time passively. When contingencies occur, the process may deviate from
its original specification in order to mitigate the effect of the contingency on

the process. CPs are triggered as a deviation from normal process execution to handle contingencies.

When a contingency occurs, a process actively searches for a contingency plan that is suitable for handling the contingency. In Fig. 6, we show how contingency plans are selected during runtime. When a change in the scheduled time of an event is detected we, first of all, determine the event's type. We do this because contingencies are specified for event types. Next, we determine the bitemporal contingency rule with the highest priority associated with that event type. If no rule is found, then a domain expert is alerted about the contingency and it is manually handled. On the other hand, if a rule is found the rule is activated and its conditions evaluated. Once the conditions are satisfied, the corresponding contingency plans are executed automatically. If the conditions for a rule are not met, then the next rule specified for the event type with the highest priority is evaluated. This is repeated until a rule satisfies the conditions. If no such rule exists, then an expert is contacted to manually handle the contingency.

6   Discussion

The distinction made between NoD_{st} and NoD_{ec} is important because of the significance of the event's scheduled time. While the scheduled time may be part of an event's content, the distinction is made to properly categorise and handle scheduled time changes. From related work, most changes occur to the scheduled time of an event [7,10,20,26]. Either NoD_{st} or NoD_{ec} may be absent in the conditions of a bitemporal ECA rule. This separation mitigates the difficulty in determining possible contingencies and designing their corresponding CPs.

This work forms part of our ongoing project which consists of (1) detecting event modification in business processes, (2) propagating event modification to other events depending on their causal relationship, (3) managing mutable event provenance and consequently, (4) managing contingencies resulting from event modification in business processes. We have handled some of these aspects in our earlier work [25].

Our implementation of this project so far handles detection, propagation, and provenance of event modification in business processes. This part of the implementation is based on Domain Specific Modelling Environment (DoME)[3]. The DoME tool-set is an extensible collection of integrated model-editing, meta-modelling, and analysis tools supporting a Model-Based Development approach to system and software engineering. DoME uses a *domain specific modelling* approach to problem-solving, i.e., the solution is specified by explicit use of the concepts in the problem domain. DoME is the only open source meta-modelling tool which allows the definition of arbitrary diagram notations using a visual meta-class notation. It also features code generation from visual meta-models, thus, simplifying the implementation of software. The implementation is done

---

[3] http://dome.ggrossmann.com/.

in Cincom's VisualWorks Smalltalk system. The bitemporal ECA rules responsible for triggering contingency plans during runtime form the back-end to our implementation. The complete prototype is being currently being developed.

7   Related Work

Most research approaches in event-driven BPM consider events to be first class citizens and the main driving component of business processes [3,5,11]. In these approaches, events are considered to be immutable once they occur. They lean towards the ideal world scenario and are not equipped to handle situations where revision of an event's content or its scheduled time is required. To the best of our knowledge, there are no approaches in BPM that support the management of bitemporal mutable event.

Substantial research has been done in BPM to enhance process flexibility and adaptability [1,15,17,22]. These approaches can be adopted for modelling contingency plans. The drawback here is that they are not typically designed to handle event correction, but only focused on the other types of triggers, such as an approaching deadline [13], a violated temporal constraint [22], adaptation for different process scenarios [2,16], or compensation in socio-technical processes [8]. We introduce new triggers for process flexibility by incorporating contingency plans that are designed specifically to handle event modification. The practicality of our approach to the designing contingency plans, therefore, relies on existing flexibility approaches in current literature.

Revising an event's content has been considered in some research outside the scope of business process management. In bitemporal database monitoring, event mutability is considered, where an event is modified for the purpose of error correction or new information augmentation [10,20]. Sripada [20] used an event calculus approach to capture event modifications. In their approach, an event recorded in a database has a belief period starting from the transaction time of the event, i.e., when the event is entered into the database. However, the belief period is terminated when a new event revises (corrects) the original event. Belief periods can be used to determine valid events at a particular time. A similar approach to [20] is discussed in [10], where they introduce the concept of multi-history (i.e., capture both the transaction and valid time) of events. In both [10,20], they consider future events (event expected to occur in the future) in their approaches. Their approaches are capable of capturing both proactive and retroactive changes to events. While their work hints towards content modification of events, they replace the entire event with a new event, i.e., the entire content of one event is replaced with the content of a new event. These approaches are outside the scope of BPM. Our work does not focus on belief periods of events as the current state of an event is the state that holds. Instead, we focus on the handling event modification in business process by synthesising contingency plans.

In the domain of web event monitoring, Furche et al. [7] provides an approach for handling web event information update. They focus on improving real world decision making by monitoring web events and their modifications. Web events (announcements) are prone to constant modifications due to frequent web updates. Decisions in the real world, based on web events, should be capable of maintaining consistency with web events after modification. Similar to other approaches that consider future events, they consider the possibility of both proactive and retroactive changes to web events. This work is outside the scope of BPM but forms the basis for our work. Our work extends these approaches in the web domain to BPM. We focus on modification of business events and how contingency plans can be synthesised to handle such changes. We introduce bitemporal ECA rules suitable for triggering contingency plans during runtime. This makes our work is significantly more suitable for BPM.

8 Conclusion

We presented an approach for handling contingencies resulting from event modification in business processes. The contingencies result from either changes in the scheduled time of an event or changes in the event's content. We focus on determining the scope of contingencies, developing contingency plans and bitemporal ECA rules for triggering the contingency plans. The rules are triggered if the scope of a contingency matches the conditions of the rule. When no rule matches the scope of a contingency, a domain expert is alerted. In the future, we hope to integrate the techniques introduced here with our existing work [25] to develop a complete prototype.

Acknowledgement. This research was partially funded by the Data to Decisions Cooperative Research Centre (D2D CRC).

References

1. Ayora, C., Torres, V., Reichert, M., Weber, B., Pelechano, V.: Towards run-time flexibility for process families: open issues and research challenges. In: Rosa, M., Soffer, P. (eds.) BPM 2012. LNBIP, vol. 132, pp. 477–488. Springer, Heidelberg (2013). doi:10.1007/978-3-642-36285-9_49
2. Bucchiarone, A., Marconi, A., Pistore, M., Raik, H.: Dynamic adaptation of fragment-based and context-aware business processes. In: Proceedings of ICWS, pp. 33–41, June 2012
3. Buchmann, A., Appel, S., Freudenreich, T., Frischbier, S., Guerrero, P.E.: From calls to events: architecting future BPM systems. In: Barros, A., Gal, A., Kindler, E. (eds.) BPM 2012. LNCS, vol. 7481, pp. 17–32. Springer, Heidelberg (2012). doi:10.1007/978-3-642-32885-5_2
4. da Costa, L.A.G., Pires, P.F., Mattoso, M.: Automatic composition of web services with contingency plans. In: Proceedings of ICWS, pp. 454–461, July 2004
5. Damaggio, E., Hull, R., Vaculín, R.: On the equivalence of incremental and fixpoint semantics for business artifacts with guard-stage-milestone lifecycles. Inf. Syst. **38**(4), 561–584 (2013)

6. Donaldson, L.: The Contingency Theory of Organizations. Sage, Thousand Oaks (2001)
7. Furche, T., Grasso, G., Huemer, M., Schallhart, C., Schrefl, M.: PeaCE-Ful web event extraction and processing as bitemporal mutable events. ACM Trans. Web **10**(3), 16:1–416:7 (2016)
8. Gou, Y., Ghose, A., Chang, C.-F., Dam, H.K., Miller, A.: Semantic monitoring and compensation in socio-technical processes. In: Indulska, M., Purao, S. (eds.) ER 2014. LNCS, vol. 8823, pp. 117–126. Springer, Cham (2014). doi:10.1007/978-3-319-12256-4_12
9. Guide, V.R., Jayaraman, V., Linton, J.D.: Building contingency planning for closed-loop supply chains with product recovery. J. Oper. Manag. **21**(3), 259–279 (2003)
10. Kim, S.-K., Chakravarthy, S.: Temporal databases with two-dimensional time: modeling and implementation of multihistory. Inf. Sci. **80**(1), 43–89 (1994)
11. Montali, M., Maggi, F.M., Chesani, F., Mello, P., van der Aalst, W.M.P.: Monitoring business constraints with the event calculus. ACM TIST **10**, 17:1–17:30 (2013)
12. Niehaves, B., Poeppelbuss, J., Plattfaut, R., Becker, J.: BPM capability development - a matter of contingencies. Bus. Process Manag. J. **20**(1), 90–106 (2014)
13. Pichler, H., Wenger, M., Eder, J.: Composing time-aware web service orchestrations. In: Eck, P., Gordijn, J., Wieringa, R. (eds.) CAiSE 2009. LNCS, vol. 5565, pp. 349–363. Springer, Heidelberg (2009). doi:10.1007/978-3-642-02144-2_29
14. Ploesser, K., Peleg, M., Soffer, P., Rosemann, M., Recker, J.C.: Learning from context to improve business processes. BPTrends **6**(1), 1–7 (2009)
15. Reichert, M., Rinderle-Ma, S., Dadam, P.: Flexibility in process-aware information systems. In: Jensen, K., van der Aalst, W.M.P. (eds.) Transactions on Petri Nets and Other Models of Concurrency II. LNCS, vol. 5460, pp. 115–135. Springer, Heidelberg (2009). doi:10.1007/978-3-642-00899-3_7
16. Rosemann, M., Recker, J.: Context-aware process design: exploring the extrinsic drivers for process flexibility. In: Proceedings of BPMDS@CAISE (2006)
17. Schonenberg, H., Mans, R., Russell, N., Mulyar, N., van der Aalst, W.: Process flexibility: a survey of contemporary approaches. In: Dietz, J.L.G., Albani, A., Barjis, J. (eds.) CIAO!/EOMAS 2008. LNBIP, vol. 10, pp. 16–30. Springer, Heidelberg (2008). doi:10.1007/978-3-540-68644-6_2
18. Snodgrass, R.T.: Temporal databases. In: Frank, A.U., Campari, I., Formentini, U. (eds.) GIS 1992. LNCS, vol. 639, pp. 22–64. Springer, Heidelberg (1992). doi:10.1007/3-540-55966-3_2
19. Sousa, R., Voss, C.A.: Contingency research in operations management practices. J. Oper. Manag. **26**(6), 697–713 (2008)
20. Sripada, S.M.: A logical framework for temporal deductive databases. In: Proceedings of VLDB, pp. 171–182 (1988)
21. Trkman, P.: The critical success factors of business process management. Int. J. Inf. Manag. **30**(2), 125–134 (2010)
22. van der Aalst, W.M.P., Rosemann, M., Dumas, M.: Deadline-based escalation in process-aware information systems. Decis. Support Syst. **43**(2), 492–511 (2007)
23. Weber, B., Reichert, M., Rinderle-Ma, S.: Change patterns and change support features - enhancing flexibility in process-aware information systems. Data Knowl. Eng. **66**(3), 438–466 (2008)

24. Wondoh, J., Grossmann, G., Gasevic, D., Reichert, M., Schrefl, M., Stumptner, M.: Bitemporal support for business process contingency management. In: Jeusfeld, M.A., Karlapalem, K. (eds.) ER 2015. LNCS, vol. 9382, pp. 109–118. Springer, Cham (2015). doi:10.1007/978-3-319-25747-1_11

25. Wondoh, J., Grossmann, G., Stumptner, M.: Propagation of event content modification in business processes. In: Sheng, Q.Z., Stroulia, E., Tata, S., Bhiri, S. (eds.) ICSOC 2016. LNCS, vol. 9936, pp. 70–84. Springer, Cham (2016). doi:10.1007/978-3-319-46295-0_5

26. Wondoh, J., Grossmann, G., Stumptner, M.: Utilising bitemporal information for business process contingency management. In: Proceedings of APCCM, pp. 45:1–45:10. ACM (2016)

Dynamic Change Propagation for Process Choreography Instances

Conrad Indiono[(✉)] and Stefanie Rinderle-Ma

Faculty of Computer Science, University of Vienna, Vienna, Austria
{conrad.indiono,stefanie.rinderle-ma}@univie.ac.at

Abstract. Business process collaborations realize value chains between different partners and can be implemented by so called process choreographies. Change has become a major driver for costly (re-)negotiations between the participants. Static a priori prediction models exist to calculate the feasibility of a change request prior to negotiation. However, the dynamic or behavioral aspect of choreography changes at the choreography instance level has not been investigated yet, i.e., the question whether a process choreography instance is compliant with the change request and hence allows for acceptance of the change request. This work takes the dynamic perspective and analyzes the impact of a single change request from one partner on the entire (distributed) choreography based on the notion of change regions and public check points. Change strategies are elaborated to ensure choreography instance state compliance. One transaction-based approach is specified using rollback regions. It identifies probabilistically the set of activity nodes to be compensated at all levels of the business collaboration to ensure state compliance. The technical evaluation enables observing the properties of the rollback region.

Keywords: Collaborative business processes · Dynamic change

1 Introduction

Process choreographies implement business process collaborations between different partners in order to reach a joint business goal. Examples stem from the manufacturing or logistics domain. "dynamism is the basis for agility" [10] presents an omnipresent challenge to handling change in process choreographies. Though some work on the static perspective of process choreography change exists, e.g., [4,12], approaches to deal with the dynamic perspective of choreography change are almost entirely missing [17]. This paper tackles this research gap by considering the actual execution state of each participating partner's process instance to estimate the impact of a change request on the global business choreography. The distributed nature of process choreographies poses particular challenges as it introduces privacy elements, disallowing full insight into direct or indirect partners' execution environments. This means that on the static model level, partners do not know what exact activities are scheduled to be

© Springer International Publishing AG 2017
H. Panetto et al. (Eds.): OTM 2017 Conferences, Part I, LNCS 10573, pp. 334–352, 2017.
https://doi.org/10.1007/978-3-319-69462-7_22

executed in between the interaction activities. Similarly on the dynamic level, the concrete current execution state as well as the historic execution log of each process instance is not fully known for partners.

This work extends already established work in the area of static change in process choreographies [5]. Here choreography change is described as a process consisting of several steps such as checking change correctness (static and dynamic) and negotiating the change request with the partners. The reason for the latter is that in a fully distributed setting changes cannot be imposed on partners, but have to be agreed on. Hence it can be beneficiary to estimate the costs of such a negotiation beforehand [8], particularly as negotiation might become a costly multi-step process [6]. Cost estimation is realized within the so called change prediction step that is executed before the negotiation takes place. The impact of the initial change request is estimated in terms of a normalized score value that allows comparison between the set of available change requests.

So far, merely the static costs of a change have been considered. The dynamic costs have only been considered in an abstracted manner, i.e., based on the choreography model level using execution probabilities [7]. In this work, the goal is to determine the *change region* of a change for different partners, i.e., the region in which a change might be critical with respect to the choreography instance state. Change regions have been proposed for business processes, but not for process choreographies. Based on the change region it can be determined whether or not a partner can apply the change right away. Further on, it can be considered whether or not the application of change compensation actions such as rollback might be meaningful in order to realize the change. In particular, the costs for such actions can be taken into consideration when estimating the change impact. This research goal is reflected in the following research questions:

- How to identify change regions for choreography changes, i.e., those regions where changing running choreography instances could lead to an inconsistent state?
- Once the change regions are identified, how to estimate the total impact of a change request?

The questions are approached following the design science method (cf. [20]). The relevance of the topic is underpinned by literature [8,17]. The created artifacts comprise definitions of fundamental concepts such as change regions as well as algorithms, i.e., algorithms that determine the rollback regions. The approach is evaluated based on a proof-of-concept implementation. Thereby the paper is structured as follows: In Sect. 2, basic terminology is introduced: Sect. 3 provides new concepts on choreography instance change. Section 4 provides change propagation strategies including compensation actions based on rollback. The evaluation is presented in Sect. 5, followed by related work in Sect. 6 and a conclusion in Sect. 7.

2   Motivating Example and Fundamentals

As an illustrative example, we study a pc case manufacturing use case consisting of these roles: pc case manufacturer, buyer, metal supplier and coloring supplier. The choreography model of this business collaboration is depicted in Fig. 1 and can be divided into the following areas: choosing a product type (F_1), checking for raw resource availability (F_2), the abstracted manufacturing subprocess (F_4) and finally the delivery process (F_3).

Consider a change introduced by the manufacturer in (F_1), which adds an option for laser etching custom designs. An associated change is also implemented in Fragment F_4, which performs the activity for the laser etching. How can we estimate the impact these changes have on the other partners? From previous work [7] we are able to estimate this impact using execution probabilities and an abstract adaptation cost. In this work, we improve the change region to determine the beginning of the change and define rollback regions to mark all possible maximal activities process instances are able to progress to. A concrete transaction based adaptation is used, which informs the cost calculation.

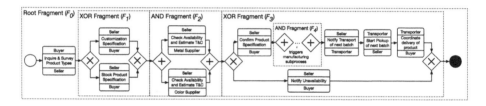

Fig. 1. Motivating example: PC case manufacturing use case

Definition 1 [Choreography]. *Adapted from [5], we define a choreography $\mathcal{C}$ as a tuple $(\mathcal{G}, \mathcal{P}, \Pi, \mathcal{L})$, where*

- *$\mathcal{G}$ is the choreography model (i.e., Fig. 1).*
- *$\mathcal{P}$ is the set of all participating partners.*
- *$\Pi = \{\pi_p\}_{p \in \mathcal{P}}$ is the set of all private models.*
- *$\mathcal{L} = \{l_p\}_{p \in \mathcal{P}}$ is the set of all public models.*

The Refined Process Structure Tree (RPST) [19] is a structured approach for representing business process models. It divides the model into several fragments, each fulfilling the single entry, single exit property (SESE). A fragment is either a trivial one (a leaf in the tree), representing a single interaction inside a sequence, or a complex one that can be either an XOR or AND fragment and contain further sub fragments. Figure 2 shows the pc case manufacturing use case as a collapsed RPST. It only shows the sequence under the root fragment (F_0), consisting of leaf nodes (trivial fragments, e.g. interactions) and sub trees representing either XOR or AND fragments (e.g., F_1, F_2, F_3). Note that the actual manufacturing process F_4 is not shown in the root sequence due to it being embedded under Fragment F_3.

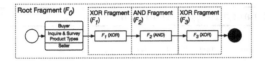

Fig. 2. Top-level RPST tree of the PC case manufacturing use case

Definition 2 [Change Patterns]. *From [5], we define the following change patterns:*

$$ChangePattern ::= REPLACE(oldFragment, newFragment)$$
$$| \ DELETE(fragment)$$
$$| \ INSERT(fragment, how, pred, succ)$$
$$how ::= Parallel \ | \ Choice \ | \ Sequence$$

Example: To implement the laser etching scenario in Fig. 1, the $REPLACE(F_0, F_{new})$ change pattern would be used, where F_{new} is based on F_0, with the following changes: Fragment F_1 has a new interaction between seller and buyer for offering the laser etching customization. Fragment F_4 is modified, where laser etching specific interactions (e.g., with the corresponding supplier) are added.

Definition 3 [Business Process Instance]. *A business process instance is a tuple (id, m, r, as), where the following holds: $m \in \Pi \wedge r \in \mathcal{P}$. as is a set of tuples each of the form (a, s), where $a \in m \wedge s \in \{inactive, activated, running, aborted, completed\}$ and id is a unique identifier that is mapped to the business process instance.*

We extend the definition of *Choreography* from Definition 1 to be: $\mathcal{C} = (\mathcal{G}, \mathcal{P}, \Pi, \mathcal{L}, \Pi^{\mathcal{I}})$, where $\Pi^{\mathcal{I}}$ is a set of business process instances (see Definition 3), which are running *instantiations* of a corresponding business process model and each activity comprising the process model having an associated state.

3 Dynamic Change Propagation Concepts

To determine the dynamic impacts of a change operation, two components are required: (1) identify the specific nodes that mark the beginning of the change: the *change region* and (2) identify the relative position of all business process instances ($\Pi^{\mathcal{I}}$) to that *change region*: state compliance. Having that relative position gives an initial indication whether there could be disproportional costs involved for the proposed change. Those process instances having their state before the *change region* are state compliant, and are thus generally non-problematic in regards to implementing the change. Those process instances whose running activities are already past the *change region* violate state compliance, and change implementation may potentially become problematic.

Algorithm 1. Change Region Algorithm

    Input:

1   δ - a change pattern (see Definition 2)

2 **Begin**

3 **if** $\delta.type \in \{INSERT, DELETE\}$ **then**

4 | **return** $\delta.fragment$

5 **else**

6        $ins \leftarrow \delta.newFragment \setminus \delta.oldFragment$

7        $del \leftarrow \delta.oldFragment \setminus \delta.newFragment$

8        $F_{root} \leftarrow \alpha(ins + del); F_{min} \leftarrow \emptyset$

9   **foreach** $node$ in $ins + del$ **do**

10   **if** $F_{min} = \emptyset \vee distance(F_{root}, node) < distance(F_{root}, F_{min})$ **then**

11 | $F_{min} \leftarrow node$

12   **return** F_{min}

3.1 Change Region

As the basis for determining the *change region*, we can start with the *smallest fragment* (see Definition 4), which returns the surrounding RPST fragment containing all supplied nodes. Example: the *smallest fragment* that contains { *Inquire & Survey Product Types* } would be the interaction itself, because a leaf node inside a RPST fragment is a fragment. The *smallest fragment* containing { *Customization Specification, Notify Transport of next batch* } is the root fragment F_0.

Definition 4 [Smallest Fragment] *(from [5]). Let σ be a public model and S be a set of nodes corresponding to σ. Then: $\alpha_\sigma(S)$ returns the smallest fragment in model σ that contains all nodes from S. Formally: $\alpha_\sigma(S) = \underset{size(F)}{\arg\min}\{F \in \sigma \mid \forall n \in S, n \in F\}$.*

While the *smallest fragment* is sufficient to determine the beginning of a change in the cases where the change pattern $\in \{INSERT, DELETE\}$, it does not hold for *REPLACE*. In the case of *INSERT*, we know a new RPST fragment is being inserted between *pred* and *succ* (see Definition 2). Thus taking the fragment itself to mark the beginning of the change is feasible. The same concept holds for *DELETE* change patterns. However, in the case where we have a *REPLACE* change pattern as in the laser etching example, the *smallest fragment* would return the root fragment F_0 as the surrounding RPST fragment. The beginning of that root fragment does not mark the real change, which is critical for determining the necessity of adaptation before implementing the change. Generally, adaptation is required once the first node marking the change enters a state past *activated* (i.e., *running, aborted, completed*). Recall that in the laser etching example we have two RPST fragments being modified inside the *REPLACE* operation: F_1 by adding an interaction *offer laser etching* and F_4 which adds new interactions with suppliers related to the new offer. To identify the concrete change region a simple *smallest fragment* call is not sufficient.

We need to further identify the fragment nearest to the start node that has been changed and set that as the beginning of the change region. To do so, we first need the fragments that have been inserted (in both F_1 and F_4) and then find the one with the shortest path from the start of the surrounding fragment: this fragment (F_1) marks the beginning of the change region. Algorithm 1 specifies such a *change region*.

3.2   State Compliance

The second problem with determining dynamic change impact is related to the following: while each partner is able to calculate the relative position of their process instances from the change region, partners are only able to estimate this location due to privacy issues. Using their own private execution log, partners are able to discern state compliance on their own [13]. For partners to estimate direct partners' state compliance, public interactions can be used as checkpoints. We know that once we have finished an interaction with another partner, the other partner has equally finished the same interaction in their own private process. Execution paths from that point on cannot not be accurately determined by direct partners as it is possible that the partner progresses in such a way that further interaction with the same partner never happens. A comprehensive monitoring approach at the cost of privacy would be required if indirect partner state tracking is required, as these would involve active state reports due to lack of direct interaction points. Estimating state compliance for direct partners can be achieved by taking a private execution log and abstracting by public activities or interactions where the direct partner is involved. The last activity can be seen as the *current state*, even though from the perspective of the private model, the actual progress might be more advanced. Only through passing public checkpoints can direct partners track state compliance.

Definition 5 *[Choreography Instance]. We extend the definition of* Choreography *from Definition 3 to include* choreography instances: $C = (\mathcal{G}, \mathcal{P}, \Pi, \mathcal{L}, \Pi^{\mathcal{I}}, \kappa, \kappa', \mathcal{G}^{\mathcal{I}})$, where

- $\kappa : corr_{id} \rightarrow \{(id, m, r, as)\}$ *is a total function that maps a unique correlation identifier to a set of business process instances working on the same business case (see Definition 3), each assigned a unique private case id, with $corr_{id} \in \mathcal{G}^{\mathcal{I}}$.*
- $\kappa' : (id, m, r, as) \rightarrow corr_{id}$, *a non-injective surjective function that maps a unique business process instance to a correlation identifier.*
- $\mathcal{G}^{\mathcal{I}}$ *is the set of all active unique correlation identifiers, each representing a single choreography instance.*

In our example, a business case is started by the buyer who intends to have a product manufactured. That buyer process starts with its own case id to track the product. The seller creates a unique case id once a product buying confirmation comes in and uses the same id when contacting suppliers. The two suppliers create their individual private case ids. Finally, the transporter is contacted by the seller, who may create a separate case id to handle the shipment. Even though many different case ids exists, the same product is being handled and referenced. The correlation id groups these unique case ids to the same business case. Thus a choreography instance groups together all partners and their private process instances working on the same business case. The known public markings are used to set the states of each public interaction activity. However, for privacy reasons, the partner who has direct connections with the most partners has the most accurate view. The choreography instance from the perspective of the buyer is more limited compared to the one of the buyer.

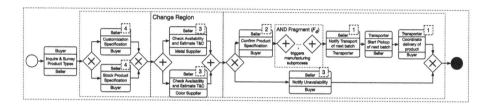

Fig. 3. Example of an aggregated choreography instance

An *aggregated choreography instance* can be created (illustrated in Fig. 3) by taking all choreography instances (in $\mathcal{G}^{\mathcal{I}}$), applying κ on each to retrieve all the relevant business process instances and abstracting on public activities of the directly involved partners. The last known public activity can be registered as the *current active node*. Such nodes can be counted and the frequency marked on the choreography model ($\mathcal{G}$). Each count represents a single choreography instance. It is now possible to determine which choreography instances violate state compliance: in Fig. 3 these are the three instances inside the change region (WITHIN), and the seven instances after the change region (AFTER).

4   Dynamic Change Propagation Strategies

Having defined the necessary concepts, we will focus our attention on the mechanics of propagating changes to partners, focusing on the question of consistency from the dynamic point of view. At the core, every process instance needs to be stopped to avoid having process instances in the BEFORE state (compliant) entering the WITHIN or AFTER state, which makes these process instances non-compliant in terms of the change requested. This is the pessimistic approach because we assume that regardless of the future path actually taken, it might lead to structural conflicts due to the changes to be committed. Activities that are still before the *change region* are free to be executed according to the previous model. The *change region* represents a critical section, and partners still need to decide on a common understanding for this region (see Change Negotiation [6]). Thus any partners proceeding the execution past the *change region* have at that point already made a choice with which change alternative to proceed. Note that choosing the old version is an alternative as well. The outcome of the change negotiation needs to match this implicitly chosen change alternative for those process instances to remain structurally compliant. Anyone diverging from the common understanding means that any work resulting from that becomes unusable and has to be discarded. Partners are tied to each other with their decisions on which change alternative to pick.

There is a significant disadvantage with this approach: all choreography instances being affected by the change are stopped and no work can be executed that falls beyond the first node in the *change region*, which means there is an opportunity cost: no productive work can be performed due to the waiting time until a change alternative has been agreed on for the contested *change region*. This waiting time ensures the collaboration stays consistent. An alternative approach is the *optimistic change strategy*, which assumes that paying the opportunity cost (of waiting) is higher than just proceeding with execution and pay the cost only if actual state violations occur, meaning a different change alternative has been agreed on than the one partners have implicitly chosen to proceed execution with. In order to accomplish this, we need an approach to estimate this actual *repair* cost, which is the cost of transforming non-compliant process instances to become compliant again. Weighing these two different costs together (pessimistic cost vs optimistic cost), allows a fair estimation of effort before change propagation is proposed to partners, and bargaining leverage once change negotiations start. In this work, the focus is placed on the technique for determining the *repair* cost, which occurs for making non-compliant process instances compliant.

4.1   Transaction-Based Optimistic Change Strategy

In this work we will focus on an optimistic change strategy that is based on transaction support and calculates the repair cost based on performing rollback, starting with the assumptions on the transaction model.

Transaction Model Assumptions. Several papers exist that explore the necessary properties a workflow transaction model should entail [3]. Here we summarize these properties and assume they are supported, independent of the concrete transaction model being used. The work in this paper builds on these assumptions.

A workflow transaction is either (i) a single activity or (ii) a sequence of activities and takes as input a consistent state transforming it into another consistent state. Furthermore, workflow transactions are hierarchically nested with the presence of subprocesses and thus sub activities. Regarding atomicity, workflow transactions relax the *nothing* property of the all-or-nothing concept of traditional transactions [3]. Whereas traditional transactions expect an opaque transformation step from one consistent state to another, workflow transactions are made transparent, where each intermediate consistent state is *opened up*. This transparency allows a more fine-grained ability to perform rollbacks to past consistent states. A rollback path can be defined, which marks the backward sequence of past activities from the current activity up to the desired past consistent state. This paper takes these eligible states as rollback target in the context of inter-organizational business processes. We assume that the past consistent states, which have been traversed so far are represented and available in the form of private and public execution logs. The visibility of which is dependent on the partner accessing it. These execution logs are one of the required inputs for performing selective rollback. Note that for all private activities, the owner of the process instance is able to decide, without coordination with the associated partners, to which past state to rollback to. But once a public activity occurs inside the rollback path, the directly associated partner is bound to the same rollback operation. A transitive effect can be observed as this partner directing its partners to rollback to the relevant public activity, due to their interaction activities in the same rollback path. A decision to rollback then, by the nature of public interactions and the message dependency, affects other partners directly as well as indirectly. Another assumption is the ability to assign compensation tasks to activities, which are executed in the case an activity needs to be rolled back. If a compensation task cannot be semantically mapped, then the activity is marked as a *critical activity*. The effects of *critical activities* are further discussed in Sect. 4.2.

Regarding consistency, we assume that each committed transaction after the execution of a single activity results in a consistent state. Accordingly, each rollback operation and thus the completion of a compensation task results in a consistent state as well.

Regarding isolation, rolling back activities should not affect concurrently running workflow transactions (assuming they are not part of the same choreography instance). In the case of sub activities we have a parent-child dependency between the parent activities and the called sub activities. Due to this parent-child relationship, whenever a decision is made to rollback the parent activity, all of the executed sub activities need to be rolled back as well. Inversely, a rollback on a sub activity does not necessarily require a rollback on the parent activity.

Regarding durability, we assume together with consistency that all committed transactions, as well as any compensation tasks executed, are made persistent.

4.2   Rollback Region

Based on the assumptions on the transaction model mentioned in the previous section, we introduce *rollback regions*. *Rollback region* is a technique that can be classified as a transaction-based optimistic change strategy. It is optimistic because it allows partners to proceed the execution even while the change request has not been committed yet. Whether or not the change request will be committed, the optimistic approach assumes that not every work result needs to be discarded. It is transaction-based due to the use of rollback, specifically compensation tasks, to bring non-compliant process instances back to compliance. *Rollback regions* determine the upper bound up to which a corresponding partner may proceed execution, and based on that calculates a cost estimation that would be required in case compliance transformation occurs. Stopping the execution of a process instance at an interaction gives us certain information about the corresponding partner we are interacting with. When we are waiting for a message then we know that the corresponding partner has not yet reached its corresponding send activity. Conversely, when there is an upcoming receive message activity with a partner and we haven't yet sent the required message for that partner to proceed, we are sure that at the worst case, that partner is waiting for us. These *checkpoints* form the basis for *rollback regions* to estimate the farthest activity direct partners are able to proceed. All possible paths that can be drawn between the current activity node up to the *checkpoints* represent the rollback paths to be traversed (i.e., by running the sequence of compensation activities). The cost of a rollback path is defined in Definition 6. The act of committing a change request is itself conducted within the context of a transaction: either all choreography instances are transformed into a consistent state, or the change commit fails and a rollback occurs which results in the consistent state before the change request was proposed. The *rollback region* captures the worst case cost in the event of a rollback. It is important to note that not all rollback paths need to be traversed, as only the actually traversed path represented by the private and public execution log needs to be compensated. In the following sections we will discuss the influencing factors that may determine the farthest *checkpoint* that builds the terminal node of a *rollback region*.

Critical Activities. The first influencing factor in determining the farthest activity is the presence of a *critical activity*. Recall that *critical activities* are those activities not associated with compensation tasks (c.f. Sect. 4.1), and thus cannot be compensated in the event of a failed transaction (e.g. rollback). Since the *rollback region* requires the presence of compensation tasks to work, *critical activities* constrain the possible execution paths for process instances in the context of a transaction-based optimistic change strategy. Concretely, from the perspective of a single process instance, if the next activity to be executed is a *critical activity*, then the optimistic change strategy becomes impossible due to not having the prerequisite compensation task to perform a rollback. This results in a hybrid change strategy where the change strategy can be optimistic up to the point of the first *critical activity*, and switching to a pessimistic change strategy from that point on. Of course, some process instances might emit a different execution log which does not entail such *critical activities*. In that case, the optimistic change strategy persists. A *critical activity* thus affects how far a process instance, under an optimistic change strategy regime, may proceed and in the same way limits the choices for the farthest activity.

Sync vs Async Message Passing in Interactions. The W3C WS Choreography Model defines two distinctive types of interaction activities[1]: The

---

[1] https://www.w3.org/TR/ws-chor-model/.

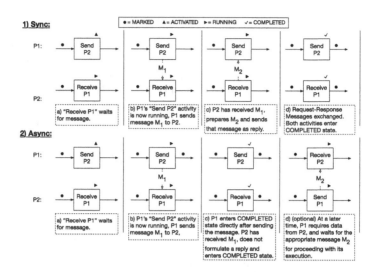

Fig. 4. Execution semantics of sync vs async interactions

(a) *one-way interaction*, for sending a single message and (b) the *request-response interaction*. In the latter case, the sender expects a response from the receiver of the initial message. In this work we define interactions to be either *async* or *sync*, corresponding to (a) and (b) respectively. The main difference between the two interaction types is the following: in the case of the asynchronous message passing style, partners are not obliged to wait for a response after sending a message to their partner (cf. Fig. 4). This sending semantic affects the boundary of the *rollback region* (i.e. the farthest activity). In a *sync* interaction, the two partners are in lock-step with each other due to the necessary message coordination in a request-response fashion. From the perspective of the sending partner p_1 in a *sync* interaction, the farthest activity for the corresponding partner p_2 is clear as long as the interaction has not completed: the current interaction node. The initial sender has two states to execute inside a *sync* interaction: the first sending of the initial message, and an implicit receive state for receiving the reply to the initial message. The latter prevents the initial sender to progress the execution of the process instance, and thus also prevent the *rollback region* from diverging with the corresponding partner.

In an *async* interaction it is not as transparent, due to the lack of a blocking receive activity that would wait for a reply. After an *async* interaction completes, the sender will come to a waiting state only if the sender has an explicit receive activity after the initial *async* send activity. Note that after a *sync* interaction completes (i.e., the receiving partner has received the reply and proceeds execution), the same non-transparency issue exists for *sync* interactions. *Rollback region* are then non-deterministic, as well as dynamic. It is non-deterministic due to not knowing which actual execution path will be taken and which blocking node reached. It is also dynamic, because the *rollback region* changes as soon as partners pass checkpoints. A *rollback region* becomes a cost estimate based on a temporary snapshot of the current state of the choreography instance, to estimate the effort and cost required to propagate a change request.

Definition 6 *[Cost of a rollback path]. Let RP be the set of private/public activities constituting a rollback path, which does not contain a critical activity. Assume further a function $cost_{comp}(a)$ that returns the cost of performing the compensation task associated with an activity a. Then the cost of a rollback path is the sum of the cost of compensation tasks to be executed: $cost_{rbp}(RP) = \sum_{a \in RP} cost_{comp}(a)$.*

Definition 7 *[Blocking Node]. A node is a* blocking node *if it is either a* sync *interaction (which includes an implicit receive activity), an explicit receive activity, a critical activity (an activity without a mapped compensation task), or an end node marking the end of the process instance.*

$$BlockingNode ::= SyncInteractionActivity \mid ReceiveActivity \mid$$
$$CriticalActivity \mid EndNode$$

Rollback Region Algorithm. Depending on the perspective of the partner applying the *rollback region*, it can have different purposes. On the one hand, from the perspective of the change initiator, the *rollback region* is used to calculate the effort required for executing compensation tasks in order to make process instances compliant according to the old process model. After initiating the change request, in an optimistic change strategy, the change initiator assumes the change request becomes accepted and proceeds execution, while the decision to commit to the change is not yet made. On the other hand, from the perspective of the change acceptor, the *rollback region* is used to calculate the effort required in order to make process instances compliant according to the new process model. This is due to the process execution maintaining the old version for running the current process instances. The effort then is to make the process instances compliant to the new process models. The *rollback region* is used as the core mechanism to determine the effort required to make process instances compliant, whether for the old process model or the new one. In both of the above cases, the *rollback region* algorithm requires the following inputs: (1) the *change region* marking the beginning of the change, (2) the private process model of the partner for which the dynamic impact is calculated, and (3) the private execution log of the same partner. Since the purpose of the *rollback region* is to calculate the cost of transforming non-compliant process instances to become compliant, we define the start node to be the first node within the *change region* (i.e., input (1) *change region*). The end node to be determined is the farthest activity as discussed earlier. In this section we will focus on the steps required to determine a *rollback region* from the perspective of a single partner. The full specification can be found in Algorithm 2.

Step 1: Determine Fragment-Local Blocking Node Probabilities. The first step of the *rollback region* algorithm is to determine for each occuring *blocking node* (cf. Definition 7) its local probability of becoming activated. The *rollback region* is not deterministic, as the actual path to be executed from the current activity is still not yet known. In the simplest case we have a single blocking node on the level of the top-most RPST fragment (inside a sequence). In this case we can determine that in all cases, landing on this node causes the execution to stop. In the case of inside an RPST fragment (XOR gateways), we require branching probabilities (e.g. based on actual historical execution data) to determine the probability of the execution engine reaching that blocking node. Thus we add another assumption: we have access to a table of branch probabilities for the private model of the partner whose impact we want to estimate. Blocking nodes inside deeply nested subfragments are determined by multiplying the branch probabilities of reaching that subfragment. Since we are only interested in the reachability of blocking nodes through branching probabilities, multiple blocking nodes inside the same subfragments are considered equal and only the first is considered. By summing up these local probabilities of reaching each blocking node, we can determine the probability of this RPST fragment becoming a terminal node: $P_{blocking}(F)$ (cf. Definition 9). The inverse probability of the execution passing through the same fragment without encountering a *blocking node* would then be $P_{pass}(F) = 1 - P_{blocking}(F)$.

Definition 8 [Reachability of Local Blocking Node]. *Given*

- *$path(F, node)$, a function that returns the shortest sequence of edges starting from the beginning of Fragment F leading to node.*
- *$edges_{xor}(F, node)$, a function that returns the edges preceeded by an XOR node from the path from the beginning of Fragment F leading to node, defined as $\alpha_{preceeded\_by\_xor}(path(F, node))$.*
- *$P_{edge}(edge)$, a function that returns the local (XOR) branching probability of that edge being traversed.*

The reachability of a local blocking node *is defined as*

$$P_{local}(F, node) = \begin{cases} \prod_{x \in edges_{xor}(F,node)} P_{edge}(F, x) & if\ is\_xor\_fragment(F) \\ 1.0 & otherwise \end{cases}$$

Figure 5 shows an example which calculates the probabilities of reaching blocking nodes $\{C, D\}$ inside a XOR RPST Fragment $F0$.

Definition 9 [Probability of a blocking RPST Fragment]. *Given the local probability of reaching a* blocking node *inside a RPST fragment F: $P_{local}(F, node)$, we can define the probability of the whole RPST fragment blocking the execution:*

$$P_{blocking}(F) = \sum\nolimits_{\forall x \in nodes(F):is\_blocking\_node(x)} P_{local}(F, x)$$

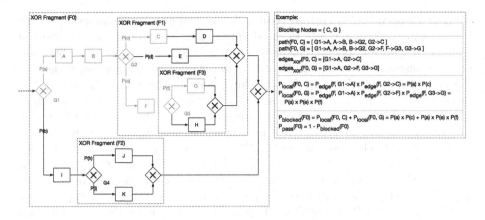

Fig. 5. Fragment-local blocking node probabilities

Algorithm 2. Local Rollback Region (LRR) Algorithm

Input:

1        $\mathcal{Q}$ - *change region*
2   π_i - *Private Process Model of partner* i
3   L_i - *Private Execution Log of partner* i
4   **Begin**
5        $\Delta \leftarrow \mathcal{Q}(\pi_i)$; $\mathcal{S} \leftarrow head(\Delta)$; $cur \leftarrow last(L_i)$
6   **if** $\mathcal{S} \notin L_i \wedge \mathcal{S}.state \notin \{running, stopped, completed\}$ **then**
7        $\lfloor$ **return** 0

8   **else**
9   // *Step 1: Determine terminal node*
10   **foreach** f *in* $nodespath_{rpst}(cur, \pi_i)$ **do**
11   **if** $P_{blocking}(f) = 1.0$ **then**
12           $node_{terminal} \leftarrow f$
13   break;

14   // *Step 2: Adjust local probabilities*
15       $residual \leftarrow 1.0$; $probs \leftarrow \emptyset$; $subfrags \leftarrow \emptyset$
16   **foreach** f *in* $nodespath_{rpst}(cur, \pi_i)$ *until* $f = node_{terminal}$ **do**
17   **foreach** n *in* $nodes(f)$ **do**
18   **if** $is\_blocking\_node(n) \wedge n \notin subfrags$ **then**
19             $probs \leftarrow probs + \{(f, P_{local}(f, n) * residual)\}$
20             $subfrags \leftarrow subfrags + \{f\}$

21       $residual \leftarrow (1 - P_{blocking}(f)) * residual$
22   assert($\sum_{\{(bn,p) \in probs\}} p = 1.0$)
23   // *Step 3: Calculate expected cost of rollback region*
24       $cost_{expected} \leftarrow 0.0$
25   **foreach** (bn, p) *in* $probs$ **do**
26       $\lfloor$ $cost_{expected} \leftarrow cost_{expected} + cost_{rbp}(nodespath(cur, bn)) * p$

27   **return** $cost_{rbp}(nodespath(\mathcal{S}, cur)) + cost_{expected}$

Step 2: Determine Terminal Node. A terminal node is the farthest activity a process instance may reach. In the simplest case, the terminal node is the *end* node of the process model. This case happens when there is no more blocking nodes to suspend the execution for a process instance, except the last node (i.e., the *end* node). Another case is a blocking node in the main sequence of the top-most RPST fragment. This becomes a terminal node due to the certainty of this node becoming activated in future executions. It could be either due to partners having to synchronize messaging (receive activity) or reaching a critical activity. The last case is an RPST fragment one level below the top-most RPST fragment, which has $P_{blocking}(F) = 1.0$ (c.f., Definition 9), meaning an absolute certainty of activating a *blocking node* once entered. In contrast, any $P_{blocking}(F) < 1.0$ will have the possibility of executions avoiding *blocking nodes* inside this RPST fragment. To generically determine the terminal node we take the top-most RPST fragment and in that sequence take each node's probability of it becoming a *blocking node*. The first $P_{blocking}(F) = 1.0$ is marked as the terminal node. An example can be seen in Fig. 6, where the end node is marked as the *terminal node*, due to F_2 being the *change region* and F_3 having a $P_{blocking}(F_3) = 0.6$.

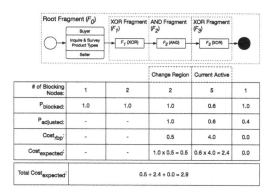

Fig. 6. Example of a *rollback region* calculation.

Step 3: Determine Absolute Probabilities of Rollback Paths Inside the Rollback Region. Having determined the *terminal node*, we can now calculate the absolute probability of reaching each *rollback path* inside the *rollback region*. The only *blocking nodes* we consider are those starting from the current activity up to the terminal node (c.f., example in Fig. 6). The *rollback paths* are built by pairing each candidate *blocking node* as the end activity with the current active node as the starting activity. For the adjustment of the probabilities we again take the sequence of the top-most RPST fragment. For each node, we accumulate the residual, which is the probability of not being blocked by the current node (starting with 1.0) and adjust the local probabilities by multiplying it with the residual. This step stops at the terminal node. The invariant to be upheld is the sum of all absolute probabilities = 1.0.

Step 4: Calculate Expected Cost of a Rollback Region. Since *rollback regions* are probabilistic and having calculated the absolute probabilities of each *rollback path*, we can calculate a final value that represents the expected cost of that *rollback region*. Using Definition 6 we can determine the cost of a single *rollback path*. This *rollback path* cost is multiplied by the absolute probability of that *rollback path* actually occuring (from the previous step). The sum of all these individual *rollback path* costs represents the expected cost of the *rollback region*. A fixed cost part exists, which is the *rollback path* cost that is accrued due to the already traversed path starting from the change region up to the current active node.

Using Rollback Region for Change Impact Analysis. Recall that the goal of this work is to have the ability to evaluate several change alternatives in terms of change impact on the whole choreography. We have tackled that challenge through the lens of the dynamic state of the individual process instances that together realize the collaboration, in relation to the position of the proposed change alternatives individually. By using *rollback regions*, it is possible to answer these questions. While the specified *rollback region* algorithm LRR works from the perspective of a single partner, and in that context for a single process instance, we can now aggregate the individual expected costs of each *rollback region* for a single choreography instance:

$$ARR(corr_{id}, q) = \sum\nolimits_{i \in \kappa(corr_{id})} LRR(q, i.m, i.as)$$

The expected cost of a change alternative over the complete collaboration would then become:

$$CRR(q) = \sum\nolimits_{corr_{id} \in \mathcal{G}^{\mathcal{I}}} ARR(corr_{id}, q)$$

With $CRR(q)$, it is now possible to compare change alternatives in terms of their impacts by estimating compensation task costs, where the individual *rollback paths* dynamically evolve as execution progresses. Note however, that the LRR expects the private model π_i of the partner for whom to calculate the impact for. The change initiator does not have access to the private models of the other partners in the collaboration. This means the aggregation conducted in $ARR(k, q)$ is a fan-out process, where the change initiator asks each partner for the results of applying the local *rollback region* algorithm (LRR) and in the end aggregates the returned individual expected costs. One approach exists to avoid the communication overhead. The change initiator could estimate the change impact of a change alternative by substituting the private model of the intended partner with the corresponding accessible public model. The required execution log can be either (i) derived through abstraction on the public nodes of the intended partner or (ii) directly requested. The result of applying the LRR on this adjusted input cannot be accurate, as no private activities are accessible and thus the complete compensation cost of those *rollback paths* unknown. What is known are the distances of these *rollback paths* through the number of *checkpoints* past the *change region*. The expected cost of change alternatives based on these inputs are only approximations. One way to increase the accuracy would be by all partners making public several metrics: branching probabilities, average number of private nodes inside RPST Fragments and average compensation cost of private activities.

5 Evaluation

As a technical evaluation we have implemented the concepts introduced in this work, mainly *rollback regions* and the dependent concepts, as a proof-of-concept. Furthermore we evaluate the output of the *rollback region* variations and ensure the critical invariants are upheld[2]. The evaluation setup follows this methodology: (1) Load a pre-defined choreography specified with BPMN 2.0 XML Format. (2) Specify a change region. (3) Randomly scale private activities for each role in the choreography with the following set as the number of private activities per fragment: $\{2, 5, 10, 30, 50\}$. (4) Randomly generate business process instances (by assigning activity states) and creating the associated choreography instances, grouping these business process instances together. (5) Calculate the expected cost using the different *rollback region* variations: (a) default LRR, (b) public *rollback region*, (c) public *rollback region* with added information (public branching probabilities, number of private activities inside each fragment, average compensation task cost of private activities inside each fragment). (6) Determine the error rate between (a) and (b) as well as (a) and (c), defined as the difference between the final costs of each respective algorithms.

The evaluation shows that the error rate is positively correlated with the number of private activities: as the number of private activities inside a fragment is scaled up, so does the error rate. Error rate reduction through the *rollback region* variant (c) can be observed. Thus it is possible to choose between variants of the *rollback region* algorithm depending on the readiness of communication overhead for determining dynamic change impact and sensitivity to the error rate.

6 Related Work

The survey presented in [17] sets out the related areas for this work. One dimension is static versus dynamic change and the other dimension process orchestrations versus process choreographies where this work sits at the intersection of dynamic change and process choreography. A plethora of approaches exists for static and dynamic change in process orchestrations [12]. Propagation strategies for process evolution have been at first defined in [2] including abort, flush, and migrate. For an efficient decision on migrating process instances change regions have been proposed in [1]. Depending on the instance state relatively to the change region the possibility to migrate this instance can be quickly made.

---

[2] The implementation can be retrieved under https://github.com/indygemma/rollback-regions.

The most interesting case are instances that are within the change regions - this holds also true for this work. Static change in process choreographies has been investigated by different approaches. The survey in [17] only names DYCHOR [14] and C^3Pro [5] to deal with change propagation, i.e., other approaches focus on structural correctness of the choreography and consistency between public and private processes. So far only [18] has provided a first approach addressing dynamic change in process choreographies according to [17]. Song et al. [18] addresses the problem of migrating instances after a choreography change by proposing strategies for handling the migration in an ordered way, i.e., by avoiding concurrent changes and by a protocol for the instances to accept or decline the migration. As opposed to [18], this work focuses on the instance states and the costs of the migration. Both approaches seem to be complementary.

Several transactional models for processes have been proposed (for an overview see [15]). Particular focus was put on how to deal with long-running transactions (i.e., instances) such as SAGAS [9] and Spheres [11]. Rolling back instances in order to reach a compliant state again was proposed for process orchestrations by [16]. This approach exploits selected ideas from transactional process management and transfers them to the context of dynamic choreography change.

7 Summary

In this work we have defined *rollback regions*, an algorithm to determine the expected cost of a change request in the context of process choreography instances. The algorithm is based on a transactional model that supports compensation tasks, the messaging semantics of interaction activities (sync vs async), as well as the actual states each single business process instance are situated in relation to the *change region*. There are several variations to the algorithm, and the evaluation shows that it is possible to choose the most appropriate one depending on sensitivity to communication overhead as well as error rate. In future work we would like to tackle the data perspective of applying change propagation requests, both in how it affects dynamic change impact analysis and state compliance, as well in the context of the change propagation algorithms themselves. *Rollback regions* can be further extended to support loops as well as studying alternative adaptations (e.g., versioning) to ensure state compliance.

Acknowledgment. This work has been funded by the Vienna Science and Technology Fund (WWTF) through project ICT15-072.

References

1. van der Aalst, W.M.P.: Exterminating the dynamic change bug: a concrete approach to support workflow change. Inf. Syst. Front. **3**(3), 297–317 (2001)
2. Casati, F., Ceri, S., Pernici, B., Pozzi, G.: Workflow evolution. Data Knowl. Eng. **24**(3), 211–238 (1998)

3. Eder, J., Liebhart, W.: Workflow transactions. In: Lawrence, P. (ed.) Workflow Handbook 1997, Handbook of the Workflow Management Coalition (WfMC), pp. 195–202. Wiley, Hoboken (1997)

4. Eshuis, R., Norta, A., Roulaux, R.: Evolving process views. Inf. Softw. Technol. **80**, 20–35 (2016)

5. Fdhila, W., Indiono, C., Rinderle-Ma, S., Reichert, M.: Dealing with change in process choreographies: design and implementation of propagation algorithms. Inf. Syst. **49**, 1–24 (2015)

6. Fdhila, W., Indiono, C., Rinderle-Ma, S., Vetschera, R.: Finding collective decisions: change negotiation in collaborative business processes. In: Debruyne, C., et al. (eds.) OTM 2015. LNCS, vol. 9415, pp. 90–108. Springer, Cham (2015). doi:10.1007/978-3-319-26148-5_6

7. Fdhila, W., Rinderle-Ma, S.: Predicting change propagation impacts in collaborative business processes. In: Symposium on Applied Computing, pp. 1378–1385 (2014)

8. Fdhila, W., Rinderle-Ma, S., Indiono, C.: Change propagation analysis and prediction in process choreographies. Int. J. Cooperative Inf. Syst. **24**(3), 1541003 (2015)

9. Garcia-Molina, H., Salem, K.: Sagas. In: Special Interest Group on Management of Data, pp. 249–259 (1987)

10. Grefen, P., Rinderle-Ma, S., Dustdar, S., Fdhila, W., Mendling, J., Schulte, S.: Charting process-based collaboration support in agile business networks. IEEE Internet Comput. **PP**, 1 (2017)

11. Guabtni, A., Charoy, F., Godart, C.: Spheres of isolation: adaptation of isolation levels to transactional workflow. In: van der Aalst, W.M.P., Benatallah, B., Casati, F., Curbera, F. (eds.) BPM 2005. LNCS, vol. 3649, pp. 458–463. Springer, Heidelberg (2005). doi:10.1007/11538394_40

12. Reichert, M., Weber, B.: Enabling Flexibility in Process-Aware Information Systems - Challenges, Methods Technologies. Springer, Heidelberg (2012)

13. Rinderle, S., Reichert, M., Dadam, P.: Correctness criteria for dynamic changes in workflow systems: a survey. Data Knowl. Eng. **50**(1), 9–34 (2004)

14. Rinderle, S., Wombacher, A., Reichert, M.: Evolution of process choreographies in DYCHOR. In: Meersman, R., Tari, Z. (eds.) OTM 2006. LNCS, vol. 4275, pp. 273–290. Springer, Heidelberg (2006). doi:10.1007/11914853_17

15. Rinderle-Ma, S., Grefen, P.: Towards flexibility in transactional service compositions. In: International Conference on Web Services, pp. 479–486 (2014)

16. Sadiq, S.W.: Handling dynamic schema change in process models. In: Australasian Database Conference, pp. 120–126 (2000)

17. Song, W., Jacobsen, H.A.: Static and dynamic process change. IEEE Trans. Serv. Comput. (2015)

18. Song, W., Zhang, G., Zou, Y., Yang, Q., Ma, X.: Towards dynamic evolution of service choreographies. In: Asia-Pacific Services Computing Conference, pp. 225–232 (2012)

19. Vanhatalo, J., Völzer, H., Koehler, J.: The refined process structure tree. In: Dumas, M., Reichert, M., Shan, M.-C. (eds.) BPM 2008. LNCS, vol. 5240, pp. 100–115. Springer, Heidelberg (2008). doi:10.1007/978-3-540-85758-7_10

20. Wieringa, R.: Design Science Methodology for Information Systems and Software Engineering. Springer, Heidelberg (2014)

A Service-Oriented Architecture Design of Decision-Aware Information Systems: Decision as a Service
(Short Paper)

Faruk Hasić[1(✉)], Johannes De Smedt[2], and Jan Vanthienen[1]

[1] Leuven Institute for Research on Information Systems (LIRIS), KU Leuven,
Leuven, Belgium
{faruk.hasic,jan.vanthienen}@kuleuven.be
[2] Management Science and Business Economics Group,
University of Edinburgh Business School, Edinburgh, Scotland
johannes.desmedt@ed.ac.uk

Abstract. Separating the decision modelling concern from the processes modelling concern has gained significant support in literature over the past few years, as incorporating both concerns into a single model has shown to impair the scalability, maintainability, flexibility and understandability of both processes and decisions. Most notably the introduction of the Decision Model and Notation (DMN) standard by the Object Management Group has provided a suitable solution for externalising decisions from processes. This paper introduces a systematic way of tackling the separation of the decision modelling concern from process modelling by providing a Decision as a Service (DaaS) layered Service-Oriented Architecture (SOA) which approaches decisions as externalised services that processes need to invoke on demand in order to obtain the decision outcome. Additionally, the benefits of the DaaS design on process-decision modelling are discussed in terms of scalability, maintainability, flexibility and understandability.

Keywords: Decision modelling · DMN · Process modelling · Integrated modelling · Separation of Concerns · Service-Oriented Architecture

1 Introduction

Business Process Management (BPM) moves towards accommodating decision management into the paradigms of *Separation of Concerns (SoC)* [1–4] and *Service-Oriented Architecture (SOA)*, by encapsulating decisions into separate models, hence viewing decisions as externalised services. This externalisation of decisions provides a plethora of advantages regarding understandability, maintainability, and flexibility of both the process and decision models [1–4]. A recent decision modelling standard, the Decision Model and Notation (DMN) [5], has

H. Panetto et al. (Eds.): OTM 2017 Conferences, Part I, LNCS 10573, pp. 353–361, 2017. .
https://doi.org/10.1007/978-3-319-69462-7_23

enjoyed significant interest in SoC literature [3,4,6,7]. Externalising decisions and setting them up as services is not the first example of adapting the SoC and SOA paradigms in the domain of BPM, as it was already applied for the interaction between rules and processes [8]. With DMN, a comparable approach can be adapted to decisions by implementing decisions as externalised services, which we call **Decision as a Service (DaaS)**. Processes, or other concerns, can invoke those decision services on demand by providing the relevant input data to the service through an interface. We call this **Decision on Demand (DoD)**. This paper proposes a **Decision as a Service (DaaS)** design for decision-aware information systems. Additionally, the implications of the DaaS design on process and decision modelling are conferred in terms of scalability, maintainability, flexibility and understandability of both the processes and the decisions.

This paper is structured as follows. Section 2 constitutes a related work section. In Sect. 3 the SOA design is established and formalised. Section 4 discusses the implications of the DaaS design in terms of advantages and disadvantages. Finally, Sect. 5 concludes.

2   Motivation and Related Work

The *SoC* paradigm is already well-established in the software modelling and design domains [1]. With the introduction of DMN, the paradigm is introduced in the BPM domain as well, effectively shifting the domain towards a *SOA*, by representing decisions as externalised services. Most modelling approaches in literature still breach the SoC between process control flow on the one hand, and data and decision aspects on the other. Consequently, issues concerning maintainability, scalability, reusability, and understandability arise [1,3,4], given the fact that decisions are hard-coded within the process flow. These issues can be avoided by externalising decision constructs to a dedicated decision model and setting decisions up as a service to be invoked by the process. Separating multi-perspective modelling tasks proves to be beneficial in multiple ways, as long as the separation and interaction between the models is conducted in a sound and consistent way [3,4]. The interaction between business rules and processes was already subject to investigation [8] Some literature on decision service platforms exists, but most works consider simple decisions and business rules, rather than holistic and intertwined decision models [9–11]. Besides, a sound and straightforward framework on how to organise decision services for processes has not been proposed yet.

3   Decision as a Service and Decision on Demand

Separating the decision modelling concern from the process modelling concern implies modelling in two separate models or layers [4]. In Fig. 1, the **DaaS** layered architecture is presented. The bottom layer depicts the *process layer*, while at the top the *decision layer* is represented. In the service-oriented approaches, the services are implemented offering a single decoupled point of entry from the top layers to the services. That way, the bottom layer, i.e. the process layer, only needs the information regarding the point of entry, or more specifically the *interface*, in order to invoke the higher level layers. This single point of entry provides a plethora of advantages, such as flexibility, maintainability, automation and scalability [4,11]. In this section, we adhere to the definitions provided in [7] and extend them to present a more profound formalisation. A decision model can be defined as follows:.

Definition 1. *A decision requirement diagram DRD is a tuple (D_{dm}, ID, IR) consisting of a finite non-empty set of decision nodes D_{dm}, a finite non-empty set of input data nodes ID, and a finite non-empty set of directed edges IR representing the information requirements such that $IR \subseteq (D_{dm} \cup ID) \times D_{dm}$, and $(D_{dm} \cup ID, IR)$ is a directed acyclic graph (DAG).*

According to the DMN standard, a decision requirement graph can be an incomplete or partial representation of the decision requirements in a decision model. The set of all DRDs in the decision model constitutes the exhaustive set of requirements, i.e. the Decision Requirement Graph (DRG).

Definition 2. *A decision requirement diagram DRD is a decision requirement graph DRG if and only if for every decision in the diagram all its modeled requirements, present in at least one diagram, are also represented in the diagram.*

According to the DMN specification, the term *decision* is the logic used to determine an output from a given input.

Definition 3. *A decision $d \in D_{dm}$ is a tuple (I_d, O_d, L), where $I \subseteq ID$ is a set of input symbols, O a set of output symbols and L the decision logic defining the relation between symbols in I_d and symbols in O_d.*

In Fig. 1 the *service layer* is implemented as the connection between the *process layer* and the *decision layer*. The communication between the *process layer* and the *service layer* is bridged by the *interface*. Consequently, the processes are only aware of the *interface* and agnostic about the underlying *service layer* and *decision layer*. Thus, to invoke the services and the decisions the processes simply need to keep information regarding the *interface* and not regarding the higher level layers. To formally define the interface and the decision services we first need to define the input requirement set of a decision as follows:

Definition 4. *The decision input requirement set $dirs_D$ of a decision D is the set of all sets of input data which are sufficient to invoke D. $dirs_D$ contains sets of input data directly or indirectly required by D. The largest set in $dirs_D$ is the set of all input data nodes for which there exists a path to D in DRG_D. The smallest set in $dirs_D$ is D's input set I_D.*

dirs_D is constructed inductively by the following rules:

- $I_D \in dirs_D$
- *For all $s \in dirs_D$ if there is an $i \in s$ such that $i \in O'_D$ for some D' in DRG_D, then $s \setminus \{i\} \cup I'_D \in dirs_D$.*

Each decision in a DRD has its own output set, as formalised in Definition 3. As seen in Fig. 1, a decision service is used to invoke a decision from the decision model.

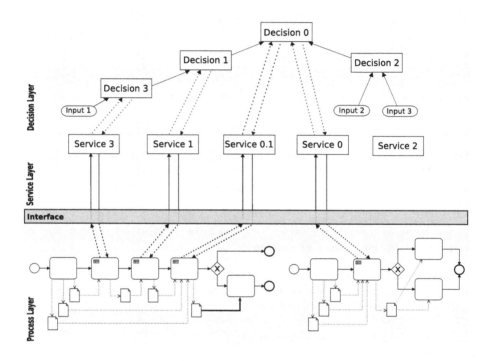

Fig. 1. Decision as a Service (DaaS) layered architecture

Definition 5. *A decision service DS_D of a decision D is a tuple (s_D, O_D), where $s_D \in dirs_D$ is a set of input data sufficient to invoke the decision D and O_D the output set representing the decision outcomes of D.*

Note that multiple decision services can be defined for a single Decision D, depending on which input data set s_D is used to access the decision layer. Consider the decision model in the decision layer of Fig. 1. For Decision D_1 two decision services can be defined: DS_{D1} with tuple (O_{D3}, O_{D1}) where O_{D3} is the output of Subdecision D_3, which serves as the input for D_1; and $DS_{D1.1}$ with tuple (id_3, O_{D1}) where id_3 is the input data object required for the decision enactment of D_3 and consequently D_1. Which decision service will be activated depends on the input data set provided through the interface by the process in the process layer. Hence, the interface will steer the information towards the suitable service that is able to invoke the required decision based on the input data set received from the process. Thus, a decision service's interface is the combination of its input requirement set and its output set:

Definition 6. *The interface* IF_D *of a decision service* DS_D *is defined as a tuple* $(dirs_D, O_D)$, *where* $dirs_D$ *is the input requirement set and* O_D *the output set of the underlying decision* D.

In summation, a process can call upon a decision by providing a data input set that is required for the enactment of the decision to the decision interface. The interface will steer the information provided by the process towards the suitable decision service. Subsequently, the decision service will invoke the requested decision from the decision model and the decision model will enact the decision, reach an outcome, and output it to the decision service. The decision service will forward the outcome back to the process layer through the interface. This mechanism is illustrated in Fig. 1. The SOA design provided in the figure makes it possible to define decisions as services, which we call **Decision as a Service (DaaS)**. These decisions and services can be invoked on demand by information systems, e.g. processes, through a well-defined interface; we call this **Decision on Demand (DoD)**. With the DaaS architecture provided in Fig. 1, the Separation of Concerns paradigm between processes and decisions is induced by design, as the process is included in the bottom layer, while the decision model is situated in the top layer.

4   Implications of DaaS for Process-Decision Modelling

The advantages of the separation of modelling concerns are emphasised in a vast array of papers [1–4]. They especially highlight the scalability, the maintainability, the flexibility, and the understandability of decisions and processes. However, they do not provide a clear design or framework on how to systematically address these issues and how to guarantee a sound SoC by design. In this section we will discuss these alleged advantages of separating the process and decision modelling concerns and relate them to the DaaS/DoD design.

4.1   Scalability

A straightforward advantage of separating the process and decision modelling concerns is scalability. Given the fact that, when separating concerns, the decision logic is not modelled together with the process, the decision logic in the decision model can be concurrently invoked by many clients. This promotes the reusability of the decisions and the underlying logic and creates scale advantages.

In the DaaS design in Fig. 1, it is clear that multiple processes from the process layer can call upon a number of services simultaneously and that all the services can access the decision model at the same time. In cases where the modelling concerns are not separated, e.g. a typical way of modelling the process and decision concerns in one model is using an intricate setup of gateways, the decisions and the decision logic are embedded in the process, thus providing no opportunities for the reuse of logic and decisions and no possibilities for parallel invocation and enactment of the decisions. The DaaS design in Fig. 1 clearly avoids these convoluted situations, as the decision logic is stored in the decision model as part of the decision layer, while the control flow is part of the process layer. The two do not convolute each other, however, the decisions can easily be invoked by any process as long as the process is able to provide the right input set to the interface of the decision service. That way, multiple decisions can access the decision model concurrently and make use of the decision logic in parallel. Hence, the DaaS design provides the suitable circumstances for the scalability of decision enactment and logic reuse.

4.2   Maintainability

Another clear advantage of separation of concerns is model maintainability. Take for example the situation of a convoluted process-decision model where decisions are hard-coded into gateways and the flow of the process: if either the process or a decision changes, the convoluted model needs to be adapted. This creates a need for constant maintenance of the convoluted process-decision model.

In the DaaS design in Fig. 1, the decisions and the processes are not convoluted as they are separated into different modules or their respective layers. If parts of a process change in the process layer, the decisions in the decision layer are not affected. This corresponds to a decision-first approach, as opposed to the process-first approach that is present a convoluted process-decision model with cascading gateways. The maintenance of the process does not affect the underlying decisions. On the other hand, if parts of the decision model change, some services and processes will need to adapt to the changes in order to be able to invoke said decisions correctly. However, not all the services and processes will need to undergo adaptations; only the ones that are directly affected by the decisions that were modified. Other services and processes that do not pertain to the adjusted decision will still function properly without any need for adaptation. Noteworthy is that in the case of the DaaS design, where the modelling concerns are separated, all the decision logic is concentrated in the decision model as part of the decision layer. If anything concerning the decisions needs to change, the adaptation happens only in one place, i.e. the decision model. However, if decisions change in the process-first approach, every process containing those decisions will need adaptations as well. Hence, the DaaS design improves the maintainability of both processes and decisions.

4.3   Flexibility

Flexibility refers to the ad-hoc reuse of decisions and decision logic, as well as the flexibility in adapting and changing the underlying decision logic. Clearly, the process-first approach where modelling concerns are not separated does not support flexibility in any way: no reuse of logic is possible, since the logic is embedded in the control flow of the process and not stored in a separate module; and adapting the decision logic is rather cumbersome, since the logic is dispersed across multiple processes and hidden in convoluted process paths. Therefore, the process-first approach shows little flexibility in general.

In the DaaS design in Fig. 1, the concerns are separated into their respective layers, and as explained earlier, the separated decision logic can be invoked ad-hoc by any client from the process layer conforming to the interface needed to invoke a decision service and consequently the necessary decision. Besides, changing the underlying decision logic is less of a burden, since the logic is concentrated in a single model, rather than dispersed across a process or even across multiple distributed or collaborating processes. Hence, the DaaS design offers a clear separation of concerns and thus a higher level of flexibility, both in terms of reusability of decisions and decision logic, as well as in terms of flexibility in logic adaptation.

4.4 Complexity and Understandability

The complexity and understandability of models is of particular interest in modelling endeavours. In the approach where modelling concerns are not separated the process quickly becomes overly complicated due to the cascading gateways. This is especially the case in knowledge-intensive processes where a lot of decisions need to be made based on a certain underlying logic. Often the term *spaghetti-like processes* is used to refer to this phenomenon of intricate and convoluted control flows.

When opting for a DaaS design, the decisions and the decision logic are externalised and encapsulated in a separate layer as part of the decision model. Therefore, the control flow of the process is alleviated from the burden of representing the different decision paths. As a consequence, the actual process becomes visible and more understandable. However, because of the fact that decisions have been externalised, the process is still burdened with data management and data propagation, i.e. the process is responsible for the collection and propagation of the data needed for the invocation of a decision service. The decision model does not concern itself with these issues of data propagation and data management. It simply transforms the input data, obtained from the process through the decision service, into a decision output which is sent back to the invoking process. Hence, in the DaaS design, process complexity in terms of control flow will decrease, however, the complexity of data management is likely to increase. Thus data management and data propagation within processes becomes of paramount importance. Besides, when separating the decisions from the process, the overall view of the entire problem might be clouded, as processes take abstraction of decisions and simply approach decisions as a black box that answers to input, without concerning themselves with the underlying decision logic. Therefore, when applying the DaaS design, the emphasis should be put on the decisions and on a decision-first approach. The process might take abstractions from

decisions and conceive them as a black box, however, everything stands or falls
with the correct definition of the decisions, as both the decision services and the
processes need to heavily rely on the decisions to function properly.

5 Conclusion and Future Work

This paper contributes a SOA design for decision-aware information systems,
enhancing the modelling of decisions and processes according to the SoC paradigm. The proposed **Decision as a Service (DaaS)** and **Decision on
Demand (DoD)** architecture improves integrated process-decision modelling
in terms of scalability, maintainability, flexibility and understandability of the
decisions.

In future work, we will investigate how the proposed architecture can aid in
solidifying the Separation of Concerns in the modelling and mining of integrated
decisions and processes. Furthermore, a guidelines framework will be set up
for integrated process-decision modelling [12]. Finally, the integration between
declarative processes and decisions will be evaluated as well.

References

1. Gordijn, J., Akkermans, H., van Vliet, H.: Business modelling is not process modelling. In: Liddle, S.W., Mayr, H.C., Thalheim, B. (eds.) ER 2000. LNCS, vol. 1921,
   pp. 40–51. Springer, Heidelberg (2000). doi:10.1007/3-540-45394-6_5
2. Vanthienen, J., Caron, F., De Smedt, J.: Business rules, decisions and processes:
   five reflections upon living apart together. In: Proceedings SIGBPS Workshop on
   Business Processes and Services (BPS 2013), pp. 76–81 (2013)
3. Janssens, L., Bazhenova, E., Smedt, J.D., Vanthienen, J., Denecker, M.: Consistent
   integration of decision (DMN) and process (BPMN) models. In: CEUR Workshop
   Proceedings of CAiSE Forum, vol. 1612, pp. 121–128. CEUR-WS.org (2016)
4. Hasić, F., Devadder, L., Dochez, M., Hanot, J., De Smedt, J., Vanthienen, J.: Challenges in refactoring processes to include decision modelling. In: Business Process
   Management Workshops. LNBIP, Springer (2017)
5. OMG: Decision Model and Notation 1.1 (2016)
6. Hasić, F., De Smedt, J., Vanthienen, J.: Towards assessing the theoretical complexity of the decision model and notation (dmn). In: Enterprise, Business-Process
   and Information Systems Modeling. Springer International Publishing (2017)
7. De Smedt, J., Hasić, F., vanden Broucke, S.K.L.M., Vanthienen, J.: Towards a
   holistic discovery of decisions in process-aware information systems. In: Carmona,
   J., Engels, G., Kumar, A. (eds.) BPM 2017. LNCS, vol. 10445, pp. 183–199.
   Springer, Cham (2017). doi:10.1007/978-3-319-65000-5_11
8. Goedertier, S., Vanthienen, J.: Compliant and flexible business processes with business rules. In: Proceedings of the CAISE Workshop on Business Process Modelling,
   Development, and Support BPMDS (2006)
9. Zarghami, A., Sapkota, B., Eslami, M.Z., van Sinderen, M.: Decision as a service:
   separating decision-making from application process logic. In: EDOC, pp. 103–112.
   IEEE Computer Society (2012)

10. Bock, A., Kattenstroth, H., Overbeek, S.: Towards a modeling method for support-
    ing the management of organizational decision processes. In: Modellierung. LNI,
    vol. 225, pp. 49–64. GI (2014)
11. Mircea, M., Ghilic-Micu, B., Stoica, M.: An agile architecture framework that
    leverages the strengths of business intelligence, decision management and service
    orientation. In: Business Intelligence-Solution for Business Development (2011)
12. Hasić, F., De Smedt, J., Vanthienen, J.: An Illustration of Five Principles for
    Integrated Process and Decision Modelling (5PDM). Technical report, KU Leuven
    (2017)

Advancing Open Innovation Capabilities Through a Flexible Integration of ICT Tools

(Short Paper)

Emmanuel Adamides and Nikos Karacapilidis[✉]

Industrial Management and Information Systems Lab, MEAD,
University of Patras, Patras, Greece
{adamides,karacap}@upatras.gr

Abstract. By adopting a capabilities perspective, we develop a flexible framework of ICTs for supporting an organization's strategy towards Open Innovation (OI). OI is associated with strategic capabilities, as well as with operational ones. ICT support in the strategic capabilities development is related to the cognitive processes of managerial staff to develop the appropriate level of absorptive capacity, whereas ICT as part of operational capabilities aims at enhancing the performance of day-to-day OI activities. Taking this into account, we develop a framework that associates specific ICTs with functionalities needed in the entire OI process. Paying much attention to the issues of collaboration and data analysis, we also comment on the integration of these technologies and their embedment in organizational processes.

Keywords: Collaborative open innovation management · ICT · Capabilities

1 Introduction

Over the last years, Open Innovation (OI) has become an established paradigm of innovation management [7]. The adoption of OI by an organization implies that the innovation management process [26] becomes porous, and ideas, concepts, designs, products, services, etc. flow in and out of its boundaries. Clearly, in large complex organizations, this is accomplished in a complex web of social processes, in which agents of different views, interests, cultures and power status [21], as well as other knowledge sources, usually being situated geographically and contextually at a distance, are part of. It has been argued that an organization that aims at an OI strategy should be able to manage these relationships, processes and agencies effectively, in order to gain from their diversity [23] by developing a set of capabilities for absorbing and assimilating knowledge from different sources and also for disseminating knowledge-containing artifacts of value to its external environment in an efficient and effective manner. These capabilities are associated with the organisation's *absorptive and desorptive capacities*, as well as with the structure of its operational processes [14]. Hence, in order to embrace OI, an organization should develop capabilities at two levels: at the *strategic level*, by preparing its members so that they are able to absorb, utilize, and transform knowledge into value to be disseminated to the society, and at the

H. Panetto et al. (Eds.): OTM 2017 Conferences, Part I, LNCS 10573, pp. 362–369, 2017.
https://doi.org/10.1007/978-3-319-69462-7_24

operational level, for connecting knowledge sources, and for supporting the transfer and transformation of knowledge. Given the complexity of these tasks, ICTs have a very important role to play in the development and appropriation of these capabilities.

The use of ICT in OI has been studied from different perspectives (e.g. [9, 10, 15]) and a number of ICT-based platforms for OI are now available [5]. Nevertheless, only limited research has addressed the role of ICT in the development of the strategic capabilities required for the adoption of OI, neither research has raised the issue of the availability of an *integrated* ICT infrastructure for OI [20]. ICT support at the strategic level of OI aims at enhancing the knowledge and social processes of managers within and at the boundaries of organization. There are a number of ICT tools that can be used outside actual operational processes to develop organizational absorptive (and desorptive) capacity in a cooperative manner. These, together with a bouquet of ICT technologies for supporting operational level capabilities, can provide the basis of a flexible technology infrastructure for OI. In this paper, adopting an *inbound OI perspective* [10], and by juxtaposing capabilities' requirements with technology characteristics, we consider the suitability of specific ICTs for supporting the entire OI process, at both strategic and operational levels. We also discuss technologies for the seamless integration of the abovementioned technologies and their embedment in organizational processes, paying particular attention to issues of collaboration and sophisticated data analysis.

2   OI Strategic Capabilities and ICTs

In an organizational capabilities perspective, strategic level capabilities for OI are linked to the notion of dynamic capabilities [25]. This concerns an organizational ability to innovate by *sensing* the environment, *seizing* opportunities and *transforming* its processes and value offerings. In the *inbound* open innovation approach, the effectiveness of these activities depends on the level of absorptive (and desorptive) capacity (ACAP) of the organization [8]. An organization's ACAP depends on the ACAP of its individual members. ACAP describes an organization's capability to recognize the value of new external information, assimilate it, and apply it to commercial ends. It depends on the richness/diversity of prior tacit and codified knowledge of the organization and is vital for its innovative capacity. In addition, the *complexity* and *centrality* of managers mental models [22] play a significant role on the level of ACAP and hence OI. Complex change- and innovation-related mental models allow firms to notice and respond to a larger number of different stimuli, thus increasing their innovation capacity.

At the strategic level, a knowledge management strategy for OI that aims at the efficient and effective creation of knowledge from different intra- and inter-organizational sources, as well as on augmenting learning capacity, should be primarily aiming at the use of ICT for the development of social capital, rather than on the installation of technology systems for the storage and distribution of codified knowledge. This suggests three main areas of intervention: intervening on the way mental model characteristics influence the accumulation of learning capacity, improving managerial participation in the development of innovation strategies and in the deployment of required operational processes, and improving the processes, methods and tools for assessing the future environment. In the following section we briefly present technologies for communicating knowledge, for exploiting existing knowledge and for facilitating the building of relations among different stakeholders in the context of their use in specific knowledge management processes. In particular, three knowledge-management-related practices for developing strategic OI capabilities for scanning, seizing and transforming knowledge with their associated processes and ICT toolsets are *boundary spanners* and *boundary objects*, *collaboration supporting methods* and *systems*, and *participative scenario planning* [2].

Boundary Spanners and Boundary Objects can be used for creating "hotspots" of creativity and innovation [13] and for increasing the complexity of managers' mental models. Boundary spanners are organization members assigned the specific role of facilitating the collaboration of different organizational and inter-organisational entities. They coordinate activities across and on the edge of boundaries where they cross or overlap. Boundary spanners can be complemented by ICT boundary objects (e.g. process maps, strategy maps, process and system dynamics simulation models, collaboratively manipulated/drawn 2D and 3D objects, as well as computer-based learning environments), which are artefacts used to create shared context among different parties in cross-module activities. Carlile [6] formed a classification of boundary objects into repositories, standardized forms and methods, and objects, models and maps. This last category is the only one that directly supports the transformation of knowledge, in addition to representation and learning, and therefore is suitable for overcoming knowledge boundaries.

As it is indicated in Sect. 4, computer-based collaboration and argumentation systems can actively support OI operational processes. Similar methods and artifacts can also be used at the strategic level to structure *problem-solving/issue-resolving processes* and the related dialoguing. The methods and systems developed over the last two decades for supporting collaboration include participative systems methodologies (problem-structuring methodologies, such as SSM, SODA, etc.), methodologies that rely on simulation modeling, methodologies that use collaborative information technology, and information systems supporting collaboration and argumentation [1]. These systems have a significant role to play as far as the democratization of innovation is concerned [27], not only by gathering ideas/concepts from diverse sources but also by implementing democratic rationales in argumentation and conflict resolution [17], thus enlarging the innovation discourse, and facilitating the re-distribution of positional power as well as preventing the abuse of rhetoric power.

Scenario planning is the most frequently used tool of the "prosessualistic", or learning, school of strategic management [24] that have penetrated the field of strategic innovation management [26]. The value of scenario planning does not stem from its outcomes, but rather from the *process* of scenario construction, which through the dynamic interaction between the organization and the environment, stimulates learning by considering possible and "impossible" future events and their consequences. Scenario methods and tools include methods of intuitive logics, which do not apply quantitative assessment and require limited IS support, methods that belong to the area of *Probabilistic Modified Trends* (PMT), and the French method of "La Prospective", supported by a number of software systems [3].

3   OI Operational Capabilities and ICT Requirements

In addition to the strategic OI capabilities associated with the development of an infrastructure for cooperative learning, the development of operational level capabilities is necessary so that OI processes are supported appropriately. After reviewing related literature, Hosseini et al. [14] have concluded that, in addition to *collaboration* (COLB), the most important capabilities for the implementation of OI are: *knowledge exploration* (KXPL), *knowledge exploitation* (KXPT), *knowledge retention* (KRTN), *decision making* (DMKG), *partner relationship management* (PREM), and *social integration* (SINT). Clearly, the importance of these capabilities varies across the spectrum of the different models of OI (*innovation markets, innovation communities, innovation contests, innovation toolkits,* and *social product development forums*), as well as on the stage of the innovation process (idea generation, selection and conceptualization of product/service/process, technical development, product/service launch, improvement of the concept, etc. [26]).

Knowledge exploration (KXPL) is a capability required in all models of OI for the idea generation phase to support the sourcing, internally and externally, of as many as possible and diverse ideas for products, concepts, etc. In addition, in the innovation community model, in the same phase, social integration (SINT) capability (capability to facilitate interaction and coordinate) is required to develop social capital and common understanding. Moreover, in the innovation toolkits model that offers a more restrictive and structured interaction process, closer relationships with external agencies require higher partner relationship management (PREM) capability.

Operational capabilities are reified in organizational processes that are dominated by activities of collaborative *interaction* (between information sources) and *transformation* (of information). In the complex and pluralistic context of modern organizations, efficient deployment of these activities requires the support of ICT. In the following section, we present such technologies (clearly, some of these technologies can be used at the strategic level too), arguing for their integration in the framework of a flexible ICT infrastructure.

4   Technologies for Operational OI Capability Development

As it has been already mentioned, collaboration is the dominant characteristic of OI processes. From the technology point of view, compared to traditional groupware applications for brainstorming, argumentation and group decision support, the diverse types of ICT-based OI systems differ in terms of scale of users and goal of the system, in that they assume a larger scale of users and a narrower goal [18]. This is largely supported by Web 2.0 technologies, which led to new knowledge sharing paradigms due to their inherent user-friendliness, intuitive character and flexibility. At the same time, it has been broadly admitted that knowledge management plays an invaluable role in innovation, in that it fosters a knowledge-driven culture and assists in creating tools, platforms and processes for tacit knowledge creation, sharing and leverage [11]. Recent research suggests that the ICT tools for facilitating such processes should (i) simplify collaboration towards knowledge co-creation, (ii) enable a formal knowledge structure for re-use and reasoning purposes, and (iii) support a sophisticated collection and analysis of associated big data [15]. Below, we present and comment on the main ICT-based practices and associated tools for building this type of OI capabilities.

Web 2.0 collaboration. The emergence of the Web 2.0 era introduced a plethora of collaboration tools, which enable engagement at a massive scale and feature novel paradigms. These tools cover a broad spectrum of needs - ranging from knowledge exchange, sharing and tagging, to social networking, group authoring, mind mapping and discussing - and enable the massive and unconstraint collaboration of users; however, this very feature is the source of the information overload problem. The amount of information produced and exchanged and the number of events generated within these tools exceeds by far the mental abilities of users to: (i) keep pace with the evolution of the collaboration in which they engage, and (ii) keep track of the outcome of past sessions. Obviously, Web 2.0 collaboration support tools can contribute directly to the development and support of processes associated with the collaboration (COLB) capability, and indirectly to those associated with knowledge exploration (KXPL), knowledge exploitation (KXPT), knowledge retention (KRTN), and social integration (SINT). In many cases, through sophisticated mechanisms for user profiling and networking, they can also support the development of the partner relationship management (PREM) capability.

Argumentative knowledge construction. Collaborative knowledge co-creation and learning processes are often based on written argumentative discourse of stakeholders, who discuss their perspectives on a problem with the goal to acquire knowledge. As far as argumentation related processes are concerned, various tools focusing on the sharing and exchange of arguments, diverse knowledge representation issues and visualization of argumentation have been developed. These tools can capture key issues and ideas, and create shared understanding in a team of knowledge workers. In any case, they are standalone applications, lacking support for interoperability and integration with other tools, and cope poorly with voluminous and complex data. Argumentative knowledge construction support tools contribute significantly to the support of knowledge exploration (KXPL), knowledge exploitation (KXPT), and knowledge retention (KRTN) capabilities. However, they have a few shortcomings that make them more suitable for use principally in OI toolkits strategies. Their emphasis on providing fixed and prescribed ways of interaction within collaboration spaces make them difficult to use in more "open" models as they constrain the expressiveness of users.

Informed decision making methods and tools. Data warehouses, on-line analytical processing, and data mining have been broadly recognized as technologies playing a prominent role in the development of current and future Decision Support Systems [16]. They may aid users make better, faster and informed decisions, and consequently develop the desired organizational decision making (DMKG) capability. However, there is still room for further developing the conceptual, methodological and application-oriented aspects of computer-supported decision making. A holistic perspective on computer-supported decision making is still missing, and there is a growing need to develop applications by following a more human-centric (and not problem-centric) view [12]. The related challenges can be addressed by adopting a knowledge-based decision-making view, while also enabling the meaningful accommodation of the results of the social knowledge and related mining processes. According to this view, which builds on bottom-up innovation models, decisions are considered as pieces of knowledge referring to an action commitment.

Social media monitoring and analytics. Social media monitoring and analytics is an evolving marketing research field that refers to the tracking or crawling of various social media content as a way to determine the volume and sentiment of online conversation about a brand or topic [4]. Their added value lies on the fact that such processes can be performed at real time and in a highly scalable way. The utilization of these tools supports the knowledge exploration (KXPL) and decision making (DMKG) capabilities. However, despite their enormous potential, so far, they have not been widely (and meaningfully) adopted, basically due to: (i) the poor quality and reliability of the existing solutions, and (ii) the difficulty of integration into the existing company structures and marketing procedures.

Opinion mining. Opinion mining (or sentiment analysis) tools employ natural language processing, machine learning, text analysis and computational linguistics to extract relevant information from the huge amounts of human communication over the Internet or other (offline) sources. In fact, the propagation of opinionated data has caused the development of web opinion mining [19] as a new concept in Web Intelligence, which deals with the issue of extracting, analyzing and aggregating data about opinions. Opinion mining methods and tools make possible for organizations to reach crowd's opinions about diverse topics of interest. As holds for the previous category, they support the development of the knowledge exploration (KXPL) and decision making (DMKG) capabilities. Generally speaking, traditional opinion mining techniques apply to social media content as well. However, there are certain factors that make Web 2.0 data more complicated and difficult to be parsed.

Integration issues. The vast majority of ICT tools presented above have been originally designed to work as standalone applications. However, in complex contexts such as that of OI processes, these tools need to be integrated and meaningfully orchestrated. In most cases, this is a hard and challenging issue, which depends on many factors, such as the type of the resources to be integrated, performance requirements, data heterogeneity and semantics, user interfaces, and middleware. At the same time, OI stakeholders are confronted with the rapidly growing problem of information overload; they need to efficiently and effectively collaborate and make decisions by appropriately assembling and analyzing enormous volumes of complex multi-faceted data residing in different sources. Admittedly, when things get complex, we need to aggregate big volumes of data, and then mine it for insights that would never emerge from manual inspection or analysis of any single data source. The above requirements can be fully addressed by an innovative web-based platform that ensures the seamless interoperability and integration of diverse components and services. The proposed solution should be able to loosely combine web services to provide an all-inclusive infrastructure ('single-access-point') for the effective and efficient support of diverse OI stakeholders. It will not only provide a working environment for hosting and indexing of services, retrieval and analysis of large-scale data sets; it will also leverage Web technologies and social networking solutions to provide stakeholders with a simple and scalable solution for targeted collaboration, resource discovery and exploitation, in a way that facilitates and boosts OI activities.

5 Conclusions

Based on a capabilities perspective, we discussed ICT as enhancer of strategic- and operational-level processes of OI. ICT support for the two levels differs significantly, as far as the human-machine relationship is concerned. Strategic processes are creative, complex, and highly subjective processes that cannot be automated and objectified. Hence, ICT support is focused on "opening" mental models and on the development of social capital through collaboration. ICT artefacts, such as computer simulation models, just play the role of "transitional objects". On the other hand, machine reasoning and data processing are required in operational processes for supporting the corresponding capabilities. Nevertheless, even at this level, flexibility and "breadth" are more important than rigid integration and "depth", for exploiting ideas from diverse sources and for switching fast between external partners.

References

1. Adamides, E.D., Karacapilidis, N.: Information technology support for the knowledge and social processes of innovation management. Technovation **26**, 50–59 (2006)
2. Adamides, E.D., Pomonis, N.: Modular organizations and strategic flexibility: the mediating role of knowledge management strategy. In: Abou-Zeid, E.-S. (ed.) Knowledge Management and Business Strategies, pp. 108–132. Idea Group Publishing, Hershey (2008)
3. Amer, M., Daim, T., Jetter, A.: A review of scenario planning. Futures **46**, 23–40 (2013)
4. Bekkers, V., Edwards, A., de Kool, D.: Social media monitoring: Responsive governance in the shadow of control? Gov. Inf. Q. **30**, 335–342 (2013)
5. Bieler, D.: The Forrester Wave: Innovation Management Solutions, Q2. Forrester (2016)
6. Carlile, P.R.: A pragmatic view of knowledge and boundaries: Boundary objects in new product development. Organ. Sci. **13**(4), 442–445 (2002)
7. Chesbrough, H.: Open innovation: a new paradigm for understanding industrial innovation. In: Chesbrough, H., Vanhaverbeke, W., West, J. (eds.) Open Innovation: Researching a New Paradigm, pp. 1–12. Oxford University Press, Oxford (2006)
8. Cohen, W.M., Levinthal, D.A.: Absorptive capacity: A new perspective on learning and innovation. Adm. Sci. Q. **35**, 128–152 (1990)
9. Corvello, V., Gitto, D., Carlsson, S., Migliarese, P.: Using information technology to manage diverse knowledge sources in open innovation processes. In: Eriksson, L.J., Wiberg, M., Hrastinski, S., Edenius, M., Ågerfalk, P. (eds.) Managing Open Innovation Technologies. Springer, Heidelberg (2013). doi:10.1007/978-3-642-31650-0_12
10. Cui, T., Ye, H., Teo, H.H., Li, J.: Information technology and open innovation: A strategic alignment perspective. Inf. Manage. **52**, 348–358 (2015)
11. du Plessis, M.: The role of knowledge management in innovation. J. Knowl. Manage. **11**(4), 20–29 (2007)

12. Evangelou, C., Karacapilidis, N.: A Multidisciplinary Approach for Supporting Knowledge-Based Decision Making in Collaborative Settings. Int. J. Artif. Intell. Tools **16**(6), 1069–1092 (2007)
13. Gratton, L.: Cooperation for innovation. In: Huff, A.S., Möslein, K.M., Reichwald, R. (eds.) Leading Open Innovation, pp. 105–116. MIT Press, Cambridge, MA (2013)
14. Hosseini, S., Kees, A., Manderscheid, J., Röglinger, J., Rosemann, M.: What does it take to implement open innovation? Towards an integrated capability framework. Bus. Process Manage. J. **23**(1), 87–107 (2017)
15. Hrastinski, S., Kviselius, N.Z., Ozan, H., Edenius, M.: A review of technologies for open innovation: characteristics and future trends. In: Proceedings of the 43rd Hawaii International Conference on System Sciences, pp. 1–10 (2010)
16. Karacapilidis, N.: An Overview of Future Challenges of Decision Support Technologies. In: Gupta, J., Forgionne, G., Mora, M. (eds.) Intelligent Decision-Making Support Systems: Foundations: Applications and Challenges, pp. 385–399. Springer, London (2006). doi:10.1007/1-84628-231-4_20
17. Karacapilidis, N., Papadias, D.: A group decision and negotiation support system for argumentation based reasoning. In: Antoniou, G., Ghose, A.K., Truszczyński, M. (eds.) PRICAI 1996. LNCS, vol. 1359, pp. 188–205. Springer, Heidelberg (1998). doi:10.1007/3-540-64413-X_36
18. Klein, M., Convertino, G.: A Roadmap for Open Innovation Systems. J. Soc. Media Organ. **2**(1), 1 (2015)
19. Taylor, E.M., Rodríguez, O.C., Velásquez, J.D., Ghosh, G., Banerjee, S.: Web opinion mining and sentimental analysis. In: Velásquez, J., Palade, V., Jain, L. (eds.) Advanced Techniques in Web Intelligence-2. SCI, vol. 452, pp. 105–126. Springer, Heidelberg (2013). doi:10.1007/978-3-642-33326-2_5
20. Malhotra, A., Majchrzak, A.: Managing crowds in innovation chanllenges. Calif. Manage. Rev. **56**(4), 103–123 (2016)
21. Mota Pedrosa, A., Välling, M., Boyd, B.: Knowledge related activities in open innovation: managers' characteristics and practices. Int. J. Technol. Manage. **61**(3/4), 254–273 (2013)
22. Nadkarni, S., Narayanan, V.K.: Validity of the structural properties of text-based causal maps: An empirical assessment. Organ. Res. Meth. **8**(1), 9–40 (2005)
23. Sarker, S., Sarker, S., Sahaym, A., Bjørn-Andersen, N.: Exploring value cocreation in relationships between an ERP vendor and its partners: a revelatory case study. MIS Q. **36**(1), 317–338 (2012)
24. Schoemaker, P.J.H.: Scenario planning: A tool for strategic thinking. Sloan Manage. Rev. **36**, 25–40 (1995)
25. Teece, D., Peteraf, M., Leih, S.: Dynamic capabilities and organizational agility: Risk, uncertainty and strategy in the innovation economy. Calif. Manage. Rev. **58**(4), 13–35 (2016)
26. Tidd, J., Bessant, J.: Strategic Innovation Management. Wiley & Sons, Chichester (2014)
27. von Hippel, E.: Democratizing Innovation. MIT Press, Cambridge (2006)

SmartGC: Online Memory Management Prediction for PaaS Cloud Models

José Simão[1,3(✉)], Sérgio Esteves[1,2], and Luís Veiga[1,2]

[1] INESC-ID Lisboa, Lisbon, Portugal
jsimao@cc.isel.ipl.pt
[2] Instituto Superior Técnico, Universidade de Lisboa, Lisbon, Portugal
sesteves@gsd.inesc-id.pt, luis.veiga@inesc-id.pt
[3] Instituto Superior de Engenharia de Lisboa, Instituto Politécnico de Lisboa,
Lisbon, Portugal

Abstract. In Platform-as-a-Service clouds (public and private) an efficient resource management of several managed runtimes involves limiting the heap size of some VMs so that extra memory can be assigned to higher priority workloads. However, this should not be done in an application-oblivious way because performance degradation must be minimized. Also, each tenant tends to repeat the execution of applications with similar memory-usage patterns, giving opportunity to reuse parameters known to work well for a given workload. This paper presents *SmartGC*, a system to determine, at runtime, the best values for critical heap management parameters of JVMs. *SmartGC* comprises two main phases: (1) a training phase where it collects, with different heap resizing policies, representative execution metrics during the lifespan of a workload; and (2) an execution phase where it matches the execution parameters of new workloads against those of already seen workloads, and enforces the best heap resizing policy. Distinctly from other works, this is done without a previous analysis of unknown workloads. Using representative applications, we show that our approach can lead to memory savings, even when compared with a state-of-the-art virtual machine - OpenJDK. Furthermore, we show that we can predict with high accuracy the best heap policy in a relatively short period of time and with a negligible runtime overhead. Although we focus on the heap resizing, this same approach could also be used to adapt other parameters or even the GC algorithm.

Keywords: Garbage collection · Machine learning · Shared execution environment · Java Virtual Machine

1 Introduction

Managed runtimes, such as the Java Virtual Machine (JVM), have been increasingly used in large-scale deployments and, particularly, in cloud environments [22]. Platform-as-a-Service providers (e.g., Heroku, AppFog and Google

© Springer International Publishing AG 2017
H. Panetto et al. (Eds.): OTM 2017 Conferences, Part I, LNCS 10573, pp. 370–388, 2017.
https://doi.org/10.1007/978-3-319-69462-7_25

App Engine) allow the deployment of workloads on high-level language virtual machines (HLL-VMs) on a multi-tenant environment. In recent years, several middleware frameworks have been developed for systems that target these runtimes. These frameworks cover several areas, such as, graph processing [9] or bio-informatics [12,14]. Furthermore, regardless of the language of choice and the development paradigm (more object-oriented or more functional), in most cases, the provider will run the resulting components in a JVM-compatible runtime, or in a runtime that has similar code generation and memory management challenges, such as the one that supports node.js - the V8 engine.[1]

Although several resources have elasticity, that is, resources can be removed or assigned without breaking the application execution, memory is one with high impact. To take advantage of the resources available in the cluster supporting the cloud, a managed runtime for these environments needs a synergy of mechanisms and allocation strategies. These mechanisms should include local adaptations to the consumption of resources (where memory must have a prominent attention) and ways to infer application progress (or progress rate).

Uninformed consolidation carries a negative impact on the application performance. With clever choices, providers can therefore consolidate more VMs on the same hardware, in terms of memory, and save costs (and reduce prices or increase revenues) without significantly worsen application performance. A policy with high-impact in the performance of managed runtimes is the one that manages memory. It has a dual effect: a direct impact on the memory allocated to the process but also a impact on the progress rate of the application. This opens the possibility to configure critical parameters of the Garbage Collector (GC) based on the application *signature* (i.e. resource usage behaviour profile, or type profile) and the correlation with a set of configurations that are previously known as favoring consolidation. The *signature* can be determined by metrics obtained from the hardware, operating system and the runtime itself. During applications execution, different access and usage patterns of memory and CPU-related structures can be identified. Previous work has reviewed the performance of programs using low-level performance metrics, namely, hardware performance counters, such as cache misses and instructions per cycle [16].

There is however no previous work that uses such information to categorize applications in terms of memory signatures, so that relevant parameters can be dynamically reconfigured. Because workloads exhibit dynamic patterns, the reconfiguration of the heap management policy should be designed as an adaptive process relying on timely collection of performance counters, identification of an application profile and the choice of the best parameters accordingly.

This paper describes *SmartGC*, a system that adapts the heap management policy of managed runtimes based on the identification of an application execution profile – a *signature* – and the relation of this signature with parameters that are known to maximize the execution yield, i.e., the relation between memory usage and execution time of the application. Although the system can classify unseen workloads, the offline signatures dataset can be easily extended with

---

[1] http://cloud.google.com/appengine.

new applications. The main contributions include: (i) a low-overhead machine learning-based classification system, based on a small set of performance counters, collected during the execution of applications that target the Java VM; (ii) an adaptive GC control for multi-tenancy to reduce memory footprint of VMs with small performance impacts, hence improving resource effectiveness and provider revenue. *SmartGC* currently supports alternative heap resizing policies with influence in the allocated memory and application performance.

The paper is organized as follows. Section 2 shows the impact of heap size management policies in memory savings and execution performance, presenting a metric that relates these observations. Section 3 describes the most effective metrics to characterize the execution of an application and how this characterization can be represented in what we call a *signature*. Section 4 describes the design of *SmartGC*, including the structure of a workload *signature* and the two distinct phases of operation. Section 5 discusses the details of techniques used in the classification adopted, and Sect. 6 makes an extensive evaluation of the major components of *SmartGC*, including the comparison with a widely deployed JVM, the OpenJDK. Section 7 makes an analysis to the state of the art and Sect. 8 closes the paper discussing future work.

2   The Case for Execution Yield

SmartGC goal is to reconfigure parameters of managed runtimes, saving memory while minimizing performance penalties. So, in our system we consider the ratio between the savings in memory, as a resource (ΔR), and the degradation in execution performance (ΔP), when compared with a large enough heap with a fixed size. We call this ratio the *execution yield*: $\frac{\Delta R}{\Delta P}$.

The parameters to be changed at runtime are JVM-specific. In this work we focus on experiences with the Jikes RVM [3]. This JVM uses a matrix to control how the heap size is modified as the application progresses. The matrix relates the percentage of live objects and the time used in GC operations to determine whether the heap will grow or shrink to a new size. The default matrix is the first presented in Fig. 1. It relates the time spent in GC activity (versus mutator activity) to the ratio of live objects. For each pair of these values, the heap will grow (positive values) or shrink (negative values) a certain percentage.

We have proposed new alternative matrices, depicted in Fig. 1 [25,26]. These new matrices explore different shrink/grow percentages across an imaginary four-quadrant space. For example, $Q2$ of M_1 is more heap conservative, meaning that, when a small percentage of time is spent on GC and live objects remain below 30%, the heap will decrease between 30% and 45%, while in M_0 the same situation implies a resizing between −10% and +10%.

Previous work typically only takes into account the heap size as seen from inside the virtual machine [7,10]. Since we target consolidated execution environments, we consider a lower-level metric, the proportional set size (PSS). This metric is deeply connected with the effective use of physical memory pages, the actual resource virtual machines compete for in cloud environments. The PSS

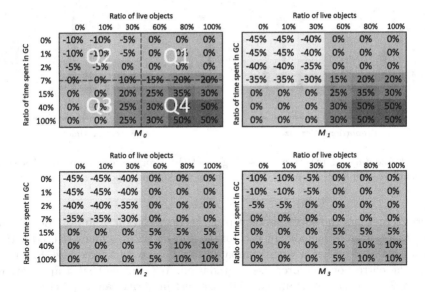

Fig. 1. Matrices options space

is reported by the virtual memory management sub-system. It is described, in Linux virtual memory documentation, as the set of pages a process has in memory, where each shared page is divided by the number of processes sharing it.

To find the best matrix, we executed a *reference set* of applications which use memory with different patterns. The goal is to correlate each of these reference applications to the matrix that maximizes the execution yield. These applications are from the Dacapo benchmark [5]. Each of these applications explore a different issue of the JIT, GC and micro-architecture, as extensively demonstrated in [5]. We expect these benchmarks represent full applications or their phases, and that the execution patterns they have regarding the use of memory are representative of other Java applications.

Figure 2a presents the memory saved by each of the four matrices, which are used within a dynamically-sized heap, when compared to a heap with a fixed size.

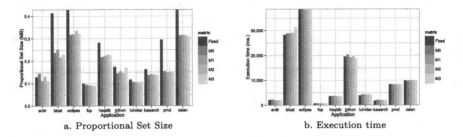

a. Proportional Set Size   b. Execution time

Fig. 2. (a) Proportional set size and (b) Execution time, of several representative applications using different heap management policies

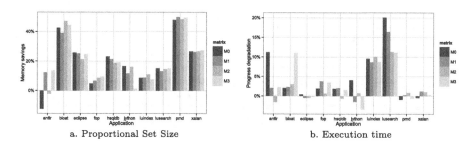

a. Proportional Set Size   b. Execution time

Fig. 3. Percentage of (a) Memory savings and (b) Progress degradation, of several representative applications using different heap management policies

Figure 2b presents the correspondent progress degradation. Figure 3a details the percentages of saved memory by each of the four matrices, which are used within a dynamically-sized heap, when compared to a heap with a fixed size. Figure 3b shows the correspondent progress degradation. We can see that 4 applications can save 20% or more of memory, while the vast majority can save between 10% and 20%. At cloud data centers scale, the potential cumulative savings are very relevant. We also note that, for example, *pmd*, a code analysis workload, can save more than 40%. Regarding progress degradation, it is limited to 20%, and in most cases it is below 5%, almost at the level of variance in execution times of several runs. In some few cases, there are a negative value, which means that the corresponding matrix actually results in a progress gain.

The ratio between saving and degradation results in a different yield for each pair of reference application and matrix. This yield is an opportunity, for the provider, to use tailored memory-saving parameters without imposing a significantly perceivable, or at all, progress penalty in each workload. Larger numbers reveal that the resources saved are several times higher than any imposed penalty. This essentially "releases" resources to other workloads that will be able to make better progress, allowing to effectively channel resources, at each moment, to where they will pay out more efficiently. The mapping between application and best heap resizing matrix is presented next: (**antlr**, M_1), (**bloat**, M_0), (**eclipse**, M_3), (**fop**, M_2), (**hsqldb**, M_0), (**jython**, M_2), (**luindex**, M_2), (**lusearch**, M_3), (**pmd**, M_1), (**xalan**, M_3). This was established by running each application with a fixed heap of 350 MiBytes (a value for which these applications exhibit low number of GCs [29]) and with each of the depicted matrices. Common practices to avoid the non-determinism inherent to the adaptive compiler were used (e.g. replay compilation).[2]

After establishing the relationship between each type of workload and the best matrix, we now need to find a set of execution characteristics to be used during the runtime identification of any given application.

---

[2] http://jikesrvm.org/Experimental+Guidelines.

3   Transparent Profiling of Workloads

There are several runtime characteristics that can be explored to build the profile of an application. Common indicators include: (i) hardware performance counters; (ii) operating system performance counters; (iii) managed runtime specific metrics; (iv) and application specific metrics. From these four indicators, application specific metrics are the less reliable. They can typically be either related to the organization of classes, or to the nature of operations performed (e.g., rate of transactions processed). Collecting these metrics at the application level is a cumbersome task and makes difficult the correlation of memory usage patterns across different applications. Following, we describe the runtime resource indicators used by *SmartGC* (Sect. 3.1) to characterize and cluster different workloads according to their memory usage patterns (Sect. 3.2).

3.1   Performance Counters

Modern CPUs support a large set of performance counters, including instructions per cycle, branch misses and L1 cache misses. Operating systems also report performance related information, such as page faults and context switches.

Using performance counters introduces some difficulties, namely: (i) selecting the appropriate number and types of performance counters for our purposes; (ii) normalizing their values across different workloads.

Regarding the first issue, we must avoid using very processor-specific hardware performance counters so that the profile to be built can be reused with new hardware. When considering Intel's© processor families, the group of *architecture performance events* ensure consistent values across different processor implementations. This group includes counters such as the number of cycles and the last level cache references. Regarding performance events supported by operating systems, it is common that counters such as page faults as exported to be easily consumed in user-space.

Regarding the second issue, normalization of performance counters across different workloads, it is necessary to capture tendencies and perform regression so that workloads can be clustered based not on the magnitude of the PCs values, but on composed relative values (e.g., growth rate).

SmartGC uses several performance counters, collected periodically, mixing both memory-related and computation-related counters, including, computation-related counters (instructions, cpu-cycles, ref-cycles); cache statistics (cache-references, cache-misses); major and minor page faults (major-faults, minor-faults) and translation lookaside buffer statistics (dTLB-stores, dTLB-loads, dTLB-load-misses, dTLB-store-misses).

3.2   Workload Signature and Mnemonics

Workloads are clustered using what we call a *signature*. A *signature* is an aggregated description computed from the considered PCs that identifies a given workload, W, in terms of its resource usage. We assume that performance counters, regardless of their nature, report a single value in each read. So, a signature contains an aggregated value for each performance counter that is computed during a period of time (e.g., lifespan of an application or a shorter period). This aggregated value represents the growth rate of a set of relevant performance counters $(pc_1 \ldots pc_N)$, between time t_i (initial time) and t_f (final time), as depicted in Eq. 1.

$$S_w(t_i, t_f) = Aggr(pc_1, t_i, t_f), \ldots, Aggr(pc_N, t_i, t_f) \qquad (1)$$

Currently, we support 2 different forms for aggregating sequences of performance counters, both taking into account the values that are available at Δt intervals. The first one is the mean of differences, as described in Eq. 2.

$$Diffs(pc_k, t_i, t_f) = \frac{1}{m} \sum_{x=t_i+\Delta t}^{t_f} pc_k(x) - p_k(x-1) \text{ where } k \in [1, N] \qquad (2)$$

The other option is a geometric mean, as described in Eq. 3.

$$GMean(pc_k, t_i, t_f) = \left(\prod_{x=t_i+1}^{t_f} pc_k(x) \right)^{1/(tf-ti)} \text{ where } k \in [1, N] \qquad (3)$$

A *mnemonic* can then be built, associating a signature to the correct best set of parameters, $\{P_1, P_2, \ldots, P_R\}$, for the executing application. Equation 4 represents a mnemonic for application W, where signature S_w is associated to the best set of parameters.

$$M_w = (S_w \to \{P_1, P_2, \ldots, P_R\}) \qquad (4)$$

An example of parameters are the matrices presented in Sect. 2. In this case, a single parameter, i.e., a matrix number, can capture a multi-dimensional associations between ratio of live objects and ratio of time spent in GC operations. The next section will detail the system design of *SmartGC*, including its two main phases – training and runtime operation.

4   System Design

Analytical modelling is a common approach to inform resource-allocation systems [18,24]. Models can be used to predict the impact of management decisions on performance, availability, and/or energy consumption. However, constructing a model of a real system is a complex task. As a consequence, not rarely models are made with over-simplified assumptions so that they can be mathematically manageable. To overcome this, researchers have developed systems for experiment-based management of virtualized data centers [18]. *SmartGC* uses this strategy, and could easily be plugged into such systems. Figure 4 shows the overall system view. *SmartGC* acts in two distinct phases: the *training* phase and the *execution* phase. In some cases, a third phase can be used to test and reinforce the quality of the training.

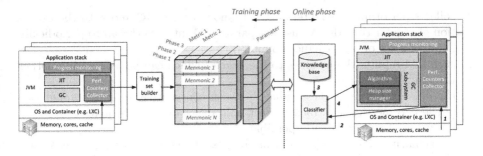

Fig. 4. System design

4.1 Training Phase

In the training phase, a set of representative applications are executed while system metrics are collected (memory and computation-related). This information is aggregated to build a training set, i.e., a set of application *signatures*, where each *signature* is associated with the heap management matrix that maximizes the *execution yield*. During the second phase, also known as online or execution phase, information about the running workload is collected to determine, using the training set, which type of application is being executed. *SmartGC* then changes the relevant parameters of the JVM according to what was predicted to be the best case for the running application, if a best parameter with a distinct confidence level is found. *SmartGC* relies on a GC system that can be instructed to change its parameters during the application runtime.

Instead of a strictly analytic model, *SmartGC* has two challenges for the training phase: (i) determine the *best* memory configuration parameters for a set of representative applications; (ii) use a set of system-related values to characterize the running application.

Regarding the first issue, resource allocation is in most cases a trade-off between two or more variables. Memory management in managed runtime is no exception. The system is designed to act on a given parameter. *SmartGC* explores the heap management policy described in Sect. 2, but structural components, such as the GC algorithm itself, can also be targeted.

The training phase can be periodically repeated when new hardware is acquired or a significant update to the runtime is made. The knowledge base built during this step can also be extended to incorporate more examples of *signatures*, so that online decisions can be made with a higher degree of confidence.

4.2 Execution Phase

After constructing a representative set of mnemonics, *SmartGC* is now ready to be plugged into a resource management middleware. It assumes that the managed runtime exposes mechanisms to dynamically reconfigure relevant systems, in our case, the garbage collector. In the online phase of the system, when a new runtime is started, *SmartGC* performs the following operations:

1. collects performance counters, asynchronously with GC and with the execution of regular threads. We use a separate VM internal thread to periodically monitor a set of performance counters;
2. each VM reports these values to the *oracle* (which can be located in another machine of the cluster);
3. the *oracle* aggregates these values between consecutive samples through one of the two metrics described in Sect. 3.2, and uses this information to find a signature, if that is already possible;
4. if a signature is found, this information is sent to the corresponding heap size manager which changes the GC-related parameter accordingly, and a GC execution is forced.

The *oracle* will make a prediction as good as the size and diversity of the applications used during the training phase. It may be the case that the *oracle* cannot predict a set of parameters with high confidence. In that case, *SmartGC* will use the default parameters and schedule a new training phase for the current application.

4.3   Online Phase

If the matrices confidence level, returned by the classifier for a given workload, are below a certain threshold (e.g., 50%), it means that the signature of that workload is not well known to the predictor. In these cases, we consider to incorporate the new workloads in our learning model, so that we are able to accurately predict the optimal matrix, should the same workload be exposed to the system at future times.

To this end, we: (i) run the (before-unseen) workload with each of the available matrices and assess which one leads to a better execution yield; (ii) include the values of program counters collected from the workload, along with the corresponding optimal matrix, in our knowledge base; and (iii) train a new learning model with an updated training-set containing the new workload footprint. This process is similar to our training phase, but it only considers executions of a single workload/application. Further, the system can be parametrized on whether to train a new workload or to use the matrix yielding the highest confidence (regardless of any defined thresholds).

5   Implementation

To achieve the best reconfiguration, *SmartGC* has to look for key runtime metrics, identify a signature and change the parameters to the appropriate values. This section describes the details of the runtime metrics collected and the machine learning algorithms used to classify a running application.

5.1   Classification Methods and Technologies

The classification process defines how the training set was built and the classification technology used. *SmartGC* classification system is organized around four classes, corresponding to the four matrices described previously. Having a known key set of workloads with the corresponding optimal matrix for each of them, we classify new workload runs, which were never seen before, and select the most appropriate matrices.

The training set is built during the so called training phase, as depicted in Fig. 4. For each execution of an application, we collect 12 performance counters (cf. Sect. 3.1) at every specified time interval (currently 100 ms). After a single run of an application we aggregate the values of the 12 counters that were observed over time thereby averaging the differences, or calculating the geometric mean, between consecutive observed values for every counter (i.e., thus having a single value for each performance counter and application execution). To stress application behavior in terms of counters variance, we run each of the 10 applications 50 times, resulting in 500 training instances. For this phase, we use the same matrix (M_0) for each execution, so that we have the same reference when obtaining counters at runtime. In the training set we also include for each instance of each application the corresponding optimal matrix (which was assessed beforehand in different trials).

We used a polynomial kernel as a parameter of a Support Vector Machine (SVM) classifier. Other kernels revealed to be less accurate. The *oracle* runs an SVM according to the Platt et al. algorithm [20]. To implement the *oracle*, we use the open source Weka framework.[3] Random Forests [8] consist of an ensemble learning, that is, a combination of various learning models to obtain better predictive performance. In this method, a large collection of tree predictors are built based on random subsets sampled from a set of training examples. The generalization capability depends on the strength of individual trees and the correlation between them. With bootstrap aggregating (also known as *bagging*) and a random selection of features to split each node it is possible to control the variance of the trees.

For pattern recognition, a classifier tries to estimate a function that, given a set with N-dimensional input data, predicts which of two possible classes form the output ($f : \mathbf{R}^N \to \pm 1$). It is also possible to classify examples in more than two classes (multiclass classification, as in our case with four classes), by using strategies like the "One-vs-Rest" that compare confidence values between pairs of binary-class classifications. The estimation of the SVM function, which corresponds to the construction of a SVM model, is based on a supplied set of training examples encompassing tuples with known correct values of input and corresponding output (i.e., supervised learning). After the model is constructed, the SVM is then able to assign new unseen examples to one of the possible multiple classes.

---

[3] http://www.cs.waikato.ac.nz/~ml/weka/index.html.

During the execution of an application, the oracle is queried every time interval with an aggregated value of each PC, at each time instance, since the application starts. This aggregated value captures the growth rate of a PC, and corresponds to the mean of the differences between every two consecutive time instances, as discussed in Sect. 3.2.

5.2   Collecting Information

The Jikes RVM can be built with support for reading performance counters.[4] It calls the PerfMon2 native Linux API.[5] In the codebase, performance events can be collected if the VM is executed in harness mode. This incurs in extra overhead so we have avoided some of the core code of the readings, and use a dedicated thread for this monitoring activity, so that the harnessing code path can be avoided. Periodically (default to $100ms$), the JVM collects and reports the performance counters to the oracle (in another physical machine of our testbed cluster) and waits for the classification response. If a new matrix is determined for the current signature, it will be used from then on. In the presence of jittering, we have a configurable threshold of n equal decisions (currently 5) necessary to make the change.

6   Evaluation

This section presents the evaluation of the three key aspects that have major impact in GC and application performance, as well as in the extent of the benefits of our approach. First it discusses the overheads of continuously monitoring the values of selected performance counters. We then focus on the classification process. The choices of a classification technology are presented followed by the confidence levels and stability of the classifier with unseen executions from applications used in the training set, and executions from applications never presented to the system. It concludes with memory savings obtained when compared with a widely used JVM, the OpenJDK.

Our execution environment is based on Jikes RVM 3.1.3, built in `production` mode (both in training and online phase), using the generational version of Immix [6]. We run on Intel Core i7-2600K processors (4 cores with hyper threading, 8M Cache and 3.40 GHz) with 12 GiBytes of RAM. The OS is Ubuntu 12.04.4 with kernel version 3.2.0-58.

We report on *SmartGC* overhead when compared with the JikeRVM 3.1.3 codebase (hereafter named as *baseline*). For each benchmark, both the *baseline* and *SmartGC* were executed 5 times to determine the run with the smallest execution. To measure this we used the record and replay infrastructure of Jikes RVM.[6] The advice files from the best run were then used in replay compilation. Each benchmark was executed 10 times, and the execution time was collected

---

[4] Using `--with-perfevents` parameter in `buildit` script.

[5] http://perfmon2.sourceforge.net/.

[6] http://jikesrvm.org/Experimental+Guidelines.

Table 1. Negligible impact on periodic collection of 12 performance counters. Mean execution time with confidence interval of 95%. Δ_{Mean} is $(SmartGC/Baseline) - 1$

	Base (ms)	CI (%)	*SmartGC* (ms)	CI (%)	Δ_{Mean}
xalan	9760.30	±0.04%	9863.7	±0.03%	1.06%
lusearch	1867.9	±0.04%	1887.6	±0.04%	1.05%
luindex	4304.8	±0.02%	4324.2	±0.01%	0.45%
fop	766.3	±0.12%	765.8	±0.04%	−0.07%
jython	19871.4	±0.08%	20030.9	±0.13%	0.80%
pmd	8459.7	±0.08%	8461.6	±0.05%	0.02%
bloat	28454.1	±0.24%	28397.0	±0.22%	−0.20%
antlr	2142.4	±0.09%	2263.1	±0.12%	5.63%
hsqldb	3646.1	±0.04%	3777.6	±0.03%	3.61%
eclipse	38110.8	±0.03%	38106.4	±0.02%	−0.01%

from the second iteration. Table 1 shows the mean of these executions along with the percentage value for a confidence interval of 95%.[7] The last column reports the percentage difference between the means of both systems. This value is very low for most cases, thus revealing that the proposed approach has a negligible impact in the execution time of these applications.

To select a good learning approach for our particular problem, we compared different widely-deployed Machine Learning algorithms thereby using the metrics: (i) accuracy; (ii) precision, proportion of instances that are truly of a class divided by the total instances classified as that class; (iii) recall, proportion of instances classified as a given class divided by the actual total in that class; (iv) F-measure, a weighted average of the precision and recall; and (v) ROC area, performance of the classifier where values approaching 1 mean optimal classifier and 0.5 being comparable to random guessing (Table 2). We used a data set with 10 applications and 500 examples of different executions, where 66% were used as the training set, and the remaining as our test set (while preserving instance order in the split). The parameters of all algorithms were tuned to get the best possible results. We decided to adopt SVM and Random Forest as our classification models to conduct all the experiences in this paper.

We have established that reading and reporting performance counters to the *oracle* (i.e., the online classifier) has a negligible impact on the performance of applications. We have also shown that, based on values collected in the training phase, the classifier which provides the best results is the one based on support vector machines. We now evaluate the *confidence* the *oracle* has that an unseen fragment of execution, either from a known or unknown application, represents one of the previously identified signatures, so that an appropriate heap resizing policy m will provide the best *execution yield* results.

---

[7] Because we have 10 final average samples we used the Students t-distribution to calculate the confidence interval [13].

Table 2. ML algorithms comparison

Algorithm	Accuracy	Precision	Recall	F-measure	ROC area
Bayes	20.58%	0.686	0.206	0.317	0.85
Logistic	52.35%	0.53	0.5224	0.475	0.806
Neuronal	99.41%	0.994	0.994	0.994	0.992
Rnd Forest	97.64%	0.978	0.976	0.976	0.990
J48 Tree	99.41%	0.994	0.994	0.994	0.992
SVM	99.41%	0.994	0.994	0.994	0.998

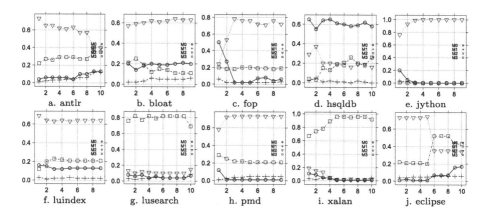

Fig. 5. Evolution of the classifiers confidence level of the unseen executions of 5 iterations of Dacapo applications

Figure 5 shows how the classification confidence of the *oracle* evolves while analyzing unseen performance counters values from the execution of 12 workloads. In each chart, the x-axis is time dependent, with a point for each classification. A classification is made based on the values of performance counters sent from a given JVM.

In the y-axis we plot the confidence value (between 0.0 and 1.0) the classifier reports for each possible matrix. This value is higher than 0.6 and there is always one classification that dominates the remaining ones. Furthermore, this high confidence levels are reached early in the execution, which is of critical importance for adaptiveness, fast change of parameters, and maximizing the overall execution yield.

For the majority of the workloads executions the *oracle* classifies them correctly, choosing the matrix that maximizes the execution yield. In cases where this is not so, like the case of `bloat`, the classifier classifies it with M_2 which is the one that represents the second best execution yield, very close to M_0, and far from M_3 which gives the worst yield for this workload.

Figure 6 also presents the confidence levels for each matrix but when running eight unseen applications, i.e., application which where not used during the

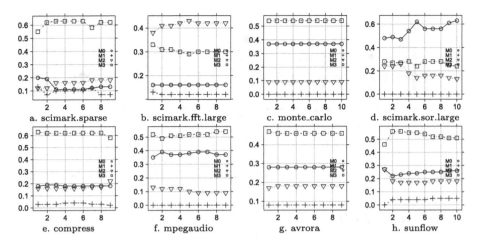

Fig. 6. Evolution of classifiers confidence level with unseen applications from SPECJVM2008 and Dacapo 2009

offline training phase. These applications are a mix of high performance computing workloads from SPECjvm2008[8] and two new benchmarks that where introduced in Dacapo 2009[9]. The results have a similar pattern to the ones discussed regarding Fig. 5, i.e., there is always a dominant matrix chosen by the classifier. Although in same cases the confidence is around 0.5, there is however a clear and stable winning choice during the execution of these unseen applications.

Finally we compare the memory saved when running reference applications with a mainstream OpenJDK 1.8.0, 64 bits, installed from the Ubuntu repository, and with *SmartGC*. OpenJDK was executed in **server** mode, the one that according to the documentation provides the best performance. OpenJDK also has a heap resizing policy, known as *Ergonomics*. This policy takes into account the application throughput, a maximum GC pause time and the minimum heap size.

Figure 7 shows the average memory used by OpenJDK and *SmartGC*. For these results, we assume that *SmartGC* chooses the correct matrix with the highest confidence. Each execution was iterated 3 times to promote the warm-up of the JVM. For 56% of the workloads there is a memory saving, which can reach 23%. This can even be underestimated because JikesRVM uses part of the heap for its own internal structures, given its meta-circular nature. This shows that our approach could be integrated with advantage in a widely deployed JVM, with limited and small changes to its codebase.

---

[8] https://www.spec.org/jvm2008.
[9] http://www.dacapobench.org.

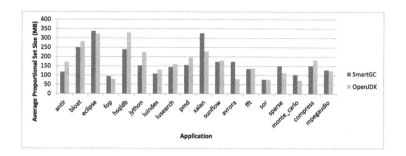

Fig. 7. Average memory used by OpenJDK and *SmartGC*. Lower is better.

7   Related Work

Memory is a relevant resource management target in cloud-like environments, typically using system-level virtual machines as in [2,27]. Recently, the need to involve the application runtime has gained more attention [7,10,22], including solutions by industry players such as VMWare's Elastic Memory for Java - EM4J.[10]

Researchers have analyzed garbage collection performance and found it to be application-dependent [11,30]. Based on these observations, several adaptation strategies have been proposed, ranging from parameters adjustments (e.g. the current size of the managed heap [15]) to changing the algorithm itself before the first execution or at runtime [30]. Other look for ways to minimize the pause time of stop-the-world garbage collectors [19]. This is an orthogonal effort that can be complemented with *SmartGC* operation because these systems are not meant to save memory that can be transferred to other tenants.

The adjustment of GC parameters to a given workload has been a topic of intense research [7,17,31], but most of them look for the parameters that give the highest throughput to a single application, regardless of memory usage. With *SmartGC* we explore a trade-off more relevant in consolidated environments - one that can reduce memory usage of applications without significantly hindering their usage of available CPU, i.e., without incurring in longer execution times.

Bobroff et al. [7] propose to investigate active memory sharing (AMS) in virtual environments. It distributes memory fairly to several running HLL-VMs. Chen et al. study the effect of consolidating HLL-VMs [11] but focus on either CPU or I/O bounded situations, leaving memory management unattained.

Only a few systems use machine learning techniques to learn the program behaviour and change the runtime algorithms or parameters. Andreasson et al. [4] used reinforcement learning techniques, to dynamically learn from GC collections. The system receives good or bad reinforcements by looking at the throughput after a GC collection. Singer et al. [28] determine, before execution, which is the best GC for a given program. It uses a J48 classification tree built from a

---

[10] https://www.vmware.com/support/pubs/vfabric-em4j.html.

long series of offline executions. Experimentations were made based just on full-heap collectors. The classification is done only offline, missing the opportunity for a finer grain adjustment to the different phases that each application might exhibit. Differently from *SmartGC*, these systems do not consider generational collectors with dynamic heap sizes or look at performance counters as a source of information.

Performance counters are typically used to detect bottlenecks and guide optimizations. Xu et al. [21] proposes to detect resource bottlenecks of multi-tier web servers, using low-level performance counters, such as, cache misses and instructions execution rate. Schneider and Gross [23] instrumented a JVM to feed the JIT compiler with performance counters information, so that more decisions can be further tailored to the underlying architecture and not only based on the program behaviour. Adl-Tabatabai et al. [1] uses performance counters to determine where to insert prefetch instructions in a JVM. This is done automatically within the same execution of the JVM. To the best of our knowledge, *SmartGC* is the first system that successfully guides the heap resizing policy of managed runtimes based on signatures learned from the observation of hardware performance counters.

8 Conclusion

We developed a low-overhead machine learning-based application classification that drives an adaptive GC control for cloud multi-tenancy. Our goal is to show that it is possible to choose at runtime, based on previous executions, the best GC parameter for a given application, obtaining reductions in the memory footprint of virtual machines with limited performance impacts, improving resource effectiveness and revenue.

The use of workload-aware policies to distribute resources among different tenants needs specific allocation mechanisms and strategies that can act upon the assets controlled by the managed runtimes, most notably memory. This paper describes the design rationale to build *SmartGC*, a system to guide the management of memory, to be used by PaaS service providers. The goal is to identify an application type, during its execution, and reconfigure relevant memory-related parameters based on the observation of current and performance counter values. We show that this can be done with low-overhead, applying parameters that can save physical memory, with minimum impact on the overall progress of applications. Future work should explore the use of *SmartGC* to target other tradeoffs in resource management, including the energy efficiency of different GC algorithms.

Acknowledgements. This work was supported by national funds through Fundação para a Ciência e a Tecnologia with reference PTDC/EEI-SCR/6945/2014, and by the ERDF through COMPETE 2020 Programme, within project POCI-01-0145-FEDER-016883. This work was partially supported by Instituto Superior de Engenharia de Lisboa and Instituto Politécnico de Lisboa. This work was supported by national funds through Fundação para a Ciência e a Tecnologia (FCT) with reference UID/CEC/50021/2013.

References

1. Adl-Tabatabai, A.R., Hudson, R.L., Serrano, M.J., Subramoney, S.: Prefetch injection based on hardware monitoring and object metadata. In: Proceedings of the ACM SIGPLAN 2004 Conference on Programming Language Design and Implementation, PLDI 2004, pp. 267–276. ACM, New York (2004)
2. Ben-Yehuda, O.A., Posener, E., Ben-Yehuda, M., Schuster, A., Mu'alem, A.: Ginseng: market-driven memory allocation. In: Proceedings of the 10th ACM SIGPLAN/SIGOPS International Conference on Virtual Execution Environments, VEE 2014, pp. 41–52. ACM, New York (2014)
3. Alpern, B., Augart, S., Blackburn, S.M., Butrico, M., Cocchi, A., Cheng, P., Dolby, J., Fink, S., Grove, D., Hind, M., McKinley, K.S., Mergen, M., Moss, J.E.B., Ngo, T., Sarkar, V.: The Jikes research virtual machine project: building an open-source research community. IBM Syst. J. **44**, 399–417 (2005)
4. Andreasson, E., Hoffmann, F., Lindholm, O.: To collect or not to collect? Machine learning for memory management. In: Proceedings of the 2nd Java Virtual Machine Research and Technology Symposium, pp. 27–39. USENIX Association, Berkeley (2002)
5. Blackburn, S.M., Garner, R., Hoffmann, C., Khang, A.M., McKinley, K.S., Bentzur, R., Diwan, A., Feinberg, D., Frampton, D., Guyer, S.Z., Hirzel, M., Hosking, A., Jump, M., Lee, H., Moss, J.E.B., Moss, B., Phansalkar, A., Stefanović, D., VanDrunen, T., von Dincklage, D., Wiedermann, B.: The DaCapo benchmarks: Java benchmarking development and analysis. In: Proceedings of the 21st Annual ACM SIGPLAN Conference on Object-Oriented Programming Systems, Languages, and Applications, OOPSLA 2006, pp. 169–190. ACM, New York (2006)
6. Blackburn, S.M., McKinley, K.S.: Immix: a mark-region garbage collector with space efficiency, fast collection, and mutator performance. In: Proceedings of the 2008 ACM SIGPLAN Conference on Programming Language Design and Implementation, PLDI 2008, pp. 22–32. ACM, New York (2008)
7. Bobroff, N., Westerink, P., Fong, L.: Active control of memory for Java virtual machines and applications. In: 11th International Conference on Autonomic Computing (ICAC 2014), pp. 97–103. USENIX Association, Philadelphia, June 2014
8. Breiman, L.: Random forests. Mach. Learn. **45**(1), 5–32 (2001)
9. Bu, Y., Borkar, V., Xu, G., Carey, M.J.: A bloat-aware design for big data applications. In: Proceedings of the 2013 International Symposium on Memory Management, ISMM 2013, pp. 119–130. ACM, New York (2013)
10. Cameron, C., Singer, J.: We are all economists now: economic utility for multiple heap sizing. In: Proceeding of Implementation, Compilation, Optimization of OO Languages, Programs and Systems (ICOOOLPS) (2014)
11. Chen, L., Serazzi, G., Ansaloni, D., Smirni, E., Binder, W.: What to expect when you are consolidating: effective prediction models of application performance on multicores. Cluster Comput. **17**(1), 19–37 (2014)
12. Francisco, A., Vaz, C., Monteiro, P., Melo-Cristino, J., Ramirez, M., Carrico, J.: Phyloviz: phylogenetic inference and data visualization for sequence based typing methods. BMC Bioinform. **13**(1), 87 (2012)
13. Georges, A., Buytaert, D., Eeckhout, L.: Statistically rigorous Java performance evaluation. In: Proceedings of the 22nd Annual ACM SIGPLAN Conference on Object-Oriented Programming Systems and Applications, OOPSLA 2007, pp. 57–76. ACM, New York (2007)

14. Gront, D., Kolinski, A.: Utility library for structural bioinformatics. Bioinformatics **24**(4), 584–585 (2008)
15. Guan, X., Srisa-an, W., Jia, C.: Investigating the effects of using different nursery sizing policies on performance. In: Proceedings of the 2009 International Symposium on Memory Management, ISMM 2009, pp. 59–68. ACM, New York (2009). http://doi.acm.org/10.1145/1542431.1542441
16. Hauswirth, M., Sweeney, P.F., Diwan, A.: Temporal vertical profiling. Softw. Pract. Exp. **40**(8), 627–654 (2010)
17. Hertz, M., Kane, S., Keudel, E., Bai, T., Ding, C., Gu, X., Bard, J.E.: Waste not, want not: resource-based garbage collection in a shared environment. In: Proceedings of the International Symposium on Memory Management, ISMM 2011, pp. 65–76. ACM, New York (2011)
18. Janakiraman, G.J., Santos, J.R., Turner, Y.: JustRunit: experiment-based management of virtualized data centers. In: Voelker, G.M., Wolman, A. (eds.) 2009 USENIX Annual Technical Conference, San Diego, CA, USA. USENIX Association, 14–19 June 2009
19. Maas, M., Asanović, K., Harris, T., Kubiatowicz, J.: Taurus: a holistic language runtime system for coordinating distributed managed-language applications. In: Proceedings of the Twenty-First International Conference on Architectural Support for Programming Languages and Operating Systems, ASPLOS 2016, pp. 457–471. ACM, New York (2016)
20. Platt, J.: Fast training of support vector machines using sequential minimal optimization. In: Schoelkopf, B., Burges, C., Smola, A. (eds.) Advances in Kernel Methods - Support Vector Learning. MIT Press, Cambridge (1998)
21. Rao, J., Xu, C.Z.: Online capacity identification of multitier websites using hardware performance counters. IEEE Trans. Parallel Distrib. Syst. **22**(3), 426–438 (2011)
22. Salomie, T.I., Alonso, G., Roscoe, T., Elphinstone, K.: Application level ballooning for efficient server consolidation. In: Proceedings of the 8th ACM European Conference on Computer Systems, EuroSys 2013, pp. 337–350. ACM, New York (2013)
23. Schneider, F.T., Payer, M., Gross, T.R.: Online optimizations driven by hardware performance monitoring. In: Proceedings of the 2007 ACM SIGPLAN Conference on Programming Language Design and Implementation, pp. 373–382 (2007)
24. Sharifi, L., Rameshan, N., Freitag, F., Veiga, L.: Energy efficiency dilemma: P2P-cloud vs. datacenter. In: IEEE 6th International Conference on Cloud Computing Technology and Science, CloudCom 2014, Singapore, pp. 611–619. IEEE, 15–18 December 2014
25. Simão, J., Veiga, L.: QoE-JVM: an adaptive and resource-aware Java runtime for cloud computing. In: Meersman, R., et al. (eds.) OTM 2012. LNCS, vol. 7566, pp. 566–583. Springer, Heidelberg (2012). doi:10.1007/978-3-642-33615-7_8
26. Simão, J., Veiga, L.: Adaptability driven by quality of execution in high level virtual machines for shared cloud environments. Int. J. Comput. Syst. Sci. Eng. **29**(6), 413–426 (2013)
27. Simão, J., Veiga, L.: A taxonomy of adaptive resource management mechanisms in virtual machines: recent progress and challenges. In: Antonopoulos, N., Gillam, L. (eds.) Cloud Computing, pp. 59–98. Springer, Cham (2017). doi:10.1007/978-3-319-54645-2_3
28. Singer, J., Brown, G., Watson, I., Cavazos, J.: Intelligent selection of application-specific garbage collectors. In: Proceedings of the 6th International Symposium on Memory Management, ISMM 2007, pp. 91–102. ACM, New York (2007)

29. Singer, J., Jones, R.E., Brown, G., Luján, M.: The economics of garbage collection. In: Proceedings of the 2010 International Symposium on Memory Management, ISMM 2010, pp. 103–112. ACM, New York (2010)
30. Soman, S., Krintz, C.: Application-specific garbage collection. J. Syst. Softw. **80**, 1037–1056 (2007)
31. White, D.R., Singer, J., Aitken, J.M., Jones, R.E.: Control theory for principled heap sizing. In: Proceedings of the 2013 International Symposium on Memory Management, ISMM 2013, pp. 27–38. ACM, New York (2013)

A Model-Driven Tool Chain for OCCI

Faiez Zalila[1,2(✉)], Stéphanie Challita[1,2], and Philippe Merle[1,2]

[1] Inria Lille - Nord Europe, Villeneuve-d'ascq, France
{faiez.zalila,stephanie.challita,philippe.merle}@inria.fr
[2] University of Lille, Villeneuve-d'ascq, France

Abstract. Open Cloud Computing Interface (OCCI) is the only open standard for managing any kinds of cloud resources, e.g., Infrastructure as a Service, Platform as a Service, and Software as a Service. However, no model-driven tooling exists to assist OCCI users in designing, editing, validating, generating, and managing OCCI artifacts (i.e., extensions that represent specific application domains and configurations that define running systems). In this paper, we propose the first model-driven tool chain for OCCI called OCCIWARE STUDIO. This tool chain is based on a metamodel defining the static semantics for the OCCI standard in Ecore and OCL. OCCIWARE STUDIO provides OCCI users facilities for designing, editing, validating, generating, and managing OCCI artifacts. We detail the tooled process to define an OCCI extension. In addition, we show how the cloud user can leverage the generated tooling for this extension to create his own OCCI configurations and manage them in the cloud. We illustrate our paper with the OCCI **Infrastructure** extensiondefining OCCI-compliant compute, network, and storage resources.

Keywords: Cloud computing · Service computing · Metamodeling · Software standards · Computer aided software engineering · Distributed information systems · Modeling environments

1 Introduction

Cloud computing has been adopted as the dominant delivery model for computing resources [3]. This model defines three well-discussed layers of services known as Infrastructure as a Service (IaaS), Platform as a Service (PaaS), and Software as a Service (SaaS) [22]. Provisioning and managing these outsourced, on-demand, pay as you go, elastic resources require cloud resource management interfaces (CRM-API) [20]. However, there is a plethora of CRM-APIs, proposed by Amazon, Eucalyptus, Microsoft, Google, OpenNebula, CloudStack, OpenStack, CloudBees, OpenShift, Cloud Foundry, to name a few. Each API is based on different concepts and/or architectures. Therefore, provisioning and managing cloud resources are faced with four main issues: *heterogeneity* of cloud offers, *interoperability* between CRM-API, *integration* of CRM-API for building multi-cloud systems, and *portability* of cloud management applications.

© Springer International Publishing AG 2017
H. Panetto et al. (Eds.): OTM 2017 Conferences, Part I, LNCS 10573, pp. 389–409, 2017.
https://doi.org/10.1007/978-3-319-69462-7_26

To tackle these issues, Open Cloud Computing Interface (OCCI) defines the first and only open standard for managing any cloud resources [16]. OCCI provides a general purpose model for cloud computing resources and a RESTful API for efficiently accessing and managing any kind of these resources. This will facilitate interoperability between clouds, as providers will be specified by the same resource-oriented model called the OCCI Core Model [33], that can be expanded through extensions and accessed by a common REST [18] API.

Today, only runtime frameworks such as erocci, rOCCI, pySSF, pyOCNI, and OCCI4Java are available, while OCCI designers/developers/users need software engineering tools to design, edit, validate, generate, and manage new kinds of OCCI resources, and the configurations of these resources. Nevertheless, OCCI lacks tools for modeling its extensions and configurations, despite the presence of a precise metamodel for OCCI [23] to be the reference of the implementation of such modeling tools.

To address this problem, the main contributions of this paper can be summarized as:

- The first model-driven tool chain for OCCI called OCCIWARE STUDIO which provides OCCI users with facilities for designing, editing, validating, generating, and managing OCCI artifacts.
- An enhanced metamodel for OCCI on which we build the OCCIWARE STUDIO. This new metamodel resolves several observed lacks in the previously proposed metamodel for OCCI [23] by introducing additional concepts such as *(i)* a mechanism to express business constraints, *(ii)* the Finite State Machine (FSM) concepts to define the behavior of OCCI resource types, *(iii)* a support for mixins, *(iv)* an own data type system for OCCI, and *(v)* a set of Ecore data types to assess the well-formedness of the OCCI artifacts.

Then, we detail the tooled process to define an OCCI extension. In addition, we show how the cloud user can leverage the generated tooling for this extension to create its own OCCI configurations and manage them in the cloud. To evaluate the efficiency of our OCCIWARE STUDIO, we elaborate a validation approach on the OCCI `Infrastructure` extension [26] defining OCCI-compliant compute, network and storage resources.

This paper is organized as follows. Section 2 explains the motivations behind our contribution. Section 3 presents our improvements made to the previous metamodel for OCCI. Then, we introduce the architecture of the OCCIWARE STUDIO and we detail the proposed approach to manage Everything as a Service with OCCI. Section 4 validates our proposed tool chain on the `Infrastructure` extension of OCCI. We position our work with related approaches in Sect. 5. Finally, Sect. 6 concludes on future work and perspectives.

2   Motivations

Today, multi-cloud computing is quite encouraged for cloud developers as a way to reduce vendor lock-in, to improve resiliency during outages and geo-presence, to boost performance and to lower costs. However, semantic differences between cloud providers, as well as their heterogeneous management interfaces, make changing from one provider to another very complex and costly. We assume for example that a developer would like to build a multi-cloud system spread over two clouds, Amazon Web Services (AWS) and Google Cloud Platform (GCP). AWS are accessible via a SOAP API, whereas GCP is based on a REST API, which leads to an incompatibility between these two different APIs. To use them, cloud consumers should be inline with the concepts and operations of each API, which is quite frustrating. The developer would like a single API for both clouds to seamlessly access their resources [9].

For this, OCCI is an open-source standard that defines a generic model for cloud resources and a RESTful API for efficiently accessing and managing resources. This will facilitate interoperability between clouds, as providers will be specified by the same resource model, and accessed by a common REST API. However, cloud developers cannot currently take advantage of this standard. Although there are several implementations of OCCI, there is no tool that allows the cloud developers to design and verify their configurations, neither to generate corresponding artifacts. This leads to several challenges:

1. Cloud users are focused on implementation details rather than cloud concerns, with the risk of misunderstandings for the concepts and the behavior that rely under cloud APIs.
2. The only way to be sure that the designed configurations will run correctly is to deploy them in the clouds. In this context, when errors occur, a correction is made and the deployment task can be repeated several times before it becomes operational. This is quite painful and expensive.
3. Cloud users need to provide various forms of documentation of their cloud configurations, as well as deployment artifacts. However, these tasks are complex and usually made in an ad-hoc manner with the effort of a human developer, which is error-prone and amplifies both development and time costs.

Recently, we are witnessing several works that take advantage of MDE for the cloud [8]. Therefore, to address the identified challenges, we believe that **there is a need for a model-driven tool chain for OCCI** in order to:

1. Enable both cloud architects and users to efficiently describe their needs at a high level of abstraction. This will be done by defining a metamodel accompanied with graphical and textual domain-specific modeling languages.
2. Allow cloud architects to define structural and behavioral constraints and validate them before any concrete deployments so they can a priori check the correctness of their systems.
3. Automatically generate *(i)* *textual documentation* to assist cloud architects, developers and users to understand the concepts and the behavior of the cloud API, and *(ii)* *HTTP scripts* that provision, modify or de-provision cloud resources.

3   OCCIWARE Tool Chain

Designing is the key activity that must be addressed to resolve other encountered challenges such as verifying, generating, etc. Therefore, in order to assist OCCI users in modeling different OCCI artifacts, we propose a metamodel for OCCI named OCCIWARE METAMODEL (cf. Sect. 3.3) extending the previous one described in Sect. 3.2. It defines the different concepts required to model OCCI extensions and configurations. In Sect. 3.4, we detail our approach to provide a tooled framework based on OCCI. It consists in mapping OCCI-WARE METAMODEL concepts into the chosen modeling framework, the Eclipse Modeling Framework (EMF). An overview of all the main OCCIWARE STUDIO features are described in Sect. 3.5. Section 3.6 presents our approach to manage Everything as a Service with OCCIWARE. It details the different interactions between the OCCIWARE STUDIO and stakeholders. Finally, Sect. 3.7 lists all the OCCI extensions currently provided with OCCIWARE STUDIO. But before that, let us start by introducing the OCCI standard.

3.1   OCCI

OCCI is an open cloud standard [16] specified by the Open Grid Forum (OGF). OCCI defines a RESTful Protocol and API for all kinds of management tasks on any kind of cloud resources, including Infrastructure as a Service (IaaS), Platform as a Service (PaaS) and Software as a Service (SaaS). In order to be modular and extensible, OCCI is delivered as a set of specification documents divided into the four following categories as illustrated in Fig. 1:

- **OCCI Core Model:** It defines the OCCI Core specification [33] proposed as a RESTful-oriented model.
- **OCCI Protocols:** Each OCCI Protocol specification describes how a particular network protocol can be used to interact with the OCCI Core Model. Currently, only the OCCI HTTP Protocol [32] has been defined.

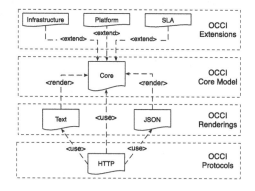

Fig. 1. OCCI specifications

- **OCCI Renderings:** Each OCCI Rendering specification describes a particular rendering of the OCCI Core Model. Currently, both OCCI Text [15] and JSON [34] renderings have been defined.
- **OCCI Extensions:** Each OCCI Extension specification describes a particular extension of the OCCI Core Model for a specific application domain, and thus defines a set of domain-specific kinds and mixins. OCCI Infrastructure [26] is dedicated to IaaS. Additional OCCI extensions are defined such as OCCI Compute Resource Templates Profile (CRTP) [13], OCCI Platform [27] and OCCI Service Level Agreements [19].

3.2 The Previous OCCIWARE METAMODEL

Figure 2 shows the OCCIWARE METAMODEL. The entry point to define it was the OCCI Core Model whose main concepts are `Resource`, `Link`, `Kind`, `Mixin`, `Attribute`, and `Action`. `Resource` is the root abstraction of any cloud resource, such as a virtual machine, a network, and an application. `Link` represents a relation between two resources, such as a virtual machine connected to a network and an application hosted by a virtual machine. Each OCCI entity (resource or link) owns zero or more `Attributes`, such as its unique identifier, the host name of a virtual machine, the Internet Protocol address of a network. As OCCI is a REST API, it gives access to cloud resources via classical CRUD operations (i.e., *Create*, *Retrieve*, *Update*, and *Delete*). In addition, each OCCI entity has zero or more `Actions` representing business specific behaviors, such as start/stop a virtual machine, and up/down a network. Each OCCI entity is strongly typed by a `Kind` and a set of `Mixin` instances. `Kind` represents the immutable type of OCCI entities and defines allowed attributes and actions. Single inheritance between `Kinds` allows us to factorize attributes and actions common to several kinds. `Mixin` represents cross-cutting attributes and actions that can be dynamically added to an OCCI entity. `Mixin` can be applied to zero or more kinds and can depend from zero or more other `Mixin` instances. The gray-colored classes in Fig. 2 show the OCCI Core Model. The latter can be interacted with over protocols/renderings and is expandable through extensions.

The OCCI Core Model does not explicitly define the notions of extension that represents a specific application domain and configuration that models a running system. For that, during a previous work [23], additional concepts, such as `Extension` and `Configuration`, are defined (the blue-colored classes in Fig. 2). This previous work has proposed the first metamodel for OCCI named, OCCI-WARE METAMODEL.

Definition 1. *Extension represents an OCCI extension, e.g., inter-cloud networking extension [21], infrastructure extension [26], platform extension [27, 39,40], application extension [40], SLA negotiation and enforcement [14], cloud monitoring extension [10], and autonomic computing extension [28–31]. Extension has a name, has a scheme, owns zero or more kinds, owns zero or more mixins, owns zero or more types, and can import zero or more extensions.*

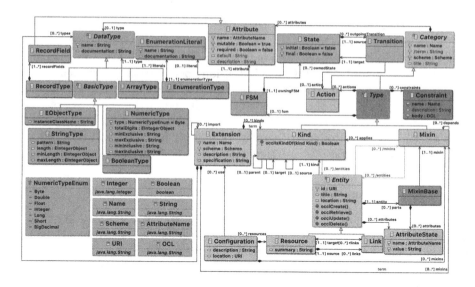

Fig. 2. Ecore diagram of OCCIWARE METAMODEL (Color figure online)

Definition 2. *Configuration represents a running OCCI system.* *Configuration owns zero or more* ***resources*** *(and transitively* ***links***)*, and* ***use*** *zero or more extensions. For a given configuration, the kind and mixins of all its entities (resources and links) must be defined by* ***used*** *extensions only. This avoids a configuration to transitively reference a type defined we do not know where.*

During the OCCIWARE research and development project[1], the proposed OCCIWARE METAMODEL [23] was used to model various OCCI extensions: OCCI Infrastructure [26], OCCI Platform [27], OCCI SLA [19], OCCI Monitoring [11] and some additional proprietary extensions such as Docker [36] and simulation [1].

3.3   The Enhanced OCCIWARE METAMODEL

We propose to extend the first version of OCCIWARE METAMODEL in order to resolve the lacks and the limitations encountered during the definition of the different OCCI artifacts.

At first, we propose to add a mechanism enabling to express constraints related to each cloud computing domain. In fact, each extension targets a concrete cloud computing domain, e.g., IaaS, PaaS, SaaS, pricing, etc. Therefore, there are certainly business constraints related to each domain which must be respected by configurations that use this extension. For example, all IP addresses of all network resources must be distinct. For that, we have extended this metamodel by adding the **Constraint** concept (the brown-colored classes of Fig. 2).

---

[1] http://www.occiware.org.

A `Constraint` has a `name`, a `description` and a `body` that can be defined with Object Constraint Language (OCL) [35]. Each kind/mixin has zero or more `constraints` inherited from the new introduced abstract `Type` class.

Then, we have extended the OCCIWARE METAMODEL by introducing a Finite State Machine (FSM) modeling language. It allows us to model the behavior of OCCI concepts such as state diagrams of OCCI `Kind` instances used in both the Infrastructure [26] and Platform [27] extensions. For that, as shown in the yellow-colored classes of Fig. 2, the FSM metamodel has three classes: `FSM`, `State`, and `Transition`. A `FSM` owns a set of states (`State`); a transition (`Transition`) is necessarily associated to a state. Additional concepts and rules related to the OCCI domain are added in the FSM metamodel such as: "a FSM is associated to a particular `attribute` of the concerned `Kind/Mixin` instance", "the identifier of a `State` is a literal (`EnumerationLiteral`)", "an `Action` is associated to a `Transition` instance", etc. In the following, we detail a subset of the static semantics of the FSM modeling language integrated in the OCCIWARE METAMODEL.

Definition 3. *The type of a FSM attribute must be EnumerationType.*

```
context FSM
inv AttributeType: attribute.type.oclIsTypeOf(EnumerationType);
```

Definition 4. *The enumerationType of a State literal is equals to the type of the attribute of the owner FSM instance.*

```
context State
inv LiteralType: owningFSM.attribute.type=self.literal.enumerationType;
```

Definition 5. *The action of a Transition instance must belong to the actions of the owner Type instance.*

```
context Transition
inv ActionMustBeDefined : self.oclContainer().oclAsType(State).oclContainer().
    oclAsType(FSM).oclContainer().oclAsType(Type).actions->includes(self.action)
```

Thereafter, we have defined an additional class named `MixinBase` (the orange-colored class) which refers to a `mixin`. This concept allows us to instantiate the attributes of the referenced `mixin` outside the owner `entity` in order to separate the `entity` attributes from the `mixin` ones.

Next, we have defined an own data type system for OCCI as shown at the left part of Fig. 2 (the green-colored classes). This data type system allows us to define primitive types, as provided in the first version of the OCCIWARE METAMODEL, such as `StringType` to model string types, `NumericType` to model numeric types and `BooleanType` to model boolean types. In addition, it allows to model a Java-based type using `EObjectType` and enumerations using

EnumerationType. It provides also the capability to model complex types like ArrayType to design array types and RecordType to design structured types.

Finally, we have defined a set of Ecore data types (the red-colored classes) such as URI, Scheme, Name, and AttributeName, etc. These string-based types are enriched with the associated regular expression in order to ensure the correct values of different attributes. For example, the following rule "*Attribute names consist of alphanumeric characters separated by dots*" has been defined in the JSON rendering specification [34]. Accordingly, the user may define non-valid attribute names. Therefore, the bugs will be detected during the last steps of the modeling process and fixing them becomes a tricky task. For that, we have defined the following regular expression pattern for the AttributeName type: value="[a-zA-Z0-9]+(\.[a-zA-Z0-9]+)+".

Once the new OCCIWARE METAMODEL is defined, we have tooled it with the EMF genmodel, the EMF model containing additional information related to the code generation, in order to generate the Java-based implementation.

3.4   Projection of OCCI to EMF

The OCCI Core Model is a simple resource-oriented model. It can be extended with several OCCI extensions.

As shown in the left part of Fig. 3, designing a new OCCI extension model consists in extending the OCCI core extension, the extension-like representation of the OCCI Core Model (as shown on the top part of Fig. 6). The OCCI core extension is composed of three kinds: a root Entity kind, and two children kinds: Resource and Link. Designing an OCCI configuration consists in defining an instance of an OCCI extension and represents a cloud architecture already deployed or to deploy.

The main goal of our work consists in introducing a tooled framework, based on OCCI, that manages any kind of resources as a service.

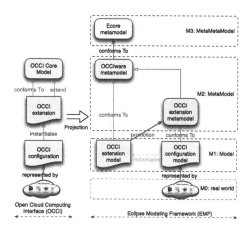

Fig. 3. Projection of OCCI to EMF

To do that, it was necessary to project different OCCI concepts into a modeling framework to benefit from the available facilities for building tools based on a metamodel (the right part of Fig. 3). To do that, EMF was chosen to embed OCCI and, thus, the OCCIWARE METAMODEL was proposed as a precise metamodel for OCCI. Therefore, we can define either an OCCI extension model or an OCCI configuration model conform to the OCCIWARE METAMODEL. However, the current tooling in EMF does not favor to encode that: an OCCI configuration is an *"instantiation"* of an OCCI extension. For that, we have proposed to promote the OCCI extension model by translating it into an Ecore metamodel, extending the OCCIWARE METAMODEL. Consequently, the OCCI configuration model becomes an instance of this generated metamodel and, thus, an instance of the OCCIWARE METAMODEL.

In the following, we detail the promotion process of OCCIWARE META-MODEL concepts into the EMF concepts:

- Each OCCI kind instance is translated into an Ecore class. If its `parent` is the `Resource` kind, the generated class extends the `Resource` Ecore class of the OCCIWARE METAMODEL. Otherwise, if its `parent` is the `Link` kind, the generated class extends the `Link` Ecore class of the OCCIWARE METAMODEL.
- Each OCCI mixin instance is translated into an Ecore class extending the `MixinBase` class of the OCCIWARE METAMODEL. Due to this added `Mixin-Base` class and this promotion rule, an instance of this generated class can now refer to the initial OCCI mixin instance.
- Each OCCI attribute instance, owned by an OCCI kind/mixin, is translated into an Ecore attribute owned by the corresponding generated Ecore class.
- Each OCCI action instance, owned by an OCCI kind/mixin, is translated into an Ecore operation owned by the corresponding generated Ecore class.
- Each OCCI constraint instance is translated into an OCL invariant.
- All Ecore data types defined in the OCCI extension are translated into the corresponding EMF concepts and/or types in the generated OCCI extension metamodel.

In order to ease the definition, edition, validation, and generation of different OCCI artifacts, we have benefited from the associated technologies to EMF in order to design and implement the OCCIWARE STUDIO.

3.5   OCCIware Studio Features

OCCIWARE STUDIO is a set of plugins for the Eclipse integrated development environment. Figure 4 shows all the main features of OCCIWARE STUDIO:

- **OCCI Designer** is a graphical modeler to create, modify, and visualize both OCCI extensions and configurations. The OCCI standard does not define any standard notation for the graphical or textual concrete syntax. This tool is implemented on top of the Eclipse Sirius framework.
- **OCCI Editor** is a textual editor for both OCCI extensions and configurations. Our OCCI textual syntax is described in [25]. This tool is implemented on top of the Eclipse Xtext framework.

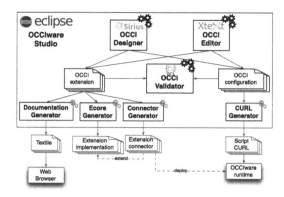

Fig. 4. OCCIWARE STUDIO features

- **OCCI Validator** is a tool to validate both OCCI extensions and configurations. This tool checks all the constraints defined in the OCCIWARE META-MODEL, *i.e.*, both Ecore and OCL ones. In addition, it checks the OCL invariants generated by the promotion process.
- **Documentation Generator** is a tool to generate a TEXTILE documentation from an OCCI extension model. TEXTILE is a Wiki-like format used for instance by Github projects. This tool is implemented on top of the Eclipse Acceleo framework.
- **Ecore Generator** is a tool to generate the promoted Ecore metamodel and its associated Java-based implementation code from an OCCI extension. This tool is directly implemented in Java.
- **Connector Generator** is a tool to generate the OCCI connector implementation associated to an OCCI extension. This generated connector code extends the generated Ecore implementation code. This connector code must be completed by cloud developers to implement concretely how OCCI CRUD operations and actions must be executed on a real cloud infrastructure. Later, this generated connector will be deployed on OCCIWARE RUNTIME [2]. This tool is implemented on top of the Eclipse Acceleo framework.
- **CURL Generator** is a tool to generate a CURL-based script from an OCCI configuration model. These generated scripts contain HTTP requests to instantiate OCCI entities into any OCCI-compliant runtime. These scripts are used for offline deployment. This tool is implemented on top of the Eclipse Acceleo framework.

3.6   Managing Everything as a Service with OCCIware

OCCIWARE STUDIO proposes a tooled framework to manage any resource in the cloud. To benefit from it, a proposed process must be followed (cf. Fig. 5). This process contains two phases: the **Design phase** and the **Use phase**.

The **Design phase** consists in defining a new OCCI extension that extends the OCCI `core` extension and/or others OCCI extensions already defined. This phase contains three steps are shown in Fig. 5.

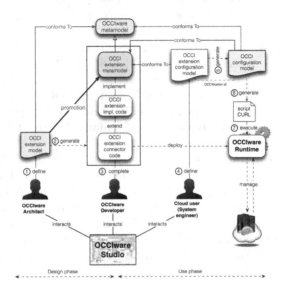

Fig. 5. OCCIWARE Approach to manage Everything as a Service

At first, an OCCIWARE architect designs his/her extension using both **OCCI Designer** and **OCCI Editor** tools (Step 1 in Fig. 5). Once the OCCI extension is defined, the generation process of the extension tooling may be triggered (Step 2 in Fig. 5). It proceeds in three steps:

1. The **Ecore Generator** tool is used to generate the extension metamodel which extends the OCCIWARE METAMODEL because:
   - the generated classes from kinds, inheriting the `Resource` kind, will inherit the `Resource` Ecore class,
   - the generated classes from kinds, inheriting the `Link` kind, will inherit the `Link` Ecore class, and
   - the generated classes from mixins will inherit the `MixinBase` Ecore class.
2. Thereafter, the EMF tooling generates the Java-based implementation of the extension metamodel.
3. Finally, the **Connector Generator** tool is used to generate the implementation of a connector for each subclass of `Resource`, `Link` and `MixinBase`.

The connector code defines an implementation of the OCCI specific callback methods (the CRUD operations) and a pseudo-code for all kind/mixin-specific actions. The code of these actions is deducted from the defined FSM.

Once the generation step (Step 2 in Fig. 5) is achieved, the OCCIWARE developer can complete the generated connector classes (Step 3 in Fig. 5). It consists in implementing the business code for each connector class. The completed connector code must be deployed on the OCCIWARE RUNTIME [2], a full OCCI-compliant model-driven server.

From now on, we can consider that the **Design phase** is achieved and the OCCI extension is completely tooled and able to be used to manage conforming configurations.

The **Use phase** can now start. In fact, thanks to OCCIWARE STUDIO enriched with the extension tooling provided during the previous design step, users can design an extension configuration model conforms to the extension metamodel (Step 4 in Fig. 5). To benefit from the OCCI-compliant tools of OCCIWARE STUDIO, this extension-specific model must be translated into an OCCI configuration model (Step 5 in Fig. 5) conforms to the OCCIWARE META-MODEL.

In order to deploy and manage the designed OCCI configuration, users interact with the cloud by sending OCCI HTTP requests to OCCI Runtime [2]. One possibility consists in generating CURL scripts (Step 6 in Fig. 5) using the **CURL Generator** tool. Then, the generated script is executed (Step 7 in Fig. 5) via the OCCIWARE RUNTIME [2] that invokes the `occiCreate()` method of each `Resource` class. This method implements how to create the considered `Resource` instance in the cloud. Finally, the created resource is deployed in the cloud.

3.7   Supported OCCI Extensions

Each OCCI extension is implemented as an Eclipse modeling project containing one extension model, which is an instance of OCCIWARE METAMODEL. Currently, OCCIWARE STUDIO supports the five OCCI extensions defined by the OGF's OCCI working group:

- **OCCI Infrastructure** [26] defines compute, storage and network resource types and associated links.
- **OCCI Compute Resource Templates Profile** [13] defines a set of pre-configured instances of the OCCI compute resource type.
- **OCCI Platform** [27] defines application and component resource types and associated links.
- **OCCI SLA** [19] defines OCCI types for modeling service level agreements.
- **OCCI Monitoring** [11] defines sensor and collector types for monitoring cloud systems.

4   OCCIware Approach Validation

In this section, we validate our approach by illustrating the proposed process on an OCCI extension with OCCIWARE STUDIO. We choose the Infrastructure extension. Then, we show how the cloud user leverages the generated tooling around this extension to create/manage his/her configuration models with OCCIWARE STUDIO and deploy them in the cloud.

4.1   Design Process of OCCI Infrastructure Extension

To design the Infrastructure extension, the OCCIWARE architect can use **OCCI Designer** and/or **OCCI Editor** tools (Step 1 in Fig. 5).

This extension defines five kinds (Network, Compute, Storage, StorageLink and NetworkInterface), six mixins (Resource_Tpl, IpNetwork, Os_Tpl, SSH_key, User_Data, and IpNetworkInterface), and around twenty data types (Vlan range, Architecture enumeration, various status enumerations, etc.), as shown in Fig. 6.

The Compute kind represents a generic information processing resource, e.g., a virtual machine or container. It inherits the Resource defined in the OCCI core extension. It has a set of OCCI attributes such as occi.compute.architecture to specify the CPU architecture of the instance, occi.compute.core to define the number of virtual CPU cores assigned to the instance, occi.compute.memory to define the maximum RAM in gigabytes allocated to the instance, etc. The Compute kind exposes five actions: start, stop, restart, save and suspend.

The Network kind is an interconnection resource and represents a Layer 2 (L2) networking resource. This is complemented by the IpNetwork mixin. It exposes two actions: up and down.

The orange-colored box in Fig. 6 illustrates the state diagram of a Network instance and describes its behavior. As shown previously, the new OCCIWARE METAMODEL provides the required concepts to describe the behavior of each kind/mixin. In addition, it allows to define extension-specific constraints. For example, the following OCL constraint specifies that each Network instance must have a unique vlan.

```
inv UniqueVlan: Network.allInstances()–>isUnique(occi.network.
    vlan)
```

In addition, we define, in the following, an additional OCL constraint in the IpNetworkInterface mixin which checks that all IP addresses must be different.

```
inv IPAddressesMustBeUnique: IpNetworkInterface.allInstances()
    –>isUnique(occi.networkinterface.address)
```

The NetworkInterface kind inherits the Link kind. It connects a Compute instance to a Network instance. The Storage kind represents data storage devices. The StorageLink kind inherits the Link kind. It connects a Compute instance to a Storage instance.

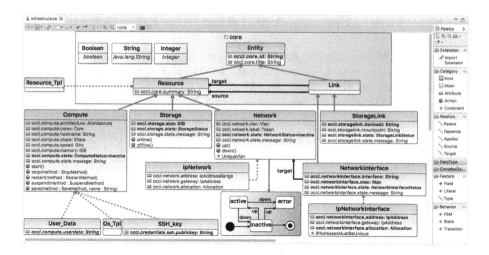

Fig. 6. OCCI Infrastructure extension model (Color figure online)

Once the `Infrastructure` extension is defined, the generation process of tooling may be triggered (Step 2 in Fig. 5). It generates three main elements: *(i)* the `Infrastructure` extension metamodel, *(ii)* the Java-based implementation of the `Infrastructure` extension metamodel, and *(iii)* the implementation of the `Infrastructure` connector.

Listing 1.1 shows a subset of the generated `Network` connector class. It extends the `NetworkImpl` class generated by the EMF tooling and contains the OCCI specific callback methods for the CRUD operations and all `Network` kind-specific actions (i.e., `up` and `down`). The generated code of specific actions is deducted from the defined FSM on the `Network` kind.

```
public class NetworkConnector extends NetworkImpl {
  NetworkConnector() {}
  // OCCI CRUD callback operations.
  public void occiCreate()    { /* TODO */ }
  ...
  // Network actions.
  ...
  public void down() {
    if(getState().equals(NetworkStatus.ACTIVE)) {
      if ( true ) {
        // TODO: Transition active -down-> inactive
        setState(NetworkStatus.INACTIVE);
      } else {
        // TODO: Transition active -down-> error
        setState(NetworkStatus.ERROR);
      }
    }
  }
}
```

Listing 1.1. The generated `Network` connector class

Once the generation step is achieved, the OCCIWARE developer can complete the generated connector classes (Step 3 in Fig. 5) by updating their methods implementations (TODO sections in Listing 1.1) with business code related to targeted API. For the NetworkConnector class, the developer completes the code to trigger that the OCCI Network resource was created (occiCreate), will be retrieved (occiRetrieve), was updated (occiUpdate) and will be deleted (occiDelete). In addition, he/she completes the generated methods (up and down) related to specific actions defined in the Network kind. The completed connector code must be deployed on the OCCIWARE RUNTIME [2], a full OCCI-compliant model-driven server.

From now on, we can consider that the OCCI Infrastructure extension is completely tooled and able to be used to manage conforming configurations.

4.2 Use Process of OCCI Infrastructure Extension

Using OCCIWARE STUDIO enriched with the Infrastructure extension tooling, users can design an OCCI Infrastructure configuration model conforms to the Infrastructure extension metamodel (Step 4 in Fig. 5). To benefit from the OCCI-compliant tools defined in the OCCIWARE STUDIO, an Infrastructure configuration model must be translated into an OCCI configuration model (Step 5 in Fig. 5) conforms to the OCCIWARE METAMODEL.

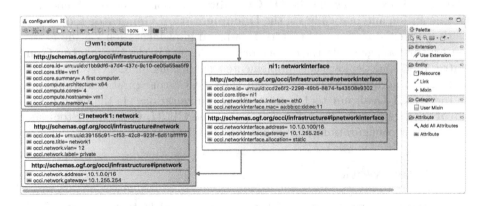

Fig. 7. An OCCI Infrastructure Configuration (Color figure online)

Figure 7 illustrates a small IaaS configuration composed of a compute (vm1) connected to a network (network1), via an OCCI link (orange-colored box), the network interface (ni1). As this configuration uses an IP-based network, the Network resource and the NetworkInterface link have an IpNetwork and IpNetworkInterface mixin, respectively. Each OCCI entity is configured by its attributes, e.g., vm1 has the vm1 hostname, an x64-based architecture, 4 cores, and 4 GiB of memory.

In order to deploy and manage the designed OCCI Infrastructure configuration, users interact with the cloud by sending OCCI HTTP requests to OCCIWARE RUNTIME [2]. These requests can be automatically generated as CURL scripts using the **CURL Generator** tool (Step 6 in Fig. 5).

```
OCCI_SERVER_URL=$1

curl $CURL_OPTS -X PUT
    $OCCI_SERVER_URL/network/39155c91-cf53-42c8-923f-6d51bffffff9
-H 'Content-Type: text/occi '
-H 'Category: network; scheme="http://schemas.ogf.org/occi/infrastructure#";
    class="kind";'
-H 'X-OCCI-Attribute: occi.core.id="39155c91-cf53-42c8-923f-6d51bffffff9"'
-H 'X-OCCI-Attribute: occi.core.title="network1"'
-H 'X-OCCI-Attribute: occi.network.vlan=12'
-H 'X-OCCI-Attribute: occi.network.label="private"'
-H 'X-OCCI-Attribute: occi.network.address="10.1.0.0/16"'
-H 'X-OCCI-Attribute: occi.network.gateway="10.1.255.254"'
```

Listing 1.2. The generated CURL script to create a `network` instance

Listing 1.2 shows the CURL script that requests OCCIWARE RUNTIME [2] via both OCCI HTTP Protocol [32] and OCCI Text Rendering [15] to create the `network1` instance. Then, OCCIWARE RUNTIME [2] invokes the `occiCreate()` method of the `NetworkConnector` class, which implements how to create the considered network instance in the cloud (Step 7 in Fig. 5). Finally, the created `Network` resource is deployed in the cloud.

In summary, thanks to OCCIWARE STUDIO, cloud architects and developers can design their OCCI extensions graphically via the **OCCI Designer** tool, validate their extensions via the **OCCI Validator** tool, generate a Web-based documentation automatically via the **Documentation Generator** tool, generate the promoted metamodel and its Java-based implementation via the **Ecore Generator** tool, generate the connector Java classes via the **Connector Generator** tool, and deploy their implemented extensions on OCCIWARE RUNTIME [2]. In addition, users can design their OCCI configurations graphically via the **OCCI Designer** tool, validate their configurations via the **OCCI Validator** tool, generate a deployment script via the **CURL Generator** tool, and manage their configurations at runtime via OCCIWARE RUNTIME [2].

5   Related Work

In this section, we present some of the cloud metamodels, tools, and standards that were recently proposed and are relevant to our contribution.

CloudML [7,17] is a cloud modeling language that allows both cloud providers and developers to describe cloud services and application components, respectively. Then, it helps to provision cloud resources by a semi-automatic matching between the defined application requirements and the cloud offerings. CloudML is exploited both at design-time to describe the application provisioning of cloud resources after performing the necessary orchestration, and at runtime to manage the deployed applications. In fact, the model at design-time is automatically handled by the Cloud Modeling Framework (CloudMF), which returns a runtime model of the provisioning resources, according to the Models@Run.Time approach [6]. CloudML is an inspiration source for future efforts in modeling the cloud, however, it lacks implementations. Unlike OCCIWARE which offers a set of tools to model, edit, validate, document, deploy, and manage any kind of cloud resources and their corresponding configurations, CloudML only provides a JSON and/or XML textual syntax to specify deployment and management concerns in IaaS and/or PaaS clouds.

SALOON [37] is an EMF graphical framework that is based on Extended Feature Models (EFMs) to represent clouds variability, as well as on ontology concepts to model the various semantics of cloud systems. It allows to translate these ontology concepts into a Constraint Satisfaction Problem (CSP) in order to select the adequate cloud environment. Although the authors state that SALOON supports the discovery and selection of multiple providers, in practice it does not. In the contrary, it selects one suitable provider at a time. SALOON targets ten IaaS/PaaS cloud environments. The objective of OCCI differs from SALOON; OCCI helps cloud users to overcome the differences between the resource kinds by defining a unified API, but it does not provide a way to select the suitable configuration.

CompatibleOne [38] is an open-source broker that offers services from several cloud providers and consists of a model called CompatibleOne Resource Description System (CORDS) and an execution platform called Advanced Capabilities for CORDS (ACCORDS). Similar to OCCIWARE METAMODEL, CORDS is based on the OCCI standard and is an object based description of cloud applications, services, and resources. However, the authors in [38] defined this model to describe only IaaS and PaaS resources managed by their broker. ACCORDS allows to handle the user's requirements, validate, and execute the provisioning plan, and to deliver the cloud services. CompatibleOne does not support the graphical design of cloud configurations nor the generation of documentations, Ecore extensions, CURL scripts, etc.

Topology and Orchestration Specification for Cloud Applications (TOSCA) [5] is a language for describing the topology/structure/architecture of a cloud application, i.e., the software components that constitute the application, the physical or virtual nodes on which the components will be deployed, and the relationships between components and nodes. The TOSCA language is only defined as a textual XML or YAML document so it is complicated to have an overview of all its concepts. Several implementations of TOSCA were developed. For example, Winery[2] provides an open source Eclipse-based graphical modeling tool for TOSCA and the OpenTOSCA [4] project provides an open source container for deploying TOSCA-based applications, hence it is responsible for translating a TOSCA description into actions to be performed in clouds. These actions are sent to the clouds through their respective APIs. While both are standards, TOSCA provides a cloud application description language when OCCI provides a cloud resource management API.

Besides OCCI and TOSCA, several elaborated and mature cloud computing standards exist. For instance, the DTMF's Cloud Infrastructure Management Interface (CIMI) standard [12] defines a RESTful API for managing IaaS resources only. OCCI Infrastructure is concurrent to CIMI because both address

---

[2] https://projects.eclipse.org/projects/soa.winery.

IaaS resource management but OCCI has a more general purpose as it can be used also for any kind of PaaS and SaaS resources.

The OASIS's Cloud Application Management for Platforms (CAMP[3]) standard targets the deployment of cloud applications on top of PaaS resources. CAMP and TOSCA can use OCCI-based IaaS/PaaS resources, so these standards are complementary.

6   Conclusion

OCCI proposes a generic model and API for managing any kind of cloud computing resources. Unfortunately, it is obvious that leading cloud providers have no interest in adopting a standard API like the one offered by OCCI to facilitate interoperability with other clouds. However, OCCI has proven its utility in several contexts. For example, the European Grid Infrastructure Federated Cloud (EGI FC), which is a hybrid cloud, is based on OCCI Infrastructure to ensure interoperability among 20 cloud providers and over 300 data centers. Furthermore, OCCI attracts several cloud brokers such as CompatibleOne that aims at ensuring seamless access to the heterogeneous resources of cloud providers.

We argue in this paper that OCCI suffers from the lack of modeling, verification, validation, documentation, deployment, and management tools for both OCCI extensions and configurations. To address this issue, we propose OCCI-WARE STUDIO, the first model-driven tool chain for OCCI. This tool chain is based on a metamodel defining the precise semantics of OCCI in Ecore and OCL. Our metamodel can be seen as a domain-specific modeling language to define and exchange OCCI extensions and configurations between end-users and resource providers. More precisely, thanks to OCCIWARE STUDIO, both cloud architects and users can encode OCCI extensions and configurations, respectively, graphically via the **OCCI Designer** tool, and textually via the **OCCI Editor** tool. They can also automatically verify the consistency of these extensions and configurations via the **OCCI Validator** tool, generate dedicated model-driven tooling via both **Ecore Generator** and **Connector Generator** tools, generate a deployment script via the **CURL Generator** tool, and manage their configurations at runtime via the generated connectors deployed in OCCIWARE RUNTIME [2]. OCCIWARE STUDIO is tightly integrated with the Java IDE, to facilitate the addition of functionality to the service skeletons and generates ready-to-deploy configurations for OCCI extensions. Our tool is validated by encoding and automatically verifying, provisioning, and managing the Compute, Network and Storage resources of the OCCI Infrastructure extension. OCCI-WARE STUDIO also succeeded in managing mobile robots in [24], which proves the ability of our tool chain in managing any kind of resources. As for previous OCCI implementations, our OCCIWARE STUDIO can be easily integrated and used with them. For example, the CURL scripts generated by OCCIWARE STUDIO can be executed on OpenStack, OCCI4Java, etc.

---

[3] https://www.oasis-open.org/committees/camp.

In the future, we target industrial validation for OCCIWARE STUDIO. Therefore, an ongoing work aims to get this tool tested and adopted within Scalair[4]. We will also validate OCCIWARE STUDIO and our improved metamodel for OCCI on all the already published OCCI extensions. In order to cover the whole cloud market, we will also continuously enrich OCCIWARE STUDIO with new extensions such as AWS, GCP, OpenStack, etc. We would also like to provide dedicated studios for these OCCI extensions as components of OCCIWARE STUDIO, i.e., DOCKER STUDIO, GCP STUDIO, etc. This will ensure a specific environment for designing configurations that conform to each promoted extension metamodel.

7 Availability

Readers can find OCCIWARE STUDIO including OCCIWARE METAMODEL and all the model-driven tools at http://github.com/occiware/OCCI-Studio. This work is supported by the OCCIWARE research and development project (www.occiware.org) funded by French Programme d'Investissements d'Avenir (PIA).

References

1. Ahmed-Nacer, M., Tata, S.: Simulation extension for cloud standard OCCIware. In: 25th IEEE International Conference on Enabling Technologies: Infrastructure for Collaborative Enterprises, WETICE, pp. 263–264 (2016)
2. Alshabani, I., Parpaillon, J., Plouzeau, N., Gibello, P.Y., Tata, S.: OCCI Core Architecture. Deliverable D4.1.1, OCCIware Project, May 2015
3. Armbrust, M., Fox, A., Griffith, R., Joseph, A.D., Katz, R., Konwinski, A., Lee, G., Patterson, D., Rabkin, A., Stoica, I., et al.: A view of cloud computing. Commun. ACM **53**(4), 50–58 (2010)
4. Binz, T., Breitenbücher, U., Haupt, F., Kopp, O., Leymann, F., Nowak, A., Wagner, S.: OpenTOSCA – a runtime for TOSCA-based cloud applications. In: Basu, S., Pautasso, C., Zhang, L., Fu, X. (eds.) ICSOC 2013. LNCS, vol. 8274, pp. 692–695. Springer, Heidelberg (2013). doi:10.1007/978-3-642-45005-1_62
5. Binz, T., Breiter, G., Leyman, F., Spatzier, T.: Portable cloud services using TOSCA. IEEE Internet Comput. **3**, 80–85 (2012)
6. Blair, G., Bencomo, N., France, R.B.: Models@run.time. Computer **42**(10), 22–27 (2009)
7. Brandtzæg, E., Mosser, S., Mohagheghi, P.: Towards CloudML, a model-based approach to provision resources in the clouds. In: 8th ECMFA, pp. 18–27 (2012)
8. Bruneliere, H., Cabot, J., Jouault, F.: Combining model-driven engineering and cloud computing. In: 4th edition of Modeling, Design, and Analysis for the Service Cloud Workshop (MDA4ServiceCloud 2010) (2010)
9. Challita, S., Paraiso, F., Merle, P.: Towards formal-based semantic interoperability in multi-clouds: the fclouds framework. In: 10th IEEE International Conference on Cloud Computing (CLOUD). IEEE (2017)

---

[4] https://www.scalair.fr.

10. Ciuffoletti, A.: A simple and generic interface for a cloud monitoring service. In: 4th International Conference on Cloud Computing and Services Science (CLOSER 2014), pp. 143–150 (2014)
11. Ciuffoletti, A.: Open Cloud Computing Interface - Monitoring Extension. Specification Document 1.2, Open Grid Forum, OCCI-WG, January 2016
12. Davis, D., Pilz, G.: Cloud Infrastructure Management Interface (CIMI) Model and REST Interface over HTTP, vol. DSP-0263, May 2012
13. Drescher, M., Parák, B., Wallom, D.: OCCI Compute Resource Templates Profile. Recommendation GFD-R-P.222, Open Grid Forum, October 2016
14. Edmonds, A., Metsch, T., Papaspyrou, A.: Open cloud computing interface in data management-related setups. In: Fiore, S., Aloisio, G. (eds.) Grid and Cloud Database Management, pp. 23–48. Springer, Heidelberg (2011). doi:10.1007/978-3-642-20045-8_2
15. Edmonds, A., Metsch, T.: Open Cloud Computing Interface - Text Rendering. Recommendation GFD-R-P.229, Open Grid Forum, October 2016
16. Edmonds, A., Metsch, T., Papaspyrou, A., Richardson, A.: Toward an open cloud standard. IEEE Internet Comput. **16**(4), 15–25 (2012)
17. Ferry, N., Rossini, A., Chauvel, F., Morin, B., Solberg, A.: Towards model-driven provisioning, deployment, monitoring, and adaptation of multi-cloud systems. In: IEEE Sixth International Conference on Cloud Computing (CLOUD 2013), pp. 887–894 (2013)
18. Fielding, R.T.: Architectural Styles and the Design of Network-based Software Architectures. Ph.D. thesis, University of California, Irvine (2000)
19. Katsaros, G.: Open Cloud Computing Interface - Service Level Agreements. Recommendation GFD-R-P.228, Open Grid Forum, October 2016
20. Martin-Flatin, J.: Challenges in cloud management. IEEE Cloud Comput. **1**(1), 66–70 (2014)
21. Medhioub, H., Msekni, B., Zeghlache, D.: OCNI - open cloud networking interface. In: 22nd International Conference on Computer Communications and Networks (ICCCN), pp. 1–8. IEEE (2013)
22. Mell, P., Grance, T.: The NIST Definition of Cloud Computing. NIST Special Publication 800(145), September 2011
23. Merle, P., Barais, O., Parpaillon, J., Plouzeau, N., Tata, S.: A precise metamodel for open cloud computing interface. In: Proceedings of the 8th IEEE International Conference on Cloud Computing (IEEE CLOUD 2015), pp. 852–859 (2015)
24. Merle, P., Gourdin, C., Mitton, N.: Mobile cloud robotics as a service with OCCIware. In: 2nd IEEE International Congress on Internet of Things (2017)
25. Merle, P., Parpaillon, J., Barais, O.: OCCI Specific Language - Structural Part. Deliverable D2.3.1, OCCIware Project, May 2015
26. Metsch, T., Edmonds, A., Parák, B.: Open Cloud Computing Interface - Infrastructure. Recommendation GFD-R-P.224, Open Grid Forum, October 2016
27. Metsch, T., Mohamed, M.: Open Cloud Computing Interface - Platform. Recommendation GFD-R-P.227, Open Grid Forum, October 2016
28. Mohamed, M.: Generic Monitoring and Reconfiguration for Service-based Applications in the Cloud. Ph.D. thesis, INT, Evry, France (2014)
29. Mohamed, M., Amziani, M., Belaid, D., Tata, S., Melliti, T.: An autonomic approach to manage elasticity of business processes in the cloud. Future Gener. Comput. Syst. **50**, 49–61 (2015)
30. Mohamed, M., Belaïd, D., Tata, S.: Monitoring and reconfiguration for OCCI resources. In: 5th IEEE International Conference on Cloud Computing Technology and Science (CloudCom 2013), vol. 1, pp. 539–546. IEEE, Bristol (2013)

31. Mohamed, M., Belaïd, D., Tata, S.: Autonomic Computing for OCCI Resources. Technical report, Telecom Sud Paris, January 2014. http://www-inf.it-sudparis. eu/SIMBAD/tools/OCCI/autonomic/AutonomicComputingForOCCIResources. html

32. Nyrén, R., Edmonds, A., Metsch, T., Parák, B.: Open Cloud Computing Interface - HTTP Protocol. Recommendation GFD-R-P.223, Open Grid Forum, October 2016

33. Nyrén, R., Edmonds, A., Papaspyrou, A., Metsch, T., Parák, B.: Open Cloud Computing Interface - Core. Recommendation GFD-R-P.221, Open Grid Forum

34. Nyrén, R., Feldhaus, F., Parák, B., Sustr, Z.: Open Cloud Computing Interface - JSON Rendering. Recommendation GFD-R-P.226, Open Grid Forum, October 2016

35. OMG: Object Constraint Language, Version 2.4. OMG Specification OMG Document Number: formal/2014-02-03, Object Management Group, February 2014

36. Paraiso, F., Challita, S., Al-Dhuraibi, Y., Merle, P.: Model-driven management of docker containers. In: 9th IEEE International Conference on Cloud Computing, CLOUD 2016, San Francisco, CA, USA, 27 June–2 July 2016, pp. 718–725 (2016)

37. Quinton, C., Haderer, N., Rouvoy, R., Duchien, L.: Towards multi-cloud configurations using feature models and ontologies. In: Proceedings of the 2013 International Workshop on Multi-cloud Applications and Federated Clouds, pp. 21–26 (2013)

38. Yangui, S., Marshall, I.J., Laisne, J.P., Tata, S.: CompatibleOne: the open source cloud broker. J. Grid Comput. **12**(1), 93–109 (2014)

39. Yangui, S., Tata, S.: CloudServ: PaaS resources provisioning for service-based applications. In: 27th IEEE International Conference on Advanced Information Networking and Applications (AINA 2013), pp. 522–529. IEEE (2013)

40. Yangui, S., Tata, S.: An OCCI compliant model for PaaS resources description and provisioning. Comput. J. **59**, 308–324 (2016)

Knowledge Is at the Edge! How to Search in Distributed Machine Learning Models

Thomas Bach[(✉)], Muhammad Adnan Tariq, Ruben Mayer,
and Kurt Rothermel

Institute of Parallel and Distributed Systems, University of Stuttgart,
Stuttgart, Germany
{thomas.bach,adnan.tariq,ruben.mayer,
kurt.rothermel}@ipvs.uni-stuttgart.de
https://www.ipvs.uni-stuttgart.de

Abstract. With the advent of the internet of things and industry 4.0 an enormous amount of data is produced at the edge of the network. Due to a lack of computing power, this data is currently send to the cloud where centralized machine learning models are trained to derive higher level knowledge. With the recent development of specialized machine learning hardware for mobile devices, a new era of distributed learning is about to begin that raises a new research question: How can we search in distributed machine learning models? Machine learning at the edge of the network has many benefits, such as low-latency inference and increased privacy. Such distributed machine learning models can also learn personalized for a human user, a specific context, or application scenario. As training data stays on the devices, control over possibly sensitive data is preserved as it is not shared with a third party. This new form of distributed learning leads to the partitioning of knowledge between many devices which makes access difficult. In this paper we tackle the problem of finding specific knowledge by forwarding a search request (query) to a device that can answer it best. To that end, we use a entropy based quality metric that takes the context of a query and the learning quality of a device into account. We show that our forwarding strategy can achieve over 95% accuracy in a urban mobility scenario where we use data from 30 000 people commuting in the city of Trento, Italy.

Keywords: Knowledge retrieval · Distributed knowledge · Query routing

1 Introduction

In many areas such as stock trading, drug design, manufacturing, and urban mobility [9,16] machine learning is the key enabler of optimization and driver of performance [9,15,21]. Besides choosing the right machine learning algorithm and applying it right, the amount of training data is key to success [9]. While the selection and application of machine learning algorithms is a research field

© Springer International Publishing AG 2017
H. Panetto et al. (Eds.): OTM 2017 Conferences, Part I, LNCS 10573, pp. 410–428, 2017.
https://doi.org/10.1007/978-3-319-69462-7_27

of its own, enough training data is needed to calibrate machine learning models, such that they can make correct predictions.

With the advent of paradigms like the Internet of Things, smart city, and Industry 4.0, data will be abundantly available [21]. Cisco, for example, estimates that the I.o.T. alone will generate over 400 ZB of data annually, by 2020 [1]. In particular, the proliferation of smart phones made training data from different sensors, such as accelerometers, cameras, microphones, and GPS units widely available [15]. Google reported that by centralizing a great amount of training data for speech recognition from Google voice search [32], it became possible to train high-quality feature-rich machine learning models for voice recognition [12].

The current approach to share such information is massive centralization. In many application scenarios, however, centralization of possibly sensitive data is not desirable as centralized data is regularly subject to breaches [14]. Today, it is well known that it is possible to derive knowledge of a user's habits, such as his home and work location from his GPS traces [4]. Many human users are thus unwilling, at least uncomfortable sharing such private information [39].

The common approach to tackle this issue is to distribute the computing infrastructure [24], and even push computing towards the edge of the network [11,26]. The upcoming trend of *fog computing* [13,23] supports this by providing computational resources close to the edge, creating a computational continuum that spans from the edge devices to the centralized cloud data centers. Sensitive data can then be processed directly on devices that are under control of the user or on fog nodes very close to them. In this respect, mobile device manufactures are building specialized machine learning hardware that enables machine learning at the edge[1]. Machine learning at the edge has many additional advantages, it allows for example to keep the user in the loop, learn personalized, and offer low latency feedback [11,26]. Google for example has recognized this trend and made approaches, where personalized learning is done directly on smart phones [25]; however, the generated local machine learning models are synchronized with a central server. Google argues, that by processing the data locally, privacy is increased compared to an entirely centralized approach.

Completely decentralized learning also holds great challenges. As training data is not centralized, each machine learning model is only trained with respect to its local experiences. In particular, such models may become local experts that are very good in predicting local phenomena. In a medical scenario this might be an advantage, as a model could learn the peculiarities of one specific patient and enable a detailed analysis. For other use cases, this is not enough. In an urban mobility scenario, for example, users are usually more interested, in traffic conditions in another part of the city which they have never seen before. Distributed learning holds the opportunity to learn about local phenomena in great detail on the one hand, on the other hand it creates the problem of locating specific knowledge.

---

[1] Mark Gurman; BloombergTechnology, Apple Is Working on a Dedicated Chip to Power AI on Devices: https://www.bloomberg.com/news/articles/2017-05-26/apple-said-to-plan-dedicated-chip-to-power-ai-on-devices.

To address this problem, in this paper, we present methods to route a query for specific knowledge through a network of nodes (local experts) that each train a local machine learning model. Our goal is to forward such a query to the node that can answer it best. In particular, we look at scenarios where knowledge in the form of machine learning models is fully distributed. Such a fully decentralized approach holds three mayor difficulties: First, we cannot assume a central index of all available knowledge. Second, the different devices (nodes) might learn based on different local observations and contexts. Third, parts of the knowledge changes or becomes outdated over time.

To this end, our contributions are: (1) We propose a decentralized routing strategy that forwards queries for specific knowledge towards nodes that can answer them best. (2) We propose methods to maintain routing tables that guide the forwarding of such a query. (3) We use entropy to evaluate how good a given query can be answered based on its context and the local machine learning model of a node. (4) We develop a modified form of the Barabasi Albert model [2] to generate a scale free topology that clusters network nodes with similar knowledge close together to deal with heterogeneous knowledge. With its scale free properties such a topology provides short paths between any two network nodes and is robust against node failures. (5) We show that we can achieve over 95% accuracy when using synthetic data and data generated by a mobility simulator where 30 000 people commute in the city of Trento, Italy, in the context of different weather conditions, times of the day, and traffic conditions.

2   System Model and Problem Formulation

We assume a distributed system of fog [13, 23] nodes that each train a machine learning model based on local observations. These nodes can join and leave the system at any time and range from user managed devices such as smart phones, laptops, and desktop computers, to infrastructure based services located in data centers, such as private clouds. All nodes communicate directly over a undirected, scale free topology, i.e. power law distributed node degree and short paths. These properties make scale free networks particularly well suited for our problem as they connect two arbitrary nodes (e.g. the source and optimal destination of a query) with a small number of hops. Furthermore, many existing networks such as social networks or the internet already show scale free properties [3]. Maintaining such a topology is also a well studied research problem [2, 17].

In order to learn, all nodes maintain a graphical machine learning model as shown in Fig. 1. Graphical models (Probabilistic Graphical Models, PGM) such as Bayesian networks or conditional random fields have a wide range of machine learning applications in computer vision, natural language processing, and bioinformatics [36]. In a PGM, random variables are represented as nodes and dependencies between them as edges of a graph. This gives them great flexibility in modeling complex dependencies.

We assume, that all network nodes maintain structural identical PGMs that are continuously evolving based on individual training data (observations).

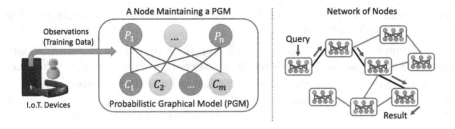

Fig. 1. System overview. (Color figure online)

This training data can be generated either by the nodes themselves, e.g. a smart phone generates GPS traces from its internal sensors, or can be received from other sensors, such as a wristband sensor that collects cardiovascular data. Furthermore, the different nodes learn about different phenomena in different contexts, leading to individual expertise of the different nodes. In an urban mobility scenario, for example, two nodes could learn about traffic conditions in different parts of the city at different times of the day. In particular, this means that the different nodes train different subsets of random variables. In consequence, not all nodes can predict all random variables equally well. A reliable prediction about the outcome of a specific random variable thus requires to search for (i.e. query) the network node that has best training.

Given any PGM, we categorize the random variables of the PGM into two groups: predicting variables and context variables. Predicting variables are the subset of random variables that we want to predict based on a certain context, modeled as context variables. In an urban mobility scenario, where we want to predict the travel time for the streets in a city with a PGM, the travel time for each street would be represented as a predicting variable. Factors that influence this travel time, such as weather or time of the day would be represented as context variables. In Fig. 1 this categorization is reflected by the color of the random variable. Predicting variables P_n in blue, context variables C_n in green and untrained random variables of both types in grey.

In this context, we define a query as a request to predict the outcome for a specific predicting variable in a certain context. In this respect, we define context as given assignments for a set of context variables. Our goal is to forward a query to a node that can answer it with the highest possible quality (we introduce quality in Sect. 3).

2.1 Formal Model and Problem Statement

More formally, we assume a set of network nodes $N = \{N_1, ..., N_n\}$ that are connected over a scale free topology. Each node N_x holds a PGM that consists of a graph $G = (V, E)$ of discrete random variables V where dependencies between random variables are modeled by the edges E. We classify the random variables into predicting variables $P_n = \{p_n^1, ..., p_n^m\}$ and context variables $C_n = \{c_n^1, ..., c_n^{m'}\}$, where p_n and c_n are possible assignments. Each random variable must be classified either as predicting variable or as context variable ($V = C \cup P$ and $C \cap P = \emptyset$).

We define a query $\overrightarrow{q} = (P_x, \{c_x^y, ..., c_{x'}^{y'}\}, H, R, Q, \overrightarrow{N})$, where P_x is the random variable that needs to be predicted in context of a given set of assignments ($\{c_x^y, ..., c_{x'}^{y'}\}$) for a subset of context variables and a limited number of hops H (number of times a query can be forwarded). Furthermore, the query contains a field to hold the prediction result R, the estimated quality of this result Q and a vector of visited nodes $\overrightarrow{N}$.

We can now define the concrete knowledge retrieval problem. Given (i) a set of nodes $\{N_1, ..., N_n\}$ holding (ii) continuously evolving, heterogeneously trained PGMs and (iii) a query $\overrightarrow{q}$ for specific knowledge, our goal is to maximize the retrieval quality of a query while forwarding it only H times.

In the following, we first establish a notion of knowledge quality in the context of PGMs and describe how to measure the quality with that a node can answer a query. (cf. Sect. 3). Based on this quality metric, we present methods to route a query towards the node that can answer it with highest quality in Sect. 4.

3   Entropy, a Measure of Training Quality

In this section we discuss how we measure the training quality of a PGM. Based on this quality, we describe how to estimate the quality with that a PGM can answer a query for specific knowledge.

As stated above, Probabilistic Graphical Models (PGM) consist of interdependent random variables. Such models are usually designed by an expert who puts his domain knowledge in the structure of the model, e.g. chooses the random variables and their conditional dependencies such that the model reflects the dependencies in the real world. Training data is then used to converge the probability distributions of the random variables from a uniform distribution to the distributions of the real world. In other words, if an increasing amount of training data is fed into the machine learning model, the uncertainty of the model decreases. In machine learning, this uncertainty (or often called surprise) of a model is measured by calculating the entropy of its random variables [20].

If, for example, we want to learn the probability of a coin flip being "Heads Up", we could use a very simplistic model that only consists of one random variable X with possible outcomes $\{0\%, ..., 100\%\}$. We now flip a fair coin several times and use the results to train the random variable as shown in Fig. 2. With an increasing number of observations (or coin tosses) the "true" probability distribution establishes and the entropy decreases.

Given a random variable X with possible assignments $\{x_1, ..., x_n\}$ we can calculate the entropy ($H(X)$) as the average surprise (or uncertainty) of the random variable (cf. Eq. 1). The logarithm of the probability of an assignment $log(P(x_n))$ represents the amount of surprise we perceive for the specific outcome [20]. The "surprises" of all possible outcomes are then weighted by their probability $P(x_n)$ and summed up to one entropy value often also called self-information [20].

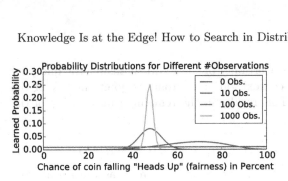
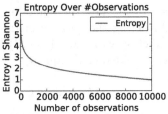

Fig. 2. Probability distribution (l) and entropy (r) of a coin flip for different number of observations.

$$H(X) = -\sum_{k=1}^{n} P(x_k) log_2 P(x_k) \qquad (1)$$

In complex machine learning models, we usually want to predict the outcome of multiple random variables. In these cases we calculate the joint entropy $H(X_0, ..., X_n)$ in order to describe their "joint uncertainty", e.g. $\{X_0, ..., X_n\}$ (cf. Eq. 2). Similar to entropy for one random variable, the idea is to calculate the uncertainty for each combination of random variables involved and weight these combinations w.r.t. their probabilities.

$$H(X_0, ..., X_n) = -\sum_{x_0 \in X_0} \cdots \sum_{x_n \in X_n} P(x_0, ..., x_n) log_2 P(x_0, ..., x_n) \qquad (2)$$

The joint entropy describes the "total uncertainty" of multiple random variables. Given a random variable X_0 that is dependent on the outcome of other random variables $\{X_1, ..., X_n\}$, the entropy $H(X_0, ..., X_n)$ denotes the uncertainty of the outcome given that we don't know anything about the outcome of $\{X_0, ..., X_n\}$. If we now gain information about the outcome of one of the variables (e.g. X_0 is a context variable and its outcome is given by a query), we can derive the remaining entropy (uncertainty) according to the chain rule of conditional entropy by subtracting the entropy $H(X_0)$ from the total entropy $H(X_0, ..., X_n)$, cf. Eq. 3. In the following, we use this chain rule to calculate the uncertainty which the PGM of a specific network node has to answer a query.

$$H(X_1, ..., X_n | X_0) = H(X_0, ..., X_n) - H(X_0) \qquad (3)$$

In Sect. 2 we divided the random variables of a PGM in two categories, predicting variables and context variables, where each predicting variable is dependent on the outcomes of a number of independent context variables. For a given predicting variable P_0 that is dependent on the outcome of context variables $\{C_0, C_1, C_2\}$ we can calculate the joint entropy $H(P_0, C_0, C_1, C_2)$ and individual entropies for the context variables $H(C_0), H(C_1), H(C_2)$. Given a query $\vec{q} = (P_0, \{c_0^1, c_1^3\}, ...)^2$ for P_0 with observed outcomes $c_0^1 \in C_0$ and $c_1^3 \in C_1$

---

[2] For better readability we do not state all fields of the query here (i.e. $H, R, Q, \vec{N}$).

we can calculate the remaining uncertainty of the PGM to answer the query by subtracting the entropy of the context variables from the joint entropy of the predicting variable (cf. Eq. 4). This results in the remaining uncertainty of the PGM to answer the query.

$$H(P_0, C_3|C_0, C_1) = H(P_0, C_0, C_1, C_2) - H(C_0) - H(C_1) \qquad (4)$$

For the rest of this paper we will also refer to entropy as the learning or training quality of a PGM.

4   Routing

Now that we have established how we can measure the training quality of the PGMs of each node, we describe how we build routing models and use them to forward queries towards the node that can answer them best. In contrast to a classic routing table, where a network address is associated with a specific port (outgoing link), each node N_i maintains a routing model $RM_{N_n}^{N_i}$ for each neighbor N_n. Each $RM_{N_n}^{N_i}$ serves as a descriptive model that cumulatively represents the knowledge available over the respective outgoing link. Keeping link-individual routing models is necessary, because we need to calculate the estimated answering quality of a query with respect to its context (cf. Sect. 3) as storing all possible combinations of contexts easily becomes too much overhead.

In order to process a query, a node first tries to improve the prediction R of a query based on its local PGM. In the next step, the node uses its routing models to determine to which neighbor the query should be forwarded. As this is an approximate routing process, we limit the number of hops (H) that a query $\vec{q}$ is forwarded before the result is returned to the sender.

In order to improve the retrieval quality we use a network topology that clusters nodes with similar knowledge (nodes that have learned about a similar set of predicting nodes). We can then optimize our routing models by maintaining context information of predicting variables only for predicting variables that have been learned by the cluster. This leads to a double-staged routing approach, where a query for a predicting variable P_n is first forwarded to a cluster of nodes that have learned about P_n. In the second stage, the query is then forwarded within a cluster to a node that has learned it in the requested context. In the following we describe how we build the routing tables, forward a query, maintain the network topology and deal with loops in the topology in detail.

4.1   Building the Routing Tables

The routing models $RM_{N_n}^{N_i}$ that each node N_i maintains for every neighbor N_n store entropy values of random variables (predicting variables and context variables) and represent the knowledge available to the respective neighbor N_n. In order to maintain them in a proactive fashion, each node sends summaries of its entropy values stored in its local PGM and its routing models to its neighbors, whenever they have changed above a certain threshold and a minimum amount of time has passed since the last update. This makes sure that all neighbors have up-to-date information about their neighbors and at the same time avoids that the network is flooded with updates.

In the following, we explain this forwarding process with respect to a set of nodes $\{N_1, N_2, N_3\}$ in detail. For better readability and without loss of generality this example is with respect to one predicting variable P_0 and context variables $\{C_1, C_2\}$. A further simplification is the use of a flat topology, i.e. all nodes are connected in a line (cf. Fig. 3). In the given example, N_3 has only one neighbor (N_2) and therefore can directly forward its set of entropy values $\{H(P_0, C_1, C_2)_{PGM},$ $H(C_1)_{PGM}, H(C_2)_{PGM}\}$ from its PGM to its neighbor N_2, where they are used as entries in the routing table $RM_{N_3}^{N_2}$ (cf. Fig. 3A). These entropy values represent the learning quality for P_0 available over the edge $N_2 \rightarrow N_3$. As described in Sect. 3 this values can be used to calculate the quality with which a query in any context (i.e. $\{\{C_0\}, \{C_1\}, \{C_0, C_1\}\}$) can be answered.

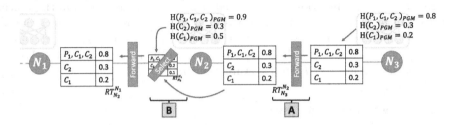

Fig. 3. Forwarding entropy values in same context.

When N_2 has received the set of entropy values from N_3, it decides to update the entropy values send to N_1. In contrast to N_3, N_2 cannot send the entropy values from its PGM directly as it has to consider the entropy received from N_3. Node N_2 needs to select which set of entropy values ($\{H(P_0, C_1, C_2), H(C_1), H(C_2)\}$) is forwarded. In order to determine this, it compares the joint entropy value of its local PGM ($H(P_0, C_1, C_2)_{PGM} = 0.9$) with the joint entropy value of its routing table $RM_{N_2}^{N_1}$ ($H(P_0, C_1, C_2) = 0.8$) As the local joint entropy is higher ($0.9 > 0.8$) it forwards the complete entropy set for P_0 received from N_3 ($\{H(P_0, C_1, C_2) = 0.8, H(C_1) = 0.2, H(C_2) = 0.3\}$) to N_1 (cf. Fig. 3B).

Generalization: If, in contrast to our example, a node has multiple neighbors, it stores the entropy value sets it receives from them in a separate routing table for each neighbor. In order to decide which entropy set (e.g. $\{H(P_0, C_1, C_2), H(C_1), H(C_2)\}$) should be forwarded, we compare all the entropy sets, including the entropy of the local PGM by their joint entropy values (i.e. $H(P_0, C_1, C_2)$) and forward the set with the lowest joint entropy. This forwarding approach makes sure that for each predicting variable (P), the lowest joint entropy (e.g. $H(P_0, C_1, C_2) = 0.8$) and the entropy values of its corresponding context variables (e.g. $H(C_1) = 0.2, H(C_2) = 0.3$) are propagated.

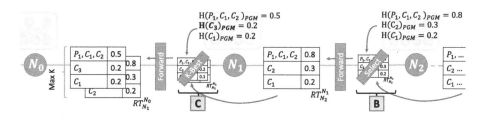

Fig. 4. Forwarding entropy values in different context.

In cases where predicting variables have been trained with respect to different context variables by different nodes, entropy values cannot simply be merged. For example, N_2 has trained P_1 with respect to context variables $\{C_1, C_2\}$ and N_1 has trained P_1 with respect to $\{C_1, C_3\}$ (cf. Fig. 4). In such cases, we store and forward up to $K \in \mathbb{N}$ different context combinations for each predicting variable P (cf. Fig. 4C). In this respect, K is a design parameter that determines how many context combinations for one predicting variable are stored in the routing tables. In general K is dependent on the number of relevant contexts in a concrete scenario. In cases where we have to limit them we can use existing dimension reduction algorithms to select the most important context variables. We will discuss the influence of K in our evaluations in Sect. 5.

4.2   Forwarding of a Query

As discussed in Sect. 2, a query $\overrightarrow{q}$ is a message issued by one node in the network, to retrieve a prediction for a specific predicting variable P_x in a given context, represented by a set of assignments $(\{c_x^y, ..., c_{x'}^{y'}\})$ for a subset of context variables $\{C_x, ..., C_{x'}\} \in C$. A query is forwarded from one node to another until the predefined number of hops, H, has been reached.

When a node N_i receives a query $\overrightarrow{q}$ it first decreases the hop counter H of the query and then determines if the query can be improved by the local PGM, by computing the entropy of answering the query (cf. Sect. 3). The resulting entropy value is then compared to the entropy value Q in the query. If the entropy value in the query is higher than the locally computed value (i.e. the node has less uncertainty cf. Sect. 3), the node predicts the outcome of the query with its PGM and updates the result field R and the quality field Q of the query $\overrightarrow{q}$ accordingly. If the hop counter H of the query is greater then zero $(H > 0)$ the node uses its local routing models to select a neighbor to which the query is forwarded.

In order to select a neighbor to send the query to, node N_i compares all routing models $RM_{N_n}^{N_i}$ by computing the conditional entropy $H(P_x|C_x, ..., C_{x'})$. This is done by subtracting the entropy values of the context variables $(\{H(C_x), ..., H(C_{x'})\})$ from the joint entropy value $H(P_x, C_x, ..., C_{x'})$ stored in the routing tables $RM_{N_n}^{N_i}$ (cf. Sect. 3). The query is then forwarded to the neighbor with the smallest conditional entropy $H(P_x|C_x, ..., C_{x'})$.

Discussion: So far, we have described how routing models are built and how they are used to forward a query. The maintenance of entropy values in multiple routing models, especially for different context combinations, produces significant overhead. The number of possible context combination grows according to the binomial coefficient. If, for example, the nodes learns w.r.t. 5 out of 10 possible context variables, there are already 252 possible combinations. In order to reduce this overhead, we cluster nodes that have learned about a similar set of predicting variables. Based on this clusters, our routing protocol uses two optimizations. First, nodes only forward entropy values from their PGM if they have a minimum level of quality (i.e. the entropy value is below a certain threshold). Second, if a node has not reached a minimum quality for a predicting variable P_n it only maintains a single joint entropy value for P_n (no entropy values for context) in its routing models. This single joint entropy value can then be used to forward a query to the next cluster that has learned about P_n, where context sensitive routing, as described above, is performed. In the following we describe how we maintain such a clustered network topology.

4.3 Topology Maintainance

As mentioned in previous sections, the topology of our network should exhibit scale free properties, such as a power law distributed node degree and short paths. Additionally we want to cluster nodes that have learned about a similar subset of predicting variables. In order to manage the overhead of maintaining routing models for each neighbor, we also need to give each node the option to limit the maximum number of neighbors. This limit can be determined node-individually, e.g., dependent on the amount of memory consumed by the routing models. In order to generate such a topology, we use a modified version of the Barabasi Albert model [2]. Our idea is to make the preferential attachment of the Barabasi Albert model dependent on node similarity. The original algorithm starts with an initial set of m_0 nodes and connects a new node N_n to an existing node N_e with a probability proportional to the edge degree of the existing nodes. This way, N_n can connect with up to $m < m_0$ existing nodes. In the original algorithm the probability of a node N_x connecting to an existing node N_y is given by $p_{x \to y} = \frac{k_y}{\sum_j k_j}$ where k_y is the degree of the existing node divided by sum of all edge degrees. We multiply this probability with a similarity factor $S(PGM_{N_x}, PGM_{N_y}) \to [0, 1]$ that describes the similarity between two PGM (e.g. PGM_{N_x} and PGM_{N_y}). If a node already has reached its individual maximum number of edges (*edgelimit*) we set the probability to zero as shown in Eq. 5.

$$p_{x \to y} = \begin{cases} \frac{k_y}{\sum_j k_j} \cdot Similarity(PGM_{N_x}, PGM_{N_y}) & \text{if } k_y \leq edgelimit \\ 0 & \text{else} \end{cases} \quad (5)$$

Let there be two nodes $\{N_1, N_2\}$ where N_1 has trained the set $A = \{P_1, P_2, P_3\}$ and N_2 the set $B = \{P_1, P_3, P_4, P_5\}$ of predicting variables of their

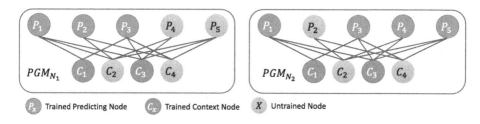

Fig. 5. Two PGMs with similar learning.

PGM as shown in Fig. 5. We define the similarity between them as the size of the intersection between A and B divided by the minimum cardinality of A and B cf. Eq. 6.

$$S(PGM_{N_1}, PGM_{N_2}) = \frac{|A \cap B|}{\min(|A|, |B|)} = \frac{2}{\min(3, 4)} = \frac{2}{3} \tag{6}$$

To demonstrate that this modified algorithm still produces a topology with power law distributed node degree, we plotted the number of edges for a network of 600 nodes using the original Barabasi Albert Model and our modified version where we limited the number of edges to 60. The major difference is, that our modified version exhibits multiple nodes with degree 60 instead of having several nodes with degree >60 (Fig. 6).

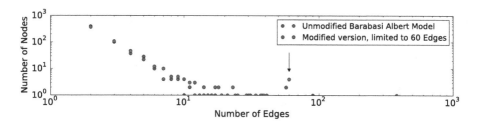

Fig. 6. Comparison of node degree between the modified and unmodified B. A. Model.

In order to deal with loops in the topology we slightly increase the forwarded entropy value with every hop. This way, propagated entropy values increase more over longer propagation paths than over shorter ones. Queries will then always be forwarded over shortest paths with lower entropy values.

5 Evaluation

In this section we evaluate the above presented aggregation based routing strategy (ABS) with respect to different network sizes, number of context nodes, context combinations, number of hops, and the context diversity parameter K (cf. Sect. 4) on synthetic training data and on data from an urban mobility scenario, in the following referred to as "Trento data". We compare our strategy to a directed random walk approach commonly used in unstructured Peer to Peer networks cf. Sect. 6.

We implemented our routing strategy (cf. Sect. 4) in the Peer to Peer simulator PeerSim [27] and performed our evaluations on an Open Stack virtual machine with 64 cores and 256 GB RAM running Ubuntu 16.04. We used Peer-Sim to instantiate up to 32 768 (2^{15}) network nodes. To represent the Probabilistic Graphical Models (PGM), each node individually trained a Bayesian network consisting of an experiment dependent number of random variables. In the following we will state for each experiment how many predicting variables (P_n) and context variables (C_n) were used to form a PGM as described in Sect. 2. The Bayesian networks on the individual nodes were then trained either on synthetic training data (Gaussian distributed observations) or on data from the Trento data set respectively. The use of synthetic data gave us the ability to flexibly generate experiment setups with any number of predicting or context variables. Based on the trained nodes, the topology between the nodes was created as described in Sect. 4.

Based on this setup, we used the cycle based engine of PeerSim to perform our evaluations. In each cycle, each node first propagated its knowledge and then issues a query that is forwarded as described in Sect. 4. In order to determine the accuracy of the result, we compared the entropy of the result with the optimal entropy which was determined by an exhaustive search over all nodes.

The "Trento data" data originates from a real world simulator for collaborative and distributed learning [29] developed at the German center for artificial intelligence (DFKI) to generate large-scale, realistic data sets for machine learning. The simulator is based on the city map of Trento in Italy. It features genuine bus tables, weather, and commuting statistics of the city. Based on this data, we used the simulator to hosts 30 000 autonomous agents that emulate the behavior of citizens, even forming traffic jams that lead to different travel times at different times of the day, days of the week under different weather conditions for different road segments of the Trento street graph. In our experiments we used weather, time of the day, and day of the week as context nodes to predict travel times for different road segments that we used as predicting variables.

In our first evaluation (Fig. 7) we compare the retrieval accuracy of our aggregation based strategy (ABS) on synthetic data and the Trento data averaged over an increasing number of cycles with the random walk approach. In this evaluation we used 1024 nodes that we trained on 100 different predicting variables and 3 context variables (weather, time of the day, and day of the week in the Trento data). We can see, that in the first cycle, the accuracy is low, as most of the routing models are empty. As knowledge gets propagated through the network, the retrieval quality increases until it settles around 90%. This evaluation already indicates the good performance of our algorithm on synthetic data and on the Trento data.

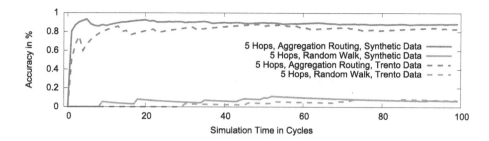

Fig. 7. Retrieval accuracy averaged over time (cycles) compared.

In Fig. 8 we evaluate the performance of ABS at different network sizes (up to 32 768 nodes) with the random walk approach using synthetic and Trento data. Just like in the previous evaluation we used 100 predicting nodes and 3 context nodes according to the Trento data. We forwarded each query $2 \cdot log(network\,size)$ times. In comparison to the synthetic data, the standard deviation (indicated by the whiskers in the graph) is a bit higher for the Trento data. The reason for this is, that the knowledge about some predicting variables is scarce and thus harder to find.

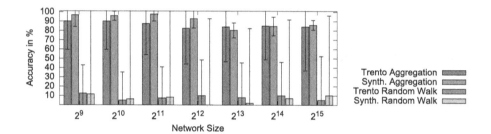

Fig. 8. Network size and retrieval quality.

Figure 9 shows the influence of the number of hops. For this evaluation we used a network of 512 nodes, 10 predicting variables, and 3 context variables on synthetic data. We can see that with an increasing number of hops not only the accuracy increases, but also the standard deviation (whiskers) decreases. In general the number of required hops grows proportional to the network diameter. As we are using a free scale topology, this is approximately logarithmic to the number of nodes (cf. Sect. 4).

In the following, we will have a closer look at the influence of the number of possible combinations of context variables and their influence on routing accuracy. We introduced the problem of different context combinations in Sect. 4 and tackled it by introducing a parameter K that defines how many different context combinations are stored in the routing models. In Fig. 10 we can see that keeping

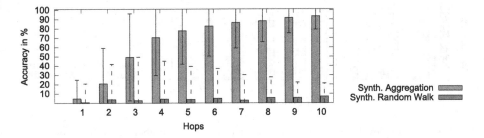

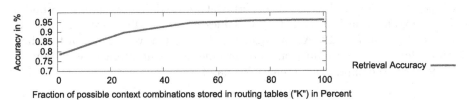

Fig. 9. Retrieval Quality w.r.t different number of hops.

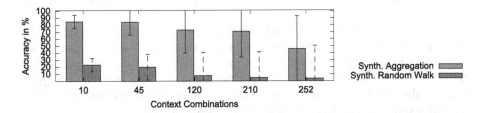

Fig. 10. Retrieval quality with respect to K.

about 50% of all possible context combinations already leads to a reasonably good retrieval accuracy.

Figure 11 shows how the accuracy degrades with an increasing amount of context combinations for a network of 512 nodes, $K = 10$, one predicting variable, and 10 context variables of which up to 4 have been trained. According to the binomial coefficient this creates up to 252 possible context combinations that could have been learned. We can see that with an increasing amount of possible context, not only the accuracy decreases, but also the standard deviation of the accuracy increases. The surprisingly good performance and low standard deviation for the random strategy for 10 context combinations can be explained when realizing that there are potentially up to 51 nodes that have learned the respective contexts.

Fig. 11. Possible context combination for $K = 10$, 512 nodes, and 3 hops.

When K is chosen around 50% of the number of context combinations, retrieval accuracy is fairly independent form the number of context variables

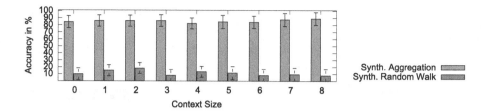

Fig. 12. Retrieval accuracy with respect to different number of context nodes

used, as shown in Fig. 12. For this experiment we used a network of 2048 nodes that were trained on 500 predicting variables in up to 10 contexts, 25 context combinations and $k = 12$. As we highlighted in Sect. 4 we can use of the shelf methods for dimension reduction to determine most important context combinations w.r.t. the application scenario at hand.

6   Related Work

Information retrieval from peer to peer (P2P) systems and machine learning are both well studied areas. Today, machine learning is often done in Big Data scenarios, where all training data is logically centralized. There exist many approaches to distribute the training data and machine learning models between several machines, for example on a cluster. These systems have the benefit of a centralized controller that actively manages how information is distributed between the different machines. In such scenarios, communication-efficient distribution of data between machines is a hard research problem of its own [22].

In this paper we argue that with the trend of decentralized computing, machine learning is coming to the edge of the network (cf. Sect. 1). On the one hand, this enables many benefits such as low latency access and the ability to maintain control over sensible information. On the other hand, without a central index structure, the problem of searching in distribute machine learning models is created. Therefore we focus on related work that tackles the problem of content sharing in P2P networks and discuss how fit these approaches are for knowledge retrieval.

First P2P systems such as Chord [35], CAN [30] and Pastry [31] tackled the problem of how to find specific data items in a distributed system. Except for CAN, most of these early work focuses on retrieval based on one unique key such as a hash value. CAN allows for multidimensional keys in euclidean space to locate data items. All approaches, however, share the draw back that they can only retrieve items that are identified by a unique index.

The second generation of P2P systems (e.g. Mercury [7], Squid [33], and Znet [34]) introduced the support for more complex, multidimensional, and range queries. This enabled searches like *Find persons age $\geq$ 10 and age $\leq$ 20 and gender = female*. These systems enable search in multidimensional space, where Data locality is usually achieved by dimension reduction techniques, such as space filling curves (e.g. [10]). A general problem is that range queries might be too restricted in cases with sparse data. For example, if there are very few results for the above mentioned query, results for persons slightly older than 20 years would also be interesting for the user.

This gap was filled by research centered around nearest neighbor queries for P2P systems, like pSearch [37] and Semantic Small World [17]. The main idea is to provide nearest neighborhood search for multidimensional queries. Most work focuses on selecting important dimensions [18,28] or methods to form an overlay network that connects nodes with similar information [38]. Just like in this paper, some of these approaches also form a small world topology [17] that has a small network diameter, which makes each node reachable with only a few hops and enables efficient routing. There also exists work that relies on a predefined similarity metric, e.g. the euclidean distance, and retrieves the k nearest data items in a large collection of high dimensional data [6,8,18,19].

All these approaches have been designed to retrieve items that are explicitly defined by matching a specific identifier (id, hash value), fall in a specific range of a set of attributes, or are close to a given query. In order to retrieve knowledge this notion has to be extended by some sort of confidence metric that can take the quality of available knowledge (information) into account. Such a confidence metric needs to express the expertise of a node, reflecting for example that it holds a lot of similar information [5] or can do reliable prediction. In our previous work [5] we have tackled this issue for knowledge modeled as N-Dimensional point-clouds. We proposed a point-cluster-based confidence metric that took the variance and number of points in each cluster as an indicator of quality into account.

To the best of our knowledge, there is no peer to peer based approach that is specifically designed so search for knowledge in graphical machine learning models.

7   Conclusion and Future Work

In this paper we have stressed the importance of machine learning at the edge of the network. We argued that with an increasing amount of fog computing devices carrying specialized machine learning hardware knowledge becomes inherently distributed. In this setting we defined and tackled the problem of finding and retrieving specific knowledge. We showed that our aggregation based routing approach can retrieve specific knowledge with over 95% accuracy even if it was learned in many different contexts.

We think that the field of distributed knowledge management is in its infancy and will rapidly gain importance. With our generic notion of predicting variables and context variables our retrieval strategy is flexible and can be adapted to many future application scenarios in health care, manufacturing, and urban mobility.

Acknowledgment. The authors would like to thank the European Union's Seventh Framework Programme for partially funding this research through the ALLOW Ensembles project (project 600792).

References

1. Cisco global cloud index : Forecast and methodology, 2013–2018. Online (2014)
2. Albert, R., Barabási, A.-L.: Statistical mechanics of complex networks. Rev. Mod. Phys. **74**(1), 47 (2002)
3. Amaral, L.A.N., Scala, A., Barthelemy, M., Stanley, H.E.: Classes of small-world networks. Proc. Natl. Acad. Sci. **97**, 11149–11152 (2000)
4. Ashbrook, D., Starner, T.: Using GPS to learn significant locations and predict movement across multiple users. Pers. Ubiquit. Comput. **7**(5), 275–286 (2003)
5. Bach, T., Tariq, M.A., Mayer, C., Rothermel, K.: Utilizing the hive mind – how to manage knowledge in fully distributed environments. In: Debruyne, C., Panetto, H., Meersman, R., Dillon, T., Weichhart, G., An, Y., Ardagna, C.A. (eds.) OTM 2015 Conferences. LNCS, vol. 9415, pp. 219–236. Springer, Cham (2015). Christophe Debruyne
6. Batko, M., Gennaro, C., Zezula, P.: A scalable nearest neighbor search in P2P systems. In: Ng, W.S., Ooi, B.-C., Ouksel, A.M., Sartori, C. (eds.) DBISP2P 2004. LNCS, vol. 3367, pp. 79–92. Springer, Heidelberg (2005). doi:10.1007/978-3-540-31838-5_6
7. Bharambe, A.R., Agrawal, M., Seshan, S.: Mercury: supporting scalable multi-attribute range queries. ACM SIGCOMM Comput. Comm. Rev. (2004)
8. Chen, D., Zhou, J., Le, J.: Reverse nearest neighbor search in peer-to-peer systems. In: Larsen, H.L., Pasi, G., Ortiz-Arroyo, D., Andreasen, T., Christiansen, H. (eds.) FQAS 2006. LNCS, vol. 4027, pp. 87–96. Springer, Heidelberg (2006). doi:10.1007/11766254_8
9. Domingos, P.: A few useful things to know about machine learning. Commun. ACM **55**(10), 78–87 (2012)
10. Ganesan, P., Yang, B., Garcia-Molina, H.: One torus to rule them all: multi-dimensional queries in P2P systems. In: Proceedings of the 7th International Workshop on the Web and Databases: Colocated with ACM SIGMOD/PODS 2004. ACM (2004)
11. Garcia Lopez, P., Montresor, A., Epema, D., Datta, A., Higashino, T., Iamnitchi, A., Barcellos, M., Felber, P., Riviere, E.: Edge-centric computing: vision and challenges. ACM SIGCOMM Comput. Commun. Rev. **45**(5), 37–42 (2015)
12. Heigold, G., Vanhoucke, V., Senior, A., Nguyen, P., Ranzato, M., Devin, M., Dean, J.: Multilingual acoustic models using distributed deep neural networks. In: IEEE ICASSP (2013)
13. Hong, K., Lillethun, D., Ramachandran, U., Ottenwälder, B., Koldehofe, B.: Mobile fog: a programming model for large-scale applications on the internet of things. In: Proceedings of the Second ACM SIGCOMM Workshop on Mobile Cloud Computing, pp. 15–20. ACM (2013)
14. World's biggest data breaches (2015). informationisbeautiful.net
15. Khan, R., Khan, S.U., Zaheer, R., Khan, S.: Future internet: the internet of things architecture, possible applications and key challenges. In: Proceedings of the FIT (2012)
16. Kienzle, M.G.: Cognitive technologies for smarter cities. In: Proceedings of the ICDCS (2016)

17. Li, M., Lee, W.-C., Sivasubramaniam, A.: Semantic small world: an overlay network for peer-to-peer search. In: Proceedings of the ICNP (2004)
18. Li, M., Lee, W.-C., Sivasubramaniam, A., Zhao, J.: Supporting k nearest neighbors query on high-dimensional data in P2P systems. FCS **2**(3), 234–247 (2008)
19. Malkov, Y., Ponomarenko, A., Logvinov, A., Krylov, V.: Approximate nearest neighbor algorithm based on navigable small world graphs. Inf. Syst. **45**, 61–68 (2014)
20. Manning, C.D., Schütze, H., et al.: Foundations of Statistical Natural Language Processing, vol. 999. MIT Press, Cambridge (1999)
21. Manyika, J., Chui, M., Brown, B., Bughin, J., Dobbs, R., Roxburgh, C., Byers, A.H.: Big data: the next frontier for innovation, competition, and productivity (2011)
22. Mayer, C., Tariq, M.A., Li, C., Rothermel, K.: GrapH: heterogeneity-aware graph computation with adaptive partitioning. In: Proceedings of the ICDCS (2016)
23. Mayer, R., Gupta, H., Saurez, E., Ramachandran, U.: The fog makes sense: enabling social sensing services with limited internet connectivity. In: Proceedings of the 2nd International Workshop on Social Sensing. ACM (2017)
24. Mayer, R., Koldehofe, B., Rothermel, K.: Predictable low-latency event detection with parallel complex event processing. IEEE Internet of Things Journal **2**(4), 274–286 (2015)
25. McMahan, B., Ramage, D.: Federated learning: Collaborative machine learning without centralized training data. Technical report, Google (2017)
26. Montresor, A.: Reflecting on the past, preparing for the future: from peer-to-peer to edge-centric computing. In: Proceedings of the ICDCS (2016)
27. Montresor, A., Jelasity, M.: Peersim: a scalable P2P simulator. In: IEEE Ninth International Conference on Peer-to-Peer Computing, P2P 2009, pp. 99–100. IEEE (2009)
28. Müller, W., Henrich, A.: Fast retrieval of high-dimensional feature vectors in P2P networks using compact peer data summaries. In: Proceedings of the ACM SIGMM International Workshop on Multimedia Information Retrieval, pp. 79–86. ACM (2003)
29. Poxrucker, A., Bahle, G., Lukowicz, P.: Towards a real-world simulator for collaborative distributed learning in the scenario of urban mobility. In: Proceedings of the SASOW 2014 (2014)
30. Ratnasamy, S., Francis, P., Handley, M., Karp, R., Shenker, S.: A scalable content-addressable network. ACM SIGCOMM Comput. Commun. Rev. **31**(4), 161–172 (2001)
31. Rowstron, A., Druschel, P.: Pastry: scalable, decentralized object location, and routing for large-scale peer-to-peer systems. In: Guerraoui, R. (ed.) Middleware 2001. LNCS, vol. 2218, pp. 329–350. Springer, Heidelberg (2001). doi:10.1007/3-540-45518-3_18
32. Schalkwyk, J., Beeferman, D., Beaufays, F., Byrne, B., Chelba, C., Cohen, M., Kamvar, M., Strope, B.: Your word is my command: Google search by voice: a case study. In: Neustein, A. (ed.) Advances in Speech Recognition, pp. 61–90. Springer, Boston (2010). doi:10.1007/978-1-4419-5951-5_4
33. Schmidt, C., Parashar, M.: Squid: enabling search in DHT-based systems. J. Parallel Distrib. Comput. **68**(7), 962–975 (2008)
34. Shu, Y., Ooi, B.C., Tan, K.-L., Zhou, A.: Supporting multi-dimensional range queries in peer-to-peer systems. In: Proceedings of the P2P. IEEE (2005)

35. Stoica, I., Morris, R., Karger, D., Kaashoek, M.F., Balakrishnan, H.: Chord: a scalable peer-to-peer lookup service for internet applications. ACM SIGCOMM Comput. Commun. Rev. **31**(4), 149–160 (2001)
36. Sutton, C., McCallum, A.: An Introduction to Conditional Random Fields for Relational Learning. Introduction to Statistical Relational Learning, vol. 2. MIT Press, Cambridge (2006)
37. Tang, C., Xu, Z., Mahalingam, M.: pSearch: information retrieval in structured overlays. ACM SIGCOMM Comput. Commun. Rev. **33**(1), 89–94 (2003)
38. Witschel, H.F.: Content-oriented topology restructuring for search in P2P networks. Technical report, University of Leipzig, Germany (2005)
39. Ziegeldorf, J.H., Morchon, O.G., Wehrle, K.: Privacy in the internet of things: threats and challenges. Secur. Commun. Netw. **7**(12), 2728–2742 (2014)

Big Data Summarisation and Relevance Evaluation for Anomaly Detection in Cyber Physical Systems

Ada Bagozi, Devis Bianchini$^{(\boxtimes)}$, Valeria De Antonellis, Alessandro Marini, and Davide Ragazzi

Department of Information Engineering, University of Brescia,
Via Branze, 38, 25123 Brescia, Italy
devis.bianchini@unibs.it

Abstract. Recent advances in the smart factory created new opportunities in industrial support, specifically in anomaly detection and asset management for maintenance purposes. Data collected from machines in operation is integrated in cyberspace with advanced cockpit and dashboard visualisation tools, as well as computerised maintenance management systems (CMMS). This paves the way to collaborative environments for Cyber Physical Systems, that assist maintenance operators in making better decisions while dealing with real time data in a dynamic context of interconnected systems. To this aim, models and techniques for data representation and treatment are highly required. In this paper, we propose a state detection service for Cyber Physical Systems, able to identify anomalies based on large amounts of data incrementally collected, organized and analysed on-the-fly. The service combines in a novel way data summarisation and data relevance techniques, to focus the computation on relevant data only, as well as a multi-dimensional model, that organises summarised data according to multiple dimensions, for flexible anomaly detection according to different analysis requirements. A pilot case study in the smart factory is also described, to demonstrate the applicability of the approach.

Keywords: Industry 4.0 · Cyber Physical Systems · State detection · Big data · Data relevance · Data summarisation · Data exploration

1 Introduction

In the new industrial evolution, as known as Industry 4.0, the strict interaction between physical spaces (embedded systems, sensors, mobile and wearable devices, RFID technology) and cyberspace (edge and cloud computing technologies), represented by Cyber Physical Systems (CPS), made the exchange of large quantity of data in (near) real-time between interconnected systems a reality. In the field of manufacturing, this created the opportunity of new applications, aimed to improve operation process performance, monitoring and control, anomaly detection and health assessment [9].

H. Panetto et al. (Eds.): OTM 2017 Conferences, Part I, LNCS 10573, pp. 429–447, 2017.
https://doi.org/10.1007/978-3-319-69462-7_28

This trend is promoting the development of collaborative applications, with the increasing inclusion of human being in the computational ambient or cyberspace [5]. For what concerns anomaly detection, data collected from machines in operation is integrated in cyberspace with visualisation tools such as cockpits and dashboards, as well as computerised maintenance management systems (CMMS). CMMS assists maintenance operators in planning activities to prevent downtime and low performance. Visualisation tools are in charge of providing different views over data, allowing for flexible anomaly detection according to different analysis requirements. Used together in a collaborative environment, these tools pave the way in making better decisions, while dealing with real time data in a dynamic context of interconnected systems. Current anomaly detection strategies generally rely on costly models built on top of experts' experience. In this context, big data and data-driven approaches are mainly unexploited [15]. Data-intensive CPS need tools and methods to deal with huge quantity of data, collected at high rate, and efficient techniques for storing and managing it. Data value declines very quickly, making organisations' wealth more and more dependent on how efficiently they can turn collected data into actionable insights [8]. The current validity of data is a critical success factor to be considered for implementing effective anomaly detection solutions. Indeed, the real value of anomaly detection using CPS concepts with big data stands in the fact that these techniques seem to be effective to identify unknown anomalies, helping industrial analysts and operators in the resolution of possible invisible problems [11].

In this paper, we propose a state detection service for Cyber Physical Systems, able to identify anomalies based on large amounts of data incrementally collected, organized and analysed on-the-fly. The aim is to demonstrate how data summarisation and data relevance techniques, proposed in [3] as ingredients to perform exploration of real time data in a dynamic context of interconnected systems, can be used in a novel way to identify anomalies in CPS. In particular, we will show how data relevance techniques, that focus the computation and monitoring on relevant data only, and a multi-dimensional model, that organises summarised data according to multiple dimensions, can be adapted for flexible anomaly detection according to different analysis requirements. In [2] we proposed IDEAaS (Interactive Data Exploration As-a-Service), a framework where innovative services are designed to enable data exploration. Here we focus our attention on the application of the multi-dimensional model, data summarisation and relevance evaluation techniques to support anomaly detection in collaborative systems in the context of Cyber Physical Systems and Industry 4.0.

The paper is organised as follows: in Sect. 2 we introduce the general idea and motivations behind a big data-driven state detection service; Sect. 3 contains the description of the multi-dimensional model; in Sect. 4, data summarisation and relevance evaluation techniques are described with reference to anomaly detection; Sect. 5 presents the implementation details and preliminary evaluation of the state detection service using a pilot study in the smart factory; cutting-edge features of our approach, compared to the literature, are discussed in Sect. 6; finally, Sect. 7 closes the paper.

2   Big Data-Driven Anomaly Detection

Motivating Collaborative Scenario. Figure 1 depicts a collaborative scenario that motivates our work on anomaly detection for CPS. In the considered scenario, an Original Equipment Manufacturer (OEM) supplies a multi-spindle machine, designed to perform flexible tasks for its clients (manufacturing enterprises in sectors like automotive, aviation, water industry). As shown in figure, the machines are equipped with three identical spindles, that work independently each others on the raw material. Each spindle is mounted on a unit moved by an electric motor to perform X, Y, Z movements. The spindle rotation is impressed by another electric motor and its rotation speed is controlled by the machine control. Spindles use different tools (that are selected according to the instructions specified within the Part Program) in order to complete different steps in the manufacturing cycle. Spindle precision, working performances, as well as minimisation of tool breaks and machine downtime are critical factors. Events to be detected and avoided regard 'spindle rolling friction torque increase' and 'tool wear'. Increase of spindle rolling friction torque may happen for lack of lubrication or other mechanical wears like bearings damage. Tool wear monitoring is referred to possible tool usage optimisation in order to balance the trade-off between the number of tools used and the risk of breaking the tool during operations that may lead to long downtimes. Data collected from the

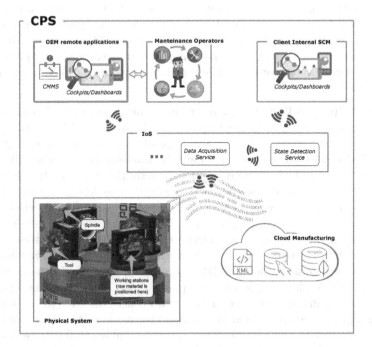

Fig. 1. A collaborative scenario for big data-driven anomaly detection of Cyber Physical Systems.

physical system is saved on the cyber side (*cloud manufacturing* component) through a *data acquisition service* and is processed by a *state detection service*. CPS is represented in the cyber-space through a set of measures, that may be collected by sensors. FMEA analysis is useful to identify possible failure modes, the relative criticality and how this can be translated as collected measures. This analysis is performed in collaboration with mechanical designers and actors at business/management level, and helps to identify the set of feature spaces of interest. In our case study, for each unit, we measure the velocity of the three axes (X, Y and Z) and the electrical current absorbed by each motor, the value of rpm for the spindle, the percentage of power absorbed by the spindle motor (charge coefficient).

Spindle rolling friction torque increase and tool wear can be monitored by observing the spindle power absorption for similar rpm. If an increased power absorption is detected disregarding the tool that is used, it is possible to identify spindle rolling friction torque increase as the possible anomaly that increases the energy request to perform the manufacturing operations. If the increase in absorbed power is related only to the usage of a particular tool, this can be recognised as a symptom of a possible exceeding tool wear. Therefore, aspects such as machine components and tools, as well as time, represent multiple perspectives to perform state detection. The state detection service we describe in this paper may interact with other modules at the application level: (i) remotely, OEM is equipped with visualisation tools such as cockpits or dashboards and computerised maintenance management systems (CMMS), to assist operators in planning activities to prevent downtime and low operation performance, according to alerts coming from the state detection service; (ii) internally, the OEM client may use service outputs to plan supply chain activities (e.g., planning of material and component orders). These collaborative needs raise a set of open issues presented in the following.

Data Modeling. Collected data must be properly stored and organised on the cloud manufacturing in order to perform efficient data-driven tasks [10]. A data-driven state detection service also depends on different analysis requirements. OEM maintenance operators might be interested on spindle rolling friction torque increase, to manage maintenance activities. On the other hand, the OEM client might monitor spindle power absorption to detect tool wear events, to better plan the supply of new tools. Therefore, anomaly detection might rely on data modeling according to "facets" (e.g., categories), evenly hierarchically organised, that represent a powerful mean to enable aggregation of data according to different perspectives, in turn related to distinct observed problems and maintenance goals, although related to overlapping data. In the motivating example, the OEM client might observe spindle power absorption during the use of multiple tools, in order to detect a tool wear event. This gives the opportunity of managing different distributed activities on the same system in a cooperative way. As described in the next sections, we will combine different technologies in order to provide a multi-dimensional representation over the large amount of collected data.

Data Volume and Velocity. In the considered pilot case study, we collected 140 millions of records from three machines, each one equipped with three spindles and different tools. The machines are identical. Records have been collected every 200 ms. The ability of providing a compact view over the huge amount of data collected from the machine is strongly required. A data summarisation approach is recommended, where data should be observed in an aggregated way, instead of monitoring each single data record, that might be not relevant given the high level of noise in the working environment (slight variations in the measured variables). Moreover, data summarisation might have positive effects on visualisation tools. At the same time, data aggregations should be observed on the fly, given the highly dynamic nature of the application domain, and efficient computation algorithms are required to summarise data.

Data Relevance. The prompt identification of anomalies by monitoring and observing collected data is one of the most important aspects to be addressed in state detection services. Data relevance evaluation techniques may help to iteratively restrict the monitoring only to the relevant measures that correspond to anomalies. This also have a positive impact on the algorithm complexity and response times, that might determine the success of an anomaly detection solution compared to the other ones. A definition of *relevance*, related to a notion of distance from an expected status, is fundamental in this case. Relevance evaluation algorithms must take into account volumes and speed of data collection phase. The dimensions of the multi-dimensional model have been considered to limit records and directions on which data summarisation and relevance detection process is applied.

Behind these open issues, heterogeneity of incoming records, as well as management of missing values have to be addressed as well, in order to prepare data for summarisation and relevance evaluation. These latter issues will be mentioned in the concluding remarks. In the following sections we will describe the models and techniques on which the state detection service relies to provide the methods described Table 1.

Table 1. Public methods of the state detection service.

Method	Inputs	Ouputs
GetData	Range of timestamps, filters (required analysis dimensions)	Sends summarised data
GetRelevantData	/	Sends summarised data that the system recognised as relevant
GetAlertStatus	Current timestamp	Reports on current status of the system
SendAlert	/	Sends alerts on detected status

3   A Multi-dimensional Data Model for State Detection

The state detection service is based on a multi-dimensional model, that organises data according to different analysis dimensions, thus allowing for flexible anomaly detection according to distinct analysis requirements. The model enables the propagation of the system status over specific dimensions. Figure 2 shows the conceptual schema of the multi-dimensional model. The core concepts of the model are *features* and *measures*, defined in the following. They are organised according to *feature spaces*, *domain-specific dimensions* and *context parameters*, that constitute the overall set of dimensions on which the multi-dimensional data model is built.

Definition 1 (Feature). *A feature represents a monitored variable that can be measured. A feature F_i is described as $\langle n_{F_i}, u_{F_i} \rangle$, where n_{F_i} is the feature name, u_{F_i} represents the unit of measure. Let's denote with $F = \{F_1, F_2 \ldots F_n\}$ the overall set of features.*

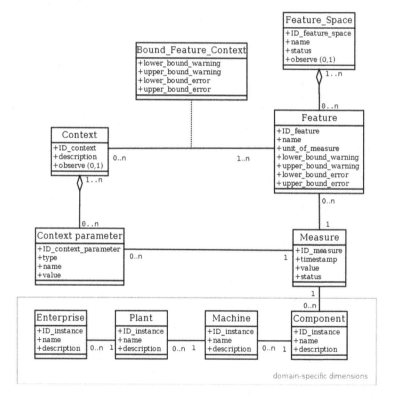

Fig. 2. Conceptual schema of the multi-dimensional model on which the state detection service relies.

In the considered case study, examples of features are the velocity of the three axes X, Y and Z, the electrical current, the value of spindle rpm and the percentage of absorbed power.

Definition 2 (Measure). *We define a measure for the feature F_i as a scalar value $X_i(t)$, expressed in terms of the unit of measure u_{F_i}, taken at the time t.*

To observe different physical phenomena of a system, multiple features can be monitored together. We call such sets of features as feature spaces.

Definition 3 (Feature space). *A feature space conceptually represents a set of related features, that are jointly measured to observe a physical phenomenon. Multiple feature spaces might be observed, and the observation of a feature might be useful to monitor more than one feature space. We denote with $FS = \{FS_1, FS_2, \ldots FS_m\}$ the set of feature spaces, where $FS_j \subseteq F$ and $m \leq n$. Given a feature space $FS_j = \{F_1, F_2, \ldots F_h\}$, we denote with the vector $\boldsymbol{X}_j(t)$ a record of measures $\langle X_1(t), X_2(t), \ldots X_h(t) \rangle$ for the features in FS_j, synchronised with respect to the timestamp t. Feature spaces can be monitored independently each others.*

In the considered case study, spindle rolling friction torque increase and tool wear can be observed by monitoring the spindle power absorption as a feature space. Feature spaces can be monitored according to different domain-specific dimensions, such as the observed machine or the tool used during manufacturing, defined as follows.

Definition 4 (Domain-specific dimension). *We denote with $\mathcal{D}$ the multi-dimensional space created by p domain-specific dimensions $\mathcal{D}_1, \ldots \mathcal{D}_p$, where $\mathcal{D} = \mathcal{D}_1 \times \cdots \times \mathcal{D}_p$. Dimensions can be organized in hierarchies, at different levels. We denote with $\mathcal{D}_j^i$ the i-th level in the hierarchy of j-th dimension and with $d_i \in \mathcal{D}_i$ a single instance of the dimension $\mathcal{D}_i$.*

For example, tools can be aggregated into tool types; in the hierarchy of the monitored physical system, components (e.g., spindles) can be aggregated into the machines they belong to, in turn organised into plants and enterprises. Furthermore, there are some characteristics that should remain constant while performing any kind of comparison between measures. For instance, spindle rolling friction torque increase or tool wear should be observed only comparing the spindle power absorption under the same working conditions of the machine, such as the *part program* that is being executed, or the *working mode* (G0, fast movement of the spindle to catch the tool, or G1, slow movement of the spindle during the manufacturing). To this aim, we introduce the concept of *context parameter*.

Definition 5 (Context parameter). *We define a context Ctx_i as a set of parameters, that identify the conditions under which the monitored system operates. Comparison between different measures makes sense only within the same context. We denote with Ctx the set of all possible contexts $\{Ctx_i\}$. A context is composed of one or more Context Parameters (e.g., part program, working mode), each one described by type, name and value.*

3.1   System Status and Alert Bounds

The goal of a state detection service is to detect anomalies and send alerts concerning the system status. We consider three different values for the *status*: (a) ok, when the system works normally; (b) warning, when the system works in anomalous conditions that may lead to breakdown or damage; (c) error, when the system works in unacceptable conditions or does not operate. Therefore, the warning status is used to perform an early detection of a potential deviation towards an error state. The migration of the system status from one value to the others raises an *alert* and occurs when one or more measures exceed a given bound, as shown in Fig. 3. We consider four types of bounds: lower bound error, lower bound warning, upper bound warning and upper bound error. In our model, bounds are further classified as feature bounds and contextual bounds.

Fig. 3. System status and alert bounds.

Definition 6 (Feature bound). *A feature bound represents a physical limit of the feature, independently from the specific context in which the monitored system is working. This bound is set as a range on the values of a feature according to the model in Fig. 2.*

Definition 7 (Contextual bound). *A contextual bound represents the limit of a feature within a specific context where the feature is measured. The rationale is that, in a specific context (e.g., the working mode, the part program), when the system works normally, a feature should assume values within a specific range, that might be different by the physical limits for the same feature. These bounds are modelled as an association with attributes between the feature and the context entities in the model in Fig. 2.*

These bounds set the ranges for the three different values of the status: ok, warning and error. Feature bounds determine the *absolute status* of a feature, while contextual bounds determine the *contextual status* of a feature. The system status (either absolute or contextual) can be propagated to the whole feature space and along the hierarchy of monitored physical system, according to the following propagation rules.

Propagation to the feature space. Consider a feature space $FS_j = \{F_1, F_2, \ldots F_h\}$, the value of the status associated to FS_j, given the status values for each feature $F_1, F_2, \ldots F_h$, is computed as follows:

   - ok, if the status of each feature F_i, $\forall i = 1 \ldots h$, is ok;
   - warning, if the status of at least one feature F_i, $\forall i = 1 \ldots h$, is warning;

- first level error, if the status of at least one feature F_i, $\forall i = 1 \ldots h$, is error;
- second level error, if the status of each feature F_i, $\forall i = 1 \ldots h$, is error.

Propagation along the hierarchy of the monitored physical system.
The value of a feature status for a component is propagated to the highest level of the hierarchy (machine, plant, enterprise) as follows: the status of the machine is

- ok, if the status for all its components is ok;
- warning, if the status of at least one component is warning;
- first level error, if the status of at least one component is error;
- second level error, if the status of all its components is error.

The same applies for the status of the plant (resp., enterprise), computed starting from the status of its machines (resp., plants).

Definition 8 (Multi-dimensional model). *We describe the multi-dimensional model as a set $\mathcal{V}$ of nodes. Each node $v \in \mathcal{V}$ is described as*

$$v = \langle \boldsymbol{X}_j(t), fs_j, d_1, d_2, \ldots d_p, Ctx_i, \sigma_j, \sigma^c \rangle \tag{1}$$

where $\boldsymbol{X}_j(t)$ represents a record of measures taken at time t for the feature space fs_j, in the context Ctx_i, for the values $d_1, d_2, \ldots d_p$ of domain-specific dimensions $\mathcal{D}_1, \ldots \mathcal{D}_p$; σ_j is the status of the feature space fs_j; σ^c is the status of the component d_k at the lowest level of the monitored physical system dimension $\mathcal{D}_k$, with $k \leq p$. The status σ^c is propagated to the other levels of $\mathcal{D}_k$ using the rules presented above.

4   Data Summarisation and Relevance Evaluation

In order to promptly detect anomalies in (near) real-time, the state detection service described in this paper relies on summarisation and relevance evaluation techniques. Details about these techniques have been presented in [3]. Here, we customise relevance evaluation techniques (Sect. 4.2), used in combination with the multi-dimensional model, to the anomaly detection problem.

4.1   Clustering-Based Data Summarisation

We apply data summarisation techniques to summarise all records of measures collected during time interval Δt in the context Ctx_i and for dimensions $d_1 \in \mathcal{D}_1$, $d_2 \in \mathcal{D}_2 \ldots d_p \in \mathcal{D}_p$, for monitoring feature space fs_j. To this aim, we denote with $\Sigma(\boldsymbol{X}_j(t), fs_j, d_1, d_2, \ldots d_p, Ctx_i)$ the application of the summarisation procedure to the records of measures $\boldsymbol{X}_j(t)$ as used in Eq. (1) of Definition 8. In our approach data summarisation is based on clustering-based techniques. When dealing with real time data, collected in CPS, we face with data streams, where not all data is available since the beginning, but is collected in an incremental way. For these reasons, an incremental data-stream clustering algorithm has been developed. The clustering algorithm produces a set of clusters aimed to summarise collected measures in a time interval Δt. The clustering algorithm is performed in two steps: (i) in the first one, a variant of Clustream algorithm [1] is applied, that incrementally processes incoming data to obtain a *set of syntheses*; (ii) in the second step, X-means algorithm is applied [14] in order to cluster syntheses obtained in the previous step. X-means does not require any a-priori knowledge on the number of output clusters. Each synthesis provides a lossless summarisation of records through five elements: the number of records included into the synthesis, the vector representing the linear sum of measures, the quadratic sum of measures, the vector representing the centroid of the synthesis and the radius of the synthesis. The second step aims to cluster syntheses. Clusters give a balanced view of the observed physical phenomenon, grouping together syntheses corresponding to close data. A cluster is represented by its centroid and the set of syntheses belonging to it. Hereafter, we denote with $SC(\boldsymbol{X}_j(t), fs_j, d_1, d_2, \ldots d_p, Ctx_i)$ the set of identified clusters, that is, the output of $\Sigma(\boldsymbol{X}_j(t), fs_j, d_1, d_2, \ldots d_p, Ctx_i)$ procedure.

4.2   Relevance-Based Anomaly Detection

Relevance-based techniques are used to detect components status over time. In literature, data relevance is defined as the *distance* from an *expected status*. The point is to define the expected status and how to compute such a distance. In our approach, the expected status corresponds to the set of clusters computed during normal working conditions for the monitored system. Let's denote with $\hat{SC}(\boldsymbol{X}_j(t), fs_j, d_1, d_2, \ldots d_p, Ctx_i)$ such cluster set. Data relevance is based on the notion of *cluster distance* between the current cluster set and $\hat{SC}(\cdot)$. This distance is defined as follows. Given a set of clusters $SC = \{C'_1, C'_2, \ldots, C'_n\}$ and the set of clusters $\hat{SC}(C_1, C_2, \ldots, C_n)$, we evaluate the distance between SC and $\hat{SC}(\cdot)$, denoted with $\Delta(SC, \hat{SC})$, by combining three factors: (i) the difference between the number of clusters, (ii) the distance between cluster centroids and (iii) the intra-cluster distances, i.e. the distance between syntheses belonging to the same cluster.

In particular, the relevance techniques allow to identify what are the clusters that changed over time. Let's denote with $\{\overline{C_i}\}$ such clusters. The distance is used to detect a *state change*. When $\Delta(SC, \hat{SC})$ exceeds a given threshold, data that is summarised in the clusters $\{\overline{C_i}\}$ is considered as relevant and, for each cluster in $\{\overline{C_i}\}$, the distance of cluster centroid from the warning and error bounds is computed. If this distance is equal or lower than the cluster radius, this means that a warning or error status has to be detected, according to the rules described in Sect. 3. Note that distance also helps to detect *potential* state changes. Consider for example Fig. 4, that shows an example of cluster evolution over time for the smart factory case study. The figure shows how the cluster C_1 doesn't changed its position, as well as its size, from time t_n to t_{n+3}. On the other hand, cluster C_2 evolves from the wealth zone to the warning and

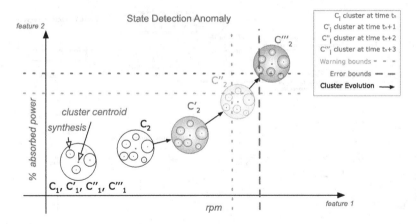

Fig. 4. Illustration of set of clusters changing over time. Distance techniques aim to detect changes in cluster set over time due to spindle rolling friction torque increase that may cause a decrease of rpm and an increase of the percentage of absorbed power.

error zones. At time t_{n+2} cluster C_2'' crosses the warning bound of rpm feature causing a warning alert, at time t_{n+3} cluster C_2''' moves into the error zone, crossing error bounds of both the features considered. At time t_{n+1} cluster C_2' still remains inside the wealth zone, however relevance techniques detected its change. Therefore cluster C_2' is recognised as relevant and monitored to detect warning or error state changes. This allows for better performance of the anomaly detection algorithm, that focuses only on potential state changes. Figure 4 also shows that it is possible to identify the feature with respect to which the warning or error bound has been exceeded (e.g., among rpm and percentage of absorbed power).

5   Implementation and Experimental Evaluation

Figure 5 depicts a collaborative scenario, where the state detection service interacts with a cockpit to visualise the status of monitored physical system and with a CMMS to plan maintenance operations. Modules are implemented as RESTful services and use JSON as data exchange format. The figure shows the data flow across interoperating systems. Data is sent from the monitored physical system as an input for the *Data Cleaning & Normalisation* module. The input files can present different formats. In our current implementation it is composed of a set of records, where each record is associated with a timestamp, a set of measures for the considered features and a set of values for the context parameters and domain-specific dimensions. An XML configuration file (*Input Data Config*) contains all meta-data about the structure of incoming data records, after the application of data acquisition and ingestion techniques to face heterogeneity issues. The cleaned and normalised data is sent to the *Data Storage* module, that is in charge of generating the JSON file for storing data in the *Summarised*

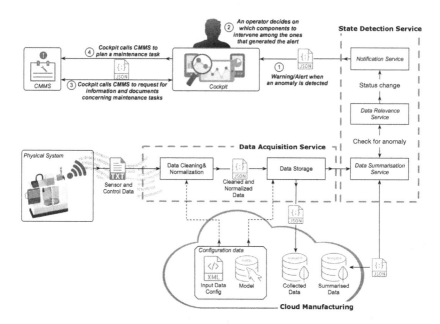

Fig. 5. Data flow in the anomaly detection collaborative scenario.

Data collection (implemented using MongoDB). Figure 6 shows the structure of JSON documents stored within this collection. Information contained in this structure, together with involved data records, are sent as output of `getData` and `getRelevantData` methods of the state detection service.

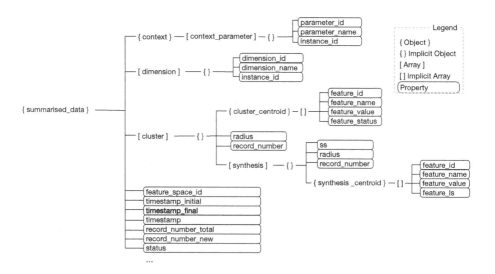

Fig. 6. The structure of the JSON document to store summarised data.

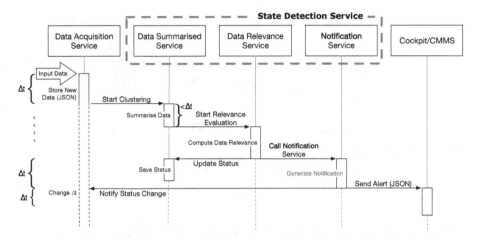

Fig. 7. Interactions between the modules of the functional architecture for the anomaly detection scenario.

The diagram in Fig. 7 shows the interactions between the functional modules. The Data Acquisition service, that includes the cleaning and storage tasks, and the State Detection service operate in parallel. This prevents the state detection procedure from being a bottleneck for the data acquisition. The Data Acquisition service activates every Δt seconds the State Detection service, that is in charge of detecting and notifying anomalies as explained in the previous sections through the execution of *Data Summarisation*, *Data Relevance* and *Notification* services. Every Δt seconds the state detection service checks the system status using clustering and relevance evaluation techniques. If the relevance evaluation detects changes in data compared to the expected status, the service computes the new status as explained in Sect. 4.2. If the status is not changed with respect to the previous check, the system simply updates the data saved in MongoDB (Summarised Data collection) with the computed status value. If the status changed, the system updates the status in MongoDB, applies propagation rules to the feature space and along all the levels of the hierarchy of monitored physical system and notifies an alert message to the cockpit containing the new status, using the `SendAlert` method to report the detected anomaly. Cockpit will handle the communication with the CMMS: it requests for information and documents regarding maintenance tasks and asks the CMMS to plan a maintenance task. It is worth to underline that, when the Data Relevance service detects a status change, it also calls the Data Acquisition module to modify the Δt value. In case of status changes from `ok` to `warning`, Δt decreases. The aim is to increment the frequency of controls when a potential breakdown might occur. This improves the performance of the system, at the cost of an increased frequency of status computation, as shown in the next section.

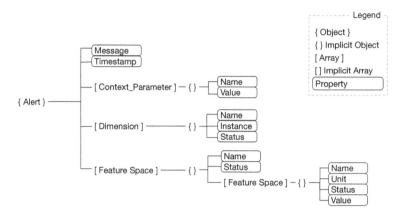

Fig. 8. Structure of JSON document as output of the `SendAlert` method.

Figure 8 shows the structure of the JSON file generated as output by the `SendAlert` method of the State Detection service, to be sent to the Cockpit. The `getAlertStatus` method returns a similar output, with the list of all current status values.

5.1 Experiments

We performed experiments on the State Detection service, in order to test its performance in terms of processing time and its effectiveness in quickly detecting anomalies. We collected 140 millions of records from the three machines considered for the case study, each one equipped with three spindles and different tools. Records have been collected every 200 ms. We run experiments on a VPS mounting Ubuntu Server 16.04, 1 vCore CPU, RAM 4 GB, hard disk 20 GB (SSD). Collected records of measures are saved within the *Collected Data* MongoDB as JSON documents.

We remark here that clustering and relevance evaluation are incrementally performed on slots of records on a time internal Δt. Also in the worst case, clustering, data extraction and cluster sets distance computation tasks were able to process $\sim162 \times 10^3$ records in ~170 s (corresponding to a processing rate of ~953 records per second), thus facing an acceptable data input rate for the considered case study. These response times must be intended also when new data arrives. As expected, clustering is the most time-consuming task. Figure 9 shows further tests, where processing time for clustering is computed by varying the number of records to which it is applied and, therefore, processing frequency $\frac{1}{\Delta t}$. As the number of new processed records every Δt seconds grows, clustering algorithm execution time presents a polynomial increment. We can conclude that the factor which impacts the most on the system performance is Δt and not the total number of processed data, thanks to the incremental approach. Therefore, when clustering processing time overtakes the data acquisition rate, a solution might be to tune the frequency

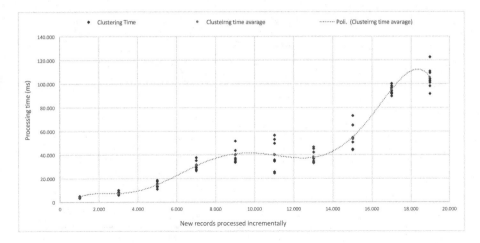

Fig. 9. Tests on the efficiency of clustering by changing the elaboration frequency ($\frac{1}{\Delta t}$).

$\frac{1}{\Delta t}$, as done during the interactions between services described above. Furthermore, higher frequency $\frac{1}{\Delta t}$ allows for faster detection of the state changes.

To test effectiveness of the state detection service, we introduced unexpected working states to simulate spindle rolling friction torque increase by increasing the spindle power absorption for similar rpm on a subset of collected data, as shown in Fig. 10. Experimental results plotted in Fig. 11 show the distance between sets of clusters calculated at time t and those calculated in case of normal working. The figure shows how the techniques proposed here allow to timely identify the unexpected situations induced in the system under observation.

Working Days	Spindle 1		Spindle 2		Spindle 3	
	rpm decrease	energy consumption increase	rpm decrease	energy consumption increase	rpm decrease	energy consumption increase
01/08/2016 (operates normally)	/	/	/	/	/	/
02/08/2016	/	/	/	/	15%	16%
03/08/2016	20%	20%	/	/	29%	29%
04/08/2016	30%	30%	/	/	/	/
05/08/2016	40%	40%	20%	20%	/	/
06/08/2016	/	/	30%	30%	/	/
07/08/2016	/	/	20%	20%	/	/

Fig. 10. Introduced variation in collected records that simulates spindle rolling friction torque increase

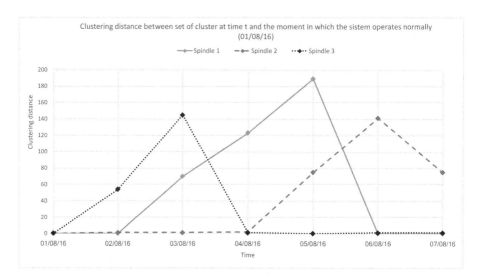

Fig. 11. Effectiveness evidences of the relevance evaluation techniques to identify changed records.

6   Related Work

Several approaches in literature have been recently proposed to address anomaly detection in presence of big data collected in (near) real time. Authors in [15] present an approach for data-driven anomaly detection in manufacturing processes. They consider joint analysis of multiple variables for detecting the anomalies. They combine real-time and historical data processing in order to increase the agility for detecting and reacting to anomalies. This approach operates in two steps: (i) learning the normal behaviour of the system (based on past data), using a clustering technique (K-means algorithm) to group together close data; (ii) detecting at real-time an anomalous behaviour when new data does not belong to previously detected clusters. Similarly, the main goal of the project and the architecture presented in [7] is to facilitate the prompt detection or prediction of failures from event data. Machine learning is used to train data collected during regular execution of the manufacturing process in order to learn a probabilistic "normal model". Furthermore, authors focused on understanding causes of failures, using structured machine learning techniques, training classifiers from labeled event sequences. The architecture includes state-of-the-art distributed cloud-based big data technologies such as NoSQL databases. In [6] an approach based on in-memory big data processing is described. Anomaly detection is performed in two phases: (i) a preparation phase is used to generate a model of the system "usual state", by applying machine learning (pre-training) on stored data; (ii) an operation phase compares real-time incoming data with the "usual state" to output an anomaly index called "anomaly score" (where 0 indicates a system close to "usual state" and 1 indicates a system close to a "state different from usual").

The framework described in [13] guides maintenance decision makers on how to model the degradation process of a condition monitored device with an indirectly observable multi-state degradation process (the status is a value in a range from "working perfectly" to "complete failure"). Then, it presents how to employ real-time condition monitoring data for online diagnostics and prognostics using some important measures. The framework is based on a flexible stochastic process for degradation modeling. In [16] statistical techniques based on the Tukey and Relative Entropy statistics are applied to provide online anomaly detection. The data is computed in streaming and doesn't rely on static profiling or limited sets of historical data. The statistical approach is implemented to exceed the limit of fixed thresholds. The approach involves measuring the changes in data distribution by windowing the data and using it to determine anomalies.

Compared to these approaches, we introduced the use of data relevance techniques to better focus the anomaly detection procedure, improving the overall performance of the approach. These techniques have been used on top of a complex multi-dimensional model, that adopts the concept of feature space to enable multi-parameter anomaly detection, as remarked in [15]. Multi-perspective anomaly detection for cooperative systems has been addressed in other application domains [4]. Nevertheless, in our model we also added the notion of domain-specific dimensions and context, to guide/filter the anomaly detection process. The application of data summarisation and relevance evaluation techniques on top of the multi-dimensional model constitutes a step forward compared to [12], where the notions of feature space and context have been introduced for the first time in the ambit of state detection services. Data summarisation techniques, combined with relevance evaluation, also prevent from using training datasets, that are often difficult to identify and select.

7 Concluding Remarks

In this paper, we proposed a state detection service for Cyber Physical Systems working on large amounts of data incrementally collected, organized and analysed on-the-fly. The service relies on novel data summarisation and data relevance techniques: the former ones enable to deal with large amount of data in a dynamic environment, by providing a synthetic view over them; the latter ones focus the computation and monitoring on relevant data only. A multi-dimensional model, that organises summarised data according to multiple dimensions, allows for flexible anomaly detection according to different analysis requirements. The adoption of Big Data technologies for data acquisition and ingestion (e.g., Kafka or Pig), in order to face heterogeneity issues and missing records, will be investigated as future work. A case study in the smart factory is also described, where we discussed the interactions of the state detection service with: (a) a cockpit that provides different views over data;

(b) a computerised maintenance management systems (CMMS), to assist maintenance operators in planning activities to prevent downtime and low operation performance. Although preliminary experiments are promising, future development will be focused on further improving the approach using technologies for streaming and parallel batch processing, such as Spark/Storm and Hadoop. Moreover, the state detection service will be enhanced by introducing pattern recognition techniques to learn from the clusters evolution. This would in principle enable the implementation of health assessment strategies, on top of the ecosystem of models and techniques described in this paper.

References

1. Aggarwal, C., Han, J., Wang, J., Yu, P.: A framework for clustering evolving data streams. In: Proceedings of 29th International Conference on Very Large Data Bases, pp. 81–92 (2003)
2. Bagozi, A., Bianchini, D., De Antonellis, V., Marini, A., Ragazzi, D.: Interactive data exploration as a service for the smart factory. In: Proceedings of IEEE International Conference on Web Services (ICWS) (2017)
3. Bagozi, A., Bianchini, D., De Antonellis, V., Marini, A., Ragazzi, D.: Summarisation and relevance evaluation techniques for big data exploration: the smart factory case study. In: Dubois, E., Pohl, K. (eds.) CAiSE 2017. LNCS, vol. 10253, pp. 264–279. Springer, Cham (2017). doi:10.1007/978-3-319-59536-8_17
4. Böhmer, K., Rinderle-Ma, S.: Multi-perspective anomaly detection in business process execution events. In: Debruyne, C., et al. (eds.) OTM 2016. LNCS, vol. 10033, pp. 80–98. Springer, Cham (2016). doi:10.1007/978-3-319-48472-3_5
5. Gorecky, D., Schmitt, M., Loskyll, M., Zuhlke, D.: Human-machine interaction in the Industry 4.0 era. In: IEEE International Conference on Industrial Informatics, pp. 289–294 (2014)
6. Hanamori, T., Nishimura, T.: Real-time monitoring solution to detect symptoms of system anomalies. FUJITSU Sci. Tech. J. **52**(4), 23–27 (2016)
7. Huber, M., Voigt, M., Ngomo, A.: Big data architecture for the semantic analysis of complex events in manufacturing. In: Proceedings of GI Jahrestagung, pp. 353–360 (2016)
8. Khalifa, S., Elshater, Y., Sundaravarathan, K., Bhat, A., Martin, P., Imam, F., Rope, D., Mcroberts, M., Statchuk, C.: The six pillars for building big data analytics ecosystems. ACM Comput. Surv. **49**(2) (2016)
9. Lee, J., Ardakani, H.D., Yang, S., Bagheri, B.: Industrial big data analytics and cyber-physical systems for future maintenance and service innovation. Procedia CIRP **38**, 3–7 (2015)
10. Lee, J., Bagheri, B., Kao, H.: A cyber-physical systems architecture for Industry 4.0-based manufacturing systems. Manuf. Lett. **3**, 18–23 (2015)
11. Lee, J., Lapira, E., Bagheri, B., Kao, H.: Recent advances and trends in predictive manufacturing systems in big data environment. Manuf. Letters **1**(1), 38–41 (2013)
12. Marini, A., Bianchini, D.: Big data as a service for monitoring cyber-physical production systems. In: Proceedings of 30th European Conference on Modelling and Simulation (ECMS), pp. 579–586 (2016)
13. Moghaddass, R., Zuo, M.J.: An integrated framework for online diagnostic and prognostic health monitoring using a multistate deterioration process. Reliabil. Eng. Syst. Saf. **124**, 92–104 (2014)

14. Pelleg, D., Moore, A.: X-means: extending k-means with efficient estimation of the number of clusters. In: Proceedings of 17th International Conference on Machine Learning (ICML), pp. 727–734 (2000)
15. Stojanovic, L., Dinic, M., Stojanovic, N., Stojadinovic, A: Big-data-driven anomaly detection in Industry (4.0): an approach and a case study. In: Proceedings of IEEE International Conference on Big Data, pp. 1647–1652 (2016)
16. Wang, F., Agrawal, G.: Effective and efficient sampling methods for deep web aggregation queries. In: Proceedings of Conference on Extending Database Technology (EDBT), pp. 425–436 (2011)

STRATModel: Elasticity Model Description Language for Evaluating Elasticity Strategies for Business Processes

Aicha Ben Jrad[1,2](✉), Sami Bhiri[1], and Samir Tata[3]

[1] OASIS, National Engineering School of Tunis, University Tunis El Manar,
2092 Tunis, Tunisia
aichabjrad@gmail.com, sami.bhiri@gmail.com
[2] SAMOVAR, Telecom SudParis, CNRS, University of Paris-Saclay,
9 rue Charles Fourier, 91011 Evry, France
[3] Almaden Research Center,
IBM Research, San Jose, CA, USA
stata@us.ibm.com

Abstract. Nowadays, Cloud Computing is receiving more and more attention from IT companies as a new computing paradigm for executing and handling their Business Processes in an efficient and cost-effective way. One of the most important features behind this attention is the Cloud Computing's elasticity which became the focus of many research works. Its management has been considered as a pivotal issue among IT community that works on finding the right tradeoffs between QoS levels and operational costs by developing novel methods and mechanisms. Elasticity controller has been used in many research works to automate the provisioning of cloud resources and control cloud applications elasticity. However, most of the previous works have been proposed based on a specific elasticity model for either vertical or horizontal elasticity. In this paper, we propose an elasticity model description language for Service-based Business processes (SBP), called STRATModel. It allows business process holders to define different elasticity models with different elasticity capabilities by providing their elasticity mechanisms through set of examples and automatically generate their associated elasticity controllers. The generated elasticity controllers are used for evaluating elasticity strategies before using them in real cloud environments. Based on STRATModel, we present our elasticity strategies evaluation framework that facilitates the description and evaluation of elasticity strategies for SBPs according to a customized elasticity model. Our contributions and developments provide Cloud tenants with facilities to choose elasticity strategies that fit to their business processes and usage behaviors using a customized elasticity controller.

Keywords: Business process · Elasticity model · Evaluation · Elasticity strategies · Cloud computing

© Springer International Publishing AG 2017
H. Panetto et al. (Eds.): OTM 2017 Conferences, Part I, LNCS 10573, pp. 448–466, 2017.
https://doi.org/10.1007/978-3-319-69462-7_29

1   Introduction

Cloud Computing is gaining more and more importance in the Information Technologies (IT) scope as an emerging computing paradigm for managing and delivering services over the internet. One of the major assets of this paradigm is its economic model based on pay-as-you-go model. It allows the delivering of computing applications as a service rather than a product by enabling ubiquitous, convenient, and on-demand network access to large pools of computing resources (*e.g.*, storage, computing, network, applications and services) that can be dynamically provisioned by increasing and decreasing services capacity to match workloads demands and usage optimization [15]. This dynamic behavior, known as Cloud Computing's Elasticity, has encouraged many companies and researchers around the world over the last decade to migrate their operations (processes) to the Cloud [8].

Elasticity is defined as the ability of a system to be adjustable to the workload change by provisioning as many resources as needed in autonomic manner in order to meet the quality of service (QoS) requirements [7]. Provisioning of resources can be made using either horizontal elasticity, vertical elasticity, or the combination of both (*i.e.*, hybrid elasticity). Horizontal elasticity consists in replicating/removing instances of system elements to balance the current workload. It is also known as replication of resources. Vertical elasticity consists in changing the characteristics/properties (*e.g.*, memory, CPU cores) of the used instances in the system by increasing or decreasing them. It is also known as resizing of resources. As a combination of horizontal and vertical elasticity, hybrid elasticity allows to add/remove instances with different characteristics. These elasticity capabilities are the main construction of an elasticity model which defines the ground terms and irrationalities that describe the elasticity of the managed system such as the elasticity actions to be undertaken, metrics to monitor to trigger the elasticity actions and properties to access and reconfigure. Elastic systems are usually managed by elasticity controllers implementing specific elasticity models. The main function of an elasticity controller is to automate the provisioning of resources by controlling the elasticity decisions according to an elasticity strategy which defines the rules for deciding when, where and how to use elasticity capabilities/actions (*e.g.*, adding or removing resources) defined in the elasticity model of the managed system.

Most of the previous works on elasticity management and evaluation have focused on either vertical elasticity [6,16] or horizontal elasticity [1,11,14,17,19]. In our previous work [11], we presented an elasticity strategies evaluation framework for Service-based Business Processes (SBP for short), named STRAT framework. It allows Business Process holders to: (1) define their elasticity strategies using STRAT, a rule-based domain specific language for describing elasticity strategies for SBPs, and (2) formally evaluate them using an elasticity controller in order to observe their behavior and choose the most effective strategy from a set of possible ones for a given SBP. STRAT language has been proposed relying on the elasticity model, presented in [2,3], that defines three main actions: (*i*) duplicate that creates a new copy of an overloaded service in order to meet its workload increase, (*ii*) `consolidate` that releases an unnecessary copy of a service in order to meet its workload decrease, and (*iii*) `routing` that controls the way a load of a service is routed over the set of its copies. The elasticity controller, used in STRAT framework, implements the same elasticity model that STRAT relies on. It is defined as a high-level petri net to allow a formal evaluation and verification of elasticity strategies on a given SBP model describing the SBP's elastic execution environment. The controller main function is to trigger fireable rules (of the elasticity strategy) and perform thereafter the corresponding elasticity actions defined in the elasticity model.

In this paper, we extend our previous work [12] by presenting a domain-specific language, called STRATModel, for describing elasticity models for SBPs which allows to generalize the use of our STRAT framework for different elasticity models. It permits business process holders to define different elasticity models, with different elasticity capabilities/mechanisms and customized monitoring metrics, and to generate their associated elasticity controllers in order to use them to evaluate elasticity strategies on a given SBP model. The elasticity capabilities in STRATModel is defined by providing their mechanisms through a set of examples which illustrate how the generated controller should operate when applying an action that changes the structure of the managed SBP model. The elasticity models, defined using STRATModel, are used to customize STRAT language and the elasticity controller in STRAT framework allowing its adaptation to business process holders needs.

The remainder of the paper is organized as follows. In Sect. 2, we present a review of related works for managing and evaluating elasticity in the cloud. The STRATModel language is thoroughly described in Sect. 3. Section 4 presents how STRATModel is integrated in the STRAT framework along with the preliminary results obtained using STRATModel. The final section presents the conclusions and proposes some lines of future work.

2   Related Works

Elasticity plays an important role in many research works that propose methods and mechanisms to harness the ability of services/processes running in the Cloud to cope with variations in workload in order to ensure the desired level of QoS while avoiding over-provisioning of resources. The provisioning of resources is usually conducted automatically by an elasticity controller which has been the focus of many research works. In [1], two adaptive horizontal elasticity controllers has been proposed to control the adding and removing of VMs to prevent QoS violations. In [14], a self-trained elasticity controller has been proposed for managing cloud-based storage services elasticity. It is defined to automatically train itself while serving workload in order to update its control model which is used to make elasticity decisions for adding/removing servers to/from the underlying storage system. In [6], the authors proposed an elasticity approach that uses control theory to synthesize a controller for vertical memory elasticity of cloud applications. Another work tackled the problem of memory elasticity of virtual machines (VMs) has been proposed in [16]. The paper introduced a framework, called CloudVAMP, for monitoring VMs and adjusting their allocated memory to adapt the current memory requirements of their running applications using a cloud vertical elasticity controller/manager. While most of these approaches are recently proposed, they focused on a specific elasticity model for constructing their elasticity controllers, which only tackle either horizontal or vertical elasticity at infrastructure scope.

Other previous works have targeted the description and evaluation of elasticity strategies, which are responsible of making decisions on the execution of elasticity mechanisms, *i.e.*, deciding when, where and how to use elasticity mechanisms. Suleiman *et al.* [19] proposed an analytical model, using queuing theory, to evaluate the influence of elasticity strategies on the performance of 3-tier applications deployed on Cloud infrastructures. The authors has simulated the logic of scale-out and scale-in actions based on CPU utilisation. In [17], Kaskos *et al.* have proposed a formal model for quantitative analysis of horizontal elasticity at the infrastructure scope using Markov Decision Processes (MDPs). The model has been defined to formally control at runtime the adding and removing of VMs to/from the managed system. However, these works have been proposed to use strategies that are limited to infrastructure metrics (such as CPU utilization) to base their decision and do not consider metrics related to deployed processes. They also focus on evaluating strategies providing horizontal elasticity decisions (*i.e.*, scale-in and scale-out). In another work, Copil *et al.* [4] have proposed a framework, named ADVISE, for the evaluation of Cloud service/application elasticity behavior based on a learning process and a clustering-based evaluation process that determines at runtime the expected elasticity behavior of Cloud service. It is proposed in order to evaluate different elasticity control processes with different elasticity capabilities exposed by cloud provided and cloud services and determine the most appropriate one regarding the considered Cloud service and a particular situation. Nevertheless, the existing works do not allow formal evaluation of elasticity strategies before investing in using them in a real Cloud environment and they do not consider the evaluation of business processes elasticity. Our interest in this work is to provide a language for describing elasticity models that allows business process holders to provide their SBPs with different elasticity mechanisms in order to ensure their elasticity in the cloud.

Another work has been presented in [20] where the authors propose a Domain-specific language called SPEEDL that simplifies the specification of elastic strategies of IaaS services. SPEEDL has been proposed to facilitate the creation of event-driven policies for resource management by leasing and releasing VMs. Triggering the elasticity of business processes and their services, we proposed in [11] a rule-based domain-specific language, called STRAT, for describing elasticity strategies for SBPs deployed in Cloud environments. STRAT has been defined relying on the elasticity model presented in [2,3], which provides operations enabling to duplicate and consolidate services. We also developed a framework enabling to define elasticity strategies using STRAT and evaluate them through simulation. The outputs of the evaluations are displayed as a set of plots showing the elastic behavior of services and processes under specific arrival laws of instantiation requests. In this work, the descriptive domain-specific language STRATModel which we define, allows business process holders to customize the evaluation framework to their needs by describing different elasticity mechanisms they want to provide to their SBPs in order to ensure their elasticity. The elasticity mechanisms/capabilities are grouped along with a set of customized metrics and properties to construct an elasticity model.

3   STRATModel: Elasticity Model Description Language

In the following, we first present STRATModel overview, followed by its grammar for describing elasticity models. Thereafter, we present the template used to generate elasticity controllers for evaluating elasticity strategies for SBPs.

3.1   STARTModel Overview

STRATModel language is proposed as a part of STRAT framework [11] for evaluating elasticity strategies for SBPs. It is designed to allow describing elasticity models which define the ground terms and functionalities that describe the elasticity of SBPs such as the elasticity actions to be undertaken, metrics to monitor to trigger the elasticity actions and properties to access and reconfigure. It is designed to allow also the generation of elasticity controller from the defined elasticity model using a pre-defined elasticity controller template. Hence, an elasticity model is the basis for specifying elasticity strategies and constructing an elasticity controller that manages and evaluates SBPs elasticity. An elasticity controller is used to monitor a SBP and analyze its performance by inspecting an elasticity strategy in order to apply some actions whenever needed which may reconfigure some properties of the managed component [12]. So, the monitoring metrics, actions, and properties are the composition of an elasticity model in STRATModel. A STRATModel metric can be defined as a basic metric, *i.e.*, obtained directly from the monitored component property, or a composed metric by introducing the composition expression. A STRATModel action is defined to make changes on an SBP model and it targets a specific type of component, *e.g.*, Service. STRATModel is designed relying on the specification of SBP model which defines the components of modeling SBPs. An SBP model depicts the elastic execution environment of an SBP. The latter can be seen as a network, of *service engines* and optionally *load balancers*. The execution of a SBP instance (*request*) is routed through *routers* over several services engines that match each of the SBP component services. The structure of SBP's execution environment evolves over time according to requests load by applying elasticity action that may reconfigure some service engines properties and/or add/remove service engine copies in order to meet its QoS requirements (*e.g.*, maximum response time). Thus, SBP model can be composed of **service engines** that may have some properties characterizing their features such as capacity, **requests** that may also have some properties related to their attached data and performance, `routers` to route requests following the defined behavior of SBP, and optionally `load balancers` to balance the load between service engines copies. The copies of `service engines` are connected with equivalence relation. In our previous works [2,3,12], we formally described a set of SBP models (*i.e.*, stateless/stateful SBP model, Timed-SBP model and Timed-Colored SBP model) based on Petri Nets where a place denotes either a service engine or (optionally) a load balancer, a transition denotes a router and a token denotes a service request/instance which may be black token, timed-token or timed-colored token.

Describing elasticity model in STRATModel depends on which type of SBP models business process holders want to provide and manage by the generated elasticity controller, and which type of elasticity strategies they want to specify. Such information is required to know whether to include or exclude some functionalities to/from the generated elasticity controller. In the following, we discuss in more details the grammar of STRATModel used to define elasticity models, followed by the procedure for generating elasticity controllers.

3.2 STRATModel Grammar

The top-level of STRATModel specification grammar is given in Grammar 1.1 using the Backus Normal Form (BNF). STRATModel documents are composed of two parts: (i) the elasticity model description part which defines the essential elements to describe an elasticity model and (ii) Business process transformation states which define the transformations occurred on the business process model when applying some defined elasticity actions. In the following, we will discuss each of the composed parts in more details. Due to space limitations, we are unable to discuss every element in STRATModel in detail.

$\langle ElasticityModel \rangle ::= \langle ModelDescription \rangle \; \langle ProcessStates \rangle$

$\langle ProcessStates \rangle ::= \langle ProcessState \rangle \; \langle ProcessStates \rangle \mid \langle empty \rangle$

Grammar 1.1. General STRATModel Grammar

Elasticity Model Description. An elasticity model in STRATModel as given in Grammar 1.2 is composed of two sets of statements, descriptive and functional, encapsulated in a block defined by *ElasticityModel* and identified by a name.

The user is allowed to firstly describe the general aspect of elasticity model such as the managed component (business process), the use of knowledge base and the frequency of monitoring, *etc.* Thereafter, the functional statements can be provided to specify the essential elements for describing the functionality of the elasticity controller implementing the elasticity model. They are splitted into actions to be undertaken, metrics to monitor to fire the elasticity actions and properties to access and to reconfigure.

⟨*ModelDescription*⟩ ::= '`ElasticityModel`' ⟨*id*⟩ '`{`' ⟨*ModelStatements*⟩ '`}`'

⟨*ModelStatements*⟩ ::= ⟨*GeneralDescription*⟩ ⟨*ItemsDefinitions*⟩

⟨*ItemsDefinitions*⟩ ::= ⟨*ItemDefinition*⟩ ⟨*ItemsDefinitions*⟩ | ⟨*empty*⟩

⟨*ItemDefintion*⟩ ::= ⟨*Action*⟩ | ⟨*Metric*⟩ | ⟨*Property*⟩

Grammar 1.2. Grammar for describing elasticity model in STRATModel

- **Action**: An `action`, as shown in Grammar 1.3, is defined by a set of statements for describing its functionality and details and for generating its implementation mechanism. It is defined by a name, a reference ID and a component. The latter is used to specify on which type of elements the action can be applied, *i.e.*, `Process`, `Service` or `Router`. For example, a Routing action can be defined for `Router` to allow to control the execution flow of requests between SBP's services. The keywords `delay` and `multiple` are used to define respectively the time delay of applying the action and whether its multiple application is allowed.

⟨*Action*⟩   ::= '`action`' '`:`' ⟨*ActionStatements*⟩ '`;`'

⟨*ActionStatements*⟩ ::= ⟨*Name*⟩   ⟨*Reference*⟩   ⟨*Component*⟩   ⟨*Delay*⟩   ⟨*Multiple*⟩
                  ⟨*Transformation*⟩

⟨*Delay*⟩   ::= '`delay`' '`:`' ⟨*int*⟩ | ⟨*empty*⟩

⟨*Multiple*⟩   ::= '`multiple`' '`:`' ⟨*Boolean*⟩ | ⟨*empty*⟩

⟨*Transformation*⟩ ::= '`cases`' '`:`' ⟨*Examples*⟩

⟨*Examples*⟩   ::= ⟨*Example*⟩ ⟨*Examples*⟩ | ⟨*empty*⟩

⟨*Example*⟩   ::= '`apply`' '`on`' ⟨*Elements*⟩ '`transform`' ⟨*id*⟩ '`to`' ⟨*id*⟩

Grammar 1.3. Grammar for describing elasticity actions in STRATModel

The execution of the defined action changes the structure of the managed SBP model. It transforms the SBP model from one state to another. This transformation can be specified in STRATModel as transformation cases. A transformation case is defined by giving an example of an initial state of the SBP model and the resulting state after applying the action (*cf.* Sect. 3.2). The idea of using examples to specify the transformation on SBP model follows the *by-example paradigm* [5] which allows the software to drive information from a set of examples specifying how things are done or what the user expects. The most prominent approaches for *by-example paradigm* are *Query by-example* [21] which has been developed for querying database systems by allowing users to give examples of query results and *Programming by-example* [13] that permits to create a program from user's actions which are recorded as replayable macros. These approaches allow the use of examples in some way to overcome the complexity of selected problems in the field of computer science. In this work, we argue that providing transformation examples is more friendly for business process holders than defining complex formal transformations instructions. So, in STRATModel, the user is allowed to give a set of examples to describe different cases of applying the action on specific elements.

- *Metric*: A *metric* in STRATModel is identified by a name and has a low-level reference. It is related to an entity that can be a process, a service, a load balancer or a requests. It also can be obtained for a specific group by allowing grouping. A metric can be either a basic metric which obtained from a low-level property or a composite metric that is denoted by the values of other base or composite metrics. For example, the metric *ExecutionTime* is a basic metric that refers to the age of a service request. When defining composite metrics, *expression* is used to specify how the value is computed. Also, a metric can be obtained for a specific group of requests by indicating *group* as true. The specification of metrics in STRATModel is given by Grammar 1.4 using also the Backus Normal Form (BNF).

$\langle Metric \rangle$ ::= 'metric' ':' $\langle MetricStatements \rangle$ ';'

$\langle MetricStatements \rangle$ ::= $\langle Name \rangle$ $\langle Reference \rangle$ $\langle Entity \rangle$ $\langle OnGroups \rangle$ $\langle Unit \rangle$ $\langle MetricExpression \rangle$

$\langle Entity \rangle$ ::= 'level' ':' $\langle MetricLevel \rangle$

$\langle MetricLevel \rangle$ ::= 'Service' | 'LoadBalancer' | 'Process' | 'Request'

$\langle OnGroups \rangle$ ::= 'group' ':' $\langle Boolean \rangle$ | $\langle empty \rangle$

$\langle Unit \rangle$ ::= 'unit' ':' $\langle string \rangle$ | $\langle empty \rangle$

$\langle MetricExpression \rangle$ ::= 'expression' ':' $\langle Expression \rangle$

Grammar 1.4. Grammar for describing metrics in STRATModel

- *Property*: In elasticity model, some actions may requires to access or modify some low-level properties of the managed SBP and its services. So, the user is allowed to define those properties and whether they are configurable or not. A *property* is primary defined in STRATModel by a name and a reference.

$\langle Property \rangle$ ::= 'property' ':' $\langle PropertyStatements \rangle$ ';'

$\langle PropertyStatements \rangle$ ::= $\langle Name \rangle$ $\langle Reference \rangle$ $\langle Config \rangle$

$\langle Config \rangle$ ::= 'configurable' ':' $\langle Boolean \rangle$

Grammar 1.5. Grammar for describing properties in STRATModel

Business Process Transformation State Definition. As previously indicated, the execution of an action transforms the SBP model from one state to another. A transformation state represents the managed SBP model at timestamp t. It is specified in a block defined by *ProcessState* and identified by a name which is used to refer to the state in the action description section (*cf.* Grammar 1.6).

Two ways are allowed to define the SBP model at timestamp t. The first is by describing the process and its components using specific notations. So, PROCESS block describes the structural representation of the SBP model at timestamp t followed by the description of requests contained in the SBP's services. The block encapsulates the process general description, the groups of requests allowed in the process, its services that are split into service engines and load balancer, the routers that connect its services, and the links between services and routers. A detailed discussion of the grammar used to describe processes is out of scope of this paper. The second way for providing the SBP model description at timestamp t, is by indicating the URL of the business process petri net model encoded in the Petri Net Markup Language (PNML).

⟨*ProcessState*⟩ ::= '*ProcessState*' ⟨*id*⟩ '{' ⟨*ProcessDefinition*⟩ '}'

⟨*ProcessDefinition*⟩ ::= '*Process*' ⟨*id*⟩ '{' ⟨*ProcessDescription*⟩ ⟨*Groups*⟩ ⟨*Services*⟩
                    ⟨*Routers*⟩ ⟨*Links*⟩ '}' ⟨*Requests*⟩
               | '*url*' ':' ⟨*string*⟩

Grammar 1.6. Grammar for defining a business process state in STRATModel

3.3   Elasticity Controller Generation

After describing an elasticity model using STRATModel syntax, we generate the elasticity controller based on a pre-defined template that groups the common functionalities of a controller. Usually, a controller is represented by a control loop to provide autonomic management which gives the system the ability to manage its resources automatically and dynamically whenever needed. This loop consists in (*i*) harvesting monitoring data, (*ii*) analyzing them using (optionally) a knowledge base and (*iii*) generating reconfiguration actions to correct violation (self-healing and self-protecting) or to target a new state of the system (self-configuring and self-optimizing) [9].

The pre-defined template as shown in Fig. 1 is modeled using high-level petri nets [10] to allow the formal evaluation of elasticity strategies on a SBP model [12]. It represents the basic construction of elasticity controllers on which the generation is based. It contains a central place *BP* of type net system that represents the managed component and surrounded with a set of transitions representing the actions to be performed on a SBP model (token in the place BP). The *Monitor* transition is used to trigger the monitoring of the SBP. It is guarded with a delay representing the frequency of monitoring. For example, if the *Monitor* transition is guarded with value 3, it means that there is three cycles between two successive monitoring actions. The firing of the transition adds new metrics values to the place *KB* representing the used knowledge base in which the history of monitoring metrics are stored. The *Check* transition is used to

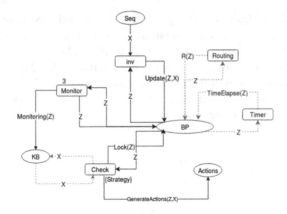

Fig. 1. Elasticity controller petri net model template

inspect an elasticity strategy and to check for QoS violations. It can optionally use information from the knowledge base component (*i.e.*, place *KB*) according to the defined elasticity model. The firing of the *Check* transition generates a set of actions to be applied according the used strategy and locks the entities on which the actions will be performed. The generated actions are stored in the place *Actions*. The *Inv* transition is used to introduce new requests to the SBP model from the place *Seq* which stores the sequence of requests arrival. According to the description of the elasticity model, two other transitions are optionally used from the template. The first one is the *Routing* transition which is responsible for routing requests between services in the SBP model. It can be omitted from the template in order to allow the user to use its customized routing action. The second one is the *Timer* transition which can be used and included in the template if the SBP model includes temporal information. It is used to increment the clocks in the SBP model.

Given an elasticity model, the elasticity controller petri net model is generated by enriching the pre-defined template with new transitions for the defined actions. Each action is translated to a transition where the name of the transition is to the name of the action. It can be associated with a time delay of applying the action. The firing of the transition executes the action mechanism on the SBP model if there is a stored action in the place *Actions* corresponding to the action of the transition where the action's reference ID is used to identify it.

Example 1. Let's consider an elasticity model for hybrid scaling that performs two main elasticity actions namely *Duplicate* and *Consolidate*. The duplication action allows to add new service copy with different configuration. The property capacity and category are reconfigurable by the action. We assume that we have a timed SBP model named '*Process1*' to manage its elasticity and that the default routing mechanisms will be used. The elasticity strategies that will be used on the latter are reactive and they do not need a knowledge base for elasticity decisions. Figure 2 illustrates the generated elasticity controller petri net model for the defined elasticity model. We added two transitions to the template named *Duplicate* and *Consolidate* corresponding respectively to the action '*Duplicate*' and the action '*Consolidate*' in the elasticity model. Since the used SBP model contains temporal information, the *Timer* transition is allowed in the final model. The elasticity model also specifies that the default routing mechanism will be used and there is no need for a knowledge base in the analysing step (*i.e.*, *Check* transition).

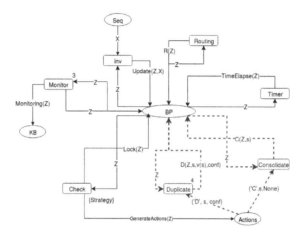

Fig. 2. Example of a generated elasticity controller Petri Net model

4   Elasticity Strategies Evaluation

Elasticity strategies govern the provisioning of necessary (to respect the agreed QoS) and sufficient (to handle the amount of requests) resources despite variations in enactment requests load. Many strategies can be defined to steer business processes elasticity. The abundance of possible strategies requires their evaluation in order to guarantee their effectiveness before using them in real Cloud environments. In our previous work [11], we presented our formal evaluation framework that is based on: (1) STRAT language for describing elasticity strategies and, (2) a specific elasticity controller implementing three actions for horizontal scaling. However, the business process holders are constrained to express their strategies using only the actions provided in the elasticity model implemented by the used controller. Here comes STRATModel to overcome this issue. In the following, we present a brief overview of the START evaluation framework and how it works using STRATModel. A detailed description is out of the scope of this paper. Therefore, we present how to adapt the STRAT language to be used across elasticity models described using STRATModel.

4.1   STRAT Evaluation Framework

START evaluation framework has been proposed to allow business process holders to define and evaluate, through simulation, elasticity strategies for a given business process model and under a given usage behavior. The strategies are described in STRAT language that allows to specify QoS requirements of a business process at different granularity levels (*i.e.*, process, service, and instance level) with taking into consideration the fundamental characteristics of SBP. They are evaluated on an business process model modelled as a petri net for a given usage behavior that represents the arrival law of clients' invocations.

The framework is composed of two main parts. The first part is the core of START language that allows to interpret and process the provided STRAT scripts. The second part is an elasticity controller that allows the execution of elasticity mechanisms. The latter is used to trigger fireable rules (of the elasticity strategy) and to perform thereafter the corresponding elasticity actions. It has been proposed implementing the elasticity model defined by Amziani et al. [3] which allows to perform three elasticity actions: (1) Routing that controls the way a load of a service is routed over the set of its copies, (2) Duplication which creates a new copy of an overloaded service in order to meet its workload increase, and (3) Consolidation that releases an unnecessary copy of a service in order to meet its workload decrease.

In order to generalize the use of the evaluation framework, we adapt it to use STARTModel which gives the description of the elasticity model that the STRAT language and the elasticity controller will be based on. So, a STRATModel script is required from the business process holders as an additional input to the framework to configure it to their needs. Using STRATModel script as input, STRAT language will be adapted to the described elasticity model (cf, Sect. 4.2) and the elasticity controller will be generated implementing the provided model as described in the previous section.

4.2   STRAT Language Adaptation

STRAT is composed of two parts: (i) STRAT grammar that specify how to describe an elasticity strategy, and (ii) STRAT Core that groups the basic functionalities of STRAT for processing the given elasticity strategies. In order to make STRAT utilisable across different elasticity models, some adaptations need to be made in both parts. In the following, we present the adaptation performed in each part separately.

STRAT Grammar. The top-level of STRAT specification grammar is given in Grammar 1.7 using the Backus Normal Form (BNF). A strategy STRAT is composed of two sections encapsulated in a block defined by Strategy (*i.e.*, indicates the beginning of the strategy) and identified by a name. The business process holder is allowed to separate the rules section identified by `Actions` from the definition of constants sets used by the rules like thresholds sets and time constrains.

$\langle ScalingPolicy \rangle ::=$ 'Strategy' $\langle name \rangle$ '{' $\langle Statements \rangle$ '}'

$\langle Statements \rangle \quad ::= \langle Initialization \rangle \langle ActionsBlock \rangle \mid \langle ActionsBlock \rangle$

$\langle Initialization \rangle ::=$ 'Sets' ':' $\langle Sets \rangle$

$\langle ActionsBlock \rangle ::=$ 'Actions' ':' $\langle Actions \rangle$

Grammar 1.7. General STRAT Grammar

In this adaptation version of STRAT, the `Actions` section is adapted to allow business process holders to specify their elasticity strategies using actions defined in a given elasticity model described using STRATModel. The actions are now referred to the actions described in the given STRATModel script. So, by changing the provided script, the actions allowed in STRAT change as well. Moreover, we provide the grammar with a syntactic validator and scope provider modules to adapt actions parameters according to their definitions in the used elasticity model by dynamically activate and deactivate parts of syntactic definition of `Action`. The parameters of an action are: (1) the elements on which the action will be applied, (2) the resulting configuration performed by the action, and (3) the multiplicity of applying the action. The first two parameters are recognized from the defined cases of the action while the multiplicity of the action is allowed if the action is defined as multiple. The configurable properties used in the second parameter are referred to the properties defined in STRATModel as configurable. The specification of actions in `Strat` is given by Grammar 1.8.

$\langle Actions \rangle \qquad ::= \langle Action \rangle$ ':' $\langle Rules \rangle$ '.'
$\qquad\qquad\quad \mid \langle Action \rangle$ ':' $\langle Rules \rangle$ '.' $\langle Actions \rangle$

$\langle Action \rangle \qquad ::= [\text{STRATModel::Action}]$ '(' $\langle id \rangle \ \langle Copies \rangle \ \langle Configuration \rangle$ ')'
$\qquad\qquad\quad \langle Multiple \rangle$

$\langle Configuration \rangle ::=$ ',' '[' $\langle ConfigItem \rangle \langle ConfigItems \rangle$ ']' $\mid \langle empty \rangle$

$\langle ConfigItems \rangle \ ::=$ ',' $\langle ConfigItem \rangle \langle ConfigItems \rangle \mid \langle empty \rangle$

$\langle ConfigItem \rangle \quad ::= [\text{STRATModel::Property}] \langle ConfigOps \rangle \langle ItemValue \rangle$

$\langle Multiple \rangle \qquad ::=$ 'by' $\langle value \rangle \mid \langle empty \rangle$

Grammar 1.8. Grammar of specifying actions in the adapted STRAT

STRAT Core. It groups the basic functionalities of STRAT that allow the processing of STRAT scripts, validating the coherence of the given rules and applying them. It provides a set of pre-defined functions used to perform certain tasks on a given business process model. Additionally, it includes the basic functions like arithmetic functions (*i.e.*, add, mul, sub, mod, div), and counter function (*i.e.*, count). The pre-defined functions are extended automatically and dynamically by a set of functions generated from the metrics definitions given in the used STRATModel script. They are used to express actions rules.

4.3   Experimental Evaluation

We present hereafter two use cases showing the definition of two elasticity models and their use to evaluate elasticity strategies using the adaptation of STRAT framework. This experiment is provided to show how the Business process holders can adapt the evaluation framework to their needs. In each use case, we define an elasticity model, a SBP, an elasticity strategy, and the arrival law of process enactment requests.

(1) First Use Case

(a) *Elasticity Model:* Our first use case is based on the elasticity model used in [11] which allows three main actions: (i) `Duplicate` which is applied on service engine components of SBP model, (ii) `Consolidate` to remove service engine copy from the managed SBP, and (iii) `Routing` which is applied on router components of SBP model to transfer requests between services. It is defined to manage an untimed SBP model. The model is defined using STRATModel and is illustrated in the Fig. 4(a). The Figure also illustrates the adaptation of STRAT language according to the defined elasticity model.

(b) *SBP model:* We use an example of SBP for online computer shopping which is composed of four services. Figure 3 presents the corresponding BPMN model of our SBP. The initial service in the SBP (*i.e*, S1) is responsible for receiving requests to purchase a computer and forwarding them to the Computer assembly service (S2) and the invoice Service (S3). Thereafter, service S2 performs the assembly of computer components based on user preferences while service S3 issues the invoice order. The last service S4 is responsible for delivering the computer along with the invoice to the requester. The elastic execution environment of this example is modeled as a Classic Petri Net where a service engine is represented by a place, a router is represented by a transition, and a service's request is represented by a black token [3].

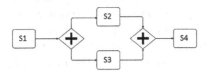

Fig. 3. BPMN model of the SBP example of the online computer shopping service.

(c) *Elasticity Strategy:* We define an elasticity strategy named `CapacityStrategy` that relies on the elasticity model `controller1` for the online computer shopping process. It defines the rules to add/remove service's copies according to the capacity threshold of each service defined in `max_c` and `min_c` sets which represent respectively the maximum and the minimum capacity of the supported elastic services. The rule for `Routing` action is also defined in the strategy to route a request if the request transfer (or router) does not cause a violation of the maximum thresholds of its post-services. It is triggered when neither of the previous actions (Duplicate and Consolidate) are allowed.

(d) _Usage behavior:_ We use in this experiment the Poisson distribution for dynamically generating the sequence of requests arrival. We set the mean of Poisson distribution to 2;

(a)   (b)

Fig. 4. (a) START adaptation for horizontal scaling (b) STRAT adaptation for horizontal and vertical scaling

(2) Second Use Case

(a) _Elasticity Model:_ The elasticity model is defined for a timed SBP model (_cf._, Fig. 4(b)). It is constructed of a set of metrics, one property and three actions. The set of actions is composed of: (i) `ScaleOut` which is defined to add a new copy of a service engine components along with a load balancer for the service engine if it is applied for the first time; (ii) `ScaleIn` which is defined to remove a copy of a service engine along with the associated load balancer if it will remain one copy of the service engine after applying the action; (iii) `Resizing` which is defined to reconfigure characteristics/properties of a service engine component (_i.e._, capacity). The metrics defined in the model are: (i) `executionTime` which is obtained by summing the waiting time and the processing time metrics of a request which are also defined in the model; (ii) `UsedCapacity` which provides the rate of capacity consumption for a service copy; and (iii) `Capacity` which is obtained from the property 'cap' that indicates the allocated capacity to a service copy. The model defines a property 'cap' as configurable which will be modified by the action `Resizing`. Figure 4(b) also illustrates the adaptation of STRAT language according to the defined elasticity model.

(b) _SBP Model:_ We use an example of SBP for molecular evolution reconstruction (MER) which is composed of eight services [12]. Figure 5 presents the corresponding BPMN model of our SBP. The initial service in the SBP (_i.e_, S1) is responsible for performing the pre-processing of FASTA file which contains a representation of nucleotide or peptide sequences. The second service consists in constructing a Multi Sequence Alignment (MSA) which is converted to PHYLIP format by the service S3. Then, the fourth service which consists in pre-processing PHYLIP file formats the input file according to the format definition and generates a second PHYLIP file. After that, the fifth service receives the PHYLIP file as input and produces a phylogenetic tree as output. The constructed phylogenetic tree is used for the MER exploration. We consider in this example that the latter is performed by two parallel services (*i.e.*, 6.1 and 6.2) which output a set of files containing evolutionary information. The last service processes the evolutionary information resulted from the previous services. The elastic execution environment of this example is modeled as a High-level petri net where service engines and load balancers are represented by places and each place characterized by a set of properties, routers correspond to transitions, and serivce's requests are represented by tokens having a set of properties [12].

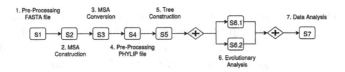

Fig. 5. BPMN model of the SBP example of molecular evolution reconstruction (MER).

(c) *Elasticity Strategy:* We define a strategy named `ResponseTimeStrategy` that relies on the elasticity model `controller2`. It defines rules for resizing the capacity of a service's copy and adding/removing copies of a service according to the response time thresholds of each service defined in `max_t` and `min_t`. The resizing action is performed also according to the allocated capacity to the service's copy. So, it can be performed only if the total allocated capacity didn't exceeded a fixed amount of capacity.

(d) *Usage behavior:* In this use case, we set the mean of Poisson distribution to 4.

Evaluation Results. Due to space limitation, we provide the results of one of the evaluation framework's outputs named capacity evolution indicator. We use it to illustrate how the evaluation of strategies is performed using different elasticity models. The capacity evaluation indicator is obtained from monitoring the given SBP models. In Fig. 6, we compare the amount of used capacity to the total provided capacity over time for the service S2 in the online computer shopping process and the service S2 in the MER process. The used capacity is computed by summing up the consumed capacity in each copy of the service engine. The provided capacity is computed by summing up the provided capacity for each service engine copy. The amount of either consumed or provided capacity is computer for the 1er use case in term of number of requests. For the

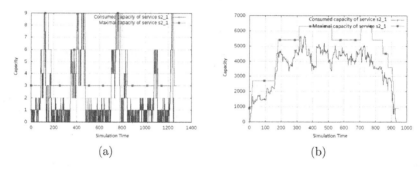

Fig. 6. The evolution of capacity consumption of (a) service S2 in online computer shopping process using `CapacityStrategy` and (b) service S2 in MER process using `ResponseTimeStrategy`

second use case, the amount of consumed and provided capacity are obtained respectively from summing up the capacity used by each request in the service and the property 'cap'. The resulted plots of the two cases show the adjustment of capacity to the change in workload over time. The business process holders can observe the elasticity behavior of their processes by analysing the difference between the allocated and the consumed capacity. This analyse allows to make decision on adjusting the defined thresholds or changing some conditions in the elasticity strategy.

5   Conclusion

In this paper, we proposed a descriptive domain specific language for expressing elasticity model for business processes, called STRATModel. STRATModel enables business process holders to define their proper elasticity models by describing customized metrics, properties and elasticity actions. The mechanism of a defined action can be provided through a set of examples illustrating how the action should be applying in certain cases. It follows the *by-example* paradigm in generating elasticity mechanisms. Moreover, STRATModel generates customized elasticity controllers and provides the basic model for STRAT to rely on which allows to define elasticity strategies describing rules for different elasticity actions using customized metrics. We also described the adaptive evaluation framework that allows to define elasticity models according to our DSL and evaluate strategies described using the adaptive version of STRAT through simulation. In our future work, we target to extend the evaluation framework and allow the evaluation of business processes as services in real cloud environments using Open Cloud Computing Interface (OCCI) [18].

References

1. Ali-Eldin, A., Tordsson, J., Elmroth, E.: An adaptive hybrid elasticity controller for cloud infrastructures. In: NOMS, pp. 204–212 (2012)
2. Amziani, M., Klai, K., Melliti, T., Tata, S.: Time-based evaluation of service-based business process elasticity in the cloud. In: CloudCom, vol. 1, pp. 573–580 (2013)
3. Amziani, M., Melliti, T., Tata, S.: Formal modeling and evaluation of stateful service-based business process elasticity in the cloud. In: Meersman, R., Panetto, H., Dillon, T., Eder, J., Bellahsene, Z., Ritter, N., De Leenheer, P., Dou, D. (eds.) OTM 2013. LNCS, vol. 8185, pp. 21–38. Springer, Heidelberg (2013). doi:10.1007/978-3-642-41030-7_3
4. Copil, G., Trihinas, D., Truong, H.-L., Moldovan, D., Pallis, G., Dustdar, S., Dikaiakos, M.D.: ADVISE – a framework for evaluating cloud service elasticity behavior. In: Franch, X., Ghose, A.K., Lewis, G.A., Bhiri, S. (eds.) ICSOC 2014. LNCS, vol. 8831, pp. 275–290. Springer, Heidelberg (2014). doi:10.1007/978-3-662-45391-9_19
5. Cypher, A. (ed.): Watch What I Do - Programming by Demonstration. MIT Press, Cambridge (1993)
6. Farokhi, S., Jamshidi, P., Lakew, E.B., Brandic, I., Elmroth, E.: A hybrid cloud controller for vertical memory elasticity: a control-theoretic approach. Future Gener. Comput. Syst. **65**, 57–72 (2016)
7. Herbst, N.R., Kounev, S., Reussner, R.: Elasticity in cloud computing: what it is, and what it is not. In: ICAC, pp. 23–27 (2013)
8. IDG: Idg enterprise cloud computing study (2014). http://www.idgenterprise.com/report/idg-enterprise-cloud-computing-study-2014
9. Jacob, B., Lanyon-Hogg, R., Nadgir, D.K., Yassin, A.F.: A Practical Guide to the IBM Autonomic Computing Toolkit. IBM Redbooks, IBM Corporation, International Technical Support Organization, Armonk (2004)
10. Jensen, K., Rozenberg, G.: High-level Petri Nets: Theory and Application. Springer-Verlag, Heidelberg (1991)
11. Jrad, A.B., Bhiri, S., Tata, S.: Description and evaluation of elasticity strategies for business processes in the cloud. In: SCC, pp. 203–210 (2016)
12. Jrad, A.B., Bhiri, S., Tata, S.: Data-aware modeling of elastic processes for elasticity strategies evaluation. In: CLOUD (2017)
13. Lieberman, H. (ed.): Your Wish is My Command: Programming by Example. Morgan Kaufmann Publishers Inc., San Francisco (2001)
14. Liu, Y., Gureya, D., Al-Shishtawy, A., Vlassov, V.: Onlineelastman: self-trained proactive elasticity manager for cloud-based storage services. In: ICCAC (2016)
15. Mell, P.M., Grance, T.: The nist definition of cloud computing. Technical report, National Institute of Standards & Technology, Gaithersburg, MD, United States (2011)
16. Molt, G., Caballer, M., de Alfonso, C.: Automatic memory-based vertical elasticity and oversubscription on cloud platforms. Future Gener. Comput. Syst. **56**, 1–10 (2016)
17. Naskos, A., Stachtiari, E., Katsaros, P., Gounaris, A.: Probabilistic model checking at runtime for the provisioning of cloud resources. In: Bartocci, E., Majumdar, R. (eds.) RV 2015. LNCS, vol. 9333, pp. 275–280. Springer, Cham (2015). doi:10.1007/978-3-319-23820-3_18

18. Nyren, R., Edmonds, A., Papaspyrou, A., Metsch, T.: Open cloud computing interface - core. Technical report, Open Grid Forum (OGF) (2011)
19. Suleiman, B., Venugopal, S.: Modeling performance of elasticity rules for cloud-based applications. In: EDOC, pp. 201–206 (2013)
20. Zabolotnyi, R., Leitner, P., Schulte, S., Dustdar, S.: SPEEDL - a declarative event-based language for cloud scaling definition. In: IEEEServices (2015)
21. Zloof, M.M.: Query by example. In: Proceedings of National Compute Conference, pp. 431–438. AFIPS Press (1975)

Boosting Aided Approaches to QoS Prediction of IT Maintenance Tickets

Raghav Sonavane[1], Suman Roy[2(✉)], and Durga Prasad Muni[2]

[1] E&ECE Department, IIT Kharagpur, Kharagpur 721302, West Bengal, India
raghavsonavane@gmail.com
[2] Infosys Ltd., # 44 Electronics City, Hosur Road, Bangalore 560100, India
{Suman_Roy,DurgaPrasad_Muni}@infosys.com

Abstract. Ticketing system is an example of a Service System (SS) which is responsible for handling huge volumes of tickets generated by large enterprise IT (Information Technology) infrastructure components, and ensuring smooth operation. The system maintains the provision of recording the time that reflects when a ticket is opened, acknowledged to user, resolved and/or closed, from which different QoS parameters could be obtained. For example, Resolution Time can be computed as the difference of resolution date and opening date of the ticket. One needs to use new technology solutions in QoS-related analysis like categorization of tickets according to their QoS, predicting QoS parameters for new tickets etc., to improve the performance of the SS. In this work we propose boosting oriented solutions to QoS prediction of tickets using crisp and fuzzy set models of QoS. In particular, we employ a two-stage analysis framework for QoS prediction for incoming tickets which includes clustering incident tickets based on QoS values and building a regression model using this categorization and the textual contents of tickets. We carry out experiments on industrial data sets using different techniques for prediction. We improve the quality of prediction by using suitable boosting techniques. We propose random forest boosting on Logistic Regression and gradient boosting on NNLS for our purpose, both of which improve the performance of prediction. We report these results and compare them.

Keywords: Quality of services (QoS) · Service System · Tickets · Crisp set · Fuzzy set · Boosting · Response Time · Resolution Time · Clustering · Logistic Regression · SVM · Non-negative least squares (NNLS)

1 Introduction

Ticketing system, an example of SS, is used as one of the inputs for Information Technology Infrastructure Library (ITIL) services such as problem management and configuration management. Dozens of tickets are raised by the users on

---

R. Sonavane—This work was done when Raghav Sonavane did his internship with Infosys during May–July'17.

H. Panetto et al. (Eds.): OTM 2017 Conferences, Part I, LNCS 10573, pp. 467–487, 2017.
https://doi.org/10.1007/978-3-319-69462-7_30

the ticketing system for the purpose of resolving their problems while using different support systems. Normally a ticket is recorded on the system with a summary which describes the related issue. The maintenance team examine the ticket, respond to the user with an acknowledgment with possibly seeking more explanation of the issue and indicating the possible solution. Subsequently on hearing back from the user, the team undertake remedial actions on the ticket, write a short note on the resolution steps and close the ticket in the end. The maintenance team also record other details about the ticket like the opening date, the response date, the resolution date and the closing date for the ticket. These parameters could be used to compute different Quality of service (QoS) parameters like Response Time (difference between response date and open date), Resolution Time (difference between resolution date and open date), Closure Time (difference between closure date and open date) etc. There is a need of introducing new QoS technologies in services that can be used for future business applications in ticketing systems. These could include categorization of tickets according to their QoS, predicting QoS parameters for new tickets etc. which are the focus of this paper.

Although exact values of a ticket can be computed for different QoS parameters it may not be appropriate to predict an exact value of a QoS parameter for a new ticket. Different persons may resolve the same ticket in different time, hence Resolution Time for a ticket can be different for the same ticket. Naturally, we choose to predict a QoS value for a new ticket using an interval of time when we model QoS values using crisp sets.

We also model different QoS parameters arising in ticketing application services using Fuzzy sets. As mentioned earlier, QoS parameters like Resolution Time could be different from person to person handling the same ticket. Also Resolution Time would depend on the description of the summary of the ticket written by different users although the issue may be more or less the same. These QoS parameters cannot be considered as noted precisely. Therefore, an appropriate way to describe the Resolution Time for a new ticket is to say that it is high, or medium, or low. These phrases 'high', 'medium' and 'low' can be regarded as fuzzy quantities. This kind of categorization of ticketing services also helps people especially those with almost no technical experience, to understand reasonable QoS values as provided by the service.

Background for the Work. Motivated by this we use a two-stage procedure for predicting QoS values for new tickets using both crisp and fuzzy sets (see Fig. 1). When we model QoS models using crisp sets we partition QoS values into different clusters which are used to label the tickets. Subsequently, we use feature vectors of tickets (using TF*IDF values of selected terms and concepts) and their labels to set up a classification problem which can be solved with Logistic Regression (LR) technique or Support Vector Machine (SVM) to predict QoS values for an incoming ticket in terms of intervals of time. In case of fuzzy modeling of QoS parameters we choose Fuzzy C-means algorithm to cluster ticket data *wrt* their QoS values and compute the membership values of tickets with respect to each cluster. Then we use these membership values and feature

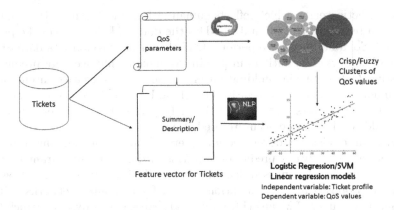

Fig. 1. A two-stage procedure for categorization and prediction of QoS parameters for tickets

vectors of tickets with TF*IDF values to set up a linear regression model. This model helps us to predict fuzzy QoS values for incoming tickets.

However, in some of these cases the accuracy of prediction turns out to be low. For improving this we resort to boosting algorithms which are ensemble of meta-algorithms for primarily reducing bias and variance in supervised learning. Boosting methods work on the principle of generating multiple predictions and majority voting (averaging) among the individual classifiers. For improving classification accuracy in case of Logistic Regression we propose Random Forest based boosting on top of it. Similarly in case of prediction of fuzzy QoS values we propose a version of gradient boosting using NNLS (non-negative least sqaure). While the accuracy of prediction witnesses improvement for Logistic Regression with the use of boosting in most of the cases, there is significant improvement in case of prediction of QoS values on using gradient boosting in all cases of NNLS. We carry out a thorough experiments on real-life ticket data sets and compare the prediction results prior to boosting and after boosting.

The paper is organized as follows. In the next section (Subsect. 1.1) we discuss the current literautre available on related work. We discuss the schema of the ticket data that we consider in Sect. 2. We describe crisp set-based approach of QoS modeling in Sect. 3. Fuzzy set modeling of QoS parameters is taken up in Sect. 4. New boosting techniques are introduced in Sect. 5. We describe our experimental effort in Sect. 6. Finally we conclude in Sect. 7.

1.1 Related Work

Fuzzy analysis has been applied to the selection of web services, which are based on different QoS parameters. Such a selection of QoS-based web services is done in [1] by applying the fundamental principles in the fuzzy set theory to model web service selection in terms of Fuzzy Multiple Criteria Decision Making (FMCDM) using a synthetic weight, which reflects human rating among others. Human preferences can be biased and often depend on the domain knowledge of the person concerned. So, a better way to perform QoS clustering is to follow a data driven approach like ours. The authors in [2] aim to make highly accurate predictions for missing QoS data via building an ensemble of non-negative latent factor (NLF) models by fitting non-negative QoS data. But the data driven association among QoS parameters and the related analysis are missing, incorporation of which could result in more accurate predictions.

Regression models are developed for different applications ranging from simulation of data to network prediction in terms of different QoS parameters like throughput, mean delay, missed deadline ratio, collision ratio etc [3,4]. Dogman *et al.* have proposed a QoS evaluation system for computer networks using a combination of Fuzzy C-Means (FCM) [5] and regression analysis by which they analyze and assess the QoS in a simulated network [4]. The authors therein consider network QoS parameters of multimedia applications like delay, jitter, loss ratio etc. and use FCM algorithms to partition those QoS paramaters into appropriate clusters, the number of which is determined using validity index. The resulting QoS parameters are given as inputs to a regression model in order to quantify the overall QoS value. In [6,7] the authors improve upon these methods are carry out a more rigorous and comprehensive analysis by using fuzzy regression model with non-negative least square method, which accepts the membership value of each ticket with respect to each category (cluster) of a QoS as response variables. However, the accuracy of QoS prediction is not up to the desired level and there is no benchmarking of results against similar techniques. We try to address these deficiencies here by way of improving the accuracy of prediction through boosting methods and compare prediction of QoS values for fuzzy modeling with crisp modeling.

2   Ticket Model

We consider incident tickets with similar schema which are frequent in IT maintenance which are described below.

2.1   Schema of Ticket Data

These tickets usually consist of two fields, fixed and free form. There is no standard value for free-form fields. These fields can contain call description or summary of ticket which capture the issue behind the creation of such tickets on the maintenance system. This can be just a sentence that summarizes the problem reported in it, or it may contain a detailed description of the incident. Normally fixed fields provide very specific information about the incidents. Fixed fields are customized and inserted in a menu-driven fashion. Example of such fields are the ticket's identifier, department of the user raising the ticket etc., priority of the ticket, the time the ticket is raised, responded to or closed on the system or, if a ticket is of incident or request in nature. Various other important information are

IncidentType	Category	SubCategory	ApplicationName	SubmitDateTi	ClosedDateTi	LastModified	Statu	Priori	Summary	Note
AS/400 - User Profile	Administration	Configuration	AS400 Legacy - Manufa	9/1/2015 17:20	9/4/2015 0:51	9/4/2015 0:51	Closed	P4 - Stan	AS/400 ID Requests - NBTY	09-0 AM -
AS/400 - Software De	Software	Application Errors	AS400 Legacy - Warehc	9/17/2015 22:23	9/21/2015 22:02	9/23/2015 19:00	Closed	P3 - Mod	Inventory/BU Location - Update error (Please see Attached Screenshot)	09-2 PM -
AS/400 - User Profile	Administration	Configuration	AS400 Legacy - Warehc	5/18/2015 19:19	5/28/2015 0:13	5/28/2015 0:13	Closed	P4 - Stan	?Need access to be authorized to use miscellaneou	05-2 AM -
AS/400 - Software De	Software	Application Functi	AS400 Legacy - Warehc	6/29/2015 21:17	7/6/2015 5:10	7/7/2015 19:00	Closed	P4 - Stan	We been trying to get a CPO to be unlocked or fixed	07-0 PM -
AS/400 - Software De	Software	Application Errors	AS400 Legacy - Warehc	8/3/2015 18:39	8/7/2015 0:34	8/10/2015 0:00	Closed	P4 - Stan	Problem with G.T.S menu - (See Attached)	08-0 AM -
AS/400 - Software De	Software	Software Enhance	AS400 Legacy - Wholes	6/10/2015 6:16	6/29/2015 5:41	6/29/2015 5:41	Closed	P4 - Stan	Tammy needs to receive emails through the winspo	06-2 AM -
AS/400 - User Profile	Administration	Configuration	AS400 Legacy - Warehc	6/11/2015 9:20	6/16/2015 9:32	6/16/2015 9:32	Closed	P4 - Stan	Can we please have Andrew's Regular Out Queue c	06-1 AM -
Oracle - ORPOS POS (	Software	Application Errors	AS400 Legacy - Retail	9/22/2015 12:14	############	############	Closed	P4 - Stan	From: Andrew Boland [mailto:AndrewBoland@hollandandbarrett.com]	10-2 PM -
AS/400 - Software De	Software	Configuration	AS400 Legacy - Retail	6/9/2015 9:26	6/11/2015 9:26	6/11/2015 9:26	Closed	P4 - Stan	Bonus Rebate Expiring Email - File Due 6/11	06-1 AM -
AS/400 - Software De	Software	Configuration	Legacy RCA	10/6/2015 12:35	############	############	Closed	P4 - Stan	Jenny Stayte Sent: Tuesday, October 06, 2015 6:34 AM	10-2 AM -

Fig. 2. A sample of ticket data

also captured through these fixed fields such as ticket's category, sub-category, application name (application under which the ticket was raised), incident type etc. A small part of ticket data is shown in Fig. 2.

2.2 Feature Vector Creation from Ticket Data

We consider the collection of summary of tickets for extracting the feature vector corresponding to a ticket. We use light natural language processing for feature vector generation. As a pre-processing we remove the tickets which do not contain any summary. In the beginning we perform lemmatization of the words in the summary of tickets. Then we use Stanford NLP tool [8] to parse the useful contents in the summary of the tickets and tag them as tokens. Then we use the tool to perform PoS tagging of the tokens. We mark only the nouns and choose them as potential keywords. Next we set up some rules for removing tokens which are stop words. We compute document frequency (DF) of each lemmatized word. We discard the words whose DF is smaller than 3. Ticket summary may contain some very rare words like name of a person (user who raised the ticket) and some noise words. By removing words with DF smaller than 3, we can remove these very rare words which do not contribute to the content of ticket summary. In this way, the feature vector size could be reduced significantly. We discard keywords whose information content [9] are among top 10% of the maximum entropy content of a word in the corpus. We denote the set of keywords extracted so far as $\mathcal{K}_1$. In the next step we choose keyphrases consisting of bigrams and trigrams. Using the PoS tagging we collect adjacent nouns (in a sentence) having length at most three. Then we choose some of the heuristics [10] like All Words Heuristic, Any Word Heuristic to select the final set $\mathcal{K}_2$ of keyphrases. The final list of keywords and keyphrases (together they will be called terms also) is $\Gamma = \mathcal{K}_1 \cup \mathcal{K}_2$. We can model a ticket as a vector $\bar{\mathbf{x}}(.,\ldots,.)$, where each element $\bar{\mathbf{x}}(t)$ represents the importance or the weight of a term t with respect to the ticket T. Here we take $\bar{\mathbf{x}}(t) = tf * idf(T,t)$ where $tf * idf(T,t)$ denotes the TF*IDF [11] of a term t wrt ticket T as its weight[1].

However, all these terms extracted from tickets fail to model the semantic aspects of ticket descriptions as they cannot capture any background information of ticket domain. For that we shall use concepts as additional features for tickets which will be extracted from the general purpose knowledge repository of WordNet. Concept extraction for domain is a key component in modeling domain knowledge. As manual concept extraction is expensive in terms of time and effort we employ heuristics for automatic concept extraction from summary of tickets, lifting some ideas from [10]. In addition to extracting unigrams as concepts we include bigrams and trigrams in the list of concepts again using heuristics based on senses and compound structures. Senses of the keywords are checked to remove the irrelevant keywords using Word Count Approach (WCA) heuristic. Individual words in bigrams and trigrams are used to find additional keyphrases as concepts which is based on WordNet sense count heuristic. The final set of concepts is Ξ. The relationship $\mathcal{R} \subseteq \Gamma \times \Xi$ between terms and concepts can be maintained through an easy bookkeeping. For details the reader is referred to [13]. Finally, a ticket T_i can be represented as a vector $\mathbf{x_i} = (\underbrace{a_{i1}, \cdots, a_{im}}_{\text{terms}}, \underbrace{b_{i1}, \cdots, b_{ih}}_{\text{concepts}})$, where a_{ip} indicates the weight (typically TF*IDF value) of term t_p in Ticket T_i and b_{iq} the weight (CF*IDF value)[2] of concept c_q in ticket T_i etc. T.

2.3   Grouping the Tickets wrt Tuples

The fixed field entries of a ticket can be represented using a relational schema. For that we shall consider only a limited number of fixed fields of a ticket by choosing attributes that reflect its main characteristics (the domain experts' comments play an important role in choosing the fixed fields), for example the attributes can be - application name, category and sub-category. They can be represented as a tuple: Ticket(application name, category, sub-category). Each of the instances of tuples corresponding to entries in the fixed fields in the ticket can be thought of an instantiation of the schema. Examples of rows of such schema can be, (AS400 - Legacy Manufacturing, Software, Application Errors), (AS400 Legacy - Retail, Software, Application Functionality Issue) etc. The relation key can vary from 1 to number of distinct tuples in the schema. One such key can

---

[1] TF*IDF is a popular metric in the data mining literature [12] and $(TF*IDF)(T,t) = TF(T,t) * IDF(t)$; $TF(T,t)$ represents the term frequency of term t wrt ticket T, that is, the number of times term t appears in T; $IDF(t) = \log(1 + \frac{N}{DF(t)})$, where $DF(t)$ is the Document Frequency of term t in the ticket corpus having N tickets, that is, the number of times t occurs in the corpus.

[2] $CF(T,c)$ denotes the concept frequency of a concept c wrt ticket T, for definition see [13].

hold several Incident IDs, that is, it can contain several tickets with different IDs. This way we can partition the tickets using their tuples.

2.4   Possible QoS Parameters for Ticket Data

Given this ticket schema one can formulate the following QoS paramaters, Response Time $\mathcal{R}$, Resolution Time $\mathcal{S}$, and Closure time $\mathcal{C}$. As the open, close, response and resolve dates are recorded for each ticket we can set the following parameters: *Response Time* $\mathcal{R}$ as the difference between response date and open date, *Resolution Time* $\mathcal{S}$ as the difference between resolve date and open date and *Closure time* $\mathcal{C}$ as the difference between closure date and open date etc. In some of the data closure date may be present while resolve date may be absent. In some cases there will be resolve date, but no closure date.

3   Crisp Prediction of QoS Values

We describe our crisp set-based method of predicting QoS values using the same 2-step approach. For each tuple in a ticket data set we cluster QoS values using K-means algorithm generating a finite set of clusters on QoS values. Tickets are modeled as features vectors of terms and concepts. We can assign a cluster to a ticket by finding the cluster with which it is the closest. We then set up a classification problem by labeling each ticket with its assigned cluster number. This classification problem is solved using any appropriate technique like Logistic Regression or Support Vector Machine (SVM).

In logistic regression, or logit regression, or logit model the dependent variable is categorical which is used for classification. For binary classification we consider a binary output variable y_i for which the conditional probability $p(y_i = 1|X = x_i)$ is modeled as a function of data point x_i; further $p(y_i = 0|X = x_i) = 1 - p(y_i = 1|X = x_i)$. One assumes $p(y_i = 1|X = x_i) = p(y_i|x_i; \beta)$, where $\beta = [\beta_0 \; \beta_1 \; \cdots \; \beta_N]^T$ and β_is are estimation parameters. One needs to estimate the unknown parameters in the function by maximizing conditional likelihood. The conditional likelihood is given by $L(\beta) = \sum_{i=1}^{N} [y_i \log p(y_i|x_i; \beta) + (1 - y_i) \log(1 - p(y_i|x_i; \beta))]$ for N data points. Assuming $\mathbf{x} = [1 \; x_1 \; \cdots \; x_N]^T$ we can write $p(\mathbf{y}|\mathbf{x}; \beta) = \frac{\exp(\beta^T \mathbf{x})}{1+\exp(\beta^T \mathbf{x})}$, by the assumption of logistic regression. Substituting the above in conditional likelihood expression we arrive at: $L(\beta) = \sum_{i=1}^{N} [y_i \beta^T x_i - \log(1 + \exp(\beta^T \mathbf{x}))]$.

The multinomial logistic regression generalizes logistic regression to multiclass problems with more than two possible discrete outcomes. Using this model it is possible to predict the probabilities of the different possible outcomes of a categorically distributed dependent variable, when a set of independent variables (which may be real-valued, binary-valued, categorical-valued, etc.) is given. In this case, one assumes the condition of independence of irrelevant alternatives (IIA) which states that the odds of preferring one class over another do not depend on the presence or absence of other "irrelevant" alternatives. With this, the choice of K alternatives can be modeled as a set of $(K - 1)$ independent binary choices, in which one alternative is chosen as a "pivot" and the other $(K-1)$ are compared against it, one at a time.

Support Vector Machines (SVMs) are learning algorithms for supervised learning models which are used for classification and regression analysis. One needs to solve a very large quadratic programming (QP) optimization problem to train a SVM learner. This can be simplified using sequential minimal optimization (SMO) algorithm due to Platt [14]. In SMO large QP problem is decomposed into a series of smaller possible QP problems which are solved analytically thus avoiding time-consuming numerical QP optimization as a whole.

4   Fuzzy Prediction of QoS Values

In this section we describe our approach of predicting QoS values when they are modeled using Fuzzy sets. First we cluster QoS values using FC-means algorithm which produces membership values of tickets with respect to each cluster for any QoS parameter. Then we set up a linear regression model for predicting QoS values where the feature vectors of tickets are taken as regressor variables and membership values of tickets are regressands. This linear regression problem is solved using Non-Negative Least Square method (NNLS). For details the reader is advised to consult [6, 7].

4.1   Clustering of QoS Values Using Fuzzy C-Means

Let us assume a QoS parameter Q modeled on ticket data, where Q can be response time or resolution time or closure time. The Fuzzy C-means (FCM) algorithm [5] partitions the ticket data of each tuple with respect to Q into κ clusters with $c_i, 1 \leq i \leq \kappa$, as cluster centers. Let us take a domain Ω for Q as the collection of cluster centers, $\Omega = \{c_1, \ldots, c_\kappa\}$. Then for a ticket T_i we can find the membership value of QoS parameter Q as μ_{ij}, for the cluster with center $c_j, 1 \leq j \leq \kappa$. That is, for each ticket we can define a fuzzy set representing QoS Q as $(\mu_{i1}/c_1 + \cdots + \mu_{i\kappa}/c_\kappa)$ on the set Ω with the constraint $\sum_{j=1}^{\kappa} \mu_{ij} = 1$.

4.2   Linear Regression on QoS Parameters

We use linear regression model for predicting QoS parameters modeled as Fuzzy sets. As the fuzzy memberships are defined on discrete sets we choose to work with separate linear regression models for these discrete fuzzy sets. Let us formulate the prediction problem for the QoS parameter Q. We have seen that each ticket T_i can be modeled using a feature vector $\mathbf{x}_i$. The responder variables can be expressed as membership values for the jth cluster:

$$\mathbf{y}^{\mathbf{j}} = \begin{pmatrix} y_{1,j}(= \mu_{1,j}) \\ \vdots \\ y_{n,j}(= \mu_{n,j}) \end{pmatrix}, 1 \leq j \leq \kappa$$

Further the TF*IDF matrix of tickets can be written as $\mathbf{X} = [\mathbf{1} \; \mathbf{x}_1^T \; \cdots \; \mathbf{x}_n^T]$. The regression equation for QoS parameter Q will look like $\mathbf{y}^j = X\beta^j, 1 \leq j \leq \kappa$ (β^j being the estimation parameters), whereas the observed response is $\tilde{\mathbf{y}}^j$. In order to estimate β, we formulate a non-negative least square (NNLS) problem:

$$\beta^* = \arg\min_{\beta \geq 0} \frac{1}{2} \left\| \left(\tilde{\mathbf{y}} - \mathbf{X}\beta \right) \right\|^2 . \tag{1}$$

Any update algorithm for solving NNLS can be used to solve this (convex programming) minimization problem. As the matrix X is sparse and skinny we use principal component analysis (PCA) to reduce its dimension before performing NNLS.

For a new ticket we need to predict its QoS parameter Q by computing its membership value *wrt* the each of the fuzzy clusters. Using NLP techniques discussed previously we can extract the feature vector of new ticket T_{in} as $\mathbf{x_{in}}^T = [1 \; x_{in,1} \; \cdots \; x_{in,p}]$. For QoS Q we predict its membership value $y_{in,j} = \mu_{in,j} = \mathbf{x_{in}}^T \beta^j$, where j ($1 \leq j \leq \kappa$) denotes the jth fuzzy cluster. Need to ensure that $\sum_{i=1}^{\kappa} \mu_{in,i} = \mu_{in,1} + \cdots + \mu_{in,j} + \cdots + \mu_{in,\kappa} = 1$. Hence we normalize the membership values as $\bar{y}_{in,j} = \frac{\mu_{in,j}}{\sum_{i=1}^{\kappa} \mu_{in,i}}$, $1 \leq j \leq \kappa$, which is finally published. We assign the likely QoS of an incoming ticket for a particular cluster if the predicted membership value is the maximum for that cluster. The cluster curve can be used to determine the range of values for this predicted QoS.

5   Boosting Techniques

For improving the performance of our prediction/classification problem we use boosting algorithms. They consist of an ensemble of meta-algorithms for primarily reducing bias and also variance [15]. High discrimination is attained in boosting by sequentially training the classifiers. The training sample on which the previous classification produces training errors, is employed to train the subsequent classifier to produce correct classification. To improve the performance of prediction we propose random forest boosting on Logistic Regression and gradient boosting for linear regression problem.

5.1   Random Forest Boosting for Logistic Regression

We shall propose a boosting for Logistic Regression based on random forest. Random forest boosting [18] is an ensemble of training algorithms which builds multiple decision trees (forest). It can remove the effect of over-fitting on the training samples by random selection of samples using similar techniques like that of bagging [19]. Splitting at nodes may be based on random selection of a subset of features which enables faster training. This creates a classifier which is robust against noise. For building the forest we adopt the classical model tree approach [16,20] which produces decision trees with linear regression function at the leaf nodes of the trees. However, in stead of using classical M5 approach

---

Algorithm 1. Boosted Random Forest for Logistic Regression

Input   : A sequence of Training samples
$\{(\mathbf{x}_1, y_1), \ldots, (\mathbf{x}_N, y_N)\}; \mathbf{x}_i$ appears in $\mathbf{X}$ (the ticket-term TF*IDF matrix),
$y_i \in \mathcal{L} (= \{0, \ldots, K\}), \ 1 \le i \le N$

1 **for** $t = 1 : \aleph$ **do**
2 | Choose a subset $\mathcal{S}_t$ of training samples randomly;
3 | **for** $c = 0 : K - 1$ **do**
4 | | Generate new labels y_i^{new} in place of y_i (using One vs not One approach);
5 | |

$$\forall i, y_i^{new} = \begin{cases} 1, & \text{if } y_i = c \\ 0, & \text{otherwise} \end{cases}$$

| | Randomly choose a subset $\mathcal{G}_c$ of features;
6 | | ModelTreeM5$(\mathcal{S}_t, \mathcal{G}_c, c)$ ▷ aka M5' algorithm [16];
7 | | **if** there are $1, \ldots, l$ leaf nodes associated with $f_1(x), \ldots, f_l(x)$ functions **then**
8 | | | $F_c(x) = \text{LBoost}(f_1(x), \ldots, f_l(x), M)$;
9 | | | ▷ Based on LogitBoost Algorithm with M iterations [17];
10 | | **end**
11 | | Set $p^t(y_i = c|\mathbf{x}_i; \beta) = \frac{1}{1+\exp(-F_c(x))}$
12 | **end**
13 | Compute error rate of the decision tree $\epsilon_t = \frac{\sum_{i=1}^N \#\{y_i \ne \hat{y}_i\}}{N}$;
14 | Weight of the decision tree $\alpha_t = \frac{1}{2}\log\frac{1-\epsilon_t}{\epsilon_t}$;
15 **end**
16 Compute $p^f(y_i = c|\mathbf{x}_i; \beta) = \frac{\sum_t(\alpha_t * p^t(y_i=c|\mathbf{x}_i;\beta))}{\sum_t \alpha_t}$;
17 **return** $p^f(y_i = 0|\mathbf{x}_i; \beta), p^f(y_i = 1|\mathbf{x}_i; \beta), \ldots, p^f(y_i = K - 1|\mathbf{x}_i; \beta), p^f(y_i = K|\mathbf{x}_i; \beta)$;

---

of building model trees [20] we use another variant M5' [16] which can generate more accurate classifiers. Below we describe our algorithm, the steps of which can be found in Algorithm 1. We also use a variant of LogitBoost [21], a boosting algorithm for forward stage fitting of additive regression models.

Training Process: We assume a training sample of tickets of size N, $\{\mathbf{x}_1, y_1\}, \ldots, \{\mathbf{x}_N, y_N\}$, where $\mathbf{x}_i s$ are independent vectors of feature variables having dimension $d = m + h$, and $y_i \in \mathcal{L}$, are the output variables $(1 \le i \le N)$ denoting class labels in $\mathcal{L} = \{0, 1, \cdots, K\}$, which has $K + 1$ labels. Using random sampling the sample sets are created from the training samples as done in random forest.

Node Splitting: A flow diagram of our proposed method is shown in Fig. 3. We consider the feature space $(\Gamma \cup \Xi)$ and select a subset $\mathcal{G}$ of features randomly from this.

Decision Tree Weighting: We use the same approach of multiclass boosting for computing the decision tree weights [22,23], by which the decision tree weight α_t for tree t is calculated by,

$$\alpha_t = \frac{1}{2}\log\frac{1 - \epsilon_t}{\epsilon_t},$$

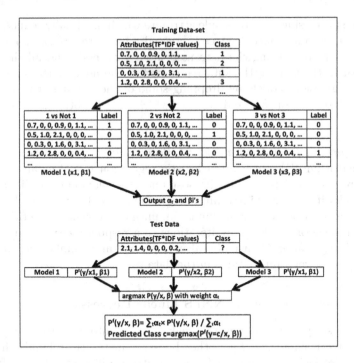

Fig. 3. Proposed training algorithm for random forest boosting on logistic regression

where ϵ_t is the error rate of the decision tree. One can classify the constructed decision tree. This error rate is computed from the weights of the incorrectly classified samples: $\epsilon_t = \frac{\sum_{i=1}^{N} \#\{y_i \neq \hat{y}_i\}}{N}$ [3].

Classification Process: When an unknown ticket is provided as an input to all the decision trees as shown in Fig. 3 the conditional probabilities $p^t(y_i = c|\mathbf{x}; \beta)$ of all classes $c \in \mathcal{L}$ in the leaf nodes for each tree t are computed. Then the outputs of the decision tree t, $p^f(y_i = c|\mathbf{x}; \beta); y_i \in \mathcal{L}$ are computed by taking the weighted average of that of each tree:

$$p^f(y_i = c|\mathbf{x}; \beta) = \frac{\sum_t (\alpha_t * p^t(y_i = c|\mathbf{x}; \beta))}{\sum_t \alpha_t}.$$

The class y^* that has the highest probability, is given as output as the classification result of the new ticket sample: $y^* = \arg \max_{c \in \mathcal{L}} p^f(y = c|\mathbf{x}; \beta)$.

Fitting Model Trees Using M5': We adopt M5' approach [16] (which improves upon the original model tree algorithm [20]) to fit a model tree for each class label. In this approach a regression tree is constructed by recursively splitting

---

[3] $\#\{y_i \neq \hat{y}_i\}$ is the number of samples where predicted value differs from the actual value.

the sample space using tests on single attribute which maximally reduce variance in the target variable. Once the tree is created a linear regression model is built for every inner node, using the data associated with node and all the features that appear in the tests in the sub-tree rooted at the node. Then one simplifies the linear regression models by dropping some of the features if it results in a lower expected error on future data after which, every subtree is considered for pruning.

LBoost Algorithm for Fitting a Logistic Regression: We use a variant of Log-itBoost algorithm [21] which fits a logistic regression model as an addition of linear regression models [17]. LogitBoost finds the maximum likelihood linear logistic model if each individual linear regression model is fit using linear least square regression when the algorithm is run until convergence. The LBoost algorithm fits regression model in M iterations to a response variable. In this way, the model learned after M (usually taken as a very small number) iterations will include the most relevant features in the data.

5.2   Gradient Boosting for NNLS Linear Regression

Gradient boosting combines several weak learners into a single strong learner in an iterative fashion [24]. Like other boosting methods, it builds the model in a stage-wise fashion by optimizing an arbitrary differentiable cost function. We follow similar principle for improving the prediction for NNLS using gradient boosting by extracting a model as a function $F(\cdot)$ to predict values in the form of $\tilde{\mathbf{y}} = F(\beta)$ and then, minimizing mean squared errors $\frac{1}{2}\|\tilde{\mathbf{y}} - \mathbf{y}\|^2$ of the observed values $\tilde{\mathbf{y}}$ to the predicted values $\mathbf{y}$. At each stage $k, 0 \leq k \leq M$ of gradient boosting, we assume some imperfect model F_k which is a very weak model that just predicts y in the training set. Our algorithm will not change F_k in any way, instead it improves on it by creating a new model with the addition of an estimator of the form $h(\beta^k)$ at each stage: $F_{k+1}(\beta^{k+1}) = F_k(\beta^k) + h(\beta^{k+1})$, where β^{k+1} and β^k denote the estimation parameters at stages $(k+1)$ and k respectively. We set $h(\beta^k) = \mathbf{X}\beta^k$. We also assume that $h(\cdot)$ should satisfy:

$$F_{k+1}(\beta^{k+1}) = F_k(\beta^k) + h(\beta^{k+1}) = \mathbf{y} \tag{2}$$

or equivalently, $h(\beta^{k+1}) = \mathbf{y} - F_k(\beta^k)$. In our case we assume β_k is updated using the rule: $\beta^{k+1} = \arg\min_{\beta^k \geq 0} \frac{1}{2}\left\|\left(\tilde{\mathbf{y}} - F_k(\beta^k)\right)\right\|^2$. We have an initial condition: $F_0(\beta^0) = \mathbf{X}\beta^0$. Now applying Eq. 2 iteratively, we get $F_M(\beta^M) = \mathbf{X}(\beta^0 + \beta^1 + \cdots + \beta^M)$.

We now present the modified algorithm for solving this regression problem in Algorithm 2 for the prediction of QoS parameters using gradient boosting in our setting. As before the aim is to estimate the parameters $\beta_0, \beta_1, \ldots, \beta_{m+h}$ (see Eq. 1).

Algorithm 2. NNLS algorithm with gradient boosting

Input   : A feature vector $\mathbf{x}_i$ for T_i (using both terms and
                    concepts) and thus the TF*IDF matrix $\mathbf{X}$ capturing ticket
                    and term-concept relationship and the observed values
                    $\tilde{y}_1, \ldots, \tilde{y}_n$

Output   : The final estimation parameters $\beta_0^M, \beta_1^M, \ldots, \beta_{m+h}^M$

Initialization: $\tilde{y}_i = y_i$, $1 \leq i \leq n$ and $\beta^0 = 0$

1 **for** $0 \leq k \leq M$ **do**

2 $\qquad \beta^{k+1} = \arg\min_{\beta^k \geq 0} \frac{1}{2} \left\| \left(\tilde{\mathbf{y}} - \mathbf{X}\beta^k \right) \right\|^2$

3 $\qquad \mathbf{y} \leftarrow \mathbf{y} - \varepsilon(\mathbf{X}\beta^{k+1})$

4 **end**

5 **return** β^M

Above ε is the learning rate of the algorithm which is taken as 1 in this work (as this is a regression problem). The final predicted value is: $\mathbf{X}(\beta^0 + \beta^1 + \cdots + \beta^M)$. Experimentally we have concluded that the value of M should be in the range of 2–6. When $M = 0$ the problem reduces to a simple NNLS regression problem. When M exceeds a threshold value it no longer affects the accuracy of prediction.

6   Experiments and Results

We conduct experiments on industrial ticket data set using both crisp and fuzzy-based approaches for the prediction of QoS values.

6.1   Datasets

We perform experiments using Infosys internal ticket data sets. These ticket data follow the similar schema described in Sect. 2.1. We have used diversified data sets, spanning three separate domains: Application Maintenance and Development (AMD), Retail Corporation (RC) and Personal Insurance Lines (PIL). RC domain has the highest number of tickets, 14379 while AMD has the lowest number of tickets, 4510. Correspondingly, RC has the maximum number of tuples having 175 of them, and AMD has the minimum having only 30 of them. Using appropriate NLP techniques discussed in Sect. 2.2 we extract terms and concepts from tickets in each domains. We produce statistics of ticket corpora in Table 1 which show term, concept and feature size of some of the tuples (containing more number of tickets) for each domain.

6.2   Evaluation Metrics

We shall use evaluation metrics for two purposes. In the first case we use them for computing the appropriate number of clusters both for crisp and fuzzy sets.

Table 1. Tuple details from different datasets

DataSet	Tuple details	# of tickets	# of terms	# of concepts	# of features
AMD	Configuration	289	355	688	1043
	Application errors	222	2018	2579	4597
	Warehouse distribution	210	2788	3972	6760
	Manufacturing packaging	180	4474	5846	10320
GLMS	Open system and associate training	3508	4671	5826	10497
	Web and mobile applications	1519	2035	3335	5370
	Talent development	669	1601	2624	4225
	Windows servers	329	144	310	454
	Management systems	184	601	1046	1647
	Benefits portal	166	67	134	201
All-State	Legacy, auto endorsement and error received	47	239	549	788
	Recent change and error received	35	170	404	574
	Legacy, auto endorsement and discount	25	134	346	480
	Recent change and activity history	24	126	330	456
	Legacy, MRP endorsement and error received	21	126	275	401
	Recent change and coverage	20	89	241	330

We use k-means algorithms for computing the clusters for crisp set analysis for which we determine the number of clusters using Silhouette index. For fuzzy QoS computation we use Xi-Beni index to compute the optimum number of clusters in FC-means clustering algorithm. In the second case we evaluate the performance of our methods using accuracy metrics like confusion matrix, accuracy of prediction etc.

Clustering Metrics. The Silhouette Index (S.I.) [25] is a measure of how similar an object is to its own cluster (cohesion) compared to other clusters (separation). The S.I. ranges from -1 to 1, where a high value of the index indicates that the object is well matched to its own cluster and poorly matched to neighboring clusters. The average value of S.I. over all data of a cluster is a measure of how tightly grouped all the data in the cluster are. In FC-means clustering normally one employs Xie-Beni (XB) validity index [5] computed as the ratio of compactness and separation degree, to determine the optimal number of clusters. The index of compactness emphasizes that the members of each cluster should be as close as possible to each other. On the other hand, separation reflects the separation of clusters from each other. Lower the value of this index, better the clustering solution is.

Accuracy Metrics. We use a couple of accuracy metrics to judge the accuracy of prediction in supervised learning. A confusion matrix is a square matrix that is often used to describe the performance of a classification model on a set of test data for which the true values are known. Using this confusion matrix one can compute the accuracy of prediction as the ratio of trace of the matrix to the sum of all elements of the matrix, - this captures the fraction of accurate results by the classifier on the test data. We also check the correlation of the accuracy of prediction with the number of terms and the number of tickets.

6.3 Evaluation and Discussion

In this section we evaluate our techniques using both crisp and fuzzy models of QoS parameters. For each tuple we can determine the appropriate QoS value for a ticket computed from the entries like Open (Submit) Date, Closure Date etc. - for these data sets we could only ascertain Closure Time as the only QoS parameter. When we use crisp sets we use Logistic Regression (LR) and Support Vector Machine (SVM) for classifying incoming tickets. Further we use Random Forest Boosting to improve the performance of LR. As SMO-SVM is an iterative algorithm for solving the optimization problem we do not use any boosting on this. In case of fuzzy modeling of QoS parameters we use NNLS to predict QoS parameters and then use Gradient Boosting on top of it to enhance its performance. While conducting th experiment we divide our data into training (80%) and test (20%) data sets. We set the regression models on training data and use these models for prediction on test data sets. In both the cases we compare the performance of prediction before boosting and after boosting using performance metrics.

Crisp Set Modeling of QoS. For grouping QoS values for each tuple we use Silhouette index for validation purposes. We compare the Silhouette indices for different values of clusters by plotting them. An optimum value of cluster is marked at the point when the curve reaches a peak and then stabilizes. We can then label each of the tickets with the appropriate cluster for the QoS on which we set up a classification problem. We use both Logistic Regression and SVM-SMO for solving them. We produce the confusion matrix for one tuple for each data set which reflects reasonable accuracy. For example, see Table 2(a), (b) and (c) for results on logistic regression prior to boosting.

For LR in some of the tuples the accuracy values are below 50% which makes it a weak learner. This is the reason for which we decide to boost LR classification and propose a Random Forest boosting for LR. The confusion matrices for all the previous tuples are shown in Table 3 after boosting. It shows appreciable improvement in the accuracy, for example, accuracy rises by almost 50% for AMD (Table 3(a)), 43% (Table 3(b)) for PIL and 20% for RC data set (Table 3(c)).

Table 2. Logistic regression without boosting

Actual Value	Accuracy 47%	Predicted Value		
		High	Med	Low
	High	10	3	6
	Med	4	5	4
	Low	7	4	14

Actual Value	Accuracy 60%	Predicted Value		
		High	Med	Low
	High	2	0	3
	Med	0	1	1
	Low	0	0	3

Actual Value	Accuracy 83%	Predicted Value		
		High	Med	Low
	High	155	1	17
	Med	2	1	0
	Low	11	1	10

(a) AMD (Warehouse Distn)

(b) PIL (Recent change and Activity history)

(c) RC (Talent Development)

Table 3. Logistic regression with random forest boosting

Actual Value	Accuracy 98%	Predicted Value		
		High	Med	Low
	High	37	0	5
	Med	0	18	2
	Low	0	0	43

Actual Value	Accuracy 86%	Predicted Value		
		High	Med	Low
	High	2	1	0
	Med	1	1	0
	Low	1	1	12

Actual Value	Accuracy 100%	Predicted Value		
		High	Med	Low
	High	234	0	1
	Med	0	4	0
	Low	0	0	47

(a) AMD (Warehouse Distn)

(b) PIL (Recent change and Activity history)

(c) RC (Talent Development)

The confusion matrix for some data sets in case of the application of SMO-SVM is shown in Table 4. In most of the cases it gives better result than LR before boosting.

Table 4. Sequential minimal optimization (SMO) SVM

Actual Value	Accuracy 70%	Predicted Value		
		High	Med	Low
	High	13	1	4
	Med	2	4	1
	Low	9	4	31

Actual Value	Accuracy 78%	Predicted Value		
		High	Med	Low
	High	0	1	0
	Med	1	2	1
	Low	1	0	8

Actual Value	Accuracy 88%	Predicted Value		
		High	Med	Low
	High	202	24	9
	Med	0	4	0
	Low	0	0	47

(a) AMD (Configuration)

(b) PIL (Recent changes and Activity history)

(c) RC (Management system)

Fuzzy Set Modeling of QoS. While modeling QOS with fuzzy sets we run FC clustering algorithm on QoS values of each tuple for all three data sets. First we validate the generated clusters using VB validity index. We choose the fuzzy weighing exponent m to be equal to 2 which is normally chosen in the objective function-based method [26]. Based on the value of this index the optimal number of clusters for Closure Time for all these tuples in all data set are set to 3. We consider 3 clusters, viz., 'high', 'medium' and 'low'. The FCM algorithm generates the cluster centers of each QoS parameters. Subsequently, these are used to produce fuzzy clustering models.

For building the regression model we use an appropriate update algorithm to solve the NNLS problem described in Sect. 4.2. As the TF*IDF matrix $\mathbf{X}$ is sparse and skinny we reduce its dimension by using principal component analysis (PCA). We carry out NNLS on the reduced matrix and obtain the regression parameters. For a new ticket we can predict its QoS parameter Q by computing its membership value *wrt* the each of the fuzzy clusters as discussed in Sect. 4.2. The confusion matrix for different (test) data sets are shown in Table 5(a), (b) and (c).

Table 5. FCM NNLS without boosting

Accuracy 67%	Predicted Value		
	High	Med	Low
High	2	4	13
Med	1	122	17
Low	1	50	52

Accuracy 68%	Predicted Value		
	High	Med	Low
High	15	0	0
Med	3	1	0
Low	5	0	1

Accuracy 75%	Predicted Value		
	High	Med	Low
High	77	111	1
Med	1	1035	0
Low	13	246	8

(a) AMD   (Configuration)

(b) PIL (Legacy - Auto endorse -discount)

(c) RC (Talent Development)

For improving the accuracy of prediction we use a gradient boosting on NNLS as described in Sect. 5.2. This appreciably improves the performance of prediction as evidenced by the confusion matrix for the same data-sets in Table 6(a), (b) and (c).

Table 6. FCM NNLS with gradient boosting

Accuracy 94%	Predicted Value		
	High	Med	Low
High	131	9	0
Med	0	98	5
Low	0	1	58

Accuracy 92%	Predicted Value		
	High	Med	Low
High	16	1	0
Med	0	2	0
Low	1	0	3

Accuracy 91%	Predicted Value		
	High	Med	Low
High	969	3	7
Med	1	212	1
Low	42	0	258

(a) AMD   (Configuration)

(b) PIL (Legacy - Auto endorse -discount)

(c) RC (Talent Development)

Accuracy of Prediction. In Table 7 we compare accuracies for QoS prediction using different techniques on different tuples for different data sets. We could see that there is significant improvement in accuracy of prediction in case of NNLS regression of fuzzy QoS values by using gradient boosting. However, in case of Logistic Regression, Random Forest Boosting improves the performance in some of the cases. Boosting has lead to most of the improvements in the domain RC, one possible reason might be that each of the tuples in it has higher number of tickets. As SMO-SVM breaks the optimization problem into a series of possible smaller sub-problems to be solved analytically, we do not propose any boosting for this mode of classification. Overall NNLS for fuzzy analysis (both pre- and post-boosting) has performed better than classification algorithms for most of the cases. It may appear that the number of clusters in each regression mode may have some bearing in accuracy values. For other data sets we have chosen the number of clusters between 3 and 5 based on Sihoutte Indices, and we have seen similar accuracy values. For fuzzy set based QoS prediction the number of fuzzy clusters are guided by XB index, which never goes beyond 3, the minimum remains obviously 2.

Table 7. Accuracy Comparisons on different techniques for QoS prediction

DataSet	# of tickets	Tuple details	FCM NNLS without boosting	FCM NNLS with gradient boosting	Logistic regression without RFB	Logistic regression with RFB	SMO SVM
AMD	262	Configuration	67.2	93.5	83.3	85.8	69.8
	222	Application errors	82.4	88.9	43.9	55.6	84.2
	204	Warehouse distribution	77.4	88.2	47.2	98.4	92.1
	179	Manufacturing packaging	81.6	95.5	57.0	100.0	94.3
PIL	46	Legacy, auto endorsement and error received	71.7	84.8	63.6	92.9	82.6
	32	Recent change and error received	71.9	93.7	62.5	71.4	87.5
	25	Legacy, auto endorsement and discount	68.0	92	50	57.1	76.0
	23	Recent change and activity history	82.6	95.6	60.0	85.7	78.3
	20	Legacy, MRP endorsement and Error received	80.0	85.0	53.0	60	85
	18	Recent change and coverage	61.1	94.4	66.7	100.0	72.2
RC	665	Talent development	74.8	91.1	83.4	100.0	84.2
	329	Windows servers	79.6	89.9	80.8	100.0	79.6
	184	Management systems	81.5	97.8	42.3	98.2	88.1
	166	Benefits portal	64.4	83.7	64.0	100.0	22.4

Correlation Analysis for Accuracy. We perform correlation analyses to test the following hypotheses: accuracy of prediction increases with the number of terms appearing in the feature vectors of tickets, but decreases with the number of tickets in a tuple.

From the analysis of Pearson coefficients analysis, we can see that accuracy for FCM NNLS without Boosting has significant correlation with total number of terms (0.471) but, not with the total number of tickets (0.163). Whereas the prediction accuracy for Logistic regression without Random Forest Boosting has positive correlation with total number of tickets (0.498) but, is negatively

Table 8. Correlations for prediction accuracy

	No of terms		No of tickets	
	Pearson correlation	t-test	Pearson correlation	t-test
FCM without boosting accuracy with PCA	0.47	1.613	0.163	0.564
FCM with boosting accuracy	−0.022	−0.076	0.03	0.1039
Logistic regression without boosting accuracy	−0.273	−0.943	0.498	1.707
Logistic regression with RF boosting accuracy	−0.295	−1.02	0.41	1.4103
SMO SVM accuracy	0.28	0.967	0.042	0.1454

correlated with the number of terms (−0.273). This analogy makes sense as for Logistic regression we are using "One v/s All" strategy for the division of the training data, which implies a need of more number of tickets for better training. On the other hand FCM NNLS training needs more number of terms for better training for obtaining higher degree of accuracy and does not depend on number of tickets. The accuracy for prediction for both the boosting algorithms are independent of all two determining factors and show almost no correlation with them, signifying the robustness which is independent of the total number of terms and tickets. FCM NNLS with gradient boosting shows correlation with terms and tickets as −0.022 and 0.030 respectively. The values for LR with RFB are −0.295 and 0.410 respectively. From correlation analysis as well we can see that gradient boosting performs better than Random Forest Boosting (Table 8).

7   Conclusion

The prediction of the performance (Resolution and Closure Time) of tickets are of paramount importance as they have high impact on the quality parameters of Service Systems. For critical service systems it is imperative to predict these QoS parameters as accurately as possible which can be achieved through application of suitable boosting techniques. In future we would like to apply these boosting algorithms to a variety of real-life problems. Also we want to modify these boosting algorithms so that they do not require any prior knowledge about the performance of the weak learning algorithm. Moreover, we would like to compare the performance of different boosting algorithms for logistic regression, *e.g.*, model trees [16], logistic model trees [17] and our random forest boosting algorithm. Similarly we would like to benchmark our experiment on the gradient descent with other NNLS boosting algorithms like functional gradient descent algorithm on our data sets.

References

1. Xiong, P., Fan, Y.: QoS-aware web service selection by a synthetic weight. In: Proceedings of the Fourth International Conference on Fuzzy Systems and Knowledge Discovery, FSKD 2007, vol. 3, pp. 632–637 (2007)
2. Luo, X., Zhou, M., Xia, Y., Zhu, Q., Ammari, A.C., Alabdulwahab, A.: Generating highly accurate predictions for missing QoS data via aggregating nonnegative latent factor models. IEEE Trans. Neural Netw. Learn. Syst. **27**(3), 524–537 (2016)
3. Ollos, G., Vida, R.: Adaptive regression algorithms for distributed dynamic clustering in wireless sensor networks. In: Proceedings of the 3rd International Symposium on Information Processing ISIP 2010, pp. 563–566 (2010)
4. Dorgman, A., Saatchi, R., AI-Khayatt, S.: Quality of service evaluation using a combination of Fuzzy C-Means and Regression Model. Int. J. Comput. Electr. Autom. Control Inf. Eng. **6**(1), 62–69 (2012)
5. Hasan, M.H., Jaafar, J., Hassan, M.F.: Development of web services fuzzy quality models using data clustering approach. In: Herawan, T., Deris, M.M., Abawajy, J. (eds.) Proceedings of the First International Conference on Advanced Data and Information Engineering (DaEng-2013). LNEE, vol. 285, pp. 631–640. Springer, Singapore (2014). doi:10.1007/978-981-4585-18-7_71
6. Roy, S., Dutta, D., Muni, D.P., Bhattacharya, A.: Fuzzy prediction of QoS for IT maintenance tickets. In: ACM CODS-COMAD Conference, Industrial Track, Chennai, March 2017 (2017)
7. Roy, S., Muni, D.P., Bhattacharya, A., Dutta, D., Budhiraja, N.: Fuzzy QoS modeling of IT maintenance tickets. In: IEEE International Conference on Web Services, ICWS 2017 (2017)
8. de Marneffe, M.C., MacCartney, B., Manning, C.D.: Generating typed dependency parses from phrase structure parses. In: International Conference on Language Resources and Evaluation (LREC 2006), pp. 449–454 (2006)
9. Li, Y., McLean, D., Bandar, Z., O'Shea, J., Crockett, K.A.: Sentence similarity based on semantic nets and corpus statistics. IEEE Trans. Knowl. Data Eng. **18**(8), 1138–1150 (2006)
10. Punuru, J., Chen, J.: Automatic acquisition of concepts from domain texts. In: IEEE International Conference on Granular Computing, GrC 2006, pp. 424–427 (2006)
11. Salton, G., Buckley, C.: Term weighing approaches in automatic text retrieval. Inf. Process. Manag. **24**, 513–523 (1988)
12. Leskovec, J., Rajaraman, A., Ullman, J.: Mining of Massive Datasets, 2nd edn. Cambridge University Press, Cambridge (2014)
13. Roy, S., Yan, J.Y.T., Budhiraja, N., Lim, A.: Recovering resolutions for application maintenance incidents. In: IEEE International Conference on Services Computing, SCC 2016, pp. 617–624 (2016)
14. Platt, J.: Fast training of support vector machines using sequential minimal optimization. In: Scholkopf, B., Burges, C., Smola, A. (eds.) Advances in Kernel Methods Support Vector Learning. MIT Press, Cambridge (1998)
15. Freund, Y., Schapire, R.E.: A decision-theoretic generalization of on-line learning and an application to boosting. J. Comput. Syst. Sci. **55**(1), 119–139 (1997)
16. Wang, Y., Witten, I.: Induction of model trees for predicting continuous classes. In: European Conference of Machine Learning, Proceedings of poster papers (1997)
17. Landwehr, N., Hall, M.A., Frank, E.: Logistic model trees. Mach. Learn. **59**(1–2), 161–205 (2005)

18. Mishina, Y., Tsuchiya, M., Fujiyoshi, H.: Boosted random forest. In: Proceedings, Internal Conference on Computer Vision Theory and Applications (VISAPP 2014). IEEE (2014)
19. Breiman, L.: Bagging predictors. Mach. Learn. **24**(2), 123–140 (1996)
20. Quinlan, J.: Learning with continuous classes. In: Proceedings of the 5th Australian Joint Conference on Artificial Intelligence, pp. 343–348. World Scientific (1992)
21. Friedman, J., Hastie, T., Tibshirani, R.: Additive logistic regression: a statistical view of boosting. Ann. Stat. **28**(2), 337–407 (2000)
22. Kim, T.K., Cipolla, R.: MCBoost: multiple classifier boosting for perceptual co-clustering of images and visual features. In: NIPS, pp. 841–856. Curran Associates, Inc. (2008)
23. Özuysal, M., Calonder, M., Lepetit, V., Fua, P.: Fast keypoint recognition using random ferns. IEEE Trans. Pattern Anal. Mach. Intell. **32**(3), 448–461 (2010)
24. Friedman, J.H.: Greedy function approximation: a gradient boosting machine. Ann. Stat. **29**, 1189–1232 (2000)
25. Arbelaitz, O., Gurrutxaga, I., Muguerza, J., Pérez, J.M., Perona, I.: An extensive comparative study of cluster validity indices. Pattern Recogn. **46**(1), 243–256 (2013)
26. Pal, N.R., Bezdek, J.C.: On cluster validity for the fuzzy C-means model. IEEE Trans. Fuzzy Syst. **3**(3), 370–379 (1995)

From VM to Container: A Linear Program for Outsourcing a Business Process to Cloud Containers

Khouloud Boukadi[✉], Rima Grati, Molka Rekik, and Hanêne Ben Abdallah

Mir@cl Laboratory, University of Sfax, Sfax, Tunisia
khouloud.boukadi@fsegs.usf.tn, rima.grati@gmail.com, molka.rekik@gmail.com,
hbenabdallah@kau.edu.sa

Abstract. Cloud computing has been recently empowered with a new service offering called Containers-as-a-Service (CaaS). This offers horizontally scalable, deployable systems and it bypasses high-performance challenges of traditional hypervisors when deploying applications. This paper assists in using CaaS for business process outsourcing to the cloud–an emerging trend that still faces several problems. In particular, this paper concentrates on the resource allocation problem from an enterprise perspective and proposes a linear program (LP) that finds out the optimal deployment of a business process on cloud containers. The herein reported experimental results show the effectiveness and performance of the LP compared to both the classic deployment (VM-based deployment) and the container First Fit strategy.

Keywords: Business process · Cloud · CaaS · Linear program · Optimal deployment

1 Introduction

The adoption of cloud computing has been gaining momentum thanks to its promised shift in the way of provisioning computing resources and expertise. Indeed, the cloud paradigm promotes access and management of physical resources by virtually leveraging storage, computing and applications. In addition, this way of provisioning software and hardware resources alleviates the users' costs of installing and managing costly information technology to support computing intensive applications.

Besides its classical three service models (IaaS, PaaS, and SaaS), cloud computing has been recently empowered with a new service offering called Containers-as-a-Service (CaaS) [17]. In fact, container-based virtualization is gaining significant acceptance because it provides for a lightweight solution that allows bundling applications and data in a simpler and more performance-oriented manner, making them runnable on different cloud infrastructures. This way of dealing with virtualization offers horizontally scalable, deployable systems while bypassing high-performance challenges of traditional hypervisors and the

© Springer International Publishing AG 2017
H. Panetto et al. (Eds.): OTM 2017 Conferences, Part I, LNCS 10573, pp. 488–504, 2017.
https://doi.org/10.1007/978-3-319-69462-7_31

overheads of managing large scale cloud infrastructures when deploying applications. Among the applications that can benefit from cloud computing in general and CaaS in particular, this paper focuses on Business processes outsourced to the cloud.

The paradigm of Business Process Outsourcing (BPO) is not a new trend. It was a common and a well-known business practice insuring among others, the business added-value. Exploiting cloud computing as a business process execution environment is, nevertheless, an emergent tendency that still faces several problems. One of these problems is how to allocate *optimally* the cloud resources to execute the business process' activities. Overall, optimality is expressed in terms of the minimal cost of the allocated cloud resources that fulfill the functional and Quality of Service (QoS) requirements of the process activities. That is, the resource allocation method must minimize the costs, minimize risk factors (*e.g.*, related to security), and enhance the business process performance (*e.g.*, reduce its execution time, increase its availability, etc.).

There are several resource management methods already available such as policy-based resource allocation and First Fit strategy [3]. Most of these methods are optimally used from a supplier's perspective. However, they fail to validate whether the provisioned resources were optimally used by a particular application. Both the cloud provider and the consumer get optimal benefits when the appropriate amount of resources for an application is correctly provisioned/unprovisioned. This paper addresses the problem of business process deployment into the cloud from the consumer's (enterprise) perspective. It proposes a new approach that relies on *(i)* a business process model with its set of functional and non-functional requirements, and *(ii)* a cloud environment infrastructure with its hosts, virtual machines, containers configurations and their related constraints. It finds out the optimal deployment under functional and QoS constraints.

More specifically, this paper has a three-fold contribution: *(i)* an extension of the *ContainerCloudSim* to consider pertinent workflow patterns to simulate business process execution in cloud containers; *(ii)* a linear program for discovering the optimal deployment of the business process in the cloud; and *(iii)* an evaluation of the proposed linear program in terms of effectiveness and performance compared to the classic deployment (VM-based deployment and First Fit strategy). To the best of our knowledge, this is the first attempt to consider both VM and container allocations for process deployment in the cloud.

The remainder of the paper is organized as follows: Sect. 2 overviews the container concept and motivates the problem with a real-world case of a petroleum company. Section 3 presents the ContainerCloudSim extension for business process simulation. Section 4 describes our approach for an optimal business process deployment in a cloud environment. Section 5 describes the implementation of the approach and discusses preliminary evaluation results. Section 6 places our work in the context of existing approaches for business process deployment in the cloud, and it outlines our ongoing research.

2   Background and Case Study

This section briefly overviews the container-based virtualization. Afterward, it presents a case study used to illustrate the proposed approach.

2.1   Container-Based Virtualization

Container-based virtualization technology is a new approach compared to the hypervisor based virtualization [11]. In this new model, the hardware resources are divided by implementing many instances with isolation properties [22]. As compare in Fig. 1, in container-based virtualization, the guest processes immediately obtain abstractions as they operate through the virtualization layer directly at the operating system (OS) level. Furthermore, in hypervisor-based approaches, there is typically one virtual machine per guest OS [22]. In contrast, one OS kernel is typically shared among virtual instances in container-based solutions. From a user's perspective, containers operate as autonomous OS, that run independently of hardware and software [20].

Nowadays, Containers as a Service (CaaS) is the new cloud service offering. The CaaS service is usually provided on top of the IaaS virtual machine (VM). A recent study [1] has outlined that VM-Container configurations obtain close to, or even better performance, than native Docker (container) [16] deployments. The users of this service submit their requests for the provisioning of the containers which run inside the virtual machines that are hosted on physical servers.

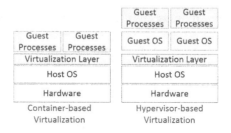

Fig. 1. Container-based *vs.* hypervisor-based approaches [11]

2.2   Running Scenario

Our research is motivated by a real business process of an offshore petroleum logistic enterprise. The business process is related to a purchasing function of an Offshore Rig Operator. As modeled with the BPMN notation [14] in Fig. 2, this process is recognized to be compute intensive. Thus, the enterprise is motivated to outsource it to a cloud environment under a set of QoS requirements. For example, the activity a_4 "Price comparison and quality assessment" must be executed in a cloud with a minimal level of security of 50% and a minimal

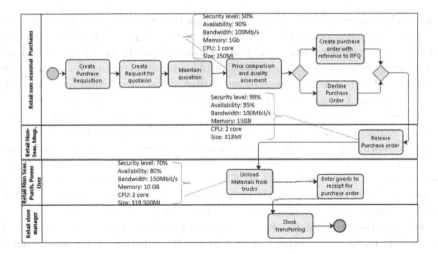

Fig. 2. Purchasing process of an Offshore Rig Operator

availability of 90%. In addition, a_4 whose size is 250 MI must be deployed on a virtual machine with a minimal memory of 1 Gb, a minimal CPU of 1 Core and a minimal bandwidth of 100 Mb/s. This process will be outsourced to a cloud whose infrastructure within a particular data center hosts a set of the virtual machines V_1, V_2, V_3,...V_p each of which has a set of properties. For instance, V_1 has a compute price of 0.05$/$hour$, a data transfer price of 0.2$/Mb, a memory of 1 Gb, 2 cores of CPU, a bandwidth of 100 Mb/s and a MIPS of 317.900 MI/s (see Fig. 3).

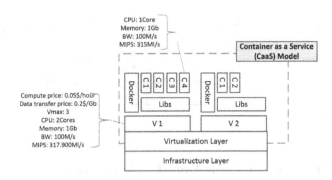

Fig. 3. CaaS model

With the emergence of the container technology, the allocation problem complexity increased. In fact, if the enterprise starts the development of a new business process from scratch, it can be committed to update it around a microservices based architecture, and consequently the use of containers is a natural choice. When it comes to monolithic business processes, the enterprise has the choice of either developing the next version of these processes using containers and micro-services or partitioning them in a set of containers. The second choice is simpler but it must identify the number of containers required for an adequate execution of the business process to avoid SLA violations. Cloud providers, like Amazon [2] and Google [7], allow an application to be deployed on containers through a declarative template. Within this template, the number of the required containers, including the memory and CPU requirements, is specified. However, defining an exact number of containers along with their requirements is not a relevant task since the enterprise needs to be assisted to identify the right number of both containers and VMs. The same observation is valid for the business process case, where different containers with different capacities (CPU, bandwidth and memory) can execute the business process activities.

One possible deployment of the presented process is illustrated in Fig. 4. It corresponds to the allocation of resources (VMs and containers) to the activities. The enterprise seeks the optimal deployment where the cost of allocated resources is minimal while respecting the QoS requirements. Finding the optimal deployment is a complex problem. Furthermore, to select the best VM offering, different VM configurations within the cloud landscape should be simulated to yield the response time as well as the cost values of the business process.

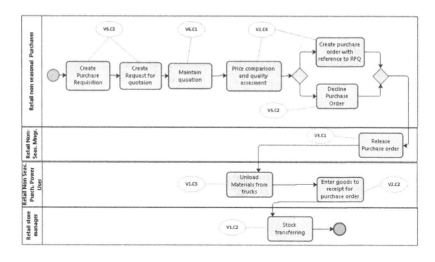

Fig. 4. One possible deployment of the purchasing process

In the following sections, we present our approach dealing with the afore-mentioned problem.

3   ContainerCloudSim Extension for Business Process Simulation

A number of simulators with various objectives were developed for the cloud environment. They mainly differ in the considered performance metrics, the supported applications, and their ability to simulate the data center power consumption. MDCSim [12], GroudSim [15], SPECI [21] and CloudSim toolkit [4] are examples of these simulators. The most used one is CloudSim which is an open source, java-based simulator that allows an easy modeling of virtualized environments and on-demand resource provisioning. The main purpose of CloudSim is to simulate the behavior of real cloud offerings and provides both the system and behavior modeling of cloud computing components. Recently, CloudSim has been extended to model a containerized cloud environment. The new extension is named ContainerCloudSim [17]. It is the first initiative that provides for modeling and simulation of containers, unlike the existing simulators whose primarily focus is the system level virtualization with virtual machine as the basic component.

To conduct business process simulations on different VM and container configurations, we propose to extend ContainerCloudSim. The proposed extension is presented in Fig. 5. In the last version of ContainerCloudSim API (4.0), the entities are parts of simulated elements and services. Our extension concerns mainly the simulated elements, which include the following classes:

- Datacenter: represents the hardware layer of the cloud infrastructure.
- Host: describes the configuration of the physical server in terms of processing capability expressed in MIPS (million instructions per second), memory, and storage.
- VM: models a Virtual Machine which is managed and hosted by a Host.
- Container: depicts a Container hosted by a VM.
- Cloudlet: refers to the applications hosted in a Container in a cloud data center.

At this level, a set of classes, such as shown in Fig. 5, are added to support the simulation of business processes:

- The *Process* class which extends the Cloudlet class and represents the business process to be executed on the cloud infrastructure. The process is composed of a set of activities.
- The *Activity* class which represents a special kind of Cloudlet. It encapsulates the minimum availability and security levels, the required memory, the CPU and bandwidth, the number of instructions to be executed, and its relationships with other activities.

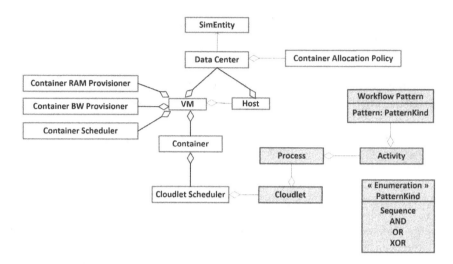

Fig. 5. Extended ContainerCloudSim class diagram

- The *Workflow patterns* class that capture common patterns required to depict the existing dependencies between the activities. For instance, a sequence is required when there is a dependency between two or more activities so that one activity cannot be started (scheduled) before another activity is finished. For example, the patterns include a Parallel Split, a Loop, etc. which help in the simulation steps compute thoroughly the business process execution time and cost.

After extending the ContainerCloudSim with the above described classes, it became possible to simulate the business process subject of outsourcing.

4   Linear Program for Optimal BP Deployment into Cloud Containers

To select the optimal set of VM instances and the containers from a variety of VM types and container configurations, we opt for an exact optimization resolution approach to obtain a global optimum solution. More precisely, we present in this section our proposed linear program to find the optimal deployment of a business process model into the cloud. Our program can discover the set of VMs and the set of containers to be allocated for an optimal business process deployment. It is defined in terms of its inputs, its decision variables as well as its objective function.

4.1   Inputs and Definitions

- We define a business process model P as a tuple (A, WP, T_{WP}, F):
  - $A \subset \mathcal{U}_A$ is a non-empty set of activities where $\mathcal{U}_A$ is the universe of activities;
  - $WP \subset \mathcal{U}_{WP}$ is the set of workflow-patterns where $\mathcal{U}_{WP}$ is the universe of workflow-patterns;
  - $T_{WP} : WP \rightarrow WPT$ is a function that assigns to each workflow-pattern $wp \in WP$ a type $t \in WPT$ where $WPT = \{AND, OR, XOR\}$; and
  - $F \subseteq (A \cup WP) \times (A \cup WP)$ is the set of edge flows representing the control flow of the process activities. For a node $n \in A \cup WP$, we denote by $\bullet\, n = \{n' \in A \cup WP \mid (n', n) \in F\}$ (respectively $n\, \bullet\, = \{n' \in A \cup WP \mid (n, n') \in F\}$) the set of predecessors (respectively successors) of n. A workflow-pattern $wp \in WP$ is a split if $|wp\,\bullet| > 1$ and it is a join if $|\bullet\, wp| > 1$. A workflow-pattern can be exclusively either a split or a join function.
    In the running example of Fig. 2, we have $\{(Sequence, \{a_1, a_2, a_3, a_4\}),$ $(XOR, \{a_5, a_6\}), (Sequence, \{a_7, a_8, a_9, a_{10}\})\}$.

- $I_A = \{1, 2, \ldots, |A|\}$ the set of indexes of the activities in A. The activity $a_i \in A$ has the index $i \in I_A$; the functional and QoS attributes of $a \in A$ are defined as a tuple (rs, ra, rc, rr, rd, mi) where $rs \in \mathbb{R} \geq 0$ is the minimum security level required by a (%), $ra \in \mathbb{R} \geq 0$ is its minimum required availability level (%), $rc \in \mathbb{R} \geq 0$ is its minimum required CPU capacity (Cores), $rr \in \mathbb{R} \geq 0$ is its minimum required memory capacity (GB), $rd \in \mathbb{R} \geq 0$ is its minimum required bandwidth (Mb/s), and $mi \in \mathbb{R} \geq 0$ is its estimated required size/length to be executed a in millions of instructions (MI). For instance, in the running example of Fig. 2, the user expresses his requirements for the deployment of the activity a_4 as $a_4 =$(50 %, 90 %, 1 Core, 1 Gb, 100 Mb/s, 250 MI).

- Virtual machines have indices in $I_V = \{1, 2, \ldots, |V|\}$. A virtual machine $V = (sec, ava, pc, pd, mc, mr, mb, vmax, mips)$ is a tuple where sec, ava is, respectively, the capability in terms of security $sec \in \mathbb{R} \geq 0$ (%) and availability $ava \in \mathbb{R} \geq 0$ (%), $pc \in \mathbb{R} \geq 0$ is the computed price ($/hour), $pd \in \mathbb{R} \geq 0$ is the data transfer price ($/Mb), $mc \in \mathbb{R} \geq 0$ is the maximum CPU capacity (Cores), $mr \in \mathbb{R} \geq 0$ is the maximum RAM capacity (Gb), $mb \in \mathbb{R} \geq 0$ is the maximum bandwidth capacity (Mb/s), $vmax \in \mathbb{N}$ is the maximal number of instances that can be deployed, and $mips \in \mathbb{R} \geq 0$ is the CPU speed measured in millions of instructions per second (MI/s). We denote by $\mathcal{U}_V$ the universe of virtual machines and $\mathbb{V} = \{V \mid V \in \mathcal{U}_V\}$ the set of virtual machines.

- $I_{\mathbb{C}} = \{1, 2, \ldots, |\mathbb{C}|\}$ the set of indexes of the containers in cloud $\mathcal{C}$. A container $\mathcal{C}_k \in \mathbb{C}$ has the index $k \in I_{\mathbb{C}}$; A container $C = (cc, cr, cb, mips)$ is a tuple where $cc \in \mathbb{R} \geq 0$ is the maximum CPU capacity (Cores), $cr \in \mathbb{R} \geq 0$ is the maximum RAM capacity (Gb), $cb \in \mathbb{R} \geq 0$ is the maximum bandwidth capacity (Mb/s) and $mips \in \mathbb{R} \geq 0$ is the CPU speed measured in millions of instructions per second (MI/s). We denote by $\mathcal{U}_C$ the universe of containers. An example of container model is shown in Fig. 3 where:

- Virtual machine $V_1 = (0.05\,\$/\text{hour},\ 0.2\,\$/\text{Gb},\ 3, 2\,\text{Cores},\ 1\,\text{Gb},\ 100\,\text{Mb}$ $317.900\,\text{MI/s}).$
- Container $C_4 = (1\,\text{Core},\ 1\,\text{Gb},\ 100\,\text{Mb},\ 315\,\text{MI/s}).$

- To execute activity a_i $(i \in I_A)$ in a container $\mathcal{C}_k \in \mathbb{C}$ $(k \in I_\mathbb{C})$ on a virtual machine $\mathbb{V}_p$ $(p \in I_\mathbb{V})$, the estimated time t_{ipk} is based on Eq. 1:

$$\forall i \in I_A, p \in I_\mathbb{V}, q \in \{1, \ldots, vmax_{V_p}\}, k \in I_\mathbb{C}$$
$$t_{ipk} = X_{ipqk} \times (mi_i/(mips_k \times mc_k)) \tag{1}$$

where X_{ipqk} is equal to 1 if activity a_i is executed by container k on instance q of V_p, otherwise it is equal to 0. Based on this equation, the execution time is estimated by dividing the activity size mi_i by the CPU speed $mips_k$ multiplied by its number of cores mc_k. Note that time is equal to 0 if the activity is not assigned to container k on instance q of V_p.

- The estimated data transfer time tt_{ij} between activities a_i and a_j is based on Eq. 2:

$$\forall i, j \in I_A, p \in I_\mathbb{V} \ \ tt_{ij} = \frac{rd_{ij}}{mb_p} \tag{2}$$

The deployment of a process into a cloud consists of allocating an adequate container within a virtual machine to each activity of the process. It is a function $\mathcal{D} : S \rightarrow \mathbb{V} \times \mathbb{C}$ that assigns the $(V, \mathcal{C})$ pair of a virtual machine and container in it $(\mathcal{C} \in V.\mathbb{C})$ to each activity $a \in A$.

In addition, a process deployment has a cost that covers the allocated cloud resources (the cost function is detailed in the following section). According to a given cost function, a process deployment $\mathcal{D}$ is **optimal** if and only if it results in a minimal cost. For example, in the process deployment in Fig. 4, the activity a_4 is deployed on the container C_4 of the virtual machine V_1, thus $\mathcal{D}(a_4) = (V_1, \mathcal{C}_4)$.

4.2 Decision Variables

The following two real decision variables are associated with our mathematical model:

- For any $i \in I_A$, $p \in I_\mathbb{V}, q \leq vmax_{VM_p}$ and $k \in I_\mathbb{C}$: X_{ipqk} expresses whether activity a_i is assigned to container $\mathcal{C}_k$ within instance q of the virtual machine V_p;
- For any $i, j \in I_A$, $p \in I_\mathbb{V}$: Z_{ij} specifies whether a_i and a_j are assigned to the same virtual machine V_p.

4.3 Objective Function

The proposed objective function is a cost function that selects the VMs and containers so as to achieve a minimum total execution cost of the business process composed by the total compute and intra-cloud communication costs. The total compute cost is the sum of VM allocation costs in order to execute activities within the BP. While, the total intra-cloud communication costs is the sum of data transfer costs between activities deployed in the different VMs. The allocation cost is computed by the multiplication of the simulated activity execution time by the VM utilization cost (per hour). While the communication cost is calculated by considering the following two scenarios: (1) If the activities are deployed in the same VM, the communication cost is null, and (2) If the activities are deployed in different VMs, the communication cost is equal to the multiplication of the transferred data size (per Mb/s) by the bandwidth utilization cost (per \$/Mb) and by the simulated data transfer time (per s).

$$MinZ = \sum_{i=1}^{|I_A|}\sum_{p=1}^{|I_V|}\sum_{q=1}^{vmax_{V_p}}\sum_{k=1}^{|I_C|} pc_p t_{ipk} X_{ipqk} + \sum_{i=1}^{|I_A|}\sum_{j=1}^{|I_A|}\sum_{p=1}^{|I_V|} pd_p rd_{ij} tt_{ij}(1 - Z_{ij}) \quad (3)$$

This optimization is *subject to* the following set of constraints:

- QoS constraints: impose the minimum security (4) and availability (5) levels required by the enterprise to outsource each activity on the cloud.

$$sec_p X_{ipqk} \geq ds_i \ \forall i \in I_A, p \in I_V, q \in \{1,\ldots, vmax_{V_p}\}, \forall k \in I_C \quad (4)$$
$$av_p X_{ipqk} \geq da_i \ \forall i \in I_A, p \in I_V, q \in \{1,\ldots, vmax_{V_p}\}, \forall k \in I_C \quad (5)$$

- Placement constraint: we propose Eq. 6 which implies that each activity should be assigned to one (and only one) container.

$$\sum_{p=1}^{|I_V|}\sum_{q=1}^{vmax_{V_p}}\sum_{k=1}^{|I_C|} X_{ipqk} = 1 \ \forall i \in I_A \quad (6)$$

- Resource constraints: the resource constraints refer to both the VM and the container related constraints.
  1. VM constraints: impose that the VM's capacities in processing, memory and network should satisfy the activity requirements (Eqs. 7, 8 and 9).

$$\sum_{p=1}^{|I_V|}\sum_{q=1}^{vmax_{V_p}} mc_p X_{ipqk} \geq rc_i \ \forall i \in I_A, \forall k \in |I_C| \quad (7)$$

$$\sum_{p=1}^{|I_V|}\sum_{q=1}^{vmax_{V_p}} mr_p X_{ipqk} \geq rr_i \ \forall i \in I_A, \forall k \in |I_C| \quad (8)$$

$$\sum_{p=1}^{|I_V|}\sum_{q=1}^{vmax_{V_p}} mb_p X_{ipqk} \geq rb_{ij} \ \forall i, j \in I_A, \forall k \in |I_C| \quad (9)$$

2. Container constraints: A container is hosted on a particular VM if its requirements in processing, memory and network are met (Eqs. 10, 11 and 12).

$$\sum_{i=1}^{|I_A|} \sum_{q=1}^{vmax_V} \sum_{k=1}^{|I_\mathbb{C}|} cc_k X_{ipqk} \leq mc_p \ \forall p \in I_\mathbb{V} \quad (10)$$

$$\sum_{i=1}^{|I_A|} \sum_{q=1}^{vmax_V} \sum_{k=1}^{|I_\mathbb{C}|} cm_k X_{ipqk} \leq mr_p \ \forall p \in I_\mathbb{V} \quad (11)$$

$$\sum_{i=1}^{|I_A|} \sum_{j=1}^{|I_A|} \sum_{q=1}^{vmax_{V_p}} \sum_{k=1}^{|I_\mathbb{C}|} cb_k X_{ipqk} \leq mb_p \ \forall p \in I_\mathbb{V} \quad (12)$$

– Binarity constraints: Because we are dealing with a linear program, Eqs. 13 and 14 are imposed to guarantee that the decision variables are 0 or 1.

$$X_{ipqk} \in \{0,1\} \ \forall i \in I_A, p \in I_\mathbb{V}, q \in \{1, \ldots, vmax_{V_p}\}, k \in I_\mathbb{C} \quad (13)$$
$$Z_{ij} \in \{0,1\} \ \forall i,j \in I_A \quad (14)$$

– Linearity constraints: imposed between the decision variables (15−17).

$$Z_{ij} \leq X_{ipqk} \ \forall i,j \in I_A, p \in I_\mathbb{V},$$
$$q \in \{1, \ldots, vmax_{V_p}\}, k \in I_\mathbb{C} \quad (15)$$
$$Z_{ij} \leq X_{jpqk} \ \forall i,j \in I_A, p \in I_\mathbb{V},$$
$$q \in \{1, \ldots, vmax_{V_p}\}, k \in I_\mathbb{C} \quad (16)$$
$$Z_{ij} \geq X_{ipqk} + X_{jpqk} - 1 \ \forall i,j \in I_A, i \neq j, p \in I_\mathbb{V},$$
$$q \in \{1, \ldots, vmax_{V_p}\}, k \in I_\mathbb{C} \quad (17)$$

5   Evaluation and Experiments

We implemented the proposed linear program using the open source Cplex 12.6[1] solver of IBM-ILOG and Microsoft Visual C++ 6.0. We conducted experiments on a laptop with a 64-bit Intel Core 2.50 GHz CPU, 6 GB RAM and Windows 8 as an operating system. We conducted three series of experiments to assess the effectiveness and performance of the linear program (LP): The first experiment compares our proposal to a non-container based deployment (VM-based deployment); the second deals with scaling-up of the container numbers; and the third compares the LP results to the classic allocation of containers based on the First-Fit strategy.

---

[1] http://www-01.ibm.com/support/docview.wss?uid=swg24036489.

5.1   Experiment Setting

To conduct our experiments, we randomly generated a testbed based on the given ranges (see Table 1) and a set of business processes provided by our industrial partner, the offshore petroleum logistic enterprise. All data input are randomly generated except for the execution time t_{ipk} taken by a resource to execute an activity, which is simulated based on Eq. 1 and using the extension of the ContainerCloudSim.

Table 1. The input ranges used to create cloud infrastructure and business process instances with different attribute' values

Information	Type	Range
VM number	Integer	[1..45]
VM' security level	Double	[0..1]
VM' availability level	Double	[0..1]
VM/Container MIPS	Double	[1..500000]
Maximum instance number of VM	Integer	[1..n/3]
CPU number of VM/container	Integer	[1..36]
RAM amount of VM/container	Double	[0.6..244]
Bandwidth amount of VM/container	Double	[0..10000]
VM' compute price	Double	[0.01..7]
Data transfer price	Double	[0.001..0.121]
Requirement in security	Double	[0..1]
Requirement in availability	Double	[0..1]
Requirement in CPU	Integer	[1..36]
Requirement in RAM	Double	[0.6..244]
Requirement in bandwidth	Double	[0..10000]
Length of activity	Double	[1..200000]
Number of activities	Double	[1..50]

Table 2 depicts the generated instances using the data inputs ranges shown in Table 1. For instance, business process P_3 contains 6 activities and the cloud infrastructure hosts 49 VMs and 21 containers.

5.2   Experiment 1: VM-Based *vs* Container-Based Deployments

The VM-based deployment consists in selecting, for all the activities in the business process model, the set of VMs that guarantee the optimal deployment (minimal deployment cost). In this experiment, we compare the VM-based deployment to the proposed linear program (container-based) for all the instances in Table 2.

Table 2. Generated instance description

BP(P)	Activities # ($|A|$)	VMs # ($|V|$)	vmax #	Containers # ($|C|$)
P_1	2	32	1	32
P_2	3	9	[1..2]	9
P_3	6	49	[1..3]	21
P_4	9	34	[1..4]	34
P_5	12	4	[1..5]	35
P_6	16	4	[1..5]	35
P_7	16	4	[1..5]	9
P_8	17	30	[1..6]	4
P_9	18	23	[1..6]	4
P_{10}	19	16	[1..6]	4
P_{11}	20	21	[1..8]	30
P_{12}	24	28	[1..8]	23
P_{13}	24	6	[1..8]	16
P_{14}	26	48	[1..7]	21
P_{15}	29	34	[1..11]	28
P_{16}	33	47	[1..12]	8
P_{17}	33	38	[1..11]	2
P_{18}	37	28	[1..14]	9
P_{19}	42	26	[1..14]	6
P_{20}	43	38	[1..16]	3

This experiment revealed an average reduction of the deployment cost by about 62.58% (see Fig. 6). The proposed LP reaches the optimal solution better (4.694 $) with an average gap of 0%. The reduction of the deployment cost can be explained by the fact that it costs much less to host an existing container on a VM than deploying a new VM to meet the requirements of the activities. Nevertheless, the average computation time increases by 10% which is explained by the search space scaling.

5.3   Experiment 2: LP with a Scaling up of the Container Numbers

We conducted this experiment while scaling up the number of containers within the data center. More containers are randomly added (scaled up) from [1.n] where n is the number of activities. As shown in Table 3, the compromise between the average gap (0.01%) and the average computation time (4.021 s) is considered as a good indicator of the performance of our proposed LP.

The LP finds solutions in a reasonable time, however it may fail shortly to provide the most optimal ones. Indeed, when the gap is zero, optimality is demonstrated. On the other hand, we know that we may get a better solution by

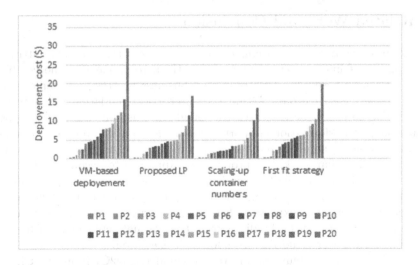

Fig. 6. Comparison between different deployment strategies

Table 3. Experimental results

	VM-based deployment	Proposed LP	LP with scaling up container numbers	First fit strategy
Average deployment cost (objective function)	7.5 $	4.694 $	3.482 $	5.867 $
Average gap	0.0%	0.0%	0.01%	-
Average CTime	3.285	3.392	4.021	3.345

working longer on our problem. Consequently, we plan to enhance the program by setting a relative tolerance on the gap or changing the CPU time limit of the CPLEX solver, which is set to 450 s, to achieve better solutions.

5.4 Experiment 3: Linear Program *vs.* First-Fit Strategy

The First-Fit strategy places the container on the first available VM once it meets the container's resource requirements. The CPLEX succeeded in finding optimal solutions for the proposed LP. The average objective function gains by 80% and the average computational effort increases by 10%. The First-Fit strategy is relatively more costly than the LP but less costly than the VM-based deployment. This proves that container-based deployment is cost effective even with a First Fit strategy.

6   Related Work and Conclusion

Over the past few years, outsourcing business processes to the cloud has attracted the attention of many researchers. Some of them recommended the combination of the cloud-based and the traditional business process management. The aim is to keep sensitive data within the enterprise boundaries and outsource the compute intensive activities to the cloud. In this context, [5,6] proposed an automated transformation support necessary for splitting business processes according to the data and activity distributions (in cloud/in premise) defined by the users. In a similar work, [18] proposed an approach to decompose the business processes between the cloud and the in-premise sides while guaranteeing the data constraints. The aforementioned approaches target the cloud service based on the enterprise's experts and propose no method that assists the enterprise with the BPO decision by discovering the optimal deployment of business process activities over cloud resources for their execution.

A Business Process Outsourcing to the cloud (BPO2C) framework in [19] encompasses several phases pertinent to the outsourcing decision; starting from the elaboration of the enterprise's motivations to identify both the implied business process in the outsourcing decision and the outsourceable process and the cloud services that minimize the business process costs, duration and mitigate the cloud risks.

In [13], the authors propose a resource allocation technique based on the execution path of the process, which is used to assist the business process owner in efficiently leasing computing resources. The technique includes three phases, namely, a process execution prediction, a resource allocation and a cost estimation. The first relies on the process model metrics and attributes in order to derive resource requirements based on the predicted process execution path, while the second minimizes the resource requirements by reusing the resources. The final phase estimates the lowest cost leasing option based on the resource allocation and pricing models offered by the cloud provider.

In [9], the authors propose a mixed integer linear programming technique for a cost-effective deployment of elastic business processes in a hybrid cloud environment. The proposed program takes into account the data transfer communication requirements. Another interesting approach focusing on QoS aspects is presented in [10]. The business process deployment into cloud environment is based on a discrete particle swarm optimization method that minimizes the business process execution time and cost and maximizes the user's satisfaction in terms of security.

In [8], the authors propose Event-B based language that specifies cloud resource allocation policies in business process deployed in a cloud environment. The proposed approach enables to check different cloud resource properties and constraints that cover both the design and the runtime requirements, analyze and check its correctness according to the user's requirements and resource capabilities.

In [23], the authors propose an approach for provisioning appropriate platform resources in order to deploy service-based processes in existing cloud platforms. This approach consists in dividing a given process to deploy into a set of elementary services through a Petri net decomposition approach. Source codes of obtained services are generated. After that, the services are packaged in their already developed service micro-containers and deployed in any target PaaS. The authors provide a realistic use case scenario in order to illustrate and show the feasibility of the proposed approach.

In summary, the proposed approaches concentrate on the deployment of the business process in the cloud environment with a main focus on the VM allocation problem. They consider different optimization techniques under different QoS requirements and resource capabilities. Similar to these approaches, our approach proposes a cost-effective business process deployment while considering the functional and QoS requirements of the process activities. Unlike the presented approaches, our linear program-based approach accounts for both VM and container allocations for process deployment in the cloud; in addition, it experimentally demonstrates an effective cost deployment reduction compared to the classic deployment (VM-based deployment and First Fit strategy). The experiments were conducted thanks to our extension of ContainerCloudSim, to consider workflow patterns during the simulation of business process execution in cloud containers.

Nonetheless, further enhancements in the proposed linear program should be elaborated to consider the inter-container communication which is a current practice used by well known cloud providers such as Google and Amazon. In addition, we are working on the scaling of the proposed linear program when the number of VMs and containers grows. Furthermore, we intend to compare the proposed linear program with other allocation strategies such as Maximum Usage, Most Correlated, etc.

References

1. Ali, Q.: Scaling web 2.0 applications using docker containers on vsphere 6.0 (2016). https://blogs.vmware.com/performance/2015/04/scaling-web-2-0-applications-using-docker-containers-vsphere-6-0.html. Accessed 13 July 2017
2. AmazonEC2: Amazon ec2 container service - docker management - aws (2017). https://aws.amazon.com/ecs/. Accessed 15 July 2017
3. Anuradha, V.P., Sumathi, D.: A survey on resource allocation strategies in cloud computing. In: International Conference on Information Communication and Embedded Systems (ICICES), pp. 1–7. IEEE (2014)
4. Calheiros, R.N., Ranjan, R., Beloglazov, A., De Rose, C.A.F., Buyya, R.: CloudSim: a toolkit for modeling and simulation of cloud computing environments and evaluation of resource provisioning algorithms. Softw. Pract. Exp. 41(1), 23–50 (2011)
5. Duipmans, E.F., Pires, L.F., da Silva Santos, L.O.B.: Towards a BPM cloud architecture with data and activity distribution. In: IEEE 16th International Enterprise Distributed Object Computing Conference Workshops (EDOCW), pp. 165–171. IEEE (2012)
6. Duipmans, E.F., Pires, L.F., da Silva Santos, L.O.B.: A transformation-based approach to business process management in the cloud. J. Grid Comput. 12(2), 191–219 (2014)
7. GoogleContainer: Google container engine (GKE) (2017). https://cloud.google.com/container-engine/. Accessed 15 July 2017

8. Graiet, M., Mammar, A., Boubaker, S., Gaaloul, W.: Towards correct cloud resource allocation in business processes. IEEE Trans. Serv. Comput. **10**(1), 23–36 (2017)
9. Hoenisch, P., Hochreiner, C., Schuller, D., Schulte, S., Mendling, J., Dustdar, S.: Cost-efficient scheduling of elastic processes in hybrid clouds. In: IEEE 8th International Conference on Cloud Computing (CLOUD), pp. 17–24. IEEE (2015)
10. Cao, J., Chen, J., Zhao, Q.: An optimized scheduling algorithm on a cloud workflow using a discrete particle swarm. Cybern. Inf. Technol. **14**(1), 25–39 (2014)
11. Kozhirbayev, Z., Sinnott, R.O.: A performance comparison of container-based technologies for the cloud. Future Gener. Comput. Syst. **68**, 175–182 (2017)
12. Lim, S.-H., Sharma, B., Nam, G., Kim, E.K., Das, C.R.: MDCSim: a multi-tier data center simulation, platform. In: IEEE International Conference on Cluster Computing and Workshops (CLUSTER 2009), pp. 1–9. IEEE (2009)
13. Mastelic, T., Fdhila, W., Brandic, I., Rinderle-Ma, S.: Predicting resource allocation and costs for business processes in the cloud. In: IEEE World Congress on Services (SERVICES), pp. 47–54. IEEE (2015)
14. OMG: Business Process Model and Notation (BPMN), Version 2.0, January 2011
15. Ostermann, S., Plankensteiner, K., Prodan, R., Fahringer, T.: GroudSim: an event-based simulation framework for computational grids and clouds. In: Guarracino, M.R., et al. (eds.) Euro-Par 2010. LNCS, vol. 6586, pp. 305–313. Springer, Heidelberg (2011). doi:10.1007/978-3-642-21878-1_38
16. Piraghaj, S.F., Dastjerdi, A.V., Calheiros, R.N., Buyya, R.: A framework and algorithm for energy efficient container consolidation in cloud data centers. In: IEEE International Conference on Data Science and Data Intensive Systems (DSDIS), pp. 368–375. IEEE (2015)
17. Piraghaj, S.F., Dastjerdi, A.V., Calheiros, R.N., Buyya, R.: ContainerCloudSim: an environment for modeling and simulation of containers in cloud data centers. Softw. Pract. Exp. **47**(4), 505–521 (2017)
18. Povoa, L.V., de Souza, W.L., Pires, L.F., do Prado, A.F.: An approach to the decomposition of business processes for execution in the cloud. In: IEEE/ACS 11th International Conference on Computer Systems and Applications (AICCSA), pp. 470–477. IEEE (2014)
19. Rekik, M., Boukadi, K., Ben-Abdallah, H.: A comprehensive framework for business process outsourcing to the cloud. In: IEEE International Conference on Services Computing (SCC), pp. 179–186. IEEE (2016)
20. Soltesz, S., Pötzl, H., Fiuczynski, M.E., Bavier, A., Peterson, L.: Container-based operating system virtualization: a scalable, high-performance alternative to hypervisors. In: ACM SIGOPS Operating Systems Review, vol. 41, pp. 275–287. ACM (2007)
21. Sriram, I.: SPECI, a simulation tool exploring cloud-scale data centres. In: Jaatun, M.G., Zhao, G., Rong, C. (eds.) CloudCom 2009. LNCS, vol. 5931, pp. 381–392. Springer, Heidelberg (2009). doi:10.1007/978-3-642-10665-1_35
22. Xavier, M.G., Neves, M.V., Rossi, F.D., Ferreto, T.C., Lange, T., De Rose, C.A.F.: Performance evaluation of container-based virtualization for high performance computing environments. In: 21st Euromicro International Conference on Parallel, Distributed and Network-Based Processing (PDP), pp. 233–240. IEEE (2013)
23. Yangui, S., Klai, K., Tata, S.: Deployment of service-based processes in the cloud using petri net decomposition. In: Meersman, R., Panetto, H., Dillon, T., Missikoff, M., Liu, L., Pastor, O., Cuzzocrea, A., Sellis, T. (eds.) OTM 2014. LNCS, vol. 8841, pp. 57–74. Springer, Heidelberg (2014). doi:10.1007/978-3-662-45563-0_4

Using Colored Petri Nets for Verifying RESTful Service Composition

Lara Kallab[1,2]([⊠]), Michael Mrissa[1], Richard Chbeir[1], and Pierre Bourreau[2]

[1] Univ Pau & Pays Adour, LIUPPA, EA3000, Anglet, France
{lara.kallab,michael.mrissa}@univ-pau.fr, rchbeir@acm.org
[2] NOBATEK/INEF4, Anglet, France
{lkallab,pbourreau}@nobatek.com

Abstract. RESTful services are an attractive technology for designing and developing web-based applications, as they facilitate reuse, interoperability, and loosely coupled interaction with generic clients (typically web browsers). Building RESTful service composition has received much interest to satisfy complex user requirements. However, verifying the correctness of a composition remains a tedious task. In this paper, we present a formal approach based on Colored Petri Nets (CPNs) to verify RESTful service composition. First, we show how CPNs are utilized for modeling the behavior of resources and their composition. Then, we present how this formal model can be used to verify relevant composition behavior properties. Our solution is illustrated with a scenario built upon an energy management web framework developed within the HIT2GAP H2020 European project (Highly Innovative building control Tools Tackling the energy performance GAP http://www.hit2gap.eu).

Keywords: REST architecture · Resource composition · Colored Petri Nets · Composition properties verification

1 Introduction

Nowadays, web-based applications provide functionalities as web services, to allow for language and platform independence and to improve interoperability with other services. The REST architectural style [4] has recently become the most adopted solution for designing and developing web services. This is due to (i) its simplicity and ease of use that make services integration cost-effective, (ii) its single approach that allows the use of generic clients to interact with any RESTful APIs, (iii) its support for different data formats (e.g., XML and JSON), and (iv) its ability to support caching for better performance and scalability. Hence, more and more, web-based applications provide functionalities as RESTful services that follow the principles of the REST architectural style, also referred to in this paper as resources. Each resource provides a well-defined functionality that meets a specific client request. However, answering some requests often requires the combination of two or more resources, forming a composition.

© Springer International Publishing AG 2017
H. Panetto et al. (Eds.): OTM 2017 Conferences, Part I, LNCS 10573, pp. 505–523, 2017.
https://doi.org/10.1007/978-3-319-69462-7_32

Although there have been many contributions related to RESTful service composition, ensuring the correctness of the composition behavior remains a challenge. In fact, verifying a composition usually relies on the formal verification of its behavioral properties [16] (e.g., Reachability, Liveness, and Persistence). Such verification typically depends on the formal modeling of the composition behavior via a modeling language with clear semantics. Several works have been carried out in this scope. Some RESTful composition approaches are based on formal languages (e.g., Petri Nets [1,3], Finite State Machine (FSM) [18], and Process Algebra [15]), and others rely on services descriptions with embedded semantics such as in [14]. Although these approaches respect the majority of REST principles, they mainly contribute in modeling and constructing RESTful services composition without verifying its correct behavior.

In this paper, we propose a formal language based on Colored Petri Nets [9], known as CPNs, to model and verify RESTful service composition. The main contribution of this work is the mapping between CPNs model and RESTful services, to allow the use of CPNs behavioral properties for verifying RESTful service composition. Such verification is necessary, in order to avoid incorrect composition behavior and unnecessary execution for an erroneous composition. CPNs were initially founded to provide design, analysis and verification techniques for concurrent processes behavior. They bring expressiveness in modeling complex systems, graphical visual notations that aid in following the process behavior simulation at each step, and more importantly, the ability to validate and verify the functional correctness of systems behavior [7]. Based on CPNs, our approach is able to ensure the correctness of a composition through the verification of the following behavior properties:

- **Interoperability**, checks if the resources involved into the composition can be linked together. This is related to data type compatibility between the linked resources where the output of a resource should be of the same type of the input of another resource.
- **Reachability**, is used to verify that the desired final composition state is reachable from the initial state.
- **Liveness**, ensures that all resources participating in the composition will be invoked during composition execution.
- **Persistence**, is used when parallel resources are executed simultaneously, and ensures that the occurrence of one resource will not disable another.

The remainder of this paper is organized as follows. Section 2 presents a scenario to motivate our work, and highlights the research problem we tackle. Section 3 gives a brief description on the principles of RESTful services and the basics of Colored Petri Nets (CPNs). Section 4 details our CPN-based approach for verifying RESTful service composition. Section 5 illustrates the proposed solution within our motivating scenario. Section 6 presents the related work and highlights the originality of our approach. Finally, Sect. 7 provides concluding remarks, and discusses perspectives for future work.

2 Motivating Scenario and Problem Statement

In this section, we present our motivating scenario illustrated in a web-based building energy management platform that is currently being developed within a H2020 European project called HIT2GAP.

Recent studies reveal that a huge gap exists between the design specifications and the operation of buildings energy performance[1]. This is due to various sources encountered in building life cycle phases (i.e., inaccuracy of the simulations tools, poor quality of the used construction equipment, inadequate verification tools, limited data analysis of the building in operation, etc.). With the aim of reducing the energy gap, HIT2GAP project provides a web-based energy management solution for managing the energy behavior of operational buildings. To do so, HIT2GAP uses mainly a 3 layered architecture as follows:

- **Field layer:** Includes heterogeneous data sensed from the building (e.g., internal temperature, energy consumptions, and other building related information such as doors and building levels), and from other sources (e.g., weather forecasts and occupants).
- **Core layer:** Contains (1) the HIT2GAP data repository to store the collected data, (2) a set of basic services (i.e., data preprocessing services used to correct erroneous data collected from the Field layer), and (3) web APIs through which third-parties modules access the framework services.
- **Management layer:** Contains third-parties modules (i.e., Forecasting, Fault Detection and Diagnosis, Occupants' Behavior Detection, etc.) developed by HIT2GAP partners.

HIT2GAP services and modules are encapsulated into RESTful services, called also resources. Each resource is dedicated to provide a specific functionality that satisfies a specific building actor request. However, some requests require the composition of several resources.

To motivate our work, let us consider the case of the teaching and research building of the National University of Ireland Galway (NUIG), one of the HIT2GAP solution pilot sites. The building manager of NUIG wants to estimate the upcoming week heating energy consumption of the corresponding building. The prediction output will help him to anticipate building energy resource needs required for the resulted consumption, and analyze building energy behavior. To do so, the building manager will have to invoke several services simultaneously, embedded as resources, through the web interface of the HIT2GAP instance deployed for NUIG and illustrated in Fig. 1. However, to satisfy his prediction demand properly, the building manager is ought to invoke the required resources and link them together correctly to reach the desired composition behavior. Figure 2 depicts the overall process to be executed for answering the building manager's request. It mainly requires the interaction with the following resources integrated into the HIT2GAP NUIG instance:

---

[1] https://www.designingbuildings.co.uk/wiki/Performance_gap_between_building_design_and_operation.

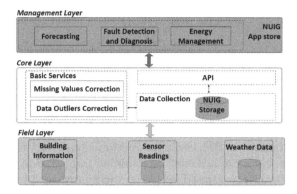

Fig. 1. HIT2GAP solution instance deployed in NUIG

1. Data collection resources to collect data required for the prediction process:
   (a) Upcoming week predicted internal temperature, which is extracted directly from the HIT2GAP database
   (b) Upcoming week predicted external temperature, which is provided by an external weather forecast RESTful service
2. Data preprocessing resources that clean the collected data from the external weather forecast service:
   (a) Resource that manages and corrects outliers values, which are data values outside the range of most of the other values
   (b) Resource that manages and corrects empty or missing values retrieved during certain timestamps
3. Resource responsible for the prediction of the NUIG heat energy consumption. This resource, embedded into the Forecasting module, uses a prediction model considered already implemented in the module.

The scenario shows the resources composition needed to satisfy the request at hand. However, building the composition properly and ensuring its correct behavior is a difficult task for the building manager. In fact, several problems may occur when building and/or executing the composition:

- **Non-interoperability:** Links between the output of a resource and the input of another may be invalid. This is due to the difference of data types that each resource handles. For example, the resource responsible for correcting outliers values only processes an array of temperature values whose data type is "Real". If it receives values from a previous resource with different data type (e.g., "Integer"), the composition will be erroneous. Such data type mismatch causes interoperability issues between resources.
- **Looping:** Starting the composition by collecting the required data (the internal and external predicted temperature), the process may not reach the final expected result, which is acquiring the predicted energy heat consumption of the NUIG building. This can be due to an end-loop occurred at a certain stage

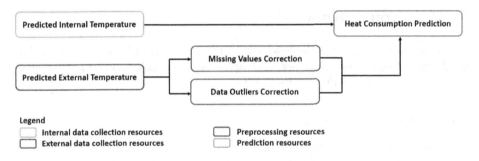

Fig. 2. NUIG resources involved in the prediction process

during composition execution, such as a loop in the preprocessing resources execution that can prevent the next resources to run.

- **Dead resources:** A resource, such as the external weather forecast, may not respond due to some technical problems on its server, and thus the next related resources involved into the composition process will not be invoked.
- **Conflict execution:** When dealing with parallel services such as preprocessing resources, a conflict in their execution may occur (deadlock).

To overcome these problems[2], we propose a formal language with clear semantics to model the behavior of RESTful services and verify the correctness of their composition. Such modeling language should cover the requirements below:

- **Support for REST principles:** Since HIT2GAP web services follow REST architecture principles, the formal language should be aligned with the main principles of REST architectural style. These principles are explained later in Sect. 3.1.
- **Data types handling:** The ability to handle the types of data flowing between services (i.e., String, Integer, etc.) allows composition syntax checking and thus a better management of the links between services.
- **Composition behavior verification:** In order to verify the correctness behavior of the composition, the modeling language should be able to verify the behavior properties considered important to HIT2GAP. These properties are: Reachability, Liveness, and Persistence.

In order to cope with the requirements mentioned above, we propose in this paper a formal language based on Colored Petri Nets (CPNs) to model Restful services and their composition. By using CPNs, our approach is able to handle data types and thus to check web services data compatibility. Moreover, with their formal syntax and semantics, they are able to validate the behavior of the built models through the execution of several verification properties embedded in open source and well known tools such as the CPN tools. The advantage of CPNs are detailed more in Sect. 3.2.

---

[2] Please note that the service discovery and selection problems are out of the scope of this paper.

3   Background Knowledge

In this section, we provide the reader with the necessary information to reach a good understanding of the paper, through a brief reminder on the principles of REST and on Colored Petri Nets (CPNs).

3.1   REST Principles

RESTful services, also called resources, are web services that follow the REpresentational State Transfer (REST) software architecture style [4]. A RESTful service provides a functionality through an abstract resource-oriented view identified by a Unique Resource Identifier (URI), and invoked via HTTP-based methods. We list below the main principles of RESTful services, inspired from [11]:

- **Resource Oriented and Addressability:** Resources are central elements addressable through their URI. They can be defined as objects with a type (e.g., JSON, XML, etc.), associated data (e.g., text files, images, etc.), relationships to other resources (i.e., Forecasting and Energy Management used in the HIT2GAP project), and a set of methods that operate on it (i.e., GET, POST, PUT and DELETE).
- **Uniform Interface:** Interacting with resources is realized through a uniform interface, which provides a set of standard operations (i.e., GET, POST, PUT and DELETE) implemented by the HTTP protocol. Each method has a well-defined semantics in terms of its effect on the state of the resource. It manipulates the resource state that can be transferred from/to clients, and represented in various types (e.g., JSON, XML, etc.).
- **Stateless Communication:** Every interaction with a resource is stateless. This means that each request is handled independently from the other, and request messages contain all the information that the server needs to process it. Stateless communication saves energy on the server side, as the state of the interaction with any client does not need to be stored in memory.
- **HATEOAS:** Known as Hypermedia As The Engine Of Application State, is a constraint of the REST paradigm used to provide directions to the client-agent regarding the next possible operations to be triggered. The principle is to include within returned server responses, the possible next resources URIs to follow, based on the current resource state. The methods used to invoke such resources can also be included.

3.2   Colored Petri Nets

Petri Nets are graphical and mathematical modeling language used to model distributed systems. They are designed to describe and study information processing systems, with concurrent, asynchronous, distributed, non-deterministic, parallel, and stochastic behaviors [10]. As shown in Fig. 3, a Petri Net is graphically represented by a number of places (represented by circles) occupied by tokens (represented by black dots), and transitions (represented by rectangles). Transitions and places are connected via arcs. A transition may fire when each of its input places has at least one token (Fig. 3(a)). When it fires, a token from each of its input places is removed, and a token is placed into every output place (Fig. 3(b)). The number and position of tokens may change during the execution of the Petri Net transitions. The assignment of tokens to places designates a state or a marking of the net.

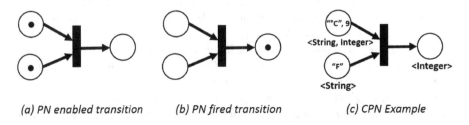

(a) PN enabled transition   (b) PN fired transition   (c) CPN Example

Fig. 3. Petri Nets vs. Colored Petri Nets

In ordinary Petri Nets, tokens cannot be distinguished and they are all identically represented as black dots. However, in more complex applications it is useful to allow the distinction between tokens and assign them some information. For these reasons, Colored Petri Nets (CPNs) combine the strengths of ordinary Petri Nets with the strengths of high-level programming languages [5], to allow handling data types and manipulating data values. In fact, within CPNs, each token can have a data type called a token color. Each color can be of a simple type (i.e., String, Integer, Boolean, etc.) or a complex type (i.e., array of String and Integer values). In addition, tokens with assigned colors can contain values. Normally, places in CPNs contain tokens of one type. An example of a CPN illustrating a temperature unit conversion is shown in Fig. 3(c). As it is illustrated, the first input place holds a record type data containing 2 variables of String and Integer type respectively (one denoting the unit of measurement of the current temperature value, and the other the actual measured temperature value). The other input place holds a String type data representing the desired temperature unit of the Integer output place value.

Using CPNs makes our modeling approach independent from the data serialization format (i.e., XML, JSON, etc.), and has the potential to handle and show explicitly data types in the composition. Moreover, CPNs are able to analyze the modeled composition and investigate its performance by making simulations of the composition process and verify several behavioral properties (e.g., Reachability, Liveness and Persistence). These analysis can lead to important insights into the behavior of the composition and help in ensuring the correctness of its design. As such, the validation process for example can aid in preventing from having a dead transition that can block the execution of other transitions (Liveness), or guaranteeing that the desired final system state will be reached (Reachability). Formally, a CPN is defined as follows [13]:

Definition 1. $CPN = (\Sigma, P, T, A, C, G, E, I)$, where:

- Σ is a finite set of non-empty types, called color sets
- $(P \cup T, A)$ forms a directed graph, where:
  - P (the set of places) and T (the set of transitions) are disjoint sets, such that $P \cap T = \emptyset$
  - $A \subseteq (P \times T) \cup (T \times P)$ is the set of arcs, such that places are only connected to transitions, and vice versa
- $C : P \rightarrow \Sigma$ is the color function that maps places to elements of Σ
- $G : T \rightarrow \mathbb{B}$ associates a precondition g (a boolean expression) to each transition. g should be evaluated to true for T execution
- $E : A \rightarrow Expr^3$ associates an expression $E(a)$ to each arc a. $E(a)$ is used to define input-output behavior of arcs, and may include variables such that:
  - Each variable in $E(a)$ has a type in Σ
  - $\forall a \in A$, $C(E(a)) = C(p)$, with p is the place connected to a
- I is the initialization function that maps each place $p \in P$ with an expression such that $I(p)$ is associated to the type $C(p)$

4   CPN-Based Approach for RESTful Service Composition

4.1   Resource Generic Interface

One of the key requirements of our modeling approach is to be aligned with REST principles. Therefore, it is essential to define the generic REST interface of a resource before presenting the interfaces of the resources involved in our motivating scenario. Generally, REST interface describes the required URI, HTTP method, query parameters[4], and responses of a resource. Responses includes:

- A list of the next resources to follow with the method used to invoke them. In this paper, the list can be empty when there are no resources to call.
- HTTP status code to indicate the query result. Such as, HTTP '200 OK' code denoting that the request has succeeded, and HTTP '201 Created' code designating that the request has been fulfilled and a new resource is created.
- The information provided by the resource, when it is available.

4.2   Interfaces of the Motivating Scenario Resources

In this paper, we restricted ourselves to the required interfaces in the motivating scenario, to ease the illustration of our approach. Table 1 lists the URIs of the composition scenario resources, and Table 2 defines their required interfaces.

---

[3] Expr is a mathematical expression that will not be detailed further here due to space limitation.

[4] Parameters are to be encoded in the URI or in the message body according to the HTTP format.

Table 1. URIs of the prediction process resources

Id	URI
1	http://www.hit2gap.eu/predictions/pred-internal-temp
2	http://www.weatherforecast.com/forecast/external-temp
3	http://www.hit2gap.eu/missing-data-manager
4	http://www.hit2gap.eu/outliers-data-manager
5	http://www.hit2gap.eu/missing-data-corrected
6	http://www.hit2gap.eu/outliers-data-corrected
7	http://www.hit2gap.eu/predictions/pred-heat-consumption
8	http://www.hit2gap.eu/predictions/heat-consumption

URI 1 is called with GET to collect the predicted internal temperature according to 2 parameters: startdate and enddate, denoting the prediction time range requested by the building manager. The required data is retrieved directly from the HIT2GAP database, and considered as preprocessed data. The array 'PrInTemp' in the responses represents the predicted internal temperatures.

URI 2 is invoked using GET to collect the predicted external temperature according to the same 2 parameters of the previous URI. The required data is retrieved from an external weather forecast resource. The array 'PrExTemp' represents the predicted external temperature. After data retrieval, URI 3 is invoked through POST to correct the missing values presented in the array 'PrExTemp', and URI 4 is called to correct the outliers values presented in the array 'PrExTemp'.

Through GET, URI 5 is called to retrieve the modifications applied on the predicted external temperature values obtained from the URI 3 (missing data manager). The '#dataset' represents the id of the preprocessed data, modified by URI 3. The array 'CorrMPrExTemp' contains the modifications applied on the predicted external temperatures.

Using GET, URI 6 is called to retrieve the modifications applied on the predicted external temperature values obtained from the URI 4 (outliers data manager). Similar to the previous step, the '#dataset' represents the id of the preprocessed data, modified by URI 4. The array 'CorrOPrExTemp' contains the modifications applied on the predicted external temperatures.

In our composition scenario, we considered that the merging of both URI 5 and URI 6 outputs, is being held on the client side to obtain the preprocessed external predicted temperatures array: [CorrExTemp{date, temp}].

URI 7 is invoked with POST to predict the energy heat consumption based on (i) the startdate and the enddate, representing the prediction period range, and (ii) the predicted internal temperatures with the preprocessed external temperature values previously collected. And finally URI, 8 is called using GET to retrieve the predicted heat energy consumptions obtained from the URI 7 and represented by '#dataset'. The array 'PrHeatEngCons' in the responses contains the predicted values of the heat energy consumption.

Table 2. Interfaces of the resources involved in the prediction process

URI	HTTP Verb	Parameters	Responses
1	GET	startdate = dd/mm/yyyy enddate = dd/mm/yyyy	200 OK [PrInTemp{date, temp}] {(POST, URI7)}
2	GET	startdate = dd/mm/yyyy enddate = dd/mm/yyyy	200 OK [PrExTemp{date, temp}] {(POST, URI3), (POST, URI4)}
3	POST	[PrExTemp{date, temp}]	201 Created {(GET, URI5)}
4	POST	[PrExTemp{date, temp}]	201 Created {(GET, URI6)}
5	GET	#dataset	200 OK CorrMPrExTemp{date, temp}] {(POST, URI7)}
6	GET	#dataset	200 OK [CorrOPrExTemp{date, temp}] {(POST, URI7)}
7	POST	startdate = dd/mm/yyyy enddate = dd/mm/yyyy [PrInTemp{date, temp}] [CorrExTemp{date, temp}]	201 Created {(GET, URI8)}
8	GET	#dataset	200 OK [PrHeatEngCons{date, HeatEngCons}] { }

4.3 Colored Petri Nets-Based Formal Composition Model

As stated in Sect. 3.1, a resource can be invoked through HTTP methods to provide a specified functionality. It is published by a service provider, and located on a specific server. An exposed resource r, has a set of inputs, a set of outputs, and a function assigned to it. In our modeling approach, a resource can be atomic or composed. In the CPN model, we define (i) an atomic resource as a single CPN with a single transition, input and output places, and (ii) a composed resource as a set of linked CPN representing linked resources.

Before we formally define a resource, we define below the following sets:

– DataType = {BasicT ∪ ExtendedT} is the date types supported by a resource, such that:
  • BasicT = {String, Integer, Real, Boolean, Date, etc.}, denotes the basic data types known in programming languages
  • ExtendedT = {Req, Status} denotes the extended data types defined to meet resources requirements

- $\mathtt{Req} = (\mathtt{HTTP} \times \mathtt{U})$ is the set of HTTP requests sent to the URIs, where:
  - $\mathtt{HTTP} = \mathtt{POST|PUT|DELETE|GET|HEAD|PATCH|CONNECT|OPTIONS|TRACE}$ is the set of HTTP methods used to invoke the URIs
  - $\mathtt{U}$ is a set of URIs based on the standard RFC3986[5]
- $\mathtt{Status} = (\text{Code}, \text{Desc})$ denoting the status of the resource response, where:
  - Code $\subseteq \mathbb{N}^*$ denotes the HTTP response status code
  - Desc represents the description of the HTTP code (e.g., 'Created', 'OK')

Definition 2. *A RESTful resource r is defined as $r = (URI, N)$, where URI is the URI associated to r, and $N = (\Sigma, P, T, A, C, G, E, I)$ is a CPN such that:*

- $\Sigma \subseteq \mathtt{DataType}$, *denoting the set of data types that the resource can process.*
- P *is a finite set of input and output places of the resource, where:*
  - $P = P_{In} \cup P_{Out}$.
  - $P_{In} = \bigcup_{i=1}^{\mathbb{N}^*} \{p_{in_i}\} \mid \bigcup_{n=1}^{\mathbb{N}^*} r_n.P_{In}$, *such that:*
    - $\bigcup_{i=1}^{\mathbb{N}^*} \{p_{in_i}\}$, *denotes the set of input places of an atomic resource. Each resource requires one input place, representing the request sent to it. Other input places can be defined according to the resources needs, such as the resources parameters.*
    - $\bigcup_{n=1}^{\mathbb{N}^*} r_n.P_{In}$, *denotes the set of input places of a composite resource.*
  - $P_{Out} = \bigcup_{i=2}^{\mathbb{N}^*} \{p_{out_i}\} \mid \bigcup_{n=1}^{\mathbb{N}^*} r_n.P_{Out}$, *such that:*
    - $\bigcup_{i=2}^{\mathbb{N}^*} \{p_{out_i}\}$, *represents the set of output places of an atomic resource. Each resource requires two output places, denoting respectively the status response code and the set of the HTTP requests that can be sent to the next possible URI resources. Other places can be defined according to the resources needs, such as resources output results.*
    - $\bigcup_{n=1}^{\mathbb{N}^*} r_n.P_{Out}$, *denotes the set of output places of a composite resource.*
- $T = t \mid \bigcup_{i=1}^{\mathbb{N}^*} r_i.T$. *$t$ represents the functionality of an atomic r, whereas the union of T sub-resources represents the functionality of a composite resource.*
- A *is a finite set of arcs linking input places to transitions and transitions to output places, such that: $P \cap T = P \cap A = T \cap A = \phi$.*
- C *is a color function. It associates a type from Σ to each place, where:*
  - $\exists p \in P_{In}$, *such that $C(p) \in \mathtt{Req}$.*
  - $\exists p_1, p_2 \in P_{Out}$, *such that $C(p_1) \in \mathtt{Status}$, and $C(p_2) \in \mathtt{Req}$.*
- G *is a guard function. It maps the transition $t \in T$ to a boolean guard expression g. The resource can only be executed if g is evaluated to true.*
- E *is an arc expression function. It maps each arc $a \in A$ into an expression that may include variables.*
- I *is an initialization function that associates places to initial values.*

Based on our defined CPN formal model for RESTful service composition, we represent formally the composed resource resulted from the HIT2GAP prediction scenario depicted in Fig. 4. Such formal language can be directly applicable to

---

[5] https://www.ietf.org/rfc/rfc3986.txt.

other scenarios and represent the corresponding web services as long as they are RESTful.

H2G_EnergyHeatPrediction = (URI, N), where URI is the address associated to the composition and $N = (\Sigma, P, T, A, C, G, E, I)$ is a CPN such that:

- $\Sigma = \bigcup_{i=1}^{8} r_i.\Sigma$ with r_i denoting the resources involved in the prediction process, and where:
  - $r_i.\Sigma \subseteq$ DataType, designates the data types handled by the resources participating into the composition.
- $P = P_{In} \cup P_{Out}$, where:
  - $P_{In} = \bigcup_{i=1}^{8} r_i.P_{In}$, such that:
    - $r_1.P_{In} = r_2.P_{In} = r_3.P_{In} = r_4.P_{In} = \{p_{in1}, p_{in2}\}$, denoting that each of these resources has 2 inputs.
    - $r_5.P_{In} = r_6.P_{In} = r_8.P_{In} = \{p_{in1}\}$, denoting that each of these resources has 1 input.
    - $r_7.P_{In} = \{p_{in1}, p_{in2}, p_{in3}\}$, denoting that this resource has 3 inputs.
  - $P_{Out} = \bigcup_{i=1}^{8} r_i.P_{Out}$, such that:
    - $r_1.P_{Out} = r_2.P_{Out} = r_5.P_{Out} = r_6.P_{Out} = r_8.P_{Out} = \{p_{out1}, p_{out2}, p_{out3}\}$, denoting that each of these resources has 3 outputs.
    - $r_3.P_{Out} = r_4.P_{Out} = r_7.P_{Out} = \{p_{out1}, p_{out2}\}$, denoting that each of these resources has 2 outputs.
- $T = \bigcup_{i=1}^{8} r_i.T$, with $r_i.T = t$ denoting the specific functionality provided by each resource.
- A is a finite set of arcs linking input places to transitions and transitions to output places, such that: $P \cap T = P \cap A = T \cap A = \phi$.
- C is a color function. It associates a type from Σ to each place, where:
  - $\exists p \in P_{In}$, such that $C(p) \in Req$.
  - $\exists p_1, p_2 \in P_{Out}$, such that $p_1 \neq p_2$, $C(p_1) \in Status$, and $C(p_2) \in Req$.
- G is a guard function. It maps the transition $t \in T$ to a boolean guard expression g.
- E is an arc expression function. It maps each arc $a \in A$ into an expression that may include variables.
- I is an initialization function that associates places to initial values.

We note that the composition URI is given after the verification of composition behavior that will be discussed in the following section.

4.4   Composition Properties

One of the main advantages of modeling RESTful services with CPNs format is the ability to analyze several behavioral properties of the composition. Mapping RESTful services and composition to CPNs with some extensions allows the execution of the algorithms related to CPNs properties. These algorithms that are used to check behavior properties in Colored Petri Nets still apply in our extended formal model, as it will be shown in this section.

To face the problems that may occur during composition execution, as it is presented in Sect. 2, four properties have been considered important to verify in the HIT2GAP project: Reachability, Liveness, Persistence [10], and Interoperability. In the Reachability, Liveness, and Persistence definitions below, we use (N, M_0) to denote a Petri Net, N, with its initial Marking, M_0. The Petri Net marking, M, designates the state of the net, which corresponds to the assignment of tokens to places. The initial marking M_0 designates the availability of some data (tokens) in the input places of one or more resources involved in the composition, before launching the composition execution.

Definition 3. Reachability - *A marking M_n is reachable from M_0 in a Petri Net N, if there exists a sequence of transitions firings from M_0 to M_n.*

One of the challenges in the composition process is to make sure that the final desired state is reachable from the initial state. In the prediction scenario, the desired final result is the predicted energy heat consumption, which is generated as an output from URI 8. To verify that the result is reached from the execution of data collection resources: URI 1 and URI 2, we use the Reachability graph.

Definition 4. Reachability Graph (RG) - *It is a set of all the reachable markings of a Petri Net represented as nodes. The nodes are connected with arcs designating the firing of a transition.*

The Reachability graph algorithm, inspired from [10], is described as follows:

Algorithm 1. Reachability Graph

1: label the initial marking M_0 as the root and tag it "new'
2: **while** "new" markings exists **do**
3:     select a new marking M
4:     if no transitions are enabled at M, tag M "dead-end'
5:     **while** there exist enabled transitions t at M **do**
6:         obtain the marking M' that results from firing t at M
7:         if M' does not appear in the graph, add M' and tag it "new'
8:         draw an arc with label t from M to M' (if not already present)

In our scenario, the Reachability property is true when: $\exists$ M_0 and $\exists$ $M \in$ RG as the end node, with M designating the final desired state.

Definition 5. Liveness - *A Petri Net (N, M_0) is considered to be L_k-live if every transition t in the net is L_k-live. t is said to be:*

- *L_0-live, if it can never be fired in any firing sequence. In this case the transition t is considered deadlocked*
- *L_1-live, if it can be fired at least once in some firing sequence*
- *L_2-live, if it can fire arbitrarily often*
- *L_3-live, if it can fire infinitely often*
- *L_4-live, if it may always fire*

Another challenge in our composition is to verify that the resources involved in the composition process will eventually be executed at least once. If for example, the preprocessing resources responsible of correcting erroneous data are not executed, the prediction resource will predict the building heat energy consumption based on inaccurate data. This will affect negatively the prediction results quality. Therefore L_1-live was our main focus, to ensure that all CPN transitions will eventually be fired by progressing through further allowed firing sequences. We note that transition firing depends on the availability of tokens (data) in all its input places. Thus, a resource can be executed only if its input places contain the required data which are: HTTP_VERB, URI, and some parameters (when it is necessary).

In our composition, a transition t is L_1-live when: $\exists\ M_0$ and $\forall\ t \in T\ in\ N$, $t \in$ RG. If not, t is considered dead.

By verifying both the Reachability and Liveness properties, we can be sure that the composition scenario contains no loop and all resources receive the required input in order to be executed.

Definition 6. *Persistence* - *A Petri Net (N, M_0) is considered persistent if for any two enabled transitions the firing of one will not disable the other.*

The notion of Persistence is useful in the context of parallel resources and is related to conflict-free nets. It is important in our composition scenario to make sure that the execution of the parallel data preprocessing resources, will be done with no conflicts. With CPN, the Persistence property is automatically guaranteed. In fact, parallel resources execution is done in any order, and no execution will disable the other.

As for the Interoperability, by definition the CPN formalism put the following as a constraint: $\forall a \in A : [C(E(a)) = C(p)]$. This means that the CPN workflow execution will not be possible unless data flowing to and from a place are of the same type. We note that data flowing to a place correspond to a transition output, while data flowing from a place denotes the input of the next transition.

5   Experimental Illustration

We illustrate our proposed CPNs-based formal composition approach in our prediction scenario using the CPN tools[6], one of the most known tools for editing, simulating, and analyzing Colored Petri Nets models. It provides a graphical user interface (GUI) with tool palettes and marking menus, to build the CPNs models. Moreover, it features syntax checking while the workflow is being constructed, and generates a standard state space report that contains information about behavioral properties of the modeled system.

Figure 4 represents the HIT2GAP scenario implemented according to our formalism. As it is illustrated, the model includes all the resources involved in the energy prediction process, with the required parameters and responses,

---

[6] http://cpntools.org.

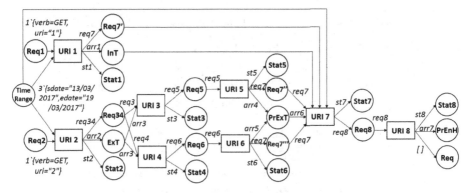

Annotations:
Req = Request, InT = Internal Temperature, ExT = External Temperature, Stat = Status, arr = Data Array, PrExtT = Processed External Temperature, PrEnH = Predicted Energy Heat, [ ]= Empty List

Fig. 4. CPN model for the resources composition relative to the prediction scenario

described in Table 2. During our tests, we extended the input places related to URI 7 to respect the HATEOAS principle. In fact, due to the existing of several resources (URI 1, URI 5 and URI 6) that point out to URI 7 in their next resources to follow, we linked each of these resources to URI 7 transition.

Using the state space tool of the CPN tools, and by implementing queries via ML code (the functional programming language of the CPN tools), we were able to verify Reachability and Liveness properties. As for the Persistence and Interoperability properties, they were verified automatically during composition construction. Below are the properties tests applied to the composition scenario:

– **Reachability Test:** Figure 5(a) shows the Reachability graph of our scenario, containing a node for each reachable state. In total we have 22 states, with node 22 representing the final state that is the prediction output results (PrEnH). Moreover, we used the 'Reachable (1,22)' boolean function (written in ML code) to test if state 22 is reached from state 1. The returned boolean value equal to "true" verifies the Reachability property.
– **Liveness Test:** Liveness property is verified through analyzing the Reachability graph arcs, which are labeled by the resources responsible of the state changing. Figure 5(b) shows examples of some Reachability graph arcs labels, appeared when clicking on the arcs. It proves that all URIs (from 1 to 8) are executed at least once. Moreover, and when generating the state space report through the state space tool of the CPN tools, several information can be retrieved including Liveness property results. Figure 6(a) for example shows the statistics representing the number of nodes and arcs of the composition scenario, and Fig. 6(b) proves that there are no dead transition instances, denoting that all the resources will be eventually executed starting from the initial state.
– **Persistence Test:** The persistence property is guaranteed automatically in the CPN tools, due to the CPN ability in modeling parallel systems. In fact,

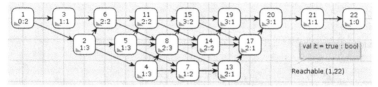

(a) Reachability graph of the prediction process

2:1->3 Heat_Consumption_Prediction'URI_1 1: {}	18:8->14 Heat_Consumption_Prediction'URI_5 1: {}
1:1->2 Heat_Consumption_Prediction'URI_2 1: {}	31:15->18 Heat_Consumption_Prediction'URI_6 1: {}
4:2->5 Heat_Consumption_Prediction'URI_3 1: {}	37:20->21 Heat_Consumption_Prediction'URI_7 1: {}
25:11->15 Heat_Consumption_Prediction'URI_4 1: {}	38:21->22 Heat_Consumption_Prediction'URI_8 1: {}

(b) Examples of the Reachability graph arcs labels

Fig. 5. Reachability graph

```
Statistics                              Liveness Properties
------------------------                ------------------------------

State Space                             Dead Transition Instances
   Nodes:   22                             None
   Arcs:    38
   Secs:    0
   Status:  Full
```

(a) State space report statistics   *(b) Dead transitions*

Fig. 6. Information retrieved from the state space report

the execution of parallel transitions can be done without interruptions, as long as each transition has the required data in all of its input places.

– **Interoperability Test:** In order to represent the flowing data types between resources, we defined the color sets as shown in Fig. 7(a). Using CPNs, our approach allows to verify the Interoperability property between the linked resources. CPN tools check the syntax of the nets during their construction where errors, such as in data types, can be visually seen through specific color indications, as shown in Fig. 7(b).

6 Related Work

6.1 Petri Net-Based Approaches

Formal languages are defined in [1,3] to describe RESTful services based on high-level Petri Nets. However, these approaches ignore internal hypermedia links, describe all tokens in XML only, and do not verify the correctness of composition behavior. In our work, we explicitly define data types as colors, within Σ that contains standard types and other defined types required for the resources. Also, in our approach we propose a unique definition of a resource as

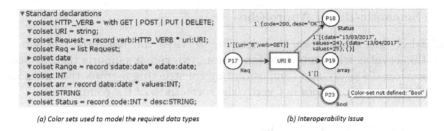

(a) Color sets used to model the required data types

(b) Interoperability Issue

Fig. 7. Defined color sets and Interoperability issue

a service, identified by a URI, which can be invoked to provide its functionality. On top of that, we focus on how CPNs properties can help verify the correctness of RESTful composition behavior.

6.2 FSM-Based Approaches

In [18], RESTful systems are modeled through a non-determinitic Finite State Machine (FSM) approach with epsilon transitions that do not need to read an input symbol in order to modify the system's state. The proposed model follows some REST design principles, i.e., uniform interface, stateless client-server operation, and code-on-demand execution. The main advantage of this work is that it supports hypermedia links between internal resources of a single web service. However, composition between different RESTful services are not modeled. Moreover, there are no composition verification properties used to ensure the correct behavior of the composition execution.

6.3 Linear Logic-Based Approaches

In [17], Intuitionistic Linear Logic (ILL) is used to model formally RESTful web services. The main contribution of this approach is web services composition modeling, and the ability to ensure composition completeness and correctness, through theorems based on propositional Linear Logic and π-calculus. However, although it respects the main principles of REST architectural style principles, linear Logic is a complicated formal language that requires extra efforts from web engineers to put it in practice.

6.4 Process Algebra-Based Approaches

In [15], RESTful services are described through the combination of process calculi format with tuple space computing, a model for managing a distributed object system. Based on this work, a semantic RESTful resource is formalized as a process associated with a triple space and a URI used for handling remote requests. However, the proposed approach does not support HATEAOS principle, nor even verification properties to check the correctness of the composition.

6.5 Semantic-Based Approaches

RESTdesc [14], is a solution that describes services' functionalities through the Notation3 syntax with embedded semantics. Although it supports the automatic web services discovery and composition by constructing a graph that links and identifies resources, it does not verify the correctness of the composition.

6.6 SOAP-Based Approaches

Before REST technology has been emerged, numerous work were conducted to compose services based on SOAP protocol [12]. In [2,6,8] SOAP web services have been translated to Petri Nets-based models. Apart from being SOAP-oriented services, none of them handle data types and they lack in analyzing composition behavior properties.

7 Conclusion

In this paper, we proposed a Colored Petri Nets-based approach for composing RESTful services, in the context of a web-based energy management solution developed in the HIT2GAP European project. We first exposed our CPN-based formalism to model the behavior of REST resources with their composition. And then, we showed how the verification of composition behavior properties (i.e., Interoperability, Reachability, Liveness, and Persistence) can be done based on our formal defined model.

Future work includes studying the limitations of our approach when the composition involves a large number of services. As well, quality of service constraints could be also modeled and taken into consideration in the verification. We also intend to extend resources descriptions with semantic annotations to automate the service orchestration task.

Acknowledgement. HIT2GAP project has received funding from the European Union's Horizon 2020 research and innovation programme under grant agreement N° 680708.

The authors acknowledge that the development work is carried out in a complementary manner with SIBEX: a French project funded by the Energy Transition Institute INEF 4.

References

1. Alarcon, R., Wilde, E., Bellido, J.: Hypermedia-driven RESTful service composition. In: Maximilien, E.M., Rossi, G., Yuan, S.-T., Ludwig, H., Fantinato, M. (eds.) ICSOC 2010. LNCS, vol. 6568, pp. 111–120. Springer, Heidelberg (2011). doi:10. 1007/978-3-642-19394-1_12
2. Chemaa, S., Elmansouri, R., Chaoui, A.: Web services modeling and composition approach using object-oriented petri nets. arXiv preprint arXiv:1304.2080 (2013)

3. Decker, G., Lüders, A., Overdick, H., Schlichting, K., Weske, M.: RESTful Petri Net execution. In: Bruni, R., Wolf, K. (eds.) WS-FM 2008. LNCS, vol. 5387, pp. 73–87. Springer, Heidelberg (2009). doi:10.1007/978-3-642-01364-5_5
4. Fielding, R.T.: Architectural styles and the design of network-based software architectures. Ph.D. thesis, University of California, Irvine (2000)
5. Gehlot, V., Nigro, C.: An introduction to systems modeling and simulation with Colored Petri Nets. In: Proceedings of the Winter Simulation Conference, pp. 104–118. Winter Simulation Conference (2010)
6. Hamadi, R., Benatallah, B.: A Petri Net-based model for web service composition. In: Proceedings of the 14th Australasian Database Conference, vol. 17, pp. 191–200. Australian Computer Society, Inc. (2003)
7. Kristensen, L.M., Christensen, S., Jensen, K.: The practitioner's guide to coloured petri nets. Int. J. Softw. Tools Technol. Transf. (STTT) 2(2), 98–132 (1998)
8. Li, B., Xu, Y., Wu, J., Zhu, J.: A Petri-Net and QoS based model for automatic web service composition. JSW 7(1), 149–155 (2012)
9. Liu, D., Wang, J., Chan, S.C., Sun, J., Zhang, L.: Modeling workflow processes with Colored Petri Nets. Comput. Ind. 49(3), 267–281 (2002)
10. Murata, T.: Petri Nets: properties, analysis and applications. Proc. IEEE 77(4), 541–580 (1989)
11. Pautasso, C.: Restful web services: principles, patterns, emerging technologies. In: Bouguettaya, A., Sheng, Q., Daniel, F. (eds.) Web Serv. Found., pp. 31–51. Springer, New York (2014)
12. Suda, B.: SOAP web services (2003). http://suda.co.uk/publications/MSc/brian.suda.thesis.pdf. Accessed 29 June 2010
13. Tian, B., Gu, Y.: Formal modeling and verification for web service composition. JSW 8(11), 2733–2737 (2013)
14. Verborgh, R., Steiner, T., Van Deursen, D., De Roo, J., Van de Walle, R., Vallés, J.G.: Description and interaction of restful services for automatic discovery and execution. In: 2011 FTRA International workshop on Advanced Future Multimedia Services (AFMS 2011). Future Technology Research Association International (FTRA) (2011)
15. Wu, X., Zhu, H.: Formalization and analysis of the REST architecture from the process algebra perspective. Future Gener. Comput. Syst. 56, 153–168 (2016). http://dx.doi.org/10.1016/j.future.2015.09.007
16. Yang, Y., Tan, Q., Xiao, Y.: Verifying web services composition based on hierarchical Colored Petri Nets. In: Proceedings of the First International Workshop on Interoperability of Heterogeneous Information Systems, pp. 47–54. ACM (2005)
17. Zhao, X.: A linear logic approach to RESTful web service modelling and composition. Ph.D. thesis, University of Bedfordshire, UK (2013). http://hdl.handle.net/10547/301103
18. Zuzak, I., Budiselic, I., Delac, G.: A finite-state machine approach for modeling and analyzing restful systems. J. Web Eng. 10(4), 353–390 (2011). http://www.rintonpress.com/xjwe10/jwe-10-4/353-390.pdf

Detecting Process Concept Drifts
from Event Logs

Canbin Zheng, Lijie Wen$^{(\boxtimes)}$, and Jianmin Wang

School of Software, Tsinghua University, Beijing 100084, China
zcb0821@outlook.com, {wenlj,jimwang}@tsinghua.edu.cn

Abstract. Traditional process discovery algorithms assume processes to be in a steady state. However, process models tend to be dynamic due to various factors, which has brought challenges such as change point detection, change localization and change process discovery. Existing techniques to identify change points are sensitive to parameters and the accuracy is not satisfactory. This paper proposes a novel approach to deal with such concept drift phenomenon. Event logs can be characterized by the relationships between activities, which motivates us to transform a log into a relation matrix. By detecting the always and never intervals in each row of the relation matrix, we obtain candidate change points for each relation. Finally, all the candidate change points are combined into an overall result. The approach is also able to localize the changes between different phases. Experiments on synthetic logs show that our approach is accurate and performs better than the state of the art in detecting sudden drift.

Keywords: Process mining · Concept drift · Process discovery · Change detection

1 Introduction

A business process refers to logically related activities or tasks conducted by people or equipments to accomplish a predetermined goal. Executions of processes are recorded in event logs including information about which activities are performed, at what time, by whom, in the context of which case, etc. Process mining techniques aim to extract knowledge from event logs commonly available in today's information systems [1]. The most crucial learning task in the process mining domain is termed process discovery, which is capable of constructing a process model (i.e. Petri nets) that accurately describes the process, as it takes place in real life. A majority of process discovery techniques like α-algorithm [18], Heuristic Miner [23], Genetic Miner [20], ILP Miner [9], etc. assume processes to be in a steady state, i.e. one event log corresponds to just one process model.

However, process models tend to be dynamic due to various factors like structure optimization, demand and supply, emergency, etc. Some changes are planned and documented, but some occur unexpectedly and may remain unnoticed. Bose et al. analyzed processes in more than 100 organizations and found

© Springer International Publishing AG 2017
H. Panetto et al. (Eds.): OTM 2017 Conferences, Part I, LNCS 10573, pp. 524–542, 2017.
https://doi.org/10.1007/978-3-319-69462-7_33

that it is very unrealistic to assume that the process being studied is in a steady state [4]. In predictive analysis and machine learning, "concept drift" means that the statistical properties of the target variable, which the model is trying to predict, change over time in unforeseen ways. Here we use concept drifts to refer to situations where the process is changing while being analyzed. An event log may correspond to multiple models serving in succession.

If we apply traditional process discovery methods to an event log with concept drifts, we will obtain a composite process model which fails to reflect the staged variation. Besides, a composite process model may be large and complex. The evolution of process models has brought challenges to process mining, such as change point detection, change localization and change process discovery. Change point detection aims to identify the time periods when changes have taken place, which is the first and most fundamental problem because change localization and change process discovery are both based on the drift detection result.

Offline analysis and online analysis are two major classes in dealing with concept drifts. Offline analysis refers to scenarios where analysts are given the entire log data from the beginning. In contrast, online analysis is one that can only deal with its input piece-by-piece in a serial fashion, i.e., in the order that the traces or events are fed to the algorithm, without having the entire log from the start.

Facing concept drifts, process analysts require methods and tools that are capable of detecting change points from event logs, helping them to analyze the evolution of processes. Literatures [2,4,5,12,13] have provided techniques to deal with concept drift problem. Some base on abstract interpretation of traces and some base on distance of activities. The mainstream method borrows ideas from *change-point detection* of which the goal is to detect changes in the generating distributions of the time-series. It performs statistical hypothesis testing on the data stream transformed from trace stream by extracting features. The performance of these techniques is not satisfactory. Parameters like window size have critical influences on the accuracy and are difficult to tune. Besides, there is a lack of theoretical illustration in their ability to recognize changes.

In this study, we focus on offline concept drift scenarios and describe a three-stage approach called TPCDD (Tsinghua Process Concept Drift Detection). First, each trace is represented by multiple relations. And then for each relation, we inspect its variation trend and partition it. Finally, we combine all change points revealed by each relation to get the final result. TPCDD achieves high accuracy and the parameter setting can fit in different scale of logs. Besides, its recognition ability is extendable by customizing relations and adding them into the relation matrix. In the meantime of identifying change points, TPCDD is also able to localize the changes.

The rest of this paper is organized as follows. Section 2 introduces basic concepts used throughout the paper. Our approach for concept drift detection (i.e., TPCDD) is described in Sect. 3, including three major steps. Section 4 provides an evaluation of TPCDD on synthetic logs and compare it with the

state of the art. Section 5 discusses related work. Section 6 concludes our work
and points out our future directions.

2   Preliminaries

This section introduces some basic concepts used throughout the paper. Petri
nets has been proved to be a powerful modeling language for business processes
[19]. We do our research on a subset of Petri nets called WF-net.

Definition 1 (Petri nets). *Let $P, T \subseteq U$ be finite and disjoint sets such that
$P \cup T \neq \emptyset$ and $P \cap T = \emptyset$, $F \subseteq (P \times T) \cup (T \times P)$. The tuple $N = (P, T, F, M_0)$
is a net system, where P is the set of places, T is the set of transitions and F
is the set of arcs. A marking M of N assigns each place a natural number $M(p)$
of tokens. M_0 is the initial marking of the net.*

We write $\bullet y = \{x | (x, y) \in F\}$ and $y \bullet = \{x | (y, x) \in F\}$ for the pre-set and
post-set of a node y respectively. The semantics of a net system N are typically
given by a set of sequential runs. A transition t of N is enabled at a marking
M_i of N iff $\forall p \in \bullet t, M_i(p) \geq 1$. If t is enabled at M_i, then t can occur and
afterwards N reaches a new marking M_{i+1} where $\forall p \in P$, $M_{i+1}(p) = M_i(p) - 1$
if $p \in (\bullet t \setminus t \bullet)$, $M_{i+1}(p) = M_i(p) + 1$ if $p \in (t \bullet \setminus \bullet t)$, and $M_{i+1}(p) = M_i(p)$
otherwise. An occurrence sequence of N is a transition sequence (e.g., $t_1 t_2 t_3 \ldots$)
in which transitions t_1, t_2, t_3, etc. occur in succession.

Definition 2 (Workflow net or WF-net). *A Petri net system $N =
(P, T, F, M_0)$ that models the control-flow dimension of a workflow is called a
workflow net (WF-net for short), which should satisfy the following requirements:*

- *There is one unique source place i such that $\bullet i = \emptyset$;*
- *There is one unique sink place o such that $o \bullet = \emptyset$;*
- *Every node $x \in P \cup T$ is on a path from i to o.*

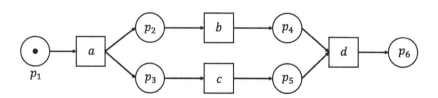

Fig. 1. A simple example of WF-net

Figure 1 shows an example of WF-net, in which $P = \{p_1, p_2, p_3, p_4, p_5, p_6\}$,
$T = \{a, b, c, d\}$. There is only one token in p_1 representing the initial marking
M_0. Taking a as example, $\bullet a = \{p_1\}$ and $a \bullet = \{p_2, p_3\}$. a is enabled at M_0. Once
a occurs, the token in place p_1 will be consumed and the numbers of tokens in

p_2 and p_3 will be increased by 1 respectively. For the WF-net in Fig. 1, there are two possible occurrence sequences $abcd$ and $acbd$.

Process mining aims to discover, monitor and improve real processes by extracting knowledge from event logs available in today's information systems. The starting point for process discovery is an event log. Each event in such a log refers to an activity (i.e., a well-defined step in a process) and is related to a particular case (i.e., a process instance). The events belonging to a case are ordered and can be seen as one run of the process. Event logs may store additional event attributes. In fact, whenever possible, process mining techniques use attributes such as the resource (i.e., person or device) executing or initiating the activity, the timestamp of the event, or data elements recorded with the event. In this paper, we abstract each process instance as a trace and each event log as a multiset of traces.

Definition 3 (Trace, Event log). *Let P_M be a process model and A be a set of activities. A^+ is the set of all nonempty finite sequence of activities from A. $\sigma \in A^+$ is called a trace when σ represents a firing activity sequence of PM. A log is a multiset of traces.*

Occurrence sequences of a net system N correspond to traces in the event log. For example, there are two possible traces $abcd$ and $acbd$ for the model in Fig. 1. The multiset $[abcd, abcd, abcd, acbd, acbd]$ is an event log with 5 cases, in which trace $abcd$ occurred 3 times while $acbd$ occurred 2 times.

We present some basic notations used in concept drift detection.

Definition 4 (Concept drift). *Let $PM_0, PM_1, \cdots, PM_n$ be $n + 1$ different process models and $T_0 < T_1 < \cdots < T_n$ be $n + 1$ moments. $PM(T_i) = PM_i$ denotes the model being used at T_i. $PM(T_0) = PM_0$ is the initial model. When moment T_i ($0 < i \leq n$) arrives, the currently used model will change into PM_i immediately. But the traces are still recorded in the same log. Such phenomenon is called concept drift. In this context, $T_1, \cdots, T_n$ are called change points.*

Note that in this paper, we focus on control flow changes which can be classified into operations such as insertion, deletion, substitution and reordering of process fragments. Data and resource perspectives are not taken into account.

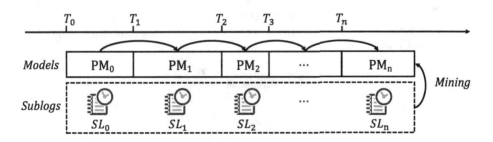

Fig. 2. Concept drift phenomenon in process mining

Besides, when model PM was changed from PM_i into PM_{i+1} at T_i, we assume that it happened in an instant (sudden drift). In other words, all traces produced at and after T_i no longer belong to model PM_i. The illustrative phenomenon for concept drifts in process mining is presented in Fig. 2.

3   Concept Drift Detection

Given a log $L = [\sigma_1, \sigma_2, \cdots, \sigma_n]$ which incorporates concept drifts, concept drift detection aims to find out the moments when the changes happened. From Definition 4, we obviously know that the model behavior is not the same before and after a change point. Take T_1 in Fig. 2 as an example. If we collect the traces between $[T_0, T_1)$ as SL_0 and the traces between $[T_1, T_2)$ as SL_1, SL_0 is different from SL_1. Therefore, a natural idea is to compare certain amount of traces before and after a candidate change point. Literatures like [4,12,13] all adopt this solution, which has two challenges. One is how to measure the differences between two sets of traces and the other is how many traces shall we collect for testing (that is to decide the window size). For the first challenge, [4,12,13] used feature extraction plus statistical hypothesis testing as a solution. For the second one, fixed and adaptive window size strategy are both introduced in [4,12,13]. However, the actual performance is not quite satisfactory because it heavily depends on the choice of window size. Once the window size is not proper, false negative and false positive are both possible to happen. To avoid such disadvantages, we proposed a totally different idea.

The topic of similarity between process models has been well studied. Existing algorithms like TAR [24], BP [22], ExRORU [21], 4C spectrum [16] offered a variety of solutions. Some can only be derived from models, some can only be derived from traces, and some can be derived from both. The principle behind

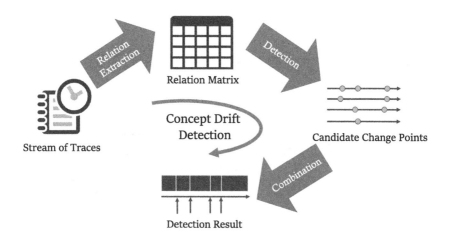

Fig. 3. Overview of our approach for detecting concept drifts

is that process models can be characterized by the relationships between activities. Motivated by similarity calculation and divide-and-conquer strategy, we proposed TPCDD to detect process concept drifts (Fig. 3).

Figure 3 provides an overview of our approach named TPCDD. First, we extract relations from each trace and transform the event log into a relation matrix. Then, for each relation, we observe its variation trend and detect candidate change points. Finally, all candidate change points are clustered to get a final result. In the following subsections, we will introduce TPCDD in detail.

3.1   Relation Extraction

Event logs can be characterized by the relationships between activities. In this section, we first introduce two basic kinds of relations and then transform a log into a relation matrix.

Definition 5 (Direct succession relation (DSR)). *Let $\sigma = t_1 t_2 \cdots t_m$ be a trace and $A = \{a_1, a_2, \cdots, a_k\}$ be a set of activities. $\forall x, y \in A$, (x, y) is a direct succession relation (denoted as $x \to y$) if and only if $\exists i \in \{1, 2, \cdots, m-1\}$ such that $t_i = x$ and $t_{i+1} = y$.*

Consider an event log $L = [abcd, acbd]$. The set of DSRs of trace $abcd$ is $\{a \to b, b \to c, c \to d\}$ and that of trace $acbd$ is $\{a \to c, c \to b, b \to d\}$. The time complexity of extracting DSRs from a trace with k activities is $O(k)$.

Definition 6 (Weak order relation (WOR)). *Let $\sigma = t_1 t_2 \cdots t_m$ be a trace and $A = \{a_1, a_2, \cdots, a_k\}$ be a set of activities. $\forall x, y \in A$, (x, y) is a weak order relation (denoted as $x \rightsquigarrow y$) if and only if $\exists i, j \in \{1, 2, \cdots, m\}, i < j$ such that $t_i = x$ and $t_j = y$.*

Consider an event log $L = [abcd, acbd]$. The set of WORs of trace $abcd$ is $\{a \rightsquigarrow b, a \rightsquigarrow c, a \rightsquigarrow d, b \rightsquigarrow c, b \rightsquigarrow d, c \rightsquigarrow d\}$ and that of trace $acbd$ is $\{a \rightsquigarrow c, a \rightsquigarrow b, a \rightsquigarrow d, b \rightsquigarrow d, c \rightsquigarrow b, c \rightsquigarrow d\}$. The time complexity of extracting WORs from a trace with k activities is $O(k^2)$.

At the first glance, DSR seems redundant as $a \to b$ leads to $a \rightsquigarrow b$. In fact, DSR and WOR are both necessary. In Fig. 4, N_1 and N_2 are distinguishable in DSR but behave the same in WOR. And the case of N_3 and N_4 are in the contrary. Accordingly, DSR and WOR complement each other in model characterization.

Definition 7 (Relation matrix). *Let L be an event log with n traces, R be a list with m relations and D be a $m \times n$ matrix. $D(i, j)$ represents the value in the i-th row and j-th column of D. Then an event log can be transformed into a relation matrix D by setting $D(i, j) = 1$ if the trace $L(j)$ contains the relation $R(i)$ and $D(i, j) = 0$ otherwise.*

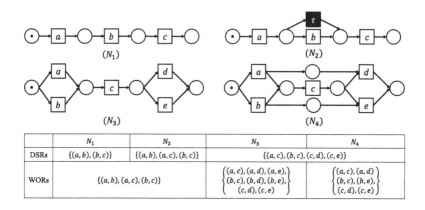

Fig. 4. DSR and WOR comparison for necessity analysis

For example, given an event log $L = [abcd, acbd, acbd, abcd, acd, abcd, acbd, ad, abd]$, Table 1 shows L's relation matrix defined over DSRs.

Table 1. The relation matrix of L defined over DSRs

	$L(1)$	$L(2)$	$L(3)$	$L(4)$	$L(5)$	$L(6)$	$L(7)$	$L(8)$	$L(9)$
$a \to b$	1	0	0	1	0	1	0	0	1
$a \to c$	0	1	1	0	1	0	1	0	0
$a \to d$	0	0	0	0	0	0	0	1	0
$b \to c$	1	0	0	1	0	1	0	0	0
$b \to d$	0	1	1	0	0	0	1	0	1
$c \to b$	0	1	1	0	0	0	1	0	0
$c \to d$	1	0	0	1	1	1	0	0	0

Let k be the average length of traces. The time complexity of converting an event log to a relation matrix is $O(nk) + O(nk^2) = O(nk^2)$.

DSR and WOR are borrowed from process similarity algorithms (i.e., TAR and BP) respectively. We use these two kinds of relations because they can be calculated from traces efficiently. Besides, they are powerful enough for change detection in general scenarios. In fact, the relation matrix is extendable. If more particular changes are required to be detected, an alternative solution is to introduce additional relations and extract them from the log.

3.2   Candidate Change Points Detection

Changes among process models can be reflected by the variation of relations between activities in the log. In this section, we observe the variation trend for each relation through its frequency level and detect candidate change points.

Definition 8 (Frequency level). *Given an integer interval $[p..q)$ and a specific relation $R(i)$, the frequency level of $R(i)$ over $[p..q)$ is*

- *always if and only if $\forall j \in [p..q), D(i,j) = 1$,*
- *never if and only if $\forall j \in [p..q), D(i,j) = 0$,*
- *sometimes otherwise.*

Correspondingly, $[p..q)$ is called as an always/never/sometimes interval.

An *always* or *never* interval implies a steady state of a certain relation and its length indicates how long the steady state lasts. If the frequency levels of a relation over two consecutive intervals are different, changes may occur at the split point. The split point is a candidate change point. In concept drift detection, we only focus on long enough (above a threshold called the minimum relation invariance distance, addr. MRID) *always* or *never* intervals because fluctuation between short intervals are regarded as normal variations inside a single model.

According to Definition 7, the variation of a relation is modelled as a row in relation matrix D. Algorithm 1 describes how to detect *always* and *never* intervals in each row of the relation matrix. The ends of these intervals form the candidate change points.

Algorithm 1. Candidate Change Points Detection

Input: the *i-th* row of relation matrix $D[i]$, MRID
Output: a set of candidate change points P

```
1  begin
2  |    P ← ∅, begin ← 0, count ← 0, n ← length of D[i]
3  |    for j ← 1 to n do
4  |    |    if  begin = 0 or D[i][j] ≠ D[i][begin] then
5  |    |    |    if count ≥ MRID then
6  |    |    |    |    P ← P ∪ {begin, j}
7  |    |    |    begin ← j
8  |    |    |    count ← 0
9  |    |    count ← count + 1
10 |    if count ≥ MRID then
11 |    |    P ← P ∪ {begin}
12 |    P = P \ {1}
```

Algorithm 1 detects *always* and *never* intervals by iterating over $D[i]$. Variable *begin* records the index of the value being observed and *count* records the number. As long as the currently visiting value is the same as $D[i][begin]$, *count* keeps increasing (line 9). Once the currently visiting value differs from $D[i][begin]$ and *count* $\geq MRID$, an *always* or *never* interval is detected (line 4–6). Afterwards, we reset *begin* and *count* for continuing detection. Let m be the number of relations and n be the number of traces. Then the time complexity of detecting all candidate change points is $O(mn)$.

Figure 5 shows an example, where $D[i] = [0,0,0,0,0,0,1,1,1,0,0,0,0,1,1,$ $1,1,1,1,1]$ and MRID $= 5$. Algorithm 1 will divide $D[i]$ into three intervals $[1..7)$, $[7..14)$ and $[14..21)$, of which the corresponding frequency levels are *never*, *sometimes* and *always*. This means that relation $R[i]$ changed over three phases, never exists, sometimes exists, always exists and the candidate change points are $\{7, 14\}$. Note that the parameter MRID only limits the minimum length of an *always* or *never* interval, which makes the algorithm adaptively fits different scales of logs.

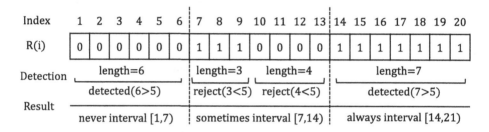

Fig. 5. An example for detecting change points of a relation

3.3   Candidate Change Points Combination

Applying Algorithm 1 to each row of a relation matrix D, we get sets of candidate change points reflecting phased variations. The procedure of combining all these candidate change points are described in Algorithm 2. Let multiset S represents all candidate change points. In concept drift context, points in S are likely to be distributed in neighborhoods of real change points, which motivates us to adopt DBSCAN[1] algorithm to partition S.

DBSCAN has two parameters: the maximum radius of a neighborhood (*eps*) and the minimum number of points required to form a dense region (*minPts*). By setting *minPts* $= 1$, no candidate change point will be dropped. *eps* is set to be a small value (like 10) or a small fraction (like 10%) of MRID. Line 9 ensures that the size of any partition of the log is no less than MRID, i.e. we keep the log size of each model to be at least MRID (consistent with Algorithm 1). More points in a cluster indicates more relations changed around the center of the cluster, which means that a real change point is more likely to lie in the cluster. Therefore, we check the clusters in descending order of size.

The clustering result not only contains change points, but also gives their corresponding weight (the size of the cluster). This provides convenience and freedom for the user to drop some less reliable points. For example, a cluster containing only one candidate change point can be treated as an exception rather

---

[1] https://en.wikipedia.org/wiki/DBSCAN.

Algorithm 2. Candidate Change Points Combination

Input: S(multiset of candidate change points), $MRID$, eps
Output: change points detected

```
1  begin
2  │   minPts ← 1
3  │   result ← [1, n]
4  │   clusters ← DBSCAN(S, minPts, eps)
5  │   sort clusters in descending order of size
6  │   for i ← 1 to |clusters| do
7  │   │   c ← average of clusters[i]
8  │   │   find change points left and right in result such that left < c < right
9  │   │   if c − left ≥ MRID and right − c ≥ MRID then
10 │   │   │   insert c into result at the position between left and right
```

a drift. By discarding clusters which has relatively few candidate change points, we can make the difference between real drifts and exceptions.

Figure 6 shows the procedure of combining candidate change points. After applying DBSCAN method, five clusters C_1, C_2, C_3, C_4, C_5 are returned with different sizes. The size indicates how many relations are changed in a cluster. Finally, we use clustering centers of C_1, C_2, C_3, C_4, C_5 as model change points.

Let m be the number of relations, n be the number of traces, s be the MRID and k be the average length of traces. Then the number of candidate change points is no more than mn/s. Therefore, the time complexity of Algorithm 2 is $O(mn/s)$. Combining relation extraction, candidate change points detection and combination, the time complexity of the overall detection is $O(nk^2) + O(mn) + O(mn/s) = O(nk^2) + O(mn)$.

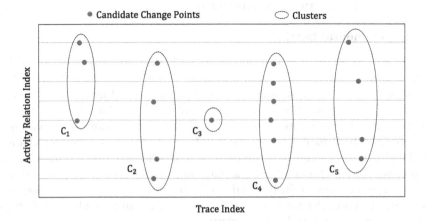

Fig. 6. The procedure of combining candidate change points

3.4   Recognition Ability

By extracting DSRs and WORs of each trace, an event log is transformed into a relation matrix and no longer used in the following steps. Therefore, we must make sure the information in the relation matrix is rich enough for distinguishing two different models. In this part, we prove that the ability of TPCDD to recognize changes is stronger than either TAR or BP.

The definition of TAR is the same as DSR. TAR algorithm distinguishes two models based on the TAR set, which contains all TARs of a model. Let S_1, S_2 be the TAR sets of model PM_1 and PM_2 respectively. If PM_1 and PM_2 are distinguishable by TAR algorithm, there exists a relation $a \rightarrow b$ such that $a \rightarrow b$ is only in one of the two TAR sets. Without loss of generality, we assume $(a \rightarrow b) \in S_1 \wedge (a \rightarrow b) \notin S_2$. Then for any log belonging to PM_2, the frequency level of $a \rightarrow b$ over any period is *never*. And for any log belonging to PM_1, the frequency level of $a \rightarrow b$ may be *always* or *sometimes*. This means that PM_1 and PM_2 are also distinguishable by TPCDD since the frequency levels of $a \rightarrow b$ are different.

BP algorithm uses the set of behavioral relations including strict order relation (denoted as SOR), exclusiveness relation (denoted as ER) and interleaving order relation (denoted as IOR) to measure the similarity of two process models. SOR, ER and IOR are all grounded on a fundamental relation called weak order relation (i.e., WOR), which is also used in our paper. If two models PM_1, PM_2 are distinguishable by BP set, then the weak order relation sets of PM_1, PM_2 are different, which means that PM_1, PM_2 are also distinguishable by TPCDD. The proof is similar to that of TAR set.

With DSRs and WORs, TPCDD possesses good capability of change recognition. Besides, it is extendable since one can customize activity relations if more particular changes are needed to be recognized.

4   Evaluations and Comparisons

4.1   Experimental Settings

To simulate the presence of concept drifts artificially, we generated mixed logs of different processes used in [12]. The base model is a textbook example of a business process for assessing loan applications (see Fig. 7), which has 15 activities, one start event and three end events, and exhibits different control-flow structures including loops, parallel and alternative branches. Maaradji et al. altered the base model by applying different types of control-flow changes, including 12 simple change patterns organized into three categories: Insertion (I), Resequentialization (R) and Optionalization (O) (shown in Table 2) and 6 composite change patterns (IOR, IRO, OIR, ORI, RIO, ROI) [12]. We modeled these processes in forms of Petri nets via PIPE[2]. The algorithm presented in this paper has been implemented[3] in Python language.

---

[2] https://sourceforge.net/projects/pipe2/.
[3] https://github.com/THUBPM/process-drift-detection.

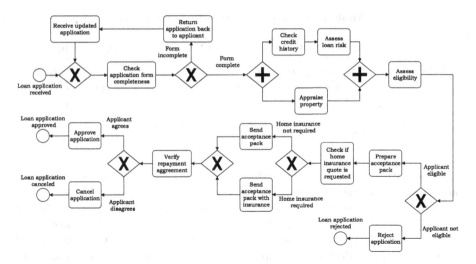

Fig. 7. A business process for assessing loan applications

Table 2. Simple control-flow change patterns

Code	Simple change pattern	Category
re	Add/remove fragment	I
cf	Make two fragments conditional/sequential	R
lp	Make fragment loopable/non-loopable	O
pl	Make two fragments parallel/sequential	R
cb	Make fragment skippable/non-skippable	O
cm	Move fragment into/out of conditional branch	I
cd	Synchronize two fragments	R
cp	Duplicate fragment	I
pm	Move fragment into/out of parallel branch	I
rp	Substitute fragment	I
sw	Swap two fragments	I
fr	Change branching frequency	O

A mixed log is determined by two parameters, the number of models (denoted as N) and the log sizes of each model (denoted as S). Given N and S, we first randomly select N adjacent different models and generate $S[i]$ traces for each model M_i ($i \in [1..N]$) and then merge these logs into one. To simulate different situations, we generated 32 mixed logs with different number of models and different sizes of logs for each model as shown in Table 3.

In Table 3, the first column indicates the log size for each model and the first row indicates the number of models in a mixed log. For example, $L_{4,500}$ is a

Table 3. Testing logs used in the experiments

	2	4	6	8	10	12	14	16
250 ± 50	$L_{2,250}$	$L_{4,250}$	$L_{6,250}$	$L_{8,250}$	$L_{10,250}$	$L_{12,250}$	$L_{14,250}$	$L_{16,250}$
500 ± 50	$L_{2,500}$	$L_{4,500}$	$L_{6,500}$	$L_{8,500}$	$L_{10,500}$	$L_{12,500}$	$L_{14,500}$	$L_{16,500}$
750 ± 50	$L_{2,750}$	$L_{4,750}$	$L_{6,750}$	$L_{8,750}$	$L_{10,750}$	$L_{12,750}$	$L_{14,750}$	$L_{16,750}$
1000 ± 50	$L_{2,1000}$	$L_{4,1000}$	$L_{6,1000}$	$L_{8,1000}$	$L_{10,1000}$	$L_{12,1000}$	$L_{14,1000}$	$L_{16,1000}$

log mixed with traces of 4 adjacent different models and each model contributes 500 ± 50 traces to the mixed log.

Knowing the number and positions of change points in logs provides a gold standard to evaluate detection algorithms. A true positive (TP)/false positive (FP) detection is defined as a detection inside/outside a neighborhood of the precise concept drift timestamp. A false negative (FN) is defined as missing a detection within the neighborhood. The detection quality is measured by $F\text{-}score = 1/(1/recall + 1/precision)$ where $recall = TP/(TP + FN)$ and $precision = TP/(TP + FP)$. We refer to the radius of the neighborhood as error tolerance (addr. ET). For TP detections, the detection error is defined as the average of their distance to precise concept drift timestamp.

4.2   Experiments on MRID

First, we evaluated the impact of MRID by executing TPCDD with 4 different mixed logs ($L_{10,250}, L_{10,500}, L_{10,750}, L_{10,1000}$) under different MRIDs. Figure 8 presents the $F\text{-}score$ with ET $= 10$.

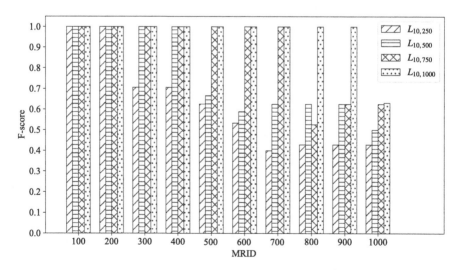

Fig. 8. F-scores under different MRIDs

As MRID grows, the *F-score* maintains steady at the beginning and then drops from a specific point. These points vary but lie around the corresponding log size of each model. This is because models of which the lifetime is less than MRID will be overlooked during detection. This experiment provides substantial evidence for the conclusion that TPCDD is not sensitive to parameter setting. As long as MRID is less than the log size of each model, the algorithm is able to fit in different log scale.

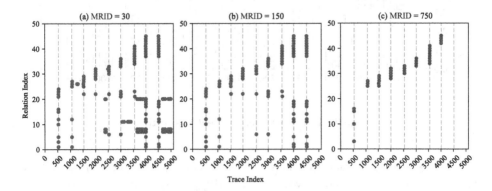

Fig. 9. Distribution of candidate change points under different MRIDs

Figure 9 provides us a visualization of the distribution of candidate change points (detected from $L_{10,1000}$). As is shown in the plot, most points lie around real change points. When MRID is too small (plot a), there exist some redundant points which are inside a model's lifetime. As MRID grows, the number of redundant points will gradually decreases. But in the main time, we may also lose some necessary points (plot c). This experiment confirms that our choice to use DBSCAN to combine candidate change points is reasonable and natural.

4.3 Experiments on *eps*

We also carried out experiments on the parameter of DBSCAN method *eps*. In this experiment, MRID was fixed to 100 and *eps* varied from 100 to 1100. Figure 10 shows the impact.

The F-score maintains steady at the beginning and drops from a specific *eps*. Take log $L_{10,500}$ as an example. When $eps \leq 400$, the F-score keeps high to 1.0. When $eps > 400$, the F-score falls off. This is because if *eps* is too large, a cluster would take in points in a wider range, resulting missing of real change points. Usually, *eps* is set to a small fraction (10% to 30%) of MRID.

4.4 Comparisons

According to Maaradji [12], their method outperforms the most mature approach [4] for process drift detection available. Since there was no algorithm better than

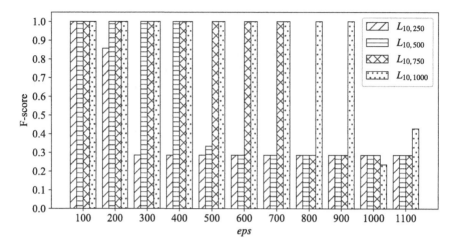

Fig. 10. F-scores under different *eps* settings

[12] at the time of writing, we compared TPCDD with [12] using synthetic logs listed in Table 3. As the adaptive method in [12] performs unstably, we chose to use only its fixed window mode here.

All 32 mixed logs were used in effectiveness experiment. We calculated the average for comparison. Figure 11 shows the result under different ETs. As is shown in Fig. 11, TPCDD performs better than Maaradji's, especially in situations where high precision (low error tolerance) is required. Maaradji's method cannot achieve a satifying F-score until ET increases to more than 100, which is unacceptable for small scale logs. Besides, the average detection error of TPCDD

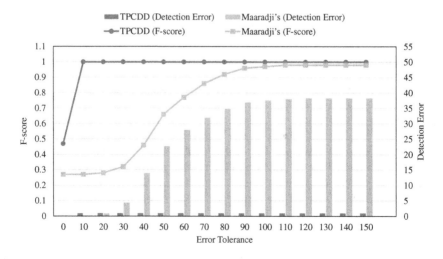

Fig. 11. F-score and detection error comparison between TPCDD and Maaradji's

is far less than Maaradji's, proving that the result of TPCDD is more close to real change points.

We also compared the time efficiency. TPCDD is implemented in Python while Maaradji's is implemented in Java. Figure 12 shows the time costs (not including file parsing time) over different mixed logs. Although Python is slower than Java in language perspective, TPCDD runs even faster than Maaradji's.

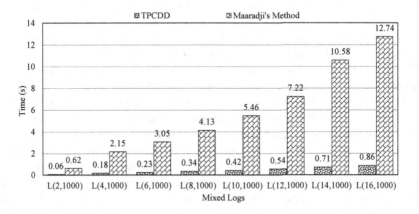

Fig. 12. Time efficiency comparison between TPCDD and Maaradji's

5   Related Works

A closely related problem to concept drift detection is change-point detection. The goal of the first is to detect changes that affect the mapping from the input space to the target concept, while the goal of the latter is to detect changes in the generating distributions of the time-series. Change-point detection is associated with homogeneity testing, in which, given two samples, one has to determine whether they were generated by the same distribution. Many attempts have been made to extend classical statistical tests for homogeneity for detection of changes in time-series [8,11]. Other methods are based on a predefined parametric model, such as the generating distribution [3,6,10], autoregressive models [17], and state-space models [7,14].

In the context of process mining, the input space refers to logs while the target concept is the underlying model. Concept drift detection is a new subfield in the process mining domain.

Carmona and Gavalda [5] first presented an online technique to detect concept drift by sequential monitoring of the logs. During each detection, an abstract interpretation in the form of polyhedron is first learned from the initial traces to represent the underlying concept. Then each subsequent trace is assessed whether it is within the polyhedron. Once the proportion of those traces inside the polyhedron drops exceed a specific threshold, a drift is detected. This method has a relation gap between polyhedron and traces. The ability of polyhedron to present the underlying model lacks explanation. It is impractical due to its inability to pinpoint the exact moment of the drift and low efficiency.

Bose et al. [4] proposed a framework for analyzing concept drifts in process mining. They transform a trace stream into a numerical stream by extracting some features in the perspective of control flow. To detect change points, they consider a series of successive populations of values and investigate if there is a significant difference between two subsequent populations via statistical hypothesis testing. The significance probabilities of the hypothesis tests can be visualized as a drift plot to analyze drifts. This method is complicated and not automated. The feature sets in the paper expand the input to a large scale. It requires human participation in feature selection and change point recognition.

Based on [4], Martjushev et al. [13] extended the approach in three aspects: automating change points identification, using adaptive window strategy and proposing techniques for detecting gradual drifts and multi-order dynamics.

Maaradji et al. [12] proposed a method similar to [4] for detecting concept drifts in business process logs. Rather than defining lots of features, [12] introduced a structure called run to compress the original event log. A stream of traces can be transformed into a stream of runs. Its core idea is to perform statistical hypothesis testing over the distribution of runs in two consecutive time windows. The underpinning assumption is that if a change occurs at or around a given time point, the distribution of runs before and after this time point will be statistically different and the difference can be revealed by statistical hypothesis testing. By adaptively resizing the window, this method also strikes a trade-off between classification accuracy and drift detection delay. But in experiments we found that the adaptive window strategy sometimes performs much worse than fixed window strategy. Building upon this technique, Ostovar et al. characterized process drifts from event streams, in terms of the behavioral relations that are modified by the drift [15]. But it is not designed to detect change points.

The core idea behind [4,12,13] is measuring the differences between two consecutive sliding windows of traces. A small window size may result in redundancy while a large one may lead to miss of change points. It is difficult to achieve one size fits all. Besides, performing statistical hypothesis testing on the data stream transformed from trace stream is not an accurate way to measuring the difference.

6   Conclusion and Future Work

In this paper, we proposed a three-staged approach to detect concept drifts from event logs. An event log is transformed into a relation matrix by extracting direct succession and weak order relations of each trace. Rather than directly comparing two sublogs, we focus on specific relations and analyze their variation trend. We partition the life cycle of each relation by detecting *always* and *never* intervals (exceed a certain size) and obtain candidate change points. Finally, all the candidate change points are clustered and combined into real change points.

In the detection framework, one can easily enhance the recognition ability by customizing relations and add them into the relation matrix. From the result, we not only learn when the changes occurred, but also distinguish successive models in the form of activity relation changes. Experiments showed that TPCDD is accurate and performs better than the state of the art in detecting sudden drift.

TPCDD currently focuses on offline scenarios. However, the framework is easy to fit in the online setting by doing little adjustment. Besides, situations where successive processes coexist for some time (denoted as gradual drift) is not supported in this paper and additional analysis for gradual drift is needed. Another avenue for future work is to make the method data and resource aware.

Acknowledgement. The work was supported by the National Key Research and Development Program of China (No. 2016YFB1001101) and the NSFC projects (No. 61472207, 61325008 and 71690231), and Tsinghua TNList Lab Key Projects.

References

1. van der Aalst, W.M.P., et al.: Process mining manifesto. In: Daniel, F., Barkaoui, K., Dustdar, S. (eds.) BPM 2011. LNBIP, vol. 99, pp. 169–194. Springer, Heidelberg (2012). doi:10.1007/978-3-642-28108-2_19

2. Accorsi, R., Stocker, T.: Discovering workflow changes with time-based trace clustering. In: Aberer, K., Damiani, E., Dillon, T. (eds.) SIMPDA 2011. LNBIP, vol. 116, pp. 154–168. Springer, Heidelberg (2012). doi:10.1007/978-3-642-34044-4_9

3. Basseville, M., Nikiforov, I.V., et al.: Detection of Abrupt Changes: Theory and Application, vol. 104. Prentice Hall, Englewood Cliffs (1993)

4. Jagadeesh, R.P., Bose, C., van der Aalst, W.M.P., Zliobaite, I., Pechenizkiy, M.: Dealing with concept drifts in process mining. IEEE Trans. Neural Netw. Learn. Syst. **25**(1), 154–171 (2014)

5. Carmona, J., Gavaldà, R.: Online techniques for dealing with concept drift in process mining. In: Hollmén, J., Klawonn, F., Tucker, A. (eds.) IDA 2012. LNCS, vol. 7619, pp. 90–102. Springer, Heidelberg (2012). doi:10.1007/978-3-642-34156-4_10

6. Gustafsson, F.: The marginalized likelihood ratio test for detecting abrupt changes. IEEE Trans. Autom. Control **41**(1), 66–78 (1996)

7. Kawahara, Y., Yairi, T., Machida, K.: Change-point detection in time-series data based on subspace identification. In: Seventh IEEE International Conference on Data Mining, ICDM 2007, pp. 559–564. IEEE (2007)

8. Kifer, D., Ben-David, S., Gehrke, J.: Detecting change in data streams. In: Proceedings of the Thirtieth International Conference on Very Large Data Bases, vol. 30, pp. 180–191. VLDB Endowment (2004)

9. Lamma, E., Mello, P., Riguzzi, F., Storari, S.: Applying inductive logic programming to process mining. In: Blockeel, H., Ramon, J., Shavlik, J., Tadepalli, P. (eds.) ILP 2007. LNCS, vol. 4894, pp. 132–146. Springer, Heidelberg (2008). doi:10.1007/978-3-540-78469-2_16

10. Lavielle, M., Teyssiere, G.: Detection of multiple change-points in multivariate time series. Lithuanian Math. J. **46**(3), 287–306 (2006)

11. Lung-Yut-Fong, A., Lévy-Leduc, C., Cappé, O.: Homogeneity and change-point detection tests for multivariate data using rank statistics. arXiv preprint arXiv:1107.1971 (2011)

12. Maaradji, A., Dumas, M., La Rosa, M., Ostovar, A.: Fast and accurate business process drift detection. In: Motahari-Nezhad, H.R., Recker, J., Weidlich, M. (eds.) BPM 2015. LNCS, vol. 9253, pp. 406–422. Springer, Cham (2015). doi:10.1007/978-3-319-23063-4_27

13. Martjushev, J., Bose, R.P.J.C., van der Aalst, W.M.P.: Change point detection and dealing with gradual and multi-order dynamics in process mining. In: Matulevičius, R., Dumas, M. (eds.) BIR 2015. LNBIP, vol. 229, pp. 161–178. Springer, Cham (2015). doi:10.1007/978-3-319-21915-8_11

14. Moskvina, V., Zhigljavsky, A.: An algorithm based on singular spectrum analysis for change-point detection. Commun. Stat.-Simul. Comput. 32(2), 319–352 (2003)

15. Ostovar, A., Maaradji, A., La Rosa, M., ter Hofstede, A.H.M.: Characterizing drift from event streams of business processes. In: Dubois, E., Pohl, K. (eds.) CAiSE 2017. LNCS, vol. 10253, pp. 210–228. Springer, Cham (2017). doi:10.1007/978-3-319-59536-8_14

16. Polyvyanyy, A., Weidlich, M., Conforti, R., La Rosa, M., ter Hofstede, A.H.M.: The 4C spectrum of fundamental behavioral relations for concurrent systems. In: Ciardo, G., Kindler, E. (eds.) PETRI NETS 2014. LNCS, vol. 8489, pp. 210–232. Springer, Cham (2014). doi:10.1007/978-3-319-07734-5_12

17. Takeuchi, J., Yamanishi, K.: A unifying framework for detecting outliers and change points from time series. IEEE Trans. Knowl. Data Eng. 18(4), 482–492 (2006)

18. van der Aalst, W.M.P., Weijters, T., Maruster, L.: Workflow mining: discovering process models from event logs. IEEE Trans. Knowl. Data Eng. 16(9), 1128–1142 (2004)

19. van der Aalst, W.M.P.: The application of petri nets to workflow management. J. Circ. Syst. Comput. 8(01), 21–66 (1998)

20. van der Aalst, W.M.P., de Medeiros, A.K.A., Weijters, A.J.M.M.: Genetic process mining. In: Ciardo, G., Darondeau, P. (eds.) ICATPN 2005. LNCS, vol. 3536, pp. 48–69. Springer, Heidelberg (2005). doi:10.1007/11494744_5

21. Wang, S., Wen, L., Kumar, A., Wang, J., Su, J.: ExRORU: a new approach to characterize the behavioral semantics of process models (short paper). In: Debruyne, C., et al. (eds.) OTM 2016. LNCS, vol. 10033, pp. 318–326. Springer, Cham (2016)

22. Weidlich, M., Mendling, J., Weske, M.: Efficient consistency measurement based on behavioral profiles of process models. IEEE Trans. Softw. Eng. 37(3), 410–429 (2011)

23. Weijters, A.J.M.M., van der Aalst, W.M.P., Alves De Medeiros, A.K.: Process mining with the heuristics miner-algorithm. Technische Universiteit Eindhoven, Technical report. WP, vol. 166, pp. 1–34 (2006)

24. Zha, H., Wang, J., Wen, L., Wang, C., Sun, J.: A workflow net similarity measure based on transition adjacency relations. Comput. Ind. 61(5), 463–471 (2010)

Multi-objective Cooperative Scheduling for Smart Grids

(Short Paper)

Khouloud Salameh[1,2(✉)], Richard Chbeir[1], and Haritza Camblong[2]

[1] University Pau & Pays Adour, LIUPPA, 64600 Anglet, France
{khouloud.salameh,richard.chbeir}@univ-pau.fr
[2] University of Basque Country, 20018 Donostia, Spain
aritza.camblong@ehu.eus

Abstract. In this work, we propose a multi-objective cooperative scheduling for Smart Grids (SG) consisting of two main modules: (1) the Preference-based Compromise Builder and (2) the Multi-objective Scheduler. The Preference-based Compromise Builder generates the best balance or what we call 'the compromise' between the preferences or associations of sellers and buyers that must exchange power simultaneously. Once done, the Multi-objective Scheduler proposes a power schedule for the associations, in order to achieve optimal benefits from different perspectives (e.g., economical by reducing the electricity costs, ecological by minimizing the toxic gas emissions, and operational by reducing the peak load of the SG and its components, and by increasing their comfort). Conducted experiments showed that the proposed algorithms provide convincing results.

Keywords: Multi-objective optimization · Smart Grid · Scheduling

1 Introduction

Nowadays, power systems are moving into a new era of reliability and efficiency. During this transition period, it is important to implement adequate techniques allowing to ensure that the benefits envisioned from Smart Grids or SG become a reality. Commonly considered as a key mechanism towards a more efficient and cost effective SG, the Demand-Side Management or DSM [10] refers to the planning and implementation of the utility companies' programs[1] designed to directly or indirectly influence the consumer consumption in the aim of reducing the system peak load and electricity costs. DSM techniques can be mainly gathered in two main categories: the load shifting [4] and the energy efficiency and conservation [2] programs. In our study, we focus on the load shifting, and more specifically on the power scheduling, since it has been observed that it is easier

---

[1] A utility company is a company that engages in the generation and the distribution of electricity for sale generally in a regulated market.

© Springer International Publishing AG 2017
H. Panetto et al. (Eds.): OTM 2017 Conferences, Part I, LNCS 10573, pp. 543–551, 2017.
https://doi.org/10.1007/978-3-319-69462-7_34

to motivate users to reschedule their needs rather than asking them to reduce their consumption [4]. Several approaches have been provided in the literature to address the power scheduling problem [1,3,7–9,11]. However, and to the best of our knowledge, none of the them fully covers the operational, economical and ecological aspects of the SG. To overcome the existing approaches limitations, we introduce here $MOCSF$, a 'Multi-Objective Cooperative Scheduling Framework' designed for the power scheduling in SG. $MOCSF$ aims at scheduling a set of couples, each consisting of a seller and buyer, working together in a mutual spirit so to ensure a better reduction of (economically, operationally and ecologically) costs and impacts within the SG. In addition, our approach presents several advantages over existing approaches, namely:

1. It provides a scheduling coverage able to consider all of the power consumption, production and storage entities of the SG,
2. It considers multiple energy sources (unlike existing approaches [1,3,7–9,11] that studied the interaction of the consumers with one energy source),
3. It takes into account individual SG components' preferences unlike existing approaches [1,8,9] that consider them partially.

2   Related Work

Many approaches have been proposed in the literature to solve the power scheduling problem. The comparison between the existing DSM approaches shows the following drawbacks:

- **Scheduling coverage**: All the existing DSM approaches [1,3,7–9,11] focused only on the power consumption scheduling, with the exception of [1,8] that addressed the storage scheduling as well. However, none of them covers the power production scheduling. Note that, all the DSM approaches [1,3,7–9,11] target the interaction of the consumers while assuming having only one utility grid and consequently one energy source.
- **Consumer satisfaction**: Few approaches [1,8,9] took into account the consumers' satisfaction. In [9], the consumers' comfort is ensured by reducing the gap between the desired and the actual hot water, and between the desired and the actual indoor temperature. However, in [1,8], the satisfaction is measured by the delay time between the desired start time and the real operation of its household appliances. Contrariwise to [3,7,11] where this aspect was completely absent.
- **Restricted goal**: Another limitation of all the existing approaches is that they do not cope with mutli-objectives (operational, ecological, economical, etc.) of a successful DSM. In almost all the approaches [1,3,7–9,11], the goal was mainly to reduce the electricity costs (economical aspect). In [9,11], the peak load reduction (operational aspect) is addressed aiming at reducing the peak hours in the power grid. However, none of the approaches considers the gas emission reduction (ecological aspect).

3   Multi-objective Cooperative Scheduling

In this section, we detail our 'Multi-Objective Cooperative Scheduling Framework' or $MOCSF$ aiming at reducing electricity bills, peak loads and environmental bad effects, while enhancing the comfort of the SG components. In order to conceive a cooperative environment, $MOCSF$ takes as input a set of couples, or what we call: seller-to-buyer associations, each consisting of a seller and buyer having mutual benefits in working together, with their individual desired schedules reflecting their operational preferences in terms of: start time, end time and power quantity (to sell or to buy). The main reason behind this choice [5,6] relies on the fact that we do not want to schedule the sellers and the buyers randomly but we rather want to maintain the power exchange between the sellers and the buyers having the biggest interest in working together (the interest can be expressed via an objective function that takes into account the ecological, economical and operational parameters).

3.1   Preference-Based Compromise Builder

As mentioned before, the input of this module is a set of seller-to-buyer associations, each composed of a seller and a buyer. Note that, each seller or buyer might belong to one or several associations. While sellers and buyers of the same association have to exchange power, each one of them has its own preferences to be respected so to establish a successful cooperative SG. Hence, the first step towards each association scheduling is to find the best balance or what we call the compromise, between the preferences of the related seller and buyer. This module aims at proposing an optimal distribution of the sellers' available power at each time t, in that it can meet its preferences and the buyers preferences. Before detailing the process, we present first some definitions used in our study.

Definition 1 (Components $[nR^+]$, $[nR^-]$, $[nR^0]$): *A seller, denoted as nR^+, has a power surplus, while a buyer, denoted as nR^-, has a power need. A self-satisfied, denoted as nR^0, has a power satisfaction*   ◆

Definition 2 (Schedule $[S]$): *A schedule S consists of the power exchanged vector $s_{\mathcal{R}} = [s_{\mathcal{R}}^1, s_{\mathcal{R}}^2, ..., s_{\mathcal{R}}^T]$, where $s_{\mathcal{R}}^t$ denotes the corresponding power quantity (in KW) that an entity $\mathcal{R}$ is willing to exchange, at a time t over a period T* ◆

Definition 3 (PurchaseGraph $[PG]$): *A PurchaseGraph PG is an oriented graph (V, E, S, EV) consisting of representing power scheduling of vertices v_i and associations e_i^j where each vertice $v_i \in V = \{nR^+\} \cup \{nR^-\}$ represents a component, each edge e_i^j connects a seller $v_i \in \{nR^+\}$ to a buyer $v_j \in \{nR^-\}$ with the total power quantity in EV exchanged between them, and each vertice v_i or edge e_i^j is associated to one desired schedule, denoted $s^{init} \in S$, and one operational schedule $s^{op} \in S$. The desired schedule designates the component operational preferences expressing its willing power quantity to exchange at each time t within a period T. The operational schedule designates the proposed schedule (provided by our algorithm). Note that, $\forall nR \in \{PG_k\} \Rightarrow nR \notin \{PG_{\neq k}\}$. To simplify in what follows,*

- $e.nR^+$ designates the edge seller,
- $e.nR^-$ designates the edge buyer,
- $e.EV$ designates the edge total power quantity,
- s_{nR}^{init} designates the component desired schedule,
- s_{nR}^{op} designates the component operational schedule,
- s_e^{init} designates the edge desired schedule, and
- s_e^{op} designates the edge operational schedule   ♦

Definition 4 (Satisfaction [$S(e, W)$]): The satisfaction of an edge e is defined as an optimization function according to several parameters. It allows to consider sellers and buyers' comfort (operational), power peak load (operational), electricity bills (economical) and environmental impacts (ecological).   ♦

The Preferences-based Compromise Builder consists of three main components: (1) Candidate components' prescheduling, (2) Final components' prescheduling, and (3) Compromise prescheduling. They are detailed below.

3.1.1   Candidate Components' Prescheduling

The aim of this module is to dissociate the desired schedule of each seller/buyer, so as to distribute the power quantity at each time t (its capacity of selling/buying) between the components with which, it must exchange, without exceeding nor being inferior to its desired capacity at time t. The pseudo-code of the candidate components' prescheduling is provided in Algorithm 1. Briefly, for each seller/buyer, we retrieve the list of edges to which the seller/buyer belongs. Then, we generate the list of all the possible permutations of the retrieved edges (Lines 7–16). For each possible permutation list of edges at a time t, we verify if the seller/buyer has enough power to sell/to buy to the buyer/from the seller of the same edge (Line 17). If there is enough power (Lines 18–20), we fill the schedule with the quantity to buy/to sell and recall the process by the next seller/buyer. If not, we fill the schedule with the quantity to buy/sell, reduce the quantity to sell/buy, and verify the quantity to sell/buy to the next buyer/from the next seller of the next edge (Lines 21–25).

3.1.2   Final Components' Prescheduling

The aim of this module is to select the candidate components' schedules that guarantee that each edge is provided with its exchanged value (EV) at each time t (e.g., at the end of the day, where t = 24 h). In other words, for each edge, the sum of the power quantity exchanged between its sellers and the buyers at T, should be equal to their exchanged value (EV) in the PG. So, the sellers sell all their power surplus and the buyers satisfy all their needs. The pseudo-code of the final components' schedules is provided in Algorithm 2. Briefly, for each candidate schedule of each seller (Lines 2–9) and for each time t of the day, we calculate the sum of the energy exchanged of the edges to which the seller/buyer belongs. The schedule is accepted if the sum is equal to the exchanged value of the edge (Lines 12–13). If the equality is verified for all the edges, we add the candidate schedule to the final components' schedules (Lines 13–16).

Algorithm 1. Candidate Components' Prescheduling

```
Input: PG[]                                                    // Set of PG forming the SG
Output: PG[]                          // Set of SG components updated with their candidate preschedules
1  S.CS = new int [][]                      // Initialize a candidate Schedule CS
2  S.e = new Edge []              // Initialize a candidate Solution S having a list of edges e
3  int RPL
4  for int i = 0; i < | SG.PG[] |; i + +                        // For each PurchaseGraph in the Smart Grid
5  do
6     E[] = GLE(SG.PG[i].e.nR)        // Retrieve the list of edges to which the seller/buyer of the edge belongs
7     PE[][] = Permutate(E[])              // Retrieve the possible permutation of the list of edges
8     for each e[] ∈ PE[][]                            // For each list of permutated lists of edges
9     do
10       S.CS = new int [| e[] |][T]                    // Initialize a candidate schedule CS for a solution S
11       S.Ce = new Couple [| e[] |]                    // Initialize a set of edges Ce for a solution S
12       RP[] = ∑_{t=1}^{T} S_{SG.PG[i].e.nR}^{sinit}[t]   // Initialize the remaining production to sell/buy to the desired
             selling/buying vector
13       for int j = 0, j < | e[] |, j + +                  // For each edge in the pemutated list of edges
14       do
15          RPL = e[j].EV        // Initialize the remaining production to buy/sell (of the linked component) with
               the valued exchanged of the couple
16          for int k = 0; k < T; k + +                       // For each time k
17          do
18             if RP[k] >= RPL                    // If there is sufficient power to sell
19             /buy then
20                S.CS[j][k] = RPL                 // Fill the schedule with the quantity to buy/sell
21                RPL = 0                          // No more power need to buy/sell
22             else
23                S.CS[j][k] = RP[k]               // Fill the schedule with the quantity to buy/sell
24                RPL− = RP[k]                     // Reduce the quantity to buy/sell
25             RP[k]− = S.CS[j][k]                 // Reduce the quantity to sell/buy
26       SG.PG[i].e.nR.S.Add(S)                    // Add S as a candidate solution of the seller/buyer
```

Algorithm 2. Final Components' Prescheduling

```
Input: PG[]                                                    // Set of PG forming the SG
Output: PG[]                          // Set of SG components updated with their final preschedules
1  for int i = 0; i < | SG.PG[] |; i + +                        // For each PurchaseGraph in the Smart Grid
2  do
3     for each s ∈ SG.PG[i].e.nR.S                        // For each possible solution of the seller/buyer
4     do
5        bool isAcceptedSolution = true
6        for int j = 0; j < | s.CS[0][] | ; j + +              // For each candidate schedule of the seller/buyer
7        do
8           int sev = 0                             // Initialize the sum of the exchange value with zero
9           for int k = 0; k < T ; k + +                         // For each time k
10          do
11             sev+ = s.CS[j][k]        // Calculate the sum of the energy exchanged during T of the edge
12          isAcceptedSolution = sev.Equals(s.Cs[j].EV)  // The solution is accepted if the sum is equal
               to the value exchanged of the edge
13       if isAcceptedSolution                      // If the equality is verified for all the edges
14       then
15          SG.PG[i].e.nR.S.Add(S)                  // Add S as an accepted solution of the seller/buyer
```

3.1.3   Compromise Prescheduling

The aim of this module is to generate every seller-to-buyer association (edge) desired schedule. It consists of selecting the best combination between the final preschedules of the sellers and buyers. This can be done by selecting the combination that ensures the minimum gap between the desired schedules of the sellers and buyers and the proposed compromise desired schedule. The pseudo-code of the final components' schedules is provided in Algorithm 3. First, we generate the combinations between the final preschedules of the sellers and the buyers (Lines 1–3). Then, for each seller/buyer of each combination, we calculate the power quantity for each edge in each candidate schedule for all combinations at a time t and fill it into a new vector ($FinalQuantity$) (Lines 4–17). After that, a similarity computation of the resulting vector and the initial desired schedule (vector) of each seller/buyer is done using the cosine similarity measure (Line 18). In fact, we adopted the commonly adopted cosine measure to calculate the distance between the proposed and the desired schedule vectors (instead of many others such the Euclidean Distance, the Pearson Correlation Coefficient, etc.) since it provides better results when there are many values in common between the two schedules to compare. Finally, the combination vector having the biggest similarity or what we call it here 'minimum delay' will be retrieved (Lines 19–27).

Algorithm 3. Compromise Prescheduling

```
Input: PG[]                                                    // Set of PG forming the SG
Output: S_e^{init}[]                                           // Edges desired schedule
1  for int i = 0; i < | SG.PG[] |; i + +                       // For each PurchaseGraph in the Smart Grid
2  do
3      Comb[] = Combination(SG.PG[i].e.nR^+.S[], SG.PG[i].e.nR^-.S[])    // Retrieve the possible
           combinations of the final sellers and buyers preschedules

4  for int i = 0; i < | Comb[] |; i + +                        // For each combination
5  do
6      for int j = 0; j < | SG.PG[] |; j + +                   // For each PurchaseGraph in the Smart Grid
7      do
8          for int k = 0; k < | Comb[i].Ce |; k + +           // For each set of edges of the selected combination
9          do
10             if Comb[i].Ce[k].nR == SG.PG[j].nR              // Check if we are verifying the schedules of the same
                   seller/buyer
11             then
12                 for int l = 0; l < | Comb[i].CS |; k + +    // For each set of candidate schedules of the
                       selected combination
13                 do
14                     Comb[i].FinalQuantityPerHour[j][l]+ = Comb[i].CS[k][l]    // Sum the power
                           quantity for each edge in each candidate schedule of each combination

15             for int x = 0; x < | Comb[i].CS |; x + +        // For each set of candidate schedules of the selected
                   combination
16             do
17                 FinalQuantity[x] = Comb[i].FinalQuantityPerHour[j][x]    // Calculate the
                       combination's power quantity for each seller/buyer

18         Comb[i].TotalDelay+ = 1 − Cosinus(S_{SG.PG[j].nR}^{init}, FinalQuantity)    // Calculate the
               similarity between the desired schedule of the seller/buyer and the combination's schedule

19  minDelay = Comb[0].TotalDelay
20  for int i = 0; i < | Comb[] |; i + +                       // Retrieve the minimum delay
21  do
22      if Comb[i].TotalDelay < minDelay then
23          minDelay = Comb[i].TotalDelat

24  for int i = 0; i < | Comb[] |; i + +                       // Retrieve the combination having the minimum delay
25  do
26      if Comb[i].TotalDelay == minDelay then
27          S_e^{init}[] = Comb[i]
```

3.2 Multi-objective Scheduler

Once done with the preferences-based combination generator that aims at extracting the desired schedules of the seller-to-buyer associations based on the sellers and buyers desired schedules given as input, it is time to schedule the resulting associations in a way to minimize costs (e.g., operational, economical, and ecological).

In our work, we adopted Particle Swarm Optimization (PSO), to search for the near-optimal scheduling for each seller-to-buyer association, thanks to its straightforward implementation and demonstrated ability of optimization. In essence, PSO is a computational method that optimizes a problem by iteratively trying to improve a candidate solution with regard to a given measure of quality using an objective function. It is considered as a powerful tool to solve complex non-linear and non-convex optimization problems. Moreover, it has several other advantages, such as fewer parameters to adjust, and easier to escape from local optimal solutions.

4 Experiments

4.1 Preferences-Based Compromise Builder Effectiveness

In this test, the efficiency of the generated desired compromise schedule is measured by its similarity with the desired schedules of the sellers and the buyers given as an input. The similarity measure used in our module is the 'Cosine Similarity Measure', which results a similarity between 0 and 1.

Figure 1a shows that the worst similarity ratio obtained is 0.72 and the best one is 0.95. This result reflects that our module ensures nice results providing an adequate compromise between the seller and the buyer preferences.

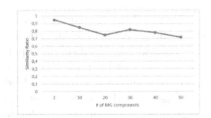

(a) Compromise Similarity w.r.t the number of *SG* Components

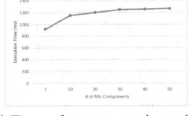

(b) Time performance w.r.t the number of *SG* Components

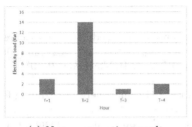

(c) Non-cooperative result

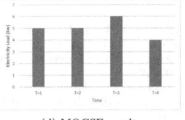

(d) MOCSF result

Fig. 1. Experimental results

4.2 Preferences-Based Compromise Builder Performance

This test consisted of measuring the necessary time to build the compromise from the sellers and buyers associations (cf. Fig. 1b).

4.3 Multi-objective Scheduler Impact on the *SG*

In this test, the objective function to optimize included economical, ecological and operational costs within two different scenarios: (1) a non-cooperative scheduling, where each association is selfish in that it only considers its desired schedule, and (2) a cooperative scheduling based on our proposed approach.

Figure 1c shows the electricity load resulting from the non-cooperative case. At $T = 2$, a peak load (Electricity load = 14 Kw) appeared having several bad effects on the (economical, ecological and the operational) costs. From the economical perspective, this peak load leads to a total electricity cost of 163 c. From the ecological perspective, and having at $T = 2$, a conventional power generator (emitting 0.26 Kg $Co2$/Kwh), the non-cooperation scheduling caused a simultaneous gas emissions of 3.64 KgCO2.

Figure 1d shows the electricity load resulting from our multi-objective scheduler. It shows how the peak loads are shaved (Max electricity load = 6 KW). The result is a trade-off between the economical, ecological and operational aspects.

5 Conclusion

In this paper, we proposed $MOCSF$, a Multi-Objective Cooperative Scheduling Framework providing a multi-type scheduling for the power generation, storage and consumption, while taking into account the ecological, economical and operational costs in a power system. Currently, we are working on implementing a privacy-by-design grid control allowing to protect the components privacy whilst preserving the advanced control and monitoring functionalities of the power systems. Further, it is interesting to apply strategy-proof techniques, in order to avoid cheating in the desired schedules.

References

1. Adika, C.O., Wang, L.: Smart charging and appliance scheduling approaches to demand side management. Int. J. Electr. Power Energy Syst. **57**, 232–240 (2014)
2. Allcott, H.: Social norms and energy conservation. J. Public Econ. **95**(9), 1082–1095 (2011)
3. Fakhrazari, A., Vakilzadian, H., Choobineh, F.F.: Optimal energy scheduling for a smart entity. IEEE Trans. Smart Grid **5**(6), 2919–2928 (2014)
4. Monteiro, J., et al.: Scheduling techniques to enable power management. In Proceedings of the 33rd annual Design Automation Conference, pp. 349–352. ACM (1996)
5. Saad, W., et al.: Coalitional game theory for cooperative micro-grid distribution networks. In: IEEE International Conference on Communications Workshops (ICC), 2011, pp. 1–5. IEEE. (2011)

6. Salameh, k., et al. Microgrid components clustering in a digital ecosystem cooperative framework. In: International Conference on Knowledge Based and Intelligent Information and Engineering Systems, vol. 112, pp. 167–176. Elsevier (2017)
7. Setlhaolo, D., Xia, X., Zhang, J.: Optimal scheduling of household appliances for demand response. Electr. Power Syst. Res. **116**, 24–28 (2014)
8. Vytelingum, P., et al.: Agent-based micro-storage management for the smart grid. In: Proceedings of the 9th International Conference on Autonomous Agents and Multiagent Systems, vol. 1, pp. 39–46 (2010)
9. Zhu, J., Lauri, F., Koukam, A., Hilaire, V.: Scheduling optimization of smart homes based on demand response. In: Chbeir, R., Manolopoulos, Y., Maglogiannis, I., Alhajj, R. (eds.) AIAI 2015. IAICT, vol. 458, pp. 223–236. Springer, Cham (2015). doi:10.1007/978-3-319-23868-5_16
10. Gellings, C.W., Chamberlin, J.H.: Demand-side management. In: Goswami, D.Y., Kreith, F. (eds.) Energy Efficiency and Renewable Energy Handbook, Chap. 15, 2nd edn, pp. 289–310. CRC Press, Boca Raton (1988)
11. Mohsenian-Rad, A.H., Wong, V.W., Jatskevich, J., Schober, R., Leon-Garcia, A.: Autonomous demand-side management based on game-theoretic energy consumption scheduling for the future smart grid. IEEE Trans. Smart Grid **1**(3), 320–331 (2010)

Privacy-Aware in the IoT Applications: A Systematic Literature Review

Faiza Loukil[1]($\boxtimes$), Chirine Ghedira-Guegan[1], Aïcha Nabila Benharkat[2], Khouloud Boukadi[3], and Zakaria Maamar[4]

[1] University of Lyon, CNRS,
IAE - University of Lyon 3, LIRIS, UMR5205, Lyon, France
{faiza.loukil,chirine.ghedira-guegan}@liris.cnrs.fr
[2] University of Lyon, CNRS, INSA Lyon, LIRIS, UMR5205,
Lyon, France
nabila.benharkat@liris.cnrs.fr
[3] Mir@cl Laboratory, Sfax University, Sfax, Tunisia
khouloud.boukadi@fsegs.usf.tn
[4] Zayed University, Dubai, UAE
zakaria.maamar@zu.ac.ae

Abstract. The Internet of Things (IoT) emerged as a paradigm in which smart things collaborate among them and with other physical and virtual objects using the Internet in order to perform high level tasks. These things appear in a variety of application domains, including smart grid, health care and smart spaces where several parties share data in order to tackle specific tasks. Data in such domains are rich in sensitive data and data owner-specific habits. Thus, IoT raises concerns about privacy and data protection. This paper reports on a systematic literature review of privacy preserving solutions used in Cooperative Information Systems (CIS) in the IoT field. To do so, and after retrieving scientific productions on the subject, we classify the results according to several facets. In this paper, we consider a subset of them: (i) data life cycle, (ii) privacy preserving techniques and (iii) ISO privacy principles. We combine the facets then express and analyze the results as bubble charts. We analyze the proposed solutions in terms of the techniques they deployed and the privacy principles they covered according to the ISO standard and the data privacy laws and regulations of the European Commission on the Protection of Personal Data. Finally, we identifies recommendations to involve privacy principle coverage and security requirement fulfillment in the IoT applications.

1 Introduction

The rapid growth of the Internet of Things (IoT) technology has resulted into different advances in the IoT field that are affecting both businesses and persons. In general, IoT applications, such as smart grid and smart cities require the collaboration of several parties in order to achieve their goals. These parties can be data owners or requesters, including an individual, a group of individuals

© Springer International Publishing AG 2017
H. Panetto et al. (Eds.): OTM 2017 Conferences, Part I, LNCS 10573, pp. 552–569, 2017.
https://doi.org/10.1007/978-3-319-69462-7_35

or an organization. For instance, the different parties can share their energy consumption in order to help energy provider to predict its energy production.

However, despite the bright side of IoT, several concerns continue to undermine its adoption. In fact, collecting data in IoT applications increases the data owner's worries about the potential uses of these data. In fact, some of the collected data can be sensitive and the data owner wish not share them with other competitor organizations without retaining some level of control. Thus, this work focuses on one non-functional requirement of IoT applications, which is privacy protection for the collaborating parties.

As part of our research agenda on privacy in the IoT era, we deem necessary conducting a comprehensive analysis on this topic. To this end, we provide an overview of existing IoT privacy preserving solutions in order to identify gaps and come up with solutions and recommendations. This overview is the result of a systematic literature review.

According to [20], a systematic literature review consists of five interdependent steps including (i) choose a research scope by defining research questions, (ii) retrieve candidate papers by querying different scientific databases, (iii) select relevant papers that can be used for answering the research questions by defining inclusion and exclusion criteria, (iv) define a classification scheme by analyzing the abstracts of the selected papers to identify the terms that will be used as categories for classifying the papers, and (v) produce a systematic literature review by sorting papers according to the classification scheme.

Our objective is to identify open issues and trends regarding privacy preserving in the IoT applications. Therefore, our classification scheme consists of six facets that stress out application domains, IoT architectures, security properties and requirements, data life cycle, and privacy preserving techniques. We also define an additional facet to identify the ISO privacy principle that the IoT solutions consider.

This paper is organized as follows. Section 2 gives a general idea about IoT, security and privacy concepts. Section 3 describes our systematic review study. Section 4 identifies the recommendations in order to involve privacy principle coverage and security requirement fulfillment in the IoT applications. Section 5 discusses existing reviews that study privacy issue in the IoT applications. Section 6 concludes the paper and presents some future endeavors.

2   Internet of Things, Security, and Privacy

In this section, we shed light on: IoT, security, and privacy by discussing their definitions, the existing IoT application domains and architectures as well as the security properties and requirements according to the ISO standard. Afterwards, we present the existing privacy legislation and the privacy preserving techniques.

2.1   Some Definitions

Commonly agreed definitions of IoT, security and privacy do not exist. Thus, we summarize those that are deemed relevant for our work.

The Internet of Things (IoT) is a network of physical objects that contain embedded technology to communicate with the external environment [12]. According to Guillemin et al. [13], IoT connects people and things at anytime, anyplace, with anything and anyone, ideally using any path or network and any service.

Security involves the application and management of appropriate measures that involve consideration of a wide range of threats. In this context, ISO standard [14] defines a set of security properties and requirements detailed in Sect. 2.3.

Privacy is *"the claim of individuals, groups, or institutions to decide for themselves when, how and to what extent information about them is communicated to others"* [25]. With the IoT applications, it is important to consider the context when dealing with privacy issue. Data privacy is about data security and while taking into account requirements from legal regulations and individual preferences [6].

2.2 Application Domains and Architectures of IoT

Different application domains in IoT exist. We categorize these applications into two domains:

1. **Personal and home:** including (i) location sharing, which the aim of providing services based on the collected location information (i.e. geographical position) of IoT terminals, (ii) health care, which consists of offering care monitoring services without necessary visiting hospitals and (iii) smart home, which automates the ability to control smart devices around the house.
2. **Government and industry:** including the smart city, which monitors all its critical infrastructures and smart grid, which allows grid monitoring in order to reduce energy consumption. Such IoT applications need collaboration between several parties in order to fulfill their ends. For instance, house, office and industry consumers should be aware of the collaboration benefits in a smart grid to reduce energy consumption. Besides, the smart grid should provide the adequate privacy protection for the collaborating members to reassure them.

Moreover, we distinguish four types of architecture that are: centralized, decentralized, third party and hybrid architecture.

1. **Centralized Architecture:** it is where all the entities in the network are passive: their only task is to provide data. The collected data will be stored, processed by a central server which is the only server that provides IoT services to the other entities [21]. The main challenge with this architecture is resilience. In fact, all the computation tasks are managed by a single server. Thus, in case of server failure, the IoT services will be unavailable.
2. **Decentralized Architecture:** each entity can process data and provide IoT services to other entities in the network. Moreover, the decentralized architecture overcomes the single point of failure issue of the centralized architecture.

In fact, a failure in one entity in the network will not affect the whole system. However, malicious entities intrusion arises because any entity can connect with any other entity at any time.

3. **Third Party Architecture:** it is where a public institution or a private corporation is responsible for data collection, transfer, storage, and/or processing. Example of ready to use platform is the Smart-Meter-Analytics (SAP) [2]. The main challenge with such architecture is that it gives full trust to the third party for the whole data management.

4. **Hybrid Architecture:** it consists of combining several architectures in order to take advantage of the existing architecture structures and overcome their disadvantages. For instance, Birman et al. [7] addressed the privacy issue in smart grid data collection phase by combining peer-to-peer communications with some elements of centralized control in order to help utilities to effectively use the collected data while preserving the consumers' privacy.

2.3 Security Properties and Requirements

According to ISO standard [14], the purpose of information security is to protect and preserve three essential properties, namely:

- **Confidentiality:** referring to data protection from unauthorized accesses, disclosures and processes.
- **Integrity:** referring to data accuracy and completeness protection from unauthorized modifications.
- **Availability:** referring to assure data accessibility and usability upon demand by an authorized entity.

Moreover, information security may also involve protecting the authenticity, the authorization and ensuring that entities can be held accountable.

- **Authentication:** referring to ensure that a claimed characteristic of an entity is correct.
- **Authorization:** referring to provide permissions towards information.
- **Accountability:** referring to the entity responsibility for its actions.

In the next subsection, we present the requirements defined by the ISO standard in order to preserve data privacy.

2.4 Privacy Legislation

In 1980, the Organization for Economic Co-operation and Development (OECD) issued Guidelines on the Protection of Privacy and Transborder Flows of Personal Data. These guidelines consist of eight principles known as Fair Information Practices (FIP) that enable individuals to express their privacy requirements and place obligations on organizations to follow those requirements.

Besides US privacy legislation, the European Union's application of a comprehensive legislation resulted in the Directive 95/46/EC [5] on the protection of individuals with regard to the processing of personal data and on the free movement of such data. The directive embeds the FIPs.

The ISO standard [14] also defines eleven privacy safeguarding requirements to protect sensitive information, namely: consent and choice; purpose legitimacy and specification; collection limitation; data minimization; use, retention and disclosure limitation; accuracy and quality; openness, transparency and notice; individual participation and access; accountability; information security and privacy compliance. We refer the readers to [14] for more information about these privacy principles.

In the next section, we propose a taxonomy of the existing privacy preserving techniques that are used to satisfy the requirements discussed above.

2.5   Privacy Preserving Techniques in IoT

Existing privacy preserving techniques are classified into: data perturbation and data restriction.

Data Perturbation Techniques. These techniques are a series of operations that modify or hide some sensitive parts on the original data to preserve privacy [10]. To this end, noise addition and anonymization techniques are adopted.

Noise Addition Techniques. These techniques transform confidential attributes by adding noise to the original data to prevent the identification of a particular individual [17]. They can be categorized into four groups: (1) data sampling techniques, which aim at releasing a new table that includes only the data of a sample for the whole population, (2) random-noise techniques, which consist of adding or multiplying the value of the sensitive attribute with a randomized number, (3) data swapping techniques, which modify a subset of the data by introducing uncertainty about the true data value [22], and (4) differential privacy techniques, which consist of adding Laplace noise to a database query result [17].

Anonymization Protection Techniques. These techniques hide a data owner's identity by removing any explicit identifier and makes the data less precise. There are three well-known privacy preserving methods: k-anonymity [23], l-diversity [16] and t-closeness [15]. The k-anonymity is a formal method that is proposed to counter the re-identification problem caused by the quasi identifier attributes. However, k-anonymity can be susceptible to background knowledge attacks. Therefore, researchers designed other versions, such as l-diversity [16], the main idea of which is that there must be at least l distinct values for the sensitive attribute in each quasi identifier group as well as t-closeness method [15], which requires the distribution of a sensitive attribute in any quasi identifier group to be close to the distribution of the attribute in the overall table.

Data Restriction Techniques. These techniques aim at limiting data use by blocking access or encrypting inputs. Data restriction methods include access control and cryptography-based techniques.

Access Control. These techniques are effective for ensuring data sharing [10]. Data owners can express their individual preferences about who can access to what data and how others manipulate their shared data. Control mechanisms include Role Based Access Control (RBAC) and Attribute Based Access Control (ABAC). RBAC assigns access permissions based on the roles whereas ABAC defines permissions based on attributes, such as subject, resource and environment attributes [10].

Cryptographic Protection. These techniques are heavily used when preserving privacy. They can be categorized into three major groups: (1) secure multiparty computation, which aggregates inputs of distributed entities to produce outputs, while preserving the privacy of inputs [22], (2) asymmetric/symmetric encryption, which uses keys to protect the data, and (3) public key infrastructure, which delivers the entity a certificate to make sure that the public key belongs to the identified entity.

In recent years, the blockchain technology emerged. In fact, this technology successfully overcomes the problem related to trusting a centralized party. The first system was Bitcoin [18], which allows users to transfer securely the currency (bitcoins) without a centralized regulator. Specific nodes in the network known as miners are responsible for collecting transactions, solving challenging computational puzzles (proof-of-work) in order to reach consensus and adding the transactions in form of blocks to a distributed public ledger known as the blockchain. Since then, other projects demonstrate how these blockchains can serve in other domains, such as the Storj project [3], which is a decentralized peer-to-peer cloud storage network, and the Onename project [1], which is a distributed and secured identity platform. Blockchain technology is also used in order to address the privacy issue in the IoT domain. However, the existing blockchain-based solutions [19,27] concentrate at addressing the access control issue in the IoT applications. In fact, they adapt the blockchain by eliminating financial bitcoin and introducing new types of transactions in order to limit unauthorized access. Moreover, the examples cited above show that the existing approaches are only concerned with one phase, and generally not address the whole data life cycle.

Based on the above overview of IoT, security and privacy, we work on a systematic literature review detailed in the following section.

3 Systematic Literature Review

To analyse privacy in the IoT applications, a systematic literature review as defined in [20], has been carried out and reported in this section. Figure 1 depicts the systematic literature review process consists of five steps.

Fig. 1. Systematic literature review process [20]

3.1 Step 1: Definition of Research Scope

This step consists of defining research questions. The main goal of our study is to (i) categorize the contributions of the research carried out on privacy preserving in the IoT applications from an end to end view, (ii) discover the limitations of the existing works from a privacy principle coverage view, and (iii) verify if using new technology, such as the blockchain can overcome the existing problems of the privacy issue.

Table 1 lists our research questions.

Table 1. Research questions for our systematic literature review

Research question	Aim
RQ1: How and what are the techniques used by published papers to preserve privacy during data life cycle in the IoT application domains?	This question aims at identifying the used techniques in order to preserve privacy from an end to end view. It will also help to understand in which phase of the data life cycle privacy should be more enhanced
RQ2: What are the privacy principles that have been supported by the proposed solutions and in which architecture and life cycle phase?	This question will help to understand the actual privacy principle coverage state by the existing solutions. It will also help to identify the least considered principles that should be addressed in the future
RQ3: What are the privacy preserving techniques that have involved the privacy principle coverage and in which architecture of the literature?	The objective of this question is to know how the chosen technique can involve the privacy principle coverage in the IoT area and its effect on the architecture choice
RQ4: What are the privacy preserving techniques that have involved the security property respect and the security requirement fulfillment on the IoT applications?	This question will help to know how the chosen technique can involve the respect of security properties in the IoT area and its impact on the fulfillment of security requirements

3.2   Step 2: Research Conducting

This step consists of collecting papers from relevant electronic databases like ACM, IEEE Xplore, Science Direct, and Web of Science. We restrict our search to papers published between 2010 and 2017. Moreover, a set of keywords is chosen and used to retrieve papers from databases. Thus, we used the following query:

Internet of Things AND privacy AND
(preserving OR principle OR blockchain)

3.3   Step 3: Paper Screening

This step consists of choosing the relevant papers that would help answer the research questions. To do so, a set of inclusion and exclusion criteria are defined. Our study is restricted to papers published in English and addressing privacy issue in the IoT applications. Publications that are table of contents, foreword or summary of conference are deleted. As a result of the filtering process, we exclude 1345 publications. Table 2 summarizes the number of papers included in and excluded from each scientific database. The outcome of this step is 90 papers to include in our study.

Table 2. Number of papers included in and excluded from each database

Database	Amount	Included	Excluded
ACM	113	25	88
IEEE Xplore	271	9	262
Science Direct	945	33	912
Web of Science	106	23	83
Total	1435	**90**	1345

3.4   Step 4: Keywording Using Abstracts

This step consists of defining a classification scheme composed of facets that group frequent relevant terms that are derived from papers' abstracts. After analyzing the abstracts of papers derived from the previous steps, we consider the frequent relevant terms as dimensions. Then, we cluster the final set of dimensions in order to form the categories (i.e., facets) for our map.

Figure 2 shows our proposed facets and dimensions for our study of the privacy issue in the IoT era.

We define seven facets for our study. These facets cover the eleven privacy principles defined by the ISO standard (see Sect. 2.4). Each IoT application should follow those principles according to the privacy legislation in order to preserve privacy.

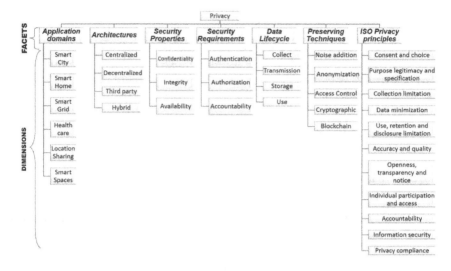

Fig. 2. Classification scheme of facets and dimensions

- **Application domains:** such as smart city, smart home, smart grid, health care, location sharing and smart space.
- **Architectures:** such as centralized, decentralized, third party and hybrid.
- **Security properties:** such as confidentiality, integrity and availability.
- **Security requirements:** such as authentication, authorization and accountability.
- **Data lifecycle:** such as collecting, transmission, storage and use phases.
- **Privacy preserving techniques:** such as noise addition, anonymization, access control, cryptography and blockchain.
- **ISO privacy principles:** such as consent and choice; purpose legitimacy; collection limitation; data minimization; use, retention and disclosure; accuracy and quality; openness, transparency and notice; individual participation and access; accountability; information security and privacy compliance.

3.5   Step 5: Data Extraction and Mapping Process

The facets are combined and the results presented in bubble charts to provide answers to our research questions. It should be noted that for the horizontal axis, the plotted values are related to the total number of publications (i.e. 90 publications) whereas the plotted values on the vertical axis are related to the horizontal axis values.

RQ1: *How and what are the techniques used by published papers to preserve privacy during data life cycle in the IoT application domains?*
To answer this question, we combine the privacy preserving techniques, the application domains, and the data life cycle facets. Figure 3 shows that cryptography (67 publications - 74.44%) is the most used privacy preserving technique in the

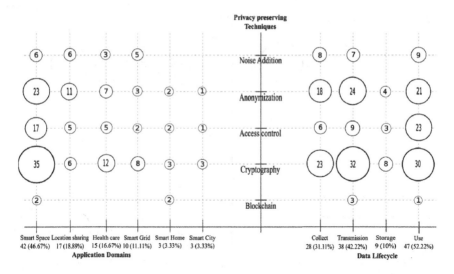

Fig. 3. Privacy preserving techniques, application domains and data lifecycle facets

current proposed solutions in each application domain. According to our study, few solutions (2.22%) are found to cover the whole data life cycle. The most addressed data phase by the studied publications is the use phase (47 publications - 52.22%) followed by the transmission phase (38 publications - 42.22%). Both of these phases are based on cryptography technique to provide data protection. The storage phase is the least addressed by publications (9 publications - 10%). It seems that with the emergence of cloud computing, solutions count on cloud data security guarantees and consider it as a trust party. Moreover, anonymization (47 publications - 52.22%) and access control techniques (32 publications - 35.56%) have also emerged in the whole data life cycle. Noise addition is the least used privacy preserving technique in the current proposed solutions for many reasons. First, noise addition leads to a significant utility loss of data [24]. Second, the obfuscated data produced by the classical obfuscation techniques are easy to be detected by an adversary [9]. Finally, noise addition requires smart devices with high storage and computation capabilities to support data storage, aggregation and communication [7].

RQ2: *What are the privacy principles that have been supported by the proposed solutions and in which architecture and life cycle phase?*
By combining the privacy principles, the architectures and the data life cycle facets (see Fig. 4), we can observe the privacy principle coverage by the different publications. In general, the existing solutions cover only six out of eleven principles. The most covered principle is the 'information security' (76 publications - 84.44%) followed by the 'use, retention and disclosure limitation' principle (57 publications - 63.33%). This can be explained by the relationship between these principles and the most addressed data life cycle phases, such as transmission and use, respectively. However, 'privacy compliance' (2 publications - 2.22%)

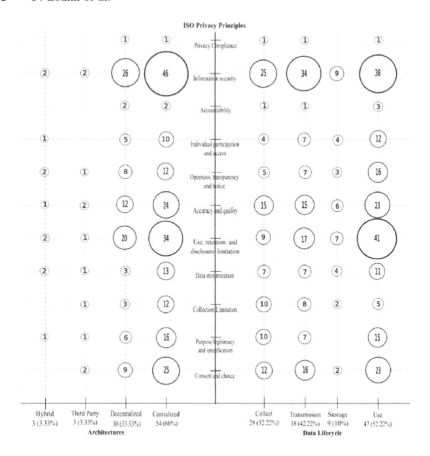

ISO Privacy Principles

Fig. 4. ISO privacy principles, architectures and data lifecycle facets

and 'accountability' (4 publications - 4.44%) are the least considered principles. The rest of privacy principles are covered by less than 50% of the publications. Moreover, both centralized and decentralized solutions involve covering all the privacy principles.

RQ3: *What are the privacy preserving techniques that have involved the privacy principle coverage and in which architecture of the literature?*

The result of combining the privacy preserving techniques, the privacy principles and the architectures facets is shown in Fig. 5. Cryptography is the dominant technique in most of the current proposed solutions that helps to cover all privacy principles. This technique is used to cover the 'information security' principle with 70% of the total publications, while the blockchain, as a decentralized solution, covers only the average of privacy principles (i.e. 6 principles). This can be explained by the life cycle coverage by this technique. In fact, blockchain is used only in transmission and use phases. It can be said that addressing the

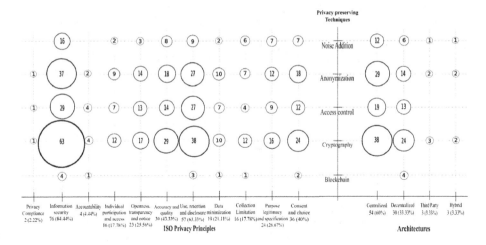

Fig. 5. Privacy preserving techniques, ISO privacy principles and architectures facets

life cycle coverage by the blockchain use can involve privacy principle coverage. Moreover, centralized is the most used architecture (54 publications - 60%) followed by the decentralized one (30 publications - 33.33%). Independently of the used techniques, these architectures suffer from several technical and legal limits, such as vulnerabilities to attacks, performance and scalability issues as well as need to purpose and data storage duration specifications.

RQ4: *What are the privacy preserving techniques that have involved the security property respect and the security requirement fulfillment on the IoT applications?*

The result of combining the privacy preserving techniques, the security properties and the security requirements facets is shown in Fig. 6. Cryptography (67 publications - 74.44%) is the most used technique in the current proposed solutions that helps to respect all security properties and fulfill all security requirements. Moreover, access control technique is used with 20% of the total publications in order to fulfill both authentication and authorization security requirements. It seems that few of the existing solutions address all the security properties. Especially availability (1 publication - 1.11%) that is the less considered property compared with confidentiality (16 publications - 17.78%) and integrity (17 publications - 18.89%).

It is worth noting that the systematic review results may have been influenced by multiple factors such as researchers' opinions, selection of databases, the used query string for search, and time constraints.

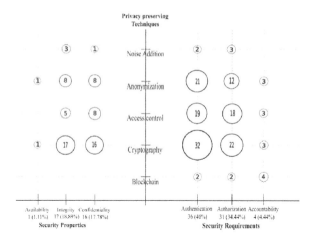

Fig. 6. Privacy preserving techniques, security properties and security requirements facets

4   Privacy and Security Concerns in the IoT Applications

According to our study, many issues in the IoT data protection and privacy preserving area are still to be dealt with. In fact, privacy should be protected in each data phase to preserve sensitive data of the data owner, who can be an individual, a group of individuals or an organization. Considering all the privacy principles defined by the ISO standard [14] is the best way to respect data privacy laws [4,5]. Nevertheless, assuring security is essential to ensure privacy.

For this purpose, we identify and suggest for IoT application consumers and designers some recommendations in order to aware them about key points for protecting data and involving privacy principle coverage from an end to end view. In our work, we distinguish two types of privacy principles: (i) general principles, such as 'accountability', 'information security' and 'privacy compliance', which need to be covered during the whole data life cycle and (ii) specific principles, including the remaining nine principles, which are bound to a particular phase.

According to our study results discussed above, addressing privacy principle coverage can be achieved by programming laws and principles into the blockchain. In fact, law enforcement can be automatically ensured by the use of smart contracts. In practice, Ethereum allows for an easy implementation of such smart contracts [8].

The next subsections identify the privacy principles for each data phase.

4.1   Privacy at Collection Time

Smart devices collect periodical data from the environment and human bodies. The sensitive information may be leaked out in case of unauthorized manipulation in these devices. For example, an attacker can reprogram a surveillance camera to gather data like the legitimate server. Thus, Privacy by Design as well as defining an appropriate authentication are important for devices that gather sensitive data in order to prevent illegal device access.

Data perturbation techniques, such as data aggregation, noise mechanism and differential privacy are the used solutions to preserve privacy in this phase.

Regarding the privacy principles, both 'consent and choice' and 'purpose legitimacy and specification' principles should be considered before beginning the collection phase. Each data owner has the right to know the reasons behind collecting each data by his smart device. Thus, the respect of these two principles can help data owner to choose his preferences about the collect frequency that can also influence privacy and data granularity he wants to disclose to third party. For instance, unlike the traditional electricity architecture in which metering data are read monthly, in the smart grid, more detailed energy data are collected. Thus, these data can expose a great amount of valuable and intimate information about the customers. Besides, specifying the reasons of collecting particular data can lead to consider the privacy principle of the 'data collection limitation'.

Preserving privacy in the collection phase is essential and can affect the whole data life cycle. Thus, privacy should be preserved before the transmission phase instead of trying to preserve it when the data are already stored in the utility data center. Thus, privacy should be preserved at the smart devices. However, smart devices generally do not support the privacy preserving techniques. For that, the use of an access point between smart devices with low memory and storage capabilities and the main data system can enable to locally store the data for pre-processing before the transmission phase. Such access point should provide a portal to the data owner to manage his smart devices and choose his preferences about how others access and manipulate his data. Thus, data owners in the IoT applications will able to keep control on their shared data and preserve their privacy. Besides, this access point should have the capability to interact with a blockchain.

4.2 Privacy at Transmission Time

After being temporally stored in the smart devices or in the broker, the collected data will be sent to external servers. Many techniques are used in order to protect the data from attacks during the transmission phase.

We notice that the most used common technique is cryptography. In our study, we distinguish three transmission types: data that are periodically transmitted, data that are transmitted as a replay to a request and the third type is the use of Publish/Subscribe system. In order to ensure data confidentiality during transmission, data encryption is absolutely required. Digital signature and certificate are also needful in order to ensure integrity and prove the identity.

In this phase, the privacy principle that needs to be covered is 'accuracy and quality'. Data should not be modified during the transmission phase. Moreover, detecting malicious entities that try to inject data in order to congest the network or influenced the analysis results is still an issue to be solved. Note that it is difficult to separate privacy and security preserving solutions in this phase. Therefore, assuring security is essential in order to ensure privacy.

The blockchain can involve privacy preserving in the IoT applications and ensure the 'accuracy and quality' principle coverage by enhancing collaboration between all entities in the network in order to verify data accuracy, integrity and reject unauthorized data access. Moreover, the blockchain offers non-repudiation principle compliance, which consists of preventing an entity from denying actions that are performed by itself since blockchain can ensure auditing functions. The main purpose of using the blockchain technology is to prevent any privacy violation attempts in the IoT applications. In fact, thanks to the immutable characteristic of the blockchain, the malicious nodes cannot modify the blockchain.

4.3   Privacy at Storage Time

After being periodically collected by devices and temporally sent through the network, data should be stored to be available for analyzing. A high storage capability is required to support the huge amount of data generated by IoT devices.

To conceal the real identity linked to the stored data, cryptography and anonymization techniques are used. Secure multiparty computation and asymmetric/symmetric are the solutions the most used by the existing approaches to preserve privacy during the storage phase. We distinguish two solutions types. The first one consists of storing encrypted data and the second solution type aims at decrypting data before the data storage. Each solution had its advantages and disadvantages. Although storing encrypted data can overcome the trust issue, data querying process will be more complicated. Contrary to the first solution, storing unencrypted data simplifies data querying by end-users (i.e., data consumers), but it necessitates giving trust to cloud computing that will not disclosure sensitive data.

Regarding the privacy principles, French and European data privacy laws state that personal data collected and stored within a European Union country territory should be stored for a reasonable time duration [4,5]. Thus, the 'use, retention and disclosure limitation' principle is to be considered in this phase. This principle aims at limiting the retention of personal information to fulfill the specified purpose as long as necessary and thereafter securely destroying data.

IoT generates a large amount of data that cannot be supported by traditional data storage solutions. For this purpose, the cloud computing seems to be the best scalable solution for storing data for many reasons. First, cloud computing offers a reduced cost. Second, it allows benefiting from scalability and high performance computing. Finally, it guarantees data security and recovery. Moreover, data cannot be altered. In fact, when storing the data hash in the blockchain, the data owner can detect any change in his stored data by comparing the hash of his data in the data center with the stored hash in his gateway. Thus, data owners can store their data without relying on a trusted Third Party Authority.

4.4   Privacy at Processing Time

Data processing is an important phase in analyzing and using the collected and stored data by end-users. The generated data in the IoT applications can be shared between multiple parties for two purposes. The first purpose is when the data owner should be known, such as billing purpose or patient's treatment. The second purpose is when the data are used for governmental programs and research. In this case, data must be anonymized.

In order to preserve privacy in query output, secure multiparty computation, anonymization, and differential privacy are used in the use phase.

According to European data privacy laws [4], data should have multiple levels of disclosure (i.e., no data sharing, restricted access, open access). Thus, without explicit acceptance of the data owner, personal data should not be disclosed to third parties. To this end, the privacy principle 'use, retention and disclosure limitation' should be considered in this phase. Furthermore, 'data minimization' principle, which aims at minimizing the processing of personal information to just fulfill the specified purpose should be more studied. Moreover, it is time that data owners be aware of their rights to have clear, complete and accessible information to correct inaccuracies. These rights are presented by 'openness, transparency and notice' and 'individual participation and access' privacy principles.

Access control techniques are necessary in order to fulfill authorization security requirement and help limit unauthorized access. However, the traditional access control models, such as RBAC and ABAC cannot support the data distribution in the IoT applications because of the lack of flexibility, scalability and usability [10]. Therefore, inspiring from the blockchain technology and proposing a permission token rather than a bitcoin can ensure a new access control solution with auditing functions and non-repudiation compliance. Thus, the blockchain-based access control can help to determine who accesses to what according to the data sensibility level (i.e., production data are less sensitive than decision support data) and under what circumstances in CIS and consequently enable data owners to own and control their shared data.

5   Related Work

Compared to security, privacy in the IoT applications has only received more attention since last year.

Fernández-Alemán et al. [11] conducted a systematic literature review concerning the security and privacy of electronic health record (EHR) systems. The authors defined and analyzed their selected articles based on several security areas. The study conclusion shows that most of the solutions defined EHR system security controls, but these are not fully deployed in actual tools.

A review of privacy threats related to the IoT applications was conducted by Ziegeldorf et al. [26]. The authors classified the evolving technologies used in the IoT applications and highlighted the most important features considered in the context of privacy. Afterwards, the authors studied and analyzed seven threat categories. Their study identified privacy-preserving approaches from related work to determine whether they could mitigate in an IoT context.

To sum up, it can be said that the existing systematic review works concerning privacy preserving issue in IoT focus on analyzing the challenges and threats of IoT in the context of entities and information flows. Our work extends the existing works by examining the IoT-specific solutions considering the security property and requirement fulfillment and studying privacy principle coverage.

6   Conclusion

Actually, IoT is considered as a promising technology that may improve collaborative working at anytime, anyplace, with anything and anyone. However, IoT opens the collaborators up to a possible loss of privacy. For this reason, both privacy and security should be carefully considered in this technology. The paper's aim is to present a detailed study about the privacy preserving solutions in the IoT applications. To achieve that, we have conducted a systematic literature review. Our analysis of the existing works helped us to identify recommendations in order to involve privacy principle coverage and security requirement fulfillment in the IoT context. We expect that our elaborated study will be an interesting contribution. In fact, researchers who want to target privacy in the IoT field can be based on our exhaustive analysis for proposing future scientific contributions that overcome existing solution limits.

In our ongoing work, we intend to propose a blockchain-based solution that takes into account our recommendations in order to preserve privacy in the IoT applications. In fact, the promising technology blockchain that successfully overcomes the problem related to trusting a centralized party in several domains, can be adapted by the IoT application designers in order to improve collaborative working in CIS and overcome privacy issue in the IoT applications.

References

1. Onename. https://www.onename.com/
2. Sap. https://www.sap.com/product/analytics/smart-meter-analytics.html
3. Storj. www.storj.io
4. Loi 78–17 du 6 janvier 1978 modifiee (1978). https://www.cnil.fr/fr/loi-78-17-du-6-janvier-1978-modifiee
5. Regulation (eu) 2016/679 of the european parliament and of the council (2016). http://eur-lex.europa.eu/eli/reg/2016/679/oj
6. Bertino, E.: Data security and privacy: concepts, approaches, and research directions. In: 2016 IEEE 40th Annual Computer Software and Applications Conference (COMPSAC), vol. 1, pp. 400–407. IEEE (2016)
7. Birman, K., Jelasity, M., Kleinberg, R., Tremel, E.: Building a secure and privacy-preserving smart grid. ACM SIGOPS Oper. Syst. Rev. **49**(1), 131–136 (2015)
8. Delmolino, K., Arnett, M., Kosba, A., Miller, A., Shi, E.: Step by step towards creating a safe smart contract: lessons and insights from a cryptocurrency lab. In: Clark, J., Meiklejohn, S., Ryan, P.Y.A., Wallach, D., Brenner, M., Rohloff, K. (eds.) FC 2016. LNCS, vol. 9604, pp. 79–94. Springer, Heidelberg (2016). doi:10.1007/978-3-662-53357-4_6

9. Elkhodr, M., Shahrestani, S., Cheung, H.: A semantic obfuscation technique for the internet of things. In: 2014 IEEE International Conference on Communications Workshops (ICC), pp. 448–453. IEEE (2014)
10. Fang, W., Wen, X.Z., Zheng, Y., Zhou, M.: A survey of big data security and privacy preserving. IETE Tech. Rev. **34**, 1–17 (2016)
11. Fernández-Alemán, J.L., Señor, I.C., Lozoya, P.Á.O., Toval, A.: Security and privacy in electronic health records: a systematic literature review. J. Biomed. Inform. **46**(3), 541–562 (2013)
12. Gartner: Internet of things (2017). http://www.gartner.com/it-glossary/internet-of-things
13. Guillemin, P., Friess, P., et al.: Internet of things strategic research roadmap. Technical report, The Cluster of European Research Projects (2009)
14. International Organization for Standardization: Information technology security techniques privacy framework, ISO/IEC 29100 (2011)
15. Li, N., Li, T., Venkatasubramanian, S.: t-closeness: privacy beyond k-anonymity and l-diversity. In: 2007 IEEE 23rd International Conference on Data Engineering, pp. 106–115. IEEE (2007)
16. Machanavajjhala, A., Kifer, D., Gehrke, J., Venkitasubramaniam, M.: l-diversity: privacy beyond k-anonymity. ACM Tran. Knowl. Disc. Data (TKDD) **1**(1), 3 (2007)
17. Mivule, K.: Utilizing noise addition for data privacy, an overview. arXiv preprint arXiv:1309.3958 (2013)
18. Nakamoto, S.: Bitcoin: a peer-to-peer electronic cash system (2008)
19. Ouaddah, A., Abou Elkalam, A., Ait Ouahman, A.: Fairaccess: a new blockchain-based access control framework for the internet of things. Secur. Commun. Netw. **9**(18), 5943–5964 (2016)
20. Petersen, K., Feldt, R., Mujtaba, S., Mattsson, M.: Systematic mapping studies in software engineering. EASE **8**, 68–77 (2008)
21. Roman, R., Zhou, J., Lopez, J.: On the features and challenges of security and privacy in distributed internet of things. Comput. Netw. **57**(10), 2266–2279 (2013). Towards a Science of Cyber Security: Security and Identity Architecture for the Future Internet
22. Sharma, M., Chaudhary, A., Mathuria, M., Chaudhary, S.: A review study on the privacy preserving data mining techniques and approaches. Int. J. Comput. Sci. Telecommun. **4**(9), 42–46 (2013)
23. Sweeney, L.: k-anonymity: a model for protecting privacy. Int. J. Uncertain. Fuzziness Knowl. Based Syst. **10**(05), 557–570 (2002)
24. Ukil, A., Bandyopadhyay, S., Pal, A.: Privacy for IoT: involuntary privacy enablement for smart energy systems. In: 2015 IEEE International Conference on Communications (ICC), pp. 536–541. IEEE (2015)
25. Westin, A.F.: Privacy and freedom. Wash. Lee Law Rev. **25**(1), 166 (1968)
26. Ziegeldorf, J.H., Morchon, O.G., Wehrle, K.: Privacy in the internet of things: threats and challenges. Secur. Commun. Netw. **7**(12), 2728–2742 (2014)
27. Zyskind, G., Nathan, O., et al.: Decentralizing privacy: using blockchain to protect personal data. In: 2015 IEEE Security and Privacy Workshops (SPW), pp. 180–184. IEEE (2015)

Self-adaptive Decision Making for the Management of Component-Based Applications

Nabila Belhaj[1]([✉]), Djamel Belaïd[1], and Hamid Mukhtar[2]

[1] SAMOVAR, Telecom SudParis, CNRS, Paris-Saclay University, Evry, France
{nabila.belhaj,djamel.belaid}@telecom-sudparis.eu
[2] National University of Sciences and Technology (NUST),
Islamabad 44000, Pakistan
hamid.mukhtar@seecs.edu.pk

Abstract. The increasing complexity of modern applications has motivated the need to automate their management functions. The applications are then able to manage themselves and meet their SLA requirements by means of autonomic MAPE-K loops based on predefined policies. However, the common use of fixed and hand-coded policies, known for being knowledge-intensive, is inadequate to dynamically changing contexts. Autonomic management should be dynamically adaptive to learn appropriate policies at runtime. Towards this direction, we propose to provide autonomic systems with learning abilities to render the decision making self-adaptive. In this paper, we propose to enrich the decision making process of an autonomic MAPE-K loop with a learning-based approach. We demonstrate the usage of learning techniques as building blocks of sophisticated and better performing autonomic systems. We have illustrated our approach with a real-world application example. The experimental results have shown a dynamic adjustment to a changing context in a shorter time as compared to existing approaches. They have also shown less frequent time spent in SLA violations during the learning phase. The approach converges faster and demonstrates higher efficiency and better learning performance.

Keywords: Autonomic computing · Self-adaptive decision making · Reinforcement Learning · Component-based applications

1 Introduction

Today's computing systems have quickly increased in size, complexity and distribution. As a result, an urgent need has developed to render the system management more automated and less dependent on human management experts. Autonomic Computing Systems (ACS) [11] have emerged to cater to the computing systems management needs. Using autonomic loops, ACS have the ability to react to changes and, therefore, manage themselves without the need of human intervention. Ideally, sophisticated autonomic systems are supposed to adapt

© Springer International Publishing AG 2017
H. Panetto et al. (Eds.): OTM 2017 Conferences, Part I, LNCS 10573, pp. 570–588, 2017.
https://doi.org/10.1007/978-3-319-69462-7_36

in dynamic contexts and learn to solve problems based on their past experiences [16]. However, due to the limited availability of past data at runtime, foreseeing system behavior for unseen events is not feasible.

In fact, when deploying ACS, experts may not have exhaustive knowledge of the context of the system and can only make use of their limited view of system states and actions for hand-coding management policies. However, achieving good performance with a decision process based on fixed and hand-coded policies is very difficult and inflexible [4]. These policies can neither envisage all the possible adaptation scenarios nor be accurate enough to reach the best possible decision. As a consequence, the approaches based on fixed policies are completely unable to adapt to a changing context, they are not able to take an appropriate decision for previously unknown situations. Such approaches entail developing accurate models of system dynamics which is typically a difficult and time-consuming task. It requires a detailed understanding of the system design to be able to predict how changes brought to resources may affect system's performance. This makes it difficult to engineer knowledge models with acceptable performance, thus, presenting a major obstacle to the widespread use of autonomic computing in deployed systems.

Reinforcement Learning (RL) [10] brings great promises in overcoming the gap in creating efficient knowledge models and offers several key advantages from ACS perspectives. First, it provides the possibility to develop optimal policies without requiring explicit model of the system. Furthermore, by its foundation in Markov Decision Processes [17], RL's underlying theory is fundamentally a sequential decision theory [13] that deals with dynamic phenomena of an environment. Therefore, RL potentially offers benefits for ACS by being part of the building blocks composing its decision process.

In the context of ACS, some research work have considered RL techniques to automate some specific management tasks (e.g., provisioning of VM resources [19]). The studies [19,24,26] have solely relied on RL-based agents to solve their management tasks in contrast to our approach where we build autonomic systems using autonomic loops combined with RL techniques. Furthermore, these work have shown a dynamic but very slow adaptation to context changes due to the usage of the one-step learning approaches. However, the usage of RL approaches remains limited in deployed systems due to the poor performance induced during the early stages of learning [4]. In fact, existing one-step approaches do not cope with large state spaces of nowadays applications. This represents a major hindrance to delivering an acceptable service quality that respects agreement terms established for an application. In our approach, not only we use RL algorithms in the decision logic of the autonomic loops, but we also use a better-performing multi-step learning approach. To the best of our knowledge, none of the existing work targeting autonomic management has proposed such approach.

In this paper, we endow traditional ACS with sophisticated and better-performing learning abilities. We propose to enhance the decision process of an autonomic loop with RL blocks to render self-adaptive the decision making for component-based applications. We demonstrate the utilization of RL algorithms by adopting a multi-step learning approach in the Analysis decision process of the autonomic loop to dynamically learn a decision policy. Through a use case study, we show how we used the multi-step learning approach in the autonomic loop to self-adapt a real-world application. In our approach, not only the system learns faster compared to one-step approaches (stated in Sect. 6), it also reduces considerably the time spent in SLA violations during learning phase.

The remaining article is structured as follows. Section 2 reviews some basic concepts related to our work. In Sect. 3, we introduce the structure of the enhanced Analysis component of the autonomic loop. We also provide details about the underlying RL theory and its mapping with our decision making problem. Through a use case study in Sect. 4, we give a description of a real-world application that we use in Sect. 5 for the evaluation of our approach. An overview of existing related work is presented in Sect. 6. Finally, we conclude the paper and provide some future work in Sect. 7.

2   Background

We give an overview of Autonomic Computing and Reinforcement Learning in order to introduce the context necessary to describe our contributions.

2.1   Autonomic Computing

Autonomic computing was introduced by IBM researchers after drawing inspiration from the human nervous system that handles unconscious reflexes (e.g., digestive functions). In 2001, IBM launched the autonomic computing initiative [7] as an attempt to intervene in computing systems in a similar fashion as its biological counterpart. The essence of this initiative is to develop systems endowed with autonomic behavior, capable of self-government and self-organization without human intervention. The ultimate aim is to render computing systems completely autonomous by means of autonomic management whereby the system's conditions remain in accordance with the established high-level objectives. An autonomic system should comply to four major self-properties [8]. *Self-reconfiguration* for an "on the fly" adaptation to the context. *Self-optimization* for an optimal performance and resource utilization. *Self-healing* for failure discovering to promote service delivery and diagnosing to prevent disruptions. *Self-protection* for identification and anticipation of unauthorized accesses and attacks protection. Among these properties, we focus on *self-reconfiguration* and *self-optimization*, as the *self-healing* and *self-protection* properties are not relevant in the current context of our work.

Bringing autonomic features to computing systems imply the usage of the so-called autonomic MAPE-K loop (Monitoring, Analysis, Planning, Execution and Knowledge) [9]. As shown in Fig. 1, the autonomic loop is associated with a system resource that is called a Managed resource. The loop is known for the classical behavior that consists of collecting monitoring data, analyzing them and in case of abnormalities, generating corrective adaptation plans then finally

Fig. 1. A standard autonomic MAPE-K loop

executing them. The Knowledge is used by the other MAPE elements to retrieve necessary information for their functioning. In most cases, this loop acts upon some hand-coded logic, embedded by management experts. Its main limitation resides in its inability to learn on its own to fit to unexpected dynamics of the systems. Consequently, we focus on *self-learning* whose aim is to reason about decision making (i.e., action selection) to provide better decision policies that fit to a changing context. It is worth noting that we raise similar issues that the Artificial Intelligence [20] field is dealing with. Such issues include autonomic decision making, learning and reasoning and are tackled by Reinforcement Learning techniques.

2.2 Reinforcement Learning

Reinforcement Learning [10] is a way of programming an agent in order to make it learn a desired behavior through trial-and-error interactions with its dynamic environment. Rewards and punishments (i.e., reinforcement signals) are distributed to the agent to guide its decision without explicitly stipulating how the task should be achieved. On each interaction step with its environment, the agent perceives an input s that indicates the current state of the environment. The agent acts by choosing an action a that changes the environment state to s'. A reinforcement signal r is communicated to the agent to indicate the utility of using the action a in the state s and is associated to the transition $(s, a) \rightarrow s'$. The agent should behave in a way that increments the long-run sum of reinforcement signal values [5].

Along the transitions, the agent actively accumulates experience about the possible states, actions, rewards and transitions of the environment which strengthen its ability to act optimally in the future. The agent's main objective is to find a decision policy π that maximizes the long-run sum of reinforcement signals, namely the optimal policy π^*. A policy π represents a function that maps states to actions whereas an optimal policy π^* maps states to the best possible actions (i.e., optimal actions).

More formally, RL draws its grounding from Markov Decision Processes (MDPs) [17]. An MDP is a mathematical formalism of decision problems that generalizes the shortest path approaches in stochastic environments. MDPs integrate the concepts of: state that summarizes the agent's situation, action (i.e., decision) that influences the state's dynamics, reward that is associated with each of the state transitions. Finally, transition probability that indicates the cognition of the context dynamics. Therefore, MDPs are Markov's chains that visit states, controlled by actions and valued with rewards. Solving an MDP is controlling the agent's behavior so that it acts optimally in order to maximize its incomes and finally optimize the system's performance [22].

3   Reinforcement Learning for Autonomic Adaptation

In this research, we associate system resources with autonomic loops whenever they needed autonomic management. A resource that is associated with a loop for its self-management is henceforth called *Managed component*. System components are heterogeneous and distributed and provide functionalities to each others to fulfill the business logic they are designed for. In this section, we propose to enhance the internal behavior of a classical autonomic MAPE-K loop. Thereby, we discuss the improvements brought to the Analysis component of the MAPE-K loop to enrich it with learning abilities. In the following, we introduce the structure of the Analysis in Sect. 3.1. Afterwards, we describe the functioning of its internal components in Sects. 3.2 and 3.3.

3.1   Structure of the Analysis Component

Classically, the Analysis component examines monitoring data and determines adaptation contexts by means of static rules it holds. Whenever an adaptation context is confirmed, the Analysis component alerts the Planning component. In the following, we propose to introduce the new Analysis structure in Fig. 2 that illustrates its internal components and their connections as well.

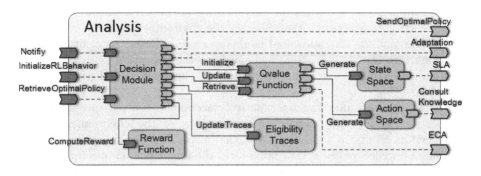

Fig. 2. Architecture of the analysis component

As shown in this figure, the Analysis is equipped with a State Space component that has the ability to generate the possible states that the system might encounter. This component requires SLA (Service Level Agreement) [2] contract to parse agreement terms for metrics extraction. The metrics and their objective values help to foresee and determine the state space values. The possible decision actions applied on the Managed component are generated by the Action Space component. To do so, it consults the Knowledge component to retrieve descriptions about the available Management services provided by the Execution component. The adaptation logic is defined by the Management services and they describe the overall self-management behavior provided by the Execution component. The Management services are about, but not limited to:

- LifeCycle: enables creation, starting, stopping, restarting and destroying a component,
- Scaling: allows insertion (i.e., scale-out) or removal (i.e., scale-in) of new instances of the Managed component,
- Adapter: permits the insertion and removal of adapters components (i.e., Load Balancer, Compressor),
- Binding: handles the functional dependencies of a component,
- Migration: manages the migration of a component from VM to another in case of deployment in Cloud environment,
- Parametric: has the ability to adjust the values of reconfigurable properties of a component.

The Reward Function component computes reinforcement signals according to the current running state of the Managed component and its improvement or degradation regarding the previous state. Its service ComputeReward is consumed by the Decision Module during learning phase at each state transition.

In fact, after state and action spaces generation, these spaces are fed to the Qvalue Function component. This latter is used during learning phase to maintain Q value function computations for each couple (s, a). Q value function represents a weighted accumulation of computed reinforcement signals. Accordingly, this component provides the services Initialize, Update and Retrieve for the Q value function. Initially, the component requires an ECA (Event Condition Action) table provided by an expert which represents an initial knowledge that may be fed to the system.

As shown in the same figure, the Analysis is enriched with a Decision Module that holds RL algorithms to learn optimal decision policies. It provides a callback Notify service to be notified from the Monitoring component about system states. It also offers InitializeRLBehavior service in order to specify the learning algorithm to be run and its related parameters as well.

Eligibility Traces component is used to enhance the performance of the Decision Module and make it learn an optimal policy faster. When activated, this component is responsible for backing up the couples (s, a) that are said to be eligible for Q value updates after a given transition (see Sect. 3.3). The suitable number of memorized couples is computed dynamically. This component offers UpdateTraces service to the Decision Module to store and update these traces.

Each time the Analysis is notified, it forwards the call to the Decision Module. This latter matches the notification to existing state in the state space then it chooses an action among the list of action space. To execute the selected action, the Decision Module requires the Adaptation service of the Execution component. During learning phase, the Decision Module requires the services ComputeReward and Update in order to use reward computations for Q value function updates. As soon as the Decision Module reaches convergence, it invokes Retrieve service to extract Q value function for each couple (s, a). Finally it deduces the optimal policy and sends it to the Planning component of the MAPE-K loop.

3.2   Mapping Decision Process to MDP Problem

We consider the decision making process of the Analysis component as a MDP problem. We formulate the decision problem as a finite MDP with a finite set of states and actions. During each state transition, the proposed decision maker obtains a reinforcement signal that indicates whether it chose the suitable action for the current state or not. In our approach, we define the relevant concepts of MDP as following:

State Space: a running state of the managed component is defined by a multi-tuples vector of monitored metric values. The state space S is a finite set of states s_i described as follows: $S = \{s_i \; / \; s_i = (v_{m_{1i}}, v_{m_{2i}}, .., v_{m_{ji}})\}$ where $v_{m_{ji}}$ is the current sensed value of metric m_j for the ith state.

A state vector should comply with some SLA requirements that are described as objective metrics for the managed component $O = \{o_{m_1}, o_{m_2}, .., o_{m_j}\}$ where o_{m_j} denotes the objective value for metric m_j explicitly required in the SLA.

Action Space: we have three types of actions:

- Elementary services A_{el}, they represent an elementary management service provided by a Management component,
- Service compositions A_{comp}, they are defined as a composition of elementary management services. They may entail local and/or remote services calls from remote autonomic loops,
- Orchestrated actions A_{BPM}, they are Business Process Management and represent an orchestration of local and/or remote elementary management services.

The action space A is a finite set of actions defined as: $A = \{A_{el} \cup A_{comp} \cup A_{BPM}\}$.

Reward Function: produces real values as feedback signals that guide the decision maker to learn the best possible behavior. Thus, it ought to explicitly punish bad action selection that leads to degraded performance, meanwhile, encouraging good action selection in the future. It also rewards good action selection that helps maintaining or reaching SLA objectives. The reward function described in Algorithm 1 computes the feedback signal based on compliance or violation degree of the new state s_t' and the system improvement or degradation degree regarding the transition from old state s_t to the new one s'_t. For a given transition t: $(s_t \rightarrow s'_t)$, each metric value $v'_{m_{kt}}$ of the new state s'_t, where $1 \leqslant k \leqslant j$ and j is the number of metrics composing a state, is compared to its value objective o_{m_k} to compute a reward distance $r_{v'_{m_{kt}}}$ (line 3). $d_{v'_{m_{kt}},o_{m_k}}$ represents a function that computes a real value as a distance between the value objective o_{m_k} and the current value $v'_{m_{kt}}$. The minimum distance of a given metric is the most significant regarding the estimation of s'_t compliance degree $C_{s'_t}$ (line 5). Compliance degrees are split into five categories to which states are assigned. A state might be: BEST, GOOD, BORDER, BAD, WORST and are listed from the most to the least desired SLA compliance degree. The value $r_{s'_t}$ is computed according to category belonging of state s'_t (line 6). We also estimate the SLA compliance degree C_{s_t} of the old state s_t (line 8). We examine the new state improvement or degradation compared to the old one by comparing the compliance degrees of transition $(s_t \rightarrow s'_t)$ to compute $r_{(s_t \rightarrow s'_t)}$ (line 9). The final reward r_t is assigned to couple (s_t, a_t) (line 11).

Algorithm 1. Reward Function

1: **Input:** s_t, s'_t

2: **for all** metrics values $v'_{m_{1t}}, v'_{m_{2t}}, .., v'_{m_{jt}}$ **do**

3: $\quad r_{v'_{m_{kt}}} = \begin{cases} 0 & \text{if } v'_{m_{kt}} = o_{m_k} \text{ (border)} \\ d_{v'_{m_{kt}},o_{m_k}} & \text{if } v'_{m_{kt}} \text{ satisfies } o_{m_k} \text{ (good or best)} \\ -d_{v'_{m_{kt}},o_{m_k}} & \text{otherwise (bad or worst)} \end{cases}$

4: **end for**

5: $C_{s'_t} \leftarrow ComplianceDegree(min\{r_{v'_{m_{1t}}}, r_{v'_{m_{2t}}}, .., r_{v'_{m_{jt}}}\})$

6: $r_{s'_t} \leftarrow ComputeReward(C_{s'_t})$

7: do loop in Step 2 for metrics values $v_{m_{1t}}, v_{m_{2t}}, .., v_{m_{jt}}$

8: $C_{s_t} \leftarrow ComplianceDegree(min\{r_{v_{m_{1t}}}, r_{v_{m_{2t}}}, .., r_{v_{m_{jt}}}\})$

9: $r_{(s_t \rightarrow s'_t)} \leftarrow CompareStates(C_{s_t}, C_{s'_t})$

10: $r_t \leftarrow r_{s'_t} + r_{(s_t \rightarrow s'_t)}$

11: *assign r_t to couple (s_t, a_t)*

12: **Output:** r_t

3.3   Online RL Decision Making

The Decision Module makes decisions at runtime with no prior knowledge about the context dynamics. This is known as online learning. More formally, at each step t, the Managed component state transits from s to s' with the unknown transition probability $P_a(s, s') = P(s'_t = s'/s_t = s, a_t = a)$ and collects the immediate reward function r_t. P_a represents the knowledge of context dynamics where $\sum_{s' \in S} p(s'/s, a) = 1$. The value function of choosing action a in state s is computed as:

$$Q(s, a) = E\left\{ \sum_{l=0}^{\infty} \gamma^l r_{t+l}/s_t = s,\ a_t = a) \right\}, \tag{1}$$

where $0 < \gamma \leqslant 1$ and denotes the discount factor that helps with the convergence of the value function. Herein, we formulate an optimization problem in which the action a picked at state s should always maximize the function $Q(s, a)$. Identifying the optimal policy π^* is similar to derive an estimation of $Q(s, a)$ that approximates its real value. Note that the real value of $Q(s, a)$ is computed when the context dynamics are known. The estimation of the value function for each couple (s_t, a_t) is then computed and updated by the end of each interaction with the equation:

$$Q(s_t, a_t) = Q(s_t, a_t) + \alpha_t(s_t, a_t)\delta_t(s_t, a_t) \tag{2}$$

For each couple (s_t, a_t), $0 \leqslant \alpha_t < 1$ and denotes the learning rate that decays asymptotically along the iterations to indicate that the Decision Module has learned enough on that couple. The temporal difference error δ_t indicates the correction of the value function estimation for this couple.

One of the recurring issues of online training is poor performance induced during the early phases of learning. The decision learner may have not acquired enough experience and may accumulate selection of random and potentially bad actions. To address this issue, we exploit the initial knowledge contained in the ECA component to determine a good value function initialization. This approach appears to hold the potential of reducing the time that the system might spend in poor performance during the first learning stages.

Algorithm 2. Learning Decision Process

1: **Input:** $A, O, ECA, \alpha_0, \gamma, \epsilon, \lambda, \epsilon_z, \xi$
2: $Initialize(Q_0) \leftarrow Parse(ECA)$
3: $z_0(s, a) \leftarrow 0, \forall(s, a)$
4: $\alpha_0(s, a) \leftarrow \alpha_0, \forall(s, a)$
5: $t \leftarrow 0$
6: **Repeat:**
7: $s_t \leftarrow MonitoreCurrentState$
8: $a_t \leftarrow ChooseAction(s_t)$
9: $ApplyAction(a_t)$
10: $s'_t \leftarrow MonitoreCurrentState$
11: $r_t \leftarrow RewardFunction(s_t, s'_t)$
12: $\delta_t(s_t, a_t) = r_t + \gamma\ max_b Q_t(s'_t, b) - Q_t(s_t, a_t)$
13: $z_t(s_t, a_t) = z_t(s_t, a_t) + 1$
14: Update $Q_t(s, a)$ and $z_t(s, a)$:
15: $(S, A)_{M_{transition}} \leftarrow (S, A)_{M_{transition}} \cup \{(s_t, a_t)\} \setminus \{argmin_{(s,a)} z_{(s,a)}\}$
16: **for all** $(s, a) \in (S, A)_{M_{transition}}$ **do**
17:     $Q_{t+1}(s, a) \leftarrow Q_t(s, a) + \alpha_t(s, a)z_t(s, a)\delta_t(s_t, a_t)$
18:     $z_{t+1}(s, a) = \gamma\lambda z_t(s, a)$
19: **end for**
20: $\alpha_t \leftarrow Decay(\alpha_t)$
21: $s_t \leftarrow s'_t$
22: $t++$
23: **Until:** All $\alpha\delta \leq \xi$
24: **Return** All $Q(s, a)$ and Retrieve π^*

As shown in Algorithm 2 (line 8), the decision making is based on function $ChooseAction(s_t)$ for action selection. This function exploits and explores for the action selection during learning phase. It exploits when it follows the best policy obtained so far, while it explores when it chooses randomly an action that has never explored before. Exploration is helpful to refine the existing policy but risky since it can compromise the system's performance. To preserve it then, this function follows a $\epsilon - greedy$ policy (i.e., best policy known so far) to control the exploration degree. It uniformly picks a random action with a small probability ϵ and follows the best policy known so far for the rest of time $1 - \epsilon$.

However, Temporal Difference methods [22], present the lack of updating the value function of the current couple (s, a) once per step. By doing so, the convergence remains very slow especially when the states and actions are large which potentially increases time spent in SLA violation. This represents a major hindrance to scale to large states and actions spaces, which limits the use of these methods in large and complex computing systems. To tackle these issues, we propose to use a combination of Temporal Difference methods and Monte Carlo methods [22]. These allow the estimation of value function of a state after a trajectory (i.e., episode of N transitions) simulation. The update of value function occurs at the end of each observed trajectory. The combination of both methods implies an update of Q value function estimation in incremental fashion but in many steps instead of a single step.

Consequently, we propose to endow the Decision Module with memory dedicated to propagation of value function, namely eligibility traces [23]. As shown in Algorithm 2, when the error $\delta_t(s_t, a_t)$ is computed, this latter is propagated back to the previous and freshly visited couples (s, a) to perform several updates on their Q value functions. Eligibility traces $(S, A)_{M_{transition}}$ indicate the set of backed up couples (s, a) eligible for update. They are computed by their coefficient trace $z_t(s, a)$ that is incremented to indicate some couple recentness. Afterwards, the Q value functions of eligible couples are updated. Their traces are then decayed by $\gamma\lambda$, where $0 < \lambda \leqslant 1$, to indicate that their freshness is reduced. All traces $z_t(s) < \epsilon_z$ are initialized to 0 where $0 < \epsilon_z < 1$. Recent visited states with the transition memory $M_{transition} = \frac{log(\epsilon_z)}{log(\gamma\lambda)}$ are maintained. Beyond this value, traces are initialized and corresponding couples are removed from memory. The algorithm converges when the correction $\alpha\delta$ brought to value function $Q(s, a)$ (line 17) for all couples reaches some convergence criterion ξ (line 23).

4   Use Case Study

In order to validate our proposal, we implemented an application that allows users to consult products on the Internet. The application is deployed in the cloud and is represented by the *Products Composite* depicted in Fig. 3. This composite allows web users to fetch and consult products through an on-line *Catalog* that brings products information from multiple inventories. The user has the ability to order their favorite products as well. Once called, the Consult

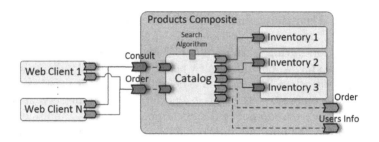

Fig. 3. Component-based description of products composite

service loads descriptions and pictures of products that match the fetched product. In order to fulfill service execution, this component runs a search algorithm to process products information required from multiple inventories as well as users' information. The search algorithm represents the property of the *Catalog* as show in Fig. 3.

The SLA terms related to this component concerns particularly Consult service, since it is heavily requested. In order to deliver good service quality to users, guarantee terms are established in a SLA, part of which is shown in Listing 1.1. We use our custom SLA parser to inspect the metrics objectives. The parsed metrics need to be monitored as they represent the most significant metric targets to determine the running state of the *Catalog* component.

Listing 1.1. SLA extract guarantee terms for Catalog component

```
1  <?xml version="1.0" encoding="UTF-8"?>
2  <wsag:AgreementOffer AgreementId="ab97c409">
3  <wsag:Name>xs:ShoppingApp</wsag:Name>
4  <wsag:AgreementContext> ... </wsag:AgreementContext>
5  <wsag:Terms>
6    <wsag:All>
7      <wsag:GuaranteeTerm wsag:Name="g1" wsag:Obligated="
           ServiceProvider">
8      <wsag:ServiceLevelObjective>
9        <wsag:KPITarget>
10       <wsag:KPIName>availability_UpTimeRatio</wsag:KPIName>
11         <wsag:CustomServiceLevel xmlns:xsi="http://www.w3.org/2001/
               XMLSchema-instance" xmlns:exp="http://www.telecom-
               sudparis.com/exp">
12           <exp:Greater><exp:Variable>minthreshold</exp:Variable> <
               exp:Value>0.9</exp:Value></exp:Greater>
13         </wsag:CustomServiceLevel>
14       <wsag:KPIName>averageResponseTime</wsag:KPIName>
15         <wsag:CustomServiceLevel xmlns:xsi="http://www.w3.org/2001/
               XMLSchema-instance" xmlns:exp="http://www.telecom-
               sudparis.com/exp">
16           <exp:Less><exp:Variable>maxthreshold</exp:Variable> <
               exp:Value>1200</exp:Value></exp:Less>
17         </wsag:CustomServiceLevel>
18       <wsag:KPIName>numberServiceCalls</wsag:KPIName>
19         <wsag:CustomServiceLevel xmlns:xsi="http://www.w3.org/2001/
               XMLSchema-instance" xmlns:exp="http://www.telecom-
               sudparis.com/exp">
20           <exp:Less><exp:Variable>maxthreshold</exp:Variable><
               exp:Value>10</exp:Value></exp:Less>
21         </wsag:CustomServiceLevel>
```

```
22            </wsag:KPITarget>
23          </wsag:ServiceLevelObjective>
24          ...
25        </wsag:GuaranteeTerm>
26      </wsag:All>
27    </wsag:Terms>
28  </wsag:AgreementOffer>
```

We define the state of the *Catalog* component as $(v_{avai}, v_{resp}, v_{calls})$, where $v_{avai}, v_{resp}, v_{calls}$ denote runtime values of service availability, average response time and number of service calls, respectively. The SLA stipulates that:

- $v'_{avai} = true$; service must be available 90% of time,
- $v'_{resp} \leqslant 1200$ ms; average response time must not exceed 1200 ms,
- $v'_{calls} \leqslant 10$; 10 simultaneous service calls at most can be handled by the component.

Parsing the SLA helps to determine all the possible values for a given metric. Subsequently, it allows the generation of state space by realizing all possible combinations of metrics values. For instance, the Availability metric is represented by boolean values while the average response time and the number of service calls metrics are represented by real values. Thus, the boolean metric values are: $\{true, false\}$ whereas the real metrics values are discretized by the parser according to an interval determined by the expert.

For the action space generation, a list of actions is generated for each available Management service provided by the Execution component. To do so, the Analysis component requires information from the Knowledge component. For instance, to generate actions related to Scaling component. The Analysis consults the Knowledge to retrieve the services provided by its Managed component (MC) *Catalog*. The Analysis generates two actions: scale-out and scale-in and both actions represent service compositions. For instance, the scale-out action is composed of the following sequence of elementary actions, executed in this order: create second instance of MC → start second instance → create loadBalancer → bind MC to loadBalancer → bind second instance to loadBalancer → bind consumers to loadBalancer. It is worth noting that in this paper, we do not deal with conflicting adaptation actions, since a single decision learning process is run at a time.

After generation of state and action spaces, the expert inserts in the Analysis, the ECA depicted in Table 1. It is parsed to identify the ECA couples (s, a). Finally the parsed couples are deployed in the Decision Module to constitute the initialization of the Q value function.

To illustrate the context dynamics and corresponding reward computation, let's take the example of transition $t = (s_t, a_t) \rightarrow s'_t$, where $s_t = s = (true, 1350, 14)$. Suppose that the Decision Module at this transition follows the current $\epsilon - greedy$ policy, its decision would be $a_t =$ scale-out MC. As shown in Fig. 4, depending on the context, the MC state might be improved (e.g., $s'_t = s'_1 = (true, 995, 8)$) or degraded (e.g., $s'_t = s'_2 = (false, 0, 0)$) in case the MC breaks down for some reasons. State s violates the SLA because of the last

Table 1. ECA table of catalog component

Events with conditions	Action
$v_{avai} = false$	Restart component
$v_{resp} > 1200$ and $v_{calls} > 10$	Scale-out component
$v_{resp} > 1200$ and $v_{calls} < 10$	Migrate component to VM with better performance

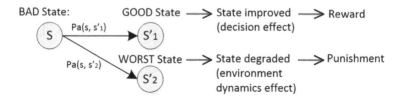

Fig. 4. Decision effect vs context dynamics during a transition

two metrics. According to Algorithm 1, the last two values give negative distances compared to the metrics objectives. Subsequently, it is assigned to BAD state category. If the state transits to s'_1 this latter is assigned to GOOD category which means that the action has improved the current state. Real reward values are shown during learning phase until convergence in Sect. 5. The Decision Module should deal with the context dynamics and learn about its decisions to dynamically compute the best possible policy.

5   Evaluation

Based on the use case study, we evaluate self-adaptivity, performance optimization and SLA guarantee of the Analysis component throughout workload variations. The workload is determined by the number of concurrent clients' calls. Three intensities of workload are mixed during training: light, medium and heavy. The context dynamically changes its workload intensity due to the concurrent clients' service demands.

We performed the experiments several times with random starting states of the *Catalog* component. The first set of experiments is depicted in Fig. 5 in which we implemented the one-step RL algorithm (i.e., without eligibility traces) used by some approaches cited in the related work section. To compare them with the multi-step RL algorithm (i.e., with eligibility traces), we realized the second set of experiments shown in Fig. 6.

Both Figs. 5 and 6 illustrate runtime values $v_{avai}, v_{resp}, v_{calls}$ at a transition number of the three metrics targeted in the use case study. They also depict the running reward signals that reflect the runtime state of the *Catalog* component. These signals are positive whenever the current state is improved and negative whenever it is degraded regarding the values required in the SLA Listing 1.1. The reward function rewards or punishes according to the degree of compliance or violation of the SLA which helps guiding the algorithms in their action selection.

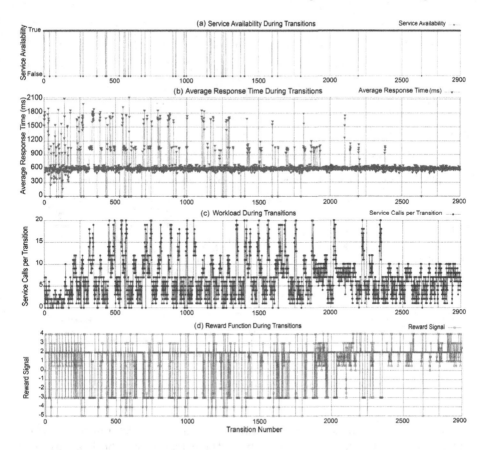

Fig. 5. Performance Metrics during one-step online learning phase

During the first learning phases, for the one-step algorithm, we note that the metrics runtime values violate the SLA terms frequently before starting to space out these violation intervals. This occurs around the 1800th transition for service availability (Fig. 5a), the 1400th transition for average response time (Fig. 5b) and the 1800th transition for service calls (Fig. 5c). In contrast, the multi-step algorithm started reducing its time spent in SLA violation around the 100th transition for service availability (Fig. 6a), the 150th transition for average response time (Fig. 6b) and the 200th transition for service calls (Fig. 6c). These results show that both algorithms start to self-adapt to the workload dynamics and the unexpected service break downs. Over the iterations, after acquiring enough experience on their environment, both algorithms learn how to avoid bad action selection.

The training session lasted for about 2420 transitions before the one-step algorithm comes to convergence. This is demonstrated by the *Catalog* runtime state that has stabilized as shown in the metrics values of Fig. 5. In contrast, the multi-step algorithm reaches convergence after about 320 iterations which

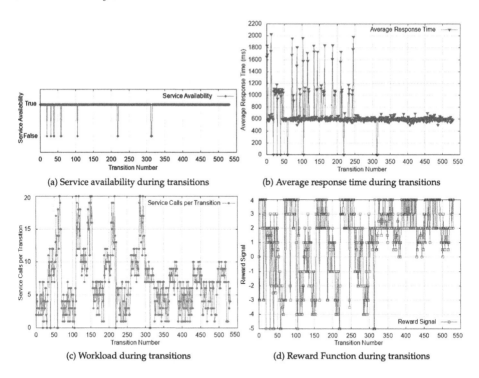

(a) Service availability during transitions

(b) Average response time during transitions

(c) Workload during transitions

(d) Reward Function during transitions

Fig. 6. Performance metrics during multi-step online learning phase

is proven by the stabilized runtime state starting from this transition. After convergence, we notice that starting from these transitions, both algorithms accumulate positive reward signals (Figs. 5d and 6d) due to the optimal decision actions they make.

The experiments were also repeated for several initial learning rates. Figure 7 shows that the number of transitions needed for convergence is largely reduced when the eligibility traces are activated. During learning phase, the average time required by the Analysis for its decision making is 0.0545 ms while the standard deviation equals 0.115 ms.

We conclude that, in either case, RL algorithms show their effectiveness regarding their self-adaptivity to the changing context. However, an efficient meeting of quality requirements is hard to achieve for one-step approaches. Spending a long time in SLA violation is not tolerated for real-world applications. As the state and action spaces are large, the one-step algorithm needs more time to visit all couples (s, a) several times before it actually learns a good quality policy.

The experimental results show clearly an efficient learning phase for the multistep algorithm. Not only our proposed algorithm can self-adapt to its context dynamics but it is also able to learn faster to optimize the application performance. Consequently, it reduces considerably the time spent in SLA violation

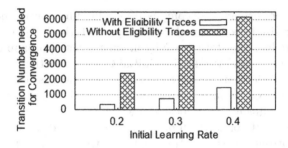

Fig. 7. Convergence speed: one-step vs. multi-step online learning

even during the learning phase. We deduce that the multi-step learning approach performs better than the one-step approaches and quickly learns an optimal policy. It allows a fast self-adaptation to context dynamics for applications especially with large state and action spaces.

6   Related Work

Many research work have devoted to system performance management. In cloud environments, SLA violation management were addressed in [14,15] to improve web servers response time by allocating resources (e.g., memory, CPU, etc.) and service-oriented applications management in [21]. These approaches rely on predefined and implemented stationary reconfiguration rules to react to SLA infractions.

There were other approaches based on feedback control theory for data center resource management. In [18], authors targeted optimization of power consumption, cooling and performance management over different resource levels (i.e., virtual machines, servers, clusters, etc.). Another work stemming from control theory for the management of component-based systems is presented in [6]. In this work, authors equip an autonomic manager with a model of its behavior described as a behavioral contract. Such approaches are based on predefined explicit model of the system which is laborious and knowledge-intensive. Stationary behavioral rules are also adopted in [3] for the management of smart environments. Similarly, in [12] rules are used to perform service compositions for the execution of applications workflow.

The aforementioned approaches tend to create controllers that use hardcoded policies known for their reactive behavior but deprived from reasoning abilities to self-adapt their behavior in highly dynamic real-world contexts. In contrast, our approach proposes a model-free and learning-based approach for self-adapting to previously unseen situations and to reason for dynamic optimal policy inference.

Recent learning-based research work show early promises in using RL techniques in autonomic contexts. Work presented in [24] addresses the problem of dynamic server allocation to applications in data centers. In order to optimize the applications performance, decomposed Q value function learners were used but with a centralized decision maker. To cope with the RL poor performance, authors had to reduce the state space by making strict approximation to represent the applications state. Pre-learned initial Q value function was also claimed to address poor performance and convergence slowness issues. Another work [19] targets virtualized resource provisioning in cloud environments. A distributed learning mechanism was proposed in which there was a dynamic tuning of CPU, memory and I/O bandwidth of VMs. Authors used the Cerebellar Model Articulation Controller (CMAC) [1] to represent the Q value function. However, their one-step learning approach remains less effective compared to the enhanced multi-step learning approach. Another study [26] showed the feasibility of RL approaches in adaptive service composition. The work targets achieving dynamic adjustment of service composition for web services based on some QoS attributes to ensure user satisfaction. Similarly, a recent work [25] targets the dynamic optimization of web service compositions. They propose a distributed learning algorithm that splits a task into sub-tasks, each one is treated by an agent to speed up the convergence. They tested their approach in a simulation web service environment which showed its effectiveness. However, they used a standard one-step learning algorithm that could have been improved with eligibility traces for better performance.

In contrast to these learning-based approaches, in our work, the decision and learning logics are both embedded in the Analysis component and are dedicated to each Managed component of the application. They are also separated among different components to allow an isolated replacement or removal to ensure more flexibility and reusability to the Analysis component. In addition, the combination of Monte Carlo and Temporal Difference methods show very promising results in accelerating convergence time, especially for large state and action spaces. Subsequently, the time that might be spent in SLA violation is reduced during learning phase. Thus the system can scale for larger state and action spaces which is hardly achievable with one-step approaches.

7   Conclusion and Future Work

This paper proposes to improve the decision making process of a traditional MAPE-K loop. Instead of using inflexible hand-coded strategies, we equipped the Analysis component with sophisticated learning blocks for a dynamic adaptation of component-based applications. We modeled the decision problem of the Analysis component as a Markov Decision Process. Thereafter, we associated each MAPE-K loop with a component of the application for its self-management. The decision making algorithm was guided by the reward signals to learn from its past experiences and be able to compute the optimal policy at runtime. In this work, we have equipped each MAPE-K loop with a multi-step learning approach for the management of component-based applications. To our knowledge, none of the existing work targeting application management has proposed such approach so far. Our research objectives are mainly focused on rebuilding traditional autonomic computing systems with sophisticated and better-performing learning blocks to render the decision making self-adaptive. Towards this objective, we extend our current work to propose a generic framework that provides management experts with tools that facilitate building autonomic systems with self-adaptive behavior.

References

1. Albus, J.S.: A New Approach to Manipulator Control: the Cerebellar Model Articulation Controller (CMAC) (1975)
2. Bianco, P., Lewis, G., Merson, P.: Service Level Agreements in Service-Oriented Architecture Environments. Technical report (2008)
3. Cano, J., Delaval, G., Rutten, E.: Coordination of ECA rules by verification and control. In: Kühn, E., Pugliese, R. (eds.) COORDINATION 2014. LNCS, vol. 8459, pp. 33–48. Springer, Heidelberg (2014). doi:10.1007/978-3-662-43376-8_3
4. Tesauro, G.: Reinforcement learning in autonomic computing: a manifesto and case studies. IEEE Internet Comput. **11**, 22–30 (2007)
5. Gerard, P.: Apprentissage par Renforcement: Apprentissage Numérique (2008)
6. Gueye, S.M.K., de Palma, N., Rutten, E.: Component-based autonomic managers for coordination control. In: De Nicola, R., Julien, C. (eds.) COORDINATION 2013. LNCS, vol. 7890, pp. 75–89. Springer, Heidelberg (2013). doi:10.1007/978-3-642-38493-6_6
7. Horn, P.: Autonomic Computing: IBM's Perspective on the State of Information Technology. Technical report (2001)
8. Huebscher, M.C., McCann, J.A.: A survey of autonomic computing - degrees, models, and applications. ACM Comput. Surv. **40**, 7 (2008)
9. IBM: An Architectural Blueprint for Autonomic Computing. Technical report (2005)
10. Kaelbling, L.P., Littman, M.L., Moore, A.P.: Reinforcement learning: a survey. J. Artif. Intell. Res. **4**, 237–285 (1996)
11. Kephart, J.O., Chess, D.M.: The vision of autonomic computing. Computer **36**, 41–50 (2003)
12. Li, Z., Parashar, M.: A decentralized agent framework for dynamic composition and coordination for autonomic applications. In: DEXA Workshops (2005)
13. LaValle, S.M.: Sequential decision theory. In: Planning Algorithms, Cambridge University Press, Cambridge (2006). Chap. 10. http://planning.cs.uiuc.edu/ch10.pdf
14. Mola, O., Bauer, M.: Collaborative policy-based autonomic management: in a hierarchical model. In: CNSM (2011)
15. Mola, O., Bauer, M.: Towards cloud management by autonomic manager collaboration. IJCNS **4**, 790 (2011)
16. Nami, M.R., Bertels, K.: A survey of autonomic computing systems. In: The 3rd International Conference on Autonomic and Autonomous Systems (2007)
17. Puterman, M.L.: Markov Decision Processes: Discrete Stochastic Dynamic Programming. Wiley, New York (1994)
18. Raghavendra, R., Ranganathan, P., Talwar, V., Wang, Z., Zhu, X.: No "power" struggles: coordinated multi-level power management for the data center. In: ASPLOS'13 (2008)
19. Rao, J., Bu, X., Wang, K., Xu, C.Z.: Self-adaptive provisioning of virtualized resources in cloud computing. In: SIGMETRICS 2011 (2011)

20. Russell, S.J., Norvig, P.: Artificial Intelligence: A Modern Approach. Prentice Hall, Upper Saddle River (2003)
21. Ruz, C., Baude, F., Sauvan, B.: Component-based generic approach for reconfigurable management of component-based SOA applications. In: 3rd International Workshop on Monitoring, Adaptation and Beyond (2010)
22. Sigaud, O., Buffet, O., Dutech, A.E.A.: Processus décisionnels de Markov en intelligence artificielle (2008)
23. Sutten, S.R., Barto, A.G.: Reinforcement Learning : An Introduction. Bradford Book, MIT Press, Cambridge (1998)
24. Tesauro, G.: Online resource allocation using decompositional reinforcement learning. In: Proceedings, AAAI 2005 (2005)
25. Wang, H., Wang, X., Hu, X., Zhang, X., Gu, M.: A multi-agent reinforcement learning approach to dynamic service composition. Inf. Sci. Int. J. **363**, 96–119 (2016)
26. Wang, H., Zhou, X., Zhou, X., Liu, W., Li, W., Bouguettaya, A.: Adaptive service composition based on reinforcement learning. In: Maglio, P.P., Weske, M., Yang, J., Fantinato, M. (eds.) ICSOC 2010. LNCS, vol. 6470, pp. 92–107. Springer, Heidelberg (2010). doi:10.1007/978-3-642-17358-5_7

On the Bitcoin Limitations
to Deliver Fairness to Users

Önder Gürcan[✉], Antonella Del Pozzo, and Sara Tucci-Piergiovanni

CEA LIST, Point Courrier 174, 91191 Gif-sur-Yvette, France
{onder.gurcan,antonella.delpozzo,sara.tucci}@cea.fr

Abstract. While current Bitcoin literature mainly focuses on miner
behaviors, little has been done to analyze user participation. Because
Bitcoins users do not benefit from any incentive, their participation in
the system is conditional upon system ability to provide a transactional
service at a reasonable cost and acceptable quality. A recent observed
trend on a growing number of unconfirmed transactions seems, however,
to substantiate that Bitcoin is facing service degradation. The objective of this paper is to shed some light on user participation in Bitcoin
against a notion of system fairness, through a utility-based approach.
We first introduce fairness to quantify the satisfaction degree of participants (both users and miners) with respect to their justified expectations
over time. We then characterize user strategies, deriving the necessary
condition for fairness, and we show Bitcoin limitations in delivering it.
The utility-based model allows to finally draw conclusions on possible
improvements for fairness to promote user participation.

Keywords: Bitcoin · Blockchain · Fairness · User expectation

1 Introduction

The blockchain protocol, introduced by Nakamoto [13], is the core of the Bitcoin-
like decentralized cryptocurrency systems. Participants following this protocol
can create together a distributed economical, social and technical system where
anyone can join (or leave) and perform transactions in-between without neither
needing to trust each other nor having a trusted third party. It is a very attractive
technology since it maintains a *public*, *immutable* and *ordered* log of transactions
which guarantees an *auditable* ledger accessible by anyone.

Technically speaking, all participants of a blockchain system store unconfirmed transactions in their memory pools and confirmed transactions in their
blockchains. Participants called users create transactions with a fee and then
broadcast them across the blockchain network for being confirmed. After receiving a certain number of transactions, participants called miners try to confirm
them as a block by solving a computational puzzle (a hash-based proof-of-work
- PoW) of pre-defined difficulty by consuming a considerable amount electricity
power. Note that, each miner tries to confirm a different block, with its own set

© Springer International Publishing AG 2017
H. Panetto et al. (Eds.): OTM 2017 Conferences, Part I, LNCS 10573, pp. 589–606, 2017.
https://doi.org/10.1007/978-3-319-69462-7_37

of transactions, put into its own order. The successful miner broadcasts its block to the network to be chained to the blockchain. This block contains the transactions, the unique hash value (the solution of the puzzle) and a block reward, which is composed of a static block reward plus the total fees of transactions, expressed as a transaction to the successful miner.

To analyze the security properties of Bitcoin-like blockchains [13], several formal studies have been conducted so far [7,8,14,18]. Garay et al. [8] showed that the number of blocks created by honest miners is proportional to their fraction of computational powers if majority of the computational power belongs to them, assuming that there are honest and Byzantine miners. Eyal and Sirer [7] and Sapirstein et al. [18] showed that even if the majority of the miners are honest, a selfish miner having enough resource and good network connectivity can increase its proportion of blocks, assuming that there are honest and selfish miners. These studies conclude that Bitcoin-like blockchains are not promoting honest participation. In such context, Pass et al. [15] provided a first definition of *fair* protocol for miners: if, in any sufficiently long window of time, honest miners create blocks proportionally to their computational powers then the protocol is fair.

However, as well as miners, the participation of users is also important. Without users, miners will have no transactions to confirm[1]. In this sense, we claim that a comprehensive definition of *fairness* is crucial for improving overall participation. Both users and miners consider worthwhile to join and stay over time in the system only if they find it fair. For instance, miners find the system fair if they are able to create blocks as they expected, and users find the system fair if their transactions are confirmed as they expected. Hence, fairness can be defined as the satisfaction of expectations of participants to a certain degree.

The contributions of this paper are as follows:

- A detailed basic model for Bitcoin blockchain;
- An elaborated rational system model where all participants (users and miners) are modeled as rational agents;
- A formal definition of *fairness* with respect to the utilities of rational agents;
- A *necessary condition* to have user fairness;
- An analysis of the Bitcoin protocol limitations in delivering fairness to *users*.

The paper is organized as follows. Section 2 gives the related work about fairness in blockchain systems. Section 3 provides a formalization of the existing Bitcoin protocol. Section 4 defines the rational agent model and proposes a definition of *fairness* related to rational behaviors for each type of participant (users and miners). Moreover, a necessary condition for fairness provided to the users is presented. Section 5 presents an analysis of the agent behaviors and finally, Sect. 6 concludes the paper.

---

[1] Technically the miners can create empty blocks and get block rewards. But this is not the purpose of blockchain systems.

2 Related Work

In the rational agents context, two concepts play a key role: *"incentive compatibility"* and *"fairness"*. Informally, a protocol is incentive compatible if rational nodes have incentives to follow it, while a protocol is fair if rational nodes are satisfied by executing the protocol actions, i.e., they have a profit acceptable for them.

Eyal et al. [7] show that the mining protocol itself is not incentive compatible: a miner that follows the protocol has a lower gain than a miner that does not. To such purpose they define a new attack strategy, the *selfish mining* in which colluding miners obtain a revenue larger than their fair share withholding new created blocks. Hereafter, other works analyzed scenarios where for miners it is more profitable to be selfish than to correctly follow the protocol (cf. [5,9,18], just to cite a few). The blockchain protocol does not rely only on the mining task but also on the information (transactions, blocks, etc.) flooding. As discussed by Babaioff et al. [2], rational agents have not incentives to forward information.

Contrarily to incentive compatibility, less research has been done concerning fairness in the Bitcoin-like systems. In [10], the authors present an inclusive protocol and discuss about its fairness as related to proportion between the miner hashing power and their rewards (a similar concept appears in [6]). Finally, Pass and Shi [15] present a formal concept of *"fairness"* and a protocol that provides it. Informally they propose a protocol where *"honest players contributing a ϕ fraction of the computational resources get a ϕ fraction of the blocks (and thus rewards) in a sufficiently long window"*.

All those works consider the strategies that rational miners can put in place with respect to the actual Bitcoin situation. Actually the Bitcoin protocol provides two incentives for a miner that solve the PoW, the static block reward and the transaction fees. Let us recall that the static block reward is periodically halving to eventually disappear. Carlsten et al. [3] analyze what would be more profitable for miners in a scenario where the transaction fees dominate the static block rewards. To do so they stated the decisions that miners can take at each time instant as: which block to extend, how many of the available transactions include in the block and for each unpublished block, whether or not to publish it. As a result, the Bitcoin will be an unstable system where selfish mining performs even better than in the current scenario advocating that block reward plays a key role for the stability of the system.

Carlsten et al. [3] is the closest to our contribution, not only for the scenarios considered, but also for the analysis on the decisions that miners can take. On our side we considered not only the miners' but also the users' actions in order to model the expectations and rewards of those participants and provide a definition of fairness general enough for both of them. At the best of our knowledge the only notion of fairness applied to the user side concerns the fair exchange in the e-commerce context [1] which is extended to the Bitcoin-Like scenario in [11] more in the sense that if there are two players performing an exchange then either both of them get what they want or none of them. All those fairness definitions apply globally to the system, contrarily, in our work we provide a local definition of fairness looking at the perceptions that users have about the system fairness.

3   Basic Blockchain System Model

In this section we provide a high-level Bitcoin protocol description[2]. To this end, we first introduce a basic model for each element involved in the protocol and then we provide a high-level detailed pseudocode.

3.1   Network Model

We model the blockchain network as a dynamic directed graph $G = (N, E)$ where N denotes the dynamic node (vertex) set, E denotes dynamic directed link (edge) set. A node n can enter and leave G by using its $join(G)$ and $leave(G)$ actions respectively. Upon joining G, n discovers neighbor nodes to connect to[3]. A link $\langle n, m \rangle \in E$ represents a directed link $n \rightarrow m$ where $n, m \in N$, n is the owner of the link and n is the neighbor of m.

A node n can communicate with a set of recipient nodes R_n (where $\forall m \in R_n | \{n, m\} \in E$) by exchanging messages of the form $\langle n, msg, d \rangle$ where n is the sender, msg is the type and d is the data contained.

3.2   Node Model

Each node $n \in N$ has a list of its neighbors N_n where $N_n \subseteq N$ and $\forall m \in N_n | \langle n, m \rangle \in E$. A node n adds and removes another node m as its neighbor using its $addNeighbour(m)$ and $removeNeighbour(m)$ actions respectively. Each neighbor $m \in N_n$ is represented as a 3-tuple $\langle m, \times_m, t_m \rangle$ where $\times_m$ is the ban score and t_m is the last communication time. t_m is updated at each message receipt and if m has not communicated for more than some time, m is removed from N_n.

Each node n has a memory pool Θ_n in which it keeps unconfirmed transactions that have input transactions, an orphan pool $\bar{\Theta}_n$ in which they keep unconfirmed transactions that have one or more missing input transactions (orphan transactions) and a blockchain ledger B_n in which they keep confirmed transactions where $\Theta_n \cap \bar{\Theta}_n = \emptyset$, $\Theta_n \cap B_n = \emptyset$ and $\bar{\Theta}_n \cap B_n = \emptyset$ always hold.

---

[2] This description is based on the Nakamoto paper [13], the Bitcoin source code (https://github.com/bitcoin/bitcoin), bitcoin.org (https://bitcoin.org/) and the Bitcoin StackExchange forum (https://bitcoin.stackexchange.com/).

[3] https://bitcoin.stackexchange.com/questions/53938/how-does-one-node-connect-to-other-nodes, last access 30 May 2017.

3.3 Miner Model

A (user) node n can turn to be a miner node if it chooses to create blocks for confirming the transactions (mining) in its memory pool Θ_m. It can start and stop mining using the actions of the form $startMining()$ and $stopMining()$ respectively, and n is said to be a miner node if it started mining but has not stopped yet. The set of miner nodes is then denoted by M where $M \subseteq N$. In order to be able to mine, $n \in M$ has to solve a cryptographic puzzle (i.e. Proof of Work) using its hashing power[4] q_n where $q_n > 0$. The first successful miner is awarded by a fix amount of reward plus the total fee of the transactions.

Algorithm 1. The actions of a miner node m.

```
1:  action startMining()          14:
2:    ω ← true                     15:  action createBlock(i, θₙ)
3:    createBlock()                16:    Σf ← calculateTotalFee(θₙ)
4:                                  17:    txc ← createTransaction(n, R + Σf)
5:  action stopMining()           18:    Ψᵢ ← createMerkleTree(txc, θₙ)
6:    ω ← false                    19:    t_bᵢ ← getTime()
7:                                  20:    hᵢ = ⟨vₙ, H(hᵢ₋₁), H(hᵢ), t_bᵢ, ηᵢ, ψᵢ⟩
8:  action createBlock()          21:    try
9:    while (ω) do                 22:      find ηᵢ such that H(hᵢ) < μ holds
10:     i ← |B*ₙ| + 1              23:      bᵢ = {hᵢ, Ψᵢ}
11:     θₙ ← selectTransactions(Θₙ) 24:      processBlock(∅, bᵢ)
12:     createBlock(i, θₙ)         25:    catch (i = |B*ₙ| ∨ ω = false)
13:   endwhile                     26:
```

The actions performed by miners are reported in details in the Algorithm 1.

3.4 Blockchain Model

We model the blockchain ledger of a node n as a dynamic append-only tree $B_n = \{b_0 \xleftarrow{r_0} b_1 \xleftarrow{r_1} \ldots \xleftarrow{r_{h-1}} b_h\}$ where each block b_i $(0 < i \le h)$ contains a cryptographic reference r_{i-1} to its previous block b_{i-1}, $h = |B_n|$ is the depth of B_n, b_0 is the root block which is also called the *genesis block* and b_h is the furthest block from the genesis block which is referred to as the *blockchain head*.

A block b_{i-1} can have multiple children blocks, which causes the situation called a *fork*. The *main branch* is then defined as the longest path h from any block to b_0 and is denoted as $B_n^\star$ where $|B_n^\star| = h$ and $B_n^\star \subseteq B$ such that $|B_n^x| < |B_n^\star|$ for all branches $B_n^x \subset B_n$ where $B_n^x \ne B_n^\star$. All branches other than the main branch are called *side branches*. If at any time, there exists more than 1

---

[4] Hashing power is proportional to computational power and nodes may change this power by time.

longest path with a depth h (i.e. there are multiple heads), the blockchain ledger B_n is said to be *inconsistent* and thus $B_n^\star = \emptyset$. This situation disappears when a new block extends one of these side branches and creates $B_n^\star$. The blocks on the other branches are discarded and referred as *stale blocks*.

3.5   Block Model

We denote a block as $b_i = \langle \mathbf{h}_i, \Psi_i \rangle$ where $\mathbf{h}_i$ is the block header and Ψ_i is the block data. The block data Ψ_i contains all the transactions organized as a Merkle tree [12]. Basically, the copies of each transaction are hashed, and the hashes are then paired, hashed, paired again, and hashed again until a single hash remains, the merkle root of a merkle tree. We denote a merkle tree and its root as Ψ_i and ψ_i respectively. A merkle tree Ψ_i is created using the action of the form $createMerkelTree(tx_c, \theta_m)$ where tx_c is the coinbase transaction[5] that rewards the miner node $m \in M$ with the block reward $R = \mathbf{F} + \Sigma f$ for its work[6] (where $\mathbf{F}$ is the static block reward, and Σf is the total fees of the transactions included in this block), $\theta_m \subseteq \Theta_m$ is the set of candidate transactions chosen for this block. Here it is important to note that, blocks have limited sizes[7] and thus the size of θ_m can not exceed this limit[8]. The set of candidate transactions θ_m are selected using the action of the form $selectTransactions(\Theta_m) : \theta_m$ where Θ_m is the memory pool of the miner. The total fee Σf is then calculated by using the action of the form $calculateTotalFee(\theta_m)$.

The block header is denoted as $\mathbf{h}_i = \{v_n, \mathcal{H}(\mathbf{h}_{i-1}), \mathcal{H}(\mathbf{h}_i), t_{b_i}, \eta_i, \psi_i\}$ where v_n is the version number of the protocol used by n, $\mathcal{H}(\cdot)$ is the cryptographic hash function, $\mathcal{H}(\mathbf{h}_{i-1})$ is the cryptographic hash code of the header of the previous block b_{i-1} ($i > 0$), $\mathcal{H}(\mathbf{h}_i)$ is the cryptographic hash code of $\mathbf{h}_i$ generated by m, t_{b_i} is the current time stamp, η_i is an integer nonce value ($\eta_i \geq 0$) to be found by the miner in order to generate the right $\mathcal{H}(\mathbf{h}_i)$ conforming to the difficulty level μ defined in the protocol version v (the cryptographic puzzle mentioned in Sect. 3.3) and ψ_i is the root of the merkle tree.

3.6   Transaction Model

We model a transaction as $tx = \langle I, O \rangle$ where I is a list of inputs ($I \neq \emptyset$) and O is a list of outputs ($O \neq \emptyset$). Each input $i \in I$ references to a previous unspent output (for spending it). Each output then waits as an Unspent Transaction Output (UTXO) until an input spends it. If an output has already been spent

---

[5] Any transaction fees collected by the miner are also sent in this transaction.

[6] Note to remember is that the coins in a coinbase transaction cannot be spent until they have received 100 confirmations in the blockchain. All things being equal, 100 confirmations should equate to roughly 16 h and 40 min.

[7] The current maximum block size in Bitcoin is 1 MB. See https://bitcoin.org/en/glossary/block-size-limit, last access on 18 July 2017.

[8] Average block size for Bitcoin is given in https://blockchain.info/charts/avg-block-size, last access on 18 July 2017.

by an input, it cannot be spent again by another input (no double spending). We model the outputs as $o_i = \langle m, \mathcal{c}_{o_i} \rangle$ where $m \in N$ is the receiver of the coins $\mathcal{c}_{o_i}$ ($\mathcal{c}_{o_i} \geq 0$). All inputs of a transaction have to be spent in that transaction and the total input coins $\mathcal{c}_I$ has to be greater than or equal to the total output coins $\mathcal{c}_O$. The fee f_{tx} of a transaction tx is then modeled as $f_{tx} = \mathcal{c}_I - \mathcal{c}_O$. The fees of transactions are not fixed and are estimated by using the action of the form $estimateFee(I, O)$ where I is the set of inputs and O is the set of outputs. Depending on the fee to be paid, if there are still some coins left to be spent, the sender can add an output that pays this remainder to itself.

Algorithm 2. The actions of a user node n.

```
 1: action makeTransaction(m,¢)           14: action processTransaction(s, tx)
 2:   tx = createTransaction(m,¢)         15:   × ← validateTransaction(tx, Θ_n)
 3:   processTransaction(n, tx)           16:   if (× = 0) then
 4:                                       17:     U ← getUnspentTransactionOutputs()
 5: action createTransaction(m,¢)         18:     if ({∀i ∈ I_tx ∧ ∃o ∈ U|o ≺ i} = ∅) then
 6:   I ← selectUnspentTransactionOutputs(B_n, ¢)  19:       Θ_n ← Θ_n ⋃{tx}
 7:   o_1 ← ⟨m,¢⟩                         20:     else acceptTransaction(tx) endif
 8:   f ← estimateFee(I, {o_1})           21:   else updateBanscore(s, ×) endif
 9:   ¢_r ← ¢_I − ¢ − f                   22:
10:   o_2 ← ⟨n,¢_r⟩                       23: action acceptTransaction(tx)
11:   return tx = ⟨I, {o_1, o_2}⟩         24:   Θ_n ← Θ_n ⋃{tx}
12:                                       25:   sendMessage(⟨n,"inv",ℋ(tx)⟩,N_n)
13:                                       26:   processTransaction(∅, ∀tx' ∈ Θ̄_n|tx' ≺ tx)
```

The coinbase transaction tx_c (see Sect. 3.5) is special transaction that collects and spends any transaction fees paid by transactions included in a block and exceptionally it does not have any input set ($I = \emptyset$). It is the first transaction in a block and can only be created by a miner.

Creating a transaction tx by a node n is modeled as the action of the form $createTransaction(m, \mathcal{c})$ where $\mathcal{c}$ is the amount of coins ($\mathcal{c} > 0$) paid to m ($n, m \in N$). Selecting the right inputs to be able spend $\mathcal{c}$ is modeled using the action of the form $selectUnspentTransactionOutputs(B_n, \mathcal{c})$ where B_n is the blockchain and $\mathcal{c}$ is the coin to be spent. All those actions are detailed in Algorithm 2.

In the next section, we define a rational system model by augmenting the basic system model for better capturing its properties formally and defining the concept of *fairness*.

4   Rational Model

In this section we augment the basic blockchain system model given in Sect. 3 by the rational agent concept. Informally speaking, such agent chooses its actions/behavior with respect to its perceptions in order to maximize its utility. In such context we introduce a concept of fairness dependent on the agent behaviors and the corresponding utilities. To this aim, the remaining part of the section characterizes both agent and minor behaviors.

4.1   Rational Agent Model

A rational agent behaves according to its local perceptions and local knowledge, models uncertainty via expected values of variables or actions, and always chooses to perform the actions with the optimal expected outcome (among all feasible actions) for maximizing its utility [17]. We model all nodes in the blockchain network as rational agents and denote the rational agent set as N. Each rational agent $n \in N$ has a set of actions A_n and a utility function $\mathcal{U}_n$. Using A_n and $\mathcal{U}_n$, n uses a decision process where it identifies the possible sequences of actions to execute. We call these sequences as rational behaviors of n and denote as β. The objective of n is to choose the behaviors that selfishly keep $\mathcal{U}_n$ as high as possible.

We model the utility function of a rational agent $n \in N$ as $\mathcal{U}_n = u_0 + \sum_{i=1}^{k} \mathcal{U}(\beta_i)$ where u_0 is the initial utility value, $k \geq 0$ is the number of behaviors executed so far and $\mathcal{U}(\beta_i)$ is the utility value of the behavior β_i. A utility value $\mathcal{U}(\beta_i)$ can also be interpreted as the *degree of satisfaction* experienced by the realization of β_i. The utility value $\mathcal{U}(\beta_i)$ is calculated as $\mathcal{R}(\beta_i) - \mathcal{C}(\beta_i)$ where $\mathcal{R}(\beta_i)$ is the overall reward gained and $\mathcal{C}(\beta_i)$ is the overall cost spent for the execution of β_i.

When an agent needs to choose a behavior for execution, it needs to calculate its expected value. The expected value $\mathcal{E}(\beta_i)$ depends on the probabilities of the possible outcomes of the execution of β_i. We model the expected value as $\mathcal{E}(\beta_i) = \sum_{j=1}^{m} p_j \cdot \mathcal{U}(\beta_i^j)$ where $m > 0$ is the number of possible outcomes, $\mathcal{U}(\beta_i^j)$ is the utility value of the possible jth outcome β_i^j and p_j is the probability of this outcome such that $\sum_{j=1}^{m} p_j = 1$. Since behaviors are chosen based on their expected values, it can be said that a utility value $\mathcal{U}(\beta_i)$ represents the *satisfaction of expectation* about β_i.

Fairness: A rational agent $n \in N$ finds a system (i.e. the blockchain network) G *fair*, if the total satisfaction of its expectations $\mathcal{U}_n$ is above a certain degree τ_n where $\tau_{n'} < u_0$.

If at any time, an agent n finds G *unfair* ($\mathcal{U}_n \leq \tau_n$), it may decide to leave G if from its points of view it will not be possible to increase its overall utility above τ_n by calculating the expected values of its possible future behaviors. In other words, n may decide to leave G if $\mathcal{U}_n + \sum_{j=k}^{m} \mathcal{E}(\beta_j) \leq \tau_n$ where $\beta_k, ..., \beta_m$ are sufficiently enough desired future behaviors of n.

Based on this model, to be able to decide if Bitcoin-like blockchains are fair, we propose behavioral models for rational user and miner agents in the next subsection.

4.2   Rational Behavior Model

The rational behaviors given in this section define the possible strategies of agents by using and improving the basic actions given in Sect. 3.

To formalize the behaviors, we model a round based approach (like Garay et al. [8]) in which miner agents start creating a new block with at the beginning of the round and a round ends when a new block is successfully created by one of the miners. Both user and miner agents make their decisions on a roundly basis. This round-based model implicitly assumes that the block sent at the end of the round is immediately delivered by all participants, i.e. communication delay is negligible with respect to block generation time.

User Agent Behaviors. The user agent in the system makes transactions. To make a transaction, she has to determine a fee to assign to the transaction. This decision depends on the probability that the assigned fee has to lead the transaction confirmed as soon as possible and the interest of the agent on the transaction. All the duration the user agent spends waiting for the transaction confirmation is said to be the waiting cost. *Such a cost is unavoidable*, the user agent has to wait as long as the transaction is confirmed since the basic protocol does not allow revoking transactions, even if they are not confirmed after a considerable time.

We model making a transaction with a specific fee f as with the action of the form $makeTransaction(m, ¢)$ where $m \in N$ is the receiver and $¢$ is the amount of coins. For simplicity, it is assumed that a users agent has an ordered set of fees $\{f_1, f_2, \ldots, f_k\}$ to use. It is also assumed that $f_i < f_{i+1}$ where $0 < i < k$ and $0 \le P(f_i) \le P(f_{i+1}) \le 1$ where $P(f)$ is the probability of a transaction with a fee f to be confirmed.

We model the interest the user agent has with respect to the transaction as I. It is assumed that the initial interest does not decrease round after round. We denote the waiting cost as $C(f)$, which is proportional to the fee f applied to the transaction. Intuitively, waiting for the confirmation of a transaction with a high fee assigned is more costly than waiting for the confirmation of a transaction with a lower fee assigned.

Along with the interest and the waiting cost, an important role is played by the fee assigned to the transaction. The fee is paid only when the transaction is confirmed in a block in the main chain. Thus, when a transaction is confirmed at round r, the expected value is given by the net difference between the interest I and the assigned fee f plus the waiting cost $C(f)$ for all the $r-1$ rounds while the user agent waited.

Let us consider a user agent that issues a transaction tx and chooses a fee f, has an interest I on tx and a waiting cost $C(f)$. The confirmation probability of tx at the first round is $P(f)$, thus if tx is confirmed during the first round the user agent gains: $(I - f) \cdot P(f)$ and no waiting cost. Let $\overline{P(f)} = 1 - P(f)$ be the unconfirmation probability of tx during a round. If tx is confirmed during the second round she gains: $\overline{P(f)} \cdot P(f) \cdot (I - f) - \overline{P(f)} \cdot C(f)$, and if confirmed during the third round she gains: $\overline{P(f)}^2 \cdot P(f) \cdot (I - f) - \overline{P(f)}^2 \cdot C(f)^2$. Generalizing for an infinite number of rounds r, the expected value $\mathcal{E}$ is:

$$\mathcal{E}(\beta) = \sum_{r=1}^{\infty} \overline{P(f)}^{r-1} \cdot P(f) \cdot (I - f) - \sum_{r=1}^{\infty} \overline{P(f)}^{r-1} \cdot C(f)^{r-1} \qquad (1)$$

where $\beta = \{makeTransaction(m, \math{c}, f)\}$. An analysis of Eq. 1, along with necessary condition for fairness, is given in Sect. 5.1.

Miner Agent Behaviors. In the decision model for the user agent, we considered a probability P(f) constant over rounds[9]. On the other hand, in this section we want to shed some light on the miner behavior, in order to have insights on how the probability changes over rounds.

To this end, we consider the miners at the beginning of a round and focus on understanding how miners choose the transactions. In particular, since the size of the block is limited, *miners could decide to deliberately exclude an unconfirmed transaction that has already been received with a lower fee.* This behavior would obviously delay the confirmation time of that transaction and affects its confirmation probability. More in detail, we consider that at the beginning of a round a miner can decide if to start immediately creating a block with the transactions it already has in its pool (Tx_1)[10] or to wait for a set of transactions with better fees (Tx_2) to arrive. If such transaction arrives a miner decides if to continue to mine Tx_1 or to start mining with Tx_2. This decision process is depicted in the state machine given in Fig. 1.

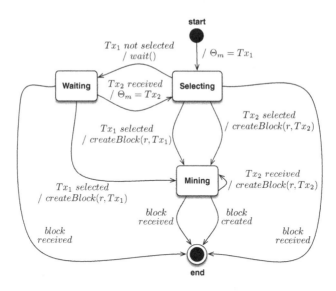

Fig. 1. The state machine of the rational mining behavior of a miner m for a round r.

Let $\{0,1\}^k$ be the range of the hash function run to solve the Proof-of-Work (PoW) and k the security parameter and let $\mathcal{D}$ be difficulty level. Let q be the

---

[9] Technically speaking this means to consider $P(f)$ modeled as a stationary process, and the blockchain system as an ergodic dynamical system [4].

[10] For simplicity, it is assumed that the sizes of both Tx_1 and Tx_2 are equal to the maximum block size.

upper bound on the number of times the hash function can be invoked during a round and each invocation is independent from the other (as independent Bernoulli trials). The probability that an attempt solves the PoW is $p = \frac{D}{2^k}$ thus the upper bound on the number of solutions that a miner can found per round is $P = p \cdot q$. If the miner solves the PoW, it gets a reward $R = \text{F} + f_{Tx}$, where f_{Tx} is the total fee associated to the transactions in the set Tx. In the following, we model how P and R changes over the time, to take into account the miner decision to wait or not before starting creating a block.

Along with the miner rewards, we consider also the power consumption cost. We consider that each attempt implies a cost that we call C. Thus, the miner gain is net between the reward and the costs multiplied by the probability to win the PoW which depends on how many attempts are performed. The reward is $R = \text{F} + f_{Tx_2}$ if the expected Tx_2 arrives and $R = \text{F} + f_{Tx_1}$ otherwise. At each round r, miners calculate the expected values of each different behavior and choose one of them.

Algorithm 3. Rational $createBlock()$ action of a rational miner agent m.

```
1: action createBlock()
2:   while (w) do
3:     i ← |B*_n| + 1
4:     ℰ ← calculate expected values
5:     Tx₁ ← selectTransactions(Θ_n)
6:     depending on ℰ either:
7:       wait until Σf_Tx₂ ≫ Σf_Tx₁ such that Tx₂ ← selectTransactions(Θ_n)
8:         if Tx₂ arrives either createBlock(i, Tx₂) or createBlock(i, Tx₁)
9:         if Tx₂ not arrives either createBlock(i, Tx₁) or continue waiting
10:    or:
11:      start createBlock(i, Tx₁)
12:        if Tx₂ arrives either createBlock(i, Tx₂) or continue
13:        if Tx₂ not arrives continue
14:  endwhile
```

Based on the state machine given in Fig. 1, the rational behaviors of miner agents can be implemented by improving the $createBlock()$ action given in the Basic System Model (Algorithm 1) as shown in Algorithm 3. In particular, we consider three behaviors: (β_1) the miner waits the arrival of Tx_2 ignoring Tx_1; (β_2) the miner starts mining with Tx_1 ignoring Tx_2 (either if it arrives or not); (β_3) the miner starts mining with Tx_1 and when Tx_2 arrives she starts mining with Tx_2. Following the same reasoning as in Sect. 4.2, for each behavior we compute the expected values:

$$\mathcal{E}(\beta_1) = \sum_{q'=1}^{q} 1/2 \cdot (p \cdot (1-p)^{q'-1} \cdot (\text{F} + f_{Tx_2}) - C) \qquad (2)$$

$$\mathcal{E}(\beta_2) = \sum_{q'=1}^{q} (p \cdot (1-p)^{q'-1} \cdot (F + f_{Tx_1}) - C) \tag{3}$$

$$\mathcal{E}(\beta_3) = \sum_{q'=1}^{q} (1/2 \cdot p \cdot (1-p)^{q'-1} \cdot (F + f_{Tx_1}) + 1/2 \cdot p \cdot (1-p)^{q'-1} \cdot (F + f_{Tx_2}) - C) \tag{4}$$

These expected values depend on the transaction the miner is considering to mine (Tx_1) and the probability that a juicy transaction (Tx_2) arrives. Having a probability distribution over such an event is far from being trivial. We consider that the user has no information about the exact distribution probability, i.e., she takes a decision under ignorance. By applying the principle of insufficient reason[11] we then assign the same probability, to the two events: "Tx_2 arrives"; "Tx_2 does not arrive".

We show the results associated with a scenario in which the two events are considered as equally as possible in Sect. 5.2.

5   Analyses and Results

We analyzed the rational behaviors of each agent proposed in Sect. 4.2 using Matlab R2017a. It is assumed that the initial utility values u_0 of all agents are the same and high enough from the threshold τ. In the following, we provide results of these analyses considering both the users and the miners employing synthetic data.

5.1   User Agent

Recalling Eq. (1), it is clear that depending on the values of $\overline{P(f)}$ and $C(f)$ the expected value $\mathcal{E}$ may or may not converge to $-\infty$.

Let us first give a graphical intuition of the series behavior (Fig. 2). For simplicity, we denote the first part of Eq. (1) as $\alpha_{gain} = \sum_{r=1}^{\infty} \overline{P(f)}^{r-1} \cdot P(f) \cdot (I - f)$ and the second part as $\alpha_{cost} = \sum_{r=1}^{\infty} \overline{P(f)}^{r-1} \cdot C(f)^{r-1}$. Consequently, $\mathcal{E} = \alpha_{gain} - \alpha_{cost}$. As can be seen, if $C(f) \geq 1/\overline{P(f)}$ then α_{cost} goes to ∞, thus $\mathcal{E}$ goes to $-\infty$ (cf. Fig. 2a). Otherwise, if $C(f) < 1/\overline{P(f)}$ the cost converge to a bounded quantity and $\mathcal{E}$ as well (cf. Fig. 2b). In the first case, it can be said that the user agent is facing an unfair situation, if she chooses the fee f her expected value $\mathcal{E}$ is $-\infty$ due to the fact she would pay an infinite waiting cost[12] with no possibility to change the strategy.

Based on this observation we state that

$$C(f) < 1/\overline{P(f)} \tag{5}$$

---

[11] The principle of insufficient reason prescribes that if one has no reason to think that one state of the world is more probable than another, then all state should be assigned equal probability [16].

[12] This is the same situation as in the Saint Petersburg Paradox [19].

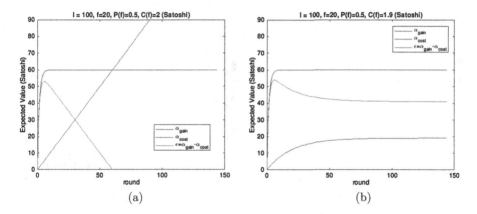

Fig. 2. Expected values during a period of 144 rounds (one day), depending on the relationship between $C(f)$ and $\overline{P(f)}$, $\mathcal{E}$ may converge to a finite quantity or to $-\infty$. $\overline{P(f)}$ and $C(f)$ are chosen such that the different behavior of $\mathcal{E}$ can be seen when (a) Equation (5) holds, and (b) not holds.

is a *necessary condition* for *fairness* since its violation leads the waiting cost to ∞ and the $\mathcal{E}$ for the transaction to issue to $-\infty$. That is, in such scenario is never profitable to try.

In the following, we consider a scenario where the necessary condition given in Eq. (5) is met and we show with a practical example why having $C(f) < 1/\overline{P(f)}$ is not sufficient to have fairness as defined in Sect. 4. We consider an user agent whose already issued some transactions, her $\mathcal{U}_n = -30Satoshi$ ($1Satoshi = 0.00000001BTC$) and her threshold is $\tau_n = 70Satoshi$. Since $\mathcal{U}_n < \tau_n$ the user finds the system unfair if the expected value of the next transaction does not allow to reach the threshold, in this case the user will leave. Let us consider that such user has a transaction to issue and can decide to play with two fees: a low and a high fee, 20 and 40 Satoshi respectively. In order to rise $\mathcal{U}_n$ above τ_n, the expected values have to be at least 100 *Satoshi*.

Figure 3 plots the expected values for two different scenarios with respect to rounds where in both cases the necessary condition (5) is met. In the first scenario (Fig. 2a),the user agent can issue a transaction with a low fee of 20 Satoshi such that $P(20) = 0.25$ and $C(20) = 0.5$. If the user agent chooses a high fee of 40 Satoshi then such probability rise to 0.75 and the waiting cost to 1.5. In the second scenario (Fig. 2b) nothing changes but the probability and the waiting cost for the low fee, $P(20) = 0.6$ and $C(20) = 1$ respectively. In both cases the user interest I is 100 Satoshi and is constant.

As can be seen, whatever fee the user decides to assign to the transaction the expected value is below 100 *Satoshi* so the user has no way to rise $\mathcal{U}_n$ above τ_n and since she will find the system unfair with no possibility to change such situation, she will leave the system. It is important to underline that *the perception of fairness is strictly local to the user since it depends to the interest she puts in the transactions issue and the waiting cost it perceives.*

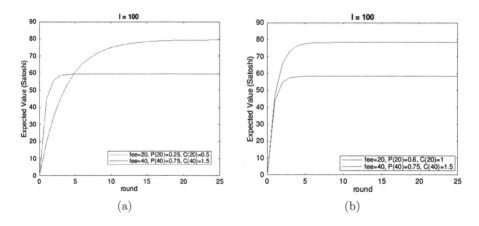

Fig. 3. Expected values of user behaviors given in Sect. 4.2 with respect to rounds.

5.2 Miner Agent

Figure 4 depicts miner expected values as defined in Sect. 4.2. We consider different scenarios depending on the reward that a miner gets solving the PoW: a fixed part and the sum of the transactions fee included in the block. For simplicity, it is assumed that each miner can insert only one transaction in a block[13] and that at the beginning of the round the miner has in its transaction pool a transaction Tx_1 and it is expecting to deliver a second transaction Tx_2 such that $Tx_2 > Tx_1$. We consider that the miner has a AntMiner S9 device[14], thus miner has $q = 504e15$ attempts to solve the PoW during a round and the probability to solve it is $p = 6.1201e - 66$ for each attempt. In this analysis we are comparing the different strategies in one single round and the power cost is common and constant but a way greater than the possible reward in one round, for this reason we do not consider the cost in the plots. In Fig. 4 on the y-axis, we have the Expected Value in Bitcoins (BTC) and on the x-axis the q attempts.

As can be seen, whenever Tx_2 arrives it is convenient to swap the two transactions and start creating a new block with Tx_2. How convenient it is depends on the difference between the two fees, *independently of the fixed part amount*. In Fig. 4a and b, $Tx_2 \gg Tx_1$ and for the miner it is clearly convenient to swap from Tx_1 to Tx_2. This this is also true if the transaction fees are close to each other, Fig. 4c and d. Let us note that in the case of a bigger block size the miner would have started to mine a block with both transactions to increase her reward.

---

[13] In the Bitcoin protocol blocks have fixed size (see Sect. 3.5).

[14] Such device hash rate is 14TH/s, which means $14e12 * 36000$ attempts to solve the PoW in a round (10 min in average).

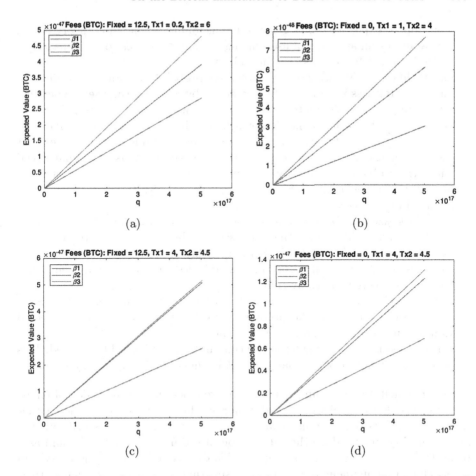

Fig. 4. Expected values of miner behaviors with respect to rounds. In cases (a) and (c) the fixed reward is 12.5 BTC, in cases (b) and (d) fixed reward is zero.

6   Discussion and Conclusion

To the best of our knowledge, this is the first study focusing on *users*. The rational system model given in Sect. 4 allows us to define fairness as the overall satisfaction of participants. However, there is still no mechanism which guarantees fairness for users. As shown in Sect. 5, the expected values for the users are always decreasing when their waiting costs are not taken into account by the miners. Moreover, we showed that the miners as rational agents have no incentive to keep the probability over rounds stationary.

Based on these findings, the main issues arisen by Bitcoin-like blockchain systems can be resumed as follows:

- There is no mechanism for a user to cancel an already issued transaction, i.e. once she decides to engage in the game, it can never abandon the game.

This implies expected values going to minus infinity. In decision theory terms, this would mean assuming a user having an infinite interest on a transaction, which is in fact hard to assume in real settings.

– Given different fees with their associated probabilities, more flexibility to guarantee fairness would have been reached by allowing the user (experiencing a long waiting time) to resend the transaction with a lower fee. This flexibility, on the other hand, is not achieved in Bitcoin-like blockchain systems mainly due to two reasons: (1) there is no deterministic guarantee that the system chooses a given copy of the transaction, it is only guaranteed with high probability that only one copy will be inserted into the blockchain, and (2) the miner behavior favors the transaction with the highest fee to get included in a block.

– The block size is a fixed parameter today. Obviously, if the volume of transactions increases over time, the block size can become a scarce resource and miners will be more apt to deliberately delaying a transaction with a low fee. This will dramatically drop the probabilities for low fees and their corresponding expected values by time.

– Fees are decided by the users, leading to a possible race among users fees. Such a situation would make the probabilities difficult to predict, and expected values of users might drop even more quickly.

– The fairness is a locally perceived concept and it must indeed be tracked. This needs further detection mechanisms not included in the current systems.

In a nutshell, we can claim that Bitcoin-like blockchain systems do not envisage any specific mechanism to avoid unfairness for the users. Unfairness situations are not caused by the users and it is the users who has to sacrifice more. Such situations may reduce the satisfaction of the users dramatically, and as a result the users may leave the system totally. Consequently, the rational system model should be augmented with mechanisms that allow agents to change their behaviors upon detecting unfair situations for balancing their overall satisfaction. Such a system will obviously encourage honest participants to stay in the blockchain system.

Besides, it is particularly difficult to devise such mechanisms while preserving the security properties of blockchain systems. This is especially true for those solutions envisaging to give rewards to users (this will change the computation of the expected value given in Sect. 4) for other protocol actions they do in the system, such as forwarding messages or validating transactions and blocks. Unfortunately, no proved solution exists on this line, to the best of our knowledge.

Pragmatically speaking, to avoid security issues and related complex analyses, we claim that a promising approach would be to stick as much as possible to the original Bitcoin protocol, introducing mechanisms to improve the degree of fairness of the system. For example, it would be interesting to consider how to give the possibility to the user to revoke a transaction, or to re-issue it again with a lower fee. It could also be interesting to allow the block size to be an adjustable parameter, as it is now for the difficulty parameter of the mathematical puzzle[15].

References

1. Asokan, N.: Fairness in electronic commerce (1998)
2. Babaioff, M., Dobzinski, S., Oren, S., Zohar, A.: On bitcoin and red balloons. In: Proceedings of the 13th ACM Conference on Electronic Commerce, pp. 56–73. ACM (2012)
3. Carlsten, M., Kalodner, H., Weinberg, S.M., Narayanan, A.: On the instability of bitcoin without the block reward. In: Proceedings of the 2016 ACM SIGSAC Conference on Computer and Communications Security, pp. 154–167. ACM (2016)
4. Coudene, Y.: Ergodic Theory and Dynamical Systems. Universitext. Springer, London (2016). http://dx.doi.org/10.1007/978-1-4471-7287-1
5. Eyal, I.: The miner's dilemma. In: 2015 IEEE Symposium on Security and Privacy (SP), pp. 89–103. IEEE (2015)
6. Eyal, I., Gencer, A.E., Sirer, E.G., Van Renesse, R.: Bitcoin-ng: a scalable blockchain protocol. In: NSDI, pp. 45–59 (2016)
7. Eyal, I., Sirer, E.G.: Majority is not enough: bitcoin mining is vulnerable. In: Christin, N., Safavi-Naini, R. (eds.) FC 2014. LNCS, vol. 8437, pp. 436–454. Springer, Heidelberg (2014). doi:10.1007/978-3-662-45472-5_28
8. Garay, J., Kiayias, A., Leonardos, N.: The bitcoin backbone protocol: analysis and applications. In: Oswald, E., Fischlin, M. (eds.) EUROCRYPT 2015. LNCS, vol. 9057, pp. 281–310. Springer, Heidelberg (2015). doi:10.1007/978-3-662-46803-6_10
9. Göbel, J., Keeler, H.P., Krzesinski, A.E., Taylor, P.G.: Bitcoin blockchain dynamics: the selfish-mine strategy in the presence of propagation delay. Perform. Eval. **104**, 23–41 (2016)
10. Lewenberg, Y., Sompolinsky, Y., Zohar, A.: Inclusive block chain protocols. In: Böhme, R., Okamoto, T. (eds.) FC 2015. LNCS, vol. 8975, pp. 528–547. Springer, Heidelberg (2015). doi:10.1007/978-3-662-47854-7_33
11. Liu, J., Li, W., Karame, G.O., Asokan, N.: Towards fairness of cryptocurrency payments. arXiv preprint arXiv:1609.07256 (2016)
12. Merkle, R.C.: A digital signature based on a conventional encryption function. In: Pomerance, C. (ed.) CRYPTO 1987. LNCS, vol. 293, pp. 369–378. Springer, Heidelberg (1988). doi:10.1007/3-540-48184-2_32
13. Nakamoto, S.: Bitcoin: a peer-to-peer electronic cash system (2008). https://bitcoin.org/bitcoin.pdf
14. Pass, R., Seeman, L., Shelat, A.: Analysis of the blockchain protocol in asynchronous networks. IACR Cryptology ePrint Archive 2016:454 (2016)
15. Pass, R., Shi, E.: Fruitchains: a fair blockchain. Cryptology ePrint Archive, Report 2016/916 (2016). http://eprint.iacr.org/2016/916.pdf
16. Resnik, M.D.: Choices: An Introduction to Decision Theory. University of Minnesota Press, Minneapolis (1987)
17. Russell, S.J., Norvig, P.: Artificial Intelligence - A Modern Approach, 3rd edn. Pearson Education, Essex (2010)

---

[15] During the submission process of this paper (August 2017), the Bitcoin (BTC) protocol, which has 1 MB of block size, has been hard forked as the Bitcoin Cash (BCC) protocol, which has 8 MB of block size. However, it is early to conclude if BCC is better than BTC for the moment.

18. Sapirshtein, A., Sompolinsky, Y., Zohar, A.: Optimal selfish mining strategies in bitcoin. In: Grossklags, J., Preneel, B. (eds.) FC 2016. LNCS, vol. 9603, pp. 515–532. Springer, Heidelberg (2017). doi:10.1007/978-3-662-54970-4_30

19. Weiss, M.: Conceptual Foundations of Risk Theory. Technical Bulletin. U.S. Department of Agriculture, Economic Research Service, Washington, D.C. (1987)

Scalable Conformance Checking of Business Processes

Daniel Reißner[1]([✉]), Raffaele Conforti[1], Marlon Dumas[2], Marcello La Rosa[1], and Abel Armas-Cervantes[1]

[1] Queensland University of Technology, Brisbane, Australia
{da.reissner,raffaele.conforti,m.larosa,a.armascervantes}@qut.edu.au
[2] University of Tartu, Tartu, Estonia
marlon.dumas@ut.ee

Abstract. Given a process model representing the expected behavior of a business process and an event log recording its actual execution, the problem of business process conformance checking is that of detecting and describing the differences between the process model and the log. A desirable feature is to produce a minimal yet complete set of behavioral differences. Existing conformance checking techniques that achieve these properties do not scale up to real-life process models and logs. This paper presents an approach that addresses this shortcoming by exploiting automata-based techniques. A log is converted into a deterministic automaton in a lossless manner, the input process model is converted into another minimal automaton, and a minimal error-correcting synchronized product of both automata is calculated using an A* heuristic. The resulting automaton is used to extract alignments between traces of the model and traces of the log, or statements describing behavior observed in the log but not captured in the model. An evaluation on synthetic and real-life models and logs shows that the proposed approach outperforms a state-of-the-art method for complete conformance checking.

Keywords: Conformance checking · Process mining · Automata · Behavioral alignment

1 Introduction

Modern information systems maintain detailed business process execution trails. For example, an enterprise resource planning system keeps records of key events related to a company's order-to-cash process, such as the receipt and confirmation of purchase orders, the delivery of products, and the creation and payment of invoices. Such records can be grouped into an *event log* consisting of sequences of events (called *traces*), each consisting of all event records pertaining to one case of a process.

Conformance checking techniques exploit such event logs in order to determine if and to what extent the actual behavior of a process conforms to a process

© Springer International Publishing AG 2017
H. Panetto et al. (Eds.): OTM 2017 Conferences, Part I, LNCS 10573, pp. 607–627, 2017.
https://doi.org/10.1007/978-3-319-69462-7_38

model capturing its expected behavior. A conformance checking technique takes as input an event log and a process model, and returns a set of differences between the model and the log. In real-life scenarios, the set of differences between an event log and a process model can be large. Hence it is necessary to represent them in a way that is compact and interpretable, yet complete, or as exhaustive as desired by the user.

State-of-the-art techniques for computing a complete set of differences include *behavioral alignment* [15] and *(all-optimal) trace alignment* [2]. The former computes a set of statements describing behavioral relations that exist in the model but not in the log. The latter computes minimal alignments between each trace in the log that cannot be parsed by the model, and a corresponding trace that can be parsed by the model. These techniques however do not scale up to large and noisy event logs. For example, our experimental evaluation (reported later) shows that the all-optimal trace alignment technique in [2] takes more than five minutes to compute an incomplete set of alignments over real-life and noisy logs and sometimes does not converge after hours. These execution times make it impractical to use these techniques in an interactive setting, e.g. when conformance checking is performed multiple times to iteratively repair a process model so as to better fit the log. Additionally, scalability issues of conformance checking techniques indirectly affect several process mining techniques, such as model repair [7,25] or process discovery [8,17], which rely on conformance checking to justify the quality of their outputs.

This paper aims to tackle this scalability issue by proposing an automata-based technique for conformance checking. In our approach, an event log is encoded as sequences of words and compressed into a minimal Deterministic Acyclic Finite State Automaton (DAFSA). Concomitantly, the process model is transformed into another automaton (its *reachability graph*). The two automata are combined into an error-correcting product automaton whose transitions correspond to either a synchronous move (on both automata) or an asynchronous move (i.e. a move on the automaton of the log that does not exist in the model or vice-versa). The produced automaton contains a minimal number of asynchronous moves. From this product automaton, we can extract either the optimal alignments of each trace in the log and a corresponding trace in the model (as in [2]) or a set of behavioral difference statements (as in [15]). Thus, our approach unifies these two previous approaches, while achieving higher scalability, as shown by an evaluation on synthetic and real-life process models and logs.

The next section discusses existing conformance checking techniques in more detail. Section 3 introduces the proposed approach, while Sect. 4 presents its evaluation. Section 5 summarizes the contributions and discusses improvement avenues.

2   Related Work

Conformance checking techniques detect two types of discrepancies between a process model and a log: behavior observed in the log that is disallowed by the model (*unfitting behavior*), and behavior allowed by the model but not observed in the log (*additional behavior*). A simple approach to detect and measure unfitting behavior is *token-based replay* [26]. The idea is to replay each trace against the model, represented as a Petri net. The transitions in the model are fired following the order dictated by a given trace. To fire, a transition needs to be enabled, i.e. it requires at least one token in each of its incoming places. When a transition cannot fire because it is not enabled, the technique determines which tokens need to be *added* to enable it. Once a trace has been replayed, if there are any tokens left in a non-sink place of the Petri net, they are labeled as *remaining tokens*. The *fitness* between the model and the log is quantified in terms of the number of added and remaining tokens (replay errors). An extended version of this approach, namely *continuous semantics fitness* [4], achieves higher performance at the expense of incompleteness. Another extension [32] decomposes the model into single-entry single-exit fragments, such that each fragment can be replayed independently. Other extensions based on model decomposition are discussed in [23].

Replay fitness methods fail to identify a minimum number of errors required to explain unfitting log behavior, thus overestimating the magnitude of differences. Trace alignment fitness [2] addresses this limitation. For each trace in the log, this technique identifies the closest trace reproducible by the model and aligns the two traces by highlighting the points where mismatches occur. This log-model alignment is achieved in several steps. The first step consists in transforming every trace in the log into a Petri net. The result is a sequence of transitions, one per event in the trace. Next, a product is computed between the Petri net of the trace and the Petri net of the model. This is done by pairing transitions of the two models that have matching labels. The product between the two Petri nets is used to create a transition system representing all possible alignments, i.e. matches and mismatches. This transition system is explored, using the A^* search algorithm, to retrieve the alignments with the minimum number of mismatches. An exhaustive and complete version of this technique, namely *all-optimal alignments*, computes all minimal alignments between each log trace and the model. *One-optimal alignment* is an alternative technique that achieves higher scalability at the expense of incompleteness. This technique computes only one alignment for each log trace, hence missing on some behavioral differences. Several heuristics-based approaches, such as sequential prefix alignments [29] or decomposing trace replay technique [27], improve on the scalability for identifying alignments. Those approaches, however, drop the guarantee to find the optimal alignments and thus trade accuracy for performance.

Approaches for identifying additional behavior include *negative-events precision* [31] and *ETC precision* [22]. The former adds negative events to the traces in the log. Given a trace, an event is negative if it is never observed after a given trace prefix. Additional behavior is identified by replaying these extended traces over the model. Whenever a negative-event is successfully replayed, the approach marks it as additional behavior. ETC precision generates a prefix automaton from the log, where each state corresponds to a distinct trace prefix in the log. The states of the prefix automaton are matched with the states of the model. When a state in the model enables a transition that it is not enabled in the matching state of the automaton, it is marked as additional behavior. The approach has been extended in [1] to handle tasks with duplicate labels and unfitting traces by means of trace alignment. The technique proposed in this paper is complementary the above ones, since it computes trace alignments that can be used for example to speed up the technique in [1].

An approach for fast approximate computation of fitness and precision metrics is presented in [18]. This technique computes these metrics over subsets of process tasks and aggregates the results at a process level. This technique has been shown to be highly scalable, however, it does not identify the behavioral differences between the model and the log, but it merely computes the fitness and precision metrics. As part of this paper, we are interested in a complete list of exact differences, hence a comparison with the approach presented in [18] is out of scope.

Another conformance checking technique, namely *behavioral alignment* [15], addresses the problems of detecting unfitting behavior and additional behavior in a unified setting. In this technique, both the input event log and the process model are transformed into event structures. A minimal error-correcting product of these two event structures is then computed. Based on this product, a set of statements are derived, which characterize all behavioral relations between tasks captured in the model but not observed in the log and vice-versa. While producing a complete set of differences, which is smaller in number than the number of trace alignments, this technique suffers from similar scalability requirements as the all-optimal alignment.

The approach herein presented uses automata as novel and memory-efficient representation for event logs and process models. By mapping the problem of conformance checking to that of synchronizing a DAFSA representing the event log, and a finite state machine (FSM) representing the model, the proposal unifies the techniques proposed in [2,15]. This allows us to extract both a set of optimal trace alignments and a set of difference statements. Thus, the paper aims at improving the efficiency of state of the art conformance checking techniques leveraging automata and memoization techniques. Unlike [15] though, we only focus on detecting unfitting behavior.

3   Approach

Figure 1 outlines the steps of the proposed method and their respective inputs and outputs. First, the input process model is expanded into a reachability graph (1). In parallel, the event log is compressed into a minimal DAFSA (2). The resulting reachability graph and DAFSA are then compared (3) to create an error-correcting synchronized product automaton (herein called a Partial Synchronized Product or PSP), wherein each state is a pair of a state in the reachability graph and a state in the DAFSA. From this result, we can directly enumerate a set of optimal trace alignments or derive a set of behavioral difference statements via further analysis (4). The rest of this section introduces some preliminary definitions, followed by a description of each of the steps.

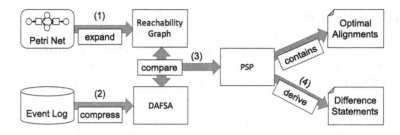

Fig. 1. Overview of the approach.

For illustration purposes, we will use the loan application process model displayed in Fig. 2. The process starts when a credit application is received, then the credit history and the income sources are checked. Then, once the application is assessed, either a credit offer is made, the application is rejected or additional information is requested (the latter leading to a re-assessment).

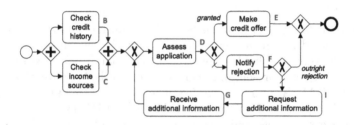

Fig. 2. Example loan application process model adapted from [15].

3.1 Preliminaries

Our approach relies on the notion of finite state machine defined as follows.

Definition 1 (Finite State Machine (FSM)). *Let L be a finite non-empty set of labels. A finite state machine is a directed graph $\mathscr{F} = (N, A, s, R)$, where N is a finite non-empty set of states, $A \subseteq N \times L \times N$ is a set of arcs, $s \in N$ is an initial state, and $R \subseteq N$ is a set of final states.*

An arc in a FSM is a triplet (n_s, l, n_t), where n_s is the *source* state, n_t is the *target* state and l is the *label* associated to the arc. The set of incoming and outgoing arcs of a state n is defined as $\blacktriangleright n = \{(n_s, l, n_t) \in A \mid n = n_t\}$ and $n \blacktriangleright = \{(n_s, l, n_t) \in A \mid n = n_s\}$, respectively. Finally, a sequence of (contiguous) arcs in a FSM is called a *path*.

3.2 From Event Log to DAFSA

Logs recording the execution of activities in a business process are called *event logs*. These logs represent the executions of process instances as *traces*

– sequences of activity occurrences (*a.k.a. events*). A trace can be represented as a sequence of labels, such that each label signifies an event. Generally speaking an event log is a multiset of traces containing several occurrences of the same trace. However, in the context of this paper, we are only interested in the distinct executions of a business process and, therefore, we define a log as a set of traces.

Definition 2 (Trace and event log). *Let L be a finite set of labels. A trace is a finite sequence of labels $\langle l_1, ..., l_n \rangle \in L^*$, such that $l_i \in L$ for any $1 \leq i \leq n$. An* event log *$\mathscr{L}$ is a set of traces.*

Event logs can be represented as *Deterministic Acyclic Finite State Automata* (DAFSA), which are acyclic and deterministic FSMs. A DAFSA can represent words, in our context *traces*, in a compact manner by exploiting prefix and suffix compression.

Definition 3 (DAFSA). *A DAFSA is an acyclic and deterministic finite state machine $\mathscr{D} = (N_{\mathscr{D}}, A_{\mathscr{D}}, s_{\mathscr{D}}, R_{\mathscr{D}})$, where $N_{\mathscr{D}}$ is a finite non-empty set of states, $A_{\mathscr{D}} \subseteq N_{\mathscr{D}} \times L \times N_{\mathscr{D}}$ is a set of arcs, $s_{\mathscr{D}} \in N_{\mathscr{D}}$ is the initial state, $R_{\mathscr{D}} \subseteq N_{\mathscr{D}}$ is a set of final states.*

Daciuk et al. [11] presents an efficient algorithm for constructing a DAFSA from a set of words, such that every word is a path from the initial state to a final state. Conversely it holds, that every path from an initial state to a final state represents a word present in the given set of words. We reuse this algorithm to construct a DAFSA from an event log, where every trace in the log represents a word. The complexity of building the DAFSA is $O(|L| \cdot \log n)$, where L is the set of distinct event labels, and n is the number of states in the DAFSA.

Given a path from the initial state to a state $n \in N_{\mathscr{D}}$, we refer to the labels associated to the arcs in the path as the *prefix* of n, and, analogously, given a path from n to a final state, we refer to the labels associated to such path as a suffix of n. Note that the prefix of the initial state is $\{\langle\rangle\}$. By abuse of notation, the set of prefixes of a state n is represented by $pref(n) = \bigcup_{(n_s, l, n_t) \in \blacktriangleright n} \{x \oplus l \mid x \in pref(n_s)\}$, where $\oplus$ denotes the concatenation operator. Similarly, the set of suffixes of n is represented by $suff(n) = \bigcup_{(n_s, l, n_t) \in n \blacktriangleright} \{l \oplus x \mid x \in suff(n_t)\}$, and if n is a final state then $\{\langle\rangle\} \in suff(n)$. Prefixes and suffixes are said to be *common* iff they are shared by more than one trace.

Definition 4 (Common prefixes and suffixes). *Let $\mathscr{D} = (N_{\mathscr{D}}, A_{\mathscr{D}}, s_{\mathscr{D}}, R_{\mathscr{D}})$ be a DAFSA. The set of common prefixes of $\mathscr{D}$ is the set $\mathscr{P} = \{pref(n) \mid n \in N_{\mathscr{D}} \wedge |n\blacktriangleright| > 1\}$. The set of common suffixes of $\mathscr{D}$ is the set $\mathscr{S} = \{suff(n) \mid n \in N_{\mathscr{D}} \wedge |\blacktriangleright n| > 1\}$.*

Figure 3 depicts an example of an log containing activities of the loan application process in Fig. 2 and its corresponding DAFSA representation. For the sake of readability, Fig. 3 uses the letters next to the each of the tasks in Fig. 2 as task labels. In this example there is only one final state f_D, and all traces in the log are paths from s to f_D. For instance, the trace $\langle B, D, E \rangle$ is represented by the path $\langle (s, B, n_1), (n_1, D, n_2), (n_2, E, f_D) \rangle$. In this example, the prefixes of state n_2 and the suffixes of state n_1 are common for all the traces.

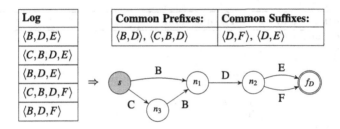

Log	Common Prefixes:	Common Suffixes:
$\langle B,D,E \rangle$	$\langle B,D \rangle$, $\langle C,B,D \rangle$	$\langle D,F \rangle$, $\langle D,E \rangle$
$\langle C,B,D,E \rangle$		
$\langle B,D,E \rangle$		
$\langle C,B,D,F \rangle$		
$\langle B,D,F \rangle$		

Fig. 3. Example log for our loan application process, and its DAFSA representation.

3.3   From a Process Model to a FSM

Process models are normative descriptions of business processes and define the expected behavior of the process. Over the years, several business process modelling languages have been proposed, such as Petri nets, BPMN and EPC. In the context of this work, business processes are modelled as (labelled) Petri nets.

Definition 5 (Labelled Petri net). *A (labelled) Petri net is the tuple* $\mathcal{N} = (P, T, F, \lambda)$, *where P and T are disjoint sets of places and transitions, respectively, $F \subseteq (P \times T) \cup (T \times P)$ is the flow relation, and $\lambda : T \to L \cup \{\tau\}$ is a labelling function mapping transitions to the set of task labels L and to a special label τ.*

Note that τ is a special label and it is used to represent invisible transitions, i.e. actions not recorded in the event log when executed. Places and transitions are conjointly referred to as nodes. A node x is in the preset of a node y if there is a transition from x to y and, conversely, a node z is in the postset of y if there is a transition from y to z. Then, the preset of a node y is the set $\bullet y = \{x \in P \cup T | (x,y) \in F\}$ and the postset of y is the set $y\bullet = \{z \in P \cup T | (y,z) \in F\}$. A marking m is a multiset of places representing a state during the execution of a system. A transition $t \in T$ is *enabled* at a marking m iff $\bullet t \subseteq m$. An enabled transition t can *fire* and yield a new marking $m' = m - \bullet t + t\bullet$. The reachability graph [21] of a Petri net $\mathcal{N}$ with an initial marking m_0 contains all possible markings of $\mathcal{N}$ – denoted as M. Intuitively, a reachability graph is a deterministic FSM where states denote markings, and arcs denote the transitions fired to go from one marking to another. The complexity of constructing a reachability graph is at worst exponential on the size of the Petri net [19], i.e. $O(2^{|P \cup T|})$.

Definition 6 (Reachability graph). *The* reachability graph *of a Petri net $\mathcal{N}$ is a deterministic finite state machine $\mathcal{R} = (M, A_{\mathcal{R}}, m_0, M_f)$, where M is the set of reachable markings, $A_{\mathcal{R}}$ is the set of arcs $A_{\mathcal{R}} = \{(m_1, \lambda(t), m_2) \in M \times L \times M \mid m_2 = m_1 - \bullet t + t\bullet\}$ and $M_f = \{m \in M \mid \nexists t \in T, \text{ such that } \bullet t \subseteq m\}$.*

Algorithm 1. Remove Tau Transitions

---

input: Reachability Graph $\mathscr{R}$

1 $\sigma \leftarrow \langle m_0 \rangle$;
2 $\Omega \leftarrow \{m_0\}$;
3 **while** $\sigma \neq \langle\rangle$ **do**
4      $m \leftarrow head\ \sigma$ [a];
5      $\sigma \leftarrow tail\ \sigma$ [b];
6      $\Psi \leftarrow \{a = (m_1, l, m) \in \blacktriangleright m \mid l = \tau \wedge m \notin M_f\}$;
7   **for** $a \in \Psi$ **do** $replaceTau(a, m, \{m\})$;
8      $A_{\mathscr{R}} \leftarrow A_{\mathscr{R}} \setminus \Psi$;
9   **for** $(m, l, m_2) \in m\blacktriangleright \mid m_2 \notin \Omega$ **do**
10         $\sigma \leftarrow \sigma \oplus m_2$;
11         $\Omega \leftarrow \Omega \cup \{m_2\}$;

12 $\Xi \leftarrow \{m \in M \mid (\blacktriangleright m = \varnothing \wedge m \neq m_0) \vee (m\blacktriangleright = \varnothing \wedge m \notin M_f)\}$;
13 **while** $\Xi \neq \varnothing$ **do**
14   **for** $m \in \Xi$ **do** $A \leftarrow A \setminus (\blacktriangleright m \cup m\blacktriangleright)$;
15     $M \leftarrow M \setminus \Xi$;
16     $\Xi \leftarrow \{m \in M \mid (\blacktriangleright m = \varnothing \wedge m \neq m_0) \vee (m\blacktriangleright = \varnothing \wedge m \notin M_f)\}$;

17 **return** $\mathscr{R}$;
18 **Function** $replaceTau((m_1, \tau, m_t) \in A, m \in M, \Theta \in 2^M)$
19   **for** $(m, l, m_2) \in m\blacktriangleright$ **do**
20   **if** $l \neq \tau \vee m_2 \in M_f$ **then** $A_{\mathscr{R}} \leftarrow A_{\mathscr{R}} \cup \{(m_1, l, m_2)\}$;
21   **else if** $m_2 \notin \Theta$ **then**
22             $\Theta \leftarrow \Theta \cup \{m_2\}$;
23             $replaceTau((m_1, \tau, m_t), m_2, \Theta)$;

---

[a] *head* in Z notation [14] to obtain the first element of a sequence.
[b] *tail* in Z notation [14] to obtain a subsequence after the first element of a sequence.

A large amount of τ-transitions in a Petri net can lead to large reachability graphs. In principle, we assume that the Petri nets have a minimal number of τ-transitions, e.g., resulting from the application of reduction rules in [24]. However, oftentimes some τ-transitions cannot be removed because they represent the "skip" or parallel execution of transitions. In this regard, we propose a further τ-reduction over the reachability graph that does not modify the underlying behavior. Algorithm 1 shows the top-down approach for the proposed reduction. Intuitively, for each arc $a = (m_1, \tau, m_2)$ referring to a τ-transition, the algorithm replaces a with $a' = (m_1, l, m_3)$ for each outgoing arc of m_2, such that $(m_2, l, m_3) \in A_{\mathscr{R}}$. This replacement is repeated until all arcs referring to τ-transitions are removed. If all incoming arcs of a state m are replaced, then m and its outgoing arcs are removed. The algorithm refrains from removing τ transitions targeting final markings to ensure proper completion. Figure 4 shows the τ-less reachability graph of the loan application process aside. Observe that the arc $[p5, p4] \rightarrow [p6]$ is replaced by $[p5, p4] \rightarrow [p7]$ with label D, and the state $[p3, p2]$ is removed.

3.4 Error-Correcting Synchronised Product

The computation of similar and deviant behavior between an event log and a process model is based on an error-correcting synchronized product (a.k.a.

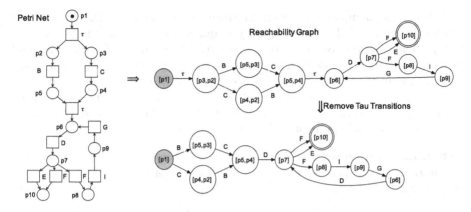

Fig. 4. Petri net obtained from the BPMN model in Fig. 2, and its tau-less reachability graph.

PSP) [5]. Intuitively, the traces represented in the DAFSA are "aligned" with the executions of the model by means of three operations: (1) synchronized move (*match*), the process model and the event log can execute the same task/event w.r.t. label; (2) log operation (*lhide*), an event observed in the log cannot occur in the model; and (3) model operation (*rhide*), a task in the model can occur, but the corresponding event is missing in the log.

Both a trace in a log and an execution represented in a reachability graph are totally ordered sets of events (sequences). Then, an alignment aims at *matching* events from both sequences that represent the same tasks w.r.t. their labels, such that the order between the matched events is preserved. For example, given a trace in a log $\langle B, D, E \rangle$ and an execution in a model $\langle D, B, E \rangle$, it is possible to match the events with label E, and either the events with label B or the events with label D, but not both. An event that is not matched has to be hidden using the operation *lhide* if it belongs to the log, or *rhide* if it belongs to an execution in the model.

In our context, the alignments are computed over a pair of finite state machines, a DAFSA and a reachability graph, therefore the three operations: *match, lhide* and *rhide*, are applied over the arcs of both FSMs. An operation applied over a pair of arcs (one in the DAFSA and one in the reachability graph) is called a *synchronization*. Note that *lhide* and *rhide* are applied only over one arc, thus we use ⊥ to denote the absence of the other element in the triplet.

Definition 7 (Synchronization). *Let $A_{\mathscr{D}}$ and $A_{\mathscr{R}}$ be the arcs in the DAFSA and in the reachability graph, respectively. A synchronization β is a triplet $\beta \subseteq op \times A_{\mathscr{D}} \times A_{\mathscr{R}}$, where $op \in \{match, lhide, rhide\}$. The set of all synchronizations is denoted as S.*

All possible alignments between the traces represented in a DAFSA and the executions represented in a reachability graph can be computed inductively as follows. The construction starts by pairing the initial states of both FSMs and then applying the three defined operations over the events that can occur in the DAFSA and in the reachability graph – each application of the operations (synchronization) yield a new pairing of states. Note that the alignments between (partial) traces and executions are implicitly computed as sequences of synchronizations. Then, an alignment is defined as follows.

Definition 8 (Alignment). *Given a set of synchronizations S, an alignment is defined as $\epsilon = \langle \beta_1, ..., \beta_n \rangle$ with $\beta_i \in S, 1 \leq i \leq n$. All the possible alignments are denoted as $\mathscr{C}$.*

Given an alignment $\epsilon = \langle \beta_1, \beta_2, \ldots, \beta_m \rangle$, we use $\hat{\epsilon}$ to denote the aligned trace in the log, i.e., $\hat{\epsilon} = \langle l_1, l_2, \ldots, l_n \rangle$, such that for any l_i, l_j, where $1 \leq i < j \leq n$, there exist $\beta_x = (op_x, (b_s, l_i, b_t), a_2)$ and $\beta_y = (op_y, (b_v, l_j, b_w), a_3)$ in ϵ, where $1 \leq x < y \leq m$, $op_x \in \{match, lhide\}$ and $op_y \in \{match, lhide\}$. In case there exists β_x but no β_y, $\hat{\epsilon} = \langle l_x \rangle$, and in case there exists no β_x, $\hat{\epsilon}$ is the empty sequence $\langle \rangle$. Thus, $\hat{\epsilon}$ is a sequence of log task labels, that have been aligned in ϵ with *match* or *lhide* operations.

All alignments can be collected in a finite state machine called PSP [5]. Every state in the PSP is a triplet (n, m, ϵ), where n is a state in the DAFSA, m is a state in the reachability graph and ϵ is the (partial) alignment of the events occurred at n and m; every arc is a synchronization; the pairing of the initial states is the initial state; and the finial states are those with no outgoing arcs.

Definition 9 (PSP). *Given a DAFSA $\mathscr{D}$ and a reachability graph $\mathscr{R}$, their PSP $\mathscr{P}$ is a finite state machine $\mathscr{P} = (N_{\mathscr{P}}, A_{\mathscr{P}}, s_{\mathscr{P}}, R_{\mathscr{P}})$, where $N_{\mathscr{P}} \subseteq N_{\mathscr{D}} \times M \times \mathscr{C}$ is the set of nodes, $A_{\mathscr{P}} = N_{\mathscr{P}} \times S \times N_{\mathscr{P}}$ is the set of arcs, $s_{\mathscr{P}} = (s_{\mathscr{D}}, m_0, \langle \rangle) \in N_{\mathscr{P}}$ is the initial node, and $R_{\mathscr{P}} = \{f \in N_{\mathscr{P}} \mid f \blacktriangleright = \varnothing\}$ is the set of final nodes.*

The PSP contains all possible alignments, however we are interested in those containing the minimum amount of hides for each trace in the log. These alignments are called *optimal*. The computation of all optimal alignments can become infeasible when the search space is too large. Thus, we use an A^* algorithm [16] to consider the most promising paths in the PSP first, i.e., those minimizing the number of hides. Given an event log $\mathscr{L}$, the resulting PSP is complete and minimal, since it contains only the optimal alignments for every trace $c \in \mathscr{L}$. The cost function for our A^* algorithm is $\rho(x, c) = g(x) + h(x, c)$, where x is a node in the PSP and c is a trace in the log.

Algorithm 2. Construct the PSP

input: Event Log $\mathscr{L}$, DAFSA $\mathscr{D}$, Reachability Graph $\mathscr{R}$

1 **for** $c \in \mathscr{L}$ **do**
2      $\sigma \leftarrow \{(s_{\mathscr{D}}, \rho(s_{\mathscr{D}}, c))\};$
3      $\rho_{max} \leftarrow |c| + \text{minModelSkips};$
4   **while** $\sigma \neq \varnothing$ **do**
5   choose a tuple $(n_{act} = (n_{\mathscr{D}}, m, \epsilon), \rho) \in \sigma$, such that
             $\nexists (n'_{\mathscr{D}}, \rho') \in \sigma : \rho > \rho';$
6          $\sigma \leftarrow \sigma \setminus \{(n_{act}, \rho)\};$
7   **if** $n_{\mathscr{D}} \in R_{\mathscr{D}} \wedge m \in R_{\mathscr{R}} \wedge \hat{\epsilon} = c$ **then**
8   **if** $\rho(n_{act}, c) < \rho_{max}$ **then**
9                  $\rho_{max} \leftarrow \rho(n_{act}, c);$
10                  $Opt \leftarrow \varnothing;$
11                  $\sigma \leftarrow \{(n, \rho(n, c)) \in \sigma \mid \rho(n, c) \leq \rho_{max}\}$
12              $Opt \leftarrow Opt \cup \{n_{act}\};$
13   **else**
14              $n_{new} \leftarrow \varnothing;$
15   **for** $\alpha_{\mathscr{D}} = (n_{\mathscr{D}}, l_{\mathscr{D}}, n_t) \in n_{\mathscr{D}} \blacktriangleright \mid l_{\mathscr{D}} =$
             $c(|\{\beta = (op, a_{\mathscr{D}}, a_{\mathscr{R}}) \in \epsilon \mid op \neq rhide\}| + 1)^a$ **do**
16                  $n_{new} \leftarrow n_{new} \cup \{(n_t, m, \epsilon \oplus (lhide, \alpha_{\mathscr{D}}, \bot))\};$
17   **for** $\alpha_{\mathscr{R}} = (m, l_{\mathscr{R}}, m_t) \in m \blacktriangleright \mid l_{\mathscr{R}} = l_{\mathscr{D}}$ **do**
18                      $n_{new} \leftarrow n_{new} \cup \{(n_t, m_t, \epsilon \oplus (match, \alpha_{\mathscr{D}}, \alpha_{\mathscr{R}}))\}$
19   **for** $\alpha_{\mathscr{R}} = (m, l_{\mathscr{R}}, m_t) \in m \blacktriangleright$ **do**
                 $n_{new} \leftarrow n_{new} \cup \{(n_{\mathscr{D}}, m_t, \epsilon \oplus (rhide, \bot, \alpha_{\mathscr{R}}))\};$
20              $\sigma \leftarrow \sigma \cup \{(n_{next}, \rho(n_{next}, c)) \mid n_{next} \in n_{new} \wedge \rho(n_{next}, c) \leq \rho_{max}\};$
21   **for** $f \in Opt$ **do** $InsertIntoPSP(f, c, \mathscr{P})$;
22 **return** $\mathscr{P}$;

---

[a] $c(i)$ is the operator in Z notation [14] to obtain the ith element in a sequence.

The current cost function of a state $x = (n, m, \epsilon)$ is $g(x) = |\{(op, a_1, a_2) \in \epsilon \mid op \neq match\} \setminus \{(rhide, \bot, (b_s, l, b_t)) \in \epsilon \mid l = \tau\}|$, i.e., the number of hide operations in an alignment without the operations over the τs. The heuristics function $h(x, c) = min\{|F_{Log}(x, c) \setminus f_{Model}| + |f_{Model} \setminus F_{Log}(x, c)|\}$, such that $f_{Model} \in F_{Model}(x)$, gives an optimistic approximation of the least amount of hide operations required to match the remaining labels in a trace c. In this formula $F_{Log}(x, c)$ represents the future task labels of a trace, such that given $x = (n, m, \epsilon)$, then $F_{Log}(x, c) = MultiSet(c) \setminus MultiSet(\hat{\epsilon})$, i.e., the multiset representation of c minus the labels of the trace matched or hidden so far.[1] The future labels in the model $F_{Model}(n)$ are computed with a bottom-up traversal on the strongly connected components of the reachability graph, where the multisets of task labels are collected and stored in each node of the graph. Observe that h assumes that all events with the same label in F_{Log} and f_{Model} are matched, this

---

[1] *MultiSet* retrieves the multiset representing the labels in a trace or the labels of a set of arcs.

is clearly an optimistic approximation, since some of the those matches might not be possible; then the optimistic approximation computed by h signifies an admissible heuristics for the A^*-search, which guarantees the optimality of the computed alignments.

Algorithm 2 shows the procedure to build the PSP, where an A^* search is applied to find the optimal alignments for each trace in a log. The algorithm chooses a node with minimal cost ρ, such that if it represents the alignment of a complete trace and the pairing of two final states (one in the DAFSA and one in the reachability graph), then it is marked as an optimal alignment. Otherwise, the search continues by applying *lhide*, *rhide* and *match*. As shown in [15], the complexity for constructing the PSP is in the order of $O(3^{|N_{\mathscr{D}}| \cdot |M|})$ where $N_{\mathscr{D}}$ is the set of states in the DAFSA and M is the set of reachable markings of the Petri net.

In order to cope with the complexity of the computation of the PSP, we propose an optimization based on two memoization tables: prefix and suffix memoization tables. Both tables store a set of partial trace alignments for common prefixes and suffixes that have been aligned previously. The tables are constructed incrementally by identifying common prefixes/suffixes after the alignment of each trace and storing the corresponding partial trace alignments. The integration of these tables requires the modification of Algorithm 2, as shown in Algorithm 3. For each trace c, the algorithm starts by checking if there is a common prefix for c in the prefix memoization table. If this is the case, the A^* starts from the nodes after all partial trace alignments for this common prefix instead of the initial node. In the case of common suffix memoization, the algorithm checks at each iteration whether the current pair of nodes and the current suffix is stored in the suffix memoization table. If this is the case, the algorithm appends nodes to the A^* search for each pair of memoized final nodes and appends all partial suffix alignments to the current alignment instead of continuing the regular search procedure. By reusing the information stored in these tables, the search space for the A^* is reduced.

Algorithm 3. Construct the PSP with Prefix- and Suffix Memoization

---

▷ replace line 2 with the following block:

> ▷ Reuse common prefix alignments
> **for** $i = 1 \rightarrow |c|$ **do** $\sigma \leftarrow \sigma \cup \{(n_{next}, \rho(n_{next}, c)) \mid n_{next} \in PrefixTable(c\ for\ i)\}$ [a]
> ;
> **if** $\sigma = \varnothing$ **then** $\sigma \leftarrow \sigma \cup \{(s_{\mathscr{D}}, \rho(s_{\mathscr{D}}, c)\}$;

▷ replace line 14 with the following block:

> ▷ Reuse common suffix alignments
> $suff_{act} \leftarrow c\ after\ |\{\beta = (op, a_{\mathscr{D}}, a_{\mathscr{R}}) \in \epsilon \mid op \neq rhide\}|$ [b];
> $n_{new} \leftarrow \{(f_{\mathscr{D}}, f_{\mathscr{R}}, \epsilon \oplus g_{suff}) \mid (f_{\mathscr{D}}, f_{\mathscr{R}}, g_{suff}) \in SuffixTable(n_{\mathscr{D}}, m, suff_{act})\}$;
> $\sigma \leftarrow \sigma \cup \{(n_{next}, \rho(n_{next}, c)) \mid n_{next} \in n_{new}\}$;
> **if** $n_{new} \neq \varnothing$ **then continue** ;

---

[a] *for* in Z notation [14] to obtain the first i elements of a sequence.
[b] *after* in Z notation [14] to obtain the elements after the first i*th* elements of a sequence.

The approach illustrated so far produces a PSP containing all optimal alignments. Nevertheless, if only one optimal alignment is required, then the algorithm can be easily modified to stop as soon as the first alignment is found. Overall, the complexity of the proposed approach consists of the construction of the DAFSA, the construction of the reachability graph and the computation of the PSP, therefore it is exponential in the worst case, i.e. $O(|L| \cdot \log n + 2^{|P \cup T|} + 3^{|N_{\mathscr{D}}| \cdot |M|})$. The technique presented in this paper does not intend to lower the complexity class for the problem of trace alignment, but rather to implement a more efficient solution within the same complexity class.

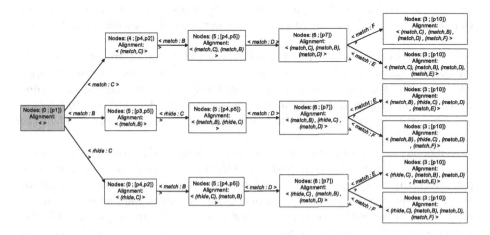

Fig. 5. The PSP for our loan application process example.

Figure 5 shows the PSP obtained by synchronizing the DAFSA of the loan application process in Fig. 3 and the τ-less reachability graph of Fig. 4, we remind the reader that a PSP represents the synchronization of the whole log. To understand its construction let us consider the sample trace $\langle B, D, E \rangle$. Starting from the source node we have $g(n) = 0$, $F_{Log}(n,c) = \{B^1, D^1, E^1\}$, and $F_{Model}(n) = \{B^1, C^1, D^1, E^1\}$. The A^* will compute the cost of performing the following possible synchronizations: $(match, B)$, $(lhide, B)$ $(rhide, B)$, and $(rhide, C)$. Out of these four possibilities it will only explore $(match, B)^2$ and $(rhide, C)$ which have a cost of one. Both $(rhide, B)^3$ and $(lhide, B)$ will never

---

[2] In case of $(match, B)$ we have a current cost of zero since it is a match (i.e. $g(n) = 0$), and a future cost of one (i.e. $h(n,c) = |\{D^1, E^1\} \setminus \{C^1, D^1, E^1\}| + |\{C^1, D^1, E^1\} \setminus \{D^1, E^1\}| = 1$).

[3] In case of $(rhide, B)$ we have a current cost of one since it is a hide (i.e. $g(n) = 1$), and a future cost of two (i.e. $h(n,c) = |\{B^1, D^1, E^1\} \setminus \{C^1, D^1, E^1\}| + |\{C^1, D^1, E^1\} \setminus \{B^1, D^1, E^1\}| = 2$).

be explored since they have a cost of three and there exist nodes with a lower cost. The A^* will continue exploring the possible synchronizations until all optimal alignments are discovered, which are found in nodes f_3 and f_5 for the trace $\langle B, D, E \rangle$.

3.5   Extracting Behavioral Mismatch Statements

In the previous section, we presented a scalable approach to discover a complete set of optimal alignments between an event log and a Petri net. While it is general practice to assess these alignments one-by-one or to aggregate them into a single metric [2], García-Bañuelos et al. [15] showed that practitioners prefer to reason in terms of natural language statements when investigating behavioral issues.

García-Bañuelos et al. [15] defined nine mismatch patterns over the PSP for the generation of natural language statements, which characterize behavior present in the log and missing in the model and vice versa. Out of these nine mismatch patterns we only support the seven patterns related to *unfitting behavior*. The detection of patterns related to *additional model behavior* is out of the scope of this paper.

Differences related to *unfitting behavior* can be divided into relation mismatch patterns and event mismatch patterns. On the one hand, relation mismatch patterns comprise cases when a pair of events in the log has a different behavior relation (sequence, concurrency, conflict) than the corresponding events in the model. E.g., it is possible to obtain statements such as: *In the log, after "A", "B" occurs before "C", while in the model they are concurrent* or *In the model, after "A", "B" occurs before "C", while in the log they are mutually exclusive*. On the other hand, event mismatch patterns characterize all other cases of unfitting behavior; e.g., *In the log, after "A", "B" is optional* or *In the model, "A" occurs after "B" instead of "C"*. In the running example, we return the statement *In the model, "C" occurs after "B" and before "D"*. A detailed description of each pattern and their verbalization can be found in [15].

Given that our technique uses the same PSP as in [15], we adapt their algorithm for the generation of *behavioral mismatch statements*. Similar to the original approach, we rely on an oracle for the computation of concurrency relations between events in a log. We use the local concurrency oracle presented in [6], however other oracles can be used, e.g., $\alpha+$ relations [13]. The approach in [6] helps to alleviate the generalisation of the behavior while computing potential concurrent behavior from a log. Roughly speaking, the local concurrency oracle delimits the scope of a concurrency relation between pairs of events to a pair of execution states. Thus, during the generation of statements, the oracle requires a pair of events, as well as an execution state, and outputs *true* if the given events can occur concurrently at that particular state, or *false* otherwise.

4   Evaluation

The presented approach was implemented as a standalone tool.[4] Given a log in XES or MXML format and a model in BPMN or PNML format, the tool returns a list of one or all-optimal alignments, and the list of behavioral mismatch statements. Other structures, such as the DAFSA, reachability graph and PSP, can be also retrieved.

The conducted set of experiments measure the quality and time performance of our approach in comparison with the trace alignment approach. Our approach was compared against the ProM [33] plugin "Replay a Log on Petri Net for Conformance Checking"[5] [3] for one-optimal trace alignment, and against "Replay a Log on Petri Net for All Optimal Alignments" [3] for the case of all-optimal. This latter plugin relies on different baseline algorithms, but only the "Tree-based state space replay for all optimal alignments" algorithm actually aims at generating all-optimal alignments; however it also returns non-optimal results. Therefore, non-optimal results were filtered out in a post-processing step, i.e., those with bigger cost than the optimal computed by the one-optimal trace alignment (this step was not included in the performance measure). The behavioral alignment approach based on event structures was not included in the evaluation, since this approach showed to be generally slower than trace alignment [15].

The performance was measured in terms of execution time (ms) and quality of the results (number of optimal alignments). The alignments are considered optimal when they have the same cost as one optimal trace alignment. Given that the computation of all-optimal trace alignments oftentimes ran for hours before running out of memory, we use two bounds in the experiments: a time bound of 5 min (after 10 min the alignment will also continue for 12 h [23]) and a state exploration bound of 100,000 states. Hence, we report on all optimal alignments found for each approach until one of the bounds is reached or until termination. The experiments were conducted on a 6-core Xeon E5-1650 3.50 Ghz with 128 GB of RAM running JVM 8.

4.1   Datasets

The experiments use three model-log pairs. The first pair is a (publicly available) dataset of a real-life Italian road fines management process (hereafter RTFMP) [12], its normative description is presented in [20], whereas its model is presented in [15].

The second dataset is the real-life log "closed problems" of the BPI Challenge 2013 (hereafter BPIC13 cp.) [28] This log originates from an IT incident and problem management system used at Volvo. From this log, the model was discovered using Structured Miner [8]. The log was preprocessed with a noise

---

[4] Available from http://apromore.org/tools.
[5] "A* Cost-based Fitness Express with ILP, assuming at most 32,767 tokens in each place".

filter [9]. The resulting model is sound (a requirement for both approaches[6]). In this case the model generated by the Structured Miner has high precision, in contrast to the model discovered by other techniques, such as Inductive Miner [17] that generates an over-generalized model, causing state space explosion when computing all alignments.

The third dataset is the SAP R/3 collection [10], a repository of 604 EPCs documenting the reference model to customize the R/3 ERP product. The models were converted into Petri nets and those with behavioral issues (i.e. unsound models) were filtered out. The event logs were generated from the remaining models using the ProM plugin "Generate Event Log from Petri Net" [30]. This plugin produces unique traces for each possible execution in the model. Next, we filtered out all logs with less than ten unique traces, since such small logs are not useful to measure scalability. This resulted in 120 pairs of real-life models and logs. Additional event logs were created with different levels of noise $(2.5\%, 5\%, 7.5\%, 10\%)$. For that, we duplicated each unique trace in the logs tenfold to maintain the original behavior and used the noise generator tool in [9] to create the noisy logs. This tool inserts events into randomly chosen traces, such that new directly-follows dependencies are created until the noise threshold is reached. The reason for inserting noise is because otherwise there is a perfect fit between the log and the model, and hence the output of the conformance checking is empty, which does not help to test for scalability.

Table 1 provides descriptive statistics of the datasets. The size of the models and of their reachability graphs ($\mathscr{R}$) correspond to the number of places and transitions, and nodes and arcs, respectively. For the SAP R/3 collection, we report the average and standard deviation for the event logs and models, for each noise level. The last column reports on the time required for constructing the reachability graph plus that for removing tau transitions for the given models.

Table 1. Descriptive statistics of the event logs and models.

Dataset	Events	Unique events	Traces	Unique traces	Model size	$\mathscr{R}$ size	$\mathscr{R}$ time (ms)
RTFMP	561,470	12	150,370	231	35	33	16
BPIC13 cp	6,660	5	1,487	183	28	9	96
SAP R/3 2.5%	37,580($\pm$116,515)	15($\pm$5)	2,795($\pm$7,897)	1,062($\pm$3,192)	49($\pm$16)	128($\pm$79)	4($\pm$3)
SAP R/3 5%	38,569($\pm$119,581)	15($\pm$5)	2,795($\pm$7,897)	1,551($\pm$4,830)	49($\pm$16)	128($\pm$79)	4($\pm$3)
SAP R/3 7.5%	39,612($\pm$122,813)	15($\pm$5)	2,795($\pm$7,897)	2,075($\pm$6,147)	49($\pm$16)	128($\pm$79)	4($\pm$3)
SAP R/3 10%	40,712($\pm$126,225)	15($\pm$5)	2,795($\pm$7,897)	2,342($\pm$6,966)	49($\pm$16)	128($\pm$79)	4($\pm$3)

4.2   Results

Table 2 reports the number of optimal alignments and execution times for each conformance checking approach (for the SAP R/3 datasets, we report on the

---

[6] Strictly speaking, trace alignment requires easy-soundness while our approach requires safeness. However both requirements are satisfied by soundness.

average and the upper bound of the 95% confidence interval for these measurements). To ensure comparability of the results we only count the alignments for our approach (shortened as DAFSA in the table) with the same cost as trace alignment, i.e. the same number of asynchronous moves. However, our approach did not detect any additional non-optimal alignments.

In the case of one-optimal, our approach always returned the same number of alignments as trace alignment. Both approaches are expected to find one-optimal alignment per unique trace of an event log, thus the number of alignments and the number of unique traces is the same. However, there is no intuitive expectation for all-optimal alignments. In this regard, our approach found many more optimal alignments than trace alignment within the same state space and time bounds. For example, on the RTFMP log, our approach found 467 alignments instead of 338 returned by trace alignment. This difference increases substantially in the other datasets: in logs with high noise levels (SAP R/3 7.5 and 10%), our approach returned up to five times the number of all-optimal alignments than trace alignment. This is due to the reuse of partial trace alignments (prefix and suffix memoization). Our approach scaled well to the number of unique traces and to the amount of unfitting behavior observed in the logs. Additionally, (all-optimal) trace alignment suffers from reporting non-optimal results that have to be filtered in a preprocessing step[7] (unfiltered results are reported in square brackets). Thus, our approach was capable of finding a more complete set of alignments.

Comparing the execution times for the all-optimal variants, our approach outperforms the tree-based trace alignment approach by 1–2 orders of magnitude. For example, our approach took 125 ms to compute all alignments for the RTFMP dataset, as opposed to 52 s for trace alignment. Additionally, trace alignment times out in 207 out of 480 cases for the SAP dataset, while our approach only timed out in two cases (trace alignment also timed out in these two cases). Our one-optimal variant performs 1.5 to nearly 40 times faster (trace alignment timed out in 2 cases for the SAP datasets, while our approach never timed out). Only in the BPIC13 cp. dataset the trace alignment outperformed our approach by nearly a factor two. The process model in this dataset contains a large state space due to the presence of nested loops, which can lead to a combinatorial state space to be explored by the A^* algorithm. Thus, when the estimation of our heuristics is imprecise due to complex loop structures, the memory and time requirements increase. Conversely, trace alignment uses a more accurate heuristic function, which leads to outperforms our approach in the BPIC13 cp. dataset. In short, the execution times positively correlate with the number of unique traces in a log, both approaches, DAFSA and trace alignment, apply an A^* algorithm for each unique trace. Our approach scales better in the case of more complex SAP R/3 logs, which exhibit a very high number of unique traces compared to the RTFMP and BPIC13 cp., as per Table 1. Our approach calculates all

---

[7] An alignment was filtered if it had a higher cost than that computed by one-optimal alignment or if it represented the swap of the label of an invisible task with that of a visible one.

optimal alignments in less than ten seconds, while trace alignment reaches the time bound of five minutes on average in every second model-log pair for the same dataset.

Table 2. Evaluation results.

Dataset	Optimal alignments (upper bound of 95% confidence interval)				Execution time (ms) (upper bound of 95% confidence interval)			
	All optimal		One optimal		All optimal		One optimal	
	DAFSA	Trace align. [#unfiltered]	DAFSA	Trace align.	DAFSA	Trace align.	DAFSA	Trace align.
RTFMP	467	338 [1,898,182]	231	231	**125**	52,041	**56**	1,844
BPIC13 cp.	28,656	22,259 [1,904,057]	183	183	**5,360**	50,160	453	**260**
SAP R/3 2.5%	4,253 (22,675)	1,233 [1,067,533] (6,470 [1,929,629])	1,062 (7,319)	1,062 (7,319)	**1,102** (7,778)	127,013 (300,000)	**814** (6,132)	1,800 (12,891)
SAP R/3 5%	7,672 (41,133)	1,751 [1,224,079] (9,178 [2,199,248])	1,551 (11,019)	1,551 (11,019)	**2,832** (28,040)	150,017 (300,000)	**1,718** (18,696)	3,415 (25,857)
SAP R/3 7.5%	11,652 (61,504)	2,154 [1,283,583] (14,207 [3,039,240])	2,075 (14,122)	2,075 (14,122)	**3,208** (19,502)	163,593 (300,000)	**2,083** (12,912)	4,967 (58,961)
SAP R/3 10%	15,754 (84,167)	2,809 [1,286,568] (22,883 [3,302,068])	2,342 (15,996)	2,342 (15,996)	**8,204** (75,643)	173,438 (300,000)	**3,480** (25,371)	7,365 (66,003)

In our approach, a trade-off between the execution time and number of alignments is observed from the comparison of the results of one-optimal versus all-optimal. It is more obvious when the amount of unfitting behavior increases. E.g., in the logs SAP R/3 with 10% noise level, our approach took, on average, five seconds longer for computing all-optimal alignments than one-optimal, but returns ten times more alignments.

The extraction of behavioral mismatch statements shows that a large number of alignments can be represented by a significantly smaller, yet interpretable, number of statements. E.g., 3,295 all-optimal alignments in the BPIC13 cp. dataset can be summarized with only 14 statements, and reduced to eight statements in the case of one-optimal alignment. In the RTFMP dataset, 120 statements were computed from 467 all-optimal alignments, whereas only 69 statements were computed for the one-optimal variant. Some example statements are:

- *In the log, after "Insert Fine Notification", "Payment" occurs before task "Add penalty", while in the model they are mutually exclusive.*
- *In the log, after "Add penalty", "Payment" is substituted by "Send Appeal to Prefecture".*

5   Conclusion

We showed that the problem of conformance checking can be mapped to that of computing a minimal error-correcting product between an automaton representing the event log (its minimal DAFSA) and an automaton representing the process model (its reachability graph). The resulting product automaton can be used to produce sets of optimal alignments between each trace in the log and a corresponding trace in the model, or even statements capturing behavioral relations (e.g. conflict relations) observed in a state of the DAFSA but not captured in the corresponding state in the model.

The use of a DAFSA to represent the event log allows us to benefit from both prefix and suffix compression of the traces in the log. This is a distinctive feature of the proposal with respect to trace alignment, which computes an alignment between each trace in the log and the model, without any reuse across traces. Due to this distinctive feature, our approach addresses some of the scalability issues of existing conformance checking techniques allowing more interactivity in redesigning process models with conformance issues. The approach can be employed to assess the quality of automated process model discovery techniques, as well as used as the cornerstone technique for process model repair.

The empirical results show that the execution times of our approach are one to two orders of magnitude faster than those of the all-optimal trace alignment method. When restricted to the problem of computing one alignment per trace (one-optimal alignment), our approach generally but not always outperforms the baseline [2]. This is attributable to the fact that the latter uses a tight heuristic function, whereas the heuristic function we use over-approximates in some cases. Designing a better heuristic function is a direction for future work. The ideas in [2] are not directly transposable as the heuristic function in our approach needs to compute a bound starting from a state of the whole log (the DAFSA), while [2] does so for one trace at a time.

In this paper, we focused on the problem of identifying unfitting log behavior. A possible avenue for future work is to extend the approach to detect additional model behavior, by adapting the ideas proposed in [15] in the context of event structures.

Finally, the empirical evaluation, while based on synthetic and real-life models and logs, is limited in that it only covers models with sizes of up to 50 tasks and logs with up to ca. 2.5K distinct traces. Conducting a more thorough evaluation with even larger process models and event logs is another avenue for future work.

Acknowledgments. This research is partly funded by the Australian Research Council (grant DP150103356) and the Estonian Research Council (grant IUT20-55).

References

1. Adriansyah, A., Muñoz-Gama, J., Carmona, J., van Dongen, B.F., van der Aalst, W.M.P.: Measuring precision of modeled behavior. IseB **13**(1), 37–67 (2015)
2. Adriansyah, A., van Dongen, B.F., van der Aalst, W.M.P.: Conformance checking using cost-based fitness analysis. In: Proceeding of EDOC, pp. 55–64. IEEE (2011)
3. Adriansyah, A., van Dongen, B.F., van der Aalst, W.M.P.: Memory-efficient alignment of observed and modeled behavior. BPM Center Report (2013)
4. Alves de Medeiros, A.K.: Genetic Process Mining. PhD thesis, TU/e (2006)
5. Armas-Cervantes, A., Baldan, P., Dumas, M., García-Bañuelos, L.: Diagnosing behavioral differences between business process models: An approach based on event structures. Inf. Syst. **56**, 304–325 (2016)

6. Armas-Cervantes, A., Dumas, M., La Rosa, M.: Discovering local concurrency relations in business process event logs. eprint # 102438, QUT (2016)
7. Armas-Cervantes, A., La Rosa, M., Dumas Menjivar, M., García-Bañuelos, L., van Beest, N.R.: Interactive and incremental business process model repair. eprint # 106611, QUT (2017)
8. Augusto, A., Conforti, R., Dumas, M., La Rosa, M., Bruno, G.: Automated discovery of structured process models: discover structured vs. discover and structure. In: Comyn-Wattiau, I., Tanaka, K., Song, I.-Y., Yamamoto, S., Saeki, M. (eds.) ER 2016. LNCS, vol. 9974, pp. 313–329. Springer, Cham (2016). doi:10.1007/978-3-319-46397-1_25
9. Conforti, R., La Rosa, M., ter Hofstede, A.H.M.: Filtering out infrequent behavior from business process event logs. IEEE TKDE **29**(2), 300–314 (2016)
10. Curran, T., Keller, G.: SAP R/3 Business Blueprint: Understanding the Business Process Reference Model. Upper Saddle River (1997)
11. Daciuk, J., Mihov, S., Watson, B.W., Watson, R.E.: Incremental construction of minimal acyclic finite-state automata. Comput. Linguist. **26**(1), 3–16 (2000)
12. de Leoni, M., Mannhardt, F.: Road traffic fine management process (2015)
13. de Medeiros, A.K.A., van der Aalst, W.M.P., Weijters, A.J.M.M.: Workflow mining: current status and future directions. In: Meersman, R., Tari, Z., Schmidt, D.C. (eds.) OTM 2003. LNCS, vol. 2888, pp. 389–406. Springer, Heidelberg (2003). doi:10.1007/978-3-540-39964-3_25
14. Diller, A.: Z: An Introduction to Formal Methods. Wiley, New York (1990)
15. García-Bañuelos, L., van Beest, N.R.T.P., Dumas, M., La Rosa, M.: Complete and interpretable conformance checking of business processes. IEEE Trans. Softw. Eng. (2017, to appear). doi:10.1109/TSE.2017.2668418. IEEE Computer Society
16. Hart, P.E., Nilsson, N.J., Raphael, B.: A formal basis for the heuristic determination of minimum cost paths. IEEE TSSC **4**(2), 100–107 (1968)
17. Leemans, S.J.J., Fahland, D., van der Aalst, W.M.P.: Discovering block-structured process models from event logs - a constructive approach. In: Colom, J.-M., Desel, J. (eds.) PETRI NETS 2013. LNCS, vol. 7927, pp. 311–329. Springer, Heidelberg (2013). doi:10.1007/978-3-642-38697-8_17
18. Leemans, S.J., Fahland, D., van der Aalst, W.M.: Scalable process discovery and conformance checking. Softw. Syst. Model. **16**, 1–33 (2016)
19. Lipton, R.: The reachability problem requires exponential space. Research Report 62, Department of Computer Science, Yale University, New Haven, Connecticut (1976)
20. Mannhardt, F., de Leoni, M., Reijers, H.A., van der Aalst, W.M.P.: Balanced multi-perspective checking of process conformance. Computing **98**, 407–437 (2016)
21. Mayr, E.W.: An algorithm for the general petri net reachability problem. SIAM J. Comput. **13**(3), 441–460 (1984)
22. Muñoz-Gama, J., Carmona, J.: A fresh look at precision in process conformance. In: Hull, R., Mendling, J., Tai, S. (eds.) BPM 2010. LNCS, vol. 6336, pp. 211–226. Springer, Heidelberg (2010). doi:10.1007/978-3-642-15618-2_16
23. Muñoz-Gama, J., Carmona, J., van der Aalst, W.M.P.: Single-entry single-exit decomposed conformance checking. Inf. Syst. **46**, 102–122 (2014)
24. Murata, T.: Petri nets: Properties, analysis and applications. Proc. IEEE **77**(4), 541–580 (1989)
25. Polyvyanyy, A., Van Der Aalst, W.M.P., Ter Hofstede, A.H.M., Wynn, M.T.: Impact-driven process model repair. ACM Trans. Softw. Eng. Methodol. (TOSEM) **25**(4), 28 (2016)

26. Rozinat, A., van der Aalst, W.M.P.: Conformance checking of processes based on monitoring real behavior. Inf. Syst. **33**(1), 64–95 (2008)
27. Song, W., Xia, X., Jacobsen, H.A., Zhang, P., Hu, H.: Efficient alignment between event logs and process models. IEEE Trans. Serv. Comput. **10**(1), 136–149 (2017)
28. Steeman, W.: Bpi challenge 2013, closed problems (2013)
29. van Dongen, B., Carmona, J., Chatain, T., Taymouri, F.: Aligning modeled and observed behavior: a compromise between computation complexity and quality. In: Dubois, E., Pohl, K. (eds.) CAiSE 2017. LNCS, vol. 10253, pp. 94–109. Springer, Cham (2017). doi:10.1007/978-3-319-59536-8_7
30. vanden Broucke, S., De Weerdt, J., Vanthienen, J., Baesens, B.: An improved process event log artificial negative event generator. Technical Report KBL_1216, KU Leuven (2012)
31. vanden Broucke, S.K.L.M., De Weerdt, J., Vanthienen, J., Baesens, B.: Determining process model precision and generalization with weighted artificial negative events. IEEE TKDE **26**(8), 1877–1889 (2014)
32. vanden Broucke, S.K.L.M., Munoz-Gama, J., Carmona, J., Baesens, B., Vanthienen, J.: Event-based real-time decomposed conformance analysis. In: Meersman, R., Panetto, H., Dillon, T., Missikoff, M., Liu, L., Pastor, O., Cuzzocrea, A., Sellis, T. (eds.) OTM 2014. LNCS, vol. 8841, pp. 345–363. Springer, Heidelberg (2014). doi:10.1007/978-3-662-45563-0_20
33. Verbeek, H.M.W., Buijs, J.C.A.M., Van Dongen, B.F., Van der Aalst, W.M.P.: Prom 6: The process mining toolkit. Proc. BPM Demonstr. Track **615**, 34–39 (2010)

Domain-Independent Monitoring and Visualization of SLA Metrics in Multi-provider Environments

(Short Paper)

Robert Engel[1]([✉]), Bryant Chen[1], Shashank Rajamoni[1], Heiko Ludwig[1], Alexander Keller[2], Mohamed Mohamed[1], and Samir Tata[1]

[1] Almaden Research Center, IBM Research, San Jose, CA, USA
engelrob@us.ibm.com
[2] IBM Global Technology Services, Chicago, IL, USA

Abstract. Over the past decade IT services have become increasingly multi-sourced, from IT outsourcing of functions such as network management and server provisioning to the widespread use of micro-services from different vendors in Cloud applications. Managing end-to-end service quality is becoming an increasing challenge. One of the key issues lies in the exchange of quality-related data between service participants in order to analyze not only SLA compliance but also root causes of failure. Formal SLA specification and management approaches today do not facilitate the specification of metric data in conjunction with a convenient way to supply this data to business analytics tools for detailed analysis. In this industry paper, we propose an extension of the rSLA DSL defining data items to be monitored and a specification-driven generator producing relational, denormalized time series tables accordingly. The resulting time series can be ingested by commonly used Business Intelligence tools such as Watson Analytics or Tableau in a straightforward and efficient manner. Using this approach, service integrators quickly identify problem areas in federated service delivery and consequently improve operational efficiency and avoid penalties related to SLA violations.

Keywords: SLA · Monitoring · Cloud computing · Measurement · DSL

1 Introduction

Service Integration and Management (SIAM) deals with the management of multiple suppliers of IT and business services and integrating them to provide a single, business-facing IT organization. Service Level Agreements (SLAs) are essential to SIAM processes since they set boundaries and expectations with respect to the services to be delivered. However, SLA management and regulatory compliance continues to be challenging and has recently been identified as the single most relevant inhibitor for future growth of the SIAM market [9].

© Springer International Publishing AG 2017
H. Panetto et al. (Eds.): OTM 2017 Conferences, Part I, LNCS 10573, pp. 628–638, 2017.
https://doi.org/10.1007/978-3-319-69462-7_39

In prior art, several frameworks for the formal specification and automated management of SLAs have been proposed (cf. Sect. 3). For instance, the rSLA domain-specific language (DSL) and the rSLA framework allow SIAM providers to declaratively specify SLAs in multi-provider environments, collect relevant metrics, evaluate Service Level Objectives (SLOs), and execute actions to remedy situations of SLA violations or near-violations [6, 10]. However, managing operations of services and maintaining SLA compliance requires that SIAM providers understand details of SLO violations and validations (e.g. the "extent" to which an SLO has been missed). Moreover, visual analysis of fine-granular time series of metrics underlying SLO evaluations can help identify root causes of phenomena that are eventually visible on the SLO level. This requires active, continuous monitoring of not only SLOs, but also underlying performance metrics.

While the rSLA framework and other approaches to SLA management tend to focus on the definition and evaluation of SLOs and the automatic execution of actions, the efficient monitoring of structured time series data of SLA-related metrics that are decoupled from and logically preceding the concept of SLOs (e.g., composite metrics) currently poses a challenge (cf. Sect. 2). In this industry paper, we address this problem by extending the rSLA language with a *Monitor* construct and provide an implementation that executes these specifications in the rSLA Service. An rSLA *Monitor* allows the SIAM provider to define metrics to be monitored and how they are to be persisted in time series tables.

The structure of the paper is as follows: In the next section we discuss in more depth the background and the motivation for our problem. Subsequently, related work is discussed. In Sects. 4 and 5 we propose our solution, an extension of the rSLA language with *Monitors* and a change in the rSLA engine to generate data for ingestion in commonly used Business Intelligence tools such as Watson Analytics or Tableau. A case study follows it to discuss the viability of our approach. Finally, we summarize and conclude.

2   Background and Motivation

Many frameworks for managing SLAs in multi-tenant environments provide formal languages to specify metrics (e.g., using an expression language), as well as means for executing actions based on condition evaluations over those metrics. These frameworks behave similarly, and we use the rSLA framework [6, 10] as a starting point for the new monitoring constructs proposed in this paper.

The rSLA framework is made up of three main components: (i) the rSLA language to formally represent SLAs, (ii) the rSLA Service, which interprets the SLAs and implements the behavior specified in them, and (iii) a set of Xlets - lightweight, dynamic adapters for monitoring and configuration.

The rSLA language includes the following concepts for describing SLAs and corresponding actions. *Base metrics* describe how metrics are to be obtained from instrumentation. *Composite metrics* aggregate values of other metrics (basic or composite). An aggregation is described using expressions over values from the metrics it depends on. *Service Level Objectives (SLOs)* define the

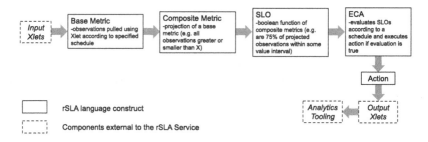

Fig. 1. Current monitoring pipeline in the rSLA Service

commitment of a service provider to its customer. This is defined in a Boolean expression over metric values. *Actions* designate operations to be taken in case of events like violation or validation of SLOs. Typically, actions are expressed in the form remote procedure calls (i.e., HTTP endpoints for calling RESTful APIs). *Event-Condition-Actions (ECAs)* describe active rules and are composed of three parts: event, condition and action. The event specifies when the rule is considered for evaluation. The condition, when satisfied, causes the execution of an action.

The rSLA framework is designed to handle all input and output (i.e., all communication with external systems) through Xlets. Input Xlets provide measurements and Output Xlets execute actions based on the results of SLO evaluations as described in corresponding ECA and Action definitions. By using an Xlet that consumes the results of SLO evaluations, rSLA users can describe how to proceed if SLOs are met or violated (e.g., changing the system configuration or auto-scaling a micro-service). This same mechanism can be used to export SLO evaluation results for reporting and analytics purposes (cf. Fig. 1). For instance, [10] describes an exemplary use case of the rSLA framework in which a *Reporting Xlet* is used to consume the results of SLO evaluations in order to generate reports on SLA compliance over discrete time periods.

2.1   Shortcomings of ECA-Based Monitoring Pipelines

Most approaches to SLA management in current literature (cf. Sect. 3) focus on the above described ECA paradigm for SLA monitoring on the level of SLOs, but for managing operations of services and maintaining their compliance with SLAs it is important for SIAM providers to understand the details of SLO violations such as the "extent" to which an SLO has been missed or the extent to which SLOs may have been fulfilled or over-fullfilled. Moreover, for determining the root cause of a particular SLO violation, validation, or its over-fullfillment, they need to be able to examine and analyze in detail the time series of various low-level metrics directly and indirectly related to that SLO.

Typical BI tools, such as IBM Watson Analytics or Tableau, lend themselves to the analysis of such time series of metrics since they allow users to compose visualizations and dashboards by dragging and dropping tables and columns onto design canvases (i.e., without any programming skills) and subsequently view and compare time series, drill down, slice the data, etc. Moreover, such BI analytics tools are optimized for processing and rendering of big data sets, as typically found in multi-tenant environments with many services, metrics, and time frames to be taken into account.

Hence, the question arises how detailed time-series data of metrics underlying SLO evaluations can be shared in an efficient manner with BI analytics tools, ideally enabling real-time analysis. The ECA-style monitoring pipelines of typical SLA management systems are unsuitable for connecting SLA management systems with BI tools for in-depth analytics for reasons including the following:

1. The need of specifying events, conditions and/or actions for every datapoint of a time series to be exported does not facilitate the "bulk" export of time series data. This is not only inconvenient for the above described analytics use case, but can also result in significant (and in many cases unacceptable) overhead in terms of performance. This holds especially true for SLA management systems in which actions are designed as remote procedure calls.
2. In some SLA management frameworks (such as rSLA without the extensions proposed in this industry paper), reporting through ECA-style monitoring pipelines may be limited to whether individual SLOs or arbitrary expressions have been validated or violated, i.e., to time series of Boolean values only.

2.2 Other Approaches

Further conceivable approaches include the use of alternative monitoring tools in parallel with an SLA management system in order to monitor metrics of interest in a more detailed fashion. However, this approach suffers from (i) potential misalignment and/or semantic heterogeneity of the monitored metrics with SLOs as defined in the SLA (e.g., different units of measurement, differences in breakdown of values with respect to time windows, different data sources, etc.), and (ii) additional overhead for configuration and operation of the additionally deployed monitoring tools. Similarly, directly accessing/exporting observations tables of the SLA management system (i.e., raw data from instrumentation) for ingestion by BI analytics tools may not provide data from downstream metrics evaluations in the context of an SLA (i.e., from metrics "closer" to the SLA's SLOs). On the other hand, simply exporting *all* metric and/or expression evaluations from an SLA management system may result in a potentially ill-structured superset of the data that is actually needed for analysis, in turn necessitating the out-of-band definition of domain-specific Extract-Transform-Load (ETL) pipelines before off-the-shelf BI analytics tools can ingest the data.

3   Related Work

The most widely used approaches for formally specifying SLAs in research and industry are WS-Agreement [1], and Web Service Level Agreement (WSLA) [5].

WSLA consists of a flexible SLA language for Web services based on XML Schema, and an architecture for interpreting the language and monitoring performance at runtime. The monitoring approach taken by WSLA is centered on notifications through *Operation Descriptions*, a mechanism for remote procedure call notifications similar to rSLA's Actions. WS-Agreement is also based on XML but is targeted at establishing and managing SLAs in the context of Grid services. Both the WSLA and WS-Agreement language specifications focus on the SLA communication protocol rather than on SLA data management aspects [11, pp. 39–40], and do not provide explicit constructs for the customization of time series data from metrics evaluations. Such explicit constructs are – best to our knowledge – neither available in other existing approaches to SLA modeling (e.g., SLA* [2], CSLA [3], SLAng [4], GXLA [12]). In [7], the authors propose to extend WS-Agreement with an event calculus based language, called EC-Assertion. While their proposed approach could be potentially used for the specification of specific time series to be monitored, individual records in time series tables would only be created as a result of a triggered event (e.g., as illustrated in [8]). For an extensive discussion of approaches for the specification and representation of SLAs in current literature the reader is referred to [10].

There is a plethora of tools and approaches in current literature for performance monitoring independent of the notion of SLAs. *Prometheus*[1] is an opensource systems monitoring and alerting toolkit featuring a multi-dimensional data model and a query language to query time series data. Graphite[2], InfluxDB[3], and OpenTSDB[4] are specialized databases for storing time series data. While these tools can be potentially used to monitor SLA-related metrics, they do not provide explicit constructs for modeling or representing SLAs.

This list is far from being exhaustive. A plethora of other proposals to provide monitoring in the cloud exists but, as far as we know, none of them successfully addresses the issue of efficient monitoring of time series based on metrics evaluations with explicit SLA semantics. This assessment is corroborated by a recent literature review study [11, pp. 39–40], which identified a number of research gaps in SLA management including *data methods for the systematic treatment of information used in [..] service level monitoring [and] auditing during/after service runtime*, and *absence of systematic query methods for SLA data*.

4   rSLA Monitors

In order to enable the efficient sharing of time series data from SLA management systems with BI analytics tools, we propose an approach based on explicit constructs in SLA management systems that define how evaluation results underlying or preceding SLO evaluations are to be exported into well-defined time series tables. We do so by extending the rSLA language by a corresponding

---

[1] http://prometheus.io.

[2] http://graphite.readthedocs.io.

[3] http://influxdata.com.

[4] http://opentsdb.net.

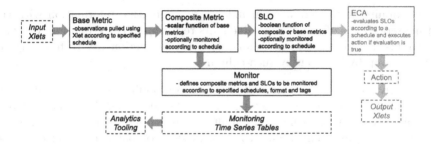

Fig. 2. Monitoring pipeline in rSLA extended by Monitors

Monitor construct and provide an implementation that executes these specifications in the rSLA Service. With the introduction of Monitors, the monitoring data pipeline of the rSLA Service (cf. Fig. 1) is modified as illustrated in Fig. 2.

4.1 Monitor Definitions

A Monitor is defined in an rSLA document according to the following grammar:
⟨*monitor*⟩ ::= monitor do
    entity ⟨*id*⟩
    name ⟨*id*⟩
    [compositemetrics ⟨*metric_id*⟩ as ⟨*column*⟩(, ⟨*metric_id*⟩ as ⟨*column*⟩)*]
    [slos ⟨*slo_id*⟩ as ⟨*column*⟩(, ⟨*slo_id*⟩ as ⟨*column*⟩)*]
    [schedule ⟨*schedule_id*⟩(, ⟨*schedule_id*⟩)*]
    [tags ⟨*key*⟩ = ⟨*value*⟩(, ⟨*key*⟩ = ⟨*value*⟩)*]
    end

- The Monitor's *entity* refers to the database table in which the time series represented by this Monitor is to be persisted.
- The *compositemetrics* and *slos* definitions are used to specify which metric values and SLO evaluation results are to be included in individual data points of the time series. The labels of the values within this vector can be specified through the *column* attribute and result in correspondingly named columns in the rendered time series table.
- The periodicity of the time series represented by the Monitor is specified by its *schedule* attribute. The periodicity also determines the time window for which observations pertaining to base metrics will be included in the computation of composite metrics and SLOs of a given Monitor.[5]

---

[5] The Monitor's schedule overrides any potentially existing schedule definitions directly placed on the referenced SLO objects, which are exclusively used for determining time windows when performing SLO evaluations in the context of ECA jobs.

– The Monitor's *tags* allow to include static contextual information with the time series in the form of key-value pairs resulting in correspondingly named columns. BI analytics tools commonly provide functionality to filter, aggregate, group by and join by such columns.

On activation of a particular SLA the corresponding time series tables are dynamically created according to the Monitors defined in the SLA. Since different Monitors can share the same *entity*, multiple time series can be persisted in the same database table. This allows the side-by-side comparison and/or combination of related time series represented by different Monitors in BI analytics tools, which typically work best with denormalized tables for performance reasons.

4.2 Example

The following Monitor captures time series data for composite metrics and SLOs of task executions of *task1* in the context of an SLA concerned with execution times of service delivery processes of a SIAM provider. The Monitor defines both weekly and monthly schedules along with contextual information specifying the task's provider and its corresponding service offering (i.e., process definition):

```
monitor do
  entity"Tasks"
  name"task1"
  compositemetrics FAST_TASK1 as Fast, SLOW_TASK1 as Slow, COUNT_TASK1 as Count
  slos SLO_TASK1 as SLO
  schedule"every1w","every1m"
  tags"Provider"="cloudauto","Offering"="ITDCCP1583"
end
```

Assuming similar Monitor definitions for *Task2* and *Task3* with a *schedule* of every1w, a corresponding time series table is rendered as illustrated in Table 1.

Table 1. Time series table example: `Tasks` table

ID	Schedule	Timestamp	Provider	Offering	SLO	Fast	Slow	Count
task1	everyweek	4/1/17 0:00	cloudauto	ITDCCP1583	1	15	5	20
task2	everyweek	4/1/17 0:00	team1	ITDCCP1873	0	7	8	15
task3	everyweek	4/1/17 0:00	resauto	ITDCCP1001	1	25	5	30
task1	everyweek	3/24/17 0:00	cloudauto	ITDCCP1583	1	12	2	14
task2	everyweek	3/24/17 0:00	team1	ITDCCP1873	1	9	1	10
task3	everyweek	3/24/17 0:00	resauto	ITDCCP1001	0	10	10	20

5 Implementation

We implemented the rSLA Service with Monitors in a Java EE based implementation, which consists of three services, namely Collector, Executor, and Lifecycle

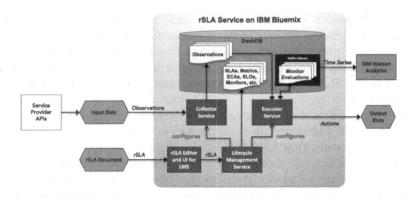

Fig. 3. Implementation of the rSLA Service with Monitors

Management, as shown in Fig. 3. These services are deployed in IBM Bluemix, a PaaS infrastructure that offers horizontal scalability and high availability.

The purpose of the Collector service is to gather observations according to schedules and base metrics specified in the SLA and persist them in the database. The Executor service performs jobs for evaluating Monitors and ECAs according to the schedules specified in the rSLA document. Evaluations for monitored metrics and SLOs are persisted in a public schema, which can be accessed by a BI analytics tool (in our case, Watson Analytics). Finally, the Lifecycle Management service parses the rSLA document and schedules the jobs in the Collector and Executor services. The rSLA language is defined using the Xtext DSL framework, which generates the visual rSLA Editor component for the rSLA Service GUI.

During execution, the time series tables specified by the rSLA Monitors are regularly imported into Watson Analytics using Data Connect. Pre-designed dashboards are then regularly updated using the imported data.

6   Case Study

In this section, we describe a case study in which we applied the rSLA framework with Monitors to an active service management engagement with an IBM client. In this case, IBM in its role as SIAM provider integrates and manages multiple IT and business services of different suppliers (i.e., "service towers") such that the client experiences a single business-facing IT organization. The SIAM provider has an SLA in place with the client that is governing the service levels of the integrated IT environment. Similarly, the SIAM provider has negotiated individual SLAs with the providers of the individual services to be integrated.

Service requests represent services delivered by the SIAM provider to the client on their request and typically carry concrete business value for the client (e.g., *"MS SQL Standard/Enterprise on Windows Virtual Server"*). The SIAM provider fulfills service requests by integrating various tasks that pertain to different service providers (e.g., *"Provision Virtual Machine in AWS"*, *"Install and*

Configure Windows OS Image", etc.) according to a *service offering*. In other words, a service offering consists of several tasks and a service request consists of several task instances (i.e., task executions). The SLO for each offering requires that 95% of service requests must be completed in the maximum duration that is defined in the SLA between the SIAM provider and the client. Similarly, the SLO for each task requires that 75% of instances must be completed by the individual service providers in the maximum duration specified in the individual SLAs between the SIAM provider and the service providers.

In order to ensure that IBM fulfills its obligations to its client, the SIAM manager needs to actively monitor each SLO and related performance metrics (e.g., the percentage of service requests that are not being completed in the specified duration, average duration of service requests, etc.). For problematic service offerings, the manager should be able to easily assess which individual tasks and which providers are responsible for poor performance. Similarly, he/she should be able to analyze the performance of various providers for different tasks.

Our dataset consists of nine months of execution times (beginning of Q4 2016 to end of Q2 2017) for service requests and their corresponding tasks from an IBM client. During this period, there were 34,660 successfully completed service requests and 176,954 task instances. To perform a rapid, live simulation using this data, we implemented an Xlet that not only reads the dataset and feeds it to the rSLA Collector service, but also projects the timestamps into the present and compresses them so that one month of data is collected in one minute.

For each service offering and task, composite metrics SLOW (number of instances whose duration was greater than the maximum duration), CT (number of instances), and SUM (sum of durations for each instance) were defined. The SUM composite metric will be used to compute averages over different time periods in Watson Analytics. Each of these metrics, as well as the SLO, were included in a Monitor entity for the service offering/task. The schedule for each Monitor was set to every minute, which translates to every month in the dataset.

Lastly, dashboards were created in Watson Analytics to visualize the data generated by the monitoring pipeline. These dashboards allow the user to visualize SLO violations over time for service offerings and the tasks that comprise them. They also allow the user to drill down into particular service offerings or providers, allowing him/her to determine which tasks and providers are responsible for poor performance at the service offering level. Conversely, they also allow the user to aggregate task SLO evaluations and performance metrics at the provider and offering level to assess the overall performance of providers. Lastly, they allow the user to visualize the extent to which offering and task SLOs are being violated or over-fulfilled.

Once the base metrics, composite metrics, SLOs, and Monitors are defined in the SLA document, it can be uploaded and activated in the rSLA Service. Once activated, the client dataset is uploaded to the Xlet, which projects the timestamps to the present. Now, the Collector will begin collecting observations from the Xlet every minute, as specified in the SLA document. Similarly, the Executor will begin evaluating monitored composite metrics and SLOs every minute, as specified in the SLA document, and persists the results in the monitoring tables. These tables are then uploaded every minute to Watson Analytics.[6]

The global delivery project executive for this client has confirmed that the rSLA framework with Monitors provides utility and value to IBM in SIAM settings and deployment is currently underway. Thus, we feel that this case study has demonstrated that our extensions to the rSLA framework successfully enable the desired monitoring and analysis of SLA performance.

7   Conclusion

In this industry paper, we addressed the problem of managing service quality and SLA performance in multi-provider environments. We provided a means for the specification of metrics relevant for root cause analysis of quality behavior in typical SLA management frameworks by introducing *Monitors*. Monitors specify vectors of time series of SLA-related metrics and their representation in domain-independent, denormalized, relational time series tables suitable for convenient and efficient import into BI analytics tools such as Watson Analytics or Tableau.

Our approach enables the real-time analysis of detailed time series data underlying SLO evaluations without the performance problems and domain-specific setup efforts faced with other approaches. Dashboards showing deep insights into the root causes of phenomena eventually visible on the SLO level can be designed without any programming skills. We demonstrated our approach in a case study on SLA-related service integration process metrics of a client in which we drilled down on specific tasks and providers that were responsible for SLO violations, thereby helping service integration managers quickly identify problem areas and consequently improve operational efficiency with respect to SLA management.

References

1. Andrieux, A., Czajkowski, K., Dan, A., Keahey, K., Ludwig, H., et al.: WS-Agreement Specification. Technical report, Global Grid Forum/GRAAP (2005)
2. Kearney, K., Torelli, F., Kotsokalis, C.: SLA*: An abstract syntax for service level agreements. In: 11th IEEE/ACM International Conference on Grid Computing, pp. 217–224 (2010)
3. Kouki, Y., de Oliveira, F., Dupont, S., Ledoux, T.: A language support for cloud elasticity management. In: 14th IEEE/ACM International Symposium on Cluster, Cloud and Grid Computing (CCGrid), pp. 206–215, May 2014
4. Lamanna, D.D., Skene, J., Emmerich, W.: SLAng: A language for defining service level agreements. In: 9th IEEE Workshop on Future Trends of Distributed Computing Systems, FTDCS 2003, p. 100. IEEE (2003)
5. Ludwig, H., Keller, A., King, R.P., Franck, R.: Web service level agreement language specification. In: IBM Germany Scientific Symposium Series (2003)

---

[6] A video demonstrating this process can be found at https://youtu.be/68sOCjAH0U4.

6. Ludwig, H., Stamou, K., Mohamed, M., Mandagere, N., Langston, B., Alatorre, G., Nakamura, H., Anya, O., Keller, A.: rSLA: Monitoring SLAs in dynamic service environments. In: Barros, A., Grigori, D., Narendra, N.C., Dam, H.K. (eds.) ICSOC 2015. LNCS, vol. 9435, pp. 139–153. Springer, Heidelberg (2015). doi:10.1007/978-3-662-48616-0_9

7. Mahbub, K., Spanoudakis, G.: Monitoring WS-agreements: an event calculus–based approach. In: Baresi, L., Nitto, E.D. (eds.) Test and Analysis of Web Services, pp. 265–306. Springer, Berlin, Heidelberg (2007). doi:10.1007/978-3-540-72912-9_10

8. Mahbub, K., Spanoudakis, G., Tsigkritis, T.: Translation of SLAs into monitoring specifications. In: Wieder, P., Butler, J., Theilmann, W., Yahyapour, R. (eds.) Service Level Agreements for Cloud Computing, pp. 79–101. Springer, New York, NY (2011). doi:10.1007/978-1-4614-1614-2_6

9. MarketsAndMarkets.com: Service Integration and Management (SIAM) Market - Global Forecast to 2021, March 2017

10. Mohamed, M., Anya, O., Sakairic, T., Tata, S., Mandagere, N., Ludwig, H.: rSLA framework: monitoring and enforcement of service level agreements for cloud services. Int. J. Coop. Inf. Syst. IJCIS **26**(02) (2017). doi:10.1142/S0218843017420035

11. Stamou, A.: Systematic SLA data management. Ph.D. thesis, University of Geneva (2014)

12. Tebbani, B., Aib, I.: GXLA a language for the specification of service level agreements. In: Gaïti, D., Pujolle, G., Al-Shaer, E., Calvert, K., Dobson, S., Leduc, G., Martikainen, O. (eds.) AN 2006. LNCS, vol. 4195, pp. 201–214. Springer, Heidelberg (2006). doi:10.1007/11880905_17

Efficient Service Variant Analysis with Markov Updates in Monte Carlo Tree Search (Short Paper)

Fuguo Wei$^{(\boxtimes)}$, Alistair Barros, Rune Rasmussen,
and Adambarage Anuruddha Chathuranga De Alwis

Queensland University of Technology, Brisbane, Australia
{f.wei,alistair.barros,r.rasmussen,adambarage.alwis}@qut.edu.au

Abstract. Static analysis techniques can be used to analyse and simplify interfaces of enterprise systems, such as those from SAP, Oracle and FedEx, which becoming more prominent on the internet and vying for new systems integration and extension opportunities. Web services of enterprise systems are notoriously complex, having hundreds of parameters per operation, multiple levels of nesting, leading to ambiguities about valid invocations of operations. To derive valid invocations, which in turn assists service users with invoking services correctly, this paper focuses on a challenging aspect of static interface analysis, namely, the identification of service variants in operations, in which the parameters are subtypes of business entities involved in a service. To efficiently search for which combinations of parameters are for a valid invocation, we have proposed a Monte Carlo method, based on likelihood-free Bayesian sampling, to identify higher probability parameters spaces, from which to test prospective invocations. A significant performance boost was found by extending Monte Carlo sampling with Markov look-up, with validation using a simulated FedEx service interface, whose structural complexity exceeds many web services of enterprise systems available on the internet.

Keywords: Web service · Markov look-up · Service variant analysis · Service interface analysis

1 Introduction

Web services, as a prominent form of application programming interface (API), providing read and write operations for software systems, have proliferated on the Internet, giving rise to the notion about the API Economy [1]. Service interfaces consist of operations together with input and output parameters and potentially behaviour of operational sequence [2], shielding software implementations from external access. In addition to internet applications, interfaces of older,

This work was sponsored by the Australian Research Council Discovery Grant DP140103788.

H. Panetto et al. (Eds.): OTM 2017 Conferences, Part I, LNCS 10573, pp. 639–647, 2017.
https://doi.org/10.1007/978-3-319-69462-7_40

enterprise systems, available through SAP, Oracle and others, are also being opened through service interfaces for new integration and software extension opportunities.

Enterprise service interfaces are often designed around a small set of mandatory parameters that lead out to optional parameters, in support of a diverse range of special invocations [3]. Such interfaces can involve large sets of parameters, even though the mandatory parameters are few. For example, the SAP Goods Movement service involves 104 parameters, of which only 12 are mandatory in every invocation [3]. The optional 84 parameters support the special variations of the service. Thus, parameters are either mandatory or optional, with the logical consequence that different parameter evaluations will lead to different optional parameters.

A specific challenge encountered with the interfaces of enterprise and other business commercial systems, is their structural complexity, as apparent in the fine-grained details of operations seen through WSDL interfaces. Due to poor design or design erosion over multiple maintenance cycles, they are overloaded. Consequently, it is not clear which parameters in operational signatures are required for valid invocations of service operations. Determining valid invocations of service operations is a tedious and time-consuming process given WSDL based service specifications together with XML schemas. For instance, the FedEx Open Shipping service interface specification has 7727 lines and more than 1000 parameters[1]. Many of these parameters are of a complex data type and hierarchically structured [4]. What FedEx has provided is just a 645-page pdf document[2], depicting the details of what each parameter means, which does not assist service users with consuming the service effectively and efficiently. As a consequence, web services integration is a costly and error-prone task, and is recognised as a key impediment for broader adoption of service-orientation in industry [3,5].

The underlying challenge for service variant derivation in operational signatures is to efficiently search for parameter sets, given the computational complexity of a brute-force approach invoking combinations of parameters to yield valid invocations based, particularly when the only input is a service interface specification (such as a WSDL file) as source codes of web services are usually inaccessible. In this paper, we have exploited the Markov assumption in a heuristic look-up that vastly improves interface search performances - thus allowing service variant analysis to be feasibly supported in static interface analysis techniques and tools contributing to API Economy applications.

2   Method

In favour of successful outcomes we have required the following assumptions for our method: 1. At least one acceptable variant is *a priori* known; 2. Detailed responses to an invocation from a service are not analysed, and the method only

---

[1] https://github.com/jzempel/fedex/blob/master/fedex/wsdls/OpenShipService_v9.wsdl.

[2] https://www.fedex.com/templates/components/apps/wpor/secure/downloads/pdf/201607/FedEx_WebServices_OpenShipping_WSDLGuide_v2016.pdf.

counts whether an invocation is accepted or rejected; 3. The acceptable parameters are clustered around common ones. The first condition simply ensures
that the search has a viable starting point. The second assumption simplifies
the search, as it is much more complex to factor in various error codes and messages returned from a service. The final precondition ensures that the acceptable
variants have some reasonable likelihood of being reached through small random
changes to a given or known variant. This precondition is a reasonable one, given
that logically associated parameters tend to be put together when a service is
designed. For example, a postal service may have its parameters constructed
out of a set of required parameters and those entailed by standard procedures
and legal requirements. Clustering around common parameters is also evident
in interfaces of web services such as SAP enterprise services [3], FedEx, UPS
and many APIs advertised through hubs such as programmableweb and Zapier.
If the final precondition does not hold, then service variant derivation may still
be possible where a partial or deprecated book version is available to guide the
search, and where this book has a near fit to the expected target book.

In our previous work [6], the distribution of acceptable variants were assumed
to be *a priori* unknown. In addition, the process of approximating a distribution for the acceptable variants involved a tree-based Monte Carlo search using
rejection sampling, where the incremental Bayesian updates of the distributions
involved importance sampling weights [7].

Given that each parameter consisted of generic values arranged in a strictly
increasing order of index, the main objective in [6] was to exploit this monotonic
pattern through stratified sampling in a tree. That is, a search tree was so defined
with the generic values in domain A assigned to the vertices, so that the paths in
the tree, which were rooted to a null ϕ generic value, had a one-to-one mapping
to the complete set of possible parameters. In this tree, a distributed sampling
method was applied, where good samples high in the tree could potentially guide
searches to the acceptable variants with great efficiency.

The Monte Carlo search was configured to expand random search paths by
sampling successors from posterior distributions, in a sequence (see Fig. 1). Since
a_n was not the only possible terminating value, these distributions required an
escape probability to facilitate terminal values other than a_n. The distributions
therefore involved a sample space $A \cup \{\omega\}$ where A was the generic values domain
and ω was a special last or terminating value. Given independent random vari-
ables $\theta_1, \theta_2, \ldots, \theta_n$ on the sample space $A \cup \{\omega\}$ and the signal T_f for variants
accepted by a server, the index for the next posterior distribution in sequence
could be determined by the values already realised along any given search path,
as shown in Fig. 1. The search path would terminate whenever the value realised
for some θ_i was either a_n or ω. At termination, a variant was computed from
those values in A already realised, and this variant was tested with the service provider. If the variant was accepted by the service provider, then T_f was
realised and each posterior distribution sampled along the path was then given
an incremental update, where the magnitude of that update was determined

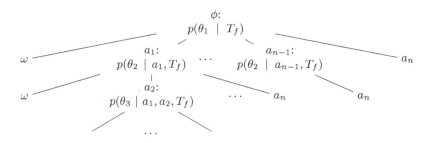

Fig. 1. The ϕ-rooted tree showing the pattern of posterior distributions associated with the interior vertices.

by computing an importance sampling weight (see [6]). However, if the service provider rejected the variant, then the search also rejected the variant.

The work herein explores the Markov property for optimising search time performances over the full Bayesian method described above and given in [6]. In this context, the Markov property would be manifested whenever the selection of a succeeding generic value depended exclusively on the last value realised, and no other preceding value. For example, the first and last name of a sender in a postal service can exhibit the Markov property, if the last name depends only on the sender's full name and not on any other postal values, which is generally the case for postal items.

Thus, the Markov property can be identified to hold whenever some distribution $p(\theta_i \mid \theta_1, \ldots, \theta_{i-2}, a, T_f)$, where $a \in A$, is invariant to the assignment of $\theta_1, \ldots, \theta_{i-2}$. That is, whenever the Markov property holds (and $i \geq 3$ in this given problem), a single distribution $p(\theta_i \mid a, T_f)$ can replace every distribution in the family $p(\theta_i \mid \theta_1, \ldots, \theta_{i-2}, a, T_f)$, for all of the possible assignments of $\theta_1, \ldots, \theta_{i-2}$.

The reader should note that given n generic values, $p(\theta_i \mid \theta_1, \ldots, \theta_{i-2}, a, T_f)$ is a family of $\frac{n!}{(n-i+3)!}$ possible valid distributions, where for sufficiently large i the update for any particular distribution in this family will be exceedingly rare. If the Markov property holds in such a case, then the entire family can be substituted with the single distribution $p(\theta_i \mid a, T_f)$, which will result in a significantly smaller search space and search time. Under such conditions, a massive boost in efficiency is possible because the substitution means that a vast set of pathways in a search tree will stop branching out, but instead, they will all converge upon a single common search node.

In exploiting the Markov property, we have assumed that the property does not initially hold for value pairs, so that our method can draw on evidence in the search to support that the Markov property does indeed hold for certain value pairs. Thus, we have avoided the non-sequitur of assuming that every Bayesian program is a type of Markov program, and have at least preserved the search performances found in our initial Bayesian program. Our method derives support for the Markov property, by estimating the probability that value pairs appear in acceptable service variants found during the search, and then by recording those probabilities into a look-up table, called a *Markov* look-up table. Thereafter, our method relies on the Markov look-up table, whenever an indexed column in the table has less information entropy than does the corresponding posterior distribution in the search.

Thus, to illustrate, let us consider here the full Bayesian method described above, which we have also described more extensively in [6]. In addition, let the Markov look-up table be a table M of probability values, so that for value set $A_\omega = A \cup \{\omega\}$, the table M has a mapping from $A_\omega \times A_\omega$ to the interval $[0, 1]$. Therefore, over all $(x, y) \in A_\omega \times A_\omega$, $M(x, y)$ is a probability distribution function. The method given in our previous work [6] concerns a search that samples values from proposal distributions along a search path. When the search arrives at either the last possible value (a_n), or the terminating value (ω), the search performs a backtrack of the search path. Such backtracks specifically involve updates of proposal distributions, whenever the search backtracks along a path that has yielded an acceptable service variant (that is, whenever T_f is true).

Let $\mathcal{P}(\theta) = p(\theta \mid \theta_1, \ldots, \theta_{i-2}, a, T_f)$. Thus, at some level i in a forward track of the search, the full Bayesian method would just sample from a proposal distribution $\mathcal{P}(\theta)$; however, by assessing the Markov property, our extension of the method requires a decision to be made between drawing from either $\mathcal{P}(\theta)$, or $M(\theta_i, a)$. Here, our new method solves the decision problem by comparing information entropy measurements, as follows:

$$\sum_{\theta \in A_\omega} M(\theta, a) \log [M(\theta, a)] < \sum_{\theta \in A_\omega} \mathcal{P}(\theta) \log [\mathcal{P}(\theta))]$$

So that, if this condition holds, the methods takes a sample from $M(\theta_i, a)$, otherwise it takes the sample from $\mathcal{P}(\theta)$. Finally, if the path yields an acceptable service variants (T_f is true), then the method performs an importance sample update of both $\mathcal{P}(\theta)$ and $M(\theta_i, a)$, on the backtrack. Here, we rewrite the condition $H(M(\cdot, a)) < H(\mathcal{P}(\cdot))$.

3   Implementation and Experiment

To validate the proposed Markov look-up table, a prototypical tool called the Service Integration Accelerator[3], part of our Service Engineering Hub toolkits, has been developed in Java 1.7.

In designing our experiments, we have gathered statistics through the structural analysis of thirteen popular service interfaces (the details of these services have been presented in our previous work [8]), to characterise their complexities. We found that the average number of parameters of enterprise services was about 200, and where the challenge has some softness given that some are under 50 and 20 parameters. Thus, for the experiments here, we have simulated services

---

[3] https://github.com/fuguowei/ServiceIntegrationAccelerator.

in stages of 20, 50 and 100 parameters, with structural complexities comparable to the services analysed.

In measuring the performance boost in our new variant derivation method, we have compared it against our full Bayesian method, which we gave in [6]. By applying the Markov assumption as prescribed herein, we have shown that our new variant derivation method can well outperform our full Bayesian method, on the problem stages.

In the simulated servers, variants were generated at random, so that we could determine the success rates of recovering those variants with our method. In each experiment, the server generated sets of twenty service variants of different lengths and deviations from one another. Experiments in the problem stage of 20 parameters, involved variants selected at random of lengths: 5, 8, 11, 14, and 17; in the problem stage of 50 parameters, the lengths of variants were: 10, 15, 20, 25, 30, 35, and 40; and for the problem stage of 100 parameters, the lengths were: 10, 20, 30, 40, 50, 60, 70, and 80. In creating statistical confidence, two-hundred experiments were conducted for each problem stage, and experiments ran for six days[4].

Figure 2 shows in box-plot form, the average recovery rates and standard deviations on those recovery rates (in % recovered). Figure 2a shows that the Markov look-up table provided a significant performance boost over our previous method, with recovery rates approaching 100% for variants selected out of twenty parameters. In the larger problem sizes, where variant selections were out of 50 and 100 parameters, the Markov look-up table gave recovery rates close to 60% for the 50 parameter case (see Fig. 2b), but approached 30% for the 100 parameter case (see Fig. 2c). These results demonstrate a method of service variant analysis that is able to sample in vast search spaces efficiently, and able to recover variants in spaces that were previously thought to be prohibitive.

FedEx Example: By emulating the "processShipment" operation of a FedEx shipment service[5], we were able demonstrate our method on a genuine server case. Here, the operation involved 1053 input parameters and 565 output parameters, from which we have derived 34 business entities [8], where a set of 43 parameters were selected to demonstrate our method. Here, our method gave a 36.34 % success rate. These variants derived were then transformed into subtypes of business entities [9] involved in the service. The subtype derivation tool[6], implemented using GoJS[7], part of our Service Engineering Hub toolkits, was able to derive 11 subtypes of Shipment (Fig. 3) - the key business entity reflected by the "processShipment" operation.

---

[4] Using the Queensland University of Technology high-performance computing lab: http://www.itservices.qut.edu.au/researchteaching/hpc/.

[5] http://www.fedex.com/templates/components/apps/wpor/secure/downloads/xml/Aug13/advanced/ShipService_v13.xml.

[6] http://anutestingfuguo.atwebpages.com/#about.

[7] http://gojs.net/latest/index.html.

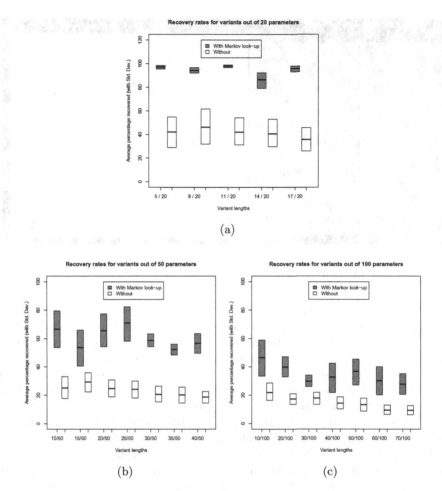

Fig. 2. The results of our new method (With Markov look-up) compared against the full Bayesian method given in [6] for variant recovery. Here, sub-figure (a) shows results for variants selected at random out of 20 parameters; sub-figure (b) shows results for variants selected out of 50 parameters; and, sub-figure (c) for variants selected out of 100 parameters.

4 Discussion and Related Work

Service variant analysis is a complex problem that can be challenging even when a number of the variants are known. This is apparently the first time that it has been possible to sample service variants in such vast search spaces, given only a service specification and informed only by whether a proposed invocation is accepted by a service provider. This is important, because our method performs well in the worse case.

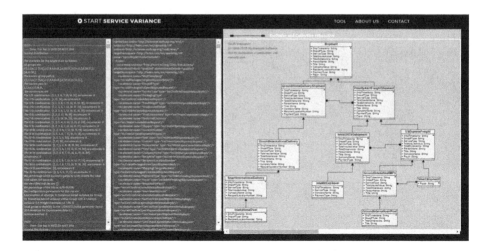

Fig. 3. A screenshot of the subtype derivation toolkit.

Our approach to service variant analysis is unique. Conventional methods approach service variant analysis by either manually annotating service interfaces, applying model-driven innovations, or both [3,10]. These approaches rely on experts to manually annotate and derive models from service interfaces. Interface annotation using semantic ontologies in [11] relies on semantically annotated descriptions, which are usually not provided as part of service delivery. Our method however derives service variants automatically with the only input - a syntactic service interface description, giving intelligence to potentially transforming service systems into fully informed "smart" systems.

Our method of service variant derivation and analysis can enable the reengineering of enterprise services from complex and overloaded ones to simple and modularised ones by extending the repertoire of the interface analysis tools that we have developed [8,12]. Specifically, big services can be split into small ones according to their variants. Since our method of service variant analysis exposes possible ways of invoking a service, it can expose variants that are common to different service interfaces. In other applications, we expect that our methods can even inform service compositions, or provide language support for service interfaces through structural insights. Our method can be used to analyse APIs, particularly legacy enterprise ones, within an organisation, where a large amount of probing is allowed. It can also be used when invoking services externally, in which case, although service providers may not allow a large number of invocations in a short period of time, services can be simulated using tools such as Smartbear and getsandbox.com.

5   Conclusion

In this paper we have demonstrated an efficient service variant analysis method based on Monte Carlo sampling with Markov look-up. The method was robust in the face of complex service interfaces and performed well in very large search spaces, which included a simulated complex FedEx process shipment case. Any errors returned from a service provider are just considered a rejection to the current proposed variant by the method. In fact, some services such as UPS Shipping return meaningful messages and error codes, for instance code 001 representing a shipper required or an invalid reference number being provided, that can be used to improve future sampling. Therefore, analysing informative responses can assist the variant analysis method to determine service variants more efficiently, which is our future work.

References

1. Zaha, J.M., Dumas, M., ter Hofstede, A.H.M., Barros, A., Decker, G.: Bridging global and local models of service-oriented systems. IEEE Trans. Syst. Man Cybern. Part C Appl. Rev. **38**(3), 302–318 (2008)
2. Barros, A., Decker, G., Dumas, M.: Multi-staged and multi-viewpoint service choreography modelling. In: Proceedings of the Workshop on Software Engineering Methods for Service Oriented Architecture (SEMSOA), Hannover, Germany, vol. 244. CEUR Workshop Proceedings (2007)
3. Stollberg, M., Muth, M.: Efficient business service consumption by customization with variability modelling. J. Syst. Integr. **1**(3), 17–32 (2010)
4. Barros, A., Duddy, K., Lawley, M., Milosevic, Z., Raymond, K., Wood, A.: Processes, Roles, and Events: UML Concepts for Enterprise Architecture. In: Evans, A., Kent, S., Selic, B. (eds.) UML 2000. LNCS, vol. 1939, pp. 62–77. Springer, Heidelberg (2000). doi:10.1007/3-540-40011-7_5
5. Autili, M., Inverardi, P., Tivoli, M.: Automated integration of service-oriented software systems. In: Dastani, M., Sirjani, M. (eds.) FSEN 2015. LNCS, vol. 9392, pp. 30–45. Springer, Cham (2015). doi:10.1007/978-3-319-24644-4_2
6. Rasmussen, R., Wei, F., Barros, A.: Technical report: Service variant discovery using a likelihood-free bayesian search method (2016)
7. Robert, C.P., Casella, G.: Monte Carlo Statistical Methods. Springer-Verlag, Secaucus (2005)
8. Wei, F., Barros, A., Ouyang, C.: Deriving artefact-centric interfaces for overloaded web services. In: Zdravkovic, J., Kirikova, M., Johannesson, P. (eds.) CAiSE 2015. LNCS, vol. 9097, pp. 501–516. Springer, Cham (2015). doi:10.1007/978-3-319-19069-3_31
9. Halpin, T., Morgan, T.: Information Modeling and Relational Databases. Morgan Kaufmann series in data management systems. Elsevier/Morgan Kaufmann Publishers, San Francisco (2008)
10. Nguyen, T., Colman, A., Han, J.: A feature-based framework for developing and provisioning customizable web services. IEEE Trans. Serv. Comput. **9**(4), 496–510 (2016)
11. Howar, F., Jonsson, B., Merten, M., Steffen, B., Cassel, S.: On handling data in automata learning. In: Margaria, T., Steffen, B. (eds.) ISoLA 2010. LNCS, vol. 6416, pp. 221–235. Springer, Heidelberg (2010). doi:10.1007/978-3-642-16561-0_24
12. Wei, F.: On the Analysis and Refactoring of Service Interfaces for Improving Service Integration Efficiency. PhD thesis, Queensland University of Technology (2016)

A Framework for Evaluating Anti Spammer Systems for Twitter

Kenny Ho[1], Veronica Liesaputra[1], Sira Yongchareon[2]([✉]),
and Mahsa Mohaghegh[2]

[1] Department of Computer Science, Unitec Institute of Technology,
Auckland, New Zealand
kenny098x@yahoo.co.nz, vliesaputra@unitec.ac.nz
[2] Department of Information Technology and Software Engineering,
Auckland University of Technology, Auckland, New Zealand
{sira.yongchareon,mahsa.mohaghegh}@aut.ac.nz

Abstract. Despite several benefits to modern communities and businesses, Twitter has attracted many spammers that have overwhelmed legitimate users with unwanted and disruptive advertising and fake information. Detecting spammers is always challenging because of the huge volume of data that needs to be analyzed while spammers continue to learn and adapt to avoid being detected by anti-spammer systems. Several spam classification systems are proposed that use various features extracted from the content and user's information from their Tweets. Nevertheless, no comprehensive study has been done to compare and evaluate the effectiveness and efficiency of these systems. It is not known what the best anti-spammer system is and why. This paper proposes an evaluation framework that allows researchers, developers, and practitioners to access existing user-based and content-based features, implement their own features, and evaluate the performance of their systems against other systems. Our framework helps identify the most effective and efficient spammer detection features, evaluate the impact of using different numbers of recent tweets, and therefore obtaining a faster and more accurate classifier model.

Keywords: Spam detection · Evaluation workbench · Feature selection · Machine learning

1 Introduction

Spams are unwanted activities such as when marketers send members unwanted advertisements, post fake reviews, or steal user information by directing users to malicious external pages [10]. As Social Network Services (SNS) becoming an important mode of communication, it attracts spammers who overwhelms users with unwanted content. Among these sites, Twitter, which was started in 2006, has grown to be one of the most popular SNS [21]. There are 500 million number of messages (called tweets) produced by 328 million active Twitter users (called twitterers) every day. Unlike other popular SNS, tweets can be read by anyone and people can follow a user without their consent. To attract users to their target websites, spammers post a

© Springer International Publishing AG 2017
H. Panetto et al. (Eds.): OTM 2017 Conferences, Part I, LNCS 10573, pp. 648–662, 2017.
https://doi.org/10.1007/978-3-319-69462-7_41

large number of coordinated messages containing specific URLs and sometimes describing them with unrelated words [25]. Because SNS helps build intrinsic trust between their users, 45% of them will click on links posted by their online friends even though they do not know those people in real life [23]. Twitterers also tend to post shortened URLs and write in abbreviated forms that rarely appear in conventional text documents or e-mails as a tweet can only contain up to 140 characters. Consequently, it is difficult for users to know the source URL and identify the content of the URL without clicking the link and loading the page. The noisy, unstructured, and informal expressions, such as "2mo is a new daaaaay!" or "TIL DC Comics stands for Detective Comics", used in the text also made it difficult for automatic spam detection system to accurately identify the semantic meaning of the tweets. Hence, social spamming is more harmful and complex than SMS, email or Web spams. It is becoming an important problem for users and service providers.

Around 83% of users of social networks have received at least one unwanted friend request or message and over 3% of tweets are spam [6]. To make Twitter a spam-free platform, Twitter enable twitterers to report spam URLs, tweets and accounts which after being verified will be included in the Twitter's blacklist. All URLs, tweets or accounts in the blacklist will be automatically filtered, suspended, or deleted by Twitter. However, due to time lag, 90% of users may visit a new spam link before it is included in the blacklist [25]. Furthermore, twitterers identify spammers manually based on experience that could lead to false positives. Therefore, it is important to have a tool that can automatically identify spammers. The approach must be scalable too, i.e. it can handle a large amount of data in a short amount of time with limited computation resources.

We can divide anti-social spammers systems into two types: tweet-level detection and account-level detection [23]. The tweet-level detection system checks each tweet for spam text content or URLs. If an account has posted a certain number of spam tweets, it is flagged as a spammer. As around 350,000 tweets are generated per minute [21], tweet-level detection consumes too many computing resources and is harder to be run in real-time. Account-level detection checks individual account's profiles and activity patterns for evidence of them sending spam tweets or being a fake account. Because there is very limited amount of imbalanced-labeled data, account-level detection systems tend to have high precision and accuracy but low recall. When they predict that an account is a spammer account, it has a high probability of it being true. However, there are many more spammer accounts out there that were not considered as spammer candidates, i.e. they are classified as legitimate. This is of course not useful to us as we are interested in detecting all spammers.

Spam detection is a never-ending game of cat and mouse. Although security companies, as well as Twitter, are working on creating systems to detect spam and spammers, spammers are always trying to avoid being detected. They deploy different techniques to post unwanted messages to users on SNS for advertisement, frauds or spreading of malware through the malicious URLs [11]. For instance, spammers create many fake accounts to post spam tweets for a specific purpose (or known as a spam campaign), send messages with different texts to convey the same meanings or pay users to follow their accounts [14]. Thus, the statistical attributes of spammers and spam tweets vary over time. Systems that rely on old samples may struggle to detect the new spammers or spam tweets.

As there are many spam detection systems proposed, it is hard for users and providers to decide which is the best one. We have also found that different work uses different evaluation metrics and datasets so it is hard to achieve a standard evaluation. This brings in a research challenge about comparing and evaluating the performance of various spam detection systems and identifying the best technique w.r.t effectiveness (i.e., accuracy, true positive rate, and precision) and efficiency (run-time execution for training and classification). This is particularly beneficial to the research community as any newly proposed techniques can be evaluated against the existing ones allowing us to know if the technique is an improvement.

In our study, we reviewed 172 content-based and user-based features from the majority of existing literature. Based on these features, we aim to develop a workbench, namely WEST (Workbench Evaluation Spammer detection system in Twitter) to evaluate their proposed features against the ones defined in WEST as well as to set the best number of recent tweets and find the best possible subset from all the features available. We have designed an evaluation method with a set of experiments to help select the optimal subset of features.

We organized the rest of this paper as follows. Section 2 introduces the background and existing work related to spam detection for Twitter. Section 3 discusses our proposed evaluation workbench and Sect. 4 discusses experimental results from our study. Lastly, a conclusion and future work are given in Sect. 5.

2   Related Work

This section provides an overview of related work with approaches and methods for SNS spammer detection.

[24] shows that machine learning methods demonstrated by [11] can be utilized with significant success in spammer detection on Twitter. Such methods are able to extract user or context-based features from user-behavioral patterns or linguistic features in tweets [1, 5]. [2] shows that supervised machine learning techniques such as Support Vector Machine [9, 23] are able to train features extracted from user profiles in order to find profiles linked to spam activity. Performance is evaluated based on precision (the percentage of correct positive prediction), recall (the percentage of positive instances that were predicted as positive), and accuracy (overall percentage of correct prediction). [6] demonstrates a method of extracting the user and context-based features from the dataset before running this through Meda et al.'s Random Forest classifier [24]. The output was evaluated based on precision and f-measure, the harmonic means of precision and recall. [16] uses information gain and relief methods to determine the five best features from the dataset. [16] uses Information Gain, and Relief methods to find the best five from features. These approaches all use different features, datasets and classifiers, and as such, we are unable to evaluate and compare their performance to ours.

Context-based features are linguistic features extracted from tweet context [5]. Twitter performs no checks on the legitimacy of shortened URLs, so spammers often exploit this by using a shortened URL service in an attempt to lure in legitimate Twitter users. [18] points out that spammers often use the same URLs in multiple tweets in order to increase the chance of it being clicked on by legitimate users. A number of researchers have utilized features related to URLs. Examples of these are the number of URLs [8], the number of URLs per word [5], and the number of unique URLs [9]. In Twitter, the #hashtag is used to describe a term, event, or emotion. If multiple tweets occur with the same #hashtag, it will become a trending topic [6]. Spammers often include a trending #hashtag with their tweets (though with unrelated content) in order to lure in legitimate users [16]. #Hashtags can be manipulated in the same way as URL features and expanded to other forms such as the number of #hashtags per work on a tweet [5]. Twitter users are able to include @username in their tweets (called a "mention"). This enables the tweet to be sent to the user in the @username, regardless of whether or not they are followers of or followed by the user who tweeted. This is a feature that spammers also exploit, enabling them to push tweets to users [16]. This feature has been explored by several authors [6, 18]. It has been noted that tweets from spammers often include a larger number of spam-related words (up to 39%) while legitimate users around 4% [1]. Because of this, some papers use a spam word feature based on spam words from sites such as Wordpress.org [1, 6]. Other methods and approaches include features such as percentage of words not contained in a dictionary in their system [7].

User-based features are derived from properties related to user behaviour [1]. Generally, spammers follow as many users as possible to gain their attention, and increase the likelihood of success with spam attacks [6]. Common user-based features include number of followers (users following the user in question), number of followings (the users the user in question is following), and reputation (determining the user's influence on Twitter). These features are used in different combinations with varying success. [6] uses the number following and number followed features, and achieve 95.7% precision whereas [13] uses number following, number followed, and reputation but only achieves 91% precision. Both [2, 11] use the followers-to-followings ratio. The reason for this feature is that while spammers attempt to follow as many accounts as possible, it is difficult to achieve "follow-backs", and the features ensures a healthy ratio of followers and followings is maintained. The approach used in [11] was able to achieve 93.6% precision. However, [14] points out followers can be purchased from certain websites, effectively reducing the reliability of the followers-to-followings ratio feature. They introduce a new feature called bi-directional links ratio. This is defined as "mutual followings" – i.e., two accounts following each other. This feature is more difficult for spammers to evade, since it results in them having to purchase more followers. The only way this is evaded is through reflexive reciprocity – when a user follows someone back out of courtesy [26]. 95% of spam tweets contain shortened malicious URLs. [15] proposes URL rate and Interaction Rate - two features to address URL-based spam attacks. Interaction rate notes the lack of normal interaction behavior in spammers, while URL rate compares the ratio of URL-based tweets to normal tweets. This is a particularly effective feature since is it almost impossible to evade.

3   Evaluation Workbench

As fundamentally formulated in the existing literature, given a set of users in a dataset, a spammer detection system is intended to build a classifier model to predict whether a user is a spammer based on a set of features extracted from the user's social activities, relations, profiles and/or textual contents gathered from the user's recent tweets. Spammers True Positive Rate, Spammers False Positive Rate, Precision, Accuracy, F_1-measure, and Time are then used to measure the efficiency and effectiveness of the model.

Unfortunately, so far none have compared the performance of their systems with the other systems. For example, both [1, 2] use Support Vector Machine model to identify spammers. However, they built the models based on different features. [2] does not include any features from the users' tweet content while [1] does. Because [1] did not use the same dataset as [2] or compare their performance with [1] it is unclear which system is better. Spammers are continually changing their strategies to fool the anti-spam systems [3]. Current effective features might not be effective in the future thus it is imperative to be able to quickly find the ineffective features and test the performance of new extracted features based on a new dataset.

The main motivation for our Workbench for Evaluating Spammers in Twitter (WEST) is to provide a comprehensive collection of social activities, relations, profiles and textual content features that researchers can quickly select and evaluate on new data sets. People can then easily determine which features or which overall system is the most effective and efficient. WEST is written in JAVA and it has extensible architecture that enables new features to be easily implemented and integrated to the workbench. It relies on WEKA [19] to perform the feature selection and classification on the dataset. Therefore, users can choose any attribute selection technique and supervised learning algorithm available in WEKA or include their own implementation to WEKA when building their anti-spammer model.

With WEST, we are able to answer the following important questions.

1. What are the most effective sets of features for identifying spammers?
2. What is the most efficient model for detecting spammers?
3. Can the number of recent tweets used significantly affect system performance?

Figure 1 illustrates the overall framework of WEST and this section outlines the set of features included in the workbench and the way researchers or practitioners can utilize and extend WEST. To avoid confusion, from now on, we will use the term *user* to signify the users of WEST and *twitterer* to represent the users of Twitter.

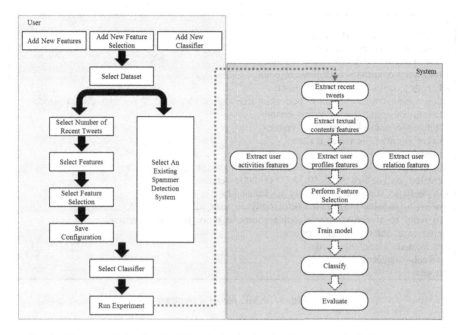

Fig. 1. Framework for the Workbench for Evaluating Spammers in Twitter (WEST)

3.1 Dataset

The input dataset must be comprised of two folders, *spam* and *ham*, containing XML files for spammers and ham twitterers respectively. Data about each twitterer is stored in a separate XML file as shown in Fig. 2. WEST then extracts features that were selected by the users from the file.

Based on the collection of features that WEST currently has, Table 1 displays the XML tags read by WEST and the list of information that can be extracted from each tag. Given a twitterer ID, we utilize Twitter4J [20] to get the age of a Twitterer's account and the list of the twitterer's followees. For each followee, we find out their name and if they are followed back by the twitterer. The result is stored in a CSV file illustrated in Table 2.

3.2 Number of Recent Tweets

It is computationally expensive to analyze every tweet that a twitterer sends to decide if the account is a spammer account. Furthermore, one solution for tackling the drift problem of twitter spams is by re-training a spammer classifier model every day based on the new spam tweets. Consequently, anti-spammer detection systems must be able to identify spammers from the smallest number of recent tweets possible. Different researchers extract features from different numbers of recent tweets. For instance, [18] used 200 tweets and achieved 81% precision while [6] extracts features from 100 recent tweets and obtained 95.7% precision. However, it is still unclear whether the difference

```
<root>
<id>187758822251704321</id>
<name>water lillies</name>
<screen_name>freshlillie</screen_name>
<followers_count>1541</followers_count>
<friends_count>1996</friends_count>
<description>ownerwaterlillies bodyskincare~so pure it's edible 32 yrs. exp.</description>
<favourites_count>1</favourites_count>
<statuses_count>3328</statuses_count>
<tweet_count>139</tweet_count>
<tweets><tweet>
<text>@thepeoplescourt marilynyou're smart i love your hair. </text>
<created_at>Tue Sep 30 05:30:32 +0000 2003</created_at>
<in_reply_to_status_id></in_reply_to_status_id>
<in_reply_to_user_id>27914673</in_reply_to_user_id>
<in_reply_to_screen_name>thepeoplescourt</in_reply_to_screen_name>
<retweet_count>0</retweet_count>
<retweeted>false</retweeted>
</tweet></tweets>
</root>
```

Fig. 2. XML file of a twitterer

Table 1. List of information extracted from each twitterer XML file

XML tags	Information extracted
Name	Twitterer's name
Screen_name	Twitterer's username
ID	Twitterer's profile id, WEST uses Twitter ID and Twitter4J to see whether a twitterer has followed another twitterer and to get the age of a twitterer's account
Followers_count	Total number of followers
Friends_count	Total number of followees/friends
Description	Twitterer's profile description
Tweet_count	Total number of tweets
Retweet_count	Total number of tweets being retweeted
Statuses_count	Total number of status updates
Retweeted	Whether the tweet has been retweeted
In_reply_to_screen_name	The screen name of the Twitterers mentioned in a tweet
Created_at	Date & time a tweet is posted
Text	The content of a tweet

in performance is due to the differing number of recent tweets or due to the set of features and dataset that they used. As shown in Sect. 4, with WEST, researchers can easily define the number of recent tweets to be included from the dataset and evaluate whether changing that number increases the model's performance.

Table 2. The table representation of the CSV file storing the twitterer's age of account and followee list

Followee ID	Followee name	Is Twitterer follows followee
3273499957	CloudCommerceCO	FALSE
1207668950	greatdeals2bid4	FALSE
81489175	tgparker2009	TRUE
Age of Twitterer Account In Days	2646	

3.3 Feature Extraction

In WEST users can either select which features they want to extract from the dataset or select the name of an existing spammer detection system. When the user selects the name of an existing anti spammer system WEST automatically selects the set of features used by that system. WEST also allows for new features to be added by the user and presently there are 17 systems [1, 2, 4, 5–18] and 173 features. Each of those features can be grouped into four types: profile, activities, relations and tweet contents.

Profile features are information obtained from the twitterer's profile page such as screen name, description, age of the account, profile's URL and reputation.

Relations features represent the twitterer's friendship status and activities like followers, followee, friends, bi-directional links (followers that are followed by the twitterer) and interaction (twitterer's reply or mention of a follower or a non-follower's name or tweet).

Content features capture all the linguistic properties of the text in a tweet such as URLs, URLs to a social media domain, hashtags, mentions, retweets, special characters (e.g. exclamation marks, question marks, blank spaces), alphanumeric characters, capital letters, consecutive words, non-dictionary words, named entity (places, organization, people), and spam words.

Activity features are acquired from the twitterer's general activities like tweets, duplicate tweets, time a tweet has been posted and the device used to post a tweet. WEST implemented three ways of determining levels of similarity between tweets: tweet cluster [17], cosine similarity [9] and minimum distance [13].

For each metric being extracted WEST obtains the sum, minimum, maximum, median, average and standard deviation of that feature or the unique instances of that feature appearing over a certain period of time (hours, days, weeks) or over a certain number of words or tweets. For example, total number of spam words on screen name, ratio of followers to following, median number of hashtags per word, average number of unique URLs on a tweet, maximum idle time between tweets, and total number of tweets posted between 3pm and 4pm.

3.4 Build and Evaluate Model

As WEST relies on WEKA to perform feature selection and classification, all features retrieved from every twitterer in the dataset is stored in an ARFF file format. This enables users to either use WEKA through WEST to perform their machine learning tasks or to use it directly on WEKA. Just like in WEKA, users can also add new feature selection and classification techniques.

4 Experimental Results

This section illustrates how we can use WEST to answer the three questions mentioned in Sect. 3. Feature selection is a step of selecting features that are more relevant to a model to improve the accuracy of a system [22]. Theoretically, we should be able to find the most effective and efficient set of features by performing feature selection. To prove this hypothesis, we need to first find the best model generated by performing feature selection and then compare it with the model generated by the existing anti-spammer model. Because we are interested in knowing the impact of the number of recent tweets on the model's performance, we will also compare the result and the attributes selected by the model generated from a various number of recent tweets. Section 4.1 describes the dataset, classifiers and evaluation criteria that we will use in our experiment. The results of the feature selection models obtained from the varying number of recent tweets are presented in Sect. 4.2 and are compared with the results of the existing spammer detection systems in Sect. 4.3.

4.1 Experimental Setup

Dataset. The dataset collected by [17] contains the profile and 100 tweets of 7,549 twitterers separated into 315 spammers and 7234 hams. Because some of those accounts are no longer available or are missing information that we need such as Age of Account or Bi-directional links, only 1729 (206 spam and 1523 ham) are usable. We split the dataset into training and test sets. 70% of the spammers and 70% of the ham twitterers are used as training set. The rest are used for testing.

Classifiers. Five most commonly used classifiers for detecting spammers are Support Vector Machine (SVM), Decision Tree (DT), Naïve Bayes (NB), k-Nearest Neighbors (KNN) and Random Forest (RF). To find the most effective and efficient model, we will classify each selected set of features with each of those classifiers.

Evaluation Criteria. To measure the performance of a model, we will use Time, Accuracy, and Spammers' True Positive Rate (TP), False Positive Rate (FP), Precision, and F_1-measure. The time here refers to the total time required to extract features, perform feature selection, build training models and classify all test instances. As mentioned by [9] and seen in our dataset, the number of ham twitterers is much greater than that of spammers, i.e. we have a class imbalance problem. Thus, the most *effective* model is the one that can identify spammers with a high true positive rate and low false positive rate. Although many researchers have proposed various systems for detecting spammers in Twitter, none have mentioned the time it requires to achieve those. Considering that on average there are 6000 tweets sent per second [21], it is essential to have an anti-spammer system that enables legitimate twitterers quickly determine if a tweet is sent by a spammer. Hence, the most *efficient* model should be able to distinguish spammers effectively in the shortest amount of time possible. ANOVA, t-test and equivalence testing will be used to help us identify whether there are significant differences in the performance of each model, and statistically speaking which one is the best.

4.2   Best Feature Selection Model (FS)

To see if the number of recent tweets used affects the performance of the spammer detection system we compared the performance of the models extracted from 20, 50, 100, 150 and 200 recent tweets. For each of them, we obtained the Top 10 attributes based on their Information Gain values and classified them using the five aforementioned classifiers.

Table 3 displays the Spammers True Positive values and the bolded value signifies the highest TP value that each subset can obtain. ANOVA result shows that there is a significant difference in 95% confidence level (p-value = 0.0016) between the number of recent tweets. Although there is no significant difference between 20, 50, 100 and 150 recent tweets, t-test shows that the TP rate value from 100 recent tweets is significantly better than 200 Recent tweets (p-value = 0.02). Through equivalence testing, we found that the model obtained from 100 recent tweets is the best, followed by the model for 20, 50, 150 and 200 recent tweets.

Table 3. Spammers True Positive Rate results obtained by using Top 10 attributes obtained from the various number of recent tweets

	Number of recent tweets				
	20	50	100	150	200
NB	**93%**	**92%**	81%	**61%**	23%
SVM	0%	0%	0%	0%	0%
KNN	69%	62%	**100%**	13%	21%
DT	67%	61%	**100%**	37%	**26%**
RF	71%	65%	**100%**	40%	22%

From Table 4, we can see that regardless of the number of recent tweets, nine out of the Top 10 attributes are the same and the same attributes are selected when we use 20 or 50 recent tweets. Furthermore the majority of the attributes in the list are idle time related features. *Idle time* is the length of time interval between two tweets while *Mean Idle time per tweet* (*total idle time/total number of tweets*) and *Max Idle time per tweet* (*maximum idle time/total number of tweets*) represent the relationship between twitterer's idle time and the total number of posts. This supports the findings by [2] that on average spammers tend to have more posts and less idle time between posts.

As we use more recent tweets more information can be captured from content related features and so they have higher information gain values. For instance, *Mean of Number of Characters* is rank 10 in 20 and 50 recent tweets, but it moves up to rank 7 in 100 recent tweets, rank 5 in 150 recent tweets, and rank 3 in 200 recent tweets.

Table 4. Top 10 subset of features extracted from 20, 50, 100, 150 and 200 recent tweets in descending order based on their Information Gain values

	20 Tweets	50 Tweets	100 Tweets	150 Tweets	200 Tweets
1	Max. idle time per tweet	Mean idle time per tweet	Mean idle time per tweet	Mean idle time per tweet	Mean idle time per tweet
2	Mean idle time	Mean idle time	Mean idle time	Mean idle time	Mean idle time
3	Mean idle time per tweet	Max. idle time per tweet	Max. idle time per tweet	Tweet similarity - Cosine	Mean number of characters
4	Std. dev. of idle time	Std. dev. of idle time	Tweet similarity - Cosine	Max. idle time per tweet	Tweet similarity – Cosine
5	Tweet similarity - Cosine	Tweet similarity - Cosine	Std. dev. of idle time	Mean number of characters	Mean number of words
6	Max. idle time	Max. idle time	Mean number of characters	Mean number of words	Max. idle time per tweet
7	Mean number of words	Mean number of words	Mean number of words	Std. dev. of idle time	Max. idle time
8	% of followers per followees	% of followers per followees	Max. idle time	Max. idle time	Std. dev. of idle time
9	Age of account	Age of account	% of followers per followees	% of followers per followees	% of followers per followees
10	Mean number of characters	Mean number of characters	Mean number of URL per word	Fraction of tweets with spam words	Mean number of numeric characters

In Table 4, we have shaded the attributes that do not appear in the list of attributes for the other number of recent tweets.

Spammer's accounts are normally banned by legitimate twitterers and Twitter [1], so their *Age of Accounts* is smaller than normal twitterers. However when we increase the number of recent tweets to 100 there are more tweets containing the URLs than 20RT and 50RT, so the *Mean number of URL per word* feature can capture more information with respect to the class target.

[4] says that spammers tend to post more URLs than normal twitterers. In fact, he found that 95% of spammers' tweets contained URLs. [15] suggests using the *URL Rate* (*total number of URLS/total number of tweets*) to identify spammers. However, this feature was not in the best Top 10 subset of features. By looking at the tweet contents in the dataset, we found that spammers usually send tweets containing a URL with similar sort of text, for example "*#FREE PDF to Excel Converter* http://t.co/XfouPlN" and "*#FREE PDF to Word Converter* http://t.co/XfouPlN". Legitimate twitterers will have different text content and a different total number of words accompanying a URL that they post in their tweets. Therefore, *URL rates* are less effective than the *Mean Number of URLs per Word* because it ignores the contextual information surrounding the tweet and just counts the number of URLs.

Although it is good to check for the occurrence of spam words in tweets, *Fraction of tweets with spam words* is not a strong feature because it is impossible to create an exhaustive list of spam words. Spammers create new spam words all the time and it takes a long time to perform string matching on each tweet [5]. Similarly, *mean number of numeric characters* in 200 recent tweets is not a good feature as the difference between the number of numeric characters used by spammers and normal twitterers is not significant—1 numeric character per tweet for hammers and 1.5 for spammers.

4.3   FS vs. Other Existing Spammer Detection Systems

We can compare the best models obtained via feature selection (FS) with the 17 systems we have implemented in WEST. However, due to space limitation in this paper, we chose to compare FS with the five representative systems including [1, 5–7, 12]. As shown in Table 5, we compare each of the systems against the six evaluation criteria including execution time (in minutes), accuracy, TP, FP, precision, and F_1-measure. ANOVA results show that there is significant difference of 95% confidence between the models in terms of TP (p-value = 0.0003), precision (p-value = 0.0070) and F_1-measure (p-value = 0.0001). There is no significant difference in terms of accuracy (p-value = 0.5326) and FP (p-value = 0.5581). T-test and equivalence testing show that FS is the best model and thus supporting our hypothesis that the most efficient and effective model is the model generated through feature selection.

Table 5. Evaluation results for FS and the selected five existing anti spammers models

Number of recent tweets	FS	[1]	[5]	[6]	[7]	[12]
	100	200	200	100	200	20
Execution time (min.)	**16**	565	35	**30**	897	158
Accuracy	**100%**	93%	93%	**94%**	92%	91%
TP	**100%**	**65%**	61%	46%	17%	42%
FP	**0%**	2%	1%	**0%**	**0%**	**0%**
Precision	**100%**	65%	76%	72%	13%	**80%**
F_1	**100%**	**60%**	55%	51%	15%	47%

Alonso et al. [7] model is the worst performing, ranked least efficient and effective. With this model, the system must check whether each tweet contains named entities, social media domains or non-dictionary words. It is not only time-consuming but also requires the system to keep exhaustive and up-to-date lists of named entities, social media domains and dictionary words. Hence, it does not perform well on our dataset.

Although the accuracy and precision of the [6, 12] systems are good, their TP and F_1-measure are low because they cannot handle the class imbalance problem, i.e. most instances are classified as normal twitterers. No content or idle-time related features are included and so they cannot distinguish spammers from legitimate twitterers.

Out of all the existing systems we have evaluated, [5] model is the most efficient. Many content related features, such as *Mean number of URL per word* and *Maximum number of words*, included in the model helped the system obtain quite good Accuracy, Spammers' TP, FP, Precision, and F_1-measure in a short amount of time. However we can improve the performance of the model even more by including *Tweet Similarity* and time-relation features such as *Mean idle time per tweet*. This is because spammers tend to produce many tweets in a short period with duplicated content.

Compared with [1, 5] system uses the same number of recent tweets and produces similar accuracy, Spammer's TP, FP and F_1-measure, nevertheless [1] system takes longer to build because it extracts more features from tweets than the other systems.

Generally, spammers will follow many accounts but almost none of them will follow them back. Nevertheless, it is not enough to calculate *number of followees*, *number of followers* and *Reputation* (*number of followers/(number of followers + number of followees)*) to identify spammers because they can just buy more followers to evade these features [14]. Replacing these features in [1, 6, 12] with *% of followers per followees* or *Total Number of Bi-Directional Link* will improve the model's performance.

Furthermore, spammers tend to post many tweets containing the same URLs to increase their chance of being clicked by legitimate twitterers and so their *Ratio Unique URL per Tweet* would be small. We can improve the performance of [1, 5–7, 12] models by replacing their URL related features with *Ratio Unique URL per Tweet*.

5 Conclusion

This paper studied a number of works in spam detection for Twitter in order to build a framework for comparing and evaluating their performance. We propose WEST as an evaluation workbench for researchers and users to measure the performance of their proposed techniques against existing ones. We have included 172 content-based and feature-based features in our study making it easier for researchers to quickly create and evaluate their models against existing models. Our experiments found that the most effective and efficient set of features for detecting spammers are idle time related activity and tweet content features and the number of recent tweets used can significantly affect the model's performance. In the future, we will consider other types of features such as graph/network based features in the tool as well as creating a user interface to make the workbench more user friendly.

References

1. Benevenuto, F., Magno, G., Rodrigues, T., Almeida, V.: Detecting spammers on Twitter. In: Proceedings of the 7th Annual Collaboration, Electronic messaging, Anti-abuse and Spam Conference (2010)
2. Gee, G., Hakson, T.: Twitter Spammer Profile Detection (2010). cs229.stanford.edu/proj2010/GeeTeh-Twitter Spammer Profile Detection.pdf
3. Chen, C., Zhang, J., Xiang, Y., Zhou, W.: Asymmetric self-learning for tackling twitter spam drift. In: Computer Communications Workshops, pp. 208–213 (2015)
4. Lee, K., Eoff, B.D., Caverlee, J.: Seven months with the devils: a long-term study of content polluters on Twitter. In: Proceedings of the 5th International AAAI Conference on Weblogs and Social Media (2011)
5. Wang, B., Zubiaga, A., Liakata, M., Procter, R.: Making the most of tweet-inherent features for social spam detection on Twitter (2015). https://arxiv.org/pdf/1503.07405.pdf

6. Mccord, M., Chuah, M.: Spam detection on twitter using traditional classifiers. In: Proceedings of the International Conference on Autonomic and Trusted Computing (2011)

7. Alonso, O., Carson, C., Gerster, D., Ji, X., Nabar, S.U.: Detecting uninteresting content in text streams. In: Proceedings of the SIGIR Crowdsourcing for Search Evaluation Workshop (2010)

8. Burnap, P., Javed, A., Rana, O.F., Awan, M.S.: Real-time classification of malicious URLs on Twitter using machine activity data. In: Proceedings of the IEEE/ACM International Conference on Advances in Social Networks Analysis and Mining (2015)

9. Lee, K., Caverlee, J., Webb, S.: Uncovering social spammers: social honeypots + machine learning. In: Proceedings of the 33rd International ACM SIGIR Conference on Research and Development in Information Retrieval (2010)

10. Lupher, A., Engle, C., Xin, R.: Feature Selection and Classification of Spam on Social Networking Sites (2012). bid.berkeley.edu/cs294-1-spring12/images/archive/6/6a/20120515031244!Spam-lupher-engle-xin.pdf

11. Amleshwaram, A.A., Reddy, N., Yadav, S., Gu, G., Yang, C.: Cats: characterizing automation of twitter spammers. In: Proceedings of the 5th International Conference on Communication Systems and Networks (2013)

12. Wang, A.H.: Don't follow me: Spam detection in Twitter. In: Proceedings of International Conference on Security and Cryptography (2010)

13. Wang, A.H.: Detecting spam bots in online social networking sites: a machine learning approach. In: Foresti, S., Jajodia, S. (eds.) DBSec 2010. LNCS, vol. 6166, pp. 335–342. Springer, Heidelberg (2010). doi:10.1007/978-3-642-13739-6_25

14. Yang, C., Harkreader, R.C., Gu, G.: Die free or live hard? empirical evaluation and new design for fighting evolving twitter spammers. In: Sommer, R., Balzarotti, D., Maier, G. (eds.) RAID 2011. LNCS, vol. 6961, pp. 318–337. Springer, Heidelberg (2011). doi:10.1007/978-3-642-23644-0_17

15. Lin, P.-C., Huang, P.-M.: A study of effective features for detecting long-surviving Twitter spam accounts. In: Proceedings of the 15th International Conference on the Advanced Communication Technology (2013)

16. Song, J., Lee, S., Kim, J.: Spam filtering in Twitter using sender-receiver relationship. In: Sommer, R., Balzarotti, D., Maier, G. (eds.) RAID 2011. LNCS, vol. 6961, pp. 301–317. Springer, Heidelberg (2011). doi:10.1007/978-3-642-23644-0_16

17. Chakraborty, A., Sundi, J., Satapathy, S.: SPAM: A Framework for Social Profile Abuse Monitoring (2012). http://www3.cs.stonybrook.edu/~aychakrabort/courses/cse508/report.pdf

18. Dhingra, A., Mittal, S.: Content based spam classification in Twitter using multi-layer perceptron learning. Int. J. Latest Trends Eng. Technol. 5(4) (2015)

19. Piatetsky-Shapiro, G.: KDnuggets news on SIGKDD service award (2005). www.kdnuggets.com/news/2005/n13/2i.html

20. Twitter4J (2007). twitter4j.org/en/

21. Sayce, D.: Number of tweets per day? (2017). http://www.dsayce.com/social-media/tweets-day/

22. Kira, K., Rendell, L.A.: A practical approach to feature selection. In: Proceedings of the 9th International Workshop on Machine Learning (1992)

23. Stringhini, G., Kruegel, C., Vigna, G.: Detecting spammers on social networks. In: Proceedings of the 26th Annual Computer Security Applications Conference (2010)

24. Meda, C., Bisio, F., Gastaldo, P., Zunino, R.: Machine learning techniques applied to Twitter spammers detection. In: International Carnahan Conference on Security Technology (2014)

25. Thomas, K., Grier, C., Song, D., Paxson, V.: Suspended accounts in retrospect: an analysis of Twitter spam. In: Proceedings of ACM SIGCOMM Conference on Internet Measurement Conference (2011)
26. Weng, J., Lim, E.-P., Jiang, J., He, Q.: TwitterRank: finding topic-sensitive influential twitterers. In: Proceedings of the 3rd ACM International Conference on Web Search and Data Mining (2010)

Formal Model and Method to Decompose Process-Aware IoT Applications

Samir Tata[1]([⊠]), Kais Klai[2], and Rakesh Jain[1]

[1] IBM Research - Almaden, 650 Harry Rd, San Jose, CA 95120, USA
{stata,rakeshj}@us.ibm.com
[2] LIPN, CNRS UMR 7030, Université Paris 13, Paris, France
kais.klai@lipn.univ-paris13.fr

Abstract. The Internet of Things (IoT) integrates a large number of pervasive things that continuously generate data about the physical world. While such generated data can be sent to the Cloud for processing, more and more scenarios are considered where IoT created data will be stored, processed, analyzed, and acted upon close to or at the Edge of the network. Consequently, IoT applications require to be deployed in distributed environments. In this paper, we propose a formal approach for the decomposition of process-aware applications to be deployed in IoT environments. We model process-aware applications in Petri nets, formally define the decomposition process and prove its correctness. As a proof of concept, we have extended the Node-RED tool to allow the modeling, deployment and distributed running of IoT applications.

Keywords: Internet of Things · Petri net · Business process · Node-RED

1 Introduction

The Internet of Things (IoT) integrates a large number of heterogeneous and pervasive things that continuously generate information about the physical world. These things refer to the network of objects, devices, machines and other physical systems with embedded sensing, computing, and communication capabilities [7]. The first adopted organization model for IoT applications consisted in collecting IoT data from sensors, devices, etc. and sending them to an ingestion tier. This tier is software and infrastructure that runs in a corporate data center or in the cloud and receives and organizes the streams of data coming from the things. Additionally, an analytics tier takes the organized data and processes it to be consumed by the end user applications.

Nowadays, an alternative organization model for IoT applications is being considered. Indeed, IDC predicts that by 2019 at least 40% of IoT created data will be stored, processed, analyzed, and acted upon close to or at the Edge of the network [10]. In early days of IoT, the simple sensors (*e.g.* collecting temperature) were the major sensing target for IoT edge devices. But now a large amount

© Springer International Publishing AG 2017
H. Panetto et al. (Eds.): OTM 2017 Conferences, Part I, LNCS 10573, pp. 663–680, 2017.
https://doi.org/10.1007/978-3-319-69462-7_42

of data in the form of image and audio is also being produced by IoT sensors. Therefore, increasingly, edge devices do not only serve as simple sensors or actors, but also provide execution environments with limited processing, memory, and storage capabilities. By using these capabilities, IoT applications offload parts of their business logic to the edge of the infrastructure to reduce communication overhead and increase application robustness [30].

Manipulating data at edge devices of IoT environment requires the applications to be distributed over these devices. We believe that the process-aware applications [15] are prominent examples of IoT applications that can be deployed and run in IoT environments. Indeed, process-aware applications are composed of clearly separated tasks that are structurally adapted to run in distributed environments. When it comes to IoT environments, process-aware applications are usually deployed individually to each compute node, where the compute node can be on the edge or in the cloud [23,24]. Compute nodes are then connected to each other in the application network for achieving the end result. Therefore the deployment of application is also, more or less, individual compute node basis. Contrary to this method, we believe that by providing a mechanism to model such a distributed IoT application in one place, decompose into a set of fragments, and deploy these fragments from the same place to all the different compute nodes in the edge, gateways and the cloud, the developers of IoT applications will benefit from faster time to market, less error and improved visibility of the application.

Nevertheless, deploying and management of applications in such a context poses several research and development issues including the decomposition of the application into a set of connected fragments, the decision about the placement of those fragments, and the provisioning of IoT infrastructure and platform facilities to host application fragments and their orchestration. Other interesting research aspects to tackle include model verifications and enforcement of good properties in process-aware applications such as timing, security enforcement, privacy protection, tolerance to unreliable communication between components etc. In this paper, we are particularly interested in proposing a formal approach for the decomposition of process-aware IoT applications to be eventually deployed in IoT environments. While we believe that the issues raised above are important to tackle, the formal decomposition of process-aware IoT applications discussed in this paper is complex enough in itself to deserve separate treatment.

The rest of this paper is organized as follows. Section 2 presents some preliminaries on Petri nets and process-aware IoT applications. Section 3 presents our approach for the decomposition of process-aware IoT applications. A proof of concept of our approach is presented in Sect. 4. Section 5 discusses related works and highlights the limitations of classical application decompositions in Cloud and IoT environments. Finally, in Sect. 6, we conclude the paper and give directions for future work.

2   Preliminaries

2.1   Petri Nets

Definition 1. *A Petri net $N = \langle P, T, F, W \rangle$ consists of the following. P is a finite set of places and T a finite set of transitions with $(P \cup T) \neq \emptyset$ and $P \cap T = \emptyset$. $F \subseteq (P \times T) \cup (T \times P)$ is a flow relation. $W : F \to \mathbb{N}^+$ is a mapping that assigns a positive weight to any arc.*

Notations

- Each node $x \in P \cup T$ of the net has a pre-set and a post-set defined respectively as follows: $^\bullet x = \{y \in P \cup T \mid (y, x) \in F\}$, and $x^\bullet = \{y \in P \cup T \mid (x, y) \in F\}$.
- Adjacent nodes are then denoted by $^\bullet x^\bullet = {}^\bullet x \cup x^\bullet$.
- By extension, for a set of nodes X, $^\bullet X = \cup_{x \in X}{}^\bullet x$, $X^\bullet = \cup_{x \in X} x^\bullet$ and $^\bullet X^\bullet = \cup_{x \in X}{}^\bullet x^\bullet$.
- For a transition t, $W^-(t)(resp.W^+(t)) \in \mathbb{N}^{|P|}$ denotes the vector where $W^-(t)(p) = W(p, t)$ (resp. $W^+(t)(p) = W(t, p)$).

Semantics. A marking is a vector $m \in \mathbb{N}^{|P|}$ that assigns a non-negative integer to each place. We denote by $m(p)$ the marking of place p. A marked PN (N, m_0) is a PN N with a given initial marking m_0. For short, a marked PN will be called PN afterward. A transition t is said to be enabled by a marking m (denoted by $m \xrightarrow{t}$) iff $W^-(t) \leq m$ (i.e., $\forall p \in P$, $W^-(t)(p) \leq m(p)$). If a transition t is enabled by a marking m, then its firing leads to a new marking m' (denoted by $m \xrightarrow{t} m'$) s.t. $m' = m - W^-(t) + W^+(t)$. For a finite sequence $\sigma = t_1 \ldots t_n$, $m_0 \xrightarrow{\sigma} m_n$ denotes the fact that σ is enabled by m_0 i.e., $m_0 \xrightarrow{t_1} m_1 \xrightarrow{t_2} m_2 \to \ldots \xrightarrow{t_n} m_n$.

A sequence of transitions $\sigma = t_1 t_2 \ldots t_k$ is fireable at marking m, if $\exists m_1, m_2, m_{k-1}$ s.t. $m \xrightarrow{t_1} m_1 \xrightarrow{t_2} \cdots m_{k-1} \xrightarrow{t_k}$. This can be denoted as $m \xrightarrow{\sigma}$. The projection of a firing sequence σ on a subset of transitions T_1 leads to a sequence σ_{T_1} obtained by erasing each transition in σ not belonging to T_1. The language generated by N is $\mathcal{L}(N) = \{\sigma \in T^* \mid m_0 \xrightarrow{\sigma}\}$. For short, we write $\mathcal{L}$ instead of $\mathcal{L}(N)$. The projected language $\mathcal{L}_{T_1}$ on a subset of transitions T_1 is defined as $\mathcal{L}_{T_1} = \{\sigma_{T_1} \mid \sigma \in \mathcal{L}\}$. A marking m is reachable in (N, m_0) iff there exists a firing sequence σ such that $m_0 \xrightarrow{\sigma} m$. The set of all markings reachable from m_0 defines the *reachability set* of (N, m_0) and is denoted $R(N, m_0)$.

Subnets. Given a PN $N = (P, T, F, W)$ and a subset of transition $T_1 \subseteq T$, the subnet of N induced by T_1 is a PN $N_1 = (P_1, T_1, F_1, W_1)$ which is defined as follows:

- $P_1 = {}^\bullet T_1{}^\bullet$,
- $F_1 = F_{(P_1 \times T_1) \cup (T_1 \times P_1)}$ i.e. the projection of the flow relation on the places and transitions of N_1,
- $W_1 = W_{F_1}$ i.e. inherited arcs have the same weight as in N.

Petri Nets with Priority. *Priority* Petri nets [8] are an extension of P/T Petri nets that offer a convenient description technique to resolve conflicts in favor of particular actions. It consists in imposing an additional relation $\rho \subseteq T \times T$ on the set of transitions. $(t, t') \in \rho$ denotes that transition t' has priority over t. Using general relations for the specification of priorities often leads to problems or simply makes no sense. Since the intended use of priorities is to resolve conflicts, reasonable relations ρ should be, of course, irreflexive, asymmetric and transitive.

Definition 2. *A priority Petri net is a pair $\langle N, \rho \rangle$ where N is a Petri net and $\rho \subseteq T \times T$ is the priority relation which is irreflexive, asymmetric and transitive.*

For a priority Petri net, enabling is defined as follows.

Definition 3. *A transition $t \in T$ is $\rho-$enabled in a marking m, denoted by $m \xrightarrow{t}_\rho$, iff $m \xrightarrow{t}$ and $\forall t' \in T : m \xrightarrow{t'} \implies (t, t') \notin \rho$.*

The concept of $\rho-$enabling restricts the firing sequences of a net. In particular, all firing sequences of the priority Petri net are also firing sequences of the underlying net but not vice versa. Thus, the reachability set of a priority net is a subset of the reachability set of the underlying net. All definitions for "ordinary Petri nets" can be carried over to priority Petri nets in the context of $\rho-$enabling.

$\langle N, \rho \rangle$ is a Petri net with static priorities. Dynamic priorities are, e.g., defined in [20] where the priority relation depends on the current marking, namely *state-controlled priority*, of the net or on the firing sequences executed starting at the initial marking, namely *event-controlled priority*. Here, we are interested in state-controlled priority: a marked Petri net $\langle N, m \rangle$ is completed with a ternary relation $\rho_d \subseteq T \times T \times 2^{\mathbb{N}^{|P|}}$ where $(t, t, S) \in \rho_d$ denotes that, when the current marking of N belongs to S then transition t' has priority over t. Otherwise, firability of a transition t at a marking m is decided as in an "ordinary" Petri net.

Labeled Petri Nets. We consider an extension of Petri nets (resp. priority Petri nets) called Labeled Petri nets (LPN) (resp. Priority Labeled Petri nets (PLPN)). We enrich a Petri net with a labeling function which associate each transition with a label representing the location where this transition will be deployed in the IoT environment. A LPN is a tuple $N_L = (N, m_0, \Sigma, \varphi)$, where (N, m_0) is a marked PN, Σ is a finite set of events (i.e., labels) and $\varphi \colon T \to \Sigma$ is the transition labeling function. φ is also extended to sequences of transitions, $\varphi \colon T^* \to \Sigma^*$. Also, one should notice that various transitions can share the same event label, i.e., φ is not bijective. We denote by T_σ the set of transitions sharing the same event σ, i.e., $T_\sigma = \{t \in T : \varphi(t) = \sigma\}$.

2.2   Example

We consider a real world scenario where trains are equipped with sensors and devices for monitoring. Sensors are attached to each bearing which supports a wheel. They measure vibration, temperature, torque, speed and other factors of wheels. The measurements from such sensors are sent to a processing unit on the train cart itself. If data from any of the sensors breaches a preset threshold range, an alarm is triggered for the local staff on the train. Figure 1 shows a description using Node-RED notation of such a system for one bearing supporting one wheel [28].

Fig. 1. Single wheel monitoring in a train monitoring system

As depicted by Fig. 2, data from all the wheels on the cart go to the processing unit on the cart, and data from all the carts go to the main computer on the train, which does filtration and aggregation of data and sends it to the cloud. This shows that the processing unit on a single cart is responsible for quite a bit of computation, analysis and action before sending the data to the central computer on the train, which receives such data from all the carts in the train. It then does its own processing and analysis before sending the data to the cloud. In cloud, data from many such trains is received, which helps the train company with maintenance and management of its operations.

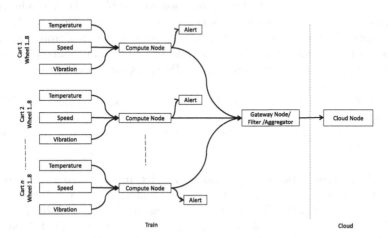

Fig. 2. Train monitoring system connected to the cloud

Figure 3 shows the Petri net representations of the example of train monitoring system connected to the cloud.

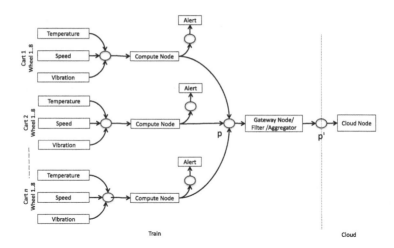

Fig. 3. Petri net representation of the train monitoring system connected to the cloud

3   Decomposition Approach

3.1   Approach Overview

Our approach takes as input a modeled process aware IoT application where its tasks are annotated by a set of IoT locations. It produces a set of remotely communicating application fragments to be deployed at different IoT locations. The modeling, decomposition and deployment proceeds as follows.

Step 1. The user would model the application in one place. We consider in our paper modeling applications as business processes (*e.g.* described in BPMN [29] or Petri nets). Without losing generality, we consider that IoT applications are modeled as Petri nets. In case of using any other modeling language, we can consider a mapping from that language to Petri net back and forth.

Step 2. The user considers a set of IoT locations that will host the different fragments of the IoT application. He/she uses these locations to annotate the modeled process by assigning a location to each task/transition of the considered IoT application.

Step 3. The resulted annotated process is then decomposed into a set of communicating fragments. The decomposition is done such that each possible execution in the original application is also a possible execution in the distributed communicating fragment and vice versa. To do so, we propose a labeled Petri net decomposition method based on the transitions' labels (*i.e.* annotations). The main idea is to split the net into as many subnets as IoT locations (transitions having the same label belong to the same subnet) so that each subnet can be executed in its location while communicating with the other subnets as needed. Then, to preserve the original net language, we propose a translation algorithm allowing to obtain, from the original LPN and its decomposition into subnets, a new "distributed" LPN (with priority) where each component can perform autonomously while communicating with the other components.

Step 4. The user uses an adequate deployment mechanism to deploy the different produced fragments of the application to the different locations.

3.2 Label-Based Decomposition of LPNs

Definition 4. *Let* $N_L = (N, m_0, \Sigma, \varphi)$ *be a labeled Petri net and let* l *be a label in* Σ. *The* l-*based subnet of* N_L, *denoted by* N_l, *is the subnet induced by* $T_l = \{t \in T : \varphi(t) = l\}$ *i.e. the set of transitions having as label* l *in* N_L.

Note that two label-based subnets have no common transitions but could have common places. Assuming that $\Sigma = \{l_1, \ldots, l_n\}$, the LPN $N_L = (N, m_0, \Sigma, \varphi)$ can be then obtained by composing N_{l_i}, for $i \in \{1, \ldots, n\}$, through the fusion of common places. Such a composition is denoted by $N_L = N_{l_1} \oplus \cdots \oplus N_{l_n}$.

3.3 DLPN: Distributed Labeled Petri Nets

Using the label-based subnets defined above, we propose in the following to translate an LPN to a priority LPN, called *Distributed LPN* (DLPN for short), where the subnets are distributed on different IoT locations and can be executed autonomously while communicating distantly with each other subnet through newly added communication transitions/activities. The DLPN is obtained by adding new communication transitions and places to the original LPN and by splitting places that are common to different label-based subnets. In order to guarantee the marking coherence of any pair of places, resulting from the same place in the original net, priority is introduced between some newly added transitions and the original ones.

To ease the definition of a DLPN, these common places are renamed according to the subnets they belong to, as follows:

Definition 5. *Let* $N_L = (\langle P, T, F, W \rangle, m_0, \Sigma, \varphi)$ *be an LPN where* $\Sigma = \{l_1, \ldots, l_n\}$ *and let* $N_{l_i} = (\langle P_{l_i}, T_{l_i}, F_{l_i}, W_{l_i} \rangle)$, *for* $i = 1 \ldots n$, *be the corresponding set of label-based subnets. The renaming function* Rn *applies on the subnet sets of places to ensure different names for common places.*

$$\forall i \in \{1, \ldots, n\}, \ \forall p \in P_{l_i}, \ Rn(p) = \begin{cases} p & if \quad \nexists j \neq i \mid p \in P_{l_j} \\ p_{l_i} & otherwise \end{cases}$$

In the following, the label-based subnets are considered such that the common places have been renamed according to Definition 5. Moreover, given a label-based subnet N_{l_i} and a place p_i in N_{l_i}, the set $Dup(p_i)$ stands for the set of places whose old name, in the original LPN, is the same as the old name of p_i. Formally,

Definition 6. *Let* $N_L = (\langle P, T, F, W \rangle, m_0, \Sigma, \varphi)$ *be an LPN where* $\Sigma = \{l_1, \ldots, l_n\}$ *and let* $N_{l_i} = (\langle P_{l_i}, T_{l_i}, F_{l_i}, W_{l_i} \rangle)$, *for* $i = 1 \ldots n$, *be the corresponding set of label-based subnets.* $\forall i \in \{1, \ldots, n\}$, $\forall p_i \in P_{l_i}$, $Dup(p_i) = \{p_j \in P_{l_j} \mid j \neq i \wedge Rn^{-1}(p_i) = Rn^{-1}(p_j)\}$.

Definition 7. *Let* $N_L = (\langle P, T, F, W \rangle, m_0, \Sigma, \varphi)$ *be an LPN where* $\Sigma = \{l_1, \ldots, l_n\}$ *and let* $N_{l_i} = (\langle P_{l_i}, T_{l_i}, F_{l_i}, W_{l_i} \rangle)$, *for* $i = 1 \ldots n$, *be the corresponding set of label-based subnets. The DLPN associated with* N_L *is a priority LPN* $N_d = (\langle P_d, T_d, F_d, W_d \rangle, m_{0_d}, \Sigma_d, \varphi_d, \rho)$ *where:*

- $P_d = \bigcup_{i=1\ldots n} P_{l_i} \cup P_a$, $T_d = \bigcup_{i=1\ldots n} T_{l_i} \cup T_a$ *and* $F_d = \bigcup_{i=1\ldots n} F_{l_i} \cup F_a$, *where* P_a, T_a *and* F_a *are newly added sets of places, transitions and arcs, respectively, satisfying the following:*
  $\forall i \in \{1, \ldots, n\}$, $\forall p_i \in P_{l_i}$, $Dup(p_i) \neq \emptyset \Leftrightarrow$ *the following holds:*
  1. $\forall t_k \in {}^{\bullet}p_i$, $\exists p_k^i \in P_a \wedge \exists t_k^i \in T_a$ *s.t.,*
     - $\{(t_k, p_k^i), (p_k^i, t_k^i)\} \subseteq F_a \wedge \forall p_j \in Dup(p_i)$, $(t_k^i, p_j) \in F_a$
     - $W_d^+(t_k, p_k^i) = W_d^-(p_k^i, t_k^i) = W_{l_i}^+(t_k, p_i) \wedge$
       $\forall p_j \in Dup(p_i)$, $W_d^+(t_k^i, p_j) = W_{l_i}^+(t_k, p_i)$
  2. $\forall t_k \in p_i{}^{\bullet}$, $\forall p_j \in Dup(p_i)$, $\exists p_k^{ij} \in P_a \wedge \exists t_k^{ij} \in T_a$ *s.t.,*
     - $\{(p_k^{ij}, t_k), (t_k^{ij}, p_k^{ij}), (p_j, t_k^{ij})\} \subseteq F_a$
     - $W_d^-(p_k^{ij}, t_k) = W_d^+(t_k^{ij}, p_k^{ij})) = W_d^-(p_j, t_k^{ij}) = W_{l_i}^-(p_i, t_k)$
  3. $\forall i \in \{1, \ldots, n\}$, $\forall p_i \in P_{l_i}$, $Dup(p_i) \neq \emptyset$ *the following priorities are introduced:*
     - $\forall t_k^i \in T_a$ *associated with* p_i *and a transition* $t_k \in {}^{\bullet}p_i$, $\forall p_j \in Dup(p_i)$, $\forall t \in {}^{\bullet}p_i{}^{\bullet} \cup {}^{\bullet}p_j{}^{\bullet}$, $(t, t_k^i) \in \rho$ *(i.e.* t_k^i *has priority over* t).
     - $\forall t_k^{ij} \in T_a$ *associated with* p_i *and a transition* $t_k \in p_i{}^{\bullet}$, $\forall t \in p_i{}^{\bullet} \cup T_a$, $(t, t_k) \in \rho$ *(i.e.* t_k *has priority over* t).
- $\forall i = 1 \ldots n$, $\forall (x, y) \in (P_{l_i} \times T_{l_i} \cup T_{l_i} \times P_{l_i})$, $(x, y) \in F_d \Leftrightarrow (x, y) \in F_{l_i}$
- $\forall i = 1 \ldots n$, $\forall (x, y) \in (P_{l_i} \times T_{l_i} \cup T_{l_i} \times P_{l_i})$, $W_d(x, y) = W_{l_i}(x, y)$
- $m_{0_{d_P}} = m_0$, *and* $\forall p_i \in P_{l_i}$, $\forall p_j \in Dup(p_i)$, $m_{0_d}(p_i) = m_{0_d}(p_j) = m_0(Rn^{-1}(p_i))$
- $\Sigma_d = \Sigma \cup T_d$
- $\varphi_d \colon T_d \to \Sigma_d$ *s.t.*

$$\varphi_d(t) = \begin{cases} \varphi(t) \; if \; t \in T \\ t \quad\; if \; t \in (T_d \setminus T) \end{cases}$$

Given an LPN, Definition 7 gives the structure of a DLPN highlighting the label-based subnets whose underlying process can be executed within its own site. First, the DLPN is obtained by composing the set of label-based subnets. Hence, the initial marking, the set of arcs (and the corresponding weights), the set of labels, and the labeling function are inherited from the composed label-based subnets. This is ensured by the five last items of the above definition. Then, in order to isolate the label-based subnets, some places, transitions and weighted arcs are added around shared places as stated in the first item of the definition as follows: for each place p_i shared by two or more subnets,

- The first point (1) states that, for each input transition t_k of p_i a new transition t_k^i and place p_k^i are added such that any change (token production) of the marking of p_i will eventually similarly change any place p_j (by the firing of t_k^i), where p_j represent the same place as p_i in the original LPN (i.e. $p_j \in Dup(p_i)$).
- The second point (2) states that, for each output transition t_k of p_i, for each place $p_j \in Dup(p_i)$, a new transition t_k^{ij} and place p_k^{ij} are added such that any change (token consumption) of the marking of p_i is conditioned by the same change for any place p_j (by the firing of t_k^{ij}).
- The third point introduces some priorities between the newly added transitions and the original ones in order to ensure coherent markings between related places (places that come from the split of the same original place). Given two related places p_i and p_j, it first states that, when new tokens are created in p_i (for instance) by the firing of a transition t_k, then these tokens must be transferred to p_j before any change of both p_i and p_j markings. The second kind of priority allows to avoid contradictory choice in case of conflict. For example, if a newly added transition t_k^{ij} is fired, this means that the sufficient number of tokens have been moved from p_j to p_i in order to fire an output transition t_k of p_i. Thus, t_k has priority over any newly added transition t that would move tokens from other direction (i.e., from p_i to p_j). Note that, without such a condition, a deadlock state can be reached.

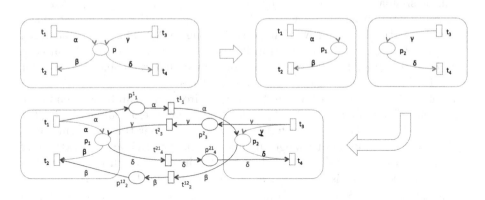

Fig. 4. From LPN to DLPN

Figure 4 illustrates the translation algorithm allowing to build a DLPN from an LPN. p_1 and p_2 are two places of two label-based subnets resulting from the renaming of the same place p in the original LPN. The added nodes (red places and red transitions) are connected to these places, following the above definition, to guarantee that, the behavior around these two places is the same in each subnet. Any token consumed (resp. produced) by a transition in one subnet is also consumed (resp. produced) in the other one. For instance, using point (1) of the first item, since t_1 is an input transition of p_1, the place p_1^1 and the transition t_1^1 are added and connected to t_1 and p_2 such that each time t_1 is fired, producing α tokens in p_1, the same number of token is eventually produced in p_2 by the firing of t_1^1. Thanks to the priority of t_1^1 over transitions t_1, t_2, t_3 and t_4, α tokens will be transferred into p_2 before any change of its marking by the original transitions. Similarly, using point (2) of the first item of the definition, since t_2 is an output transition of p_1, the place p_2^{12} and the transition t_2^{12} are added and connected to p_2 and t_2 such that each time t_2 is fired, consuming β tokens from p_1, the same number of token is necessarily consumed from p_2 by the firing of t_2^{12}. If t_2^{12} is fired, and in case p_1 contains β tokens, then t_2 has priority over t_4^{21} (for instance) which will move tokens in other direction (from p_1 to p_2). Without such a priority, the firing sequence $t_2^{12}.t_4^{21}$ (or $t_4^{21}.t_2^{12}$) is possible from the initial marking where both p_1 and p_2 contain one token, and leads to a deadlock state.

The following Theorem states that the LPN and the corresponding DLPN have the same language projected on the common transitions.

Theorem 1. *Let* $N_L = (\langle P, T, F, W \rangle, m_0, \Sigma, \varphi)$ *be an LPN and let* $N_d = (\langle P_d, T_d, F_d, W_d \rangle, m_{0_d}, \Sigma_d, \varphi_d)$ *be the corresponding DLPN. Then,* $\mathcal{L}_T(N_d) = \mathcal{L}(N_L)$.

Proof. The proof of Theorem 1 is directly guaranteed by construction.

- In fact, any firing sequence $m_0 \xrightarrow{t_0} m_1 \ldots \xrightarrow{t_n} m_n \ldots$ in N_L, starting from m_0 can be simulated by a firing sequence $m_{0_d} \xrightarrow{t_0} m_1' \xrightarrow{\sigma_1} m_1'' \ldots \xrightarrow{t_n} m_n' \xrightarrow{\sigma_n} m_n'' \ldots$ in N_d, starting from m_{0_d}, where σ_i (for $i = 1 \ldots n$) are the maximal sequences containing transitions of N_d that do not belong to N_L. This can be achieved by prioritizing the firing of the transitions that are in N_d but not N_L at each reached marking. By doing this, we guarantee that each marking m_i'' (for $i = 1 \ldots n$) coincides with m_i for common places.
- Inversely, any firing sequence, in N_d, starting from m_{0_d}, can be written in the form $m_{0_d} \xrightarrow{\alpha_1} m_1' \xrightarrow{\sigma_1} m_1'' \ldots \xrightarrow{\alpha_n} m_n' \xrightarrow{\sigma_n} m_n'' \ldots$, where $\alpha_i \in T^*$ and $\sigma_i \in (T_d \backslash T)^*$ (for $i = 1 \ldots n$). Given the definition of m_{0_d}, it is clear that the firing of σ_i (for $i = 1 \ldots n$) does not change the marking of the original places (those of N_L). Thus $m''_{i_P} = m'_{i_P}$ for $i = 1 \ldots n$, and since α_i, which is composed of N_L transitions only, is fireable at m''_i, it is also fireable by m'_{i_P}. To conclude, the behavior $m_0 \xrightarrow{\alpha_{1_T}} m'_{1_P} \ldots \xrightarrow{\alpha_{n_T}} m'_{n_P} \ldots$ is possible, in N_L, from the initial marking m_0.

3.4   Example

Figure 5 illustrates the DLPN construction for the Train monitoring system example. Place p is split into $n + 1$ places p_i (for $i = 1 \ldots n$) and place p_g, while place p' is split into two places p_1' and p_2'. For each place p_i (for $i = 1 \ldots n$), a new transition t_{CN}^i and a new place p_{CN}^i are added to the original model in order to guarantee that each firing of transition *Compute Node*, producing a token in p_i, will eventually produce a token p_g as well as in any place p_i by the firing

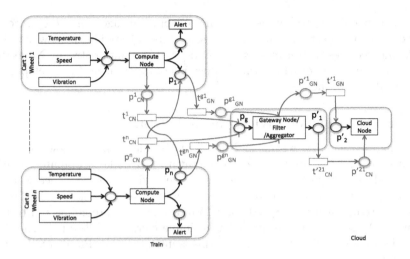

Fig. 5. From LPN to DLPN: an illustration

of t^i_{CN}. The issue is considered for place p'_1 and its input transition *Gateway Node/Filter/Aggregator*. Then, the firing of *Gateway Node/Filter/Aggregator* (resp. *Cloud Node*) is not conditioned by the marking of p_g (resp. p'_1) only, but, indirectly, by the marking of all places p_i (for $i = 1 \dots n$). This is ensured by the new added transitions t^{gi}_{GN} and places p^{gi}_{GN} (for $i = 1 \dots n$) (resp. by the newly added transition t'^{21}_{CN} and place p'^{21}_{CN}).

4 Implementation

We have enhanced and extended the open source Node-RED [28] tool as a proof of concept of our proposed approach. In particular, we have developed mechanisms to define locations and used them to annotate IoT applications. We have also implemented the decomposition process and a distributed deployment of decomposed fragments.

4.1 Node-RED Overview

The open source tool Node-RED offers capability to graphically design, deploy and run process-aware IoT applications. Node-RED can be run on a variety of platforms, from edge of network devices to the cloud. It is a web browser based flow editing tool which makes it easy to wire together flows using the wide range nodes in the palette.

The *nodes* in Node-RED are individual components of the application which perform some function. The connection mechanism, called wiring, depends on the output of a node and the input of connecting node. It could be simple text message transfer from one node to another locally, or a message transfer from one node to a remote node using HTTP or MQTT protocol, involving nodes developed specifically to handle input and output messages for these protocols. For example, an Inject node allows a message to be injected in the flow, a twitter input node receives twitter messages from defined account or stream. Node-RED is open source and is provided with thousands of different nodes developed and made available by the community.

4.2   Semantics of Node Wiring in Node-RED

The semantics of node wiring in Node-RED is informal. As shown in Fig. 1, the direct connection from node "Temperature" to node "Compute Node" is interpreted as a sequence execution, *i.e.* as defined by the Sequence workflow pattern [31]. When several nodes are connected to one node, as it is the case of nodes "Temperature", "Speed" and "Vibration" that are connected to node "Compute Node", this is interpreted as a simple merge workflow pattern. When one node is connected to several nodes this is interpreted as Parallel Split workflow pattern.

Consequently, among the basic workflow patters, Node-RED only supports Sequence, Parallel Split and Simple Merge patterns. It does not natively support Synchronization and Exclusive Choice workflow patterns.

4.3   Supporting Node Annotation in Node-RED

Node-RED has a special type of node called *Config* node, which is used to store configuration information of entities like MQTT broker, HTTP address, credentials, etc. These *Config* nodes are used within the actual nodes where such information is needed. We developed a new *Config* node to store location information, representing the type of location and other details, like IP address, port number, userid, password, etc. as shown in Fig. 6.

Multiple different locations can be defined in this manner. Other nodes are enhanced to include *Location* as an attribute as shown in Fig. 6. All the defined locations show up as a list in the node configuration, where user can select one location for that node, which is what we refer to as annotating the node with location attribute.

Fig. 6. Location config node and node annotation in Node-RED

4.4 Node-RED to/from Petri Net Transformation

Once the user creates full design of the application in Node-RED and then annotates each node in the model with its corresponding *Config* location node, the model is ready to be translated to Petri net, decomposed, translated back to node-RED and deployed.

As per the semantics of node-RED, the transformation to Petri net is straightforward. As for the transformation from Petri net to Node-RED, it consists of the following: Shared places that model remote communication between fragments (*e.g.* p_{GN}^{g1}) are transformed into a MQTT broker node. The incoming arcs are transformed into *mqtt.out* nodes (*e.g.* arc from t_{GN}^{g1} to p_{GN}^{g1}) and outgoing arcs are transformed into *mqtt.in* nodes. Transitions added by the decomposition (that do not belong to Synchronization and Exclusive Choice workflow patterns) are transformed into function nodes that output the incoming payload as is.

In addition to the above cases, our decomposition approach may generate Synchronization and Exclusive Choice workflow patterns that are not natively supported in Node-RED (as the one illustrated in Fig. 4). To deal with this issue, we have developed two new function nodes that implement Synchronization and Exclusive Choice. These function nodes use Node-RED context variable to store state of incoming completed branches and place marking consumption. Petri net places that are only shared by one incoming transition and one outgoing transition that belong to two different locations are modeled by an MQTT *Config* node along with a *mqtt.in* node and a *mqtt.out* node.

Figure 7 shows an application fragment result of the decomposition of example presented in Fig. 2 where we considered two wheels.

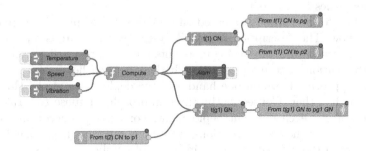

Fig. 7. IoT application fragment with generated nodes

4.5 Distributed Deployment

Node-RED comes with three different options for deploying the application. These three options are: *Full*-where it deploys everything in the editor workspace, *Modified Flows*-where it deploys only flows that contain changed nodes, and *Modified Nodes*-where it deploys only the nodes that have changed. We have added a new option for deployment, named *Distributed*, which is used for deploying the application to multiple different remote Node-RED servers based on the location attribute of the nodes.

Once the network of nodes with local and remote communication nodes is complete, each group of nodes is deployed at its specified location using the REST interface already available in Node-RED to deploy the flows remotely. Figure 7 shows how the deployed fragment for one such Wheel looks like, where additional nodes are generated during decomposition process which are responsible for remote communication and synchronization. When the node flows are deployed in each platform location, they are automatically started in running mode by Node-RED, making the whole IoT application in operation. This way, we achieve designing of the distributed IoT application in one place, usually *Cloud*, and then deploying it across different locations comprising *Edge* locations as well as *Cloud* locations.

5   State of the Art

We present in the following existing approaches for the design and deployment of applications in IoT and the Cloud as well as formal methods for process decomposition and show how these approaches are suitable to the context we consider in this paper.

5.1   Designing and Deployment of IoT and Cloud Applications

Designing and deployment of IoT applications has been tackled from two different point of views: approaches based on mashups (e.g. [16,26,28]) and model-based approaches (e.g. [3,6,11]).

According to the approaches based on mashups, IoT applications are defined as process flow. The description of an IoT system architecture is represented as connection between IoT physical components and the process flow representing an IoT application. The main issues with these approaches is the coupling of applications/process flows on one hand and the deployment model (IoT architecture) on the other. Consequently, these approaches impose to write/design a specific app for each different deployment. For example, according to these approaches, smart home applications would be different from one home to another, even if the difference is only in terms of number of rooms.

For model-driven approaches, an IoT application is modeled in IoT component-independent model. Then, its corresponding device-specific, native code is generated and possibly compiled. Finally, the generated code is deployed. These approaches impose the use of different models at different levels of abstractions and described using different domain-specific languages.

Several approaches focused on deploying business processes and choreography-based process in the Cloud. In [5], the authors draw up an inventory of the different delivery models available to execute a business processes described in BPEL process at IaaS, PaaS and SaaS layers. In addition to that, some works have proposed to transform and decompose business processes to equivalent processes in order to be able to deploy them (or a part of them) in the Cloud [5,14,32].

In addition to that, several Cloud providers begin to integrate features to design, deploy and execute business processes. For example, Amazon Web Service proposes the Amazon Simple Workflow Service (Amazon SWF) which assists developers to coordinate the various processing steps in the process to deploy and to manage distributed execution state [4]. Force.com introduced a "Visual Process Manager" [4] and WSO2, which is based on Apache ODE process engine, provides a variant of its business process server as-a-Service [4]. In addition, Cloud Foundry, proposes a feature for deploying the required framework and/or specific platform resources (e.g. a proprietary service container instance) before deploying the application [17].

While the approaches developed to deploy business processes in the Cloud tackle the principle of deployment and decomposition we address in this paper, these approaches are not adequate for IoT environments since they rely on centralized process engines to orchestrated the deployed applications fragments.

5.2 Decomposition Approaches

Decomposition approaches can be roughly divided into two categories: horizontal and vertical decomposition. In the first category, the model is sliced into different parts that can be analyzed locally, while in the second, Decomposition approaches have mainly be explored in the context of model checking [12], in order to tackle the underlying state space combinatory explosion problem. They can roughly divided into two categories: horizontal and vertical decomposition.

In order to avoid to build the whole state space of the original model, horizontal approaches (e.g. [13,18,19,22,33]) use a "divide and conquer" strategies to, first, split the global model into several pieces, then to compute and check each piece's behavior separately, and finally deduce the properties of the original model from its components ones. Since, most properties (reachability, deadlock freeness, invariants, ...) are not preserved by decomposition (neither by composition), the common way is to constraint the components communication pattern. Such approaches consider that each part of the system can run independently from the others, leading to an over-approximation of the possible behavior. These behaviors are then avoided by composition while, in our approach, the components are built in such a way that each component can run autonomously but communicate with its environment. Using the DPN, the communication between these component ensures that no extra behavior can take place within the distributed execution of the whole system.

More specifically based on Petri nets, numerous other approaches, proposing vertical decomposition techniques exist (e.g. [21,25,27]). They consist in replacing parts of the model by a unique place/transition abstracting a subnet whose reachability graph can be analyzed separately or deduced from the abstract level. For example, the authors of [9] introduced the notion of regions i.e. parts of the net that will be substituted by a less detailed representation in an abstraction operation (using the P-invariants of the original model). This leads to a hierar-chical Petri net whose reachability graph is reduced compare to the original one. However, this method can not be used in our context for many reasons: first, the region computation algorithm can not terminate (termination is guaranteed in case the model is covered by a P-invariant). Second, a region could contain two transitions belonging to two different locations, which is not appropriate in the context we consider.

In [1], the authors present some the public-to-private approach to inter-enterprise business process cooperation. The proposed approach is organized in three steps. First, a global/public business process is defined and serves as a contract between the cooperating enterprises. Second, each process fragment of the public business process is mapped onto one of the domains (*i.e.*, enterprise). Each domain is responsible for a part of the public business process, referred to as its public part. Third, each domain can make use of its autonomy to create a private business process. To satisfy the correctness of the overall inter-enterprise business process, however, each domain may only choose a private business process which is a subclass of its public part in the sense of business process inheritance [2]. The main peculiarity of the public-to-private approach is that we do not split the public business process model but we split it on the base of an arbitrary partition of its set of transitions/locations.

6   Conclusion and Future Work

To allow the modeling of IoT applications in one place and deploy it over multi-ple IoT applications, we have proposed an approach that consists in decompos-ing process-aware applications into a set of communication fragments. We have adopted a formal approach to prove the correctness of the decomposition with respect to language preservation. We have also implemented our approach as an extension to Node-RED tool.

As future work in this field, we want to extend this distributed deployment mechanism such that if there are a bunch of edge nodes running same application logic, the designer will have the option to just design with one such application fragment, and with a special type of tagging, we will determine all the compute nodes where the same fragment of the application needs to be deployed, and as part of distributed deployment, we would perform such a replicated deployment as well.

References

1. van der Aalst, W.M.P., Weske, M.: The P2P approach to interorganizational work-flows. In: Bubenko, J., Krogstie, J., Pastor, O., Pernici, B., Rolland, C., Sølvberg, A. (eds.) Seminal Contributions to Information Systems Engineering, 25 Years of CAiSE, pp. 289–305. Springer, Heidelberg (2013)
2. van der Aalst, W., Basten, T.: Inheritance of workflows: an approach to tackling problems related to change. Theoret. Comput. Sci. **270**(1), 125–203 (2002)

3. Alferez, M., Tessier, P., Janssens, C., Roubekas, P., Pascual, G., Nicolas Fauvergue, V.L., Damus, C.W., Gurcan, O., Adam, M., Radermacher, A., Tatibouet, J., Geoffroy, J., Maggi, B., Peretokin, V., Dumoulin, C., Letavernier, C., Hafsteinn, Schnekenburger, R., Landre, T., Benois, J.: Papyrus Modeling environment, 1 September 2016. https://eclipse.org/papyrus. Accessed 15 Jan 2017

4. Amazon: The Amazon Simple Workflow Service developer guide (2016). http://docs.aws.amazon.com/amazonswf/latest/developerguide/swf-welcome.html

5. Anstett, T., Leymann, F., Mietzner, R., Strauch, S.: Towards BPEL in the cloud: exploiting different delivery models for the execution of business processes. In: 2009 Congress on Services - I, pp. 670–677, July 2009

6. Barais, O., Tricoire, M., Dartois, J.E., Bourcier, J., Morin, B., Nain, G., Plouzeau, N., Sunye, G., Jezequel, J.M.: The Kevoree Book (2013). http://www.kevoree.org. Accessed 15 Jan 2017

7. Barcelo, M., Correa, A., Llorca, J., Tulino, A.M., Vicario, J.L., Morell, A.: IoT-cloud service optimization in next generation smart environments. IEEE J. Sel. Areas Commun. **34**(12), 4077–4090 (2016)

8. Best, E., Koutny, M.: Petri net semantics of priority systems. Theor. Comput. Sci. **96**(1), 175–215 (1992)

9. Buchholz, P., Kemper, P.: Hierarchical reachability graph generation for petri nets. Form. Methods Syst. Des. **21**(3), 281–315 (2002)

10. MacGillivray, C., et al.: IDC FutureScape: worldwide Internet of Things 2017 predictions. https://www.idc.com/getdoc.jsp?containerId=US40755816

11. Ciccozzi, F., Spalazzese, R.: MDE4IoT: supporting the Internet of Things with model-driven engineering. In: Badica, F., et al. (eds.) Intelligent Distributed Computing X. SCI, vol. 678, pp. 67–76. Springer, Cham (2017). doi:10.1007/978-3-319-48829-5_7

12. Clarke, E.M., Grumberg, O., Peled, D.A.: Model Checking. The MIT Press, Cambridge (2000)

13. Cobleigh, J.M., Giannakopoulou, D., PǍsǍreanu, C.S.: Learning assumptions for compositional verification. In: Garavel, H., Hatcliff, J. (eds.) TACAS 2003. LNCS, vol. 2619, pp. 331–346. Springer, Heidelberg (2003). doi:10.1007/3-540-36577-X_24

14. Dornemann, T., Juhnke, E., Freisleben, B.: On-demand resource provisioning for BPEL workflows using Amazon's elastic compute cloud. In: Proceedings of the 2009 9th IEEE/ACM International Symposium on Cluster Computing and the Grid, CCGRID 2009, pp. 140–147. IEEE Computer Society, Washington, D.C. (2009)

15. Dumas, M., van der Aalst, W.M., ter Hofstede, A.H.: Process-aware Information Systems: Bridging People and Software Through Process Technology. Wiley Inc., New York (2005)

16. Fleurey, F., Morin, B.: ThingML: a modeling language for embedded and distributed systems, 20 October 2014. http://thingml.org. Accessed 15 Jan 2017

17. Foundry, C.: Cloud Foundry official blog. Deploying a service container on CF using the standalone framework (2016). http://blog.cloudfoundry.com/2012/06/18/deploying-tomcat-7-using-the-standalone-framework/

18. Klai, K., Petrucci, L.: Modular construction of the symbolic observation graph. In: ACSD, pp. 88–97. IEEE (2008)

19. Klai, K., Tata, S., Desel, J.: Symbolic abstraction and deadlock-freeness verification of inter-enterprise processes. In: Dayal, U., Eder, J., Koehler, J., Reijers, H.A. (eds.) BPM 2009. LNCS, vol. 5701, pp. 294–309. Springer, Heidelberg (2009). doi:10.1007/978-3-642-03848-8_20

20. Koutny, M.: Modelling systems with dynamic priorities. In: Rozenberg, G. (ed.) Advances in Petri Nets 1992. LNCS, vol. 609, pp. 251–266. Springer, Heidelberg (1992). doi:10.1007/3-540-55610-9_174

21. Le Cornec, Y.S.: Compositional analysis of modular petri nets using hierarchical state space abstraction. In: Joint 5th International Workshop on Logics, Agents, and Mobility, LAM 2012, the 1st International Workshop on Petri Net-Based Security, WooPS 2012 and the 2nd International Workshop on Petri Nets Compositions, CompoNet 2012, vol. 853 (2012)

22. McMillan, K.L., Qadeer, S., Saxe, J.B.: Induction in compositional model checking. In: Proceedings of the 12th International Conference on Computer Aided Verification, CAV 2000, Chicago, IL, USA, 15–19 July 2000, pp. 312–327 (2000)

23. Meyer, S., Ruppen, A., Hilty, L.: The things of the Internet of Things in BPMN. In: Persson, A., Stirna, J. (eds.) CAiSE 2015. LNBIP, vol. 215, pp. 285–297. Springer, Cham (2015). doi:10.1007/978-3-319-19243-7_27

24. Meyer, S., Ruppen, A., Magerkurth, C.: Internet of Things-aware process modeling: integrating IoT devices as business process resources. In: Salinesi, C., Norrie, M.C., Pastor, Ó. (eds.) CAiSE 2013. LNCS, vol. 7908, pp. 84–98. Springer, Heidelberg (2013). doi:10.1007/978-3-642-38709-8_6

25. Miczulski, P.: State space calculation algorithm of hierarchical petri nets with application of decision diagrams. In: DESDes 2001, p. 67 (2001)

26. Naef, L.: ClickScript, 29 May 2014. https://github.com/lnaef/ClickScript. Accessed 15 Jan 2017

27. Notomi, M., Murata, T.: Hierarchical reachability graph of bounded petri nets for concurrent-software analysis. IEEE Trans. Softw. Eng. **20**(5), 325–336 (1994)

28. O'Leary, N., Conway-Jones, D.: Node-RED: a visual tool for wiring the Internet of Things, 11 January 2017. https://nodered.org/. Accessed 15 Jan 2017

29. Object Management Group: Business process model and notation (BPMN) version 2.0. Technical report formal/2011-01-03. Object Management Group (OMG), January 2011. http://taval.de/publications/BPMN20

30. Shi, W., Dustdar, S.: The promise of edge computing. Computer **49**(5), 78–81 (2016)

31. Van Der Aalst, W.M.P., Ter Hofstede, A.H.M., Kiepuszewski, B., Barros, A.P.: Workflow patterns. Distrib. Parallel Databases **14**(1), 5–51 (2003)

32. Wagner, S., Kopp, O., Leymann, F.: Towards choreography-based process distribution in the cloud. In: 2011 IEEE International Conference on Cloud Computing and Intelligence Systems, pp. 490–494, September 2011

33. Xiong, P., Fan, Y., Zhou, M.: A petri net approach to analysis and composition of web services. IEEE Trans. Syst. Man Cybern. Part A **40**, 376–387 (2010)

Towards an Automatic Enrichment of Semantic Web Services Descriptions

Mohamed Lamine Mouhoub[(⊠)], Daniela Grigori, and Maude Manouvrier

Université Paris-Dauphine, PSL Research University,
CNRS, UMR [7243], LAMSADE, 75016 Paris, France
{mohamed.mouhoub,daniela.grigori,maude.manouvrier}@dauphine.fr

Abstract. Web service discovery consists in identifying suitable existing services that satisfy specific goals or user requirements. Web service discovery can be a hard task because of the lack of explicit and semantic information for apprehending what services really do. In this article, we propose a service description enrichment approach based on I/O relations for improving Web service discovery. Our process is divided into two parallel steps. The first step consists in extracting existing relations between I/O of the services from the underlying ontologies using SPARQL, while the second step concerns the extraction of services' I/O relations from the text descriptions of services using NLP techniques. Matching the I/O relations extracted by the two steps is applied in order to enrich the initial service description, allowing a more accurate automatic service discovery. This article presents our approach that uses dependency grammar and *word2vec* as well as our experimental results on OWLS-TC.

Keywords: Semantic web services · Natural Language Processing · Semantic annotation · Service description enrichment · I/O relationship extraction

1 Introduction

Web service discovery consists in identifying suitable existing services in a repository that satisfy a specific goal or user query [1,2]. Automatic Web service discovery can be a tricky task today because the standard service descriptions are not sufficient for an automatic understanding of what a service does. Moreover, the inputs and outputs of a service, which are the basis for expressing its functionality, cannot provide a full understanding of what a service does with the inputs and how the outputs are related to the inputs. In other words, two identical web services in terms of I/O elements might be totally different in terms of functionality. Hopefully, such functionality of a service and the relations between its inputs and outputs can be explicitly detailed in its documentation or in the textual part of its formal description (OWL-S [3], etc.). These relations, so called *text I/O relations*, can be taken advantage of to improve the understanding of service functionality but require sophisticated tools and methods given that they are informal and written in natural language.

© Springer International Publishing AG 2017
H. Panetto et al. (Eds.): OTM 2017 Conferences, Part I, LNCS 10573, pp. 681–697, 2017.
https://doi.org/10.1007/978-3-319-69462-7_43

On the other hand, semantic web services [4] use ontologies to define their inputs and outputs. These ontologies can roughly be seen as an underlying general schema for web services that use them. Obviously, given the connected-graph nature of ontologies, if the inputs and outputs of a service refer to some nodes in an ontology, there would be at least one path (called a relation in the following) that links these nodes. Therefore, there can be a lot of ontological relations between the I/O of a service in the ontologies that do not correspond to its functionality. The challenge is to find for each service the *ontology I/O relations* that best describe its functionality.

In this paper we propose an approach that exploits the I/O relations found in textual descriptions of semantic web services to promote ontology I/O relations that can be used to enrich service descriptions in order to remedy the functionality ambiguity problems. The semantic web layer on top of the service descriptions offers the possibility to formally and explicitly express the relations between the inputs and outputs of a service in the form of a basic graph pattern (BGP, a set of triple patterns, (subject, predicate, object)) as proposed in [5]. Therefore, our approach is solely focused on semantic web services but can be extended to classical web services and APIs if coupled with a semantic annotation tool for web services like WebKarma [6].

Figure 1a illustrates a an example of ambiguous service functionality. Service #1 returns the list of books written by a given author while Service #2 returns the books for which he has written a preface. When looking only at the I/O, the two services consume and produce the same I/O and one may think that they do the same thing. However, when reading their textual descriptions, these services do completely different things.

(a) Service description #1

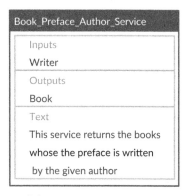

(b) Service description #2

Fig. 1. An example of two web services with identical I/O types but with totally different functionality

The objective of our approach is to resolve the functionality ambiguity problem by enriching the semantic service description with BGPs that represent the I/O relations. Achieving such a goal would have many benefits in different domains like: (a) improving the accuracy of automatic service discovery, (b) facilitating the automatic service composition and automatic service replacement in faulty compositions, (c) facilitating data integration when using data from data-providing web services (DPS). etc.

The rest of this paper is structured as follows: In Sect. 2, we briefly present some of the recent literature related to our work. Section 3 is dedicated to the presentation of our proposed approach. In Sect. 4 we evaluate different aspects of our work and prove its feasibility. We discuss some limitations and perspectives in the conclusion in Sect. 5.

2   Related Works

A lot of research works have been conducted on semantic annotation of web services. A systematic literature review is available in [7]. A richer service description allows better automatic service discovery, matchmaking, recommendation or composition.

Authors of [8] propose an automatic approach for semantically annotating I/O of Web services, based on DBpedia.[1] In this approach, I/O of services are associated with the appropriate concepts in the DBpedia ontology, using DBpedia Spotlight.[2] In [9], a framework is proposed to enrich the functional description of services written by service providers. This approach is also based on semantic annotations from DBpedia. In [10], the descriptions of API (e.g. from ProgrammableWeb[3]) are enriched by semantic annotations, linking API properties to concepts of shared vocabulary. In [11], a semantic similarity measure applied to service description and service signature (I/O) is defined, using relational distance among concepts of WordNet [12] and other ontologies. All these approaches are evaluated using OWLS-TC.[4] Speiser and Harth [5] implicitly tackles the ontological I/O relations by proposing manual semantic annotation templates that describe the I/O of a service with Basic Graph Patterns linking them to web ontologies and eventually to each other.

In contrast to our work, the aforementioned approaches do no extract relations between I/O and do not apply any matching of these relations with the service description. Our work is complementary to these approaches, aiming to enrich such semantic annotations by adding the I/O relations.

On the other hand, there have been many efforts to enrich ontologies by extracting new semantic relations from text using Natural Language Processing (NLP) techniques [13–15]. Authors of [13] propose an approach based on dependency grammar to extract new RDF properties for DBpedia from the English Wikipedia[5] using a scalable MapReduce-based algorithm. We apply a similar

---

[1] http://dbpedia.org.

[2] https://github.com/dbpedia-spotlight/dbpedia-spotlight/wiki.

[3] https://www.programmableweb.com/.

[4] http://projects.semwebcentral.org/projects/owls-tc/.

[5] https://www.wikipedia.org.

approach based on dependency grammar to extract I/O relations from the text but instead of adding them as new semantic relations to the service descriptions, we use them as a reference to choose the fittest amongst the existing relations in the ontologies that best describe the I/O relations.

3   Proposed Approach

To achieve the desired service description enrichment previously mentioned, our system extracts parallelly two types of I/O relations. The first are the existing formal relations between the I/O of services from their underlying ontologies using SPARQL.[6] The other type is the informal I/O relations depicted in natural language in the text descriptions and documentation of services which requires Natural Language Processing techniques to be extracted. The two extraction processes are computationally independent and therefore are executed in parallel. At the end, the two types of extracted relations are matched against each other and the ontological relations that have the best matches in the textual relations are considered convenient for enriching the service description. They can later be added to the semantic service description after validation from the user. The overall process is illustrated in Fig. 2.

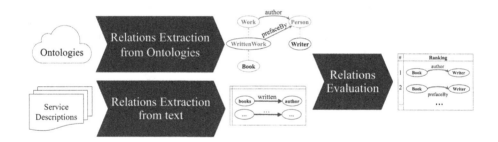

Fig. 2. I/O relations extraction process

3.1   Extracting I/O Relations from Ontologies

The inputs and outputs of a semantic web service describe what a service needs in order to be invoked (the type of data that it consumes) and what it produces after its invocation while referring to concepts or nodes in one or more ontologies. Let $I_i \in I$ be an input of a semantic service referring to a node N_i in some ontology and $O_j \in O$ be an output referring to another node N_j. We denote this reference by $I_i = N_i$ and $O_j = N_j$. Given the connected-graph nature of an ontology, there can be at least one path (relation) that links N_i and N_j. Obviously, exceptions may apply if the two nodes are from different but not interlinked ontologies.

---

[6] https://www.w3.org/TR/sparql11-query/.

The first step of our approach aims to extract all the possible relations between the input and output elements of the service within a predefined maximum distance. The extraction operates in a pairwise fashion (for each combination pair of an input and an output) but the extracted relations can all be combined and expressed in the form of a basic graph pattern (i.e. a set of RDF triples).

Figure 3 illustrates the relations between two example I/O nodes: $I_1 =$ Writer and $O_1 =$ Book. As seen in this figure, there is no edge (i.e. Property) directly linking the two nodes at a distance $d = 0$, i.e. without intermediate nodes. However, there are multiple paths at different distances $(d = 2, d = 3)$ including different nodes and properties that link I_1 and O_1.

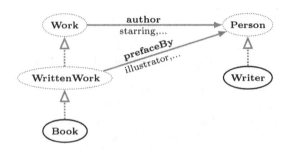

Fig. 3. BGP depicting the links between the input Book and the output Writer.

The relations graph can be extracted from the ontologies referenced in the service description to describe the I/O. For such an extraction, we can either apply graph search algorithms to the ontology dumps using graph theory tools or use a SPARQL based approach. In our process we use SPARQL because it works on-the-fly and allows to search in multiple ontologies in the Linked Data Cloud (LOD) in a federated fashion.

SPARQL-Based Extraction. We use SPARQL to query the endpoints hosting ontologies for the existing relations between all the Input and Output nodes in a pairwise fashion. Each combination of an input and an output requires its own SPARQL query consisting of a more or less complex template depending on the nature of the searched paths.

As depicted earlier, two nodes can have one or more relations that go through different paths. Each relation corresponds to a path of a certain length based on the distance between the two nodes. Depending on its length/distance $d = n$, a path involves one or more properties $p_k \in \{p_0, ..., p_n\}$ or reverse properties $q_k \in \{q_0, ..., q_n\}$ as well as zero or more intermediate nodes $M_k \in \{M_1, ..., M_n\}$.

The properties and reverse properties are owl:ObjectProperty that link two source and target owl:Class nodes representing either the I/O nodes or the intermediate nodes. This link is established in the ontology following a specific pattern. Listings 1 and 2 depict two example property extraction patterns in DBpedia and OWLS-TCv4 respectively between two nodes `?source` and `?target`. The latter is not straightforward as the first because it is meant to be inferred by an inference engine.

```
# ?source-->?target
{?p   rdfs:domain   ?source;
      rdfs:range    ?target .}
# ?source<--?target
{?q   rdfs:domain   ?target;
      rdfs:range    ?source .}
```

Listing 1. Property and reverse property extraction patterns in DBpedia.

```
# ?source-->?target
{?source           rdfs:subClassOf  ?x  .
?x                 owl:onProperty   ?p  .
?x                 ?y               ?target .
?target            rdf:type         owl:Class .}
```

Listing 2. Property extraction pattern in OWLS-TCv4 ontology.

To obtain all the existing relations in the ontology between an input node I_i and an output node O_j, we need a SPARQL query for each possible path. For a maximum user-defined length/distance, there are $2^{n+1} - 2$ combinations of paths (see Fig. 4). We use an algorithm that generates all the queries in an incremental fashion. To reduce querying costs of the ontology servers, we merge all the queries of all path combinations per Input/Output pair as sub-queries into a single SPARQL query using the `UNION` operator.

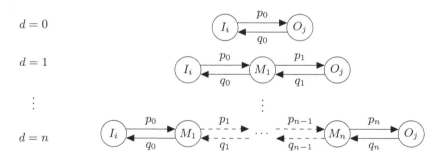

Fig. 4. Combinations of paths between input and output nodes.

An example of a relation extraction query is shown in Listing 3. It contains all the possible combinations in a union fashion to find the relations between `:Writer` and `:Book` within a maximum distance of $n = 1$. The properties are extracted using the generic pattern given in Listing 1.

```
SELECT ?p0 ?q0 ?o1 ?p1 ?q1 WHERE {
# distance=0
{?p0     rdfs:domain :Book;
    rdfs:range   :Writer .}
UNION {
?q0 rdfs:domain :Writer;
    rdfs:range   :Book .}
# distance=1
UNION {
?p0 rdfs:domain :Book;
    rdfs:range   ?o1 .
?p1 rdfs:domain :?o1;
    rdfs:range   :Writer .}
UNION {
```

```
?p0 rdfs:domain :Book;
    rdfs:range   ?o1 .
?q1 rdfs:domain :Writer;
    rdfs:range   ?o1 .}
UNION {
?q0 rdfs:domain ?o1;
    rdfs:range   :Book .
?p1 rdfs:domain ?o1;
    rdfs:range   :Writer .}
UNION {
?q0 rdfs:domain ?o1;
    rdfs:range   :Book .
?q1 rdfs:domain :Writer;
    rdfs:range   ?o1 .}}
```

Listing 3. Example I/O relation extraction query

Extraction Enhancements. When observing the most recurrent patterns found in the relation paths between concept nodes from many domains in ontology schemas (not instances), one comes immediately under the light spot: `?x rdfs:subClassOf ?x`. This hierarchical pattern is very common in DBpedia. Figure 3 gives a perfect example of these hierarchical patterns. In fact, the `author` property is not exclusive to Books and Writers but is a general relation between any `Work` and `Person`.

Such a phenomenon would make the relation paths longer and the queries more complex. Therefore, it is important to bind some parts of the paths between nodes with this pattern in order to narrow the search space and reduce the complexity. Instead of searching for all the properties and intermediate nodes, the extraction algorithm applies hierarchy patterns for up to a user-defined maximum hierarchical depth $h = m$ before looking for general properties and intermediate nodes at a distance $d = n$. Therefore, the number of unknown intermediate nodes in the path is decreased by the introduction of super-class nodes. In addition, the maximum distance value $d = n$ can be decreased as well by the user because most of the path is covered by super-classes (see Fig. 5). This improves the overall performance of the system by reducing the number of property variables as well as the number of required joins during the execution of the SPARQL relation extraction query.

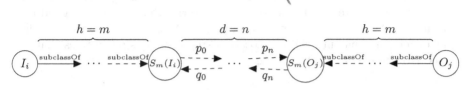

Fig. 5. Combinations of hierarchical patterns on top of path combinations.

3.2   I/O Relations Extraction from Service Descriptions

The textual descriptions of web services as well as their documentations tend to tell more about the service functionality by revealing some of its underlying data model and the conceptual relations between its inputs and outputs. Based on this assumption, the relations extraction process from text descriptions aims to discover these implicit relations in the text in order to match them later with the existing explicit relations found in the ontologies.

This process, executed in parallel to the first one, consists in three sequential tasks: text pre-processing, I/O words recognition and relations extraction. Natural language processing (NLP) techniques are used here for annotating the text and extracting the relations. *word2vec*[7] [16], on the other hand, is used in this step for detecting I/O words in the text. The following subsections provide more details about the three tasks of this process.

Service Description's Text Pre-processing. The textual part of a service description is written in an "informal" natural language. It might consist of a few sentences (a single sentence in OWLS-TC) up to many paragraphs. The pre-processing task is a pipeline that aims to annotate the text in order to highlight some useful features that help understanding the text and extracting I/O relations. It consists in:

1. *Tokenization:* Split the text into sentences, then split the sentences into tokens to generate a matrix indexing all the tokens within their sentences. In the English language, words are mostly atomic except some compound words that should be left as whole. Tokens include inter alia punctuation, numerals and symbols that are taken into account in the indexing matrix and are needed in the dependency parsing.
2. *Part-Of-Speech Tagging* (or POS tagging): to determine the nature of each word in the sentence based on its definition and its context (eg. Nouns, Verbs, etc.).
3. *Dependency parsing:* Dependency grammar is at the core of the relations extraction process. Dependency parsing approaches and tools such as Stanford CoreNLP[8] [17] exist since several years and are based on statistical models for the most part. The recent advances in deep learning have allowed the appearance of more sophisticated parsers like [18] that shows very good performance. We use the latter work as a part of the Stanford CoreNLP toolkit for dependency parsing. This annotation step generates parse trees that are used later to extract I/O relations.

I/O Recognition in Text Descriptions. Before starting to search for relations between I/O in the dependency trees, we first need to recognize their corresponding tokens in the text. The I/O recognition is a twofold task that consists in:

---

[7] https://code.google.com/archive/p/word2vec/.
[8] https://stanfordnlp.github.io/CoreNLP/.

1. Extracting the I/O words from the Inputs/Outputs declaration in the service description. Given the nature of the semantic service descriptions, in which the I/O elements are declared as URIs, there are two possibilities for I/O words: (1) They have labels that can be fetched from their reference ontologies through the `rdfs:label` property or (2) in the absence of the labels, the local names in their URIs are often meaningful words that require a particular treatment. For instance, local names can contain compound words or concatenated words in CamelCase (as in OWLS-TC), they can be dash-separated, etc. A custom pattern can be added manually to the enrichment system to deal with each case individually. Obviously, exceptions may apply here as well.

2. Recognizing the I/O tokens. After knowing what to search for from the previous step, we apply a string similarity algorithm based on *word2vec* to find the best matches for each I/O word in the text. Each Input or Output can be mentioned multiple times in different sentences and contexts, or can be composed of different words.

   Word2Vec allows to represent words in a vector space model and provides interesting features such as cosine similarity between two words. The advantage of using *word2vec* is its syntax-independent and dictionary-free representation as well as its ability to find semantic similarities based on the context. This fits perfectly in our use-case because the text can refer to I/O words with their synonyms, subclasses, super-classes, etc.

   For our recognition purposes, we calculate the cosine similarity between each I/O and each token in the text and keep all matches above some threshold value. For compound I/O elements composed of multiple words, each word is matched individually in the text. The outcome of this task is a mapping of all I/O words with the indexes of their matches and the similarity value.

   Figure 6 illustrates the extracted I/O from the example service of Fig. 1a. The input and output words are highlighted in the sentence after being matched with the I/O of the service. Table 1a shows the mappings of I/O tokens to the indexes of their matches in the example text. Note that the word *"author"* is considered as a match for *"Writer"*, thanks to `word2vec`.

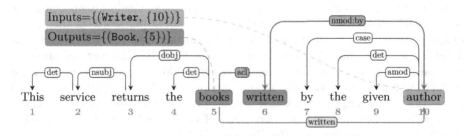

Fig. 6. I/O and relations extraction example

Table 1. Recognized I/O in Fig. 6

(a) Recognized I/O in Fig.6

Inputs	Outputs
(Writer,{10})	(Book,{5})

(b) Extracted text relations in Fig.6

	Book
Writer	[Book, written, Writer]

Relations Extraction. After the two previous tasks, extracting I/O relations is now straightforward. Starting from the generated dependency tree, a graph search algorithm finds all the dependency paths between the Inputs and Outputs in a pairwise fashion.

The extracted paths are represented as lists of strings containing only the involved words in the path without the dependency tags. The I/O tokens are replaced by their ontology labels extracted previously to help the matching process focus only on the intermediate (relation) words in the path.

Figure 6 illustrates the extracted relation in the form of a triple from the example service $S\#1$ in Fig. 1a. The only dependency path between the I/O [(author ← written ← books)] is converted into a relation array as in Table 1b where the ontology tokens for I/O are substituted for the ones from text.

3.3   Relation Evaluation

The final step of the service enrichment process consists in evaluating the extracted relations before passing them to the user to validate and add to the service description. More precisely, this step aims to evaluate the ontology relations based on their matching with the text relations.

At this point, we have two relations matrices, $L = (Orel_{ij})$ for the ontology relations and $T = (Trel_{ij})$ for the text relations where each cell contains a list of ontology and text relations respectively (see Table 1b). The evaluation algorithm iterates over all the pairs of ontology and text relations and calculates the similarities between each pair.

Calculating similarities between I/O relations of different type is not a trivial task. First, we need to homogenize the format of text and ontology relations before comparing them. For our chosen similarity calculation approach, we require each relation to be represented as an array of all the words in their order of appearance in the relation path. The text relations are already represented as an array (for each relation) as in the example in Table 1b. However, the ontology relations have a different format upon extraction. We create an array form for each relation by substituting the local names or labels for the URIs of nodes and properties except `rdfs:subClassOf` and we preserve the order of the relation's path. For the example in Fig. 3, an array of an example extracted relation is $[Book, WrittenWork, prefaceBy, Person, Writer]$.

Because relations can have different lengths, the similarity function used for matching them should be size-independent. For calculating similarities between relations, we take advantage of *word2vec*'s cosine similarity and vector space model. First, we aggregate the *word2vec* vectors of each word in each relation using an aggregation formula then we calculate the cosine similarity between the aggregated vectors. The aggregation of *word2vec* vectors is guaranteed to be size-independent, because it is only impacted by the semantics of words and not by their number (Fig. 7).

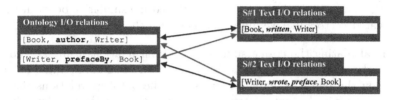

Fig. 7. Pairwise ontology (left) and text (right) relations matching. Matches are highlighted in green (Color figure online)

To calculate the similarity between relations, we create an aggregated *word2vec* vector for each relation by aggregating all its words' vectors into a single vector representing the whole relation. We use three aggregation formulae and we consider the maximum similarity value given by any of the three:

1. Average of Maxima:
   Let r_1 and r_2 be two arrays for an ontology and a text relation respectively and let $n = |r_1|$ and $m = |r_2|$ be their respective cardinals. To calculate their similarity, first we calculate the word-word similarity matrix $n * m$ of all the possible pairs of words (w_i, w_j) from the two relations r_1 and r_2. We define $sim(w_i, w_j)$ as the *Word2vec* cosine similarity function between a pair of words.

   After that, we calculate the maximum similarity values for each row (i.e. we find the best match in the text relation of each ontology relation word). Finally we calculate the average of maximums as the similarity value between r_1 and r_2.

$$AvgMaxSim(s_1, s_2) = \frac{\sum_{i=1}^{n} Max_{j=1}^{m}(sim(w_i, w_j))}{n} \tag{1}$$

2. Element-wise Vector sum method:
   First, we calculate the element-wise vector sum of all the *word2vec* vectors for each relation r as follows:

$$relVec^{\oplus}(r) = \sum_{i=0}^{n} wordVec_{w_i} \tag{2}$$

   Then the relation similarity is given by the cosine similarity between the two aggregated vectors.

3. Hadamard product method:

We first calculate the element-wise vector product (aka Hadamard product) of all the *word2vec* vectors for each relation r as follows:

$$sentVec^{\odot}(s) = \prod_{i=0}^{n} wdVec_{w_i} \tag{3}$$

then the sentence similarity is given by the cosine similarity between the two aggregated vectors like we did previously with the vector sum method.

After calculating the cosine similarity between relations using all the three formulae, the final similarity value is given by the maximum value of all the three. Upon the pairwise matching, we select the ontology relations that have the best matches in the text relations and suggest them to the user for validation before using them to enrich the semantic service description. Figure 8 illustrates the enriched service descriptions of the example services in Fig. 1. Obviously, the extracted I/O relations are represented here in a simple way for illustration purposes only.

(a) Enriched Service description #1   (b) Enriched Service description #2

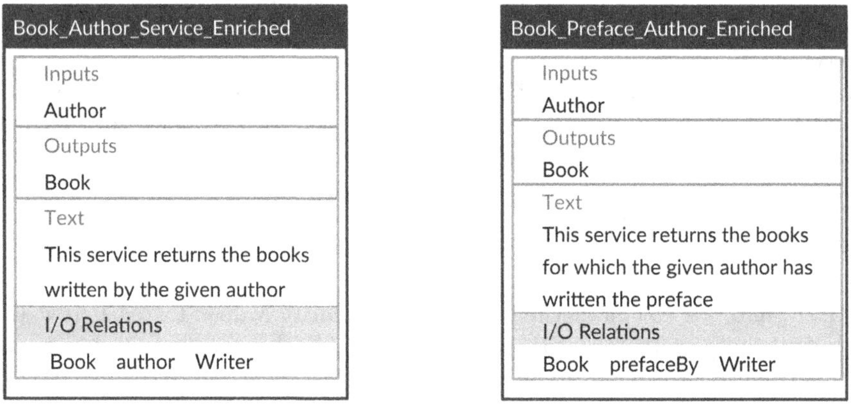

Fig. 8. An example of two web services (see Fig. 1a and 1b) after the description enrichment with I/O relations

4 Evaluation and Experimental Results

4.1 Evaluation Setup

In this section, we evaluate the feasibility and effectiveness of our proposed approach. The underlying process of our approach consists of three main tasks: extracting I/O relations from ontologies, extracting them from text and evaluating them.

We use OWLS-TCv4 as a data-set for evaluating our approach because it provides a fair number of services (1008) and a number of ontologies (48) used by these services with more than 55900 triples. Table 2a depicts the number of inputs and outputs in OWLS-TC. We host the ontologies and the services in a local non-distributed SPARQL endpoint provided by Apache Jena-Fuseki.[9]

For our *word2vec* string similarity approach, we use the *GoogleNews* pre-trained model[10] provided with the original *word2vec* paper [16]. We choose this model for its general purpose corpus and vocabulary. Obviously, another model could have a better impact on our results, but again our purpose is to prove the feasibility rather than to obtain the best possible score. We use a separate module for string similarity written in JAVA based on the *deeplearning4j*[11] toolkit. It is deployed as a RESTful API on a distant fast server for a better performance given the size of the used model and it provides interfaces for all the proposed aggregated methods presented above.

4.2   Experimental Results

Ontology I/O Relations Extraction. This first task aims to maximize its recall value by extracting all the *existing* relations in order to evaluate them later. Therefore, and given the nature of this graph search problem, the outcome has a boolean nature where relations either exist or do not. However, the number of extracted relations depends on the user-defined maximum relation length and hierarchy depth.

The outcome of our proposed approach depends on the outcome of this first task which itself depends on the quality of OWLS-TC ontologies and services. For I/O ontology relations to be extracted, a service must have at least one input and one output and all its I/O must refer to the same ontology (since OWLS-TC is not interlinked). Table 2b shows some statistics about the I/O and ontology usage in OWLS-TC where: (a) S^{all} is the set of all OWLS-TC services. S^{min} is the set of services that have the minimum requirements of at least one input and one output. S^{intra} are services whose all inputs and outputs refer to the same ontology (intra-ontology I/O). Therefore, a usable service that satisfies all the previously mentioned requirements must belong to $S^{usable} = S^{min} \bigcap S^{intra}$, the set of usable services.

Running our system on the set of usable services shows that the hierarchy depths have more impact on the extraction than the maximum lengths (number of intermediate nodes) as shown in the results in Table 3. Therefore, the ontology I/O extraction enhancement has proven to be very efficient. The last two cells of the table are missing on purpose to show the increased algorithmic complexity at this point (depth and distance). The execution times become less reasonable.

Amongst the list of usable services S^{usable}, only 64 (31% services made it through the ontology relations extraction experiments. This number is relatively

---

[9] https://jena.apache.org/documentation/fuseki2/index.html.
[10] Available at https://github.com/mmihaltz/word2vec-GoogleNews-vectors.
[11] http://deeplearning4j.org.

Table 2. I/O of services in OWLS-TC and their ontology usage

(a) Total I/O in OWLS-TC

	Services	Inputs	Outputs
#	1083	1540	1615

(b) I/O and ontology usage

	S^{all}	S^{min}	S^{intra}	S^{usable}
#	1083	996	274	205
%	100	91,96	25,30	18,92

Table 3. I/O ontology relations extraction results in OWLS-TC

Max depth	Max length							
	$d = 0$		$d = 1$		$d = 2$		$d = 3$	
	#r	#s	#r	#s	#r	#s	#r	#s
$h = 0$	12	11	12	11	258	40	307	40
$h = 1$	39	34	57	38	2385	59	17611	59
$h = 2$	64	43	117	48	3852	64		
$h = 3$	72	47	195	52	4945	64		

$\#r$: extracted I/O relations
$\#s$: services with ≥ 1 extracted relation

low compared to the 1083 services of OWLS-TC. This is due to the fact that the OWLS-TC ontologies were not designed for such a use-case and often lack relations even between its close nodes.

I/O Recognition. Recognizing the I/O tokens in the text descriptions is the basis for every textual I/O relation extraction. It depends on two major factors: (a) the quality of the textual description and (b) the quality of the used word2vec model and its sparseness. Table 4a shows the results of I/O recognition in OWLS-TC using a similarity threshold = 0.5. The overall results are fairly good and can be improved by taking into account some particular cases in OWLS-TC. We have manually checked all the services to make sure the threshold value is adequate for most if not all services.

Table 4b shows the absolute frequency of the most frequent cases of non-recognized I/O tokens as well as their percentage amongst the total cases of non-recognition. The most frequent case is the usage of compound words in the text that correspond to variable names without camelCasing (eg. maxprice, taxedprice, etc.) while the variable names themselves are written in camelCase. This requires using a more sophisticated token splitting algorithm or using an edit-distance string similarity algorithm as an alternative to word2vec similarity. Another particular case is the use of unusual token separators such as $(/, -, +,$ hidden characters, etc.). The third case is the absence of any mention to the I/O or their synonyms or similar words in the text which unfortunately cannot be dealt with automatically unless the text description is manually enriched.

Table 4. Evaluation of I/O recognition in textual descriptions

(a) I/O recognition results

	Inputs	Outputs
#	1236	1410
%	80,25	87,30

(b) Top cases of non-recognized I/O

Case	#	%
compound words	123	24,16
absent mention	98	19,25
special separators	42	8,25

Extracted I/O Relations Evaluation. To evaluate this last step, we run the whole process on the list of usable services mentioned above. For the evaluation of I/O relations, we manually evaluate the results given by our three proposed aggregated similarity functions. For this experimentation, we let the system pick the max similarity value amongst the three functions.

First of all, amongst the usable services set, only 32 of them had both text and ontology relations, 5 of which didn't have any valid I/O relations in the ontologies (absent relations). Table 5a shows the obtained results for the maximum number of usable services. Our evaluation system has an excellent precision value (see Table 5b). However, the recall is still to be optimized. For the most part, the low recall value is due to the quality of text descriptions sentences that do not have nice dependency trees. It is mainly because of a recurrent pattern of text descriptions like "This service returns ..." and because of the usage of pronouns that should be dealt with using Coreference resolution.

Table 5. Evaluation of extracted I/O relations

(a) Validity of the extracted relations

	Extracted	Valid	Invalid	Reference
#	19	18	1	28
%	67,85	64,2	3,5	100

(b) Recall and precision

Precision	Recall
94,73	64,28

5   Conclusion and Future Works

In this paper, we proposed an approach for enriching service descriptions to remedy to the problem of ambiguous service functionality. Our enrichment approach extracts existing relations between the inputs and outputs of the services from the underlying ontologies, using SPARQL, and from the text descriptions of services, using NLP techniques. We aimed to prove the feasibility of solving this problem in a way that has not been tackled before to our knowledge. We were motivated by the recent advances in deep-learning and NLP techniques that seemed to be very promising, which was proven true in our case.

Ontology I/O relations extraction with SPARQL quickly shows its limits beyond some relation lengths because of the important number of joins required for all the possible combinations of paths. Even tough the most relevant relations are found at shorter distances, some ontologies like life-science ontologies have a fine granularity level with longer paths for I/O relations. Applying heuristics to overcome this problem is very important to solve this issue. Ontology profiling data can be used to automatically generate tailored relation extraction queries specific to each ontology.

We believe that the obtained results in terms of I/O tokens recognition or in relations evaluation can be significantly improved, either by optimizing the word/sentence similarity parameters, by training other word2vec models, or by adopting other alternatives, like GloVe[12], WordNet, etc.

Since OWLS-TC ontologies and services were mainly designed for evaluating service matchmaking, they lack the interlinking of ontologies and many "real-world" relations between concepts are missing. These issues reduced the amount of usable web services for our experiments. Manually interlinking OWLS-TC ontologies to real world ontologies such as DBpedia would have permitted to extract more I/O relations. We have tried this on some OWLS-TC ontologies partially on several concepts and were able to extract more relations but we haven't included this in the experiments.

In the future, it would be interesting to extend our proposed approach to Web APIs by integrating semi-automatic semantic annotation tools for web APIs like WebKarma. This will allow us to apply our approach to real world services and ontologies but it will indeed be more challenging to go through the documentations of web APIs because of their heterogeneous HTML page structures.

Acknowledgements. The research reported on this paper was supported by the French Research Agency (ANR-14-CE23-0006).

References

1. Sbodio, M.L., Martin, D., Moulin, C.: Discovering semantic web services using SPARQL and intelligent agents. Web Semant. Sci. Serv. Agents World Wide Web **8**(4), 310–328 (2010)
2. Ngan, L.D., Kanagasabai, R.: Semantic web service discovery: state-of-the-art and research challenges. Pers. Ubiquit. Comput. **17**(8), 1741–1752 (2013)
3. Martin, D., Burstein, M., Hobbs, J., Lassila, O., McDermott, D., McIlraith, S., Narayanan, S., Paolucci, M., Parsia, B., Payne, T., et al.: OWL-S: Semantic markup for web services. W3C member submission 22, 2007–04 (2004)
4. Studer, R., Grimm, S., Abecker, A.: Semantic Web Services: Concepts, Technologies, and Applications. Springer-Verlag, New York, Secaucus, NJ, USA (2007)
5. Speiser, S., Harth, A.: Integrating linked data and services with linked data services. In: Antoniou, G., Grobelnik, M., Simperl, E., Parsia, B., Plexousakis, D., De Leenheer, P., Pan, J. (eds.) ESWC 2011. LNCS, vol. 6643, pp. 170–184. Springer, Heidelberg (2011). doi:10.1007/978-3-642-21034-1_12
6. Taheriyan, M., Knoblock, C.A., Szekely, P., Ambite, J.L.: Rapidly integrating services into the linked data cloud. In: Cudré-Mauroux, P., Heflin, J., Sirin, E., Tudorache, T., Euzenat, J., Hauswirth, M., Parreira, J.X., Hendler, J., Schreiber, G., Bernstein, A., Blomqvist, E. (eds.) ISWC 2012. LNCS, vol. 7649, pp. 559–574. Springer, Heidelberg (2012). doi:10.1007/978-3-642-35176-1_35

---

[12] https://nlp.stanford.edu/projects/glove/.

7. Tosi, D., Morasca, S.: Supporting the semi-automatic semantic annotation of web services: a systematic literature review. Inf. Softw. Technol. **61**, 16–32 (2015)
8. Zhang, Z., Chen, S., Feng, Z.: Semantic annotation for web services based on DBpedia. In: IEEE 7th International Symposium on Service Oriented System Engineering (SOSE), pp. 280–285. IEEE (2015)
9. Cheniki, N., Belkhir, A., Atif, Y.: Mobile services discovery framework using DBpedia and non-monotonic rules. Comput. Electr. Eng. **52**, 49–64 (2016)
10. Lucky, M.N., Cremaschi, M., Lodigiani, B., Menolascina, A., De Paoli, F.: Enriching API descriptions by adding API profiles through semantic annotation. In: Sheng, Q.Z., Stroulia, E., Tata, S., Bhiri, S. (eds.) ICSOC 2016. LNCS, vol. 9936, pp. 780–794. Springer, Cham (2016). doi:10.1007/978-3-319-46295-0_55
11. Chen, F., Lu, C., Wu, H., Li, M.: A semantic similarity measure integrating multiple conceptual relationships for web service discovery. Expert Syst. Appl. **67**, 19–31 (2017)
12. Miller, G.A.: Wordnet: a lexical database for english. Commun. ACM **38**(11), 39–41 (1995)
13. Nakashole, N., Weikum, G., Suchanek, F.: Patty: A taxonomy of relational patterns with semantic types. In: Proceedings of the 2012 Joint Conference on Empirical Methods in Natural Language Processing and Computational Natural Language Learning. EMNLP-CoNLL 2012, Stroudsburg, PA, USA, 1135–1145. Association for Computational Linguistics (2012)
14. Arnold, P., Rahm, E.: Extracting semantic concept relations from wikipedia. In: Proceedings of the 4th International Conference on Web Intelligence, Mining and Semantics, WIMS 2014, New York, NY, USA, 26:1–26:11. ACM (2014)
15. Angeli, G., Premkumar, M.J.J., Manning, C.D.: Leveraging linguistic structure for open domain information extraction. In: Proceedings of the 53rd Annual Meeting of the Association for Computational Linguistics and the 7th International Joint Conference onNatural Language Processing of the Asian Federation of Natural Language Processing, ACL 2015, pp. 344–354, 26–31 July 2015, Beijing, China, Volume 1: Long Papers (2015)
16. Mikolov, T., Sutskever, I., Chen, K., Corrado, G.S., Dean, J.: Distributed representations of words and phrases and their compositionality. In: Advances in neural information processing systems, pp. 3111–3119 (2013)
17. Manning, C.D., Surdeanu, M., Bauer, J., Finkel, J.R., Bethard, S., McClosky, D.: The stanford CoreNLP natural language processing toolkit. In: ACL (System Demonstrations), pp. 55–60 (2014)
18. Chen, D., Manning, C.D.: A fast and accurate dependency parser using neural networks. In: Empirical Methods in Natural Language Processing (EMNLP) (2014)

ProLoD: An Efficient Framework for Processing Logistics Data

Mohammad AlShaer[1,4], Yehia Taher[2], Rafiqul Haque[3(✉)],
Mohand-Saïd Hacid[1], and Mohamed Dbouk[4]

[1] Université Claude Bernard Lyon 1,
43 Boulevard du 11 Novembre 1918, 69100 Villeurbanne, France
{mohammad.alshaer,mohand-said.hacid}@univ-lyon1.fr
[2] Université de Versailles Saint-Quentin-en-Yvelines (UVSQ),
55 Avenue de Paris, 78000 Versailles, France
yehia.taher@uvsq.fr
[3] Cognitus SAS, 5 Rue Lacharrire, 75011 Paris, France
rafiqul.haque@cognitus.fr
[4] Lebanese University, Beirut, Lebanon
mdbouk@ul.edu.br

Abstract. Logistics is a data-intensive industry. The information systems used by logistics companies generate massive volume of data which the companies store to perform different types of analysis. In addition, the advent of Big Data technologies and Internet of Things paradigm have given logistics companies an opportunity to use external data stemming from a wide variety of sources including sensors (*e.g.*, GPS), social media and traffic controlling systems. The logistics companies aim to leverage the power of these external data and perform rigorous analysis in real-time to discover intelligence such as unpredictable delay. However, there are different challenges involved. One of the core challenges is integrating and processing a wide variety of data coming from heterogeneous sources. To the best of our knowledge, there is no off-the shelf solution which can address this challenge. In this paper, we present a framework called ProLoD which performs pre-processing and processing tasks with different types of data. Our framework relies on machine learning algorithms, for processing data; however, we found that the ready to use algorithms are not adequate to guarantee processing efficiency. Therefore, we extended an algorithm called *Hierarchical Clustering Algorithm*. We evaluated ProLoD by comparing its performance with the HCL algorithm found in the widely-adopted machine learning tool called WEKA. We found that ProLoD is performing reasonably better than WEKA in terms of producing optimal number of clusters.

Keywords: Big Data · Internet of Things · Hierarchical clustering algorithm · Unstructured data · Text analytics

© Springer International Publishing AG 2017
H. Panetto et al. (Eds.): OTM 2017 Conferences, Part I, LNCS 10573, pp. 698–715, 2017.
https://doi.org/10.1007/978-3-319-69462-7_44

1   Introduction

Internet of Things (IoT) has been emerging as a *de facto* approach for building smart applications for various industries including healthcare, manufacturing, and logistics. It is driven by sensors which create a connected ecosystem to automate business functions. Lately, the IoT-enabled logistics management system is shifting conventional semi-automated order shipment process to a smart device enabled approach. Due to various promises, the logistics companies such as DHL[1], UPS[2], and FedEx[3] have started adopting IoT in their application environment. However, the sensor-driven logistics systems are heavily data-intensive *i.e.* they produce massive scale data. In addition, there are conventional logistics information systems that are built on various business processes composed of hundreds of activities. These activities emit an enormous amount of data (*e.g.*, process logs) during executions. Furthermore, the social medias (*e.g.*, Facebook[4]), and microblog (Twitter[5]) are becoming critically important for companies including logistics because, the data stemming from these sources contain useful business information [38].

However, these data are unstructured mainly *texts* which – unlike relational data – have no well-defined structure. Thus, processing textual data is *nontrivial*. In addition, the logistics information systems produce structured data. Processing textual data with structured data within the same ecosystem may promote enormous challenge [32]. Also, it significantly increases integration challenges. Two other challenging issues involved in processing logistics data include: large *volume* and *high frequency* (better known as *velocity*). The advent of Big Data engendered several technologies in the last few years. Various data processing solutions are available such as, [9,25,42,44]. Unfortunately, these solutions focus more on optimizing system performance while data processing accuracy has been largely ignored in these solutions. In summary, an efficient solution for processing large scale logistics data with variety is missing in state of the art. However, such as solution is critically important to perform analysis effectively to discover knowledge such as *predicting delivery delay*.

Our objective is to address this shortcoming by investigating a solution. To that end, in this paper, we develop a system called ProLoD (Processor of Logistics Data) that allows to perform pre-processing functions include *collect, clean, filter, integrate*, and *store* data from different sources. Furthermore, our solution performs data processing that is, clustering data into self-explanatory groups in order to support unwanted events prediction such as predicting delivery delay. ProLoD relies on machine learning techniques. In this paper, we extended the hierarchical clustering algorithm [39] for efficient clustering of integrated data.

The remainder of the paper is organized as follows. We describe the problems more in detail in Sect. 2. The proposed solution is discussed in Sect. 3.

---

[1] DHL: http://www.dhl.com/en.html.

[2] UPS: http://www.uplonline.com.

[3] FedEx: http://www.fedex.com/us/.

[4] Facebook: https://www.facebook.com.

[5] Twitter: https://twitter.com/?lang=en.

We reported an evaluation in Sect. 4. Section 5 presents the related works. We conclude in Sect. 6.

2   Problem Description

In time delivery is one of the *key performance indicators* (KPIs) of logistics services. Delay of a scheduled (expected) delivery increases customer dissatisfaction. In order to prevent delay, logistics service providers heavily rely on automated solutions. Business intelligence is a widely used solution which enables to perform different types of *cycle time analytics* [40] that analyzes delay for different combinations of *goods, routes, modes, weather condition*. However, traditional business intelligence specially, BI&A 1.0 and BI&A 2.0 use only internal data which stem from different information systems and legacy systems [7]. Also, the process mining tool PRoM [16] lacks the ability to exploit external data. As a consequence, the analysis is performed without external data such as sensor data and global positioning systems (GPS) data that are critical to prevent delay.

The advent of Big Data technologies created wide opportunities to exploit such external data which enhance the predictability of analytics. More specifically, these data are effective to forecast potential delivery delay as they contain important information such as *high traffic, weather report, political events* such as protest, and other events such as act of God (*e.g.*, Earthquake). However, there is a data processing challenge due to variety of data model and enormous size of datasets. To be more specific, pre-processing tasks that include collecting, cleaning, filtering, integrating, and storing data with volume and variety from heterogeneous sources are highly complex tasks which need efficient approaches that are currently missing. Particularly, a seamless integration of unstructured text sourcing from for instance, Twitter with structured business process data is not possible by existing logistics solution frameworks [15]. Furthermore, although there are several techniques for processing data however according to our study there is a scope to improve these techniques such as the *clustering algorithms*.

In this paper, we aim to address two problems discussed in the above: data pre-processing and data processing. The goal is to provide an efficiently processed dataset which are employed to perform effective analysis. To that end, we developed ProLod framework that enables to fetch data from various sources and enables performing pre-processing and processing tasks efficiently.

3   ProLoD - Logistics Data Processor

In this section, we briefly introduce ProLoD system and provide a comprehensive description of pre-processing and processing functionalities of ProLoD.

3.1   Solution Overview

ProLoD is a framework for processing of logistics data stemming from multiple heterogeneous sensors (that include vehicle sensor, weather sensor, *etc.*), logistics applications, microblog (*e.g.*, Twitter), and social media (*e.g.*, Facebook).

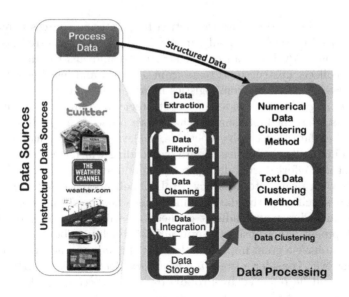

Fig. 1. The high-level functional architecture of ProLoD

ProLoD comprises two phases: *pre-processing phase* and *processing*. The former consists of data extraction (collection), data filtering, data integration, and data storage. In the last latter phase, pre-processed data are clustered. Figure 1 shows different functionalities of these phases. ProLoD relies on different machine learning techniques specifically the clustering techniques for data processing. In this section, we describe the high-level architecture of ProLoD and the underlying data processing model.

ProLoD includes five components: data extractor, data cleaner, data filter, data integrator, and data storage for performing pre-processing functions. It has a data processor which performs clustering in the second phase. Figure 1 shows the high-level functional architecture of ProLoD.

3.2 Data Pre-processing

Data Extraction is the systematic approach to gathering and measuring information from a variety of sources to get a complete and accurate picture of an area of interest. ProLoD's *data extractor* works with both *internal* and *external* sources of data. The internal data sources are typically the information systems used by the users. Consider an user has an information system consisting of a supply chain management (SCM), a customer relationship management (CRM), a logistics management system, and a account management system (AMS). These systems produce a large amount of data. ProLoD's *data extractor* collects data from these sources. It also fetches data from external sources such as Twitter, traffic sensors, weather sensors, Facebook and other social medias. In addition, ProLoD extracts archived sensor data of completed logistics processes. In most of the cases, we found that data extraction from internal sources is more trivial than external ones. Additionally, internal data were transferred faster than the external ones. ProLoD can collect structured and unstructured data. For instance, it collects unstructured texts from Twitter, and Facebook and structured business process data from logistics information system.

Data Filtering refers to a wide range of strategies or solutions for refining data sets. Datasets are refined into simply what a user (or set of users) needs, without including other data that can be repetitive, irrelevant or even sensitive. ProLoD aims to eliminate all possibilities of data overloading which can increase computational cost and effort during data processing and analysis and may jeopardies the analysis with regards to accuracy. ProLoD collects data that are related to logistics and specifically the data chunks whose *hashtags* (the words prefixed by #) determine direct and indirect connections with transportation, delivery, logistics, shipment, *etc.* Consider a term *protest* which may be a *political protest* or else but can have a great impact on delivery of goods and hence can delay the delivery. However, consider a tweet *the new York stock prices are extremely high today* which will be removed by ProLoD data filter because it does not carry any information related to logistics processes.

Data Cleaning (aka *data scrubbing*) is the process of detecting and correcting (or removing) corrupt or inaccurate records from a record set, table, or database. ProLoD cleans data from all unwanted symbols, numbers, stopping words, hashtags, and any other data items that might lead to noise and cause inaccuracy. Figure 2 shows an example of cleaning Twitter data using ProLoD.

Data Integration is performed in two steps. In the first step, the data are transformed from source to target serialization format. Currently, the target format is CSV. The second step is merging the transformed data. After preparing the integrated datasets PeoLoD *store data* into the storage.

Six car #accident in #westsacramento on WB I-80 & Capitol Ave. Reports of people injured & lanes blocked #traffic https://t.co/S29obp3ByL
Police now on scene with an #accident reported on Baird Rd Penfield Rd in #Penfield #traffic #ROC
#accident reported on Salmon Creek Rd Colby St in #Sweden #traffic #ROC
#traffic 06:47: #A4 - #accident between LATISANA S.GIORGIO towards TRIESTE
4 car smash on m4, right lane 500m after Merrylands on ramp. Avoid right lane.
#traffic #accident @channelten @Channel7 @9NewsSyd
#traffic 09:56: #A4 - #queuing traffic between PORTOGRUARO   S.STINO towards VENEZIA due #accident
#traffic 09:55: #A4 - #accident   449.4 between PORTOGRUARO S.STINO towards VENEZIA
#traffic #A4 - #accident   449.4 between PORTOGRUARO   S.STINO   towards   VENEZIA https://t.co/8zF9I8j0ft https://t.co/XeFNpGoMA8

Fig. 2. An example of cleaning data with ProLoD.

3.3   Data Processing Model

ProLoD relies on a data processing model which we developed in this paper. While choosing appropriate technique for developing the model, we considered the nature of data and operation styles. PreLoD's data processing model relies on *unsupervised learning techniques* [28]. Unsupervised learning is a machine learning approach in which a system only receives input $(x_1, x_2,..., x_n)$ without any corresponding (supervised) output (which is also called *labelled output*). *Clustering* and *dimensionality reduction* are the two most well-known unsupervised learning techniques. We choose *clustering* for our model because the objective function is expected to produce a clustered dataset which facilitates efficient analysis in prediction of *delivery delay*. Clustering is a process of grouping or segmenting data items that are *similar* between them in a cluster and *dissimilar* to the data items that belong to another cluster [28].

There are different types of cluster models which are grouped into *Connectivity models*, *Centroid model* and *Distribution models*, *Density models*, *Subspace model*, *Group model*, and *Graph-based models* [17]. We are interested in techniques used for building connectivity model which fits to our objective more than the others. *Hierarchical clustering* is a widely used approach for building connectivity model based on distance connectivity between the data items. It is a process of producing a sequence of nested cluster ranging from *singleton clusters* of individual points to an all-inclusive cluster [8]. The hierarchy of the clusters are graphically represented by a dendogram [19]. There are two approaches to develop a hierarchical cluster model:

- *Agglomeration* refers to an approach that start with the points as individual clusters and, at each step, merge the closest pair of clusters. It is also known as *Bottom-Up approach*.
- *Divisive* refers to an approach that starts with one, all-inclusive cluster and, at each step, split a cluster until only singleton clusters of individual points remain. It is also known as *Top-Down approach*.

We choose agglomerative hierarchical clustering approach for our solution because the bottom up approach is more flexible than the others in terms of choosing the number of clusters. The algorithm groups data one by one based on the nearest distance measure of all the pairwise distance between the data points. The distance between the data points is recalculated iteratively. However, which distance to consider when the groups have been formed – is a critical question. Several methods are available to address this question. These methods – found in [18] – are summarised in the following:

Definition 1. *Single-linkage:* $d(C_i, C_j) = min_{x \in C_i, x' \in C_j} d(x, x')$. *It is equivalent to the minimum spanning tree algorithm* [22]. *One can set a threshold and stop clustering once the distance between clusters is above the threshold. Single-linkage tends to produce long and skinny clusters.*

Definition 2. *Complete-linkage:* $d(C_i, C_j) = max_{x \in C_i, x' \in C_j} d(x, x')$. *Clusters tend to be compact and roughly equal in diameter.*

Definition 3. *Average distance:* $d(C_i, C_j) = \frac{\sum_{x \in C_i, x' \in C_j} d(x, x')}{|C_i| \cdot |C_j|}$.

Definition 4. *Wards method* $d_{ij} = d(\{X_i\}, \{X_j\}) = \|X_i - X_j\|^2$ *is the sum of squared Euclidean distance is minimized.*

The iteration is continued by grouping data items until a cluster is formed. As mentioned earlier the clusters are presented graphically by a dendogram which allows to calculate the number of clusters which should be produced, at the end. There are several variants of agglomerative hierarchical clustering algorithm. Below we present the steps involved in performing a hierarchical clustering. Consider a set of data points $S = (x_1, x_2, x_3, ..., x_n)$ as input. The algorithm performs the following steps:

- *Step 1*: Disjoint cluster (C) of level $\mathcal{L}(0) = 0$ and sequence number $\mathcal{M} = 0$
- *Step 2*: Calculate the least distance $(\mathcal{D})$ pair of clusters in the current C, say pair $\mathcal{P}(r, s)$, according to $\mathcal{D}(r,s) = Min(\ \mathcal{D}(i, j))$ where the minimum is over all pairs of clusters in the current clustering
- *Step 3*: Increment the sequence number, $\mathcal{M} = \mathcal{M} + 1$
- *Step 4*: Merge $C(r)$ and $C(s) \rightarrow C(z)$ which is a new cluster. Set the level of thisclustering to $\mathcal{L}(z) = \mathcal{D}(r),(s)$
- *Step 5*: Update the distance matrix Ψ, (delete the rows and columns corresponding to clusters $C(r)$ and $C(s)$ and add a row and column-corresponding to $C(z)$. The distance between the new cluster, denoted (r, s) and the old cluster(k) is defined as follows: $\mathcal{D}((k), (r, s)) = Min(\mathcal{D}((k),(r)], \mathcal{D}((k),(s))$
- *Step 6*: Repeat until ONLY one cluster remains.

We identified several disadvantages of the basic agglomerative clustering algorithm. In particular, undoing is not allowed and the time complexity is $\mathcal{O}l(n^2 logn)$ where n denotes the number of data points. For a large dataset, the performance with respect to processing time may not be satisfactory. Based on the type of distance matrix chosen for merging, different algorithms can suffer with one or more of the following drawbacks: (i) sensitivity to noise and outlier, (ii) breaking large cluster, (iii) difficulty in handling different size of clusters and handling convex shape. In this algorithm, no objective function is directly minimized. Furthermore, in some cases identifying the correct number of clusters by the dendogram can be very difficult. Therefore, we could not adopt the algorithm *as-is*. We developed our model by slightly tailoring the basic agglomerative algorithm. Below, we present our model that our solution ProLoD relies on for processing data.

- **Step 1**: Read new data streams.
- **Step 2**: Put the *unique* items in the vector format.
- **Step 3**: Fill a matrix of items absence and presence.
- **Step 4**: Calculate hamming distance.
- **Step 5**: Update the distance matrix.
- **Step 6**: Create Cluster using minimum distance.
- **Step 7**: Repeat **until** only one cluster remains.

We explain an example how the ProLoD data processing model works. It begins with reading record from file. Since data is read from the first row, thus the attribute names do not exist; we expressed them here just to make the data set meaningful to the reader.

- Start with each record as a cluster on its own.
- Read new data streams. Given below is the list of attributes.

Naval A   Milk   Low

- A unique item is added in the vector format.
- Fill in the matrix of items absence and presence. Currently, there is one record in the matrix.

	Naval A		Milk	Low
Rec1	1	1	1	1

- Build the similarity matrix using hamming distance.
  - The algorithm reads new record and adds in the matrix.

Naval	A	Milk	Low
QueenMarry	B	Cigar	Low

  - Place the new unique items.

    Naval A Milk Low QueenMarry B Cigar

  - Update the matrix.

	Naval	A	Milk	Low	Queen	B	Cigar
Rec1	1	1	1	1	0	0	0
Rec2	0	0	0	1	1	1	1

  - Build similarity matrix using Hamming distance
    * The Hamming distance can only be calculated between two strings of equal length. String 1: 1111000 String 2: 0001111.
    * Compare the bits of each string with the other.
    * If they are the same, record a "0" for that bit.
    * If they are different, record a "1" for that bit.
    * Compare each bit in succession and record either "1" or "0" as appropriate.
    * Add all the ones and zeros in the record together to obtain the Hamming distance. Hamming distance $= 1 + 1 + 1 + 0 + 1 + 1 + 1 = 6$.

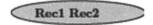

Fig. 3. The initial cluster

- Update the distance matrix (Fig. 3).

	Rec1	Rec2
Rec1	0	6
Rec2	6	0

  – Create a cluster with minimum distance
  – The systems read new records and the previous steps are repeated. At the end a new cluster is created (Fig. 4).

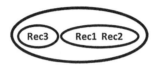

Fig. 4. The newly created cluster

The iteration stops at this step because the execution loop produces a single cluster and no cluster can be created any further. We discuss the implementation of ProLoD in the next section.

3.4 Implementation

We studied various technologies for implementing ProLoD. We investigated existing libraries for data extraction, filtering, and transformation. Our goal was to reuse existing ones instead of developing the new ones. Also, we studied machine learning libraries including DatumBox[6], SPMF[7], Massive Online Analysis (MOA), and Spark MLib[8] to implement our data processing model. We began with Datumbox. Datumbox reads the data from file stored in CSV format. The data must have attribute names in order to form the *dataframe* that will be used as the data structure in this implementation; it uses a linked hash map. Moreover, the algorithm can handle different data types like numerical, categorical, dates and so on. The reason we did not select Datumbox is its *inability* to read data by lines. The dataframe can handle data *as a batch* that can be read *in one go*. Changing the data structure for an already implemented library means changing the core of their implementation as if writing one from scratch. Therefore, we decided to investigate Spark's MLib library. However, we found MLib does not have implementation of hierarchical clustering. Additionally, MLib works only with numerical data.

---

[6] http://www.datumbox.com.
[7] http://www.philippe-fournier-viger.com/spmf/.
[8] http://spark.apache.org/mllib/.

SPMF is another potential machine learning library but we found that it suffers the same problem as Spark does. It works only with numerical data and this was explicitly mentioned in the documentation. MOA (Massive Online Analysis) was another potential candidate. It is developed by the same team who developed the most popular WEKA machine learning library. However, this library is locked-in to specific file format which is ARFF and hence it is unable to read other data format. An ARFF format needs to have attributes, types, and data explicitly mentioned within the file.[9] Nevertheless, in our case the data – are streamed and fed into the algorithm – does not necessarily have an attribute. Rather, data could be a set of records each of which is made up of different text words. From our study, we found that existing libraries could not be used to implement our model (discussed in the previous section). Therefore, we decided to implement the solution by ourselves. For implementation, we used C-Sharp language on Microsoft Visual Studio framework. The example code snippet below shows the implementation of our model using Datumbox.

```
import java.io.FileWriter;
import java.io.IOException;
import java.io.PrintWriter;
import java.util.ArrayList;
import java.util.Date;
import twitter4j.Twitter;
import twitter4j.TwitterException;
import twitter4j.TwitterFactory;
import twitter4j.conf.ConfigurationBuilder;
.......................

public class GetData {
public static void main(String[] args) throws Exception
{
ConfigurationBuilder cb = new ConfigurationBuilder();
cb.setDebugEnabled(true)
.setOAuthConsumerKey("vfjOBcYfLtHXl11DIGwJEm4k8")
.setOAuthConsumerSecret("i2Uvr8LcGVLBaD4ksY6ECpFavyLhImRRALcW9MyFMbu5rGXVZw")
.setOAuthAccessToken("720269683723673601-2KUvyUXcjoziVyBbPrruFACadGtv8uF")
.setOAuthAccessTokenSecret("bUoL9ZJ4WSk2oERaVpdFSwHEN66t0s4ZAA87KcQo8GZUE");
Twitter twitter = new TwitterFactory(cb.build()).getInstance();
Query query = new Query("#traffic AND #accident");
int numberOfTweets = 20;
long lastID = Long.MAX_VALUE;
ArrayList<Status> tweets = new ArrayList<Status>();
while (tweets.size () < numberOfTweets) {
if (numberOfTweets - tweets.size() > 100)
query.setCount(100);
-------------------------------------

}
```

---

[9] We contacted Dr. Albert Bifet the author of the library for assistance because it was only running for a specific number of data points then starts throwing errors but the problem was not solved.

The example code snippet below shows the implementation of our clustering model using Datumbox.

```
import com.datumbox.framework.common.
dataobjects.Dataframe;
import com.datumbox.framework.core.
machinelearning.
clustering.HierarchicalAgglomerative;
.......................

public class HAClustering {
public static void main(String[] args) throws
URISyntaxException {
RandomGenerator.setGlobalSeed(42L);
Configuration conf = Configuration.
getConfiguration();
```

We investigated different data processing frameworks including Apache Spark[10], and Apache Storm[11]. To the best of our understanding Storm is more potential computation system for our ProLoD. It is fast, can over a million tuples processed per second per node. It is scalable, fault-tolerant, guarantees our data will be processed, and is easy to set up and operate. Storm integrates with the queueing and database technologies we already use. A Storm topology consumes streams of data and processes those streams in arbitrarily complex ways, repartitioning the streams between each stage of the computation is however needed.

To sum up, ProLoD is integrated framework integrates which three APIs for extracting external data from different sources including Twitter API, Facebook API, and Open Weather API. It uses an open source parser. Also, it includes tools for cleaning and transforming incoming data.

4 Evaluation

In this section, we briefly discuss the results produced through experiments with ProLoD. We evaluate the performance of ProLoD over the metric *execution time*. Given below is the specification of the machine we used for our experiments:

- Processor: 2.40 GHz
- Memory: 4 GB
- HDD: 500 GB
- Operating System: Windows 10(64 bit)

We used three different data sources include Twitter, business process, and traffic data from smart city. We collected data using ProLoD and then preprocessed them. The dataset produced after pre-processing are clustered using

---

[10] http://spark.apache.org.

[11] http://storm.apache.org.

our extended HCL algorithm and the HCL algorithm deployed in WEKA. We compared our extended HCL algorithm with the hierarchical clustering algorithm provided in WEKA. The WEKA algorithm produced 7 cluster with 8 records (Shown in Table 1).

Table 1. The clusters created by WEKA

Clusters	Records
Cluster 1	Record 1 and Record 4
Cluster 2	Record 2
Cluster 3	Record 3
Cluster 4	Record 5
Cluster 5	Record 6
Cluster 6	Record 7
Cluster 7	Record 8

ProLoD produced consistent and representable clusters that will assist in exploring data, discovering insights, and supporting predictive analytics when the data distribution is observed. Table 2 shows the result.

Table 2. The clusters created by ProLoD

Clusters	Records and Clusters
Cluster 1	Record 1 and Record 5
Cluster 2	Record 2 and Record 8
Cluster 3	Record 3 and Record 6
Cluster 4	Record 7 and Cluster 3
Cluster 5	Record 4 Cluster 1
Cluster 6	Cluster 2 and Cluster 5
Cluster 7	Cluster 4 and Cluster 6

According to our analysis any labeling process using any of the attributes from the dataset will produce mresults that are representable and understandable (By representable we meant a reasonable number of clusters produced by the algorithm. For instance, all the clusters but one produced by WEKA contain only one record, which is not representable according to our view) If we try interpreting the data using *Cargo* attribute we will find that the clusters contain very similar records. If we try *Quantity* attribute, we will find similar ranges grouped together. Once the data can be visualized in this simple manner, it will be easier for experts to analyze it. Moreover, automated systems can reap the benefits from categorizing data before applying analytics, so in further steps analytics might be done on each cluster on its own. Moreover, our self-explanatory clusters that were obtained by ProLoD explicitly proved to be better than Wekas hierarchical clustering algorithm. The results obtained by

Weka were a set of seven clusters placing each record in its own cluster without any meaningful insight; nevertheless, only two records were clustered together because they had an identical word. On the other hand, our clustering results showed that food and cargos were clustered together, combustibles and cargos created a cluster, and smoking and cargos created a cluster. Thus, predictive analytics can be applied on such self-explanatory clusters instead of data points.

5   Related Work

Clustering data in real time has drawn a huge research interests in the recent past, especially with an extensive demand of using analytics on streaming data. We chose clustering because we needed to work with unlabeled data which means no model training data set is available and we know nothing about the data previously, thus we needed an unsupervised learning model. A few initiatives have been taken in the field where BIRCH (balanced iterative reducing and clustering using hierarchies) was one of the basic methods for data stream clustering [46]. Essentially, BIRCH introduced micro and macro clustering as two new concepts, and was fabricated to work with traditional data mining techniques but not with voluminous amounts of data sets like data streams. Later, the STREAM algorithm proposed by Guha et al. [20,21] was an extension of classical K-median and the first algorithm known with the ability to perform clustering on entire data streams.

In [4], Babcock et al. suggested the sliding window model as an extension to STREAM and thus they changed the concept from one single pass over the data, to the concept of receiving data points as a stream and taking into consideration the points that fall within a specific range representing the sliding window. The CluStream framework was suggested by [1] and it was considered effective in handling data streams; It divides the clustering process into two components: online and offline. The former periodically uses micro clusters to store detailed summary statistics, and the latter uses the summary statistics to produce clusters. Later, Aggarwal et al. [2] suggested HPStream that works on data streams high dimensionality reduction by means of data projection prior to clustering. Denstream algorithm was proposed by [10] as an extension for DBSCAN where they combined micro clustering concept to the density based connectivity search.

Another density-based extension is the D-Stream proposed in [11]. The proposed solution maps each new data point to a specific grid upon its arrival; the density information is stored and then clustering is applied to the density data grids. Khalilian et al. [29] suggested an improvement for well-known K-Means algorithm. They applied the widely-known divide and conquer method that is capable of clustering objects with high quality and efficiency. Specifically, the solution is suitable for analyzing high dimensional data, yet for realtime data streams it is not. EStream [30] is a data stream clustering technique, which supports five types of evolution in streaming data. They are as follows: Appearance of new cluster, Disappearance of an old cluster, split of a large cluster, merging of two similar clusters and changes in the behavior of cluster itself. It uses a decaying cluster structure with a histogram to approximate the streaming data. Although the algorithm has the disadvantage of needing an expert intervention to specify many parameters before it works, its performance is better than HPStream algorithm [41].

In [12], a multi-level unordered sampling technique was suggested to boost the time performance of fuzzy c means (make it faster). The technique is double phased. In the first phase, the random sampling is applied to estimate centroids and then fuzzy C-means "FCM" is performed on the full data with the previously initialized centroids. Fuzzy C-means together with probabilistic clustering were then extended to work on huge data sets by the sampling based proposal of Richards and James [31]. In [13] an algorithm called AFCM was suggested to speed up FCM. This is done using lookup table. In [45], the authors proposed several efficient and scalable parallel algorithms for a special purpose architecture description of a modified FCM algorithm known as 2rFCM. A fast FCM algorithm was proposed in [3]. They employed the concept of decreasing the number of distance calculations by checking the membership value for each point.

Furthermore, many machine-learning libraries are used to implement the algorithms discussed above. Since we were interested mainly in hierarchical clustering, we studied the machine learning libraries within the scope of this algorithm. Datumbox [33] is a robust framework that provides different functions like Sentiment Analysis, Twitter Sentiment Analysis, Subjectivity Analysis, Topic Classification, Spam Detection, Adult Content Detection, Readability Assessment, Language Detection, Commercial Detection, Educational Detection, Gender Detection, Keyword Extraction, Text Extraction and Document Similarity. At low-level, the basic machine learning algorithms such as K-means, hierarchical clustering, and classification algorithms perform the above functionalities. Although this library is very powerful in handling different data types (categorical, numerical, etc.), it was not implemented to work in the environment of streaming data. Therefore, it can read bulk data. Apache Spark [34] is a fast-general-purpose cluster computing system. It supports a rich set of higher-level tools including Spark SQL for SQL and structured data processing, MLlib for machine learning, GraphX for graph processing, and Spark Streaming for clustering. Spark offers limited features in particular; it supports few algorithms such as K-Means. The library is missing hierarchical clustering algorithm, which we found suitable for our research project.

Furthermore, the library offers the streaming K-means; it is applicable on numerical data only which is a limitation for Big Data where data variety is major challenge. SPMF [35] offers implementations of 120 data mining algorithms for association rule mining, item set mining, sequential pattern mining, and of course clustering and classification but not for clustering. The input is a set of vectors containing double values only, a parameter "max-distance" and a distance function. This implies the same limitation as Spark has. To the best of our understanding, this shortcoming is obvious because, the clustering is usually done according to Euclidean or Manhattan distance functions that need numerical data to be applied.

Weka [37] is a widely-known library. It is an integrated system which consists of a collection of machine learning algorithms for data mining tasks. The algorithms can either be employed directly for a dataset or called from within a Java code. Weka contains tools for data pre-processing, classification, regression, clustering, association rules, and visualization. It is a well-suited solution for developing new machine learning schemes. It is very efficient and can be used with big data analytics but needs to work only on data at rest and with a specific file format called ARFF. Recently, an initiative has been taken to extend Weka to be used for mining data Streams. MOA [5] is an open source library for data stream mining. It includes a collection of machine learning algorithms (classification, regression, clustering, outlier detection, and concept drift detection and recommender systems) and tools for evaluation. MOA provides approximately all clustering algorithms for streaming data but it is confined to a specific format just like Weka (ARFF file only). It enables to generate synthetic data streams and allows users to visualize data clustering in real-time.

Discussion

The solutions and tools discussed in the above provide a wide variety of machine learning algorithms that can be used for predictive analytics tasks, such as feature selection, parameter optimization and result validation. However, they do not offer any interactive means for manipulation, feature selection or model refinement; instead, these systems often opt to show baseline models or simple statistical measures for result validation, working as more of a black-box system. SPMF and Spark worked only on numerical data, Datum box had a specific structure for storing and processing data and it was not suitable for real time environments. Weka has been extended to MOA which we found contain bugs that did not allow us to benefit from it. Therefore, we decided to write our own implementation.

In summary, considering the evaluation of data sources, most of the existing solutions are confined to one data source for analytics and prediction. Additionally, for real-time systems with continuous improvement, most of the research used large static historical datasets for their testing while our approach does not depend on historical data only.

6   Conclusion

In this paper, we presented a framework called ProLoD for processing logistics data that are stemming from various internal and external sources within logistics domain. The proposed framework is able to perform pre-processing and processing functions which eventually produce a clustered dataset that are used by the analytics engine for an efficient predictive analytics to forecast delivery delay. In order to develop our data processing model, we studied different machine learning algorithms; eventually, we choose Johnsons hierarchical clustering algorithm. However, we modified the algorithm so that it can support incremental grouping of text messages according to similar characteristics. We presented the modified version of the algorithm in this paper. The solution architecture of ProLoD was explained and its components were explained. The implementation of ProLoD was discussed. We discussed our evaluation of ProLoD with the most widely used machine learning library called WEKA. According to our observation, ProLoD produced better representative results. However, the performance real-time version of ProLoD was worse than the batch style - which was unexpected.

Several works are lined up. We planned to use a distributed real-time computation framework such as Apache Storm. Also, we will develop *time window* in our algorithm so that ProLoD can store data for a specific period of for processing; this is an effective approach specifically for real-time processing of data. Also, we planned to develop methods to enable ProLoD to deal with numerical data.

References

1. Aggarwal, C.C., Han, J., Wang, J., Yu, P.S.: A framework for clustering evolving data streams. In: Proceedings of the 29th International Conference on Very Large Data Bases (VLDB) (2003)
2. Aggarwal, C.C., Han, J., Wang, J., Yu, P.S.: On High Dimensional Projected Clustering of Data Streams. Data Min. Knowl. Disc. **10**, 251–273 (2005)
3. AL-Zoubi, M.B., Hudaib, A., Al-Shboul, B.: A fast fuzzy clustering algorithm. In: Proceedings of the 6th WSEAS International Conference on Artificial Intelligence, Knowledge Engineering and Data Bases, pp. 28–32 (2007)
4. Babcock, B., Datar, M., O'Callaghan, R.M.L.: Maintaining variance and k-medians over data stream windows. In Proceedings of the 22nd ACM Symposium on Principles of Databases Systems (2003)
5. Bifet, A., Holmes, G., Kirkby, G., Pfahringer, B.: MOA: massive online analysis. J. Mach. Learn. Res. **11**, 1601–1604 (2010)
6. Chen, J.G., Wiener, L.J., Iyer, S., Jaiswal, A., Lei, R., Simha, N., Wang, W., Wilfong, K., Williamson, T., Yilmaz, S.: Realtime data processing at Facebook. In: Proceedings of the 2016 International Conference on Management of Data (SIGMOD 2016), pp. 1087–1098. ACM, New York (2016). doi:10.1145/2882903. 2904441
7. Chen, H., Chiang, R.H., Storey, V.C.: Business intelligence and analytics: from big data to big impact. MIS Q. **36**(4), 1165–1188 (2012)
8. Cios, K.J., Pedrycz, W., Swiniarski, R.W.: Data mining and knowledge discovery. Data Mining Methods for Knowledge Discovery. The Springer International Series in Engineering and Computer Science, vol. 458, pp. 1–26. Springer, Boston (1998). doi:10.1007/978-1-4615-5589-6_1
9. Chennamangalam, J., Karastergiou, A., Armour, W., Williams, C., Giles, M.: ARTEMIS: a real-time data processing pipeline for the detection of fast transients. In: 2015 1st URSI Atlantic Radio Science Conference (URSI AT-RASC), Gran Canaria, Spain, p. 1 (2015). doi:10.1109/URSI-AT-RASC.2015.7303171
10. Cao, F., Ester, M., Qian, W., Zhou, A.: Density-based clustering over an evolving data stream with noise. In: Proceedings of the Sixth SIAM International Conference on Data Mining (2006)

11. Chen, Y., Tu, L.: Density-based clustering for real-time stream data. In: Proceedings of the 13th ACM SIGKDD International Conference on Knowledge Discovery and Data Mining (2007)

12. Cheng, W.T., Goldgof, B.D., Hall, O.L.: Fast fuzzy clustering. In: Fuzzy Sets and Systems, pp. 49–56 (1998)

13. Cannon, R., Dave, V.J., Bezdek, C.J.: Efficient implementation of the fuzzy c-means clustering algorithms. IEEE Trans. Pattern Anal. Mach. Intell. **8**(2), 248–255 (1986)

14. Dhamankar, R. and Gade, K.: Realtime analytics @ twitter. In: Proceedings of the Fifth International Workshop on Cloud Data Management (CloudDB 2013), pp. 1–2. ACM, New York (2013). doi:10.1145/2516588.2516593

15. Dong, X.L., Srivastava, D.: Big data integration. In: 2013 IEEE 29th International Conference on Data Engineering (ICDE), pp. 1245–1248. IEEE (2013)

16. van Dongen, B.F., de Medeiros, A.K.A., Verbeek, H.M.W., Weijters, A.J.M.M., van der Aalst, W.M.P.: The ProM framework: a new era in process mining tool support. In: Ciardo, G., Darondeau, P. (eds.) ICATPN 2005. LNCS, vol. 3536, pp. 444–454. Springer, Heidelberg (2005). doi:10.1007/11494744_25

17. Estivill-Castro, V.: Why so many clustering algorithms: a position paper. SIGKDD Explor. Newsl. **4**(1), 65–75 (2002). doi:10.1145/568574.568575

18. Everitt, B.S., Landau, S., Leese, M., : Cluster Analysis Arnold. A Member of the Hodder Headline Group, London (2002)

19. Galili, T.: dendextend: an R package for visualizing, adjusting, and comparing trees of hierarchical clustering. Bioinformatics, btv428 (2015)

20. Guha, S., Meyerson, A., Mishra, N., Motwani, R., O'Callaghan, L.: Clustering data streams: theory and practice. IEEE Trans. Knowl. Data Eng. **15**(3), 515–528 (2003)

21. Guha, S., Mishra, N., Motwani, R., O'Callaghan, L.: Clustering data streams. In: Proceedings of the 41st Annual IEEE Symposium on Foundations of Computer Science (2000)

22. Graham, R.L., Hell, P.: On the history of the minimum spanning tree problem. Ann. Hist. Comput. **7**(1), 43–57 (1985)

23. Hastie, T., Tibshirani, R., Friedman, J.: Unsupervised learning. In: The Elements of Statistical Learning, pp. 485–585. Springer, New York (2009). doi:10.1007/978-0-387-84858-7_14

24. Kimball, R., Ross, M.: The Data Warehouse Toolkit: The Definitive Guide to Dimensional Modeling. Wiley, Hoboken (2013)

25. Kochhar, A.: Distributed real time data processing for manufacturing organizations. IEEE Trans. Eng. Manage. **24**(4), 119–124 (1977). doi:10.1109/TEM.1977.6447256

26. Jeseke, M., Gruner, M., Wei, F.: Big Data in Logistics - A DHL Perspective on How to Move Beyond the Hype. DHL Customer Solution and Innovation (2015)

27. Von Luxburg, U.: A tutorial on spectral clustering. Stat. Comput. **17**(4), 395–416 (2007)

28. Michalski, R.S., Carbonell, J.G., Mitchell, T.M. (eds.): Machine Learning: An Artificial Intelligence Approach. Springer, Heidelberg (2013)

29. Khalilian, M., Mustapha, N., Sulaiman, N.M., Boroujeni, Z.F.: KMeans divide and conquer clustering. Presented at ICCAE, Thiland, Bangkok (2009)

30. Udommanetanakit, K., Rakthanmanon, T., Waiyamai, K.: E-stream: evolution-based technique for stream clustering. In: Proceedings of the 3rd International Conference on Advanced Data Mining and Applications. ADMA (2007)

31. Hathaway, J.R., Bezdek, C.J.: Extending fuzzy and probabilistic clustering to very large data sets. J. Comput. Stat. Data Anal. **51**(1), 215–234 (2006)
32. Pferd, W.J.: The Challenges of Integrating Structured and Unstructured Data. Technical report. PNEC Conference (2010)
33. Vryniotis, V.: DatumBox machine learning framework. http://www.datumbox.com/
34. Meng, X., Bradley, J., Yavuz, B.: MLlib: machine learning in apache spark. J. Mach. Learn. Res. **17**, 1–7 (2016)
35. Fournier-Viger, P., Gomariz, A., Gueniche, T., Soltani, A., Wu, C., Tseng, V.S.: SPMF: a Java open-source pattern mining library. J. Mach. Learn. Res. (JMLR) **15**, 3389–3393 (2014)
36. Top Logistics Challenges Facing Shippers Today. http://www.logisticsplus.net/top-logistics-challenges-facing-shippers-today/. Date accessed: 30 Mar 2016
37. Hall, M., Frank, E., Holmes, G., Pfahringer, B., Reutemann, P., Witten, H.I.: The WEKA data mining software: an update. SIGKDD Explor. **11**(1), 10–18 (2009)
38. Stelzner A.M.: Social Media marketing Industry Report - How Marketers are using Social Media to Grow Their Business. Social Media Examiners (2016)
39. Trevor, H., Robert, T., Jerome, F.: Hierarchical clustering. In: The Elements of Statistical Learning (PDF), 2nd edn., pp. 520–528. Springer, New York (2009). ISBN 0-387-84857-6
40. Nwaubani, J.: Business intelligence and logistics. In: Proceedings of the 1st Olympus International Conference on Supply Chain, Katerini, Greece
41. Mahobiya, C., Kumar, M.: Performance comparison of two streaming data clustering algorithms. Int. J. Comput. Trends Technol. (IJCTT) **12**(2) (2014)
42. Perera, S., Suhothayan, S.: Solution patterns for realtime streaming analytics. In: Proceedings of the 9th ACM International Conference on Distributed Event-Based Systems (DEBS 2015), pp. 247–255. ACM, New York (2015). doi:10.1145/2675743.2774214
43. Taxidou, I., Fischer, F.: Realtime analysis of information diffusion in social media. Proc. VLDB Endow. **6**, 416–1421 (2013). http://dx.doi.org/10.14778/2536274.2536328
44. Vadrevu, S., Hui, C., Suju R.T., Punera, K., Dom, B., Smola, J.A., Chang, Y., Zheng, Z.: Scalable clustering of news search results. In: Proceedings of the Fourth ACM International Conference on Web Search and Data Mining (WSDM 2011), pp. 675–684. ACM, New York (2011). doi:10.1145/1935826.1935918
45. Wu, C.H., Horng, S.J., Chen, Y.W., Lee, W.Y.: Designing scalable and efficient parallel clustering algorithms on arrays with reconfigurable optical buses. Image Vis. Comput. **18**(13), 1033–1043 (2000)
46. Zhang, L., Ramakrishnan, M.: BIRCH: an efficient data clustering method for very large databases. Presented at ACM SIGMOD Conference on Management of Data (1996)

Ubiquity: Extensible Persistence as a Service for Heterogeneous Container-Based Frameworks

Mohamed Mohamed[1]($\boxtimes$), Amit Warke[1], Dean Hildebrand[1], Robert Engel[1],
Heiko Ludwig[1], and Nagapramod Mandagere[2]

[1] Almaden Research Center, IBM Research, San Jose, CA, USA
{mmohamed,aswarke,dhildeb,engelrob,hludwig}@us.ibm.com
[2] Commulytics Inc, Mountain View, CA, USA
pramod@commulytics.com

Abstract. Over the last few years, micro-services are becoming the dominant architectural approach in designing software applications environments. This new paradigm is based on developing small, independently deployable services communicating using lightweight mechanisms and being consumed by other services or end user applications. Containers have proven to be a convenient runtime platform with container frameworks enabling elastic horizontal scaling if container workloads are designed as so-called 12 factor applications (https://12factor.net/), foregoing file or block storage use for persistence. Persistence services such as databases are typically deployed in traditional runtime platforms such as virtual machines or bare metal servers. To use container frameworks to manage stateful services in a scalable way is not well supported, in particular in scenarios where state is accessed from workloads in different deployment platforms. In this paper, we present the Ubiquity framework, which provides seamless access to persistent storage across different container orchestrators (CloudFoundry, Openshift, Kubernetes, Docker and Mesos). Ubiquity is extensible to other container frameworks and different types of file and block storage systems, which can be managed independently of the container orchestrator. Ubiquity makes it easy to onboard stateful services in containerized and hybrid environments, extending the efficiency gains of containerization.

Keywords: Persistence · CloudFoundry · Kubernetes · Docker

1 Introduction

In the last few years micro-services became the new trend for designing software applications. This paradigm consists of developing software applications as a set of independent components that focus on small functionalities, can be deployed separately, and use some lightweight communication mechanism such as REST, gRPC, etc. This trend changed how IT system developers perceive the software life-cycle, from the design and going forward to the testing, deployment and release management. These different phases of the software life-cycle have shrunk

© Springer International Publishing AG 2017
H. Panetto et al. (Eds.): OTM 2017 Conferences, Part I, LNCS 10573, pp. 716–731, 2017.
https://doi.org/10.1007/978-3-319-69462-7_45

from months to weeks, days or hours giving more opportunities to innovate. The designers, architects and developers can now focus on smaller code bases and make the changes faster. They can easily isolate any fault that might occur in the overall system, fix it and redeploy only the affected part.

Another trend that goes hand-in-hand with micro-services adoption is the usage of containers. Containers are a lightweight operating system level virtualization mechanism where the application running inside the container shares the kernel with the host operating system but has its own root file system [1]. Containers consume less resources than operating systems based on hardware virtualization [20]. They exploit the *cgroup* concept[1] in order to manage resources quota (CPU, Memory, Storage and Network) and offer the needed isolation to run the application without interfering with other containers. Many platforms and orchestration systems are based on container concepts to offer an agile way of building micro-service based software such as Cloudfoundry, Docker, Kubernetes, Mesos, etc. While using these container orchestrators (COs) is beneficial for large scale deployments, there is still a lively discussion as to which type of applications they might be best suited for. Most of these COs favor stateless micro-services due to the challenges of managing state in concurrency situations.

Onboarding stateful applications into these COs in a scalable way is not well supported in terms of manageability, security and performance, in particular in scenarios where state is accessed from workloads in different deployment platforms. In these scenarios, even though we can have the same persistence backend used to maintain the state, the frameworks that are being used have different ways of consuming persistence resources. These resources are offered to the platform as volumes or persistent volumes that are specific to each CO. For example, CloudFoundry uses persistent volumes through its persistence drivers and service brokers, whereas docker uses docker plugins to create and manage these volumes. Kubernetes has a third way of doing it by having a dynamic provisioner responsible of the creation and deletion of volumes and a volume plugin responsible of the other consumption functions (e.g., attach/detach, mount/unmount). One other challenge in this context, is adding new storage backends to offer their resources as persistent volumes to be consumed by containers within these different COs. In order to do that, the storage provider needs to understand the details of every CO and how it consumes persistent storage as well as the different deployment models that it supports. This is a time consuming task for each CO that comes with adding all the management facilities for security, multi-tenancy, high-availability, etc.

In this paper, we propose the Ubiquity framework that enables seamless access to storage for heterogeneous COs. Ubiquity is agnostic to the CO since it has the needed mechanisms to integrate with the widely used orchestrators (CloudFoundry, Kubernetes, Docker, Mesos, Openshift). Moreover, the design of Ubiquity makes it easy to be extended to support new COs while maintaining all the advantages that it already offers (Security, High Availability, Scalability). It also offers an easy way of adding new storage backends without the need to

---

[1] http://man7.org/linux/man-pages/man7/cgroups.7.html.

understand the specificities of the COs that are supported. Ubiquity supports also different deployment models to accommodate the different users preferences. Our close collaboration with the leaders of most of the COs storage teams allowed us to continuously adapt Ubiquity to the newly supported functionalities in an agile way. Hereafter, we will detail how Ubiquity can be integrated and used with the different COs and how a new storage backend can be added easily.

The remainder of this paper is structured as follows: in Sect. 2 we give some motivating examples and detail the different problems that we address. In Sect. 3, we present Ubiquity and show how it solves these problems. Subsequently, we present the implementation details of Ubiquity in Sect. 4. Then, we give an overview of the state of the art in Sect. 5. Finally, we conclude the paper and give some future directions in Sect. 6.

2   Motivation and Problem Description

The motivating applications that drove the adoption of micro-services had few storage requirements. For example, Web and application servers, load balancers, and monitoring services are mostly stateless and the little state they do retain can typically be lost without any major consequence. Recently though, the COs that enabled micro-services are being used for a much broader set of applications, many of which have extensive persistent storage requirements. Key-value stores, SQL and NoSQL databases, OLTP applications, machine and deep learning analytic applications, and just about every other enterprise application is now being proposed to run inside CO environments. This demand for storage has exposed a gap in the COs that desire to run these applications.

With the rise of CO in enterprise data centers, there is a desire to shift day-to-day storage management over to the CO administrator. This avoids the need to interact with the storage administrator for each and every application deployment. But there is an ever expanding set of on-premise and cloud-based file and block storage solutions, each with their own set of unique features (e.g., encryption, storage tiering, disaster recovery). It is impractical for every CO to support each and every storage system (and all of their unique features). Other similar cloud middleware frameworks such as OpenStack simply limited the set of supported features through standardized APIs (e.g., Cinder), but this presented problems for storage vendors by limiting their ability to provide value. COs chose instead to allow storage vendors to expose their features directly to the CO administrator. This additional flexibility comes at the price of additional complexity. Given that each CO has a different set of APIs and storage features (e.g., Kubernetes is the only CO that supports storage lifecycle management), the burden is on the storage vendor to integrate into each CO. No middleware exists to glue together the diverse set of CO and storage systems.

Ubiquity resolves this complexity by providing a pluggable framework that implements each of the CO APIs and provides a simple API that allows any storage vendor to integrate without loss of differentiation. Given that most enterprise data centers use a variety of storage systems, and increasingly the use of more than one CO, Ubiquity greatly simplifies adoption CO in diverse environments.

When a CO uses Ubiquity to access file or block-based storage systems, an applications dataset can now be referred to by its Volume Name (VN). Developers can simply refer to the VN in all uses of their application, avoiding the need for remembering complex block volume ids or long file paths.

3   Ubiquity

In this section we will go into the details of the Ubiquity framework. We will start with giving an overview of our framework in Sect. 3.1. Then, we will detail how Ubiquity is able to support the existing COs in Sect. 3.2. Afterwards, in Sect. 3.3 we will describe the different deployment models. Finally, we will talk about the future of the storage plugins in Sect. 3.4.

3.1   Overview of Ubiquity Framework

As discussed in Sect. 2, accessing persistent storage is a necessity for many scenarios. For the discussed scenarios and many others, the absence of persistent storage support in container frameworks is hindering the applications from moving beyond a monolithic architecture and moving to a more scalable and agile life-cycle based on containers. There are parallel efforts going on in order to add support for efficiently accessing persistent volumes and offering them to applications running inside containers. Meanwhile, the continuous change of APIs used by the different COs makes it difficult to cope with their support for storage. More specifically, if applications need to span different COs, it is complicated to manage all changes at once. The Ubiquity framework enables the use of persistent storage by one or more COs without any need to understand the intrinsic details and mechanisms of each targeted CO.

As shown in Fig. 1, Ubiquity is made up of different loosely coupled components that are easy to extend or replace. These components are the following:

- Ubiquity Volume Service: a scalable service that links COs to storage systems.
- CO Plugins: a peace of software that allows a CO to offer a persistent volume to a container inside a host (i.e., a node).
- Storage System: a module per provider that maps the CO requirements to storage resources.
- HA Database Service: a database that maintains the state of the system and back it with high resiliency and availability.

Ubiquity Volume Service: This service is the main component of Ubiquity playing the role of the mediator between the COs and the storage backends. It is offering southbound interfaces to be consumed by COs to allow them to create persistent volumes and manage them. The management operations are continuously evolving based on the functionalities supported by the COs. These operations include attaching and detaching volumes to and from nodes, setting quotas of volumes, maintaining the coherence of the attachment of volumes. These operations come

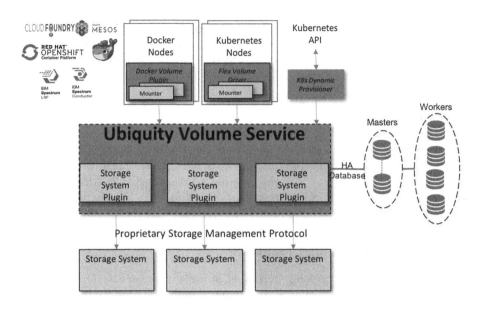

Fig. 1. Ubiquity architecture overview

with multi-tenancy support by maintaining the meta-data about the context of
the creation of a volume and prohibiting the unauthorized access to data in volumes. Other operations are being progressively included like snapshotting and
dynamic resizing of volumes. Ubiquity Volume Service is respecting the 12 factor
application recommendations. It can be easily scaled up or down when needed
to handle more queries. It supports also concurrent access by implementing a
rigorous locking mechanism. It is noteworthy that this service is maintaining
its state in a highly available database service to get more resiliency and high
availability.

Container Orchestrators Plugins: The Ubiquity framework has support for different COs. Most of these latter have a specific way of consuming storage via
storage plugins. These plugins generally live in all the nodes managed by the
CO and they allow to make host side operations to make the persistent volume
ready to consumption by the containers. These operations include mounting and
unmounting the volume to the container. In the context of containers, mounting means making the persistent volume visible in a specific location inside the
container file system and ready to be used. In some cases, the mount operation
requires other preparations before presenting the volume to the container. More
specifically, if the persistent volume is based on a block device, there is a need
to create a file system on top of that device in order to be able to mount it. If
the persistent volume is based on a shared file system, there is no need to do
any further operations other then mounting. We will give more details on the
supported COs in Sect. 3.2 where we describe the different ways COs use the
plugins in order to create and manage persistent storage.

Storage Backends: These are the mechanisms needed from the storage provider perspective to make their storage portfolio ready to COs. These components implement the needed mapping between the Ubiquity API and the specific provider API to allow the creation and management of persistence storage. Since the APIs might change from one provider to another, the mapping is not always straightforward. In some cases, we needed to integrate one or two functionalities from the provider side in order to map one functionality in Ubiquity. In other cases, we made a functionality as a no operation one because it is not relevant in the context of a particular provider.

Database Service: This component is playing a critical role in our architecture. In fact, it is critical for any 12 factor application to externalize its status to an external service to facilitate its scalability. Moreover, the Database Service is used in our locking mechanism, since the locks are created within the database and accessible from the different instances of Ubiquity. Whenever we have a concurrent access to a resource, Ubiquity consults this service to check the status on the lock granted to the needed resource. This service is also our tool to accomplish high availability and consistency. When we have different instances of Ubiquity running with a load balancer in front of them, all instances check if there is at least one that is currently playing the role of the leader in processing the calls. If the leader is not active (i.e., the leader did not update his status for a given period in the database), a new leader is elected and takes over the leadership responsibility. Ubiquity leverages the existing database solutions used to ensure high availability, leader election and locking mechanisms in different open source projects.

3.2   Supported Container Orchestrators

The Ubiquity framework plays the role of a mediator between COs and storage providers. For instance, the framework has support for any CO based on the open source PaaS CloudFoundry (e.g., IBM Bluemix, Pivotal CloudFoundry, SAP Cloud Platform, etc.). It has also support for any solution based on Kubernetes (IBM Container Service, Google Container Engine, Openshift, etc.). Finally, Ubiquity integrates very well with solutions based on the Docker Plugin API (Docker, Mesos, etc.).

CloudFoundry (CF): This Platform as a Service (PaaS) is among the most successful PaaS offerings. Support for persistent storage was introduced to the platform late 2016. The persistent storage support is based on three major components namely the service broker, the volume driver and the Diego Rep volume manager (volman):

- Service Broker: This service is compliant with the Open Service Broker API [15]. It is responsible for creating persistent volumes based on the backend offered by the storage provider. It sends back a blob of information determining how to access the created volume.

- Volume Driver: this driver is compliant with Docker Volume API [9]. It allows attaching volumes to the host and mounting them to the application container.
- Volman: is the manager that consumes the blob of information sent back by the service broker and uses it to communicate with the volume driver in order to attach and mount volumes.

The service broker and volume driver are the responsibility of the storage provider while volman is already embedded in the platform itself. Ubiquity has support for CF by implementing the docker API for the driver and by offering an implementation of the service broker with different storage backends (Spectrum Scale, OpenStack Manila, Softlayer Storage).

Kubernetes: It is *'an open-source system for automating deployment, scaling, and management of containerized applications'* [19]. This CO is one of the most popular open source COs with a huge base of contributors from all over the world. Kubernetes have support for persistent volumes using two main components: Dynamic Provisioner and Volume Plugins.

- Dynamic Provisioner: it is a unique feature for kubernetes that *allows storage volumes to be created on-demand* [17]. This feature is based on defining storage classes to present different flavors or configurations of persistence storage. Each storage class needs to specify a dynamic provisioner responsible for creating and deleting persistent volumes. An end user can specify his requirements in term of persistent volumes in a so called Persistent Volume Claim (PVC). Whenever a new PVC is created, the dynamic provisioner will take it, create a persistent volume and bind it to it.
- Volume Plugins: they are pieces of software needed to consume an existing persistent volume [18]. They offer attach/detach and mount/unmount functionalities in each node of the kubernetes cluster. These plugins can be in-tree (released with kubernetes base code) or out-of-tree (released as a separate code that can be used with kubernetes). These plugins work with the assumption that the persistent volume was already created (manually or through a dynamic provisioner). If an application requires a given type of persistent volumes, its associated volume plugin should be available on each and every host in order to be able to schedule containers on this host.

Ubiquity offers a dynamic provisioner and an out-of-tree volume plugin that are agnostic to the provided storage backend. In order to consume any persistent storage for the list that we are supporting today (Spectrum Scale, Openstack Manila and Softlayer Storage), one can specify the backend name in the storage class and Ubiquity will take care of the rest.

Docker: Since 2013, Docker [8] became the *de facto* way of deploying applications on containers. *It is an open platform for developers and sysadmins to build, ship, and run distributed applications, whether on laptops, data center*

VMs, or the cloud. Docker uses volume plugins [9] to integrate with external storage backends and allow containers to persist their data beyond the lifetime of a container. Any provider can integrate with docker by implementing a volume plugin compliant with the defined API that allows the creation of volumes and making them available to the container. Ubiquity comes with a docker plugin that is agnostic to the storage backend. A Docker user can create volumes using this plugin by specifying the backend that he desires. The plugin will just map the Docker calls to Ubiquity calls and get back the volume to be consumed.

3.3   Deployment Model

The Ubiquity framework is composed of different components that could be deployed in different ways:

Ubiquity Service: Since this service is stateless, we have different options to deploy it. The easiest way to do that and since we are working in container-based environments, is to run the service as a container. This option works for all COs as far as we have the right plugin and right endpoint of the container running ubiquity. A second option takes advantage of the concept of *deployment* proposed by Kubernetes where we can describe our deployment as a pod of containers and we can specify how many instances of that container should be running at any given time. Using this option, we can benefit from kubernetes strategy of offering high availability and fault tolerance. We can also scale up or down the number of running instances as much as we need. The third option is to run the code as a standalone binary application. Finally, we can run the service as a *systemd* script that starts with the startup of our machine.

CO plugins: These components are specific to each CO, so we are limited in term of deployment options. For CF, the plugin should be deployed as part of the runtime (diego release [3]) using the so called bosh system [2]. However, for Docker, the plugin needs just to publish a JSON file in a given directory discoverable by docker daemon. This file indicates how to consume the plugin API. Kubernetes comes with a different deployment model where the plugins must be in a given location on each node of the cluster while the dynamic provisioner can run as a container or a standalone web server. The dynamic provisioner can also run as *systemd* script as well.

Storage Backends: these latter are the responsibility of the storage providers. Some providers offer a kubernetes friendly deployment of their systems, some others offer cloud native storage and a third category have their own specific tools to deploy and manage their storage systems.

Database Service: No matter how this is deployed, it must respect the constraints that we cited before. The solution that we are using is based on kubernetes cluster that ensures high availability and fault tolerance as well as a high resiliency.

3.4 Towards a Unified Plugin Interface

Recently, there is a commitment among the different CO stakeholders to develop a unified way to consume storage by having a common API for the plugins in COs. The effort is called Container Storage Interface (CSI) under governance of the Cloud Native Computing Foundation, which has most organizations providing Cloud software among its members. The objective of this effort is to specify an industry standard that allows storage providers to develop a plugin once and use it across the different COs [23]. Using Ubiquity will enable storage providers to quickly implement SCI support for all backends they support.

4   Implementation

In this section we will provide the implementation details of each of the components of Ubiquity Framework. These components are shown in Fig. 2. We will start by describing Ubiquity Service in Sect. 4.1. Afterwards, we will detail the implementations of each CO plugin that we support in Sect. 4.2. Then, we will talk about the different backends that we implemented in Sect. 4.3. Finally we will give some indications about the database service that we are using in Sect. 4.4.

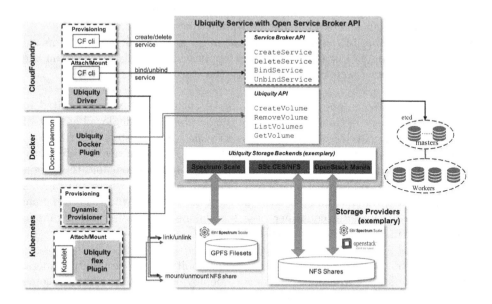

Fig. 2. Ubiquity support for CloudFoundry/Docker/Kubernetes: overview

4.1 Ubiquity Service

The Ubiquity Service[2] is implemented as go[3] language project. The project has a web server that offers REST interfaces to manage persistent volumes. As shown in Fig. 2, the framework allows creation, deletion and listing of volumes. The basic interfaces offered by the project are depicted in Fig. 3 where the Storage Client interface describes the different functions that need to be executed from the server side. These functions do not need to be executed in the node/ host that will host the container. The needed functions to be executed from the host side are described in the Mounter interface. These functions allow presenting the persistent volume and making it available inside the container to be consumed by the application.

```
//StorageClient interface to create/delete  and attach/detach volumes
type StorageClient interface {
Activate(ActivateRequest) error
CreateVolume(CreateVolumeRequest) error
RemoveVolume(RemoveVolumeRequest) error
ListVolumes(ListVolumesRequest) ([]Volume, error)
GetVolume(GetVolumeRequest) (Volume, error)
GetVolumeConfig(GetVolumeConfigRequest) (Config, error)
Attach(AttachRequest) (string, error)
Detach(DetachRequest) error
}
//Mounter interface to do host specific operations mount/unmount
type Mounter interface {
Mount(MountRequest) (string, error)
Unmount(UnmountRequest) error
}
```

Fig. 3. Ubiquity StorageClient and Mounter interfaces

The Ubiquity Service is extensible and can support new COs and new storage backends. In order to add a new CO, we can easily reuse the remote client code that we implemented. The remote client is playing the role of a proxy to invoke the Ubiquity interfaces. Moreover, in order to add a new storage backend, we need just to implement the two interfaces shown in Fig. 3 to be able to manage the specific type of storage provided by the related backend.

4.2 Supported CO Plugins

CloudFoundry Support: Ubiquity provides persistent filesystem storage to CloudFoundry (CF) applications through its implementation of the Open Service Broker API [15] and the CF Volume Services API specification [5], as well

---

[2] https://github.com/IBM/ubiquity.
[3] https://golang.org/.

as by providing a volume driver plugin for CF cells (hosts) to be used by the CF runtime called Diego. The available storage backends are exposed as CF volume services through the Ubiquity Service Broker's service catalog to support provisioning of volumes. Attaching and mounting of volumes to CF cells is supported through the Ubiquity Service Broker in conjunction with the Ubiquity CF Volume Driver Plugin. The relationships between the various involved components are illustrated in Fig. 2 and described in detail in the following.

Provisioning: Ubiquity volumes can be created for consumption by CF applications by creating instances of the services exposed by the Ubiquity Service Broker API (i.e., through `cf create-service-broker` and `cf create service`). Internally the Ubiquity Service translates these Service Broker calls into "native" Ubiquity API calls. Ubiquity/CF administrators can configure these volumes (e.g., with regard to mounting method, quotas, QoS parameters, etc.) either by (i) configuring a number of *service plans* in the Ubiquity Service Broker services catalog that encapsulate different preset configuration settings; or by (ii) passing arbitrary parameters through the `cf create-service` interface which are internally passed on by Ubiquity to the storage backend associated with the CF service. When a CF service representing a Ubiquity volume is instantiated through the Service Broker, the corresponding storage backend is eventually responsible for creating the volume.

Attaching/Mounting: Attaching to and mounting Ubiquity volumes on individual CF cells is arranged for by the Diego Rep Volume Manager component of the CF installation in conjunction with the Ubiquity CF Volume Driver Plugin residing on the individual CF cells. On binding a Ubiquity service instance to a CF application (i.e., through `cf bind-service`), the Diego Rep volume manager (volman) instructs the Ubiquity CF Volume Driver Plugin on all affected cells to attach to the Ubiquity volume represented by the service that is being bound to the application, and subsequently mounts the volume. The mounting method (e.g., NFS, native Spectrum Scale mount, etc.) and mounting details (e.g., remote IP address of an NFS share) are chosen based on configuration details provided by the Ubiquity Service Broker in the service binding response and using pluggable logic in the Ubiquity CF Volume Driver Plugin for different storage backends. Note that if NFS is chosen as a mounting method, the CF community NFS plugin [4] may be employed in lieu of the volume driver plugin provided by Ubiquity.

Kubernetes Support: Ubiquity implements support for Kubernetes persistent volumes[4] following the community recommendations. The Ubiquity framework offers an implementation of the dynamic provisioner API [17] and an implementation of the flexvolume plugin [14].

---

[4] https://github.com/IBM/ubiquity-k8s.

Provisioning: For provisioning functions, Ubiquity implements a dynamic provisioner as go language project that runs as a web server. In the context of kubernetes, a dynamic provisioner must advertise its name. This name will be used by a storage class that represents a storage flavor for kubernetes. For example, since we have support for Spectrum Scale storage, we defined a storage class named *goldFileset* that refers to filesets created within the gold filesystem from Spectrum Scale. The storage class refers to our dynamic provisioner to specify that any query related to this storage class should be forwarded to our running dynamic provisioner server. The dynamic provisioner will then handle the request, create a new Spectrum Scale fileset and bind it to the request (PVC).

Attaching/Mounting: Attaching and Mounting a volume is a host side action performed by the Kubelet component of Kubernetes. Ubiquity implements a flexvolume plugin that communicates with Ubiquity Service in order to run its functionalities. Whenever a container needing any volume provisioned by Ubiquity is to be deployed, the Kubernetes Manager will check which nodes dispose of a Ubiquity flexvolume plugin. Then it will schedule the container to run on a specific node. An attach call is sent to ubiquity in order to make the volume available to that specific host. Afterwards, the Kubelet will call the flexvolume plugin to mount the volume and make it available to the container.

Docker Support: The Ubiquity Docker Plugin[5] is an external volume plugin that works in conjunction with the Ubiquity Service to provide persistent storage to containers orchestrated by Docker. Docker communicates with the Ubiquity Docker Plugin via Dockers Volume Plugin Protocol [9], a REST based API, to provide persistence storage as Docker volumes to containers orchestrated by Docker. Ubiquity Docker plugin must be run on the same host as the Docker engine. The Plugin registers itself with the Docker daemon by placing a JSON specification file in the plugin directory, which the daemon scans to discover plugins. The plugin is activated when the Docker daemon discovers the specification file in the plugin directory and activates it with a handshake API. The Docker daemon will refer to the plugin whenever a user or container refers to the plugin by name.

The Ubiquity Docker plugin communicates with Ubiquity Volume Service using a REST based API, to perform Volume Management operations (Create, Remove, Path, Mount, Unmount, Activate, Get, List, Capabilities) requested by Docker daemon for a specific storage backend supported by Ubiquity Volume Service.

The plugin also implements the Mounter interface for each storage backend supported by the Ubiquity Volume Service. The Mounter interface performs storage backend specific host-side operations (Mount/Unmount) so that volume created using Ubiquity volume service can be made available as a path on the Docker hosts filesystem. For example, an IBM Spectrum Scale Mounter would set user permissions on a volume made available on the host if the volume was

---

[5] https://github.com/IBM/ubiquity-docker-plugin.

created with user, group permissions. Similarly, an NFS mounter would mount and unmount NFS shares on the host. Docker daemon makes these volumes available to the containers by bind-mounting this path into the right path inside the containers.

In the Docker Swarm Mode, which consists of a cluster of Docker hosts, Ubiquity Docker Plugin must be running on all the worker nodes and all of instances of the plugin must be configured to communicate with the running Ubiquity Service.

4.3 Supported Backends

As shown in Fig. 2, Ubiquity supports different storage backends. The initial reference backend is the Spectrum Scale file system, which offers two types of volumes. The first one is based on Spectrum Scale filesets while the second one is based on Spectrum Scale lightweight volumes. Fileset based volumes benefit from the possibility of fixing quota on them while lightweight based ones benefit from the fact that they don't have any number limit. A user can choose the most suitable type for his or her use case. Ubiquity also supports Spectrum Scale NFS volumes that could be consumed using any default NFS plugin developed by the community. Moreover, Ubiquity supports Openstack Manila to enable creating persistent volumes out of the offered NFS shares. All these backends are based on shared filesystems. The implementation of block storage support is ongoing.

4.4 Database Service

In our implementation we are using etcd key value store [21]. This is an open-source distributed key value store. It allows to share configurations and service discovery for Container Linux clusters. We used etcd to enforce the high availability of Ubiquity by handling the leader election and the loss of the current leader. We are also using it for all locking mechanisms and concurrency management scenarios.

5 Related Work

Persistence for COs is relatively a new topic that goes back to the last three years. The issues and challenges have not been addressed widely by the research community. In [10], the author proposed an approach to use Byte addressable Non Volatile Memory in containers to boost the performance of containerized applications while benefiting from the isolation guarantees that containers offer. In another direction, the authors of [11] propose Slacker as a new storage driver for docker to reduce the start time of a container by pulling not all the layers of the container image but only a subpart of these layers needed to start the container. And during runtime the authors proposed a lazy approach to fetch the needed layers from the registry which requires a continuous connectivity to the registry side. This paper does not target persistent storage, but instead targets the storage driver to be used by docker itself. In [13], the authors proposed a peer-to-peer solution to save image layers in order to enhance the start time of docker images. But this solution also does not target the persistent volume drivers as well. So far, these are the only research papers related to container storage that we could identify. There are other research targeting storage for virtual environments such as [7,22]. These works are focusing on virtual machines rather than containers. Consequently, they are not dealing with the same challenges and not working on the same granularity and isolation layer.

Meanwhile, in the industrial context we see that there are different efforts to resolve this problem. These efforts are mainly led by different storage system providers who want to support the trend of their customers moving towards containers. For example, EMC proposes Rex-Ray [6] as a storage integration framework for Container Orchestrator and provides plugins for their storage backends. Rex-Ray focuses on Docker, Kubernetes and Mesos, as publicly available at the time of writing. In the same context, NetApp proposes Trident [16] as a solution for Kubernetes-based systems. CoreOS proposes their solution called Torus [12], which is targeting persistent storage to containers orchestrated by Kubernetes.

Compared to the existing efforts, Ubiquity presents an easy way to bring together the different existing COs and storage providers in a open way. It also supports the paradigm of PaaS as well as regular container orchestration by providing service-oriented pathways into storage life-cycle management, as articulated in the implementation of the Cloud Foundry service broker API. This support enables the integration not only with systems based on Docker and Kubernetes but also with PaaS like CloudFoundry. Ubiquity comes also with an easy way to add support for new storage providers without the need to understand all the specificities of the existing COs. These storage providers can be on-site systems or plugins to Cloud storage providers where this makes sense.

6   Conclusions

With the rise of Container Orchestrators in enterprise data centers, demand arises to run not only stateless applications but also stateful applications in containers. This demand has exposed a gap in these COs that desire to run these applications in term of persistent storage support. Moreover, given that each CO has a different set of APIs and storage features, the burden is on the storage vendor to integrate into each CO. So far there are few efforts to integrate this diverse set of COs and storage systems. In this paper, we propose the Ubiquity framework that addresses the complexity of bringing together the different COs and Storage providers in the context of sharing persistent state across COs. Ubiquity offers an easy way to plug in the support of new - versions of - COs on the one hand and, on the other hand, to enable the support for new storage backends independently of the specific CO. Ubiquity is an open source project that is continuously adding support of new functionalities, new COs and new storage backends. While multiple storage vendors offer persistence solutions for some COs for their storage systems, Ubiquity supports a large spectrum as well as multiple orchestration paradigms, including Platform-as-a-Service, providing additional flexibility for managing the storage life-cycle for a wider range of systems. Due to its open source nature, the Ubiquity community edition is also an ideal platform for academic experimentation.

In future work we will look at emerging container interface standards such as the Container Storage Interface being discussed by the Cloud Native Computing Foundation and adding new, storage related functionality. In addition, we are evaluating storage management performance of COs and storage system combinations in the context of different workloads.

References

1. Azab, A.: Enabling docker containers for high-performance and many-task computing. In: 2017 IEEE International Conference on Cloud Engineering (IC2E), pp. 279–285, April 2017
2. bosh.io: bosh (2017). https://bosh.io/docs. Accessed 19 June 2017
3. CloudFoundry Foundation: Diego architecture (2017). https://docs.cloudfoundry.org/concepts/diego/diego-architecture.html. Accessed 19 June 2017
4. CloudFoundry Foundation: Nfsv3 volume driver (2017). https://github.com/cloudfoundry/nfsv3driver. Accessed 18 June 2017
5. cloudfoundry.org: Volume services, June 2017. http://docs.cloudfoundry.org/services/volume-services.html. Accessed 18 June 2017
6. Dell EMC: Rex-ray openly serious about storage (2017). https://rexray.readthedocs.io/en/stable/. Accessed 20 June 2017
7. Diaz, J., von Laszewski, G., Wang, F., Younge, A.J., Fox, G.: Futuregrid image repository: a generic catalog and storage system for heterogeneous virtual machine images. In: 2011 IEEE Third International Conference on Cloud Computing Technology and Science, pp. 560–564, November 2011
8. Docker Inc.: A better way to build apps (2017). https://www.docker.com/. Accessed 19 June 2017
9. Docker Inc.: Volume plugins (2017). https://docs.docker.com/engine/extend/plugins_volume/. Accessed 18 June 2017
10. Giles, E.R.: Container-based virtualization for byte-addressable NVM data storage. In: 2016 IEEE International Conference on Big Data (Big Data), pp. 2754–2763, December 2016
11. Harter, T., Salmon, B., Liu, R., Arpaci-Dusseau, A.C., Arpaci-Dusseau, R.H.: Slacker: fast distribution with lazy docker containers. In: 14th USENIX Conference on File and Storage Technologies (FAST 2016), pp. 181–195. USENIX Association, Santa Clara (2016). https://www.usenix.org/conference/fast16/technical-sessions/presentation/harter
12. Michener, B.: Presenting torus: a modern distributed storage system by coreos (2017). https://coreos.com/blog/torus-distributed-storage-by-coreos.html. Accessed 20 June 2017
13. Nathan, S., Ghosh, R., Mukherjee, T., Narayanan, K.: Comicon: a co-operative management system for docker container images. In: 2017 IEEE International Conference on Cloud Engineering (IC2E), pp. 116–126, April 2017
14. Nelluri, C.: FlexVolume explored (2017). https://www.diamanti.com/blog/flexvolume-explored/. Accessed 20 June 2017

15. Open Service Broker: Open service broker API specification (2017). https://github.com/openservicebrokerapi/servicebroker/blob/v2.12/spec.md. Accessed 18 June 2017

16. Sullivan, A.: Introducing trident: a dynamic persistent volume provisioner for kubernetes (2017). http://netapp.io/2016/12/23/introducing-trident-dynamic-persistent-volume-provisioner-kubernetes/. Accessed 20 June 2017

17. The Kubernetes Authors: Dynamic provisioning and storage classes in kubernetes (2017). http://blog.kubernetes.io/2016/10/dynamic-provisioning-and-storage-in-kubernetes.html. Accessed 19 June 2017

18. The Kubernetes Authors: Persistent volumes (2017). https://kubernetes.io/docs/concepts/storage/persistent-volumes/. Accessed 19 June 2017

19. The Kubernetes Authors: Production-grade container orchestration (2017). https://kubernetes.io/. Accessed 19 June 2017

20. Uehara, M.: Performance evaluations of LXC based educational cloud in Amazon EC2. In: 2016 30th International Conference on Advanced Information Networking and Applications Workshops (WAINA), pp. 638–643, March 2016

21. Walker-Morgan, D.: etcd introduced (2017). https://www.compose.com/articles/etcd-introduced/. Accessed 20 June 2017

22. Xie, W., Zhou, J., Reyes, M., Noble, J., Chen, Y.: Two-mode data distribution scheme for heterogeneous storage in data centers. In: 2015 IEEE International Conference on Big Data (Big Data), pp. 327–332, October 2015

23. Yu, J., Ali, S., DeFelice, J.: Container storage interface (CSI) (2017). https://github.com/container-storage-interface/spec/blob/master/spec.md. Accessed 19 June 2017

A Formal Approach for Correct Elastic Package-Based Free and Open Source Software Composition in Cloud

Imed Abbassi[1]([✉]), Mohamed Graiet[2], Sindyana Jlassi[1], Abir Elkhalfa[1], and Layth Sliman[3]

[1] ENIT, University of Tunis El Manar, Tunis, Tunisia
abbassi.imed@gmail.com, sindyana.jlassi@gmail.com, abir.elkhalfa@gmail.com
[2] INSA Centre-Val de Loire, Blois, France
mohamed.graiet@insa-cvl.fr
[3] Efrei, Paris, France
layth.sliman@efrei.fr

Abstract. Cloud environments have been increasingly used to deploy and run software while providing a high level performance with a low operating cost. Most of the existing software applications are nowadays distributed as Package-based Free and Open Source (PFOS) applications. Different requirements must be considered while configuring PFOS software. These requirements can be classified into two classes: dependency and capacity requirements.

In this paper, we proposed a novel approach to ensure the correctness of elastic composite PFOS applications. Our approach is based on Event-B and combines proof-based models with model checking to provide a more complete verification. It starts by abstractly specifying the main concepts of PFOS software, and then refining them through multiple steps to model the elastic composite PFOS software and its correctness requirements. The consistency of each model and the relationship between an abstract model and its refinements are obtained by formal proofs. Finally, we used the ProB model-checker to trace possible design errors.

Keywords: Free and open source software · Composition · Cloud · Elasticity · Event-B · Verification

1 Introduction

Cloud environments are used nowadays to deploy and run software applications [1]. These software applications use cloud components (databases, containers, Virtual machines, and so on) which in turn use cloud resources (CPU, memory, network, and so on). One of the most interesting properties provided by Cloud environments is the horizontal and vertical elasticity at different levels and particularly the software as a service (SaaS) level [2,3]. Vertical elasticity allows to

© Springer International Publishing AG 2017
H. Panetto et al. (Eds.): OTM 2017 Conferences, Part I, LNCS 10573, pp. 732–750, 2017.
https://doi.org/10.1007/978-3-319-69462-7_46

increase or decrease the software resources, while the horizontal elasticity is used to replicate or remove software instances [3,4].

Most of the existing software are nowadays distributed as Package-based Free and Open Source (PFOS) applications [5]. In this work, we focus on this type of software. The PFOS software can be installed, run or uninstalled automatically. They provide a collection of functional capabilities [6]. A functional capability is defined either as a required port, or as a provided port. Reusability is a central concept of software design [7]. For instance, it allows the construction of composite PFOS software. These composite PFOS software can be automatically extended by adding new components or removing others [8]. For example, the Rodin [9] and Eclipse platforms [10,11] are well-known extensible composite PFOS applications.

Different requirements must be considered while configuring composite PFOS software [5]. These requirements can be classified into two classes: dependency and capacity requirements. The dependency requirements are defined as a set of constraints between components [5]. An example of dependency requirement is that the installation of some components requires the installation of others. The capacity requirements are defined as a set of upper and lower bound constraints on the number of ports' bindings [5]. For instance, the provided ports shall satisfy a maximum number of bindings [5]. The required ports shall satisfy a minimum number of bindings [5].

The main objective of our work concerns mainly the correctness verification problem of the horizontal and vertical elasticity properties for composite PFOS applications. To meet such an objective, we propose a novel approach for the formal modeling and verification of such applications based on the Event-B formal method [12] and combining proof-based models with model checking to provide a more complete verification. Event-B is chosen because of its capability to effectively master the complexity of the system design using stepwise refinement. The step-wise refinement method produces a correct specification by construction since at each step the different properties of the system are formally proven. Moreover, Event-B tools can be used to perform an incremental verification by checking the defined properties and constraints at each PFOS software execution step. These execution steps are ensured by events. In order to guarantee the invariants/constraints preservation by these events, Event-B defines proof obligations. Hence, using our approach we check and prove the correctness of the PFOS software composition and its requirements and elasticity properties. Then, once the composite PFOS applications are verified, our formal model guarantees that their execution does not face failures and inconsistencies related to the violation of dependency and capacity requirements.

The rest of this paper is organized as follows: Sect. 2 presents a motivating example and the basic concept of the Event-B method and its proof procedure. In Sect. 3, we propose an Event-B formal model for the composite PFOS software. In Sect. 4, we refine this model through several steps to model the (horizontal and vertical) properties and their correctness requirements. The consistency verification process of such a model is described in Sect. 5. Section 7 concludes this paper and presents some future works.

2   Motivation and Background

In this section, we first introduce a motivating example that will be used throughout the paper. Finally, we describe the basic knowledge of Event-B.

2.1   Motivating Example

Let's consider a Wordpress application to illustrate our approach. This application is composed of the following three software components:

- apache2: a PFOS software providing an *httpd* port.
- mysql: a PFOS software providing *mysql_up* and *mysql* ports.
- wordpress: a PFOS software providing *wordpress* port, while requiring *httpd* and *mysql_up* ports.

The internal behavior of these software components is described as a finite state automaton composed of the following three states, namely installed, uninstalled and running (see Fig. 1).

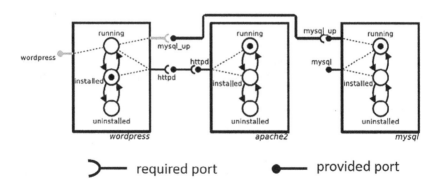

Fig. 1. The Wordpress application [5].

All of the Wordpress application components are initially set to the uninstalled state. Then, its execution state can be changed according a set of dependency requirements. Indeed, wordpress can be installed only when apache2 is already installed or is running. The running or installation process of the apache2 is required in order to activate the provided *httpd* port that is required by wordpress software. Once installed, wordpress can either be run when mysql software is running, or be uninstalled. When running, it can be stopped and rolled back to the installed state. When in use by wordpress, apache2 and mysql can neither be stopped nor uninstalled. The only way to stop them is to release their dependency with the wordpress component. This is done by stopping and uninstalling the wordpress component. When all their dependencies are correctly released, apache2 and mysql can be securely stopped or uninstalled.

2.2   Event-B Method

Event-B [12] is a formal method based on set theory and first order logic. It provides a correct-by-construction design methodology and covers the whole software's life-cycle. Indeed, the designers start by abstractly specifying the requirements of the whole system and then refining them through several steps to reach a detailed description of the system that can be translated into executable code (C, JAVA, etc.) via additional tools such as EB2C [13] and EB2J [14].

An Event-B model includes two types of entities to describe a system: machines and contexts. A machine represents the dynamic parts of a model. It may contain variables, invariants, theorems, variants and events. On the other hand, a context represents the static parts of a model. It may contain carrier/enumerate sets, constants, axioms and theorems. Those constructs appear on Fig. 2.

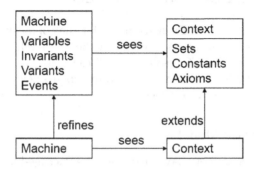

Fig. 2. Machine and context relationships

A machine is organized in clauses: the first clause, VARIABLES, represents the defined variables of the model. The INVARIANTS clause represents the invariant properties of the system and must at least include the typing of the variables declared in the VARIABLES clause. The EVENTS clause contains the list of events of the model. An event is modeled with a guarded substitution and fired when its guards is evaluated to true. The events occurring in an Event-B model affect the state described in the VARIABLES clause. An Event-B model may refer to a context or machine. A context consists of the following clauses: the SETS clause describes a set of abstract and enumerated types. The CONSTANTS clause represents the constants of the model, whereas the AXIOMS clause contains all the properties of the constants and their types.

The concept of refinement is the main feature of Event-B. It allows an incremental design of systems. At each level of abstraction we introduce further details of the modeled system. Correctness of Event-B machines is ensured by establishing proof obligations (POs); they are generated by a RODIN platform tool called *proof obligations generator* to check the consistency of the model. Let M be an Event-B model with v as variables. The invariants are denoted by $I(v)$

and E is an event with an input parameter p, a guard $G(v,p)$ and a before-after predicate $R(v,p,v\prime)$. The initialization event is a generalized substitution of the form $v : INIT(v)$. The initial proof obligation guarantees the satisfaction of the initialization invariants: $INIT(v) \Rightarrow I(v)$. The second proof obligation is related to events. The event E should preserve the invariants after its triggering. The Feasibility statement (FIS) and the invariant preservation (INV) are given in the following predicates:

- FIS: $I(v) \wedge G(v,p) \Rightarrow \exists v\prime \cdot R(v,p,v\prime)$
- INV: $I(v) \wedge G(v,p) \wedge R(v,p,v\prime) \Rightarrow I(v\prime)$

An Event-B model M with invariants I is well-formed, denoted by $M \vdash I$, only if M satisfies all of the proof obligations.

3 Composite PFOS Software Model

In this section, we propose an Event-B model for the composite PFOS software. This model explores two abstraction levels. At the first level, we abstractly specify the basic concepts of PFOS software. At the second level, we refine such an abstract model by introducing the software dependency concept in order to formally define the composition process.

3.1 PFOS Software Model

Software applications are nowadays distributed as Package-based Free and Open Source (PFOS) applications [6,15]. The package includes the current software artifact and its default configuration.

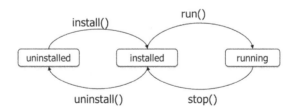

Fig. 3. State transition diagram of PFOS software

A PFOS software can be associated to a life-cycle statechart. A set of states (uninstalled, installed, running) and a set of transitions (*install()*, *uninstall()*, *run()*, *stop()*) are used to describe the PFOS software status and behavior (see Fig. 3). When being created, a software is initially set to the *uninstalled* state. Then, it can transit to the *installed* state. Once installed, it can either be uninstalled, or transited to the running state.

```
CONTEXT PFOSScontext
SETS
SOFTWARE STATE                      MACHINE PFOSSmachine
CONSTANTS                           SEES PFOSScontext
uninstalled installed running       VARIABLES
AXIOMS                              states
axm1: partition(STATE,{ uninstalled} ,  INVARIANTS
{ installed} ,{ running} )          type1: states ∈ SOFTWARE ⇸ STATE
END                                 EVENTS
                                    . . .
```

Fig. 4. The PFOSScontext and PFOSSmachine description

```
install ≙ ANY s                 run ≙ ANY s
WHERE                           WHERE
grd1: s ∈ dom(states)          grd1: s ∈ dom(states)
grd2: states(s)=uninstalled    grd2: states(s)=installed
THEN                           THEN
act1: states(s):=installed     act1: states(s):=running
END                           END
uninstall ≙ ANY s              stop ≙ ANY s
WHERE                          WHERE
grd1: s ∈ dom(states)         grd1: s ∈ dom(states)
grd2: states(s)=installed     grd2: states(s)=running
THEN                          THEN
act1: states(s):=uninstalled  act1: states(s):=installed
END                          END
```

Fig. 5. The formalization of the installation, uninstallation, run and stop processes of software using Event-B notation

We formalize the PFOS software using an Event-B model that includes a machine and a context. The context, *PFOSScontext* (see Fig. 4), describes the static part of PFOS software. By *SOFTWARE*, we denote the set of PFOS software. The enumerated set *STATE* represents all possible states of PFOS components, namely *uninstalled*, *installed* and *running*. The machine, *PFOSSmachine* (see Fig. 4), formally models the behavior of PFOS software using variables and a set of events. The variable, *states* (formalized as a partial function from the set *SOFTWARE* to *STATE*), specifies the current state of each PFOS software. Figure 5 presents the formalization of the *install*, *uninstall*, *run* and *stop* processes of a software s. The guard $grd2$ of each event defines the current state of the input parameter. The events *install* and *uninstall* are behaviorally opposite. Similarly, the *run* event has the inverse behavior of *stop*.

3.2   Composite PFOS Software Model

Reusability constitutes the main feature of software design [7]. It allows the construction of a new value-added software by combining a set of existing software. The composite PFOS software are the results of the combination of a set of existing PFOS software [5]. They describe the preconditions for installing, running and uninstalling components. These preconditions express dependency relations. The dependencies between the components are defined through ports as a set of bindings [5]. A port represents a functional capability that can be either provided or required by software components. A binding is a relation defined through a port between two software components. This port shall be provided by one of them and required by the other. For the sake of simplicity, we assume two components can be bound through only one port. Different requirements shall be satisfied while executing composite PFOS software. In this paper, we consider the following requirements:

- REQ1: the installation of a software requires the installation of all the required software [5].
- REQ2: the software that are running or having provided ports under use by others cannot be uninstalled [5].
- REQ3: running a software requires running all of the required software [5].
- REQ4: the provided ports shall satisfy a maximum number of bindings [5].
- REQ5: the required ports shall satisfy a minimum number of bindings [5].

These requirements shall always be satisfied by the composite PFOS software. The first three requirements define dependency constraints between software components. The requirement $REQ4$ defines a set of upper-bound capacity constraints on the provided software ports. The last one defines a set of lower-bound capacity constraints on the required software ports. According to our motivating example, the installation of the Wordpress component requires the installation of the apache2 and mysql components. The $mysql$ component provides two ports, namely $mysql\_up$ and $mysql$ (see Fig. 1). The $mysql$ port, is provided in the running and installed states. The $mysql\_up$ port is required by the $Wordpress$ component in the running state.

```
CONTEXT CPFOSScontext
EXTENDS PFOSScontext
SETS
PORT
CONSTANTS
requiredPorts providedPorts portsCapacity
AXIOMS
axm1: providedPorts ∈ SOFTWARE × STATE ↛ ℙ(PORT)
axm2: requiredPorts ∈ SOFTWARE × STATE ↛ ℙ(PORT)
axm3: portsCapacity ∈ SOFTWARE × PORT ↛ℕ
...
END
```

Fig. 6. The CPFOSScontext description

To formally model composite PFOS software, we first extend the *PFOSScontext* (see Fig. 6). We introduced a finite set, *PORT*, to denote all the ports that are provided or required by software. The *requiredPorts* function defines the set of ports that are required by a component in a given state. The *providedPorts* function defines the set of ports that are provided by a component in a given state. The *portsCapacity* function defines a set of capacity constraints on the required and provided ports of components.

```
MACHINE CPFOSSmachine REFINES PFOSSmachine
SEES CPFOSScontext
VARIABLES
CS states SB ports
INVARIANTS
type1: CS ∈ ℙ (SOFTWARE)
type2: ports ∈ ℙ (PORT)
type3: SB ∈ CS × CS ⇸ ports
type4: states ∈ CS ⇸ STATE
...
END
```

Fig. 7. The CPFOSSmachine description

Second, we refine the *PFOSSmachine* (see Fig. 7) in order to specify the dynamic aspect of composite PFOS software such as behavior and composition requirements (REQ1, REQ2, REQ3, REQ4, REQ5). The *CS* variable represents the set of PFOS software components. We use the *SB* variable to define the set of binding relations between software components. The *ports* variable denotes the set of ports with which the software components are bound. The types of these variables are defined using two invariants, namely *type2* and *type3* (see Fig. 7). For the sake of clarity, the invariants, defined in *CPFOSSmachine* and modeling the PFOS software composition requirements, are separately presented in Fig. 8.

```
REQ1: ∀c1, p·c1 ∈ CS ∧ states(c1) = installed ∧ p ∈ requiredPorts(c1 ↦ installed)
⇒ (∀c2·c2 ∈ CS ∧ SB(c2 ↦ c1) = p ⇒ states(c2) ∈ {installed, running})
REQ2: ∀c1·c1 ∈ CS ∧ (∃p, c2·p ∈ ports ∧ c2 ∈ CS ∧ SB(c1 ↦ c2) = p)
⇒ states(c1) ∈ {installed, running}
REQ3: ∀c1, p·c1 ∈ CS ∧ states(c1) = running ∧ p ∈ requiredPorts(c1 ↦ running)
⇒(∀c2·c2 ∈ CS ∧ SB(c2 ↦ c1) = p ⇒ states(c2) = running)
REQ4: ∀c1, p, s·c1 ∈ CS ∧ s ∈ STATE ∧ p ∈ providedPorts(c1 ↦ s)
⇒portsCapacity(c1 ↦ p) ≤ card({c2·c2 ∈ CS ∧ (c1 ↦ c2) ↦ p ∈ SB|c2})
REQ5: ∀c1, p, s·c1 ∈ CS ∧ s ∈ STATE ∧ p ∈ requiredPorts(c1 ↦ s)
⇒portsCapacity(c1 ↦ p) ≥ card({c2·c2 ∈ CS ∧ (c2 ↦ c1) ↦ p ∈ SB|c2})
```

Fig. 8. The Event-B invariants modeling the PFOS software composition requirements

4   Extending the Composite PFOS Software Model with Elasticity Properties

As we stated in the introduction, the main goal of this paper is the correctness verification of the (vertical and horizontal) elasticity properties of composite PFOS software. To achieve this goal, we extend the composite PFOS software model described in the previous section. This extension comprises two steps:

- Step 1: Modeling the vertical elasticity property
- Step 2: Modeling the horizontal elasticity property

In the following subsections, we present a detailed description of these steps.

4.1   Modeling the Vertical Elasticity Property

As stated in [4], the vertical elasticity is a rename of scalability. In fact, it is used to scale-up or down software applications deployed in Cloud by adding new resources and removing already used ones. The resources can be deplored in the Cloud such as (PFOS) software, hardware, database, and so on. In this work, we focus on the software resources. As claimed in [8], composite PFOS software can be scaled up and down automatically using the following operations:

- *create:* an operation for the creation of a new PFOS software.
- *destroy:* an operation for the deletion of an existing PFOS software.
- *bind:* an operation for the definition of a binding between two exiting PFOS software.
- *unbind:* an operation for the deletion of an existing binding.

The *bind* operation is used to define the binding relation between created software. The main purpose behind the use of the *unbind* operation is to release (delete) all of the binding relations of an uninstalled software to be removed later.

```
MACHINE VElasticityMachine REFINES CPFOSSmachine
SEES CPFOSScontext
VARIABLES
CS1 states1 SB1 ports1
INVARIANTS
type1: CS1 ∈ ℙ (SOFTWARE)
type2: states1 ∈ CS1 → STATE
type3: ports1 ∈ ℙ (PORT)
type4: SB1 ∈ CS1 × CS1 ⇸ ports1
glu1: CS = CS1
glu2: states = states1
glu3: ports = ports1
glu4: SB = SB1
. . .
END
```

Fig. 9. The VElasticityMachine description

To formally specify the vertical elasticity property for composite PFOS software, we created the new Event-B machine VElasticityMachine. This new machine sees the *CPFOSScontext* context. It refines the *CPFOSSmachine* machine by introducing a set of new events and variables (see Fig. 9). The abstract variables of *CPFOSSmachine*, CS, states, SB and ports, are respectively replaced in the refined *VElasticityMachine* machine with the following ones: CS1, states, SB1 and ports1. This replacement process is strongly required because the newly created events of the *VElasticityMachine* machine are not able to modify the variables of *CPFOSSmachine*. In addition to typing invariants (used to specify the type of each variable in the VElasticityMachine), a set of gluing invariants are also defined. These particular gluing invariants define an

```
create ≙ ANY ToBeCreated
WHERE
grd1: ToBeCreated ∈ SOFTWARE ∧ ToBeCreated ∉ CS1
THEN
act1: states1(ToBeCreated):=uninstalled
act2: CS1:=CS1∪ {ToBeCreated}
END
destroy ≙ ANY ToBeRemoved
WHERE
grd1: ToBeRemoved ∈ CS1 ∧ states1(ToBeRemoved)=uninstalled
grd2: ¬(∃c·c ∈ CS1 ∧ (ToBeRemoved ↦ c ∈ dom(SB1) ∨ c ↦ ToBeRemoved ∈ dom(SB1)))
THEN
act1: states1:={ToBeRemoved} ◁states1
act2: CS1:=CS1\ {ToBeRemoved}
END
bind ≙ ANY c1, c2, p
WHERE
grd1: {c1,c2} ⊆ dom(states1) ∧ c1 ≠ c2 ∧ p∈ PORT
grd2: ∃ s · s∈ STATE ∧ p∈ providedPorts(c1↦ s) ∩ requiredPorts(c2↦ s)
grd3: (c1 ↦ c2) ↦ p ∉ SB1
THEN
act1: ports1:=ports1∪ {p}
act2: SB1:=SB1∪ {(c1 ↦ c2) ↦ p}
END
unbind ≙ ANY c1, c2, p
WHERE
grd1: {c1,c2} ⊆ dom(states1) ∧ c1 ≠ c2 ∧ p∈ PORT
grd2: (c1 ↦ c2) ↦ p ∈ SB1 ∧states1(c2) = uninstalled
THEN
act1: ports1:=ran({c1↦ c2} ◁SB1)
act2: SB1:=SB1\ {(c1 ↦ c2) ↦ p}
END
```

Fig. 10. The formalization of the vertical elasticity operations with Event-B.

equality relation between the abstract and concrete variables (glu1, glu2, glu3, and glu4).

The vertical elasticity operations are formally specified as events (see Fig. 10). These events are defined while refining the *CPFOSSmachine*. The *create* event allows to add new components to the composite software. The added components are initially set to the *uninstalled* state. The *bind* event is used to define the required bindings between the new added components and some existing ones. In some cases, it is necessary to remove some components and their associated bindings using the *unbind* and the *destroy* events. The guards *grd*1 and *grd*2 of *destroy* event state that the only way to destroy a software is to uninstall it while releasing all of its binding relations using the *unbind* event.

4.2   Modeling the Horizontal Elasticity Property

As we explained in the introduction, the horizontal elasticity is another fundamental property provided by the Cloud paradigm. Such a property is used basically to increase or decrease the capacity of the deployed software applications by adding new copies of existing components, while removing unnecessary ones. More precisely, it is implemented using two operations: duplication and consolidation.

The duplication operation allows to increase the capacity of the composite softwares by duplicating some of its components. It can be considered as a special form of creating a component that is a copy of an existing one. The duplication of a given component implies the automatic duplication of all its bindings. The consolidation operation allows to decrease the capacity of the composite softwares through the deletion of unnecessary components copies. It can be considered as a special form of destroying a component, which is a copy of an existing one. The consolidation of a given component implies the consolidation of all its bindings.

```
duplicate ≙ REFINES create
ANY ToBeCreated, component
WHERE
. . .
grd2: component ∈ CS1
grd3:∀s·s ∈ STATE ∧ ToBeCreated ↦ s ∈ dom(requiredPorts) ∧ component ↦ s ∈
dom(requiredPorts)
  ⇒ requiredPorts(ToBeCreated ↦ s) = requiredPorts(component ↦ s)
grd4: ∀s·s ∈ STATE ∧ ToBeCreated ↦ s ∈ dom(providedPorts) ∧ component ↦ s ∈
dom(providedPorts)
  ⇒ providedPorts(ToBeCreated ↦ s) = providedPorts(component ↦ s)
THEN
. . .
act3: copies(component):=copies(component)∪ {ToBeCreated}
act4: SB1:=SB1 ∪ {c·c ∈ CS1 ∧ c ↦ component ∈ dom(SB1)|(c ↦ ToBeCreated) ↦
SB1(c ↦ component)} ∪ {c·c ∈ CS1 ∧ component ↦ c ∈ dom(SB1)|(ToBeCreated ↦
c) ↦ SB1(component ↦ c)}
END
```

Fig. 11. The formalization of the duplication operation.

To formally specify the horizontal elasticity property for composite PFOS software, we created the new Event-B machine $HElasticityMachine$. This new machine sees the $CPFOSScontext$ context. It refines the $VElasticityMachine$ machine by introducing a variable named $copies$ and refining the $create$ and $destroy$ events. This new variable is defined as a function from the set $CS1$ in $\mathbb{P}(CS1)$. It determines the set of all the copies of the exiting components. The purpose behind the refinement of the $create$ and $destroy$ events is to model the duplication and consolidation operations. It should be emphasized that the events create and destroy are already defined in the $HElasticityMachine$. The $grd3$ and $grd4$ guards of the duplicate event (see Fig. 11) express that the component to be duplicated ($component$) and the one to be created ($ToBeCreated$) shall provide and require exactly the same ports. The substitution action $act3$ of the duplication event adds a new copy ($ToBeCreated$) of $component$. The binding relations of such an added copy are those of duplicated component (see act4, Fig. 11). The $grd3$ guard of the $consolidate$ event (see Fig. 12) states that the component (specified by the input parameter $component$) to be removed is a copy of another existing one (denoted by the parameter $original$). The substitution actions of the consolidation event are those of the $destroy$ event.

```
consolidate ≙REFINES destroy
ANY ToBeRemoved, component
WHERE
· · ·
grd3: component ∈ CS1
grd4: ToBeRemoved ∈ copies(component)
THEN
· · ·
END
```

Fig. 12. The formalization of the consolidation operation.

5   Correctness Verification and Validation

In the previous two sections, we first proposed an Event-B model for verifying PFOS software composition correctness. Then, we have extended this model to cover the correctness of elasticity properties. The consistency of our PFOS software composition model and its extended version is mandatory in order to meet the attended correctness verification needs of the PFOS software composition and its elasticity properties. The adopted verification approach comprises two steps:

– Step 1: we use a proof-based verification.
– Step 2: we use an animation-based verification using ProB model-checker.

The following subsections expose a detailed description of these steps.

5.1   Proof-Based Verification

We have succeeded to combine the modeling and the proof-based verification to ensure the consistency of the Event-B models of PFOS software composition and the elasticity properties. The verification activity is based on the integrated proof of the Rodin platform. The proof-based verification consists in discharging a set of proof obligations (PO). It guarantees that:

– The initialization of a machine leads to a state where the invariant is valid.
– Assuming that the machine is in a state where the invariants are preserved, every enabled event leads to a state where the invariants are still preserved.

Rodin generates a set of proof obligations for every invariant that can be affected by an event, i.e. the invariant contains variables that can be changed by an event. The name of the proof obligation is then "evt/label/INV". It can be either automatically/interactively discharged (marked with ✓^{A}), or undischarged (marked with ✎^{A}). The symbol "A" means that the PO is automatically discharged. The goal of such a proof is to assert that when all affected variables are replaced by new values from the actions, the invariant still holds. Figure 13 reports the POs that are generated while proving the consistency of the $CPFOSSmachine$. All of these POs are discharged using the Rodin provers.

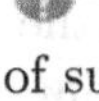

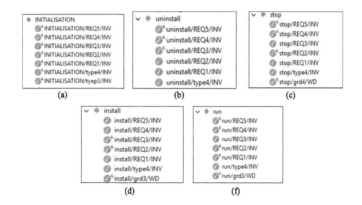

Fig. 13. All of the proof obligations of the *CPFOSSmachine* machine are discharged

We used the concept of refinement to model the elasticity properties by refining the composite PFOS software formalization. The concept of refinement allowed us to gradually introduce more details into a model. In order to ensure a correct refinement of the *CPFOSSmachine*, we must prove two things:

– The concrete events can only occur when the abstract ones occur.
– If a concrete event occurs, the abstract event can occur in such a way that the resulting states correspond again, i.e. the gluing invariant remains true.

The first condition concerns the guard strengthening of the *VElasticity Machine*. The goal of such a condition is to check if the *VElasticity Machine* behaves in a way that corresponds to the behavior of the *CPFOSSmachine*. Consequently, several POs are generated with the following form: *cEvent/aGrd/GRD* where *aGrd* is an abstract guard. For instance, the POs shown in Fig. 14a state that the concrete events *install* and *uninstall* of the *VElasticityMachine* can only occur when the abstract ones occur.

The second condition is related to the gluing invariant preservation by concretes. The goal of such a condition is to prove that the invariant of the concrete machine is valid when each occurrence of a modified variable is replaced by its new value. The resulting proof obligations have the following label form: *cEvt/cInv/INV* where *cEvt* is a concrete event and *cInv* is a concrete invariant (defined on the concrete variables). For instance, the proof obligations "destroy/glu1/INV" and "destroy/glu2/INV" (Fig. 14b) state that when the concrete event "destroy" of the *VElasticityMachine* machine occurs, the gluing invariants "glu1" and "glu2" can remain false. Similarly, the gluing invariants "glu1" and "glu2" can remain false when the create event of the *VElasticityMachine* is executed. This is justified by the fact that the proof obligations "create/glu1/INV" and "create/glu2/INV" are discharged neither automatically nor interactively (see Fig. 14b). This means that the *VElasticityMachine* is not correct.

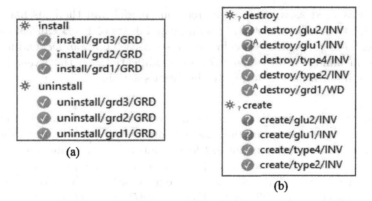

Fig. 14. Some proof obligations of the *VElasticityMachine* machine

Table 1. Summary of the results of the proof-based verification

Machines/contexts	POs	Proved	Unproved
PFOSSmachine	9	9	0
PFOSScontext	0	0	0
CPFOSSmachine	39	39	0
CPFOSScontext	0	0	0
VElasticityMachine	51	47	4
HElasticityMachine	11	11	0

Summarizing, the provers generate 110 POs (see Table 1). We notice that the work on POs is in progress: at this time, 95 of them are automatically or interactively discharged. The proof of 4 POs remains to be done. That does not mean that they are false, but that the Rodin provers simply don't succeed in demonstrating the rule: it may be due to the fact that some events are not handled in an efficient manner by the provers, or due to the fact that the heuristics used for the proof are not efficient enough in this case. An effort remains to be done to manually demonstrate the unproved POs.

5.2   Animation-Based Verification

As shown in the previous section, the provers fail to discharge automatically and interactively several POs. For this purpose, we complement the proof-based verification with an animation process by using ProB [16].

Overview of ProB. ProB is basically a PFOS software that can be composed with the Rodin platform to assist the designers in tracing possible specification errors causing the failure of the provers and then prove those POs. The counter-example generated by ProB that shows situations where the invariant is not satisfied was used as a guide to correct our model and then discharge some proof obligations. The constraint-solving capabilities of ProB can also be used for model finding, deadlock checking and test-case generation. Depending on the situation in hand, we have made several modifications on the Event-B model related to the invariant, guard or to the event's action.

Animation. The animation is performed on a concrete Event-B model. For this purpose, we substantiate our abstract model with the Wordpress application. This is done by refining the $HElasticityMachine$ machine and extending the $CPFOSScontext$ context in order to specify the main elements of such a particular application. The extended context, WordPressContext, is shown in Fig. 15. It is seen by the refined Machine ($WordPressMachine$).

```
CONTEXT WordPressContext
EXTENDS CPFOSScontext
CONSTANTS
Apache2 Mysql Wordpress httpd mysql_up mysql wordpress
AXIOMS
axm1 partition(SOFTWARE,{Apache2 },{ Mysql },{Wordpress })
axm2 partition(PORT,{httpd},{mysql_up }, {mysql }, {wordpress })
axm3 partition(requiredPorts,{(Wordpress↦installed)↦{httpd}},
{(Wordpress↦running)↦{httpd,mysql_up}})
. . .
END
```

Fig. 15. The WordPressContext description

We start the animation step by executing the INITIALISATION event. Afterward, we can interact with the model by triggering events. This is done by double-clicking on an enabled event or by right-clicking it and selecting a set of parameters, if applicable. We first trigger the event *create* with the parameter: Apache2. Figure 16 reports the results of triggering creation event. These results indicate that gluing invariants glu1 and glu2 are violated. This confirms the failure of the Rodin Provers while discharging the following POs: "create/glu1/INV" and "create/glu2/INV". To repair these specification errors, we modify the gluing invariants as follows:

- glu1: $CS \subseteq CS1$
- glu2: $states \subseteq states1$
- glu3: $ports \subseteq ports1$
- glu4: $SB \subseteq SB1$

With these redefinitions, we succeeded to discharge the four undischarged POs related to the refinement of the $CPFOSSmachine$.

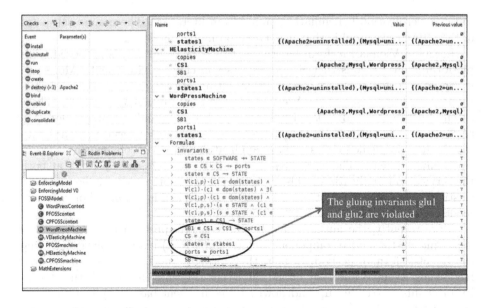

Fig. 16. ProB signals invariants violation

6   Related Work

Different approaches have been proposed to solve the configuration problem of software. The authors of [17] propose empirical data for the real power of open source style of software development. They mainly investigate the benefits of the structural code analysis. As claimed in [17], the quality of the code produced by an open source is lower than that expected from an industrial standard.

In [6], the authors developed a modular package management allowing for pluggable dependency solvers. They claim that this is the path that leads to the next generation of package managers that will deliver better results, accept more expressive input languages, and can be easily adaptable to new platforms. In [5,8,18], the authors focus on studying the deployment problem and the automatic reconfiguration of software in cloud environment. First, they propose a component-based model for cloud, called Aeolus, slightly inspired from the Fractal model. Then, they propose an algorithm for the verification and the construction of a correct configuration of PFOS software to be deployed in the cloud environment. The verification strategy conducted by such an algorithm is based on constructing all possible correct configurations of the PFOS software by applying a set of operators, namely create, destroy, bind, unbind and stateChange (an operation used to change the state of software). Then, it verifies whether the configuration to be checked is one of the constructed configurations. If so, the configuration is valid, otherwise it is invalid. In [19], the authors propose an open-source toolkit tool, called CPAchecker, for the verification of configurable software. They evaluated the efficiency of the CPAchecker tool for software-verification benchmarks coming from literature. The main weak point of the CPAchecker tool is that it is based on model-checking.

In [20], a Vienna Platform for Elastic Processes (ViePEP) is proposed. This platform allows to construct elastic processes in cloud environments. Moreover, it can be used for resources consumption optimization by using a set of elasticity actions. Extensions of the ViePEP platform are proposed by [21,22]. These extensions are knowledge-based prediction, scheduling and resources-allocation algorithms. The authors of [4] proposed the elasticity as a service (ElaaS) concept that is implemented as a SaaS application. This concept can be used in cloud environments. The work presented in [23] considers the elasticity at both the service and application levels. Nevertheless, the correctness of the proposed elasticity mechanisms is not proved since the approach is not based on a formal model.

More recently, the authors of [24] proposed an approach for verifying and deploying elastic service component architecture (SCA)-based applications using the Event-B formal method. In fact, they formally model the component artifacts using Event-B and they define the Event-B events that model the elasticity mechanisms (scaling up and down) for component-based applications. The main shortcoming of [24] is that it basically does not support the port capacity requirements.

The work we proposed in this paper is new in the sense that it (1) tackles the problem of elasticity of the PFOS software applications at the SaaS level, (2) is based on a formal model and (3) proposes a formal verification of the elasticity mechanisms.

7   Conclusion

In this paper, we proposed a novel approach to ensure the correctness of elastic composite PFOS applications. The approach consists in a correct-by-construction elastic composite PFOS applications model built on Event-B. This formal model explores four abstraction levels. At a first level, we defined an abstract model for the PFOS software. At a second level, we refined this abstract model for the definition of the PFOS software composition process and its correctness requirements (capacity and dependency requirements). At a third level, we formally defined the vertical elasticity mechanisms by refining the composite PFOS software model further. The vertical elasticity mechanisms allow to scale-up and down a composite PFOS application by adding and removing components and bindings. At a fourth level, we extended the formalization of vertical elasticity capabilities with horizontal elasticity mechanisms that consist in duplication and consolidation operations. We have succeeded to combine the modeling and verification activities to ensure the consistency of the Event-B models of PFOS software composition and the elasticity properties. The verification activity is based on the integrated proof of the Rodin platform. We complemented this verification activity with an animation process using the ProB model-checker to trace possible modeling errors.

We are currently working on developing a tool that automates the Event-B based verification process and an automatic deployment framework of elastic composite PFOS software applications. The proposal of a PFOS software selection method could be interesting extension to the proposed approach.

References

1. Fox, A., Griffith, R., Joseph, A., Katz, R., Konwinski, A., Lee, G., Patterson, D., Rabkin, A., Stoica, I.: Above the clouds: a Berkeley view of cloud computing. Report UCB/EECS 28(13) 2009, Department of Electrical Engineering and Computer Science, University of California, Berkeley (2009)
2. Rimal, B.P., Choi, E., Lumb, I.: A taxonomy and survey of cloud computing systems. NCM **9**, 44–51 (2009)
3. Dustdar, S., Guo, Y., Satzger, B., Truong, H.L.: Principles of elastic processes. IEEE Internet Comput. **15**(5), 66–71 (2011)
4. Kranas, P., Anagnostopoulos, V., Menychtas, A., Varvarigou, T.: ElaaS: an innovative elasticity as a service framework for dynamic management across the cloud stack layers. In: 2012 Sixth International Conference on Complex, Intelligent and Software Intensive Systems (CISIS), pp. 1042–1049. IEEE (2012)
5. Di Cosmo, R., Mauro, J., Zacchiroli, S., Zavattaro, G.: Aeolus: a component model for the cloud. Inf. Comput. **239**, 100–121 (2014)
6. Abate, P., Di Cosmo, R., Treinen, R., Zacchiroli, S.: MPM: a modular package manager. In: 14th International ACM SIGSOFT Symposium on Component Based Software Engineering (CBSE 2011) (2011)
7. Krueger, C.W.: Software reuse. ACM Comput. Surv. **24**(2), 131–183 (1992)
8. Di Cosmo, R., Zacchiroli, S., Zavattaro, G.: Towards a formal component model for the cloud. In: Eleftherakis, G., Hinchey, M., Holcombe, M. (eds.) SEFM 2012. LNCS, vol. 7504, pp. 156–171. Springer, Heidelberg (2012). doi:10.1007/978-3-642-33826-7_11
9. Abrial, J.-R., Butler, M., Hallerstede, S., Voisin, L.: An open extensible tool environment for Event-B. In: Liu, Z., He, J. (eds.) ICFEM 2006. LNCS, vol. 4260, pp. 588–605. Springer, Heidelberg (2006). doi:10.1007/11901433_32
10. Smith, D., Milinkovich, M.: Eclipse: a premier open source community. Open Source Business Resource, July 2007
11. Wiegand, J., et al.: Eclipse: a platform for integrating development tools. IBM Syst. J. **43**(2), 371–383 (2004)
12. Abrial, J.: Modeling in Event-B: System and Software Engineering. Cambridge University Press, Cambridge (2010)
13. Méry, D., Singh, N.K.: EB2C: a tool for Event-B to C conversion support. In: 8th IEEE International Conference on Software Engineering and Formal Methods (SEFM) Poster and Tool Demo submission. Published in a CNR Technical Report (2010)
14. Méry, D., Singh, N.K.: EB2J: code generation from Event-b to Java. In: SBMF - Brazilian Symposium on Formal Methods, CBSoft - Brazilian Conference on Software: Theory and Practice, Sao Paulo, Brazil (2011)
15. Di Cosmo, R., Zacchiroli, S., Trezentos, P.: Package upgrades in FOSS distributions: details and challenges. In: Proceedings of the 1st International Workshop on Hot Topics in Software Upgrades, p. 7. ACM (2008)

16. Leuschel, M., Butler, M.: ProB: a model checker for B. In: Araki, K., Gnesi, S., Mandrioli, D. (eds.) FME 2003. LNCS, vol. 2805, pp. 855–874. Springer, Heidelberg (2003). doi:10.1007/978-3-540-45236-2_46
17. Stamelos, I., Angelis, L., Oikonomou, A., Bleris, G.L.: Code quality analysis in open source software development. Inf. Syst. J. 12(1), 43–60 (2002)
18. Di Cosmo, R., Mauro, J., Zacchiroli, S., Zavattaro, G.: Component reconfiguration in the presence of conflicts. In: Fomin, F.V., Freivalds, R., Kwiatkowska, M., Peleg, D. (eds.) ICALP 2013. LNCS, vol. 7966, pp. 187–198. Springer, Heidelberg (2013). doi:10.1007/978-3-642-39212-2_19
19. Beyer, D., Keremoglu, M.E.: CPACHECKER: a tool for configurable software verification. In: Gopalakrishnan, G., Qadeer, S. (eds.) CAV 2011. LNCS, vol. 6806, pp. 184–190. Springer, Heidelberg (2011). doi:10.1007/978-3-642-22110-1_16
20. Schulte, S., Hoenisch, P., Venugopal, S., Dustdar, S.: Introducing the Vienna platform for elastic processes. In: Ghose, A., Zhu, H., Yu, Q., Delis, A., Sheng, Q.Z., Perrin, O., Wang, J., Wang, Y. (eds.) ICSOC 2012. LNCS, vol. 7759, pp. 179–190. Springer, Heidelberg (2013). doi:10.1007/978-3-642-37804-1_19
21. Hoenisch, P., Schulte, S., Dustdar, S., Venugopal, S.: Self-adaptive resource allocation for elastic process execution. In: 2013 IEEE sixth International Conference on Cloud Computing (CLOUD), pp. 220–227. IEEE (2013)
22. Hoenisch, P., Schulte, S., Dustdar, S.: Workflow scheduling and resource allocation for cloud-based execution of elastic processes. In: 2013 IEEE 6th International Conference on Service-Oriented Computing and Applications (SOCA), pp. 1–8. IEEE (2013)
23. Tsai, W.T., Sun, X., Shao, Q., Qi, G.: Two-tier multi-tenancy scaling and load balancing. In: 2010 IEEE 7th International Conference on e-Business Engineering (ICEBE), pp. 484–489. IEEE (2010)
24. Graiet, M., Hamel, L., Mammar, A., Tata, S.: A verification and deployment approach for elastic component-based applications. Formal Aspects Comput. 1–25 (2017)

An Optimization-Based Approach for Cloud Solution Design

Aly Megahed[(⊠)], Ahmed Nazeem, Peifeng Yin, Samir Tata,
Hamid Reza Motahari Nezhad, and Taiga Nakamura

IBM Research - Almaden, 650 Harry Rd, San Jose, CA 95120, USA
{aly.megahed,stata,motahari,taiga}@us.ibm.com, Ahmed.Nazeem@ibm.com

Abstract. Over the last few years, the rate of adoption of cloud computing and specifically managed cloud hosting and migration has accelerated significantly. Consequently, many enterprises/clients choose to migrate their local IT workload to cloud platforms. Such a migration consists in considering a cloud offering and come up with a cloud solution. In industrial settings, cloud designer may spend one day or so to come up with an acceptable cloud solution with a reduced cost/price. It is obvious that such a process is error prone, is time consuming and does not guarantee an optimal solution, e.g. a solution with a minimum cost/price. In this paper, we propose an approach and an optimization approach for cloud solution design that satisfy client requirements and cloud offering constraints and producing a solution, if there is any, with a minimum cost in a reasonable time. In addition, we present the implementation of our optimization method as well as two baseline methods and compare the results of applying all three methods on realistic data.

Keywords: Cloud solution design · Cloud computing · Optimization · Integer programming · Operations research

1 Introduction

Over the last few years, the rate of adoption of cloud computing and specifically managed cloud hosting and migration among enterprise customers has accelerated significantly. In the enterprise marketplace, managed cloud services are common in which an enterprise migrates some or all their applications from their own data centers to cloud and expects the cloud providers to manage it for them [11].

The problem of hosting workload of an enterprise customer (whether new workload or migration) into an enterprise cloud service provider cloud is a complex and challenging problem. This is because the hosting entails provision of compute resources (e.g., virtual machines), storage, and network, and running client applications with desired service levels on potentially shared infrastructure components. The managed cloud hosting calls for monitoring and management tools at various levels from infrastructure, compute, storage, networking and application levels to provide transparency and monitoring of such service levels.

© Springer International Publishing AG 2017
H. Panetto et al. (Eds.): OTM 2017 Conferences, Part I, LNCS 10573, pp. 751–764, 2017.
https://doi.org/10.1007/978-3-319-69462-7_47

The migration job, in addition to new workload hosting, requires analyzing the existing customer workload, and infrastructure needs and configurations and map, upgrade or optimize those in the service provide environment which adds to the complexity. Last but not least, the price of cloud migration or hosting job for a customer, and cost of running such a workload for the service provider, is a key factor and driver of any cloud hosting or migration task.

Any cloud hosting or migration project entails a technical solution design phase. The proposed technical solution to an enterprise customer details the required compute, storage, networking, application, monitoring and management applications and resources along with associated service levels for each. The solution design process starts with capturing and documenting client requirements. The requirements include IT requirements, i.e. the specification of the needed IT resources (at all level of stack from infrastructure to application), or existing client environment (in case of a migration project), and business-level requirements and objectives. While there are multiple business level objectives, the most common is cost saving as the result of migrating to or adopting cloud, as opposed to on-premise data center operation.

The problem that we tackle in this paper is that of automatically computing a cost optimized cloud solution for a given client IT requirements (including the delivery locations, and all levels of infrastructure and applications needs) by finding the optimized combination of solution components, from a cost point of view, offered by a service provider that meets client's requirements. We model IT requirements of a client as a set of constraints expressed over a generic model of functional and non-functional requirements of IT resources and applications. The solution elements of the service provider are expressed over a generic cloud IT resources model. In this model, various capabilities are modeled as objects with their variations in terms of different acceptable values for any given object's attribute. There are also solutioning rules that constrain value selection and enforce generation of valid solution combinations.

In large client cloud hosting or migration projects, solving the solution design problem is a complex and tedious task. This is because given a specific client IT requirement, and different solution components and cloud delivery locations of a service provider, it is often possible to generate multiple solution alternatives. Generating a detailed technical solution design for a cloud project can take a team of IT architects days or weeks (depending on the scope and complexity of requirements). Also, it is not guaranteed that the solution architects can find the optimal (or the best alternative that is available among multiple possible solutions) after such an exercise, both in terms of price for the customer and cost of solution delivery for the client.

In the literature, while there are multiple work (e.g., [13,17,19]) that comes up with a price optimized solution for a client in leveraging a public cloud resources (some by considering the mix of on-demand and reserved resources (e.g. [17]), they all tackle the problem from a cloud client or consumer point of view. In this paper, we propose a novel cloud solution design approach that models and solves a constraint-satisfaction problem that produces an cost/price optimized cloud solution, from a service provider point of view. The key advantage of our method is proposing a generic model for IT requirements and cloud offerings and solution elements over which customized client requirements can be expressed, and formulate the problem of cost/price optimized solution design as an integer programming problem. We have implemented the proposed approach, and present the result of experiments and evaluation that shows the practicality of the proposed solution.

The rest of paper is structured as follows. Section 2 presents the state of the art related to the issue of migrating local IT environment to cloud platforms. Section 3 details our optimization-based approach for cloud solution design. Section 4 presents the implementation of our approach and its experimentation with respect to two baseline solution methods that we have also implemented. Finally, Sect. 5 concludes our paper and presents our future works.

2 Literature Review

Many enterprises choose to migrate their local IT environment to a cloud platform due to its advantage on flexible scalability and low cost. The migration process also attracts researchers' attention and many works focus on solving different research issues.

Early works are usually a report or case study, analyzing the whole migration process of a particular application/environment. In [9], Khajeh-Hosseini et al. studied the migration of an IT system in oil & gas industry in terms of the benefit and risk. In [3], the authors reported the experience of migrating Hackystat, an Open Source Software (OSS) framework, to the cloud.

Considering both benefit and risk, many works develop diversified tools and frameworks to facilitate the decision making process. Saripalli and Pingali [18] apply the multiple attribute decision methodology to help decision making with respect to different cloud decision objectives. In [2], a general decision process, CloudStep, is presented to support legacy application migration to cloud. Other works focus on a particular factor such as cost of deployment [8,10], network [1], security and privacy [12,14,16].

Besides high-level analysis, there are works aiming at reconfiguration of existing systems when migrated to cloud. In [7], Frey et al. show the unmodified system has scalability issues (either under-provision or over-provision) after migrating to cloud. They propose several heuristic rules to improve resource efficiency. Trummer et al. [19] model the application deployment as a constraint-satisfying optimization problem and rely on the constraint solver to get optimal solution in terms of cost. Aniceto et al. [17] aim at optimizing the mixture of on-demand and reserved instance in cloud to cover variable computation tasks at minimum cost. In [13], an auto-scaling mechanism is proposed for VM start-up and shutdown activities in order to complete scientific computation tasks within time and budget constraints.

Other works adopt diversified techniques such as evolutionary optimization [5,20,21], particle swarm optimization [15], multi-goal genetic search algorithm [6] and so on.

In this work, we also focus on the optimization problem. However, different from existing works that solve the problem from the user angle, we solve it from the aspect of a cloud provider, who aims to offer customized cloud solutions for different user requirements at low cost. Such difference requires a unique way of problem modeling. By analyzing real business data, we abstract it into a general attribute-value combination problem and takes advantage of powerful integer programming to solve it. The general form of the optimization model allows variant definitions of customer requirements as well as cloud offerings. Evaluation shows promising results and practical potential.

3   Approach for Cloud Solution Design

3.1   Approach Overview

Cloud providers, such as IBM and Amazon, deliver multiple Cloud offerings that provide users and companies with access to an integrated set of managed IT resources including infrastructure, platform and applications. Managed IT infrastructure resources include virtual machines and network. Managed platform resources include middleware and database.

The general process a cloud provider follows to deliver a cloud solution is presented in Fig. 1. It consists in three steps: Requirement capturing, Solution design and Delivery specification.

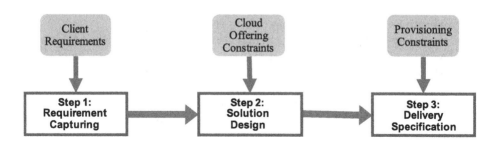

Fig. 1. Approach overview of cloud solutioning

Step 1: Requirement Capturing. The first step in a cloud sales deal is capturing the client requirements including application hosting, infrastructure needs, service level requirements, the need for disaster recovery, database resiliency, backup, etc. This consists in considering a set of attributes that characterize cloud offerings. The requirement consists in a set of constraints on the values those attributes can take. For example, if we consider a VMware offering in IBM Cloud, we may consider a set of attributes to formally describes the offering. These attributes may include:

- the *data center* selected by the client to host his/her workloads,
- the VMware *offering type*, with values such as *vCenter* and *vSphere*,

- the *number of clusters* requested by the client,
- the *number of virtual machines* per cluster to migrate and their characteristics in terms of CPU, RAM and storage,
- the *server size*, with values such as *Small*, *Standard*, *Medium*, and *Large*,
- the *server total cores* of a server,
- the *storage type*, with values such as *VSAN*, *Endurance* and *ISA*,
- the *operation System*, with values such as *AIX*, RHL and Win,
- the *Disaster Recovery*, with values such as *Yes* and *No*.
- etc.

Step 2: Solution Design. The second step, within cloud solutioning, is solution design. This step consists in determining the set of accepted/possible values of the list of attributes that describe the considered cloud offering.

Beside client requirement that define attribute values to be included or to be excluded in the solution, there are cloud offering constraints that should be also satisfied by the solution. There are mainly two types of cloud offering constraints. The first type concerns the constraints that define sets of combinations of attributes and their corresponding values that must be either included together or excluded together in the solution. For example, let's consider the attributes *server size* and *server total cores* with the corresponding values *Small* and *12*. This combination of attributes and values are either included together or excluded together in the solution. The second type of cloud offering constraints concerns constraints that define sets of combinations of attributes and their values are never included together in the solution. For example, not all VMware offerings are available in all possible data centers, not all server sizes are available with all VMware offerings.

Step 3: Delivery Specification. The third step, consists in specifying the delivery information necessary to provision the designed cloud solution. This specification includes the point of delivery, or PoD, where the servers will be provisioned, the IP addresses of servers, etc.

3.2 Need for Optimization

In the general case, solution design may come up with multiple possible solutions that satisfy client requirements and cloud offering constraints. The difference between these solutions would be the cost/price. In industrial settings, cloud designers may spend approximatively one day to come up with a solution with a reduced cost/price. It is obvious that such a process is error prone, is time consuming and does not guarantee an optimal solution, e.g. a solution with a minimum cost/price.

The objective of this paper is to tackle these issues, by providing an optimization model for cloud solution design that satisfy client requirements and cloud offering constraints and producing a solution, if there is any, with a minimum cost in a reasonable time.

3.3   Problem Abstraction and Optimization Model

In this subsection, we first abstract our problem, then we present the notation of our optimization model, and finally we formulate that model.

Problem Abstraction. There are multiple solution attributes. We denote the set of solution attributes as the set S. For each solution attribute $s \in S$, there are multiple possible values that can be chosen. However, in the solution to be proposed to the client, only one of these values (or none) is chosen for each attribute. Let the set V_s be the set of possible values for attribute $s \in S$.

There is an associated cost codes for multiple combinations of values for a subset of the attributes. Let the set of cost codes be F. Each cost code $f \in F$ has a cost $cost_f$ and the total solution cost is the sum of the costs of all cost codes that are enabled in the solution. A cost code is enabled whenever all of the combinations it is defined upon are included in the solution. For each cost code $f \in F$, we define a set C_f for all combinations of attributes and their values that enable it. That is, an element $(v, s) \in C_f$ is a tuple of the attribute $s \in S$ and its corresponding value $v \in V_s$ that form one of the enablers of that cost code $f \in F$.

We define the set NA as the set of combinations of all attributes $s \in S$ and their corresponding values $v \in V_S$ that are not allowed to be in our solution. Similarly, we define the set oMI as the set of combinations of attributes $s \in S$ and their corresponding values $v \in V_s$ that must be included in the current solution. Set IT is the set of combinations of attributes and their corresponding values that must be either included together or discluded together in the solution. That is, for each element (s, v, s', v') of this set, if solution attribute $s \in S$ and its corresponding value $v \in V_s$ is included in the solution, then attribute $s' \in S \setminus \{s\}$, with its corresponding value $v' \in V'_s$, must be included in the solution, while if the former is not included, the latter must be excluded, too. Lastly, set NT is the set of combinations of attributes and their corresponding values that are not allowed to be included together in a solution, i.e., for each element (s, v, s', v') of that set, if solution attribute $s \in S$ and its corresponding value $v \in V_s$ is included in the solution, then attribute $s' \in S \setminus \{s\}$, with its corresponding value $v' \in V'_s$, cannot be included in the solution, and vice versa. Table 1 summarizes the sets used in our model. Apart from these aforementioned sets, the only other parameter/data input to the model is the cost code values, $cost_f$, for each cost code $f \in F$.

With the aforementioned dynamics, our problem becomes: which attributes and their values should be included in the solution in order to minimize the total solution cost while satisfying all the given solution constraints? To solve this problem, we formulate and solve an integer programming (IP) optimization model. We next provide the notation for our model then present its mathematical formulation.

Model Notation. We define the following two sets of decision variables before formulation our optimization model to solve our problem: variable Y_{vs} is 1,

Table 1. Sets of our optimization model

Set Name	Set description
S	Set of solution attributes
V_s	Set of values for solution attribute $s \in S$
F	Set of cost codes
C_f	Set of combinations of all attributes and their corresponding values that enable cost code $f \in F$
NA	Set of combinations of all attributes and their corresponding values that are not allowed to be in our solution
MI	Set of combinations of all attributes and their corresponding values that must be included in our solution
IT	Set of combinations of attributes and their corresponding values that must be either included together or discluded together in the solution
NT	Set of combinations of attributes and their corresponding values that are not allowed to be included together in a solution

if value $v \in V_s$ for attribute $s \in S$ is included in our solution, and zero, otherwise. Variable X_f is 1, if cost code $f \in F$ is enabled in our solution, and zero, otherwise. Table 2 summarizes the decision variables of our solutionn model.

Model Formulation. We now present the formulation of our IP optimization model as follows:

$$Min \sum_{f \in F} cost_f . X_f \tag{1}$$

$$s.t. \sum_{v \in V_s} Y_{vs} = 1, \quad \forall s \in S \tag{2}$$

$$X_f \geq \sum_{(v,s) \in C_f} Y_{vs} - (|C_f| - 1), \quad \forall f \in F \tag{3}$$

$$Y_{vs} = 0, \quad \forall (v,s) \in NA \tag{4}$$

$$Y_{vs} = 1, \quad \forall (v,s) \in MI \tag{5}$$

$$Y_{vs} = Y_{v's'}, \quad \forall (v,s,v',s') \in IT \tag{6}$$

$$Y_{vs} + Y_{v's'} \leq 1, \quad \forall (s,v,s',v') \in NT \tag{7}$$

$$X_f \in \{0,1\}, \quad \forall f \in F \tag{8}$$

$$Y_{vs} \in \{0,1\}, \quad \forall s \in S, \quad \forall v \in V_s \tag{9}$$

Where objective function 1 minimizes the total cost of the solution, which is the sum of the costs of all cost codes included in that solution. Constraint 2 ensures that exactly one value is chosen for each solution attribute. Constraint 3 sets the logic of enabling a cost code. That is, a cost code is enabled if and only if all combinations of attributes and their values, that enable that cost

Table 2. Decision variables of our optimization model

Variable name	Variable description
Y_{vs}	$\begin{cases} 1, & \textit{if value } v \in V_s \textit{ for attribute } s \in S \textit{ is included in our solution, and} \\ 0, & \textit{otherwise} \end{cases}$
X_f	$\begin{cases} 1, & \textit{if cost code } f \in F \textit{ is enabled in our solution, and} \\ 0, & \textit{otherwise} \end{cases}$

code, are included in the solution. Constraint 4 ensures that attributes and their values, that are not wanted by the client, are not included in the solution. Similarly, constraint 5 forces attribute values, requested by the client, to be included in the solution. Constraint 6 ensures that for each combination in set IT, the corresponding attributes and values are either included together or discluded together. Constraint 7 ensures that for each combination in set NT, the corresponding attributes and their values are never included together in the solution. Lastly, constraints 8 and 9 are the binary constraints for our decision variables.

4 Implementation and Experimentation

In this section, we first describe the data, we used in our experimentation, we collected from a real cloud solutioning application of one of the world's largest cloud providers in Sect. 4.1. We then describe two baseline solution methods for our problem in Sect. 4.2. Lastly, we provide the implementation of our optimization method as well as the two baseline methods and compare the results of applying all three methods on the aforementioned realistic data in Sect. 4.3.

4.1 Data Collection

For experiment, we collect three types of data: (i) attribute and value domain data for defining solution, (ii) constraint to define a valid solution and (iii) combination of attribute-value pairs to obtain the price.

For the first one, we analyze real business data records and summarize 39 attributes. On average each attribute has a domain of 9.36 value choices. Thus in total there are about $\sim 10^{28}$ combinations, which is a huge search space. The second and third parts can be combined, since the illegal combinations would not have price. After analyzing existing pricing document, we finally obtain 209 price possibilities, of which each one is associated with 4.44 pairs of solution attribute & value.

Tables 3 and 4 respectively show examples of attribute-domain and its associated pricing rule. For confidentiality issue, we replace the real price number with different letters, representing different price.

4.2 Baseline Solution Methods

In this section, we describe two baseline methods that we compare to our optimization method in the next section. The first baseline method is simply choosing

Table 3. Example of attribute domain

Attribute	Domain
Operation System (OS)	AIX 7, RHL 6, Win 2012
Data Center Type (DCT)	Type I, Type II
Disaster Recovery (DR)	Y, N

Table 4. Example of price

OS	DCT	DR	Price
Win 2012	Type I	N	x
AIX 7	Type I	Y	y
RHL 6	Type II	Y	y

feasible attribute values at random from every attribute, putting in consideration not to choose the values that are not allowed to be chosen and choosing those that we must choose. Algorithm 1 illustrates this method. Note that we use the same notation in this section as the one we used in Sect. 3.3.

In this algorithm, we first choose the values for the attributes that have required values to be chosen (see set MI). Then, we eliminate, from our choices for the remaining attributes, the values for the attributes that have values that have to be eliminated (see set NA). Finally, we choose at random the values for the attributes that we have not chosen values for yet. To calculate the cost of our solution, we iterate over all cost codes, check whether a cost code is enabled, and add its cost, if it is.

The second baseline method that we present next is a greedy heuristic for choosing cheap solution. This heuristic is illustrated in Algorithm 2, where the idea is after choosing the values that must be chosen in the set MI, we sort the cost codes in an ascending order of their cost value and iterate on these codes enabling them one by one until all attributes have chosen values for them. That is, whenever any cost code is enabled, all of its value/attribute pairs are chosen. That is why we enable it only if all of the attributes in it have not been assigned values earlier and also if all the values of its attributes are not in the forbidden set NA.

Note that in these two methods, we respect all the different constraints, except for those in the sets NT and IT above, which do not exist in our realistic data anyway and thus we ignored them in our baseline methods. Also, it is obvious that it is not trivial to construct similar baseline methods while putting sets NT and IT in consideration. This is, therefore, one of the strong aspects of our optimization model, that it puts all these sets in consideration while giving the optimal solution at minimum cost. Further, such optimal solution could be significantly lower than those obtained by the *limited* baseline methods that we are presenting here, as we will show in the next subsection.

4.3 Implementation and Experiments

We implemented our optimization method as well as the two baseline methods. All implementations were done in the *python* programming language. We solve the optimization model using the commercial solver CPLEX [4]. We then constructed some experiments to examine the performance of our optimization method compared to the two baseline method.

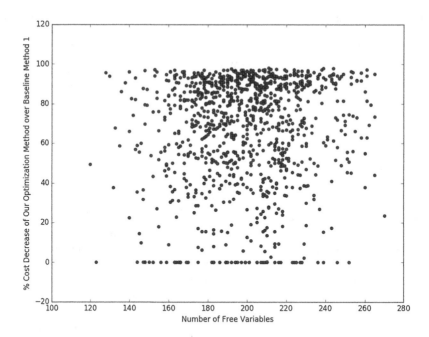

Fig. 2. % Cost decrease of our optimization method over baseline method 1

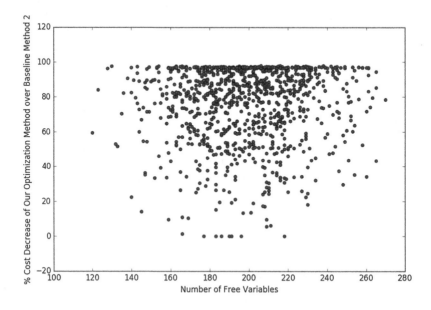

Fig. 3. % Cost decrease of our optimization method over baseline method 2

Algorithm 1. Baseline Solution Method 1

1: *Initialize* :
2:     $RemainingAttributes = S$
3:     $OurSolution = \phi$
4:     $TotalCost = 0$
5: **for** Attribute $s' \in RemainingAttributes$ **do**
6:     **for** $v' \in V_{s'}$ **do**
7:         **if** $((v', s') \in MI)$ **then**
8:             $OurSolution = OurSolution \cup \{(v', s')\}$
9:             **Break**
10:        **end if**
11:    **end for**
12: **end for**
13: $RemainingAttributes = RemainingAttributes \backslash \{s'\}$
14: **while** $RemainingAttributes \neq \phi$ **do**
15:    Choose attribute $s' \in RemainingAttributes$
16:    $V_{s'} = V_{s'} \backslash \{v' : (v', s) \in NA\}$
17:    Choose a random value $v'' \in V_{s'}$ for attribute s'
18:    $OurSolution = OurSolution \cup \{(v'', s')\}$
19:    $RemainingAttributes = RemainingAttributes \backslash \{s'\}$
20: **end while**
21: **for** Cost code $f \in F$ **do**
22:    **if** $(\forall (v, s) \in C_f, (v, s) \in OurSolution)$ **then**
23:        $TotalCost = TotalCost + cost_f$
24:    **end if**
25: **end for**

We explain our experimental design as follows: We constructed 1000 problem instances using our aforementioned realistic data for Cloud solutioning attributes and their values, where in each instance, we restrict the values that can be chosen for each attribute to a random subset of these values. Therefore, the number of free variables that the model chooses from vary from one instance to another. All of our experiments were run on a Macbook Pro computer with 16 GB of RAM and a 2.5 GHz Intel core i7 processor.

We found that our optimization model is robust; it solves any of these instances in less than 1 s. We also found that the optimal cost could be significantly lower than that obtained by each of the two baseline methods.

Figures 2 and 3 show a scatter plot for the percentage of cost decrease of our optimization method over the first and second baseline methods, respectively as a function of the free variables. Table 5 summarizes the statistics of these results. Looking at these results, one can obviously tell the effectiveness of our optimization method given how huge of a cost saving it could provide for the solution provider when used instead of any of the baseline methods. Also, given how fast our model gets solved in, the efficiency of our method becomes clear, especially compared to the hours (or few days) it used to take human solutioners to get a feasible solution for the problem.

Algorithm 2. Baseline Solution Method 2

1: *Initialize* :
2:     $RemainingCostCodes = F$
3:     $RemainingAttributes = S$
4:     $OurSolution = \phi$
5:     $TotalCost = 0$
6: **for** Attribute $s' \in RemainingAttributes$ **do**
7:     **for** $v' \in V_{s'}$ **do**
8:         **if** $((v', s') \in MI)$ **then**
9:             $OurSolution = OurSolution \cup \{(v', s')\}$
10:            **Break**
11:        **end if**
12:    **end for**
13: **end for**
14: Sort the cost codes in set F in ascending order of cost in $cost_f$
15: **for** $f \in RemainingCostCodes$ (where we iterate on the sorted set F) **do**
16:    **if** $(RemainingAttributes = \phi)$ **then**
17:        **Break**
18:    **else if** ($(v, s) \notin OurSolution\ \forall v \in V_s \land \forall s : (v, s) \in C_f) \land ((v, s) \notin NA\ \forall (v, s) \in C_f)$) **then**
19:        **for** $(v, s) \in C_f$ **do**
20:            $OurSolution = OurSolution \cup \{(v, s)\}$
21:            $RemainingAttributes = RemainingAttributes \backslash \{s\}$
22:            $RemainingCostCodes = RemainingCostCodes \backslash \{f\}$
23:        **end for**
24:        $TotalCost = TotalCost + cost_f$
25:    **end if**
26: **end for**
27: **while** $RemainingAttributes \neq \phi$ **do**
28:    Choose at random attribute $s'' \in RemainingAttributes$
29:    Choose at random value $v'' \in V_s'' : (v'', s'') \notin NA$
30:    $OurSolution = OurSolution \cup \{(v'', s'')\}$
31:    $RemainingAttributes = RemainingAttributes \backslash \{s\}$
32: **end while**

Table 5. Statistics of comparisons of our optimization method versus the two baseline methods

	% Cost decrease of our optimization method over baseline method 1	% Cost decrease of our optimization method over baseline method 2
Minimum	0.00	0.00
Maximum	0.98	0.98
Average	0.66	0.76
Standard deviation	0.27	0.22

5   Conclusions and Future Work

We tackled in this paper the issue of cost optimization of a cloud solution for a given client IT requirements. Our contribution consists in a novel approach that consists in finding the optimized combination of solution components, from a cost point of view, offered by a service provider that meets client's requirements. We have implemented our optimization method as well as two baseline methods and compared the results of applying all three methods on realistic data. Experimentation results show the effectiveness of our optimization method that provide how huge cost savings for the solution provider compared to the baseline methods. In addition, the experimentations show the efficiency of our approach that executes in less than one second compared to the hours (to few days) it used to take human solutioners to get a feasible solution for the considered problem.

We solved the problem of solution cost optimization, in which we were trying to find the minimum possible solution that satisfied the client requirements. After solution providers solve such costing minimization problem, they need to "price" their solution. That is, they need to add some gross profit on top of the cost in order to reach the price that they will offer to clients. Obviously, the higher the price, the lower their chances of selling their solution versus other competitors. Thus, a research question, that is a natural extension of our work, is: what is the optimal price (or added gross profit on top of the optimal cost) that would increase the chance of successfully winning the deal of selling the cloud service to clients?

References

1. Ahmed, E., Akhunzada, A., Whaiduzzaman, M., Gani, A., Ab Hamid, S.H., Buyya, R.: Network-centric performance analysis of runtime application migration in mobile cloud computing. Simul. Model. Pract. Theory **50**, 42–56 (2015)
2. Beserra, P.V., Camara, A., Ximenes, R., Albuquerque, A.B., Mendonça, N.C.: Cloudstep: a step-by-step decision process to support legacy application migration to the cloud. In: 2012 IEEE 6th International Workshop on the Maintenance and Evolution of Service-Oriented and Cloud-Based Systems (MESOCA), pp. 7–16. IEEE (2012)
3. Chauhan, M.A., Babar, M.A.: Migrating service-oriented system to cloud computing: an experience report. In: 2011 IEEE International Conference on Cloud Computing (CLOUD), pp. 404–411. IEEE (2011)
4. IBMI CPLEX: V12. 1: user's manual for CPLEX. Int. Bus. Mach. Corp. **46**(53), 157 (2009)
5. Csorba, M.J., Meling, H., Heegaard, P.E.: Ant system for service deployment in private and public clouds. In: Proceedings of the 2nd Workshop on Bio-inspired Algorithms for Distributed Systems, pp. 19–28. ACM (2010)
6. Frey, S., Fittkau, F., Hasselbring, W.: Search-based genetic optimization for deployment and reconfiguration of software in the cloud. In: Proceedings of the 2013 International Conference on Software Engineering, pp. 512–521. IEEE Press (2013)
7. Frey, S., Hasselbring, W.: The cloudmig approach: model-based migration of software systems to cloud-optimized applications. Int. J. Adv. Softw. **4**(3 and 4), 342–353 (2011)

8. Khajeh-Hosseini, A., Greenwood, D., Smith, J.W., Sommerville, I.: The cloud adoption toolkit: supporting cloud adoption decisions in the enterprise. Softw. Pract. Exp. **42**(4), 447–465 (2012)
9. Khajeh-Hosseini, A., Greenwood, D., Sommerville, I.: Cloud migration: a case study of migrating an enterprise it system to IAAS. In: 2010 IEEE 3rd International Conference on Cloud Computing (CLOUD), pp. 450–457. IEEE (2010)
10. Khajeh-Hosseini, A., Sommerville, I., Bogaerts, J., Teregowda, P.: Decision support tools for cloud migration in the enterprise. In: 2011 IEEE International Conference on Cloud Computing (CLOUD), pp. 541–548. IEEE (2011)
11. Linthicum, D.: The case for managed service providers in your cloud strategy, 19 May 2015. http://www.infoworld.com/article/2923441/cloud-computing/the-case-for-managed-service-providers-in-your-cloud-strategy.html. Accessed 1 June 2017
12. Ma, H., Hu, Z., Li, K., Zhang, H.: Toward trustworthy cloud service selection: a time-aware approach using interval neutrosophic set. J. Parallel Distrib. Comput. **96**, 75–94 (2016)
13. Mao, M., Li, J., Humphrey, M.: Cloud auto-scaling with deadline and budget constraints. In: 2010 11th IEEE/ACM International Conference on Grid Computing (GRID), pp. 41–48. IEEE (2010)
14. Mouratidis, H., Islam, S., Kalloniatis, C., Gritzalis, S.: A framework to support selection of cloud providers based on security and privacy requirements. J. Syst. Softw. **86**(9), 2276–2293 (2013)
15. Pandey, S., Wu, L., Guru, S.M., Buyya, R.: A particle swarm optimization-based heuristic for scheduling workflow applications in cloud computing environments. In: 2010 24th IEEE international conference on Advanced information networking and applications (AINA), pp. 400–407. IEEE (2010)
16. Pavlidis, M., Mouratidis, H., Kalloniatis, C., Islam, S., Gritzalis, S.: Trustworthy selection of cloud providers based on security and privacy requirements: justifying trust assumptions. In: Furnell, S., Lambrinoudakis, C., Lopez, J. (eds.) TrustBus 2013. LNCS, vol. 8058, pp. 185–198. Springer, Heidelberg (2013). doi:10.1007/978-3-642-40343-9_16
17. San Aniceto, I., Moreno-Vozmediano, R., Montero, R.S., Llorente, I.M.: Cloud capacity reservation for optimal service deployment. In: Second International Conference on Cloud Computing, GRIDs, and Virtualization, pp. 52–59 (2011)
18. Saripalli, P., Pingali, G.: Madmac: multiple attribute decision methodology for adoption of clouds. In: 2011 IEEE International Conference on Cloud Computing (CLOUD), pp. 316–323. IEEE (2011)
19. Trummer, I., Leymann, F., Mietzner, R., Binder, W.: Cost-optimal outsourcing of applications into the clouds. In: 2010 IEEE Second International Conference on Cloud Computing Technology and Science (CloudCom), pp. 135–142. IEEE (2010)
20. Wada, H., Suzuki, J., Yamano, Y., Oba, K.: Evolutionary deployment optimization for service-oriented clouds. Softw. Pract. Exp. **41**(5), 469–493 (2011)
21. Yusoh, Z.I.M., Tang, M.: Composite SAAS placement and resource optimization in cloud computing using evolutionary algorithms. In: 2012 IEEE 5th International Conference on Cloud Computing (CLOUD), pp. 590–597. IEEE (2012)

Author Index

Printed in the United States
By Bookmasters